Lecture Notes in Electrical Engineering

Volume 926

The book series *Lecture Notes in Electrical Engineering* (LNEE) publishes the latest developments in Electrical Engineering—quickly, informally and in high quality. While original research reported in proceedings and monographs has traditionally formed the core of LNEE, we also encourage authors to submit books devoted to supporting student education and professional training in the various fields and applications areas of electrical engineering. The series cover classical and emerging topics concerning:

- Communication Engineering, Information Theory and Networks
- Electronics Engineering and Microelectronics
- Signal, Image and Speech Processing
- Wireless and Mobile Communication
- Circuits and Systems
- Energy Systems, Power Electronics and Electrical Machines
- Electro-optical Engineering
- Instrumentation Engineering
- Avionics Engineering
- Control Systems
- Internet-of-Things and Cybersecurity
- Biomedical Devices, MEMS and NEMS

For general information about this book series, comments or suggestions, please contact leontina.dicecco@springer.com.

To submit a proposal or request further information, please contact the Publishing Editor in your country:

China

Jasmine Dou, Editor (jasmine.dou@springer.com)

India, Japan, Rest of Asia

Swati Meherishi, Editorial Director (Swati.Meherishi@springer.com)

Southeast Asia, Australia, New Zealand

Ramesh Nath Premnath, Editor (ramesh.premnath@springernature.com)

USA, Canada

Michael Luby, Senior Editor (michael.luby@springer.com)

All other Countries

Leontina Di Cecco, Senior Editor (leontina.dicecco@springer.com)

**** This series is indexed by EI Compendex and Scopus databases. ****

Kumari Namrata · Neeraj Priyadarshi ·
Ramesh C. Bansal · Jitendra Kumar
Editors

Smart Energy and Advancement in Power Technologies

Select Proceedings of ICSEAPT 2021
Volume 1

Springer

Editors
Kumari Namrata
Department of Electrical Engineering
National Institute of Technology
Jamshedpur
Jamshedpur, Jharkhand, India

Ramesh C. Bansal
Department of Electrical Engineering
University of Sharjah
Sharjah, UAE

Department of Electrical, Electronic
and Computer Engineering
University of Pretoria
Pretoria, South Africa

Neeraj Priyadarshi
Department of Energy Technology
Aalborg University
Esbjerg, Denmark

Jitendra Kumar
Department of Electrical Engineering
National Institute of Technology
Jamshedpur
Jamshedpur, Jharkhand, India

ISSN 1876-1100 ISSN 1876-1119 (electronic)
Lecture Notes in Electrical Engineering
ISBN 978-981-19-4970-8 ISBN 978-981-19-4971-5 (eBook)
https://doi.org/10.1007/978-981-19-4971-5

This Springer imprint is published by the registered company Springer Nature Singapore Pte Ltd.
The registered company address is: 152 Beach Road, #21-01/04 Gateway East, Singapore 189721, Singapore

Contents

Performance Analysis of PV Array Under Partial Shading Conditions with Bypass Diode and Static Array Configuration

Sanat Kumar Patro and R. P. Saini

Abstract Photovoltaic (PV) energy has made significant progress and nowadays considered as an affordable alternative towards meeting electrical power demands. Apart from this fossil fuels limitations, carbon emission and global warming encourage the development of PV technology. In the present study, the performance analysis of photovoltaic system has been performed under different shading conditions. A polycrystalline (RSM-50) PV module is considered to constitute SP, TCT, BL and HC 6 × 4 static array configuration, and their performance is investigated under different shading scenarios. Total-cross-tied (TCT) configured static PV array has been found to enhance the output power to 1.14 kW with bypass diode. However, without bypass diode, HC static array configuration inhabits a 1.05 kW global peak, which is found as maximum without bypass diode.

Keywords Photovoltaic · Solar irradiance · Partially shading conditions · Reconfiguration

Nomenclature

MPP	Maximum power point
HC	Honey comb
SP	Series-parallel
MPPT	Maximum power point tracking
SPV	Solar photovoltaic
PSC	Partial shading conditions
TCT	Total-cross-tied
BL	Bridge link

S. K. Patro · R. P. Saini (✉)
Hydro and Renewable Energy Department, IIT Roorkee, Roorkee, Uttarakhand 247667, India
e-mail: saini.rajeshwer@gmail.com

K. Namrata et al. (eds.), *Smart Energy and Advancement in Power Technologies*,
Lecture Notes in Electrical Engineering 926,
https://doi.org/10.1007/978-981-19-4971-5_1

1 Introduction

Photovoltaic (PV) energy has made significant progress and nowadays considered as an affordable alternative towards meeting electrical power demands in isolated and grid-connected systems. Apart from this fossil fuels limitations, carbon emission and global warming encourage the development of PV technology. The solar energy can be used with solar photovoltaic (SPV) systems. SPV systems are affordable, eco-friendly and require low maintenance. To achieve required voltage and current, photovoltaic (PV) modules can be placed in two types of combinations: parallel and series. However, the performance of SPV systems is significantly influenced with the variations in weather conditions and its surroundings. Partial shading is one of such influencing factors that can affect the performance of SPV systems. It is produced due to clouds, neighbouring buildings, trees, birds litter, dusts and other unavoidable circumstances. Due to partial shading, unequal solar radiations fall on different modules of an array at same time, and hence, unequal power is generated by modules.

The PV module under PSC operates in the reverse biased mode and acts as a load. Operating in this conditions, the module emits heat, if it operates beyond its breakdown voltage for certain time, then the module becomes permanently damaged, and this is called hot spot phenomenon [1]. Figure 1 shows the operating point of a typical module during shading and unshaded conditions [2].

The losses due to partial shedding can be reduced by using different array configurations apart from conventional array configurations. These configurations are pre-configured, and some of the typical configurations are total-cross-tied (TCT), bridge link (BL), honey comb (HC) and series–parallel (SP) (d). It is also possible to employ the dynamic switching array approach to decrease the effect of partial shading [3–5]. However, the scheme involves switching matrix and control unit which are integrated in the solar arrays. The dynamic switching array is another technique which uses a switching matrix and control unit. The switching matrix reconfigures the array during partial shading that is physical change in the location of modules. The array is separated into two parts, one of which is fixed and the other of which is adaptable. Generally, the adaptive part moves during the operation. However, these techniques do not perform outstandingly over different shading scenarios and fast-changing

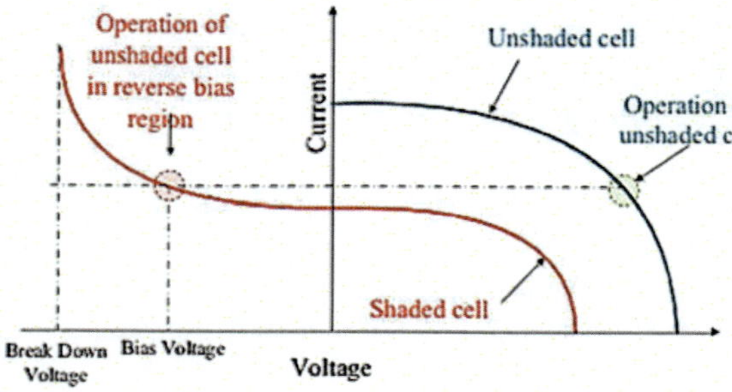

Fig. 1 Operating point of the module during shaded and unshaded conditions [2]

irradiances reconfiguration. Furthermore, these solutions necessitate the use of a monitoring system to track shading and faulty situations, a reconfiguration algorithm to select the best configuration and a switching matrix to connect PV modules. As a result, the cost of dynamic reconfiguration increases and the system complexity also increases [6–9]. The array arrangement, shading pattern and physical location of the PV modules together influence the power loss caused by partial shading. As a result, an appropriate reconfigurable pattern for putting PV modules is required to distribute shading effects across the entire array and explore its performance under various shading circumstances.

The present study is carried out under variable shading scenario with different static array configurations, i.e. TCT, BL, HC and SP. The main contribution of this paper is that MATLAB/SIMULINK PV double-diode model has been implemented, which is modelled to simulate PV static array configuration with and without bypass diode in the previous study of the authors [10].

2 PV Double-Diode Model

The accurate modelling is very crucial for simulation study of large PV arrays, evaluation, optimal sizing and maximum power point tracking (MPPT). With equivalent electrical circuit models and their parameters, all the physical properties of a PV module can be described. Figure 1 shows the circuit diagram of a standard double-diode seven-parameter electrical system [10–15]. During low isolations, the double-diode lumped model yields correct results [16]. The lumped model of double-diode PV module has been adopted to be able to precisely forecast the performance of a photovoltaic system. The double-diode model can be built by extracting the unknown parameters with analytical and numerical evolutionary algorithms [17, 18].

Figure 2 shows the complete physical characteristics of PV module which is presented with equivalent electrical circuit system and its parameters such as photocurrent (I_{ph}), saturation current (I_0), series resistance (R_S), shunt resistance (R_P) and ideality factors (a_1) and (a_2). The analysis of equivalent circuit parameters to simulate the output characteristics of a PV system at varied irradiance and temperature is included in the identification of PV system parameters [18].

Determination of parameters from the PV lumped model helps to model an accurate, fast and reliable PV model. The developed PV model can be implemented to realize different PV arrays configurations using various partial shading conditions. The optimum array configuration in most of the shading scenario can be found out based on performance investigation of different static array configurations.

P–V and I–V characteristics are investigated from PV double-diode model whose parameters are obtained using analytical method from data sheet information as given in Table 1. The PV modelling algorithm carried out by Patro and Saini [10] has been adopted to find the seven parameters (a_1, a_2, R_p, R_s, I_{01}, I_{02} *and* I_{ph}), and the values

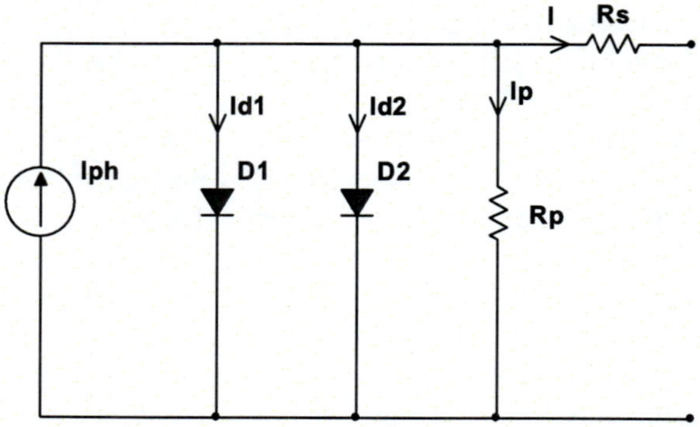

Fig. 2 Double-diode PV model

Table 1 Data sheet information and extracted seven parameters of double-diode model

Data sheet information RSM-50		Parameters obtained	
I_{sc} (A)	3.1	a_1	2
V_{oc} (V)	21.7	a_2	1.0222
I_{mp} (A)	2.82	R_S (Ω)	0.5868
V_{mp} (V)	17	R_P (Ω)	402.724
K_i (mA/°C)	1	I_{01} (A)	3.34E−09
K_v (mV/°C)	−78	I_{02} (A)	1.10E−09
N_s	36	I_{ph} (A)	3.1044

are given in Table 1 for accurate modelling of all the different configurations used in the present study. The MATLAB/SIMULINK block diagram of TCT arrangement for shading scenario 1 is shown in Fig. 3.

3 Results and Discussion

In the present study, RSM-50 poly crystalline modules are arranged in 6 × 4 with SP, TCT, BL and HC static configurations. There are four different partial shading scenarios considered for the 6 × 4 PV array as shown in Fig. 4. The values shown in Fig. 3 represent the intensity of solar radiations and are in W/m^2. The effect of bypass diode across each panel is also evaluated. I–V and P–V characteristics curves of PV array obtained for the four different considered array designs with bypass diode are shown in Fig. 5. It can be seen in the figure that the presence of bypass diode enhances the power in all types of configurations. Table 2 shows the effect of a bypass diode on various array layouts.

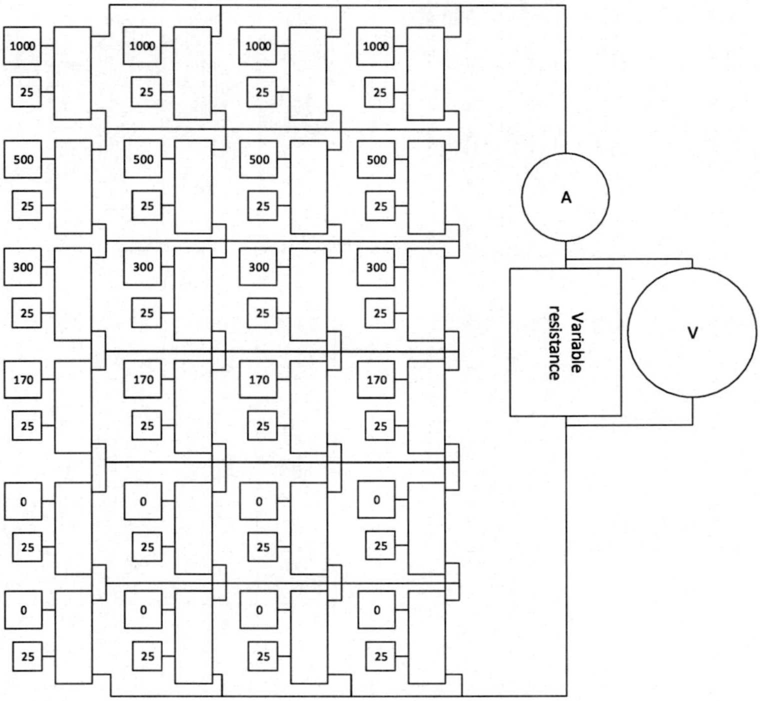

Fig. 3 SIMULINK block diagram of TCT connection for shading scenario 1

Based on the obtained results, it is found that TCT provides better power enhancement of 1.14 kW as compared to other configurations with bypass diode. However, HC configuration is found suitable for the condition when number of columns receives same irradiance is more than number of rows receives same irradiance. It is also found that HC exhibits a highest power of 1.05 kW without bypass diode. TCT configuration without bypass diode exhibits the least power of 0.927 kW. However, the TCT configuration shows superior results with bypass diode. A combination of bypass diode and TCT static array configuration strategy is found to increase the power of PV configuration using various partial shading conditions.

4 Conclusions

Under the present study, the effects of bypass diode and static array configurations using various partially shading conditions are investigated. A comparative investigation is also conducted on various static array configurations. Based on the static array configuration strategy, it is found that TCT reconfiguration shows improved performance under different shading scenarios. A global peak of 1.14 kW is obtained

1000	1000	1000	1000
500	500	500	500
300	300	300	300
170	170	170	170
0	0	0	0
0	0	0	0

(a)

1000	1000	1000	1000
220	220	220	220
150	150	150	150
0	0	0	0
0	0	0	0
0	0	0	0

(b)

1000	1000	1000	1000
700	700	700	700
500	500	500	500
150	150	150	150
100	100	100	100
0	0	0	0

(c)

1000	1000	1000	1000
500	500	500	500
500	500	500	500
300	300	300	300
250	250	250	250
250	250	250	250

(d)

Fig. 4 Various considered partial shading scenarios for 6 × 4 array **a** scenario 1, **b** scenario 2, **c** scenario 3 and **d** scenario 4

with bypass diode for the TCT configuration. However, global peak of 0.927 kW is attained without bypass diode. The BL static array configuration shows the highest global beak of 1.05 kW when compared with SP, TCT and HC without bypass diode. The use of bypass diode along with TCT configuration is found to increase 10% improvement in power output. A combination of bypass diode and TCT static configuration has been found to be a suitable strategy for power enhancement using various partial shading conditions. The results of the present study may be useful to effectively utilize the solar photovoltaic capacity for maximum energy extraction under variable operating conditions. The proposed strategy may also be adopted in future for performance investigation of SPV systems integrated with hybrid systems and concepts like nearly zero-energy building.

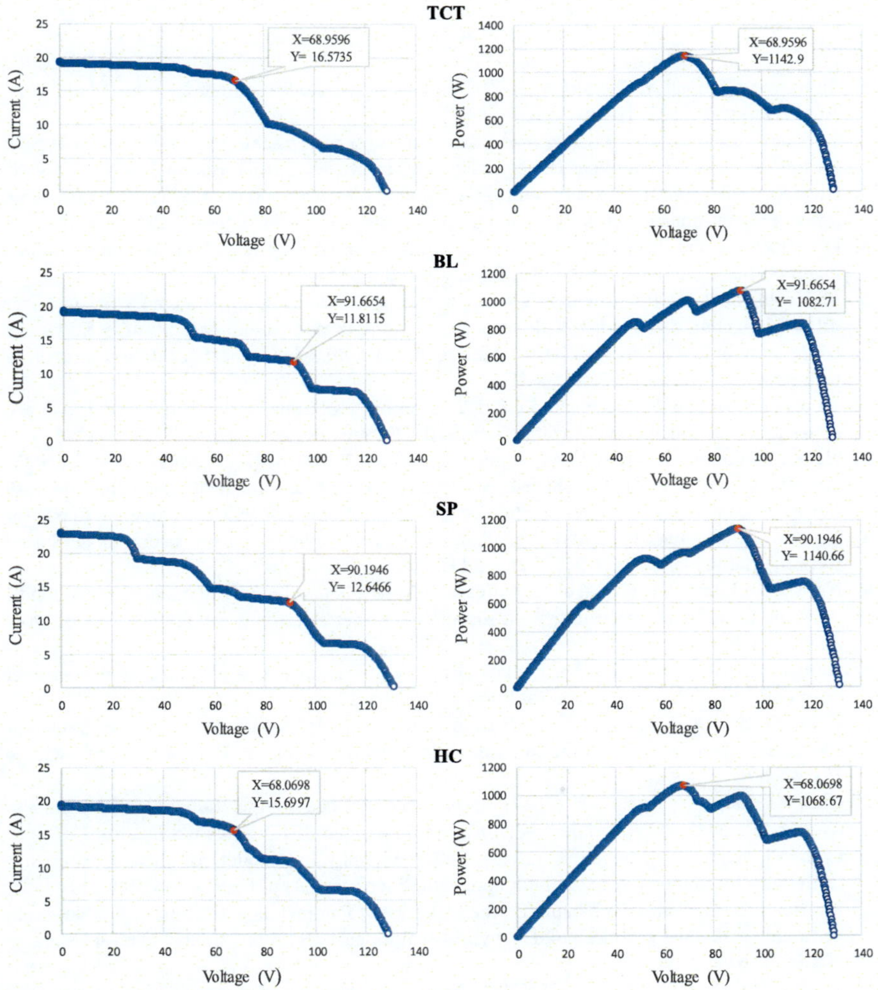

Fig. 5 I–V and P–V curves of different configurations with bypass diode

Table 2 Effect of bypass diode on different array configurations

Array size	Configuration type	Maximum power without bypass diode (kW)	Maximum power with bypass diode (kW)
6 × 4	SP	0.96	1.08
6 × 4	TCT	0.927	1.14
6 × 4	BL	1.003	1.13
6 × 4	HC	1.05	1.069

References

1. Silvestre S, Boronat A, Chouder A (2009) Study of bypass diodes configuration on PV modules. Appl Energy 86:1632–1640
2. Ishaque K, Salam Z (2013) A review of maximum power point tracking techniques of PV system for uniform insolation and partial shading condition. Renew Sustain Energy Rev 19:475–488
3. Bana S, Saini RP (2017) Experimental investigation on power output of different photovoltaic array configurations under uniform and partial shading scenarios. Energy 127:438–453
4. Bingöl O, Özkaya B (2018) Analysis and comparison of different PV array configurations under partial shading conditions. Sol Energy 160:336–343
5. Yadav AS, Pachauri RK, Chauhan YK (2016) Comprehensive investigation of PV arrays under different shading patterns by shade dispersion using puzzled pattern based Su-Do-Ku puzzle configuration. In: Proceedings 2015 1st international conference on next generation computing technologies NGCT 2015, pp 824–30
6. Storey JP, Wilson PR, Bagnall D (2012) Improved optimization strategy for irradiance equalization in dynamic photovoltaic arrays. IEEE Trans Power Electron 28:2946–2956
7. Nanno I, Haraoka R, Ikemoto M, Ahmed T, Nishida K (2016) New configuration for high-efficient operation of partially shaded PV system using an electromagnetic relay. In: 2016 18th European conference on power electronics and applications EPE ECCE Europe 2016, pp 1–9
8. Nguyen D, Lehman B (2008) An adaptive solar photovoltaic array using model-based reconfiguration algorithm. IEEE Trans Ind Electron 55:2644–2654
9. Nguyen D, Lehman B (2008) A reconfigurable solar photovoltaic array under shadow conditions. In: Conference proceedings—IEEE applied power electronics conference exposition—APEC, pp 980–6
10. Kumar PS, Saini RP (2020) Mathematical modeling framework of a PV model using novel differential evolution algorithm. Sol Energy 211:210–226
11. Chin VJ, Salam Z, Ishaque K (2016) An accurate modelling of the two-diode model of PV module using a hybrid solution based on differential evolution. Energy Convers Manage 124:42–50
12. Chin VJ, Salam Z, Ishaque K (2017) An accurate and fast computational algorithm for the two-diode model of PV module based on a hybrid method. IEEE Trans Ind Electron
13. Chin VJ, Salam Z, Ishaque K (2015) Cell modelling and model parameters estimation techniques for photovoltaic simulator application: a review. Appl Energy 154:500–519
14. Muhsen DH, Ghazali AB, Khatib T, Abed IA (2015) Parameters extraction of double diode photovoltaic module's model based on hybrid evolutionary algorithm. Energy Convers Manage 105:552–561
15. Sandrolini L, Artioli M, Reggiani U (2010) Numerical method for the extraction of photovoltaic module double-diode model parameters through cluster analysis. Appl Energy 87:442–451
16. Kumari PA, Geethanjali P (2018) Parameter estimation for photovoltaic system under normal and partial shading conditions: a survey. Renew Sustain Energy Rev 84:1–11
17. Phang JCH, Chan DSH, Phillips JR (2007) Accurate analytical method for the extraction of solar cell model parameters. Electron Lett 20:406
18. Chan DSH, Phang JCH (1987) Analytical methods for the extraction of solar-cell single- and double-diode model parameters from I–V characteristics. IEEE Trans Electron Devices 34:286–293

Performance Analysis of a Microgrid System Connected to a Grid Using EHO Technique

Monika Gupta, P. M. Tiwari, R. K. Viral, and Ashish Shrivastava

Abstract Distributed energy resources (DERs) have become more appealing to feed local loads under the abstraction of microgrids (MG). The concept of microgrid (MG) is presented for better renewable energy penetration into the utility grid. Renewable sources being highly intermittent in nature, feeder power flow at substation bus bar is required to be controlled and coordinated. One of the foremost elements to amalgamate distributed generation (DG) unit in a MG is battery energy storage system (BESS). A MG embodying single-phase photovoltaic (PV) arrays and wind turbines (WT) that functions as a foremost distributed generation elements and the "battery energy storage system" to augment the spasmodic photovoltaic as well as wind power generation and the demand variation in the microgrid is presented in this work. Although there are many heuristic algorithms for solving this problem, however, we have considered a swarm-based heuristic algorithm elephant herding optimization (EHO). This algorithm has been utilized to eradicate the error in power flow for keeping up the balance of demand and generation by adjusting the attained values of all the PI controllers optimally. Simulink and MATLAB environment has been used to study the achievement of this propound system.

Keywords Distributed energy resources · Microgrid · Photovoltaic and wind turbines · Elephant herd optimization

1 Introduction

With revolutionization, energy system of the world is focused to be low on carbon, should be green, as well as a sustainable one [1, 2]. Some of the foremost sources in renewable energy resource (RESs) are the photovoltaic (PV), wind turbines (WTs),

M. Gupta (✉) · P. M. Tiwari · R. K. Viral
Department of Electrical and Electronics Engineering, Amity University, Noida, India
e-mail: monika_gupta2208@yahoo.com

A. Shrivastava
Department of Electrical and Electronics Engineering, Manipal University Jaipur, Jaipur, India

K. Namrata et al. (eds.), *Smart Energy and Advancement in Power Technologies*,
Lecture Notes in Electrical Engineering 926,
https://doi.org/10.1007/978-981-19-4971-5_2

tidal power, biomass, and hydro which should alleviate emissions from greenhouse gas and provide a sustainable development by providing clean energy. Nevertheless, sporadic and unstable characteristic of PV and WT origins sometimes causes major issues with regard to reliability and stability [3, 4]. Therefore, for an intelligent strategy, smart power grids [5] are required to amalgamate with high number of RESs to a utility grid in order to improve performance and reduce system uncertainties [6]. Further, in order to trade-off power with the primary grid and to assist the local load during the power failure, microgrids are operated in an on-grid mode [7] and in an off-grid mode. For exponential growth in power converters, the RESs (photovoltaic and wind turbine systems) became the major distributed generation sources in MG [8, 9]. Battery energy storage system (BESS) is an indispensable and efficacious element to amalgamate a DG unit to a MG and to enhance its flexibility and controllability. Hence, to assure stored energy balance within BESSs and RESs, a coordinated control is required to amplify the steadiness and consistency of the MG system [10]. Figure 1 depicts a block of the PV-wind-BESS MG system.

"Elephant herd optimization (EHO)" technique is tendered for worldwide accretion which is stimulated by the behavioral action of elephants in nature [11–13].

EHO [14, 15], a recently propound "swarm intelligence-based algorithm", is among the most illustrative work w.r.t. "swarm intelligence" and is reviewed and summarized.

An improved EHO algorithm adopted to decipher multi-faceted "distributed energy resources (DERs)" arrangement complications of distribution systems has been impersonated by Meena et al. [16]. The same is done by taking combination of a "technique for order of preference by similarity to an ideal solution (TOPSIS)".

Exploratory results depicted that the EHO has a refinement for about 10.7% on "Arsikere Taluk" and 6.63% on "NITK campus" above "support vector machine" as Jayanth et al. [17] utilized EHO to resolve the predicament.

Succession of experiments is to comprehend controlled parameter's impact and to propound unlike approaches in order to intensify accomplishments of the EHO.

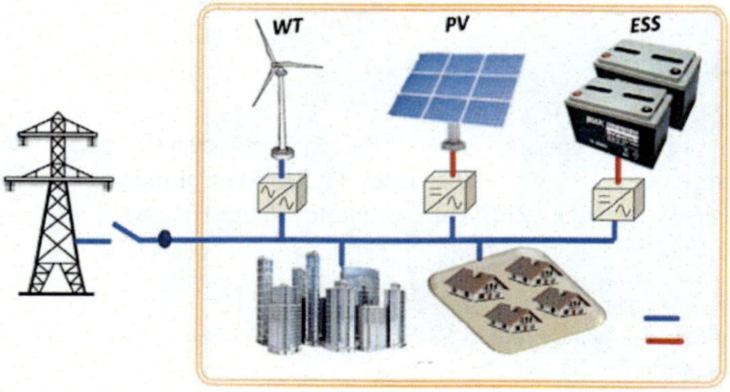

Fig. 1 Block of PV-wind-BESS MG system [10]

Anamnesis ranging between the current testing complications of "Congress on Evolutionary Computation (CEC)" 2016, to an admired engineering complications of a "gear train", a "welded beam", a "three-bar truss design complication", a "continuous stirred tank reactor", and a "fed-batch fermentor", is being applied for ratifying and examining accomplishments of said propound EHOs against the prevailing techniques/methods.

To inscribe the energy bound source confinement issue in wireless sensored networks, a "metaheuristic algorithm", EHO was used by Correia et al. [18]. The simulation results manifested that the new viewpoint outstandingly surpassed the current issues in noisy environments and examining of "metaheuristic methods".

A fresh "hybrid algorithm" which was assumed from "EHO" and "cultural algorithm (CA)", named "elephant herding optimization cultural (EHOC) algorithm", has been propound by Jafari et al. [19], wherein a conviction space delimited by a conventional algorithm was accustomed to ameliorate the regular EHO and to ameliorate its search capability and also to develop an algorithm with a suitable exploration–exploitation balance.

Thus, in this paper, "elephant herd optimization (EHO) algorithm" has been utilized to diminish inaccuracy by suitably regulating gain values in the feeder power flow of all the proportional and integral controllers [20–23] used within the MG operation. The optimal values of these controllers would assist in maintaining the balance of generated power at the source and demand power at the consumers end.

Concludingly, MATLAB/Simulink used simulation studies that are piloted on the propound microgrid structure, and its "control system" is derived as per the below test scenarios:

1. When there is low speed in the WT and high input in the PV
2. When there is high speed in the WT and low input in the PV.

This paper is sorted as follows. System structure and its explanation are discussed in Sect. 2. The propound methodology and its discussion are iterated in Sect. 3. Section 4 depicts the results of simulation and their discussions, and Sect. 5 collects the concluding remarks.

2 System Structure and Description

The MG structure comprises of renewable sources which are "photovoltaic (PV) array", a "wind turbine (WT)", and the "battery energy storage system". Wind turbine and photovoltaic is used to generate required power and to cope up with the load demand of the consumers. A "storage system" is necessary for taking care of the "load" and "function" requirements of the applications. Deploying relevant methods of storage system, the production of energy may be made autonomous from the demand. This approach powers up the sharing thereby keeping up the equilibrium within the power generation and its demand.

Figure 2 depicts the configuration of the propound MG Simulink model that is operating in grid-connected way of operation, and Fig. 3 displays the control system employed in the Simulink model.

The propound microgrid includes 66 single-phased parallel 305 W photovoltaic arrays, a 6.6Ah battery backup, and 3 single-phased residential loads. Power output of every "photovoltaic array" has been allowed to differ inside an agreeable scope. A battery backup has been affixed to the "point of common coupling (PCC)" of MG with 3 single-phased "full-bridge" inverters for managing output current and output voltage. Photovoltaic arrays and wind turbines work as the major distributed generation units of MG, and the battery backups are installed as a support that shall supply power during shortage and momentarily stores the residual power of MG. Microgrid is also coupled with a utility grid through a 400 KVA, 260/25 kV distribution transformer.

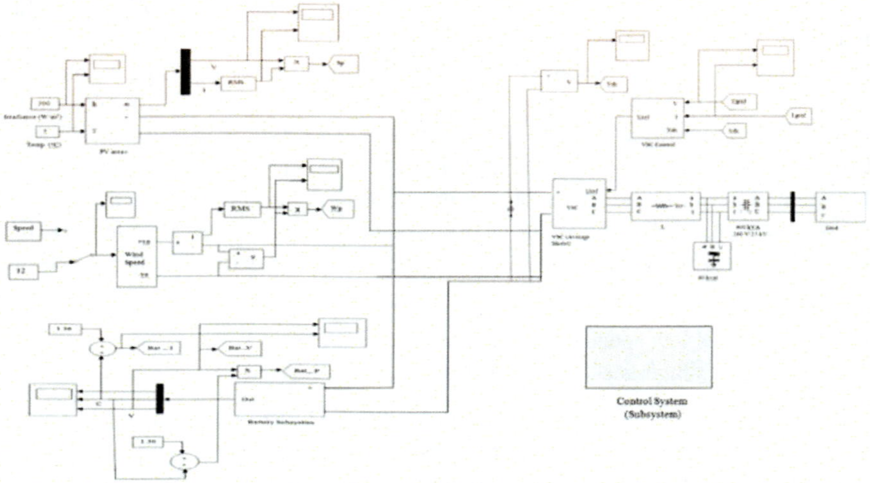

Fig. 2 Simulink model of propound MG system

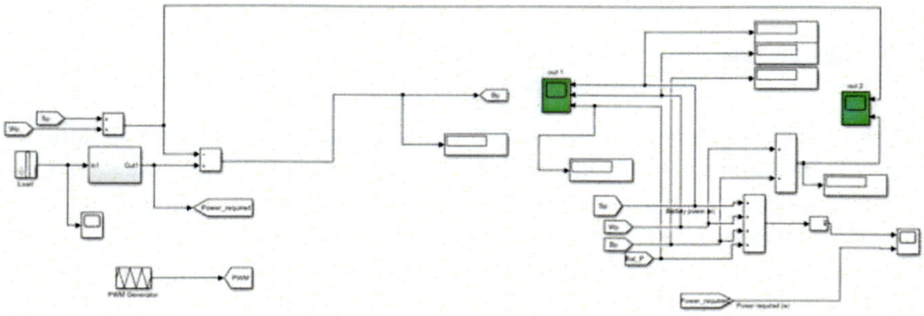

Fig. 3 Control system diagram employed in the Simulink model

In a MG, functioning of every inverter is deliberated by an upper centralized MG energy management system that formulates on total power of system, i.e., "P_{net}" and the "battery state of charge (SOC)" limits. "P_{net}" is termed as net power generation subtracted from the net load demand.

$$P_{net} = (P_{DG} - P_L)$$

Here, P_{DG} is the power dispatched by primary distributed generation units, and P_L is the power essential by load.

In MG, "energy storage system" backs up voltage of MG, and RESs injects the utmost power to the MG in usual operation. For "energy storage system", the "proportional-resonant controller" is employed, and for RESs, the "voltage-modulated direct power control (VMDPC)" is utilized to have a good performance as well as a better steady-state behavior. In this approach, the VMDPC is used for PV/WT to reach maximum grid power and a conventional controller is casted for battery energy storage system (BESS) to sustain "voltage" and "frequency" in MG.

3 Proposed Methodology

An efficient elephant herd optimization (EHO) technique-based coordinated control is presented in this paper for the performance enhancement of the hybrid renewable energy sources-connected smart grid system that consists of wind turbine (WT), photovoltaic (PV), battery energy storage system (BESS), and controllable loads.

The algorithms iterated in this paper use "hourly meteorological data (irradiance and temperature, wind speed)", then load the profile to adjust PI controller parameters by adopting a "performance function" which is known as "fitness function" for evaluating the accomplishments of controller for the cost and energy worth.

3.1 Objective Function of the Proposed Method

Augmenting of the static power from microgrid system is assumed as an essential aim of this propound formulation. Generated power equals the required power in the propound microgrid system model which is an objective function.

Maximize

$$F(\text{obj}) = \text{Max}\big(P_G = P^*\big) \tag{1}$$

Here, "$F(\text{obj})$" is a function of power to be augmented, whereas "P_G" is power generated, and "P^*" is power required.

3.2 Elephant Herd Optimization (EHO) Algorithm

"EHO algorithm" is the recent optimization algorithm which is developed [24, 25]. It apprehends the herding behavior of the elephants. Elephant being taken as a social animal and their herding comprise of various clans of female elephants and its respective calves [20, 23]. Respective clan moves in the influence of a matron or the leader elephant. The female elephant lives with its respective family group, whereas male elephant gets separated as soon as it grow up and lives in touch with its respective family group through "low-frequency vibrations". This behavior has been moldered mathematically in two different types of operators—updating operator as well as separating operator [15], which is exhibited in (2) and (5).

Elephant herd optimization is mathematically described using the following algorithm steps.

In every clan, the position of the elephants is updated as per the below equation,

$$E^i_{\text{new,cj}} = E^i_{\text{cj}} + A * \left(E_{\text{best,cj}} - E^i_{\text{cj}} \right) * r \tag{2}$$

Post that, arrangement of the fittest elephant is updated in every clan, which may be described in Eqs. (3) and (4).

$$E^i_{\text{new,cj}} = B^* E_{\text{center,cj}} \tag{3}$$

$$E_{\text{center,cj}} = \sum_{i=1}^{n} E^{i,d}_{\text{cj}} / \eta e \tag{4}$$

Thereafter, elephants which are the worst in the respective clans are separated as per the below equation,

$$E^i_{\text{worst,cj}} = E_{\min} + (E_{\max} - E_{\min} + 1)^* \text{rand}() \tag{5}$$

On the basis of the above equations, respective elephants and their respective positions are then updated.

Initially, PV power, irradiance, temperature, wind speed, wind power, and battery power are given as the input of the EHO algorithm. After that, the "DC-link voltage", "generated power", "grid current", and "voltage" are determined. The flow diagram of the EHO algorithm working is depicted in Fig. 4.

4 Simulation Results and Discussions

Intel(R) i7 core (TM), 4 GB RAM, and MATLAB/Simulink 7.10.0 (R2018a) engine are used for the simulation of the propound MG system. The block diagram of this

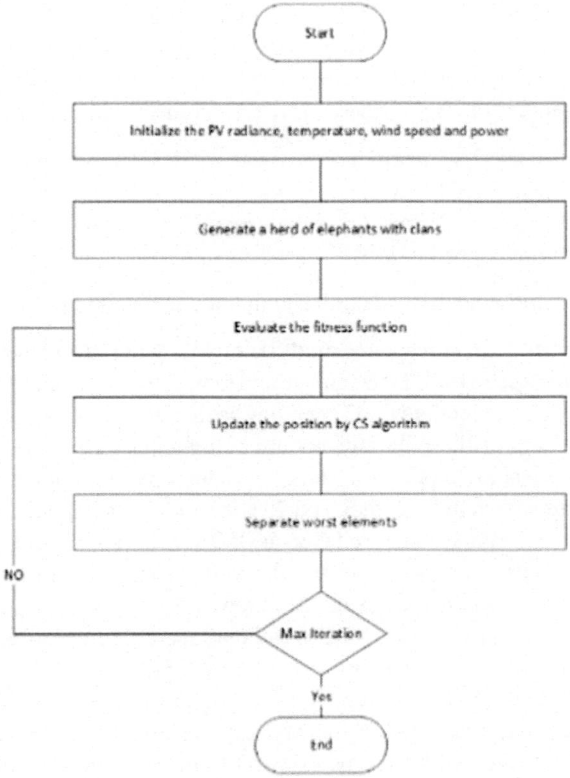

Fig. 4 Flow diagram of EHO algorithm

propound system is already shown in Fig. 2. The propound technique improves the stability of the MG system considered through balance of generation and demand power by minimizing the error by effectively adjusting the gain values of the proportional and integral controllers. Controlling parameters, i.e., voltage, current and power are then evaluated from the MG system.

Table 1 presents the implementation parameters of MG system considered while performing simulation in MATLAB using the propound optimization technique.

Table 1 Implementation parameters of microgrid system

Description of parameters	Values
Wind speed	12 (m/s)
Generator speed	1.2 (rpm)
Battery rated capacity	6.6 (Ah)
Shunt resistance	269.5934 (Ω)
Series resistance	0.37152 (Ω)
Nominal mechanical output power	10 (KW)
DC-link voltage	140 V

The EHO-based system is used to control the DC-link voltage at microgrid and enhance the power management. EHO technique-based controller is inspected in two distinct cases because of the spasmodic nature of photovoltaic and wind power generation as provided under the given simulated results [26].

Case 1: Performance analysis of MG when there is high input in PV system and low speed in WT
Case 2: Performance analysis of MG when there is low input in PV system and high speed in WT

In-depth analysis of EHO method is detailed in the below section.

Case 1: Performance analysis of MG when there is high input in PV system and low speed in WT. Irradiance and temperature are the input parameters of photovoltaic system, and the wind speed has been considered for the wind turbines. The standard values of these parameters are being taken as per the standard research papers that were studied [27, 28]. The graphs depicting irradiance, temperature, and wind speed are shown in Figs. 5, 6, and 7. The values of irradiance, temperature, and wind speed in the MG system are taken as 750 W/m2, 10 °C, and 2 m/s, respectively.

Based on this analysis, the voltage and the current in photovoltaic, wind, and battery are measured and depicted in Figs. 8, 9, 10, 11, 12, and 13. As per the graphs, at $t = 1$ s, the values of "PV voltage" and "PV current" are 138 V and 900 A. The wind and battery current and voltages are 0.57 A, 0 A, 138 V, and 30.85 V, respectively.

Numerous iterations are performed to the propound Simulink model. The EHO technique applied to the MG system optimizes the value of K_p and K_i of the PI controller, respectively. For an optimized proportional and integral controller, the DC-link voltage is kept persistent as per Fig. 14. DC-link voltage is 138 V at 1 s at steady state but at 130 V, between 0.4 and 0.5 s, some small oscillations are observed.

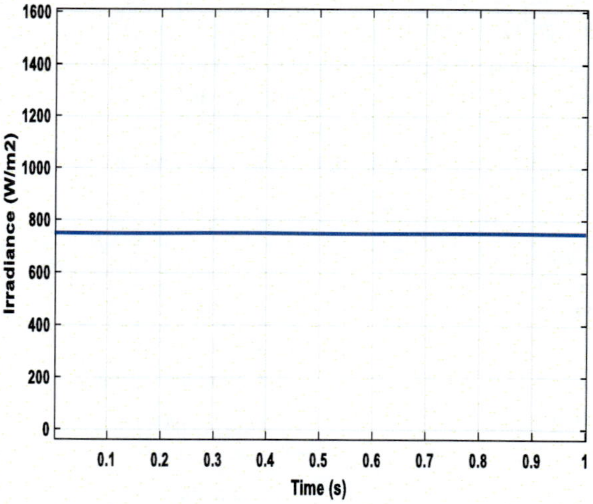

Fig. 5 Input parameters—PV irradiance—case 1

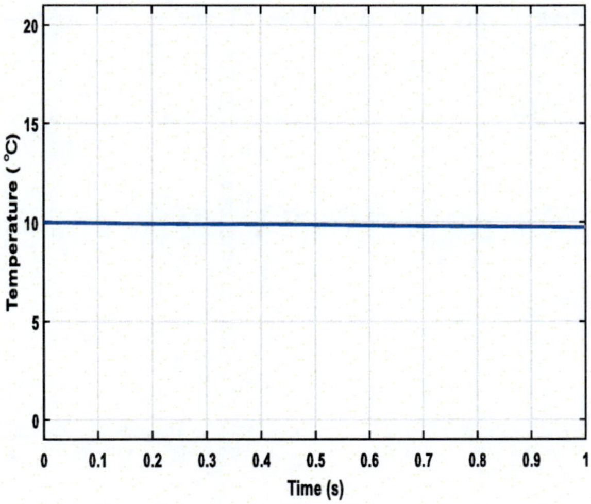

Fig. 6 Input parameters—PV temperature—case 1

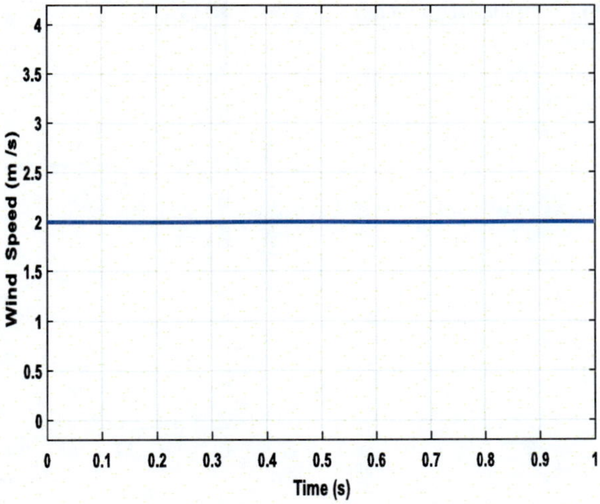

Fig. 7 Input parameters—wind speed—case 1

Figures 15 and 16 depict the grid voltage (which varies from −2000 V to 2000 V) and grid current (which varies from −32 A to 32 A) from 0 to 1 s, respectively. As per the grid current graph, a dip is observed in the value of current between 0.4 and 0.5 s.

Figures 17, 18, and 19 demonstrate the performance analysis of "photovoltaic", "wind", and the "battery" power, respectively. The "PV", "wind", and the "battery" are utilized for compensating and meeting the required load. "Battery" works to meet the load demand if it is not compensated with the "photovoltaic" and "wind" generation power.

Fig. 8 Battery voltage at $t = 1$ s in case 1

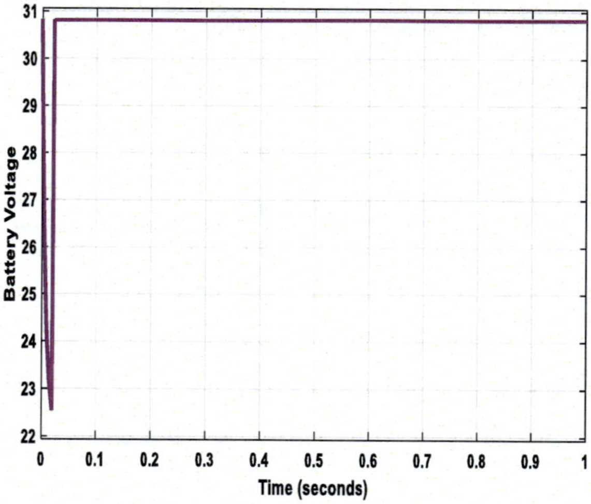

Fig. 9 Battery current at $t = 1$ s in case 1

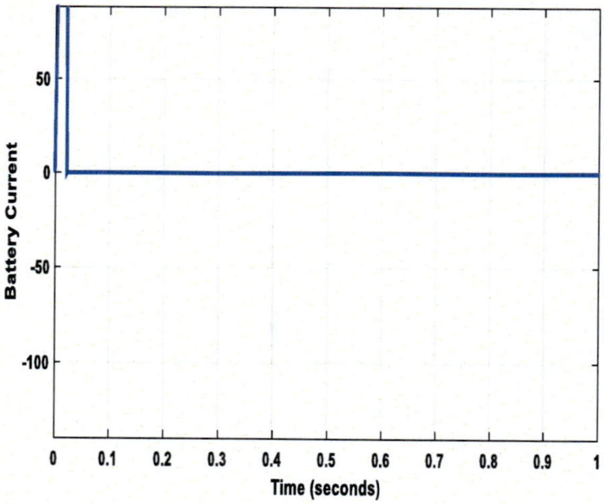

Figures 20 and 21 show that required and generated power are stabilized at the appropriate reference values to maintain the load power value at its required value of 10KW and equal to the generated power.

Case 2: Performance analysis of MG when there is low input in PV system and high speed in WT. In this case, values of irradiance, temperature, and wind speed in the MG system are taken to be as 500 W/m^2, 5 °C, and 12 m/s, respectively, and the graphs depicting the same are shown in Figs. 22, 23, and 24.

On the basis of this analysis, the voltage and the current in "photovoltaic", "wind", and "battery" are measured and depicted in Figs. 25, 26, 27, 28, 29, and 30. As per the graphs, at $t = 1$ s, the values of "PV voltage" and "PV current" are 140 V and

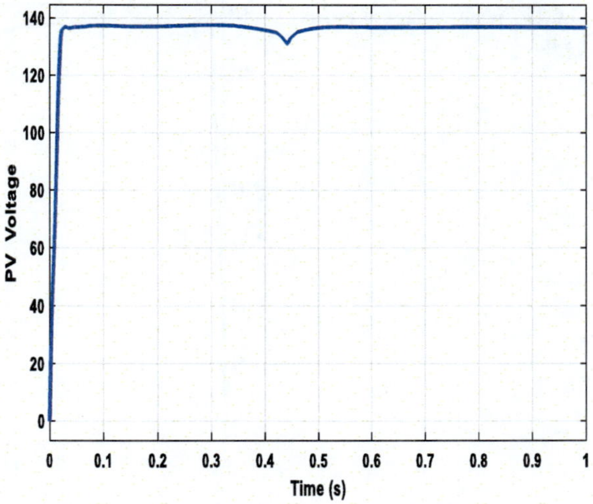

Fig. 10 PV voltage at $t =$ 1 s in case 1

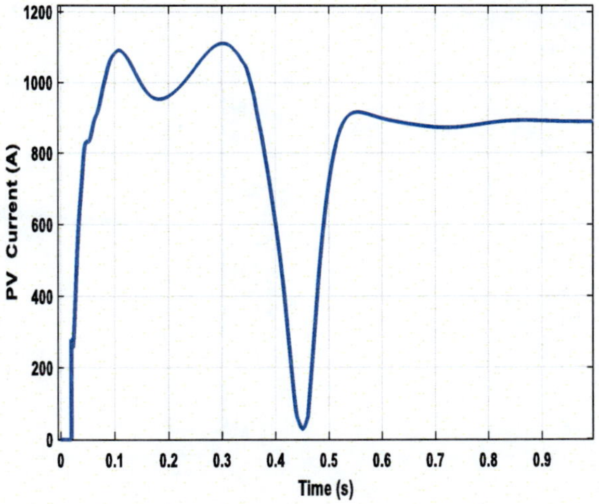

Fig. 11 PV current at $t =$ 1 s in case 1

800 A. The wind and battery current and voltages are 9 A, 0 A, 140 V, and 30.85 V, respectively.

DC-link voltage of Case 2 as depicted in Fig. 31 shows that the DC-link voltage of the microgrid arrangement and voltage is maximum at 230 V at time instant $t =$ 0–1 s.

The generated power from the power generation system is illustrated in Figs. 32, 33, and 34, which show the power generated from the PV, wind, and battery. The maximum power of wind production is 1100 W. The variation of the power required at the load side is controlled by utilizing battery storage system.

Figures 35, 36, 37, and 38 depict the performance analysis of the "grid voltage", the "grid current", the "required power", and the "generated power". These depicted

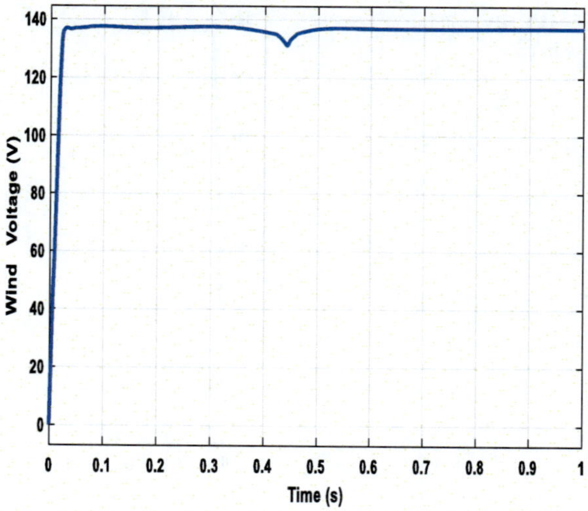

Fig. 12 Wind voltage at $t = 1$ s in case 1

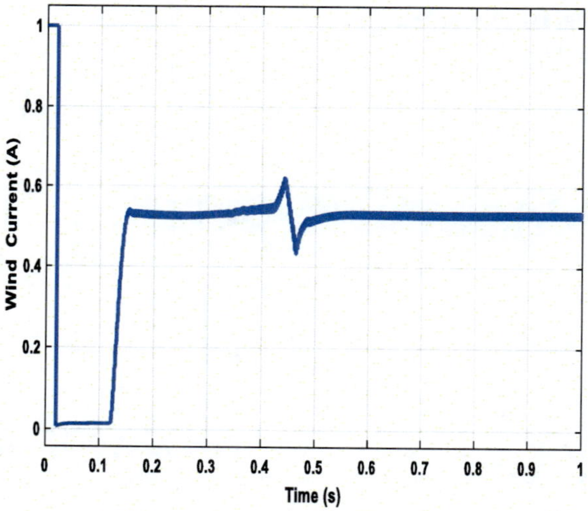

Fig. 13 Wind current at $t = 1$ s in case 1

waveforms are a repercussion of power organization in a microgrid at stable "wind speed" and diverse "load deviation". The necessary power 12 KW at 0.5 s, 11 KW at 0.7 s, and 10 KW at 1 s is achieved and equals the generated power.

Also, the optimal values of gains of PI controller obtained using the proposed EHO technique are compared with other existing optimization techniques such as particle swarm optimization (PSO) and cuckoo search (CS) [23] which exhibit a better performance of the proposed technique. It is depicted in Table 2.

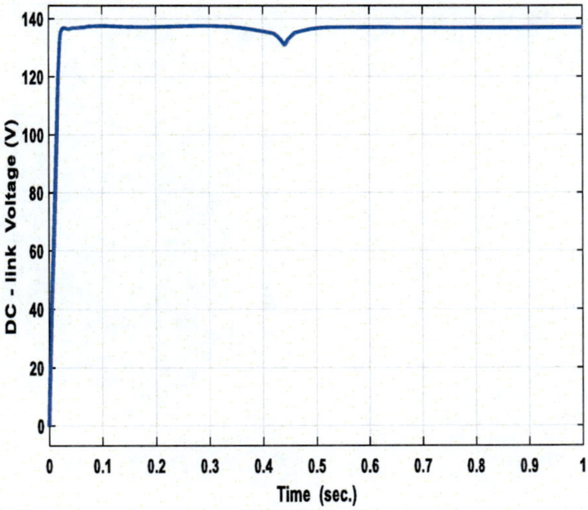

Fig. 14 DC—link voltage

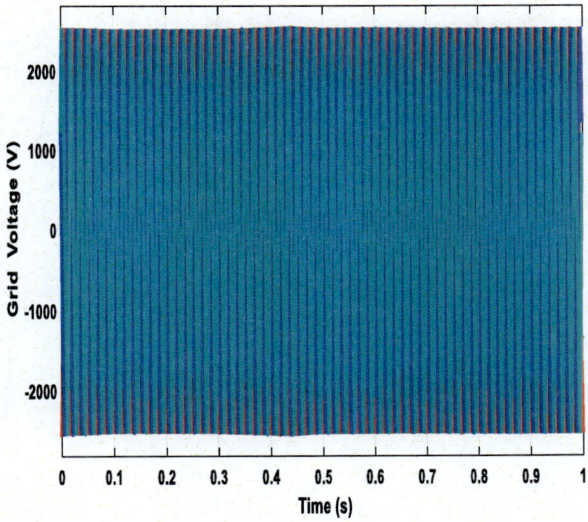

Fig. 15 Grid voltage

5 Conclusion

World is revolutionized by the concept of microgrid providing power systems with the plasticity of power distribution on a wide ranging scale. Microgrids can function in a secluded mode if the central supply collapses which is one of the prime aspects for the great evolution in this discipline in order to elude major power failures. Using the allocated control of "energy storage devices" for utmost exploitation of energy assets, this can be achieved. Analysis of strategies to improve power quality and its elaborated methods for microgrid control is explained. Here, the photovoltaic, the

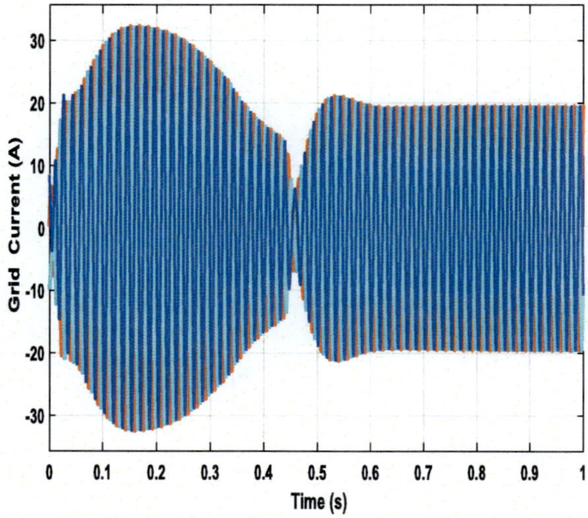

Fig. 16 Grid current

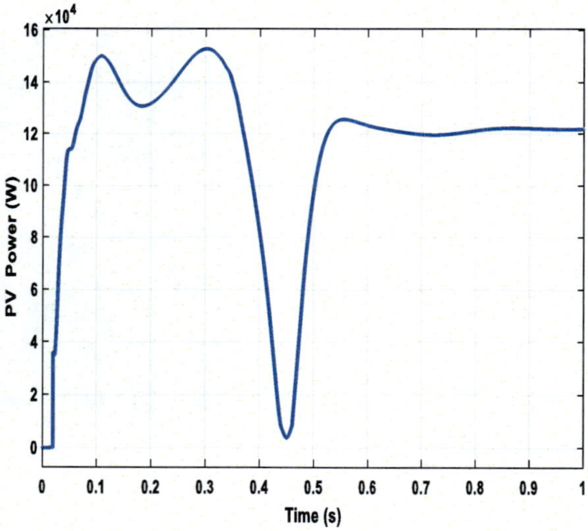

Fig. 17 Performance analysis of PV power

wind, and the battery are appraised to cope up with the required energy level, and improving stability was accomplished using the propound technique. This managed and controlled procedure was simulated in MATLAB/Simulink execution software, and its results were analyzed. The process of the photovoltaic power, the wind power, the battery power, the voltage, the current, the DC-link voltage, the generated power, and the grid current and voltage was examined by exploitation of the propound technique. The results obtained demonstrate that the propound technique has achieved desirable results. The attained optimal values of the parameters of PI controller are

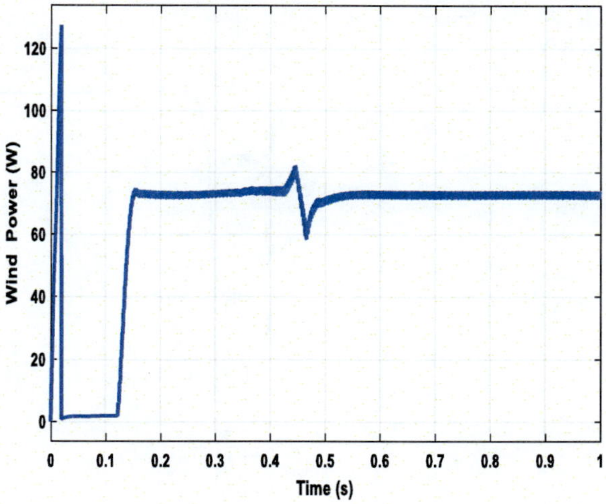

Fig. 18 Performance analysis of wind power

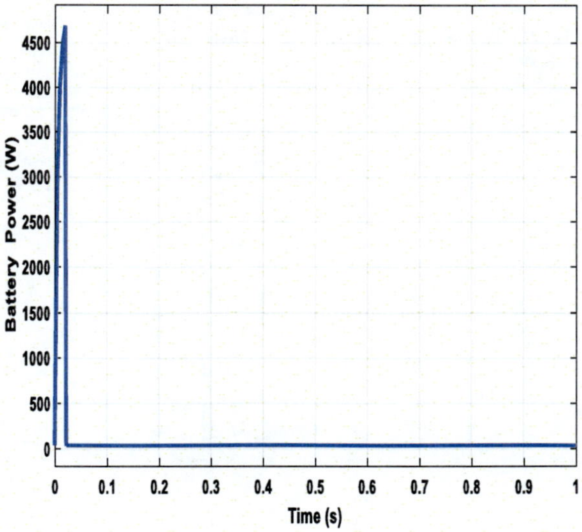

Fig. 19 Performance analysis of battery power

much better as compared to the conventional controllers using this new optimization technique. Also, the required power matches with the generated power in the microgrid system. The above-mentioned strategies confirm system's effectiveness, its constancy, its certainty, and its merchandising under variable operating circumstances for microgrids of its allocated nature. The altogether benefaction of the investigation is to come up with a sturdy foundation for developing a structured and well-grounded "distributed control techniques" for smart microgrids.

Although EHO exhibits a good performance, like other "metaheuristic methods", it does not make use of the best available details from earlier elephant individuals to escort present and the later searches. The recess shall be identified, because earlier

Fig. 20 Required power at 10 KW

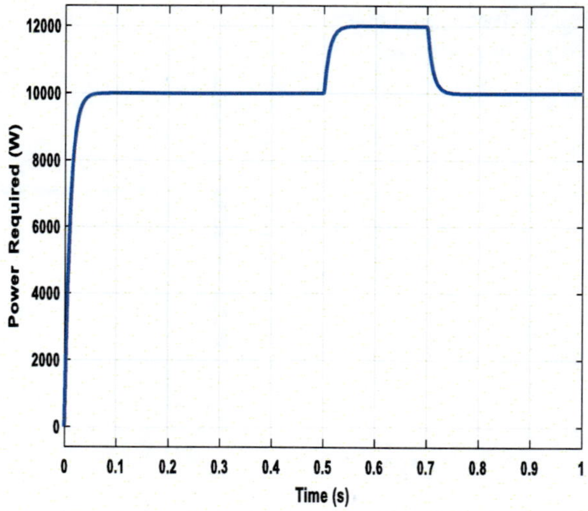

Fig. 21 Power generated at 10 KW

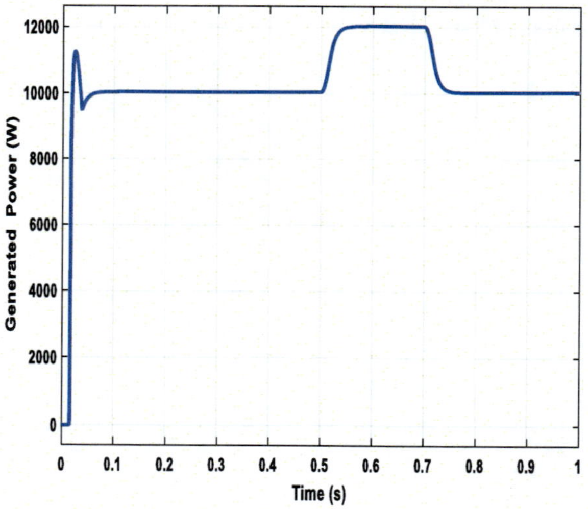

researchers may cater a diversified handy facts and figures. All this information could be completely explored, utilized, and practiced in later refurbishment processes; the performance of elephant herd optimization may be improvised outstandingly which would be taken care of in the future prospects.

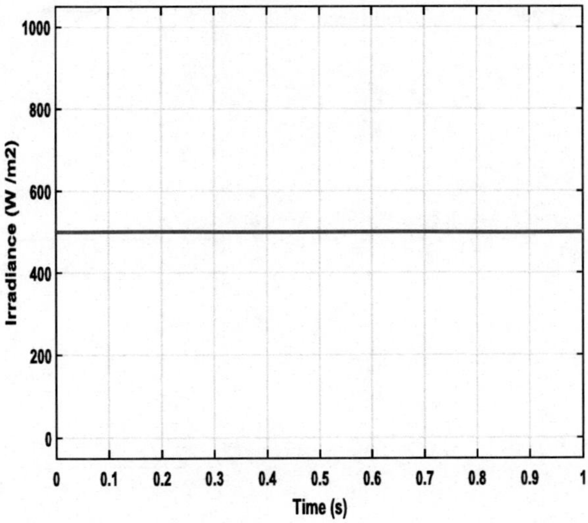

Fig. 22 Value of irradiance

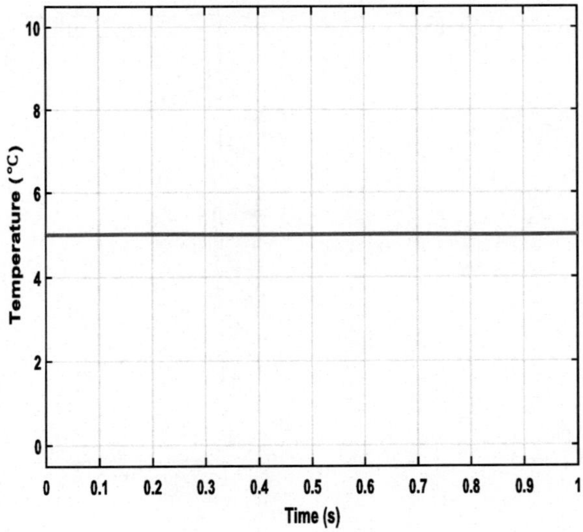

Fig. 23 Value of temperature

Fig. 24 Value of wind speed

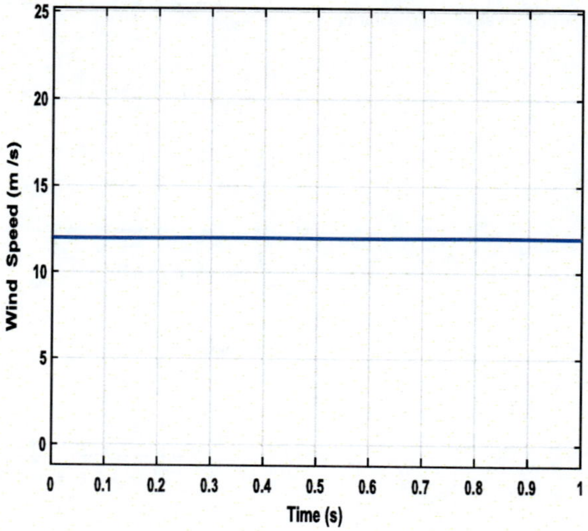

Fig. 25 PV voltage

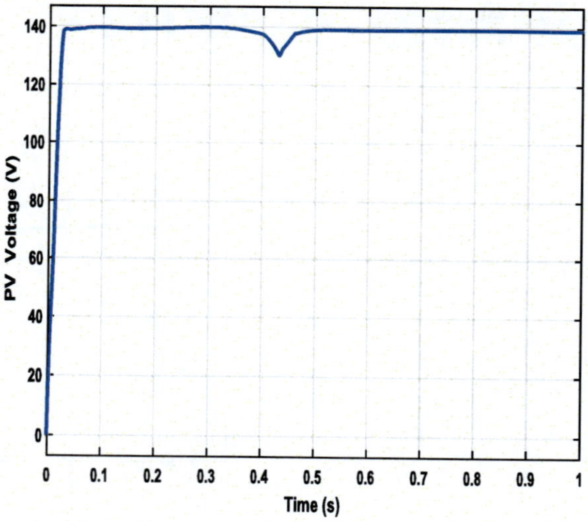

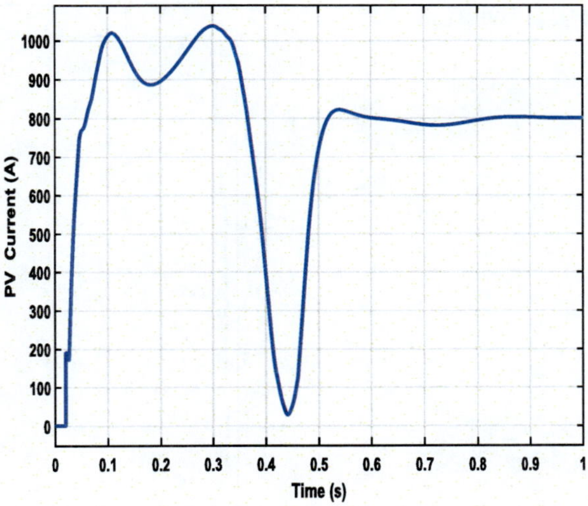

Fig. 26 PV current

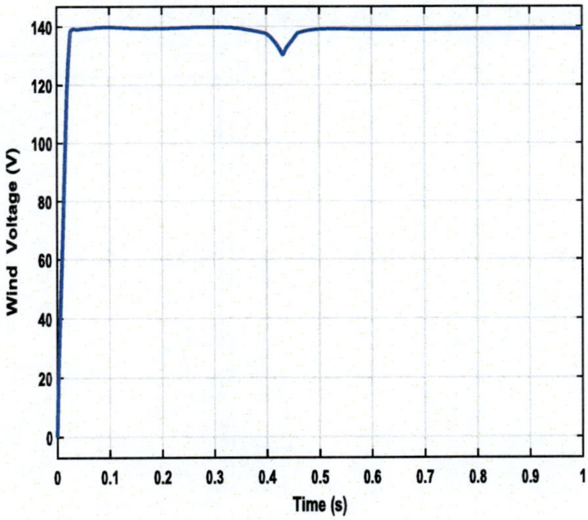

Fig. 27 Wind voltage

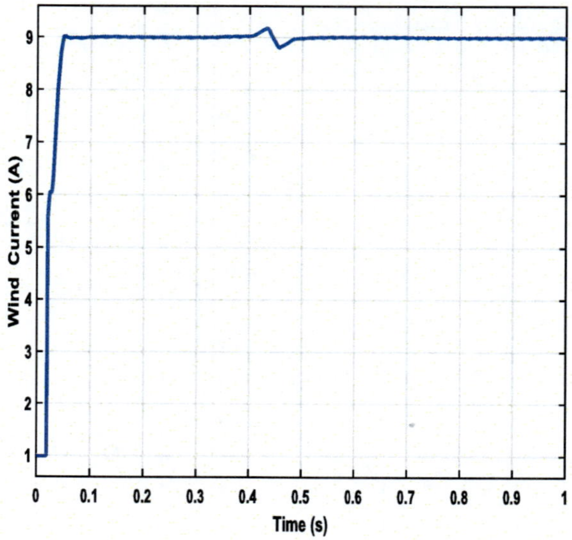

Fig. 28 Wind current

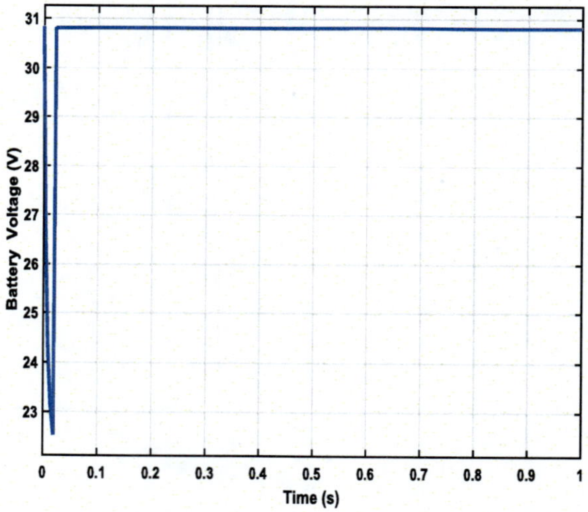

Fig. 29 Battery voltage

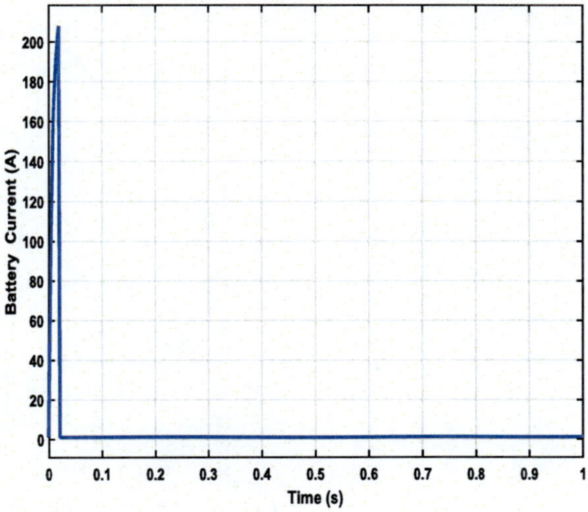

Fig. 30 Battery current

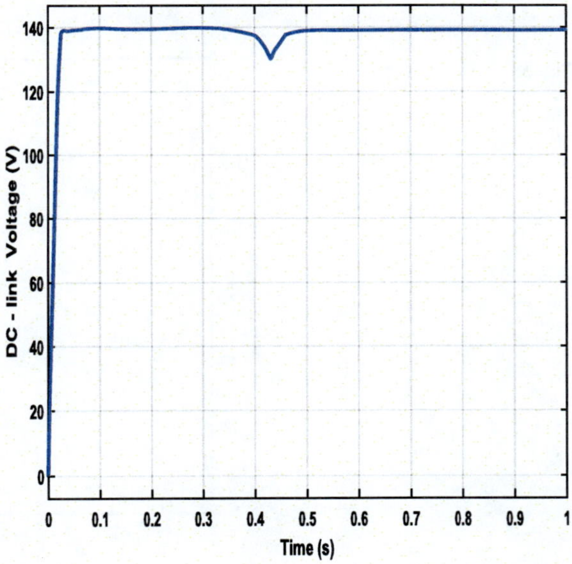

Fig. 31 DC-link voltage

Fig. 32 PV power

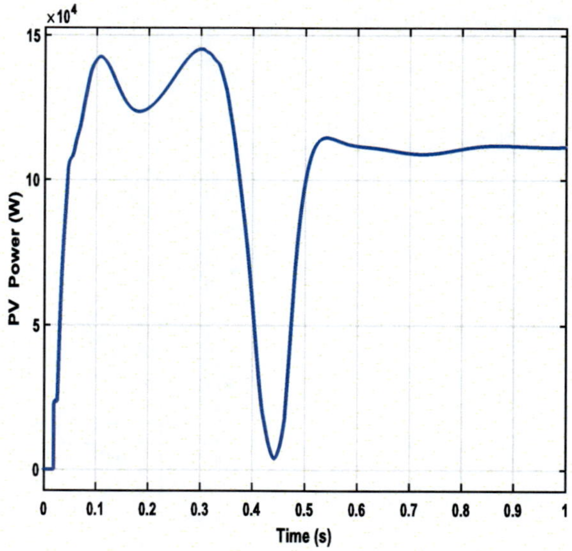

Fig. 33 Wind power

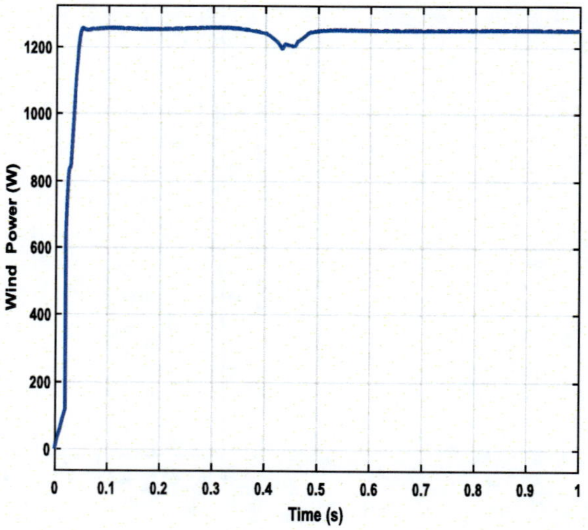

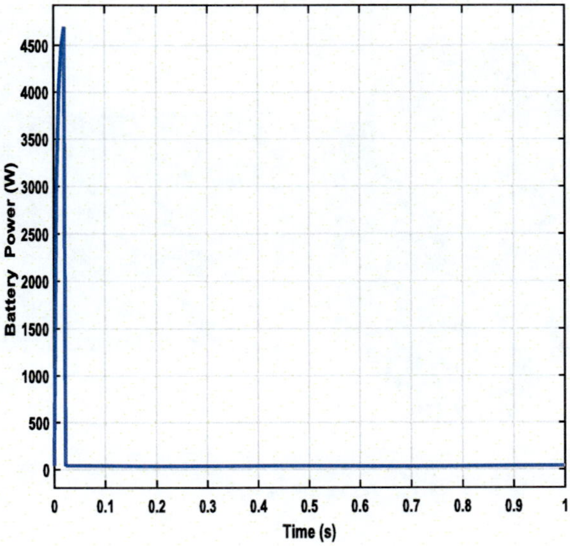

Fig. 34 Battery power

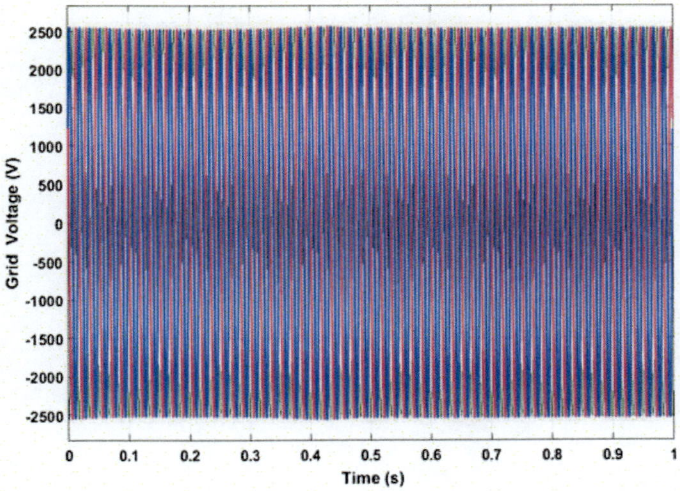

Fig. 35 Grid voltage

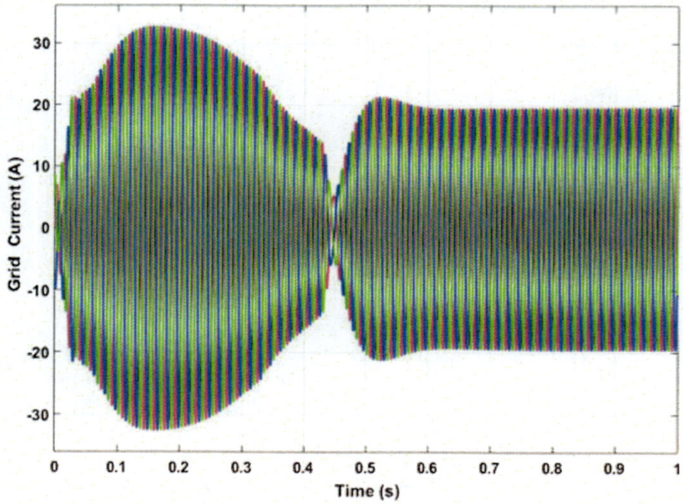

Fig. 36 Grid current

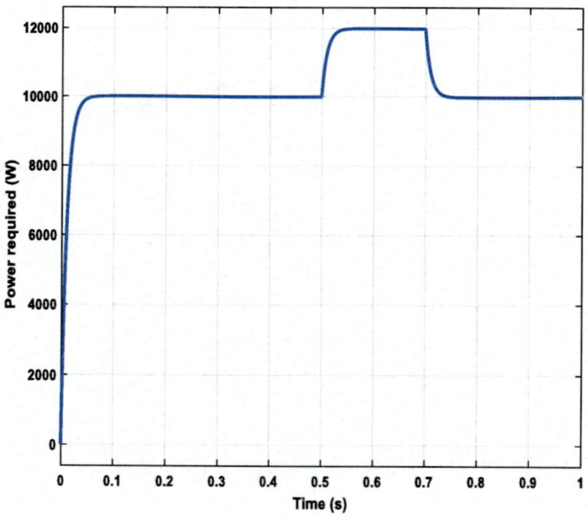

Fig. 37 Power required

Fig. 38 Generated power

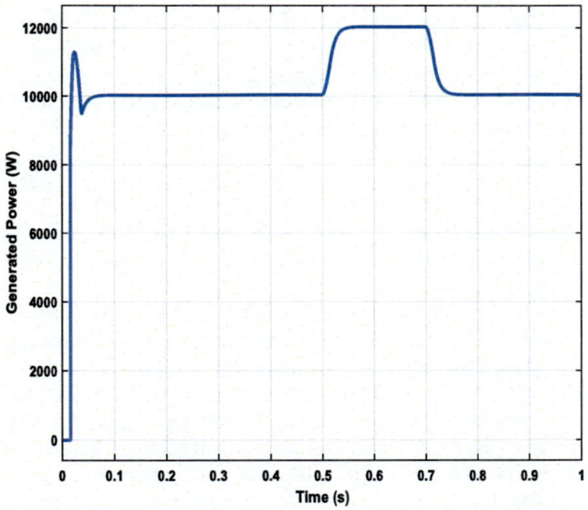

Table 2 Optimal gains of PI controller

S. No.	Optimization method	*Kp*	*Ki*
1	PSO	33.6	6.9
2	CS	43.7	10.8
3	EHO	31.3	7.9

References

1. Yildiz YE, Topal AO (2019) Large Scale Continuous global optimization based on micro differential evolution with local directional search. Inf Sci 477:533–544
2. Delghavi MB, Yazdani A (2019) Sliding-mode control of ac voltages and currents of dispatchable distributed energy resources in master-slave-organized inverter-based microgrids. IEEE Trans Smart Grid 10(1):980–991. https://doi.org/10.1109/TSG.2017.2756935
3. Ghafouri A, Milimonfared J, Gharehpetian GB (2015) Coordinated control of distributed energy resources and conventional power plants for frequency control of power systems. IEEE Trans Smart Grid 6(1):104–114. https://doi.org/10.1109/TSG.2014.2336793
4. Wang Y, Tan KT, So PL (2013) Coordinated control of battery energy storage system in a microgrid. In: 2013 IEEE PES Asia-Pacific power and energy engineering conference (APPEEC), pp 1–6. https://doi.org/10.1109/APPEEC.2013.6837211
5. Tan KT, So PL, Chu YC, Chen MZQ (2013) Coordinated control and energy management of distributed generation inverters in a microgrid. IEEE Trans Power Delivery 28(2):704–713. https://doi.org/10.1109/TPWRD.2013.2242495
6. Dou C, Lv M, Zhao T, Ji Y, Li H (2015) Decentralised coordinated control of microgrid based on multi-agent system. IET Gener Transm Distrib 9(16):2474–2484. https://doi.org/10.1049/iet-gtd.2015.0397
7. Rahman MS, Oo AMT (2017) Distributed multi-agent based coordinated power management and control strategy for microgrids with distributed energy resources. Energy Convers Manage 139:20–32, ISSN 0196-8904. https://doi.org/10.1016/j.enconman.2017.02.021
8. Baghaee HR, Mirsalim M, Gharehpetian GB, Talebi HA (2017) A decentralized power management and sliding mode control strategy for hybrid AC/DC microgrids including renewable energy resources. IEEE Trans Indus Inform. https://doi.org/10.1109/TII.2017.2677943

9. Mahmoud MS, Alyazidi NM, Abouheaf Mi (2017) Adaptive intelligent techniques for micro-grid control systems: a survey. Int J Electric Power Energy Syst 90:292–305, ISSN 0142-0615. https://doi.org/10.1016/j.ijepes.2017.02.008
10. Gui Y, Wei B, Li M, Guerrero JM, Vasquez JC (2018) Passivity-based coordinated control for islanded AC microgrid. Appl Energy 229:551–561, ISSN 0306-2619. https://doi.org/10.1016/j.apenergy.2018.07.115
11. Sabu NA, Batri K (2020) Power and area-efficient register designs involving EHO algorithm. Circuit World 46(2):93–105. https://doi.org/10.1108/CW-07-2019-0077
12. Soleimanian Gharehchopogh F, Haggi S (2020) An optimization K-modes clustering algorithm with elephant herding optimization algorithm for crime clustering. J Adv Comput Eng Technol 6(2):79–90
13. Li W, Wang GG, Alavi AH (2020) Learning-based elephant herding optimization algorithm for solving numerical optimization problems. Knowl Based Syst 195:105675, ISSN 0950-7051. https://doi.org/10.1016/j.knosys.2020.105675
14. Houssein EH, Saad MR, Hashim FA, Shaban H, Hassaballah M (2020) Lévy flight distribution: a new metaheuristic algorithm for solving engineering optimization problems. Eng Appl Artif Intell 94:103731, ISSN 0952-1976. https://doi.org/10.1016/j.engappai.2020.103731
15. Li J, Guo L, Li Y, Liu C (2019) Enhancing elephant herding optimization with novel individual updating strategies for large-scale optimization problems. Mathematics 7:395. https://doi.org/10.3390/math7050395
16. Meena NK, Parashar S, Swarnkar A, Gupta N, Niazi KR (2018) Improved elephant herding optimization for multiobjective DER accommodation in distribution systems. IEEE Trans Ind Inform 14:1029–1039
17. Jayanth J, Shalini VS, Ashok Kumar T, Koliwad S (2019) Land-use/land-cover classification using elephant herding algorithm. J Indian Soc Remote Sens. https://doi.org/10.1007/s12524-018-00935-x
18. Correia SD, Beko M, da Silva Cruz LA, Tomic S (2018) Elephant herding optimization for energy-based localization. Sensors 18:2849
19. Jafari M, Salajegheh E, Salajegheh J (2018) An efficient hybrid of elephant herding optimization and cultural algorithm for optimal design of trusses. Eng Comput Ger. https://doi.org/10.1007/s00366-018-0631-5
20. Sarwar MA, Amin B, Ayub N, Faraz SH, Khan SUR, Javaid N (2018) Scheduling of appliances in home energy management system using elephant herding optimization and enhanced differential evolution. In: Barolli L, Woungang I, Hussain O (eds) Advances in intelligent networking and collaborative systems. INCoS 2017. Lecture notes on data engineering and communications technologies, vol 8. Springer, Cham. https://doi.org/10.1007/978-3-319-65636-6_12
21. Rahim S, Javaid N, Ahmad A, Khan SA, Khan ZA, Alrajeh N, Qasim U (2016) Exploiting heuristic algorithms to efficiently utilize energy management controllers with renewable energy sources. Energy Build 129:452–470
22. Kanojiya RG, Meshram PM (2012) Optimal tuning of PI controller for speed control of DC motor drive using particle swarm optimization. Int Conf Adv Power Convers Energy Technol (APCET) 2012:1–6. https://doi.org/10.1109/APCET.2012.6302000
23. El-Naggar MF, Mosaad MI, Hasanien HM, AbdulFattah TA, Bendary AF (2021) Elephant herding algorithm-based optimal PI controller for LVRT enhancement of wind energy conversion systems. Ain Shams Eng J 12(1):599–608, ISSN 2090-4479. https://doi.org/10.1016/j.asej.2020.07.013
24. Wang G-G, Tan Y (2019) Improving metaheuristic algorithms with information feedback models. IEEE Trans Cybern 49:542–555
25. Al-Saedi W, Lachowicz SW, Habibi D, Bass O (2012) Power quality enhancement in autonomous microgrid operation using particle swarm optimization. Int J Electri Power Energy Syst 42(1):139–149, ISSN 0142-0615. https://doi.org/10.1016/j.ijepes.2012.04.007
26. Gupta M, Tiwari PM, Viral RK, Shrivastava A (2019) Performance enhancement of a grid-connected micro grid system using PSO optimisation technique. In: 2019 international conference on computing, communication, and intelligent systems (ICCCIS), pp 110–115. https://doi.org/10.1109/ICCCIS48478.2019.8974480

27. Kumar M, Sandhu KS, Kumar A (2014) Simulation analysis and THD measurements of amalgamated PV and wind as hybrid system connected to grid. In: 2014 IEEE 6th India international conference on power electronics (IICPE):1–6. https://doi.org/10.1109/IICPE.2014.7115779
28. Ibrahim H, Anani N (2017) Variations of PV module parameters with irradiance and temperature. Energy Procedia 134:276–285, ISSN 1876-6102. https://doi.org/10.1016/j.egypro.2017.09.617

Coordinated Strategy of Ultra-Capacitors and UPFC for LFC of Dual Area Conventional System Having Classical PID Controller with Set Point Filter

CH. Naga Sai Kalyan, Chintalapudi V. Suresh, U. Ramanaiah, B. Srikanth Goud, and C. H. Rami Reddy

Abstract This paper introduces the design of classical PID with set point filter (*N*) PIDN controller making use of seagull optimization algorithm (SOA) subjected to squared error over integral (ISE) function for power system load frequency control (LFC). For this purpose, a realistic system of dual area conventional units is conceived and analyzed by laying load disturbance in area-1 of 1% step load (1% SLP). However, the efficacy of different time domain objective functions is tested and ISE is proven as the best among them. Moreover, the devices of ultra-capacitors (Ucs) are placed in dual areas, and unified power flow controller (UPFC) is incorporated with tie-line to further diminish the deviations in system responses.

Keywords Seagull optimization · PIDN controller · 1%SLP · Ucs · UPFC

1 Introduction

The development of the country is measured in terms of gross domestic product (GDP). Directly or indirectly this GDP relates to the setup of industrialization thereby necessary of more electrical energy in every country. The need of more electrical

CH. Naga Sai Kalyan (✉) · C. V. Suresh
Electrical and Electronics Engineering Department, Vasireddy Venkatadri Institute of Technology, Namburu 522508, India
e-mail: kalyanchallapalli@gmail.com

U. Ramanaiah
Electrical and Electronics Engineering Department, NIT Jamshedpur, Jamshedpur 831014, India

B. Srikanth Goud
Electrical and Electronics Engineering, Anurag University, Venkatapur, Ghatkesar, Medchal, Telangana 500 088, India
e-mail: srikanth.b@anuraghyd.ac.in

C. H. Rami Reddy
Electrical and Electronics Engineering, Malla Reddy Engineering College (A), Maisammaguda, Telangana 500100, India

K. Namrata et al. (eds.), *Smart Energy and Advancement in Power Technologies*, Lecture Notes in Electrical Engineering 926,
https://doi.org/10.1007/978-981-19-4971-5_3

energy always strives the power generating utilities to incorporate new plants with the existing grid becoming the system more and more complex. Operation of the most complex system is not easy task. However, the efficient and effective operation of the interconnected system are attained up on maintaining the real power generation with the demand. Otherwise, this mismatch will directly affect the speed of the alternator in the powerhouse thereby system frequency. This task will be automatically handled by the load frequency controller.

Confessed from the literature survey that, control techniques presented by the researchers are tested on several test system models that are mentioned in [1] in detail. Different conventional, intelligent, combination of intelligent and traditional control schemes are presented in the LFC study and some of them are provided here. Conventional PID/PIDN/PIDD [2], degree of freedom (DOF) 2DOFPID/3DOFPID [3], neural network (NN)-based schemes [4], fractional order (FO) FOPI/FOPID [5], fuzzy and modified fuzzy controllers are reported in literature [6]. However, designing of NN and fuzzy controllers requires and involves many assumptions and approximations may pretend controller efficacy [7]. So, a classical controller with set point filter PIDN is chosen in this work as regulator which is simple to design and can exhibit efficacy in performance with robustness.

Moreover, different nature-inspired algorithms are employed by the researchers to optimize the regulator parameters and some of them are particle swarm optimization (PSO) [2], Ant-lion optimizer (ALO) [8], imperialist competitive approach (ICA) [6], Harris-hawks optimizer (HHO) [9], simulated annealing (SA) [10], hybrid artificial field (HAEFA) [11], water cycle algorithm (WCA) [12], sine–cosine algorithm (SCA) [5], differential evolution (DE) [13], whale optimizer (WO), grasshopper optimization (GO) [3], gray wolf optimizer (GWO) [14], other combined algorithms like DE-AEFA [15, 16] and so on. However, each approach has its own merits and drawbacks which open the scope for implementation of new algorithms for optimizing the regulators in LFC study. This inspires the researchers to implement a new nature-inspired SOA approach in this work which is a maiden attempt in the power system optimization to the best of knowledge.

The contributions of this work are:

(a) Different time domain indices performances are assessed.
(b) Implementing SOA optimization for power system control is a maiden attempt.
(c) SOA tuned PIDN is implemented successfully for LFC.
(d) UPFC and UCs coordinated strategy are enforced with the system to attain boost up in performance.

2 Test System Model

Model of power system furnished in Fig. 1 is a conventional realistic plant considered in this paper for investigation. Both the thermal and hydro units have been incorporated with constraints of the rate at which power generation is limiting and is considered to conduct analysis close to realistic practice. Total investigation is

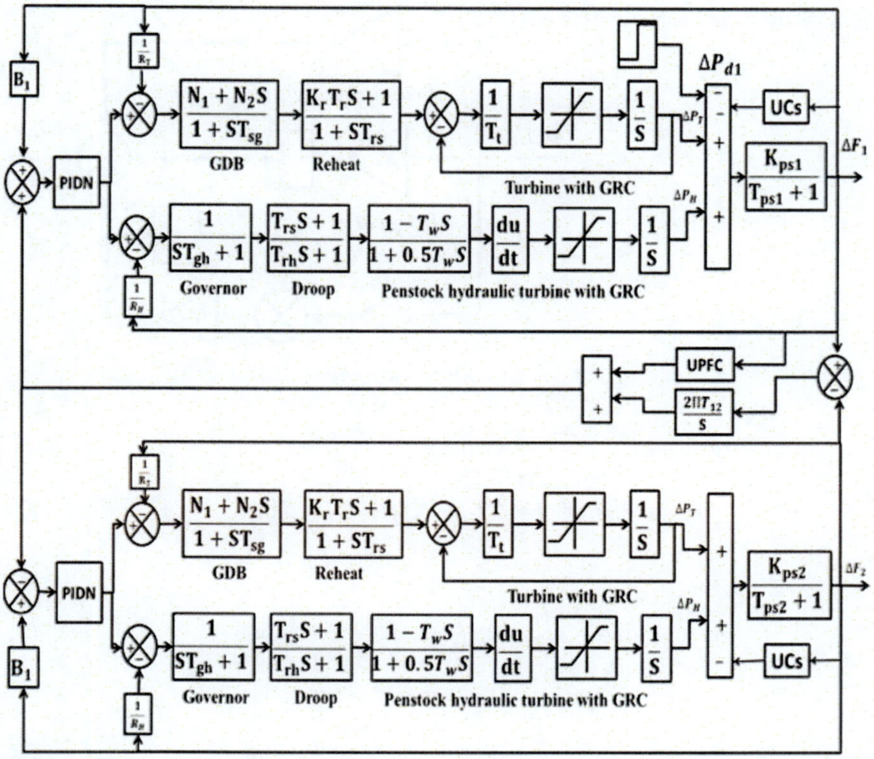

Fig. 1 Dual area conventional system with UCs and UPFC devices

carried up on laying area-1 with a disturbance of 1%SLP. Model shown in Fig. 1 is constructed in the environment of MATLAB/SIMULINK (R2016a) version. The required parameters are chosen directly from [14].

3 Objective Function

Selection of appropriate objective function as a constraint while tuning the regulator parameters to obtain optimal values is having a lot of weightage in designing the controller. The optimization in this work is the minimization function. In general, a better index will definitely escalate the performance of the regulator. Various time domain functions that are evaluated in this present work are modeled in Eqs. (1–4) [17]. PIDN regulator structure employed in this work is depicted in Fig. 2.

$$J_{\mathrm{ISE}} = \int_{0}^{T_{\mathrm{sim}}} (\Delta f_1^2 + \Delta f_2^2 + \Delta P_{\mathrm{tie}_{12}}^2)\mathrm{dt} \tag{1}$$

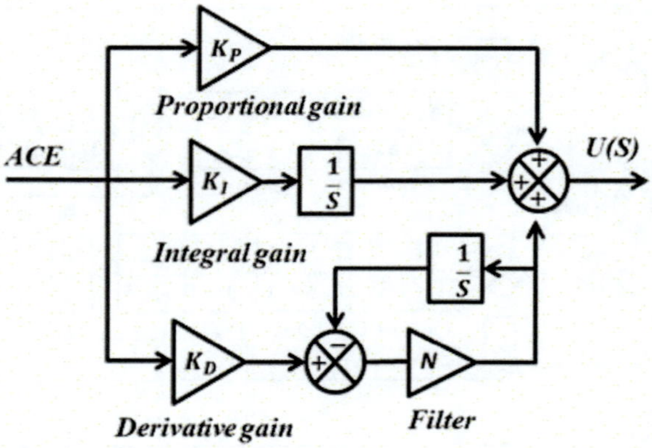

Fig. 2 Structure of PIDN

$$J_{\mathrm{IAE}} = \int_{0}^{T_{\mathrm{sim}}} (\Delta f_1 + \Delta f_2 + \Delta P_{\mathrm{tie}_{12}})\mathrm{dt} \tag{2}$$

$$J_{\mathrm{ITSE}} = \int_{0}^{T_{\mathrm{sim}}} t.(\Delta f_1^2 + \Delta f_2^2 + \Delta P_{\mathrm{tie}_{12}}^2)\mathrm{dt} \tag{3}$$

$$J_{\mathrm{ITAE}} = \int_{0}^{T_{\mathrm{sim}}} t.(\Delta f_1 + \Delta f_2 + \Delta P_{\mathrm{tie}_{12}})\mathrm{dt} \tag{4}$$

4 Coordinated Strategy

In this paper, UPFC and UCs coordinated strategy are implemented to further enhance system performance under sudden peak disturbances. UC's are also named as super capacitors having capability to store bulk power because of their greater plate surface area. Since, UCs possess the characteristics of lengthy cycle time, greater efficacy and high power density makes them comfortable for applications of peak demands. Even though these are expensive, the benefit of low maintenance makes them wide spread of application during peak disturbances. The transfer function model of UC implemented is given in Eq. (5) [13].

$$G_{\mathrm{UC}} = \frac{K_{\mathrm{UC}}}{1 + \mathrm{ST}_{\mathrm{UC}}} \tag{5}$$

UPFC is one of the multivariable comprehensive FACTS devices and best suitable for maintaining stability by controlling power flow. It is capable of both absorbing

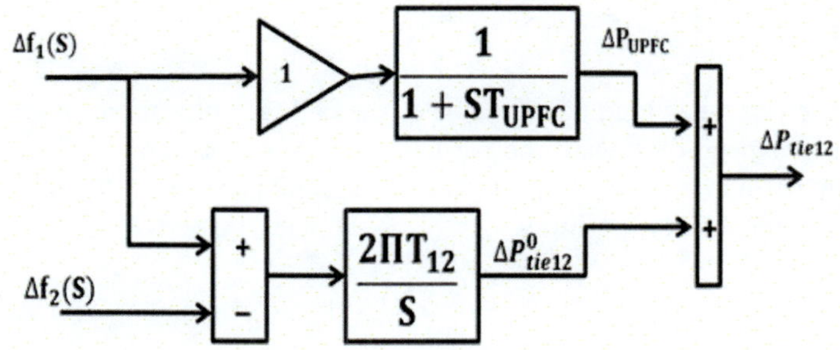

Fig. 3 UPFC damping controller

and injecting power so that the mismatch in real power with system can be minimized. The modeling of device UPFC is given detailed in [14] and its structure implemented in this paper is shown in Fig. 3.

5 SOA Algorithm

Seagulls are the birds usually live at the sea shores can be widely found all over the world, which are scientifically called as Laridae. Seagulls are likely to be the omnivorous, which feed on fishes, insects and earthworms. The species of the seagulls is differentiated in wide ranges depending up on their length and weight. These are very highly intelligent birds, and their body is covered with white feathers. The unique feature of the seagulls is they can drink salt water that no other birds can do. In contrast, they will flush out the salt through the glands beneath the eyes from the body. Seagulls attracts the fish and other insects by sprinkling the bread crumbs which are collected from environment and can also make the sound that exactly resembles to that coming during rain to attract the earthworms which are hiding in the soil.

In general, like other birds, seagulls are also live in colonies and making use of their behavioral intelligence in attacking the prey. This SOA algorithm was proposed by Gaurav [18] in the year 2019, based on the migration and attacking behavior of the seagulls. Seagulls exhibits solely behavior at the time of migration from one place to another to avoid the chances of collision with another birds in the colony. Moreover, during migration period, these birds regularly attack other sea birds by making use of the spiral shape movement strategy. This behavior of seagulls is implemented mathematically in minimizing the objective function in this present work. The following are the mathematical equations that are necessary to design the algorithm in (.m file) format.

5.1 Exploration Phase

During this phase, searching agents satisfies the collision obviations at the time of moving to new position toward the particles with best position. Collision among neighboring particle is avoided by inserting the special parameter 'N' in calculating the searching agent's new position.

$$\vec{F}_S = N\,\chi\,\vec{D}_S(k) \tag{6}$$

Search agents new position is represented with $\vec{F}_S$, and the current position is designated with $\vec{D}_S$ and 'k' represents the current iteration. The special parameter 'N' that facilitating the collision avoidance be modeled as follows:

$$N = E_c - (k * (E_c/\text{Max.Iter})) \tag{7}$$

Parameter 'N' is chosen as '2' in this paper to regulate the variable frequency that can be lessened from E_c to 0 linearly. Thereby, the search agents are attempting to move toward the direction of best particles by using

$$\vec{M}_S = A\chi\,(\vec{P}_{\text{bs}}(k) - \vec{D}_S(k)) \tag{8}$$

Each search agent position on the way toward position of best agent $\vec{P}_{\text{bs}}$ is designated with $\vec{M}_S$. The parameter 'A' is to be initialized randomly in order to insert the balance among the phases of exploration and exploitation be calculated as

$$A = 2 * N^2 * \text{rand}() \tag{9}$$

Finally, the search agent's position will be updated toward the best position is calculated as

$$\vec{R}_S = \left|\vec{F}_S + \vec{M}_S\right| \tag{10}$$

5.2 Exploitation Phase

Seagulls regularly alter their attacking angle and speed at the time of migration based on their experience. In three-dimensional plane, these behaviors can be represented as

$$P' = r * \text{Cos}(x) \tag{11}$$

$$Q' = r * \mathrm{Sin}(x) \tag{12}$$

$$R' = r * x \tag{13}$$

Spiral movement radius is indicated with 'r' and 'x' is random number from (0–2). Up on saving the best solution, the position of remaining search agents will be updated as

$$\vec{D}_S(k) = (\vec{M}_S * P' * Q' * R') + \vec{P}_{bs}(k) \tag{14}$$

Keeping these intelligent searching strategy of seagulls in mind, motivates the authors to implement for the study of LFC for the first time as this SOA approach is already implemented in other complex engineering optimization issues and proved to be one of the best suitable.

6 Simulation Results

6.1 *Scenario-1: System Performance with SOA Tuned PIDN Regulator Based on Performance Indices*

In this case, the performances of different time domain indices are evaluated up on optimizing the PIDN regulator using SOA algorithm. The responses for this case are visualized in Fig. 4, by applying area-1 with 1%SLP. The parameters of PIDN based on SOA approach are noted in Table 1 and the time taken by the responses to reach steady state is mentioned in Table 2.

It is pointed out from Fig. 4 and Table 2 that, out of different time domain, objective functions ISE plays a salient role balancing the mitigation of peak deviations and bringing responses to steady condition.

6.2 *Scenario-2: System Performance with UPFC and UCs Coordinated Strategy*

From the previous case, it is exonerated that the ISE shows the predominant performance compared to other indices urges to implement this function further. Later, the UCs are installed in both areas and the system responses under same load disturbances are retrieved and rendered in Fig. 5. Further, the FACTS device of UPFC is installed with the line as the UCs are in still incorporation regulator and the responses

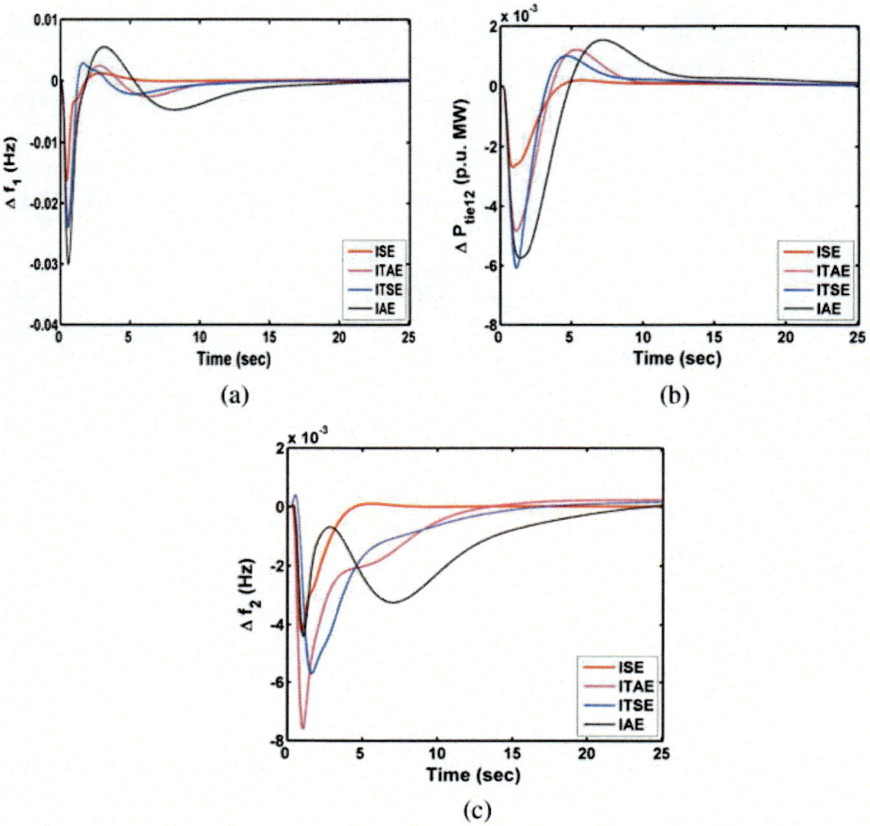

Fig. 4 Responses for scenario-1 **a** Δf_1, **b** ΔP_{tie12}, **c** Δf_2

Table 1 Optimal values of PIDN regulator tuned with SOA

Indices	IAE	ITSE	ITAE	ISE
Area-1	$K_P = 2.366$ $K_I = 0.098$ $K_D = 1.2057$ $N = 147.05$	$K_P = 2.399$ $K_I = 0.126$ $K_D = 1.093$ $N = 146.98$	$K_P = 2.558$ $K_I = 0.266$ $K_D = 0.963$ $N = 147.23$	$K_P = 2.416$ $K_I = 0.188$ $K_D = 0.980$ $N = 144.78$
Area-2	$K_P = 2.317$ $K_I = 0.124$ $K_D = 0.906$ $N = 148.04$	$K_P = 2.142$ $K_I = 0.077$ $K_D = 0.899$ $N = 145.71$	$K_P = 2.204$ $K_I = 0.152$ $K_D = 0.635$ $N = 146.32$	$K_P = 1.927$ $K_I = 0.106$ $K_D = 0.155$ $N = 147.26$
Function value * 10^{-3}	11.637	6.204	2.967	1.645

Table 2 Settling time of responses for scenario-1 and scenario-2

Parameters	Scenario-1				Scenario-2	
	IAE	ITSE	ITAE	ISE	With UCs only	With UCs and UPFC
Δf_1	24.57	19.9	17.48	9.846	8.341	4.982
ΔP_{tie12}	23.75	21.18	17.84	10.340	8.574	7.703
Δf_2	26.10	24.92	23.64	8.794	7.733	6.847

are compared in Fig. 5, under the supervision of SOA tuned PIDN, and the corresponding numerical results are noted in Table 2. Finally, it is concluded that by incorporating the UCs and UPFC devices that there will be a remarkable enhancement in the power system dynamic behavior under disturbing conditions.

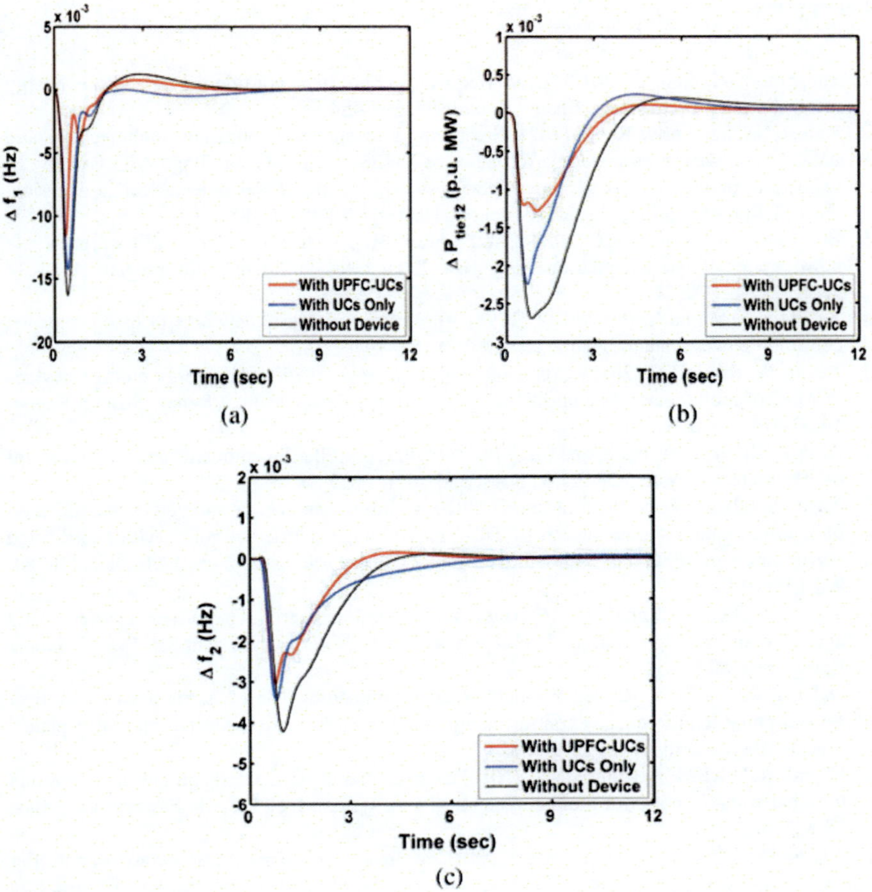

Fig. 5 Responses for scenario-2 **a** Δf_1, **b** ΔP_{tie12}, **c** Δf_2

7 Conclusion

This paper assessed the performance of different time domain functions like ITSE, ITAE, IAE and ISE in tuning the parameters of PIDN controller with SOA algorithm for LFC of realistic conventional system. Analysis revealed the efficacy of ISE function over others. Dynamical behavior of the system is analyzed for a change in load of 1% in area-1. Further, the coordinated strategy of UPFC and UCs is implemented as territorial control approach under the monitoring of SOA-based PIDN regulation. Simulation results reveals that, this territorial control strategy boosts up the system performance to some extent by diminishing the peak deviations and acquiring the responses to steady condition in less time.

References

1. Tungadio DH, Sun Y (2019) Load frequency controllers considering renewable energy integration in power system. Energy Reports 05:436–454
2. Kalyan NS, Sambasiva Rao G (2021) Stabilizing frequency and voltage in combined LFC and AVR system with coordinated performance of SMES and TCSC. In: Singh AK, Tripathy M (eds) Control applications in modern power system. Lecture notes in electrical engineering, vol 710. Springer, Singapore. https://doi.org/10.1007/978-981-15-8815-0_4
3. Biswas S, Roy PK, Kalyan C (2020) FACTS-based 3DOF-PID controller for LFC of renewable power system under deregulation using GOA. IETE J Res (2020). https://doi.org/10.1080/03772063.2020.1870874
4. Beaufays F, Magid Y, Bernard W (1994) Application of neural networks to load frequency control in power systems. Neural Netw 7(01):183–194
5. Tasnin W, Saikia LC (2018) Comparative performance of different energy storage devices in AGC of multi-source system including geothermal power plant. J Renew Sustain Energy 10:024101
6. Arya Y (2019) A new optimized fuzzy FOPI-FOPD controller for automatic generation control of electric power systems. J Franklin Inst 356:5611–5629
7. Naga sai kalyan CH, Sambasiva Rao G (2020) Combined frequency and voltage stabilization of multi-area multisource system by DE-AEFA optimized PID controller with coordinated performance of IPFC and RFBs. Int J Ambient Energy. https://doi.org/10.1080/01430750.2020.1860130
8. Raju M, Saikia LC, Sinha N (2016) Automatic generation control of a multi-area system using ant lion optimizer algorithm based PID plus second order derivative controller. Electric Power Energy Syst 80:52–63
9. Dalia Y, Sudhakar B, Ahmed F (2020) Recent methodology based harris hawks optimizer for designing load frequency control incorporated in multi interconnected renewable plants. Sustain Energy Grids Netw 22:100352
10. Chandrakala KRM, Balamurugan S (2016) Simulated annealing based optimal frequency and terminal voltage control of multi-source multi area system. Int J Electric Power Energy Syst 78:823–829
11. Naga sai kalyan Ch, Sambasiva rao G (2020) Coordinated SMES and TCSC damping controller for load frequency control of multi area power system with diverse sources. Int J Electric Eng Inform 12(04):747–769
12. Mohit J, Asha R, Nikhil P, Vijander S, Alok P (2019) Design of fractional order 2-DOF PI controller for real-time control of heat flow experiment. Eng Sci Technol Int J 22(01):215–228

13. Naga sai kalyan Ch, Sambasiva rao G (2020) Performance comparison of various energy storage devices in combined LFC and AVR of multi area system with renewable energy integration. Int J Renew Energy Res 10(02):933–944
14. Naga sai kalyan Ch (2021) UPFC and SMES based coordinated control strategy for simultaneous frequency and voltage stability of an interconnected powers system. In: 1st international conference on power electronics and energy. https://doi.org/10.1109/ICPEE50452.2021.9358576
15. Naga sai kalyan Ch, Sambasiva rao G (2020) Frequency and voltage stabilization in combined load frequency control and automatic voltage regulation of multi area system with hybrid generation utilities by AC/DC links. Int J Sustain Energy 39(10):1009–1029. https://doi.org/10.1080/14786451.2020.1797740
16. Naga sai kalyan Ch, Sambasiva rao G (2021) Impact of communication time delays on combined LFC and AVR of a multi-area hybrid system with IPFC-RFBs coordinated control strategy. Prot Control Mod Power Syst 6, 7. https://doi.org/10.1186/s41601-021-00185-z
17. Naga sai kalyan C, Sambasiva rao G (2021) Performance index based coordinated control strategy for simultaneous frequency and voltage stabilization of multi-area interconnected system. In: Singh AK, Tripathy M (eds) Control applications in modern power system. Lecture notes in electrical engineering, vol 710. Springer, Singapore. https://doi.org/10.1007/978-981-15-8815-0_4
18. Dhiman G, Kumar V (2019) Seagull optimization algorithm: theory and its applications for large-scale industrial engineering problems. Knowl Based Syst 165:169–196

Implementation of Peak Shaving Algorithm in an Islanded Microgrid for Economic Power Consumption

Viswanathan Ganesh, S. Senthilmurugan, and Rathinam Ananthanarayanan

Abstract The objective and purpose of this study is to implement a peak shaving algorithm in an islanded microgrid for economic power consumption. The simulation of this study has been created through MATLAB/Simulink. This system works for different peak demands and can easily be manipulated, providing us a dynamic approach to the system for maximum efficiency and maximum power generation for the required demand. Added to this, we have also included grid-tied solar panels which are renewable and eco-friendly. The results from the simulation show us clearly that the peak shaving algorithm gets activated as soon as the supply surpasses the peak demand. When this activation happens, the grid is islanded and it goes into self-generation mode. The current use of the grids is aged and is not able to support the current population from time to time. This paper will prove to be a viable, cost-effective and simple solution for the power stations. The simulation, functioning, observation, inference, advantages and disadvantages have been completely provided in this paper with the help of reference papers which have been mentioned towards the end.

Keywords Intentional islanding · Economic solution · Peak shaving algorithm

1 Introduction

With the substantial increase in the world of technology and with equal but negative increase in non-renewable resources (like coal, crude oil, etc.), people are trying to

V. Ganesh (✉)
Chalmers University of Technology, Gothenburg, Sweden
e-mail: viswanathanganesh1999@gmail.com

S. Senthilmurugan · R. Ananthanarayanan
Department of Electrical and Electronics Engineering, SRM Institute of Science and Technology, Kattankulathur, Tamil Nadu, India
e-mail: senthils8@srmist.edu.in

R. Ananthanarayanan
e-mail: rathinaa@srmist.edu.in

K. Namrata et al. (eds.), *Smart Energy and Advancement in Power Technologies*,
Lecture Notes in Electrical Engineering 926,
https://doi.org/10.1007/978-981-19-4971-5_4

find a quick but a firmly effective renewable solution for this problem. One of the key notes we have to keep in mind is that it is nearly impossible to totally changeover the existing topology. So, we have to make do with the current topology and find an almost perfect solution [1]. The current scenario of the electrical board has been highly towards the point of maximum stress over generation of power. This is because the increasing population of this country has increased the demand of power towards all households and industries. This will result in constant occurrence of faults over the grids and will result in a lot of chaos for the consumers. Considering all of these factors, we have decided to provide an economical and optimal solution [2]. Given a variety of grids at our disposal, we have chosen microgrids as it is the most efficient and economical type of grids which can operate in ranges of megawatts [3]. Microgrids are designed in such a way that they can operate for both base and peak load. Even though we operate in the peak load for a very short time, this may introduce higher stress to the consumer as it might be more than the demand level and may result in a penalty for the consumer [4, 13, 14].

2 Microgrid

A microgrid is an energy transmission and distribution network that relies on the loads that are present in it. It consists of AC or DC or AC/DC generating systems for generating the power for the consumers, circuit breakers as safety devices and to island the circuit whenever the trip signal is given, buses to transmit the required current and several other components [5]. This makes the microgrid independent of other external components and can function on its own. Further below we have given the classification of microgrids with respect to their type of power source and have also listed the advantages and disadvantages of each [6].

2.1 Types of Microgrid

AC Microgrid According to the types of microgrids, AC microgrid is the dominant of the lot. It directly connects to the distribution network. It uses an AC bus to connect to the distribution network, and the AC bus controls the connection and disconnection of the distribution network through the circuit breaker at the Power Control Centre. Since the AC microgrid is connected directly to the AC bus, there is no necessity of an inverter in the circuit. This also reduces the losses which occur while inversion of the supply. On the other hand, it is quite hard to control and operate during operation mode. Also due to dynamic changes in load, the frequency of the overall microgrid can be affected which may require additional control operations to maintain a constant frequency.

DC Microgrid The DC bus of a DC microgrid supplies to various DC loads of low power rating like cell phones, laptops, routers, etc. [15]. Even though the AC grid is much vastly used than DC, it has its own advantages over the AC type. This substantially increases the efficiency of the microgrid. While we need to do frequency stabilization in AC grids when faults arise, that procedure is not necessary in the DC grid since it operates with a DC waveform.

AC/DC Microgrid From the name itself, we might be able to deduce what it is. The AC/DC hybrid microgrid consists of both AC and DC subgrids interlinked with a bidirectional interlinking converter. Hybrid AC/DC microgrid enables the interconnection of different AC- and DC-based renewable energy resources and eliminates the usage of several converters due to the existence of individual AC and DC grids. In such a microgrid, where they co-exist. Though AC/DC microgrids can blend easily into the current scenario, they also produce several problems which are not easy to overcome.

- The presence of a global variable is not present in this microgrid to control the sharing of power, voltage and frequency modulation. This complicates the microgrid and hence makes it tough for the operators to handle.
- There is also the problem of finding a balance between voltage regulation and reactive power-sharing.

3 Solar Panels

Devices that take the input as photons (e.g. sunlight) and produce a DC current through the usage of photovoltaic cells are known as solar panels. The phenomenon that makes this possible is called as photoelectric effect. With respect to the microgrid, the solar panels can be connected in two ways:

3.1 Stand-Alone Solar Panels

A stand-alone solar panel operates individually and is not linked with any other circuit or grid. It is self-functioning and can produce power in the night when there is no source of natural illuminance. A battery is connected to the solar panel to store and provide the energy whenever needed. An electric system including one or more PV modules, cables, electrical components and a load range is a stand-alone photovoltaic system (PV). There are many applications of stand-alone solar panels. Some of them are:

- It is a perfect solution for rural areas and places where we can find the seldom usage of cable electricity

- It does not have to be attached to any roof or to the top of buildings for its usage. This makes it transportable and can be used in camps and other remote locations.

The space allocation of the energy storage system is 40% of the total solar system. Hence, there are losses in batteries due to the limited availability of recharge time and energy. Hence, a charge controller is required to control the over-charging and over-discharging of the batteries. In addition to that, an inverter is required to give the desired AC output.

3.2 Grid-Tied Solar Panels

It is a solar system that is directly connected to the grid with the help of a grid-tied inverter. This inverter helps us transmit power between the consumer and the grid to which the solar panel is connected and vice versa [7]. One of the main advantages of using a grid-tied solar is that it doesn't need any battery for storing energy and directly uses the main supply from the grid as a storage source. Using this as a plus point, the surplus power isn't wasted and is able to produce the required power too. Since the grid and solar are interconnected, they can support each other when the problem of supply shortage arises. For example, when the supply from the solar is short and is unable to produce the necessary power, it can import the shortfall from the grid.

Advantages of Grid Tie Solar

- Since there is no requirement of batteries and other equipment, it tends to be much more economically
- It is also an economical solution for the consumers because they don't have to be totally dependent on either of them.

Disadvantages of Grid Tie Solar

- Transportation of grid-tied solar is not possible making it hard for us to communicate
- The system will shut off when the grid goes down and leaves us without electricity. If the grid goes down, the grid-mounted equipment will immediately stop and immediately switch on when power is restored.

4 Simulation

4.1 Grid Isolation

When a grid or a part of a grid isolates itself with the help of circuit breakers, it is said to be in islanded mode [8]. This method is used in all power stations. When there

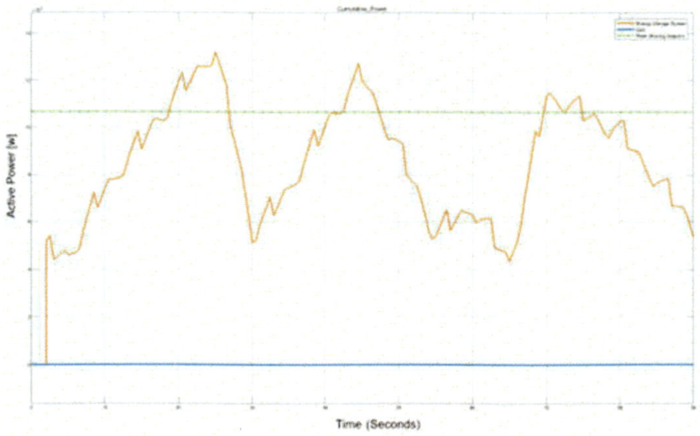

Fig. 1 Working of grid in islanded mode

is a fault or an overload in the supply, it may damage the components of the grid and may result in an electrical blackout which will also lead to a lot of unnecessary expenditures. If we sense this overload or fault and island the system, it can prevent the high current from reaching the components and will stay safe [9]. This small method will increase the efficiency and also make it cost-efficient. Figure 1 provides the load profile of the island while the grid is isolated, and the entire power is being diverted from the battery energy storage system (BESS).

4.2 Peak Shaving Algorithm

Every person has his/her own approach to the peak saving algorithm because it may vary for the application they might be using for. It is a very efficient and cost-saving method for the consumers who have high peak demands. The working of the peak shaving algorithm is quite simple. A consumer will always have a base demand and a peak demand. During the course of the day, the load will reach the peak demand maybe for only a few hours. But the problem arises when it crosses the peak demand. This will result in a penalty for the mass consumer and he will end up paying for the excess supply he utilized [10]. To avoid this, we use the peak shaving algorithm. Once the demand crosses the peak load, the main supply will be turned off and the battery with the help of the grid-tied solar panel starts to provide the power [11]. This makes it better for the consumer and can exploit the resources up to a large extent without paying any tariff. This is a huge relief for the consumer and can work much efficiently [12].

Figure 2 displays the effect of peak shaving algorithm during the time period 20 & 45 s. The amount of power required to be supplied to the consumer is being diverted from the BESS as peak limit power as been attained.

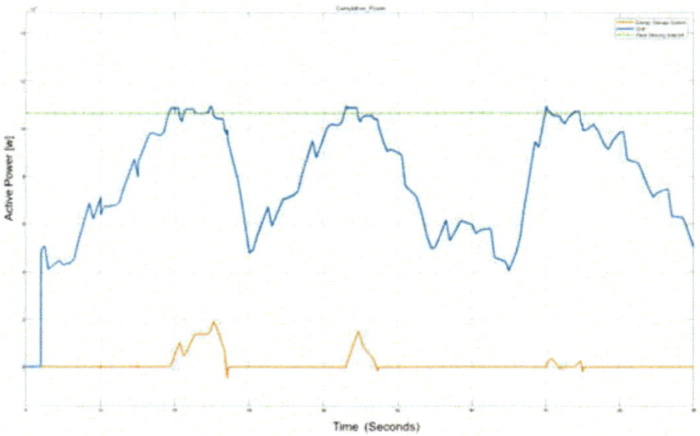

Fig. 2 Working of grid-connected mode

4.3 Circuit Breaker Working on Availability of Microgrid

To constantly maintain a high level of efficiency and monitor the functioning of the grids, we constantly have to take a set of fault analysis. In planning, protection, selection of equipment, and overall system reliability analysis, fault analysis is an extremely significant aspect. Below given is a table that explains the various combinations of faults which may occur in each microgrid. It also specifies the activation of the respective circuit breakers using the trip signal. When a fault occurs in a microgrid, it sends a signal called the trip signal to the circuit breaker. The circuit breaker in turn responds to this signal and isolates the microgrid from the rest of the system. This saves the components from the fault and also gives the station workers a safe base to work on the fault. Analysis of circuit breaker activation:

The below equations can be framed with the help of Table 1

$$\sum \mathrm{CB}_1 = (1, 5, 6, 7, 11, 12, 13, 15, 16, 20, 21, 22, 26, 27, 28, 30)$$

$$\sum \mathrm{CB2} = (2, 5, 8, 9, 11, 13, 14, 15, 17, 20, 23, 24, 26, 28, 29, 30)$$

$$\sum \mathrm{CB}_3 = (3, 6, 8, 11, 12, 14, 15, 18, 21, 23, 26, 27, 29, 30)$$

$$\sum \mathrm{CB}_4 = (4, 7, 9, 10, 12, 13, 14, 15, 19, 22, 24, 25, 27, 28, 29, 30)$$

$$\sum \mathrm{MCB} = (1, 2, 3, 4, 5, 6, 7, 8, 9, 10, 11, 12, 13, 14, 15)$$

Table 1 Combination of circuit breaker

S. No.	Main availability	MG availability	CB1	CB2	CB3	CB4	MCB
1	Yes	1	1	0	0	0	1
2	Yes	2	0	1	0	0	1
3	Yes	3	0	0	1	0	1
4	Yes	4	0	0	0	1	1
5	Yes	1,2	1	1	0	0	1
6	Yes	1,3	1	0	1	0	1
7	Yes	1,4	1	0	0	1	1
8	Yes	2,3	0	1	1	0	1
9	Yes	2,4	0	1	0	1	1
10	Yes	3,4	0	0	0	1	1
11	Yes	1,2,3	1	1	1	0	1
12	Yes	1,3,4	1	0	1	1	1
13	Yes	1,2,4	1	1	0	1	1
14	Yes	2,3,4	0	1	1	1	1
15	Yes	1,2,3,4	1	1	1	1	1
16	No	1	1	0	0	0	0
17	No	2	0	1	0	0	0
18	No	3	0	0	1	0	0
19	No	4	0	0	0	1	0
20	No	1,2	1	1	0	0	0
21	No	1,3	1	0	1	0	0
22	No	1,4	1	0	0	1	0
23	No	2,3	0	1	1	0	0
24	No	2,4	0	1	0	1	0
25	No	3,4	0	0	0	1	0
26	No	1,2,3	1	1	1	0	0
27	No	1,3,4	1	0	1	1	0
28	No	1,2,4	1	1	0	1	0
29	No	2,3,4	0	1	1	1	0
30	No	1,2,3,4	1	1	1	1	0

Figure 3 provides the normal working conditions of all the four MG. Hence, all the circuit breakers are closed. We can observe the equal amount of load being shared AC cross all the four MG's. Figure 6 provides the circuitry connection of the various MG's and main grid with the peak shaving algorithm placed between the main grid and island.

$$\sum CB_1 = \sum CB_2 = \sum CB_3 = \sum CB_4 = \sum MCB = (15)$$

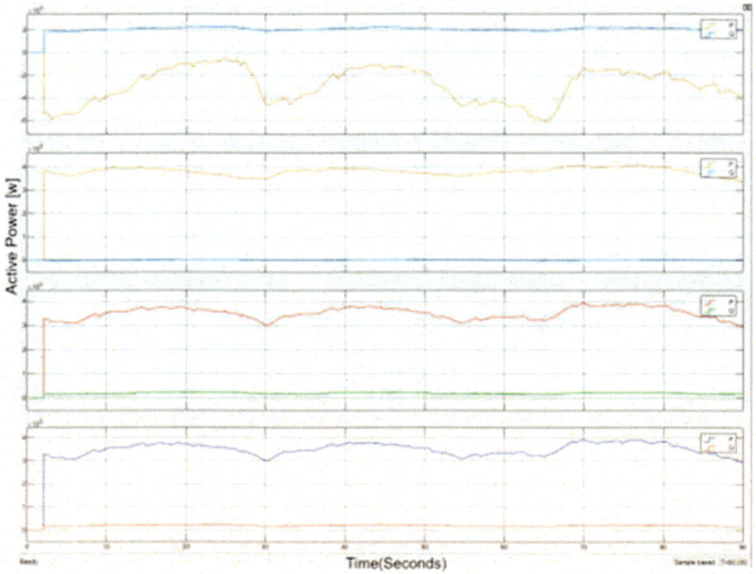

Fig. 3 Normal functioning of the main grid

Figure 4 provides a graph with load sharing, when MG1 has been isolated due to the unavailability/fault issues. Therefore, the CB1 has been opened to isolate the respective MG. Upon successful isolation of MG1, we can observe the other MGs have equally shared the load due to the property of equal load allocation.

$$\sum CB_1 \neq (15) \quad \& \quad \sum CB_2 = \sum CB_3 = \sum CB_4 = \sum MCB = (15)$$

Figure 5 provides a graph with load sharing, when MG1 and MG2 have been isolated due to the unavailability/fault issues. Therefore, the CB1 and CB2 have been opened to isolate the respective MG. Upon successful isolation of MG1, we can observe the other MGs have equally shared the load due to the property of equal load allocation (Fig. 6).

$$\sum CB_1 = \sum CB_2 \neq (10) \quad \& \quad \sum CB_3 = \sum CB_4 = \sum MCB = (10)$$

4.4 *Circuit Breaker Working on Availability of Microgrid*

The designed topology has a combination of main three phase supply and a grid-tied solar with a combination of grid-tied solar panels. On the island side, the load parameters are designed for a dynamic live load profile and a load side solar including

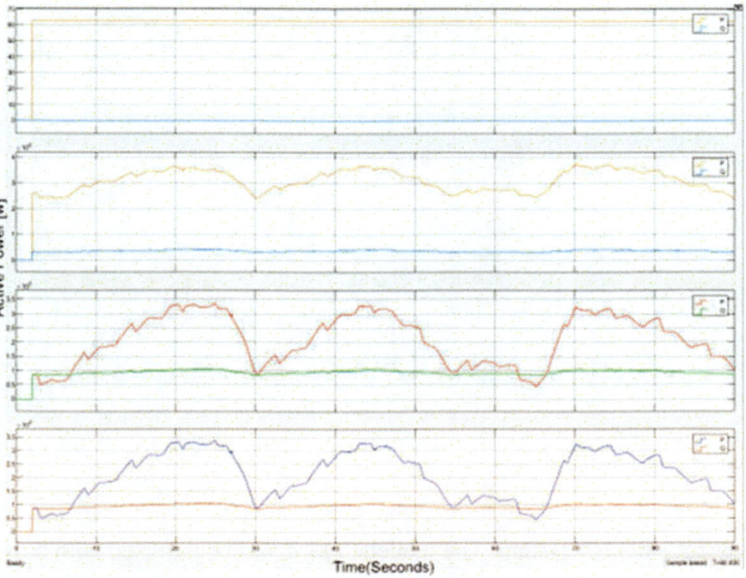

Fig. 4 Functioning of the grid when fault occurs in MG1

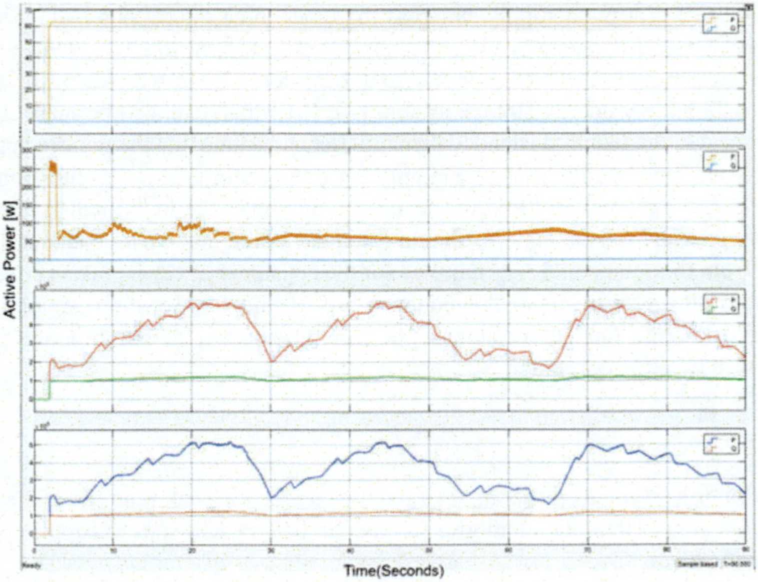

Fig. 5 Functioning of the grid when fault occurs in two grids

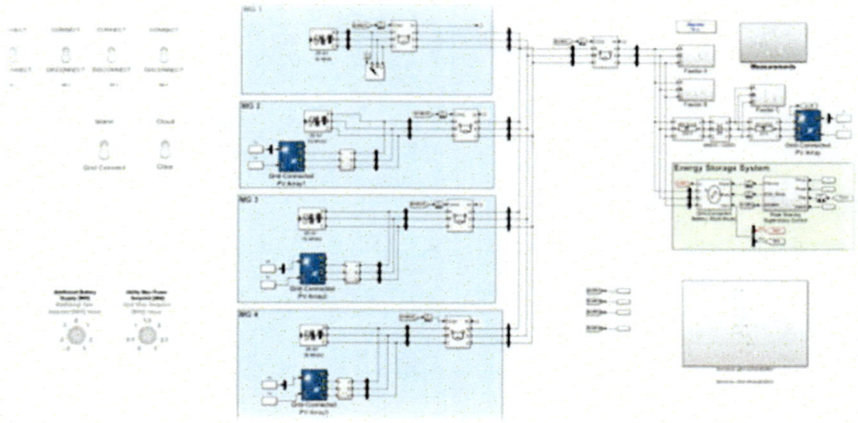

Fig. 6 Working simulation diagram

battery energy storage system. The modelled load was constructed with a dynamic peak load so that it is easier for us to examine the working of the peak shaving algorithm for different values of the peak demand. Moreover, the additional advantage of the dynamic peak load is that we can manipulate it even while running the simulation. MG1 consists of a main supply, whereas the other three microgrids (MG2, MG3, MG4) have a grid-tied solar system. This has been designed to also give us a view on how the simulation will work in a normal basis too. The island side of the simulation has a battery energy storage system (BESS) which can also be manoeuvred with respect to the required islanded demand. Let us assume that the required power, which is the load profile, is P, the supply from the main grid is $P1$ and from the battery is $P2$. The load profile is the scalar summation of the individual loads from the main grid and the battery. We should also take a note that we are neglecting the supply from the solar panel. The demand for the consumer always remains constant. So, the varying parameters of this equation are $P1$ and $P2$. So, if the supply from the grid decreases, the battery must support the grid supply to maintain a constant power supply to the consumer.

$$P = P_1 + P_2$$

Peak Shaving Algorithm Let us set a fixed value for the peak limit, K. If the peak demand is greater than the main grid supply, the battery will start to supply for the remaining power.

$$\begin{gathered}\text{If } P_1 \geq K \\ p_1 = K \& P_2 \text{ will supply the rest} \\ \text{else} \\ P_1 \text{ will supply till } K\end{gathered}$$

During islanded mode, the islanded side is now fully dependent on the battery. So, the battery will supply the required power. Since the battery is made with a set of super capacitors, it takes very less time to charge the battery and exploit it.

$$P = P_2$$

We island the grid to examine and correct the fault in the main side. So, when we connect the grid back to the islanded side, we have to examine a few parameters. So, to examine those parameters, we have given a delay of two seconds. Before we connect it back to the grid, we have to synchronize the frequency and the voltage to prevent frequency mismatching and phasor mismatching that may lead to asynchronized power supply which may ruin the components and may also result in an electrical blackout. Once the synchronization is done, we can connect the grid supply back to the consumer while the battery will go into charging mode.

5 Conclusion

From the simulation performed, we have come to a formidable conclusion. The utilization of peak shaving algorithm has quite benefited us and has successfully managed to reduce the consumption cost for the consumer. We have also managed to reduce the faults that occur in the topology used currently. From the various faults which have arisen, we have provided the respective circuit breakers to activate to make it easier for the station people to handle and examine the fault. The common needs for any human asset for the country are saving money for the future. Meanwhile, the main aim of the energy sector is to restrict the usage of non-renewable sources. With the increasing depletion of the non-renewable resources, the cost of electricity has gone over the roof. This has made the previous statements not achievable. With the help of microgrids and energy saving algorithms, the world of energy management can become a much better place to interact with the common people. It is cost-efficient, much more effective, eco-friendly, and provides a simpler platform for the humans to communicate with.

References

1. Lange C, Rueß A, Nuß A, Öchsner R, März M (2020) Dimensioning battery energy storage systems for peak shaving based on a real-time control algorithm. Appl Energy 280:115993
2. Papadopoulos V, Knockaert J, Develder C, Desmet J (2020) Peak shaving through battery storage for low-voltage enterprises with peak demand pricing. Energies 13(5):1183
3. Shamblin SM, Beeks J (2020) On-site renewable energy storage at san diego gas & electric's century park campus
4. Salles RS, Souza AC, Ribeiro PF (2020) Energy storage for peak shaving in a microgrid in the context of Brazilian time-of-use rate. Multi Digit Publishing Inst Proc 58(1):16

5. Neagu BC, Grigoraș G (2020) A fair load sharing approach based on microgrid clusters and transactive energy concept. In: 2020 12th international conference on electronics, computers and artificial intelligence (ECAI), pp 1–4. IEEE
6. Zaery M, Wang P, Wang W, Xu D (2020) Distributed global economical load sharing for a cluster of DC microgrids. IEEE Trans Power Syst 35(5):3410–3420
7. Neagu BC, Ivanov O, Grigoras G, Gavrilas M, Istrate DM (2020) New methods for improved prosumer trading and energy poverty mitigation in microgrids using crowdsourcing concepts and blockchain technologies
8. Ekanayake UN, Navaratne US (2020) A survey on microgrid control techniques in islanded mode. J Electric Comput Eng
9. Ambia MN, Meng K, Xiao W, Dong ZY (2020) Nested formation approach for networked microgrid self-healing in islanded mode. IEEE Trans Power Delivery 36(1):452–464
10. Hosseinimoghadam SMS, Roghanian H, Dashtdar M, Razavi SM (2020) Power-sharing control in an islanded microgrid using virtual impedance. In: 2020 8th international conference on smart grid (icSmartGrid), pp 73–77. IEEE
11. Chen B, Wang J, Lu X, Chen C, Zhao S (2020) Networked microgrids for grid resilience, robustness, and efficiency: a review. IEEE Trans Smart Grid 12(1):18–32
12. Aziz AS, Tajuddin MFN, Adzman MR, Mohammed MF, Ramli MA (2020) Feasibility analysis of grid-connected and islanded operation of a solar PV microgrid system: a case study of Iraq. Energy 191:116591
13. Ganesh V, Senthilmurugan S, Ajit Ram RR, Prabhu A Smart grid-meters and communications-design, challenges, issues, opportunities and applications. In: 2020 IEEE international conference on advances and developments in electrical and electronics engineering (ICADEE), pp 1–5. IEEE
14. Ganesh V, Ajay Krishna VM, Ajit Ram RR (2021) Safety feature in electric vehicle at public charging station. In: 2021 7th international conference on electrical energy systems (ICEES), pp 156–161. IEEE
15. Vijayalakshmi S, Ganesh V (2020) Design of a 60W power supply unit for a single cell cycler. Solid State Technol 63(3):503–514

A Study of the Installed Turbine, Generator, and Control Equipment of the Indira Sagar Hydropower Plant

Shreya Malaviya, Atma Ram Gupta, and Jitendra Singh Bhadoriya

Abstract Hydropower is a pollution-free alternative to thermal power facilities. Numerous hydroelectric projects contribute to the sustainable production of energy. Indira Sagar is a hydroelectric project, one in which natural resources are utilized to generate energy with a capacity of 1000 MW installed. The project's execution was not planned on a turnkey basis; instead, several agencies were assigned to various E&M equipment/systems. This article discusses different pieces of equipment that have been provided and built by various agencies. The design elements of the types of equipment are discussed in detail to familiarize the reader with the different system/E&M pieces of equipment. Additionally, this article discusses different restrictions. The packages were created in groups for ease of deployment, but attempts were made to purchase equipment directly from the source/manufacturer to keep costs low. Because water is plentiful, hydroelectric projects are more efficient at producing power. This paper provides an overview of the turbines, generators, and control systems utilized in the ISP hydroelectric project and their benefits.

Keywords Generator · Stator · Rotor · Turbine · Generator bearing · Shaft

Abbreviations

ISP	Indira Sagar Project
RTDs	Resistance Temperature Detectors
OPU	Oil Pressure Unit
NGT	National Green Tribunal
MMC	Man Machine Communication
UCB	Unit Control Board
CCR	Central Control Room

S. Malaviya (✉) · A. R. Gupta · J. S. Bhadoriya
National Institute of Technology Kurukshetra, Kurukshetra, India
e-mail: shreyamalaviya24@gmail.com

K. Namrata et al. (eds.), *Smart Energy and Advancement in Power Technologies*, Lecture Notes in Electrical Engineering 926,
https://doi.org/10.1007/978-981-19-4971-5_5

1 Introduction

Generation, transmission, and distribution make up the three main sectors of the electrical industry. The process of creating electricity utilizing various fuels and production methods is known as the generation process (generation plants). Bulk electricity from the generating plants is carried to the distribution substations on the grid using high voltages transmission line facilities. Distribution utilities connect energy users at the substations to the distribution system that distributes power to homes and businesses. Distribution occurs at lower voltages than the retail stage.

Various kind of renewable as well as conventional non renewable energy sources are utilized to produce electricity in India. The renewable energy sources utilized of generation of power includes the sun, wind, and hydroelectric (solar, wind, small hydro, biomass). The electricity generating capacity of India is at 386.88 GW as of 31 July 2021 as per central electricity authority (CEA) report. India's generating capacity has grown significantly during the past two decades. This boost may be traced to the fact that in 2003, electricity-generating was delicensed, which allowed the involvement of private sector businesses to operate without limitations. Coal provides over half of the nation's electricity production, which means that coal represents a majority of the total. The generation of thermal electricity (including coal and natural gas) accounts for 60.9% of the total. More than 25% of total generating capacity is made up of renewable energy. Beyond 12% of all energy is provided by hydroelectric power, while nuclear power provides nearly 2% of all generations. Coal is still a significant component of power-sector. A total of 576 million tons of coal was delivered to the electricity industry in 2017–2018 (84% of the total coal dispatched in the country) [1].

More than 54% of India's energy is generated by coal. Coal is also the dirtiest, if it has been mined. The current state of the Indian power grid poses a significant danger to the environment in the future years due to its continued use of fossil fuels. Hydroelectric power plants built after the project's development have zero direct waste and a much-reduced level of carbon dioxide emissions compared to fossil fuel power plants [2].

Hydropower is a fuel source that's powered by water, making it an environmental friendly option. Power stations that burn fossil fuels, such as coal or natural gas, contaminate the air. As hydroelectric power sources rely on the water cycle, they are renewable. Engineers may regulate water flow via turbines to generate energy as per the requirement [3]. There are other advantages like no fuel requirement, cheapest operation and maintenance, highly reliable, rapidly changing loads can be met without any difficult, robust and have got longer life in addition to clean energy from hydropower plants. In constructing reservoirs, impoundment hydropower produces lakes with an abundance of leisure possibilities, notably fishing, swimming, and boating. To enable the public to take advantage of these possibilities, many hydroelectric projects are obliged to offer public access to the reservoir. Flood management and water supply are among the possible advantages of this proposal [4].

Indira Sagar project (ISP) utilizes the vast but variable monsoon flows of Narmada River flowing from east to west [5]. The Omkareshwar project will generate 1167 million units of energy annually [6]. The gross storage of Indira Sagar project is 9.9 million-acre feet and line storage is 7.9 million-acre feet. As per Narmada Water Distribution Award, 8.12 million-acre feet of water is to be released for Sardar Sarovar project. For utilizing this flow, 8 units each of 125 MW are installed at Indira Sagar project which will operate as peaking power station [7]. However, during monsoon, the spill of the reservoir will be utilized to generate secondary power and powerhouse will operate during off peak hours as well. The turbine procured from M/s BHEL which have been mainly manufactured at Bhopal. The stator and rotor were assembled on site. Looking to the transportation limitation, the electronic control equipment has been supplied by M/s BHEL Bangalore. In order to select type of turbine various factors are considered like efficiency and maintenance, and many others [8]. For a different environment, different types of turbine can be more effective, but for most of hydro power plant, Francis turbine is the one that has been used [9]. The Francis type turbine is the one that has used the most in any of hydro project because of its efficiency is dependent on the water flow (which is very important in case of hydro power plant) [10]. In a hydropower plant, water is stored in dams and reservoirs and in order to get this water, small rivers (as reservoir) are better option as compared to large rivers because their energy potential is found to be much more than that of large rivers [11]. The hydropower energy production technique is one of the least expensive in terms of electricity generation expenses, perhaps because the fuel (falling water) is free. For large-scale projects, the levelized cost of energy production varies between 0.02 and 0.19 US dollars per kilowatt-hour. The comparatively cheap cost of electricity production may be one of the reasons why hydroelectricity is suggested as a base load source for most electric utility providers.

2 Hydro Generator

Hydrogenerators are used to change mechanical energy supplied by turbine to electrical energy. Typically, large-scale hydropower facilities are utilized to generate energy. Figure 1 illustrates the fundamental schematic design for a hydroelectric power generating system. The turbine output shaft is connected to the generator to generate energy. The generator is primarily composed of an electromagnetic rotor that is housed within a cylinder (referred to as the stator) that contains a coil of electric wires (known as a conductor). During operation, the stator's rotor rotates and produces energy through the electromagnetic induction process. The produced power is transported to load locations through a transmission system comprised of switchyards, transformers, and transmission lines.

The detailed description of turbine, generator and power plant control equipment of ISP given sequentially in next sections.

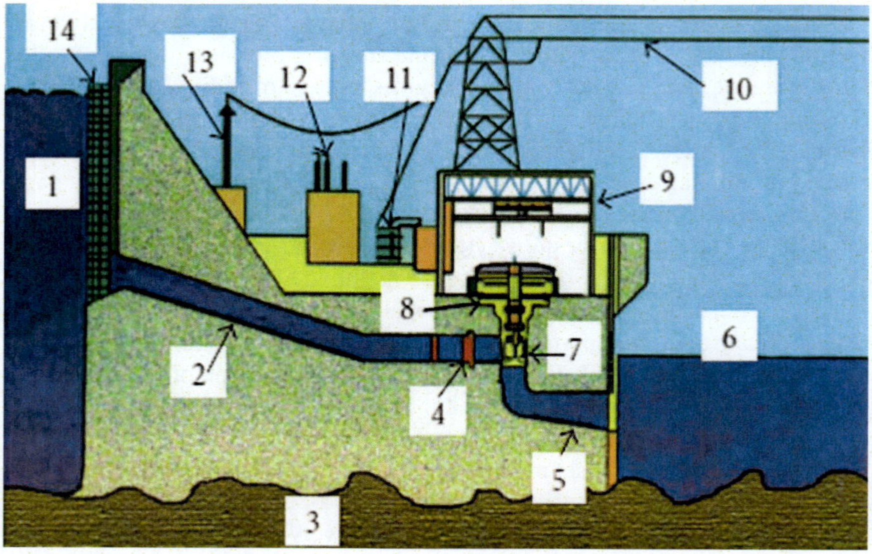

(1) Reservoir
(2) Penstock
(3) Bed rock
(4) Valve
(5) Draft tube
(6) Tailrace water
(7) Turbine
(8) Generator
(9) Powerhouse
(10) Transmission lines
(11) Transformer
(12) Insulators
(13) Transmission tower
(14) Trash rack

Fig. 1 Schematic view of a hydropower station and its basic parts [12]

3 Turbine Guide Bearing

The salient features are as follows: Eight identical hydraulic turbines are installed in power house. Each turbine is vertical Francis reaction type with turbine shaft coupled to generator shaft. Here, Francis turbine is used because it's efficiency is dependent on water flow [13]. When viewed from the generator with rotating mass, rotation is seen to be in clockwise direction. Each turbine assembly consists of the following main components:

3.1 Runner

The runner is a stainless-steel casting with coupling flange for joining to the shaft. The upper labyrinth and lower labyrinth are shrunk fitted on the runner and welded. The runner is provided with a fabricated runner and welded. The runner is provided with a fabricated runner cone.

3.2 Shaft

The shaft is forged out of 1.5% manganese. Steel it includes an integrally forged collar for the guide bearing. One end of the shaft is coupled to the runner and other end with the generator shaft [14]. Mounted on the shaft is a stainless-steel shaft sleeve, which acts as a bearing surface for the shaft seal.

The type of the turbine guide bearing is of pivoted pad type consisting of self-contained oil bath lubrication and internally mounted oil coolers. It comprises of eight babbitt lined pads organized along coolers of the shaft. The bearing housing is of the type of split construction and is gotten on the upper side of top cover with the studs and nuts. Temperature of guide bearing pads is monitored by a set of resistance temperature detectors (RTDs) and dial type thermometers. Out of eight pads, temperature of four pads is measured by RTDs for recording and indication. The bearing cover includes four transparent windows for visual observation during operation of the bearing.

3.3 Spiral Casing and Stay Ring

The spiral casing is manufactured from high-tensile steel plates. The spiral casing is of logarithmic structure and generously roundabout cross-area for keeping a uniform flow all through its entry. Stay ring is made in two parts of wieldable steel and is welded with spiral casing.

3.4 Draft Tube

Water after striking the runner blades passes through the draft tube and discharges to the tailrace channel. The draft tube is constructed from welded steel and it also consists of cone and an elbow liner embedded in concrete.

3.5 Guide Apparatus

The guide apparatus regulates the flow of water with change in load and also servers as a closing device. It has inclusion of top cover, pivot ring, guide vanes, and turning mechanism. The fabricated top cover supports the regulating ring guide bearing and houses shaft seal upper bush housing upper stationary labyrinth and liner plates.

The guide vanes are made of stainless-steel castings. The bearings are shrunk in upper bush housing, are bolted on the top cover. The lower journals of guide vanes are supported in self-lubricated brushes shrunk in pivot ring. Two servomotors housed inside the pit liner, actuate the regulating ring, which in turn operates the guide vanes through regulating gears.

3.6 Guide Vane Servomotors

Two servomotors are provided for turbine regulating ring during regulation of load on turbine and closing/opening of the guide apparatus. The servomotors are identified as LH (left hand) and RH (right hand).

3.7 Shaft Steel

The shaft sealing prevents leakage of water through clearance between top cover and the sleeve. It is located below turbine guide bearing and is fastened top cover by studs and nuts. The sealing elements consist of two numbers of rubber sealing rings and it is water-lubricating type.

3.8 Governor Oil Pressure Unit (OPU)

It consists of a PP set sump lank in which oil is kept and a pressure vessel in which air and oil are accumulated. Oil sump tank is made of electrically welded steel and of sufficient capacity to supply the oil to the governor, governor servomotors, and pressure accumulators. Two pressure-pumping sets (one main and other standby) are mounted on each sump tank.

3.9 Hydro Mechanical Control (HMC)

The hydro mechanical control is mechanical governing system to control the start/stop RPM of the turbine. The HMC gets oil at 50 kg/cm^2 pressures from the OPU pressure vessel and sends controlled oil supply to servomotors for desired opening of guide vanes.

4 Generator

Eight identical hydraulic generators are installed in powerhouse. Each generator is of the type of centrally vertical shaft type consisting of salient poles with closed air circuit ventilation. Hydraulic generator are used to convert mechanical energy received from the turbine to electricity [15]. Below the rotor, it has a combined thrust and bottom guide bearing and an upper guide bearing above it. Each turbine assembly consists of the following main parts:

4.1 Stator

4.1.1 Stator Frame

The stator frame has been developed of steel plates that are weldable and have satisfactory depth to forestall distortion during transport or under any working conditions [6]. The joints between the segments of the frame have been vigorously flanged inside and coupled by various short bolts.

4.1.2 Stator Core

The stator core is built up of varnish-insulated stampings of high-grade non-aging cold-rolled silicon alloy steel. Ventilation dusts have been provided at intervals along the stator core, these being formed by non-magnetic steel spacers welded to the adjacent stampings. The core has been securely clamped by a large number of studs outside the core, extending over its full length.

4.1.3 Stator Windings

The stator winding has class '*F*' insulation system. The winding is of double layer wave connected bar type wound in open slots. In order to minimize eddy current losses each bar consists of a number of glass-covered polyester made varnish bonded with rectangular copper conductors. The bars have *ROEBLE* type transposition in the slot portion to reduce the circulating current losses.

The bars have been tightly wedged in the slots with non-shrinkable wedges, which are restrained against any downward movement. The overhang portion of the bars have been adequately braced with packing blocks and securely lashed by varnished glass cord/tape to molded glass-fiber support rings carried on insulated brackets fixed on core end plates.

Both ends of each phase have been brought out as insulated bus bar terminals, arranged outside the generator frame. The stator winding is having four parallel paths in each phase. It is having three-phase star winding, one end connected to split-phase bus duct and another end connected to NGT.

4.1.4 Neutral Grounding Cubicle

Neutral grounding cubicle is fabricated out of sheet steel of suitable thickness and is of freestanding type and completely covered with suitable ventilation. The cubicle is complete with interior illumination lamp, spacer heater, terminal boards, insulating barrier between transformer and resistor compartment, earthing stud, common marshalling boxes, and other accessories for proper operation/functioning of the cubicle.

4.2 Rotor

The design of the rotor is such that it can withstand all mechanical stresses imposed by the maximum runaway speed safely. The static and dynamic adjusting of the rotor was completed as a piece of pre commissioning tests at site and values of rotor vibrations had been inside suitable limits as indicated by IS/BS norms.

4.2.1 Shaft and Thrust Bearing Collar

The integral shaft and thrust bearing collar have been forged from high quality manganese steel. It has been accurately machined and has a hole bored through the center for inspection and other purposes. In order to connect to the turbine, flange an integrally forged half coupling has been provided at the bottom of the shaft. Top guide bearing journal has been provided on top shaft, which will be bolted on to the main shaft.

4.2.2 Spider and Rotor Rim

The rotor spider is a fabricated structure. The lower disc of hub has been fitted on to the shaft with studs and radial keys. The spider has weighty steel bars welded on to the external part in order to support the rotor rim. The rotor rim gathered around rotor spider has been developed from thin sheet covers each covering two pole pitches and progressive layers of covers to give satisfactory solidarity to the rim.

4.2.3 Poles with Field Windings

The poles are of covered construction comprising of thin sheet steel pole punching, braced between heavy steel end plates through steel studs. The post core has number of T shaped tails to draw in with comparing T molded openings in rotor rim. The pole face profile is circular and so chosen as to ensure suitable grading of air gap. The damper winding bars are of circular section copper embedded in pole face. The ends of damper bars are short circled together by copper punching. The field coils square ended, being fabricated from rectangular copper strips, dovetailed and brazed at the ends.

4.2.4 Slip Rings and Brush Gear

The collector (slip rings) is of gentle steel and mounted on the highest point of generator tube shaft. The brush gear for collector is mounted on insulated studs upheld on the top section and is orchestrated to allow advantageous access for maintenance and review.

The type of insulation used for slip rings and their connection is non-hygroscopic and oil resistant. Leads from collector to the generator field poles have been taken along the tube shaft and have suitable joints for dismantling of rotor. The polarity of the field is reversed when required and provision has been made for this purpose in the field suppression cubicle. Insulated brush holders and copper brushes have been provided for measuring the voltage across the slip rings.

4.3 Generator Bearing

4.3.1 Thrust Bearing

The thrust bearing is of the spring mattress type, the stationary parts in this type consist of a set of white metal segmental pads which are supported on mattress of helical springs. The thrust pad is of stress relived mild steel and is faced with high quality white metal. The underside of the thrust collar has been machined so as to be accurately perpendicular to the axis of the shaft and has been polished to fine limits. The thrust bearing pads have been completely immersed in an oil bath and are cooled by means of oil cooler units. The bearing has been designed for the axial load, which is summation of the weights of all rotating parts of generator and turbine and maximum hydraulic axial thrust. It is possible to remove the bearing pads in radial direction after lifting the rotor.

4.3.2 Generator Guide Bearing

The upper and lower guide bearings are of the turned pad type comprising of a row of white metal pads organized in a support ring to bear on a diary surface. A pivot bar has been rushed to the rear of each guide pad to empower the pad to rock slightly to take up an appropriate position and work with arrangement of the oil film when running.

4.3.3 Bearing Insulation

To prevent shaft current through the guide bearing pads the upper guide bearing pads have been insulated from the rotating system by providing insulation either at the back of the pads or under the guide bearing support rings. Bearing capability rotor braking and jacking. The thrust and guide bearing housing and oil bath have been formed in the center of the bracket. The bearing will be capable of safe operation without any damage under the following conditions:

- Continuous operation at any speed from 90 to 110% of rated speed.
- If it is operated for a period of minimum 15 min at rated load and speed after the failure of bearing cooling water supply.
- If it is operated for a period of minimum 15 min under maximum runaway speed conditions with cooling water ON.
- If it is operated for a period of minimum 30 min at low speed of 4%–5% of rated speed.

4.3.4 Top Bracket

The top bracket is of a fabricated steel structure. It supports the weights of the stationary parts of brush gear is supported by top bracket and over speed device are also supported by top bracket [1]. It has been placed on stator frame wit shims for level adjustment. Jacking screws have been provided for centering. It also supports weights of the brush gear dome and partly that of the flooring.

4.3.5 Bottom Bracket

The bottom bracket is of a fabricated steel structure. It is intended to help the hydraulic thrust from the turbine notwithstanding the heaviness of rotating of the generator and turbine. Jacking screws have been provided for adjusting the bracket during leveling and centering of the bracket. It is thus possible to lift the bracket thought the stator bore. Brake and jack units have been mounted upon the bracket for rotor braking and jacking.

4.3.6 Aeration Valves

A runner-aerating valve is provided on the top of the generator shaft. Normally, this valve remains closed against runner valve seat due to spring property of rubber. Whenever pressure below the runner is lower than atmospheric, this valve allows automatically the atmospheric air to pass through. The valve supplies the air through the generator and turbine shaft central hole, to the runner cone to prevent cavitation of the runner.

5 Power Plant Control Equipment

Mosaic control panels desks and computerized power plant control equipment are installed for control and monitoring of eight number of generating units, associated unit auxiliaries common station auxiliaries and 400 kV switchgear.

The unit control boards are located in n the machine hall near each unit, whereas the mosaic control panels/desks are located in the control room. The computerized control equipments are mainly positioned at two locations, viz, machine hall and the control room. Unit process control computer along with its interface is a part of the unit control board and is being used during testing and commissioning of generating units and maintenance of process control computers. The balance computerized control equipment comprising MMC computers, data acquisition system, station process control computer are located in the control room. Auxiliary relays to release CO2, starting of HP lubrication oil pump during stopping of machine, cutting in/out of creep detector in circuit and to control all other units auxiliaries and associated station auxiliaries (included in generator manufacturer's scope supply) are also provided necessary input output devices, transducers, safety devices associated with turbine generator governor exciter, etc., electrical protective relays required for generator, unit transformer unit auxiliary transformer and other equipment are also installed. The protective relays are microprocessor based.

5.1 Control, Supervision, and Monitoring Philosophy

All eight generating units are designed to operate entirely independent of each other. Each unit has been equipped with its own auxiliary system. This concept also applies to the designs of the control system too. The following hierarchically arranged controls are also provided:

- Distributed local control for individual unit equipment/ auxiliaries with control and monitoring features on the local control cubicles located near the equipment
- Centralized manual control on the UCB at machine floor level for manual the operational parameters will be controlled manually from the central control room and the dispatch center.
- Automatic control for operational sequences to be activated from the unit control board central control room and the dispatch center.
- Centralized control of all the eight units, switchgears and auxiliary systems of the power plant from the central control room and from the dispatch center.

The control equipment and functions provided at different hierarchically arranged control levels are described below:

5.1.1 Process Controls

The process level acts as an interface between the automation system and the process. The components at the process level perform the following tasks:

- Limit switches, pressure monitors, and level monitors shall provide the feed-back signals to the automation unit.
- Electrically or hydraulically activated valves for lubrication, compressed air systems: pumps and a compressor shall provide the required medium.
- Monitors, sensors and/or other devices involved in the process shall provide the inputs for control supervision and monitoring.

5.1.2 Local Controls

This includes local control stations unit and penstock gate control panels etc. that has been installed for all field devices and also permit manual control there from.

5.1.3 Unit Controls

Each unit is having its own computer (controller) which performs the following operations:

- Control automatic operation
- Transfer of single units from one operational condition to another such as startup, synchronization, generation, stopping, and shut down with all their possible variants.

5.1.4 Plant Controls

The statistic and management station work and operating station and supervisory and control board (mosaic board) have been installed in the central control room (CCR) of the power plant. The communication process enables operator communication and monitoring of all generating units and of the overall plant via a keyboard visual display unit and peripheral printer. Facility for the following controls are also provided:

- Automatic control of generating units individually/power plant completely.
- Manual Control of 400 kV switchgear line bays.

5.2 *Starting and Stopping of Generating Unit*

The starting of the generating unit shall be planned both from the Unit Control Board/Mosaic Control Panel/Dispatch Center. The closing operation in normal, abnormal, and emergency conditions is possible both from the unit control board as well as automatically through unit shut down devices. Facility for starting to stop of the generating units is also available through keyboards of the computerized control equipment.

6 Advantages of Hydro Generator

- Hydro-electricity is a very dependable source of energy. They do not contribute to the pollution of the atmosphere. There are relatively few variations in electrical power unless a different output is specified, the electricity generated by the plants. Countries with significant hydropower resources use hydroelectricity as a baseload source of energy.
- Once a dam is built, energy may be generated continuously. If no power is required, the sluice gates may be closed, ending energy production. When dam systems are operational, their energy does not contribute to the production of greenhouse gases.
- The water from the lake may be utilized for agriculture. The accumulation of water in the lake implies that energy may be captured to be held until the water is required, at which point it is released to create electricity.
- Water may be conserved for a variety of purposes utilize a different time period when power demand is high. Dams are built to endure for decades and may contribute to energy production over an extended duration. The lake that develops behind the dam may be utilized for water sports and recreational activities.

7 Conclusions

This article presents a study of the ISP hydropower project, including the requirements for turbine generator equipment that were installed in 1993, and even after so many years, the design elements remain consistent with current trends. The machines worked at an elevation of 238 m, much lower than the usual minimum draw drawn level of 243.23 m, resulting in early activation of the powerhouse. This study contributes to the advancement of knowledge about hydroelectric power plants and its operating, the equipment used in generators and turbines, and the advantages associated with them.

Acknowledgements With profound respect, Author Shreya Malaviya would like to thank Er. Hitesh Purohit Senior Manager Electrical, NHDC Limited Bhopal for providing insight knowledge and expertise that greatly helped in writing this paper.

References

1. Raja J, Ajay-D-Vimal Raj P, Rajasekar S (2018) Overview of power sector scenario in India. Power Syst:1–21. https://doi.org/10.1007/978-981-10-2972-1_1
2. Tiewsoh LS, Jirásek J, Sivek M (2019) Electricity generation in India: present state, future outlook and policy implications. Energies 12(7):1361. https://doi.org/10.3390/en12071361
3. Kaunda CS, Kimambo CZ, Nielsen TK (2012) Hydropower in the context of sustainable energy supply: a review of technologies and challenges. ISRN Renew Energy 2012:1–15. https://doi.org/10.5402/2012/730631
4. Maradin D (2021) Advantages and disadvantages of renewable energy sources utilization. Int J Energy Econ Policy 11(3):176–183. https://doi.org/10.32479/ijeep.11027
5. Kang C (2010) Preface: special issue on Asia power and energy systems. Int J Power Energy Syst 30(4). https://doi.org/10.2316/journal.203.2010.4.203-0001
6. Station planning and design (1992) Modern power station practice, pp 205–208. https://doi.org/10.1016/b978-0-08-040735-7.50009-0
7. Gitleman L (2017) Thirty seventh annual report of Narmada control authority of Indore
8. Ransford G, Rottner J (1959) The optimization of hydraulic governor performance taking account of the grid inherent stability factor and elastic water hammer effects. La Houille Blanche 45(1):23–46. https://doi.org/10.1051/lhb/1959022
9. Murty MS, Hariharan MV (1984) Analysis and improvement of the stability of a hydro-turbine generating unit with long penstock. IEEE Power Eng Rev PER-4(2):31–32. https://doi.org/10.1109/mper.1984.5525488
10. Magureanu R, Albu M, Bostan V, Dumitrescu AM, Pelizza M, Andreea F, Dimu G, Popa F, Rotaru M (2008) Optimal operation of Francis small hydro turbines with VARIABIABLE FLOW. In: 2008 IEEE international symposium on industrial electronics. https://doi.org/10.1109/isie.2008.4677281
11. Aminov D, Gandzha S, Kosimov B (2020) Technology of using a combined excitation valve hydrogenerator as an alternative source of energy use for small rivers. In: 2020 Russian workshop on power engineering and automation of metallurgy industry: research & practice (PEAMI). https://doi.org/10.1109/peami49900.2020.9234384
12. Hydropower (2012) IEA technology roadmaps. https://doi.org/10.1787/9789264189201-en
13. Eliud K, Musembi NM, Kindole D, Mukama A, Kosgei SK, Nemoto Y, Nakajo Y (2017) Study of performance characteristics of small submersible pump run as hydro turbine generator. Sustain Energy 5(1):1–5. https://doi.org/10.12691/rse-5-1-1
14. Schleif F, Wilbor A (1966) The coordination of hydraulic turbine governors for power system operation. IEEE Trans Power Apparatus Syst PAS-85(7):750–758. https://doi.org/10.1109/tpas.1966.291702
15. Lansberry JE, Wozniak L, Goldberg DE (1992) Optimal hydrogenerator governor tuning with a genetic algorithm. IEEE Trans Energy Convers 7(4):623–630. https://doi.org/10.1109/60.182643

FACTS Devices Injection in Electrical Network for Reactive Power Compensation

Kailash Kumar, Atma Ram Gupta, and Jitendra Singh Bhadoriya

Abstract With the growth of urban and rural infrastructure, the power system is pushed to run near capacity, resulting in overloaded power grids, higher losses, peak generation, security threats, and network instability. The performance of the legacy electric network may be improved by supplying sufficient reactive power through grid voltage tuning, transformer tap setting, capacitor bank sizing, and the use of computational intelligence technologies. Voltage control, power factor enhancement, and voltage supply have become simple with these power electronic devices. These FACTS devices outperform the previous cumbersome equipment employed for the same purpose. FACTS controllers with harmonic filters are the best solution to improve supply voltage. In this vein, this article offers a thorough bibliography of different FACTS devices intended to improve power system performance by compensating reactive power. Finally, novel FACTS device combinations are proposed to magnify power system current and voltage. This study will be immensely beneficial to researchers for fast reference on the FACTS controllers.

Keywords FACTS · Reactive power · SVC · TCR

1 Introduction

Inductive or capacitive components cause the current waveform to be out of phase with the voltage waveform. Active power is generated when the power network current is in phase with the voltage. Capacitors and inductors need reactive power to generate magnetic and electric fields. Transmission lines are capacitive and inductive in nature based on the load characteristics. Reactive power is required to build

K. Kumar (✉) · A. R. Gupta · J. S. Bhadoriya
Department of Electrical Engineering, NIT Kurukshetra, Haryana, India
e-mail: kailash_11814059@nitkkr.ac.in

J. S. Bhadoriya
e-mail: jitendra_61900077@nitkkr.ac.in

K. Namrata et al. (eds.), *Smart Energy and Advancement in Power Technologies*,
Lecture Notes in Electrical Engineering 926,
https://doi.org/10.1007/978-981-19-4971-5_6

up the electric field in equipment and consumers, such as rectifiers with capacitive smoothing, CFLs, capacitors, generators with leading power factors, and no-load or low-load overhead transmission lines and cables. Unlike active power, reactive power varies between the supply (reactive power compensation device) and the drain (motor and transformer). Asynchronous motors and electrical transformers use approximately 60 and 25% of the installation's total reactive power, respectively, by generating alternating magnetic fields. The weight for industrial customers is approximately 70% for asynchronous motors and 20% for transformers. For asynchronous motors and transformers, the difference in reactive power consumption is because the magnetization reactive power, which is the essential component of reactive power, depends on the volume of the magnetic circuit, to which is added for asynchronous motors. Reactive power depending on load is another component of reactive power for asynchronous motors and transformers. The apparent current is raised by adding the active and reactive currents in proportion to their quantity and phase angle [1].

Reactive power circulation in power systems has substantial impacts on interconnected providers and consumers. The increased active power losses in the passive components of the installation, the increased voltage losses, the increased installation size, and the decreased electric energy transfer capacity are all consequences of increased reactive power consumption. Transmission lines' reactive power consumption rises with current squared. It is thus challenging to transfer reactive power. Condensers, capacitors, and static compensators can all produce reactive power. The ability of a power plant to generate reactive power is limited, and this enormous power consumes transmission lines, transformers, and causes system losses. The availability of reactive power sources near the consumer reduces costs and increases transmission line capacity [2].

Voltage limitations usually demand that a transmission system's voltage be kept within a specific range, around 5% of the nominal voltage [3]. Changes in load or faults in transmission and distribution lines or other equipment may affect the voltage in transmission lines. In these situations, the dynamic and transient voltages should be kept within a range. A line voltage that is too high may create a short circuit and damage transformers and other equipment in substations. The voltage of an AC transmission line is nearly directly linked to the wire's reactive current and reactance. Capacitors and reactors may be used to regulate voltage variations on lines. A flat voltage profile comes from balancing consumption and generation of reactive power at a certain loading level. The inability of the system to generate necessary reactive power is one of the major reasons that contribute to voltage instability. Voltage collapse occurs when the buses' voltage gradually and unpredictably drops. Reactive power controllers and series and shunt capacitors help avoid voltage instability. This correction injects reactive power to keep bus voltages near to nominal while reducing line currents and system losses [4].

The FACTS devices are used to improve transient stability of a power system, also enhancing overall power quality, and increasing power transfer capability in transmission lines.

In Fig. 1, two ABB 140 MVAR inductive to 140 MVAR capacitive static VAR compensators were installed at the Kanpur 400 kV substation of the Power Grid

Fig. 1 Showing an installed SVC in Kanpur 400 kV substation

Corporation of India in Uttar Pradesh in 1992, which is a FACTS device and contains system energy components. In the past years, thyristor current ratings also increased which help us in taking its use in our power system in which power flow in the range of MW. Due to use of modern power electronics components, FACTS devices have a greater speed and flexibility in operation, and along with these they also provide other improvements like: better power flow, increased power flow capability in transmission lines, improved transient stability, damping out low-frequency oscillations generated due to rotor swing in generator installed, and an overall control and stability over system voltage [5]. Depending on the system, a proper modification is done to our device that would work perfect there, and with these ideas we can achieve an increase in power transfer capability capacity of up to 40–50%. Compared to previously heavy machine operated controllers, FACTS controllers require less area and require less maintenance. But, in the point of view of *Transmission System Operator (TSOs)* the cost, complexity, and reliability of FACTS controllers seem to be a challenge to our modern time power system [1]. Also, with increased demand, our losses are increasing simultaneously with a restriction to not build new power lines, and these parameters are very difficult to handle. However, many high-power transmission line systems operate at a lower limit less than their actual power transfer and durability limits due to these constraints. All these conditions show that for both a reliable and an economic system a new advanced technology is required which can make a better use of existing power system [6]. In conclusion, reactive power generating devices compensate for all of the above stability issues while allowing the transmission system to operate closer to its thermal limit.

2 Advantages of FACTS Devices

When a minor interruption occurs in the power network, electromechanical oscillations occur at a low frequency. Power system variables such as line frequency, bus voltages, line currents, and generator speeds also become oscillatory as a result of these oscillations. The synchronous generators are linked to improve the dependability of the power network and to increase generating capacity, resulting in fluctuation in power system oscillations. Thus, electricity networks seem to be less safeguarded and constantly vulnerable to voltage instability, which has resulted in many significant network failures worldwide [7]. To ensure the security of power systems, it is critical to propose appropriate solutions to increase power system security and voltage stability. Preventive methods include rescheduling generation and energy transmission, putting backup generators online, load shedding, and VAR control using series or shunt capacitors. After all, most of them are electromechanical, which has drawbacks like slowness and wear and tear [8]. However, FACTS may achieve the same goals without significant modifications in the power system architecture. The benefits of using FACTS devices in transmission systems are described below [9].

- Increased power transfer capability of the existing transmission line
- Improvement in power factor
- Increase in the overall reliability of system
- Enhanced system stability
- Increased quality of supply for large industries
- Benefits the environment
- Power quality improvement
- Flicker mitigation
- Voltage profile improvement.

3 FACTS Devices

The FACTS is a static device used to increase the controllability and power transfer capacity of alternating current transmission of electrical power. FACTS' primary function is to manage congestion by improving power transfer capacity, dependability, regulating power flow, strengthening security, and enhancing overall system performance. The FACTS devices are capable of controlling the transmission system's various electrical characteristics. FACTS are solid-state flexible devices that are used to improve the capability of a transmission line to handle electricity, its transient stability, system reliability, load management, and power flow control. The following are the various FACTS devices listed:

- Thyristor-controlled series compensator (TCSC)
- Thyristor switched series reactor (TSSR)
- Thyristor-controlled series reactor (TCSR)
- Thyristor switched series capacitor (TSSC)
- Static var compensator (SVC)
- Static synchronous compensator (STATCOM)
- Static synchronous series compensator (SSSC)
- Thyristor-controlled phase shifting transformer (TCPST)

- Unified power flow controller (UPFC)

Figure 2a, b illustrates the uses of FACTS devices and equivalent circuits. As can be seen, FACTS devices provide many advantages, motivating the research community to investigate their effect on the power system network. FACTS offers a significant answer to various practical issues, including power flow management, system damping, voltage regulation, and power factor enhancement. The inverter-based power systems integrating FACTS devices has developed into an increasing area of interest for the safe, dependable, and affordable power system operation.

SVC is a shunt-connected static VAR generator and also an excellent absorber of reactive power. Its generation and absorption of reactive power can be controlled by changing its capacitive or inductive current. Static var compensators (SVC) is a shunt-connected FACTS device which has greater accuracy and quick response that makes it possible to provide a better control over power, but SVC also faces a drawback which is dynamic voltage problem [11]. The SVC is a combination of different components which are shown in Fig. 2b. Along with the use of capacitors and inductors SVC also contains a harmonic filter which is used to damp out the harmonics generated due to use of capacitors and inductors in same circuit [12]. Unlike to a synchronous condenser (a rotating electric motor used for reactive power generation), an SVC does not contain any moving parts other than an internal switchgear which make it more efficient because of less losses. SVC works in two modes. If load of the power system is leading, SVC will switch to thyristor-controlled reactors to consume the extra VARs generated and overall reducing the system voltage. In opposite conditions when load is lagging SVC switches its capacitors to generate var. If extra var is generated by our capacitor, then that can be absorbed by thyristor-controlled reactors, so in this way a better power control and a better voltage can be achieved with a perfect combined working of reactor and capacitors. In industries, SVCs are connected near to fast changing load such as arc furnaces, paper mills to smoothen the voltage flicker. An SVC mainly consists up of capacitor banks and reactors both connected in parallel. Air core reactors are connected in series with thyristors, this combination is also called thyristor-controlled reactors (TCR), and it is used to absorb the extra reactive power generated in our system. An equivalent electrical representation of SVC is shown in Fig. 2b. Amount of power absorbed by TCR depends on current flowing through inductor. Current in our inductor can be controlled by using firing angle delay method. SVC is a good source of reactive power as it has capacitors which generate vars and the surplus amount of var can be absorbed with help of TCR by varying the firing angle of TCR [13]. By generating and absorbing reactive power, SVC helps a system to achieve unity power factor.

Presence of current harmonics during the operation of the thyristor-controlled reactors is almost certain. In the linear range of control SVC voltage is constant, but in real life applications, a slope of 5% is given in voltage to enable the parallel operation of more than one SVC connected at same bus. Also, this slope prevents SVC from hitting its limits frequently [14]. Also, SVC current is considered positive when susceptance of SVC is inductive. Thus,

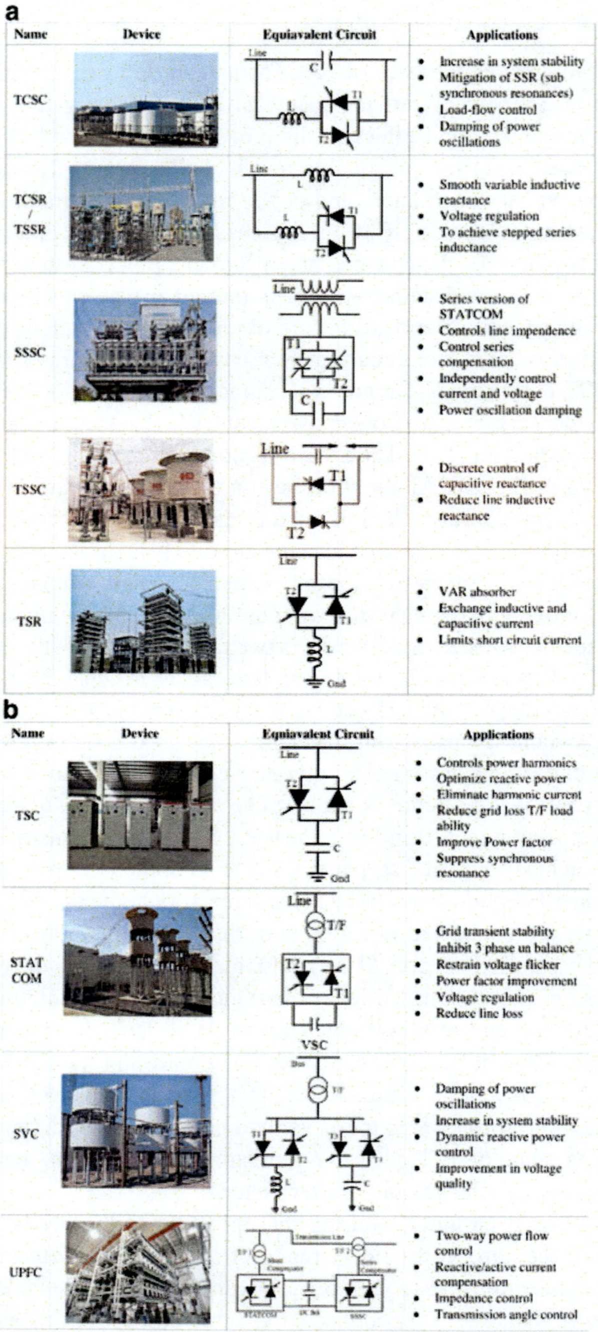

Fig. 2 **a** Types of FACTS with equivalent circuit and applications, **b** Types of FACTS with equivalent circuit and applications. *Courtesy* ABB, Siemens, electronic hub [10]

$$\text{Isvc} = -\text{BsvcVsvc} \quad (1)$$

Vref is the voltage at which current of SVC is zero or we can say at this point SVC neither absorb nor deliver energy. Reference voltage can vary between high and low limits—Vref max and Vref min. Operating range of SVC decided by current and voltage ratings of different components such as thyristors, capacitors, and inductors.

To use synchronous voltage source (SVS) for reactive power compensation DC energy source replacement is done with a small dc capacitor. For reactive power compensation capacitor connected in SVS charge itself up to required voltage level. Capacitor can be charged with the help of output voltage which is generally in lagging phase with respect to system voltage. During this charging process of capacitor, converter absorbs a small amount of real power which compensates for losses of converter. Performance characteristics of SVS in this controller are same as synchronous condenser that is why it is called static synchronous compensator. STATCOM can provide output current of both natures capacitive and inductive independent of AC system voltages. That is, STATCOM can provide maximum output current even at low system voltages. The ability of STATCOM to generate maximum output current even at low system voltages makes it way better than SVC. Another advantage of STATCOM is that it requires less area than SVC.

As the name of SSSC says it is series connected device, so in transmission line, it is connected with the help of transformer as shown in Fig. 2a. For compensation purpose, SSSC injects a 90° lagging or leading voltage with respect to current [15]. It means SSSC has control over complete inductive and capacitive compensation [16]. If we provide or connect an energy storing element with SSSC, it can control flow of power also. When SSSC is connected to an energy storage capacitor, then whole system now can only deliver and absorb reactive power to and from the system. Further, without this capacitor, SSSC works as a series compensator whose output voltage is controlled. Reactance which is changing with time will cause disturbances in electric power flow in line. There are mainly three components of SSSC which are voltage source converter (VSC), transformer and energy source. VSC is the main component of SSSC. Second component which is transformer connect SSSC with the main transmission line. Third component is energy source which compensates for losses in converter and also provides voltage across DC capacitor.

TCSC is a series connected first generator var compensator that consists of an inductor with an antiparallel combination thyristor in series with it, and this inductor with thyristor circuit in parallel to a capacitor is for var generation and inductor is for var absorption of extra var generated in power system. TCSC can work in three different modes.

I. Bypass-thyristor mode,

   In this mode, thyristors are fully on for whole 180° period. In this mode, TCSC looks like a parallel connection of just inductor and capacitors.

II. Blocked-thyristor mode,

   The triggering circuit of thyristors is removed, and thyristors will get turn. So, overall impedance of TCSC will be only capacitive.

III. Vernier mode,

In this mode, TCSC behave as continuously controllable capacitive or inductive reactance. This can be done with the help of varying firing angle in a certain range [17]. Basic principle used for compensation in TCSC is to increase the magnitude of voltage across the fixed capacitor. The circuit of TCSC with a variable value of inductor connected in parallel to FC. Equivalent impedance, Zeq, of this LC circuit can be given as [18]:

$$\mathrm{Zeq} = [(1/\mathrm{jwC})//\mathrm{jwL}] \quad (2)$$

The total reactance of TCR, XL (α) varies from its low value XL to infinity. Resonance occurs when XL (α) = XC, and at this point, net reactance of system reaches up to infinity theoretically. Further increase in firing angle would result in capacitive reactance which is decreasing with increase in firing angle. So, TCSC reactance shows both inductive and capacitive behaviour for different values of firing angle [19]. While choosing or connecting inductor in circuit we should take note that value of XL should not be equal or greater than reactance of capacitor for any value of firing angle. This is to be done to avoid occurrence of resonance. If we take reactance of inductor more than capacitor, then there is a possibility that for some firing angle both capacitive and inductive reactance become equal and cause a resonance.

4 Results and Simulations of TCR, TSC, FC-TCR, and TSC-TCR

TCR is a component of SVC which is used to absorb the extra reactive power generated into our power system either by capacitor banks (TSC) or by generators. Amount of reactive power absorbed is totally dependent on amount of current flowing through our reactor. And this value of current can be varied with the help of firing angle delay method. Simulation diagram for the same is shown in Fig. 3, and a visual representation of final waveform of current flowing through TCR is shown in Fig. 4 along with a waveform of input voltage shown in same window. All the simulation results are obtained using parameter setting given in Table 1.

TSC is a component of SVC which is used to generate extra reactive power for our power system which is consumed by inductive load connected at the distribution end. Simulation diagram of TSC is shown in Fig. 5. Here, reactive power generated when capacitor is charged with input voltage. A major disadvantage of TSC is that it cannot be triggered at required moments. Its triggering should be done at proper instants when difference between capacitive voltage and input voltage is either zero or minimum. That is why a TSC can only produce output in a step as shown in Fig. 6. Triggering at unusual times will result into high content of transients or harmonics.

FC-TCR is a combination of a fixed capacitor parallelly connected with a TCR. This combination now can do both works like absorption and generation of reactive

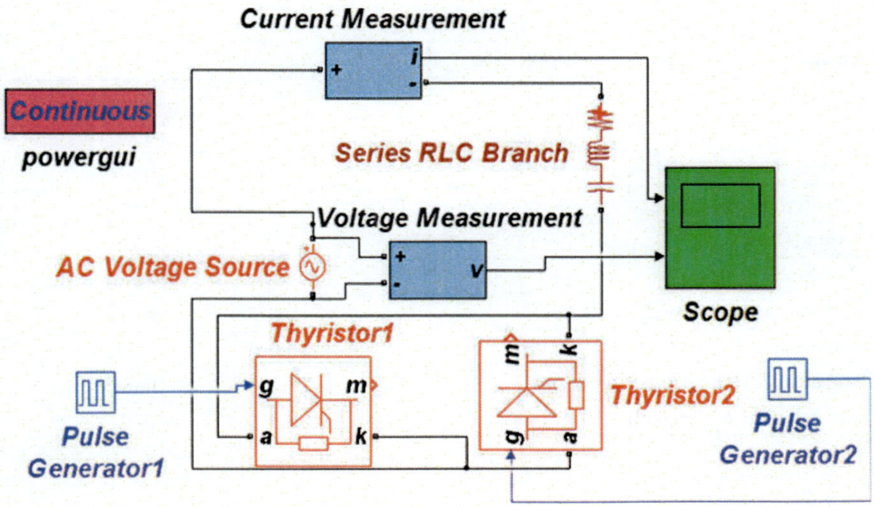

Fig. 3 Simulation diagram for TCR

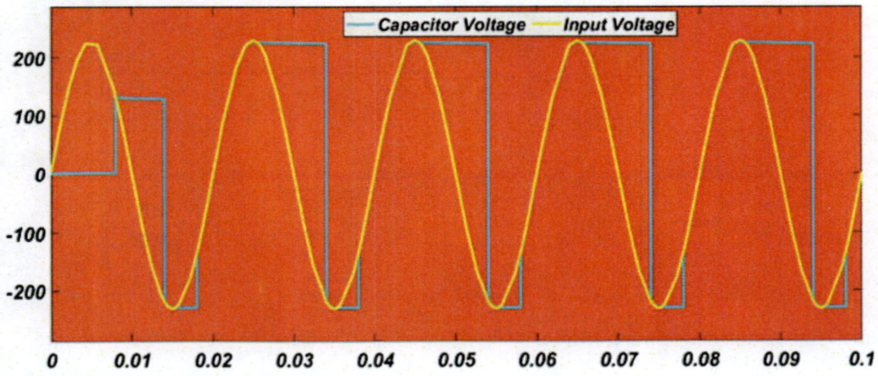

Fig. 4 Current and voltage waveforms of TCR

power as per requirement. A small value resistor is also connected in series with capacitor as shown in Fig. 7 to avoid problem of harmonics generation because of presence of inductor and capacitor in a same circuit. Reactor used in this circuit should be able to generate more reactive power than capacitor because during low-load or no-load conditions system will have extra reactive power, so reactor needs to absorb that. And in every case whether we need enough reactive power or not our capacitor is of fixed value so it will generate same power which is a disadvantage for this circuit. Current through TCR and voltage across TSC are shown in waveforms of Fig. 8.

TSC-TCR combination is same as FC-TCR with one replacement, and triggering is provided to both capacitor and reactor as in Fig. 9. And output for a specific set of

Table 1 Simulation parameters of various FACT devices

Parameters	TCR	TSC	FC-TCR	TSC-TCR
Input supply (V)	230	230	230	230
Supply frequency (Hz)	50	50	50	50
Inductor (Henry)	0.01e−1	–	0.01e−1	0.01e−1
Capacitor (Farad)	–	0.0001	0.0001	0.0001
Resistor (Ohm)	–	–	1	100
Pulse amplitude	10	10	10	10
Pulse time period (second)	0.01	0.01	0.01	0.01
Pulse width (second)	0.001	0.001	0.001	0.001
Phase delay for $T1$ (second)	0.008	0.008	0.006	0.006
Phase delay for $T2$ (second)	0.018	0.014	0.016	0.016

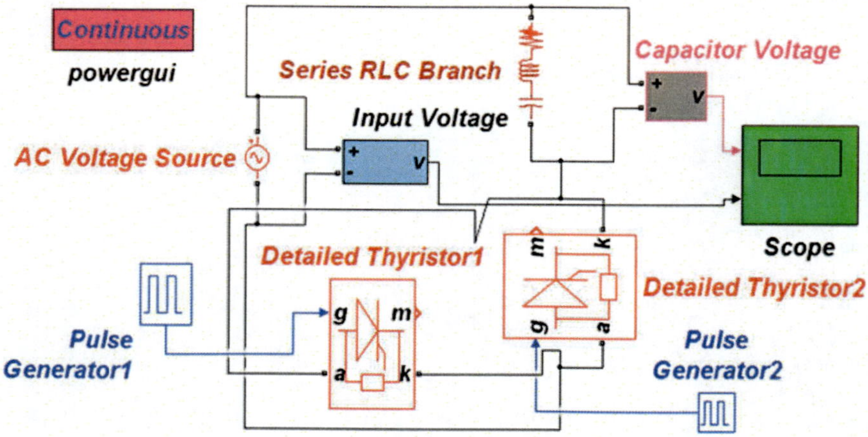

Fig. 5 Simulation diagram for TSC

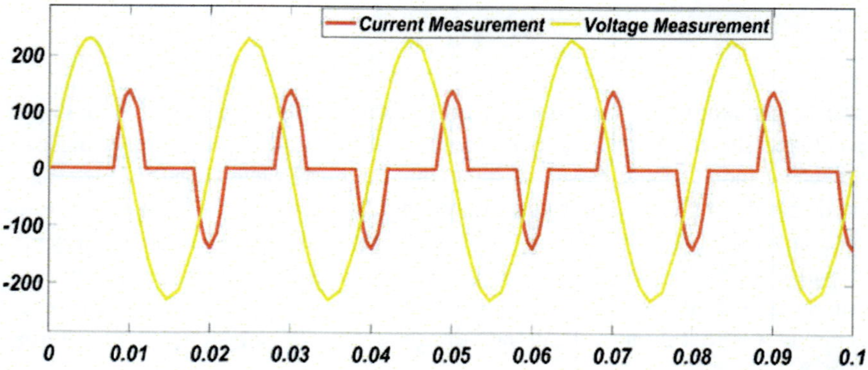

Fig. 6 Input voltage and capacitive voltage waveforms of TSC

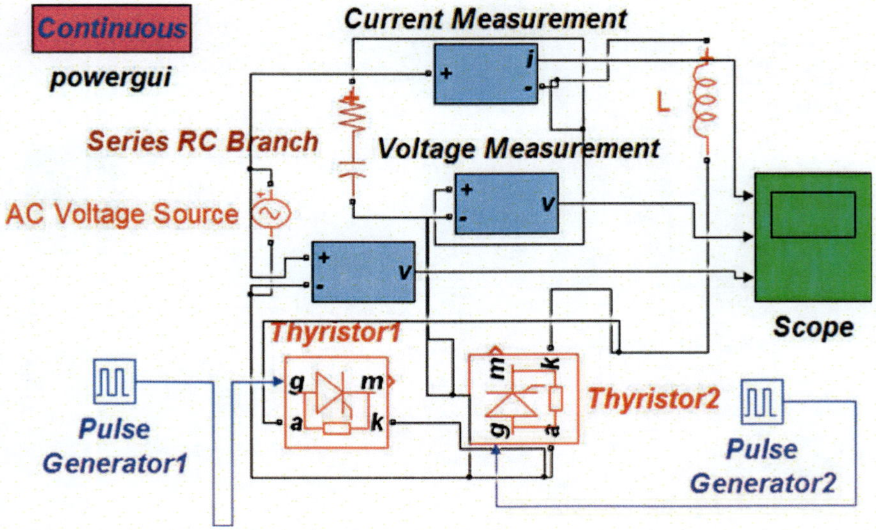

Fig. 7 Simulation diagram for FC-TCR

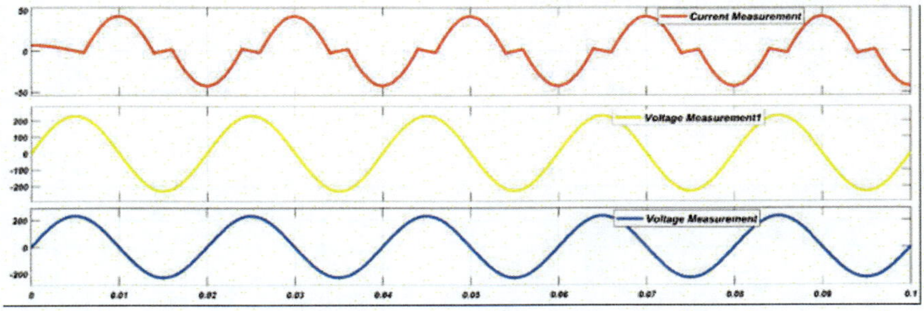

Fig. 8 Input voltage and capacitive voltage waveforms of FC-TCR

values is shown in Fig. 10. In this case only disadvantage of FC-TCR is removed as we have control over capacitor also. So, whenever we do not require extra reactive power, we can turn off capacitor by removing gate pulse. Inductor used in this circuit can have lower rating as that of capacitor. And for larger requirement we can also connect another TSCs and TCRs in parallel.

5 Conclusions

This study compares several FACTS devices TCSC, TSSR, TSSSC, SVC, STATCOM, SSSC, TCPST, and UPFC. The simulation results of integrated FACTS

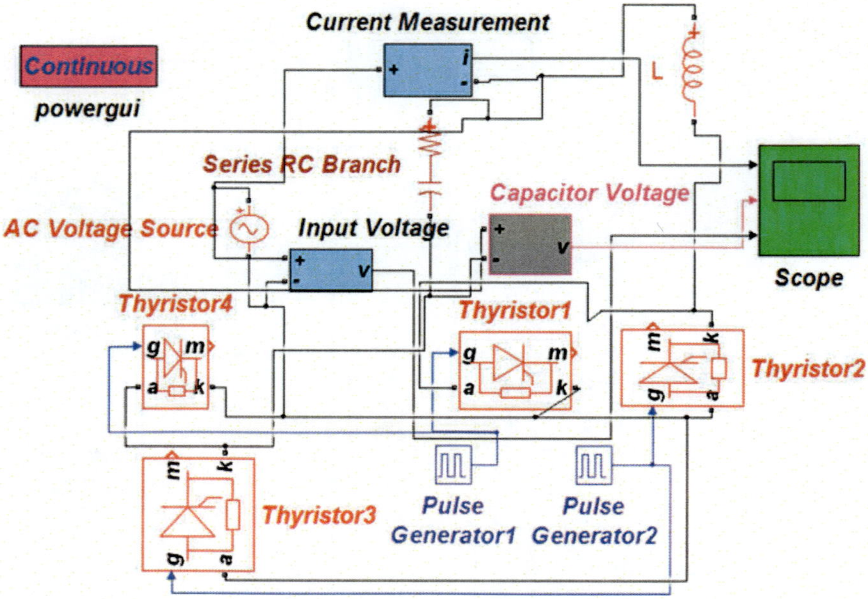

Fig. 9 Simulation diagram for TSC-TCR

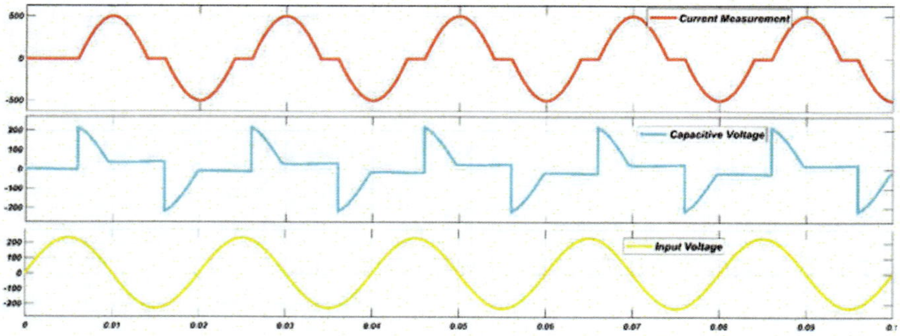

Fig. 10 Current, capacitive voltage, and input voltage waveforms of TSC-TCR

controller TCR, TSC, FC-TCR, TSC-TCR for transient state stability improvement are also presented. A FACT device improves AC grid reliability and lowers power delivery costs. Additionally, it conveys their features and mode of functioning after seeing their unique functionality. It is discovered that each gadget is unique in some manner. Each device has specific benefits. Furthermore, this power electronics converter-based controller will be very beneficial for dampening low-frequency power system oscillations and boosting power transfer capacity while also improving power quality. These FACTS devices may be deployed in the distribution network in the future using a stochastic optimization algorithm control method to enhance power stability and reliability.

References

1. Ismail B, Abdul Wahab NI, Othman ML, Radzi MA, Naidu Vijyakumar K, Mat Naain MN (2020) A comprehensive review on optimal location and sizing of reactive power compensation using hybrid-based approaches for power loss reduction, voltage stability improvement, voltage profile enhancement and loadability enhancement. IEEE Access 8:222733–222765. https://doi.org/10.1109/access.2020.3043297
2. Bhadoriya JS, Gupta AR (2021) Optimal share of DG and DSTATCOM in distribution network using firefly algorithm. In: Gupta OH, Sood VK (eds) Recent advances in power systems. Lecture notes in electrical engineering 699. Springer, Singapore. https://doi.org/10.1007/978-981-15-7994-3_46
3. Asano H, Bando S (2006) Load fluctuation analysis of commercial and residential customers for operation planning of a hybrid photovoltaic and cogeneration system. In: 2006 IEEE power engineering society general meeting. https://doi.org/10.1109/pes.2006.1709111.
4. Gupta AR, Kumar A (2019) Deployment of distributed generation with d-facts in distribution system: a comprehensive analytical review. IETE J Res:1–18. https://doi.org/10.1080/03772063.2019.1644206
5. Coates D (1998) Facts: A transmission utility perspective. IEE Colloquium Flex AC Transm Syst FACTS. https://doi.org/10.1049/ic:19980968
6. Wang L, Pourbeik P (2012) Assessment of power system stability and dynamic security performance. power system stability and control, 3rd edn, pp 1–20. https://doi.org/10.1201/b12113-19
7. Qi W, Jun Y, Liping L (2015) Optimal control strategy of a unified power flow controller incorporated with an active power source. In: 2015 IEEE international conference on cyber technology in automation, control, and intelligent systems (CYBER). https://doi.org/10.1109/cyber.2015.7287946
8. Selvarasu R, Kalavathi MS (2014) Self-adaptive firefly algorithm based transmission loss minimization using multi type facts devices. In: 2014 international conference on circuits, power and computing technologies [ICCPCT-2014]. https://doi.org/10.1109/iccpct.2014.7054761
9. Faridi M, Maeiiat H, Karimi M, Farhadi P, Mosleh H (2011) Power system stability enhancement using STATIC synchronous series COMPENSATOR (SSSC). In: 2011 3rd international conference on computer research and development. https://doi.org/10.1109/iccrd.2011.5764220
10. Wien M, Schwarz H, Oelbaum T (2007) Performance analysis of svc. IEEE Trans Circuits Syst Video Technol 17(9):1194–1203. https://doi.org/10.1109/tcsvt.2007.905530
11. Siddique A, Xu Y, Aslam W, Rasheed M (2019) A comprehensive study on facts devices to improve the stability and power flow capability in power system. In: 2019 IEEE Asia power and energy engineering conference (APEEC). https://doi.org/10.1109/apeec.2019.8720685
12. Muhammad Y, Khan R, Raja MA, Ullah F, Chaudhary NI, He Y (2020) Solution of optimal reactive power dispatch with facts devices: a survey. Energy Rep 6:2211–2229. https://doi.org/10.1016/j.egyr.2020.07.030
13. Li GJ, Lie TT, Shrestha GB, Lo KL (2000) Implementation of coordinated multiple facts controllers for damping oscillations. Int J Electr Power Energy Syst 22(2):79–92. https://doi.org/10.1016/s0142-0615(99)00039-3
14. Mithulananthan N, Canizares CA, Reeve J, Rogers GJ (2003) Comparison of PSS, SVC, and STATCOM controllers for damping power system oscillations. IEEE Trans Power Syst 18(2):786–792. https://doi.org/10.1109/tpwrs.2003.811181

15. Canizares CA, Faur ZT (1999) Analysis of SVC and TCSC controllers in voltage collapse. IEEE Trans Power Syst 14(1):158–165. https://doi.org/10.1109/59.744508
16. Gelen A, Yalcinoz T (2008) The behaviour of TSR-based SVC and TCR-based SVC installed in an infinite bus system. In: 2008 IEEE 25th convention of electrical and electronics engineers in Israel. https://doi.org/10.1109/eeei.2008.4736670
17. Chang Y (2006) Design of HVDC and SVC coordinate damping controller based on wide area signal. Int J Emerg Electr Power Syst 7(4). https://doi.org/10.2202/1553-779x.1304
18. Kundur P (2007) Power system stability. Power Syst Stab Control. https://doi.org/10.1201/9781420009248.sec2
19. Bruno S, De Carne G, La Scala M (2020) Distributed facts for power system transient stability control. Energies 13(11):2901. https://doi.org/10.3390/en13112901

IOT Enabled Smart Grid Coordinated Control Using Hierarchical PLC Architecture

Kumari Namrata, Abhishek Dayal, Ayush Ranjan, K. Arun, Dhanesh Tolia, and Akshit Samadhiya

Abstract The Smart Grid, an upgrade to the older ways of allot power fragments to industries together with regulations at home feeders. Solution is to network it out considering different heavy load applainces on the power system. As the power consumption on the system increases, shutdown cannot be avoided. Total shutdowns are unwilling main reason being the heavy-consuming devices. The level information transferred from the regional grid to the distribution substation and then to home level. Handy in manuplating heavy-consuming applainces in homes whereas low load devices like a light and fan continue to work. Customers wantings are recorded when overall expences is calculated. Recording of graphs showing their monthly power utilization using the application. Load deviation if the customers daily limit increases their limit, an pop up is sent through sms.

Keywords NodeMCU · PLC · Internet of things · Power distribution · Adafruit cloud

---

K. Namrata · A. Dayal (✉) · A. Ranjan · A. Samadhiya
Department of Electrical Engineering, NIT Jamshedpur, Jamshedpur, Jharkhand, India
e-mail: adayal487@gmail.com

K. Namrata
e-mail: namrata.ee@nitjsr.ac.in

A. Ranjan
e-mail: ayushranjan9911@gmail.com

A. Samadhiya
e-mail: akspinnacle1@gmail.com

K. Arun · D. Tolia
Department of ECE, NIT Jamshedpur, Jamshedpur, Jharkhand, India
e-mail: kakarun1908@gmail.com

D. Tolia
e-mail: dstolia07@gmail.com

K. Namrata et al. (eds.), *Smart Energy and Advancement in Power Technologies*, Lecture Notes in Electrical Engineering 926,
https://doi.org/10.1007/978-981-19-4971-5_7

1 Introduction

1.1 The Problem Background, Existing Algorithms and Issues

Power transmission is a core technology of the energy generation and consumption and value of this energy is null till the energy reaches the destination. Energy is produced in Mega and Giga Watts and transferred over hundreds of miles to arrive at load distribution points transferred to consumers through overhead power lines. Lack of intelligent and efficient energy transmission framework in a country can cause pitfalls in supplying energy to every locality even if it has surplus energy to feed its states.

Lack of efficient transmission infrastructure results in depleted energy supply to a major part of the population even through our country being energy surplus and irregular load patterns which causes mismatch in the supply and load side power. The magnitude of power consumed by the grid should be equal to magnitude of power fed into it, else it will result in a shutdown. Various factors result in the variation of renewable electrictricity production, being highly complex. These constant fluctuations need to be compensated by traditional power plants, since large magnitude of electricity cannot be stored efficiently over a long time duration. Smart Grids are a new technology for this era. But it turns out, this is not a topic of research. Let us have a look at few of these:

Intelligent Residential Energy Management System (IREMS) put together demand side management (DSM) and an exclusive setup used programming being linear as well as mixed integer major use of which is depletion in energy expense for basic home consumers [1]. In 2017, A Machine-Learning maintained home energy management system of low energy based on Bluetooth was proposed. This technology was made by superimposing communication technology over electrical network [2]. Similarly, BMS which stands for building management system was proposed in 2017 which monitored and coordinated an electric and self-movable stature of a building through Internet of Things (IoT), and PoE [3]. By conducting different bands of combined air-conditioners in 2017, technology of delivery and load formulation was introduced. Synchronizing of compound load by using VPSSS, i.e., Virtual power storage space scheme [4].

Another proposed technology utilizing active power management by coordinated multi point (CoMP) [5]. We also saw in 2016, a technology developed to keeping ecosystem for smart microgrid, known as Intelligent dynamic energy management system (IDEMS). Put forward technology combines whole relentless to the point load appeals and sends energy to load so that the microgrid can be consistent, ecologically sound and self-sustainable [6] But in 2016, Cyber-Physical Systems (CPSs) was developed using game theory based on coalition and energy-efficient smart grid [7]. In 2018, Vehicle-to-Grid (V2G)'s uses were made practical for PEV's that is plung in electric vehicles by 2 strategies to manage power consumption along with prevention of energy imbalance in corresponding microgrid [8].

Now with arrival of 2015, Distributed Location Marginal Pricing (DLMP) based on a Unified Energy Management System (UEMS) was found [9]. In 2017, one more idea was proposed aiming to reduce net production cost while keeping restorative system and caloric constraints by control concluding the good transmitting microgrids contain lesser voltage [10]. In the year 2017, (HoMeS) Home Energy Management System including storage was proposed [11].

We also found a system that had a Home Plug that could answer to PLC for overviewing a solar electric base [12]. In 2014, a setup was introduced called smart home energy management system (SHEMS) used to overview energy generation and utilization simultaneously [13]. Also, for research and development of solar electric power pump constant a Maximum power point tracking (MPPT) was introduced [14].

In 2013, in a smart home situation differentiability of heating devices and customer maintenance, excellent heating forcasting method and air conditioning was evaluated using Home energy management system (HEMS) [15]. Another method helped in protecting the formation hardships of power systems and gave two estimated resolutions using Markov Decision Process (MDP) [16]. In 2013, Functionalities had technique memorizing algorithms, announcement technology and sense using Smart Home Energy Management System (SHEMS) [17].

The major drawback of the approach adopted today is that a major part of the population has to suffer power outage because industrial loads cannot be interrupted.

2 Suggested Solution

Taking into account every loophole, are suggesting setup of an Adafruit (IoT) setup in between every high-consuming devices together with Regional load system. Taking into account the steps offered by the grid we use PLC automation to manipulate the voltage of devices. Therefore, ensuring the distribution of load correctly to all preventing a total load-shutdown in every district.

3 Conceptual Setup

Works in three parts:

(i) State Load Dispatch Center or Substation (ii) District Load Dispatch Center or Distributing Substation (iii) Locality Level or feeder or Home level,

The State Load Dispatch Center will capture data as the demand rises, NodeMCU (master) is introduced here. NodeMCU (2nd part) id in the stature of the substation. NodeMCU for both of the Mid and Master level will be connected by a series of regulations known as Message Queue Telemetary Transport (MQTT) protocol, all mid-level NodeMCU will follow directions as per the unique signals generated by the Master. This is presented in Fig. 1.

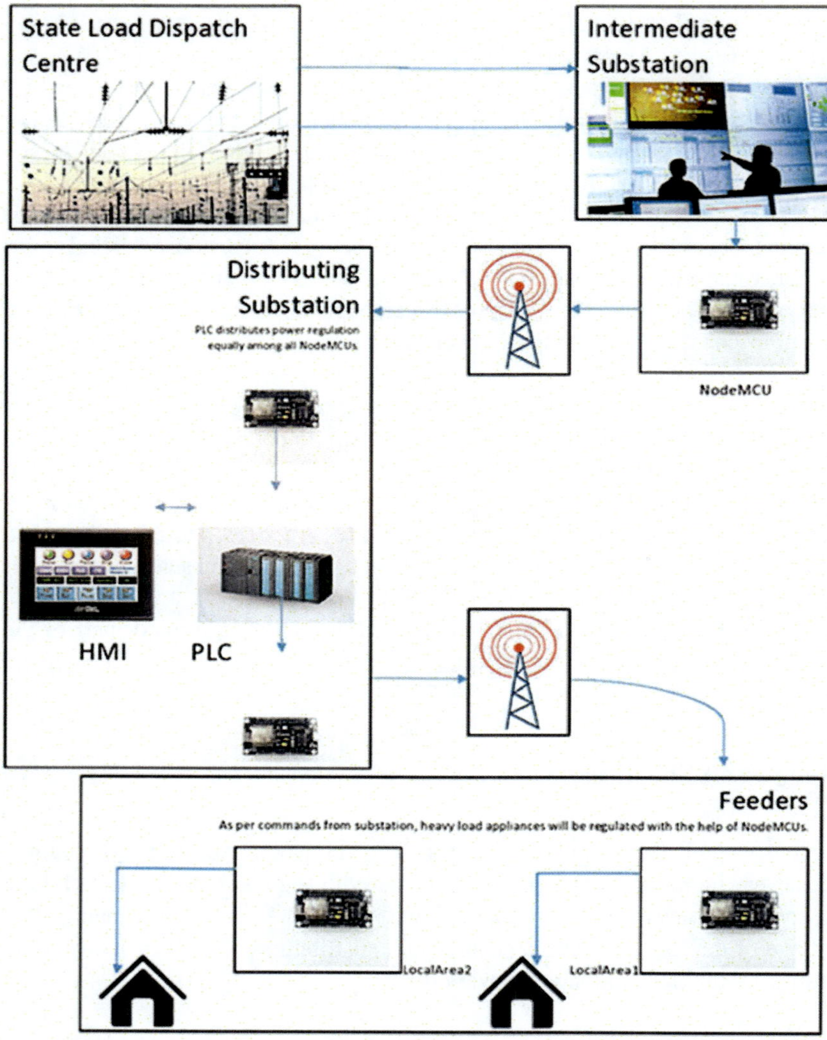

Fig. 1 Conceptual diagram

When the Master NodeMCU posts a heavy load signal on the grid, this signal is sent to the cloud as an event. Every middle layer element or mid-level circuit following this unique event will start the control and regulation of their activities and events under the Programmable block, i.e., PLC.

In the District Load Dispatch Center, 2nd part or mid-level NodeMCU transmits contrast signals through the Arduino UNO paired with relay based interface setup that sends the signal values into Programmable Logic Controller input pins.

PLC's output ports will send 3.30 V dc output which acts as an input mode to the next NodeMCU (Mid Layer-Level). This will send unique named signals under an event name which is captured by NodeMCU (slave) at the feeder or home level.

In the feeder level, NodeMCU (slave) will control and regulate the electrical appliances under them. Before control and regulation, a warning will be issued to the customer. These all NodeMCUs which are at slave level will also record power consumption of these electrical appliances which will be available for the consumer when demanded.

4 Hardware Setup

We divide the setup into three parts:

(A) State Load Dispatch Center
(B) District Load Dispatch Center
(C) Home level

4.1 Intermediate Substation

The power demand of a customer against time recorded on daily basis is known as a 'load curve'. These curves are fruitful in visualizing diagrams of consumption, maxima and dips and energy use prediction in a part or in a distribution network as the case may favor. We can provide better prediction of loads/demand, we can improve the daily demand curves. While we already have the everyday data of incoming supply for a region and we also have a regulation mechanism, we can compare them instantaneously and with the regulation, we can provide 24 * 7 basic electricity supply to everyone at every instant.

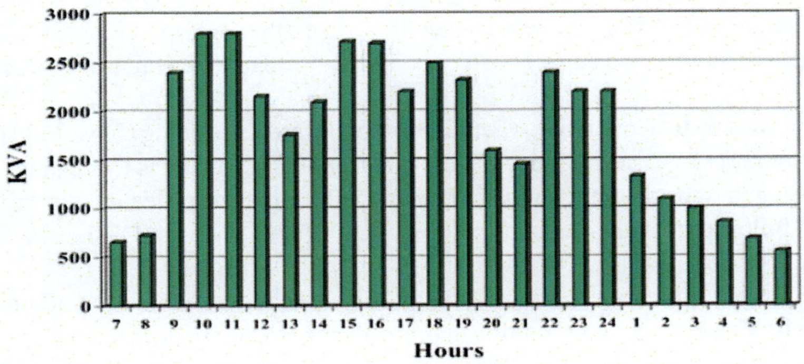

Prediction Mechanism:

We will have such graphs of everyday. From the state load dispatch centers and dispatch centers at district level, we will collect data and graphs. We will put all of them together and by using a Machine-Learning model, we will have unique load patterns through data-driven models.

This will enable us to have demand graphs which will be generated in advance for each single day. Now, according to the predicted demands and peak load for a day, the generating stations can be directed prior to each single day about the capacity at which they will work for the weekdays. So, we will have demand graphs and supply units which will be near and equal to each other. These two overlapping graphs will make the grid stable and reduce fluctuations in power.

We assume, any model cannot be perfect at the beginning. So, we will have deflections on site. But, that deflections will be small and we will reach to closer predictions with every passing year and also adding more complexity in the algorithm setting it according to the different needs of the data. Small fluctuations between supply and demand can be handled, by local equipment which can instantaneously add or store power for certain period of time.

We will also use regulation mechanisms which is exclusively designed to regulate the power consumptions in high load devices. We will use comparison tables for activating the grid under certain mechanism, if and only if the grid does not have the supply to fulfill the demands during a day. Otherwise, if there is enough power in the grid then the grid will work at normal. One typical table is shown below.

Comparison Table:

Demand level	Difference in power (in MW)
Very-high	>7500
High	5000–7500
Medium	2500–5000
Low	1000–2500
Very low	0–1000

So, from the above table, the regulation mechanism is followed. And, after following this table, if the power balance is not met, then we will employ instantaneous power storage and supply devices, which can add or extract power in/from the grid.

So, according to above table, whenever power difference is going above 7500 MW, the personnel at local substation will press the "Very-High" button in the regional grid and the regulation according to that command will start at every feeder. Similarly, it will be followed for other regulation levels viz. "High", "Medium", "Low", "Very Low".

The power limit chosen here is typical value chosen to exemplify the purpose of the levels. There is enough room for improvement based on real-time data.

The demand levels during different seasons, months, days, events, festivals, etc. is studied under the lens of Machine-Learning algorithm which is quite new and currently in development phase. The predictions depend on the volume of data supplied to it. With every passing year, the predictions get better which is boon for the generating units.

4.2 At Distributing Substation

NodeMCU (Medium-mode) gather specific consumption mode through state load dispatch center. Among 5 digital ports get the readings from adafruit io mechanism ensuring that every port specifies a deviated mode of dispatch center.

Having provided 5 input ports in automated PLC. The respective ports in PLC are switched on by pulses transferred from MCUs Adafruit cloud at dispatch center powered by a Relay-Arduino medium from 3.3–24 V.

Dividing the system into 4 sub parts, each is regulated through 1 NodeMCU, setup of four NodeMCUs regulating 5 devices (Fig. 2).

The voltage is classified into 5 modes deviating from 0 to 3.3 V (Table 1).

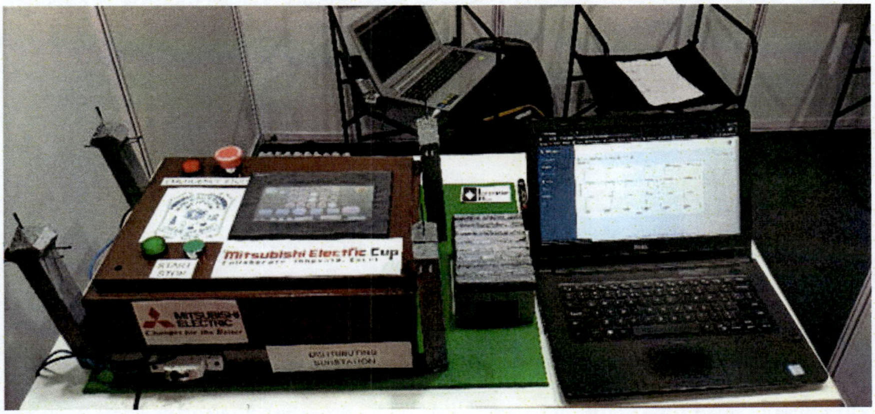

Fig. 2 Assembled apparatus of dispatch center with our monitoring website. *Note* Human Machine Interface for monitoring at the state load dispatch center

Table 1 Ports and voltages according to given load

Consumption mode	Ports of NodeMCU	Respective voltages (V)
Very low	*D*6	3.300
Low	*D*5	2.475
Medium	*D*2	1.650
High	*D*1	0.825
Very high	*D*0	0.000

Fig. 3 End model of home level and the output applainces which functions we regulate

Output ports of the PLC transfers the signals from dispatch center level to feeders via [MCU (Substation) to adafruit MCU (Feeder)] stature. Each NodeMCU attached to output ports of plc as:

- NodeMCU4: Y16-Y20
- NodeMCU3: Y11Y15
- NodeMCU2: Y06-Y10
- NodeMCU1: Y1-Y5

4.3 Home Level

NodeMCU (feeder level) on getting their mode begins regulating the appliances' function (see Fig. 3). Before power shutdown or other limitations, there would be a mail generated for the customer by the NodeMCU module. The software will gather power usage particulars of every high power consuming appliances and store them to the adafruit io. Graphs will be shown to users based on data which is compiled throughout the operations, available for customers on demand.

It should be noted that our technology would only function for "Devices that are Smart", i.e., devices containing a microcontroller or regulating device already fixed in them and also be able to connect to the Internet.

4.4 IoT Security

Security in the IoT domain has acquired prime stage as more and more attacks on IoT networks occur around the world. Many small and large scale IoT devices have become victim of hacking and data leak. Thus, putting some provisions for security in the project is of utmost importance. With this belief, all NodeMCU microcontrollers have been programmed to send and receive data over Secure Socket Layer (SSL) from the Adafruit IO cloud. The SSL helps establish authenticated and encrypted links between networked devices. This in turn prevents unauthorized personnel with malicious intent from accessing the information while it is in transit between micro-controller and cloud computer. The Man-in-the-Middle attack is one of the most common ways an IoT network is compromised.

4.5 Proposed PLC Algorithm

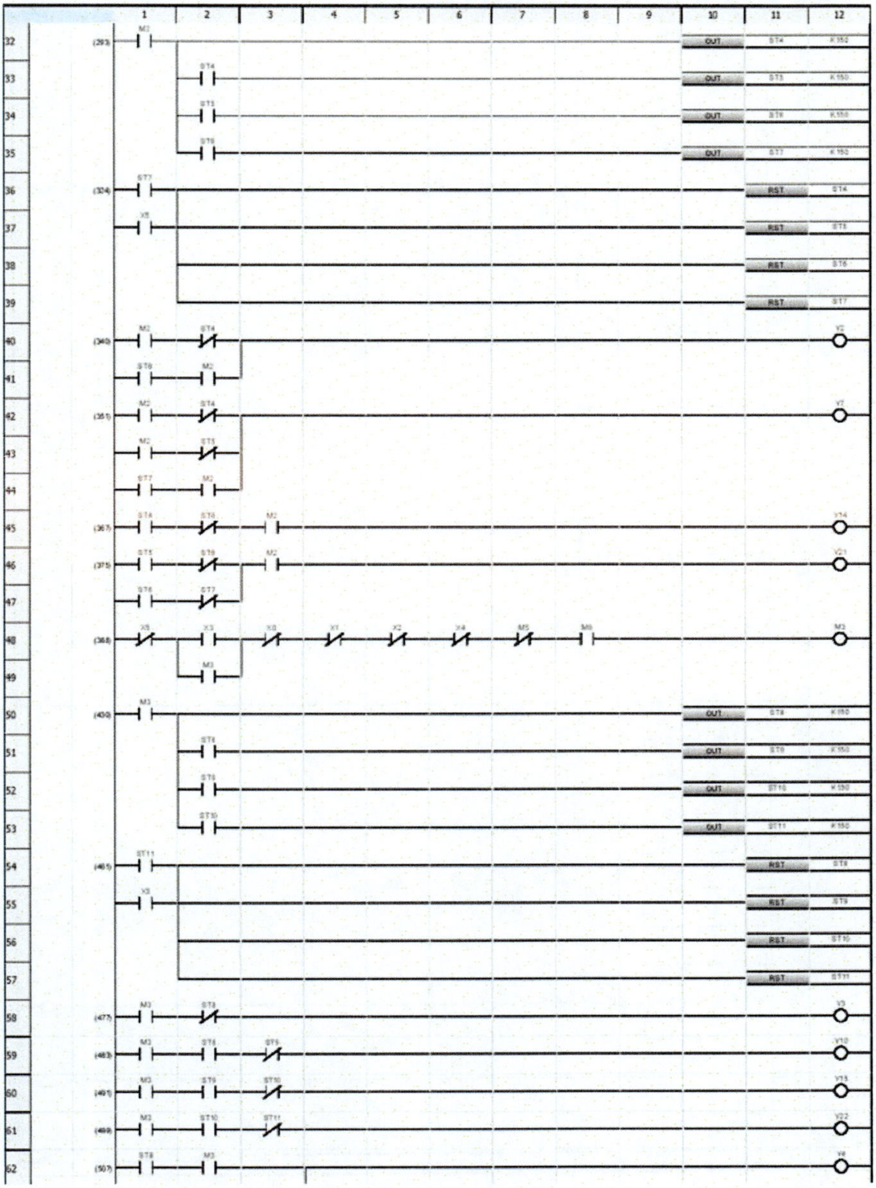

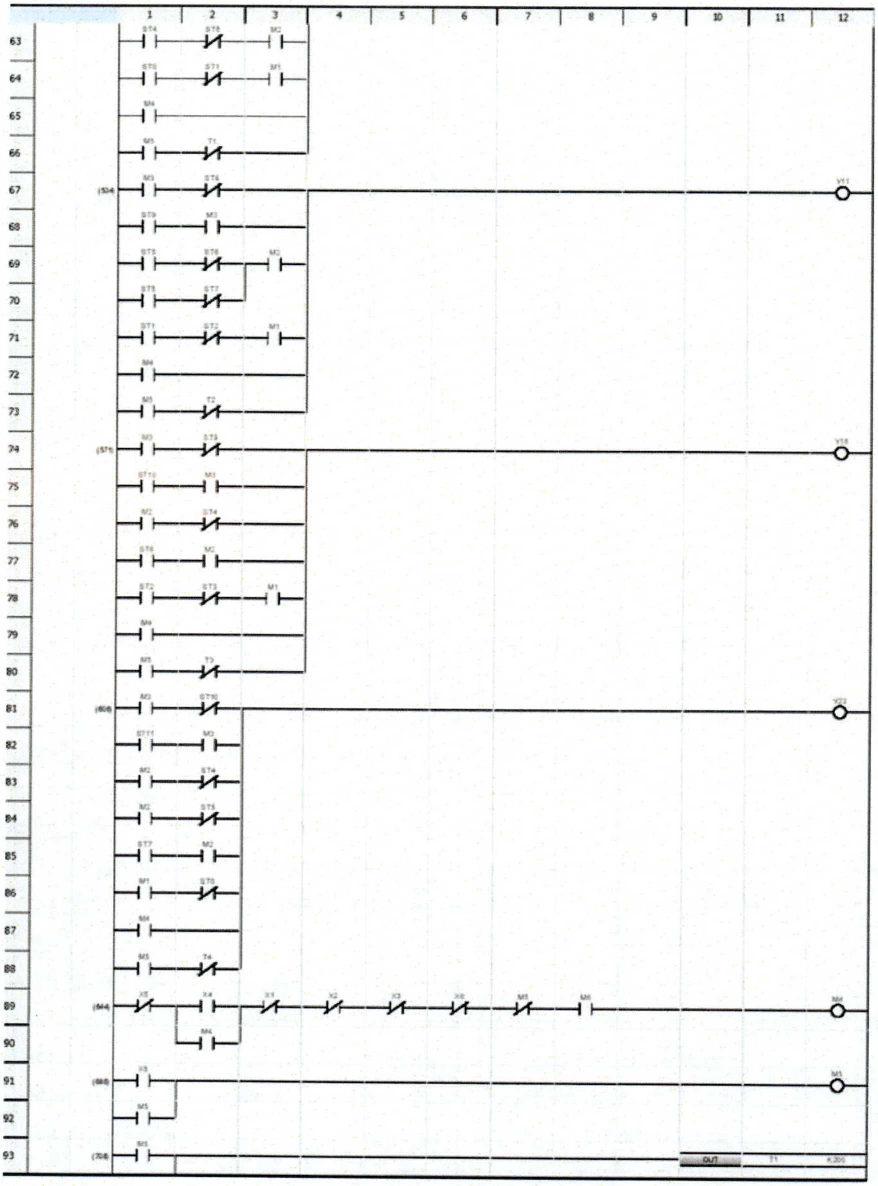

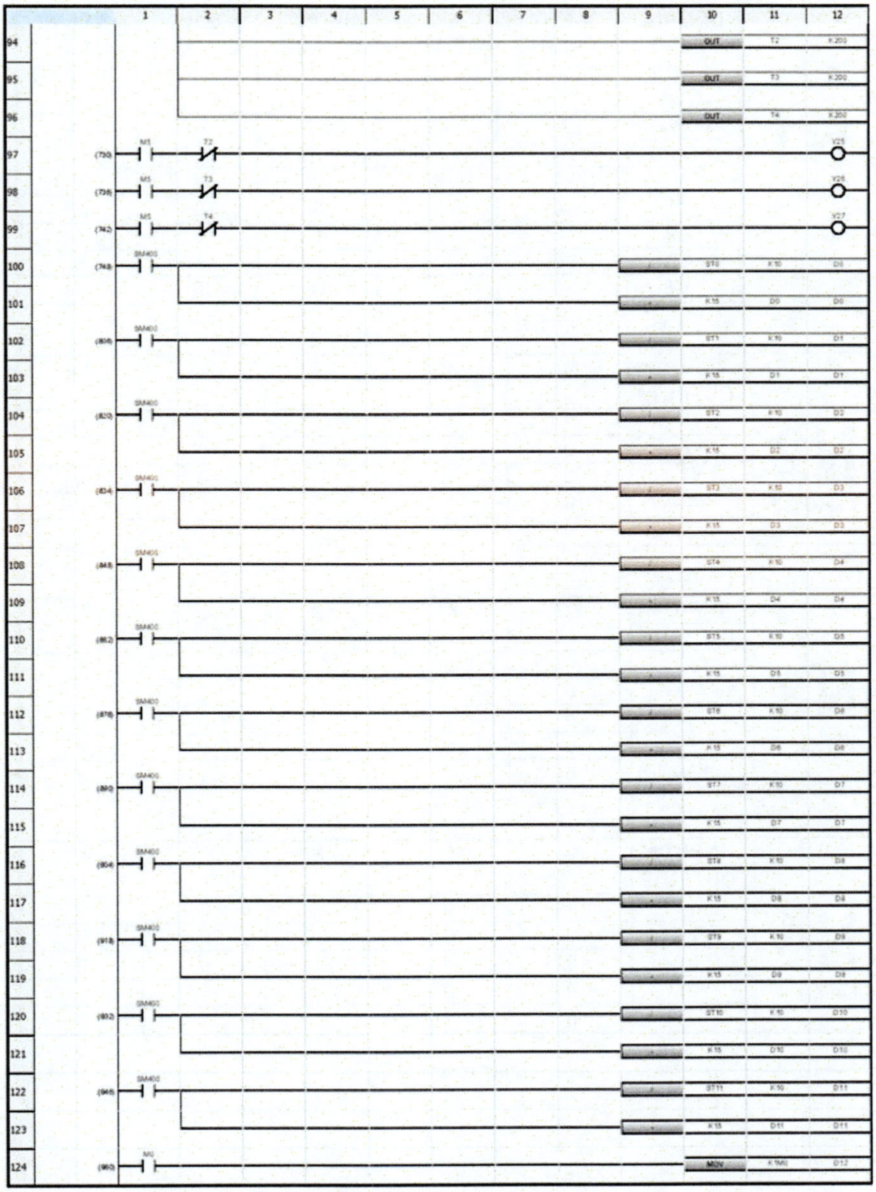

	DEMAND FEED FROM ITSS	VERY HIGH DEMAND	HIGH DEMAND	MEDIUM DEMAND	LOW DEMAND	VERY LOW DEMAND
N1 Feeder 1	INPUT→ OUTPUT↓	X0	X1	X2	X3	X4
	Y0	1	0	0	0	0
	Y1	0	1	0	0	0
	Y2	0	0	1	0	0
	Y3	0	0	0	1	0
	Y4	0	0	0	0	1

	INPUT→ OUTPUT↓	X0	X1	X2	X3	X4
N2 Feeder 2	Y5	1	0	0	0	0
	Y6	0	1	0	0	0
	Y7	0	0	1	0	0
	Y10	0	0	0	0	0
	Y11	0	0	0	1	1

	INPUT→ OUTPUT↓	X0	X1	X2	X3	X4
N3 Feeder 3	Y12	1	0	0	0	0
	Y13	0	1	0	0	0
	Y14	0	0	0	0	0
	Y15	0	0	0	0	0
	Y16	0	0	1	1	1

	INPUT→ OUTPUT↓	X0	X1	X2	X3	X4
N4 Feeder 4	Y17	1	0	0	0	0
	Y20	0	0	0	0	0
	Y21	0	0	0	0	0
	Y23	0	0	0	0	0
	Y24	0	1	1	1	1

Input output table

5 Results

Table 2 briefs the results for the recommended solution when different orders are generated from the State Load Dispatch Center or Intermediate Substation (Figs. 4 and 5).

Table 3 projects an air coditioner temperature being controlled by a drop in controlling voltage of respective NodeMCU's.

Table 2 Activated modes and voltage values generated out by various NodeMCUs

NodeMCU/Voltage when different pins are high	*D*0 (V)	*D*1 (V)	*D*2 (V)	*D*5 (V)	*D*6 (V)
1st NodeMCU	0	0.825	1.65	2.475	3.3000
2nd NodeMCU	0	0.825	1.65	3.300	3.3000
3rd NodeMCU	0	0.825	3.30	3.300	3.3000
4th NodeMCU	0	3.300	3.30	3.300	3.3000

Fig. 4 Graph projecting signal values by home level NodeMCU's

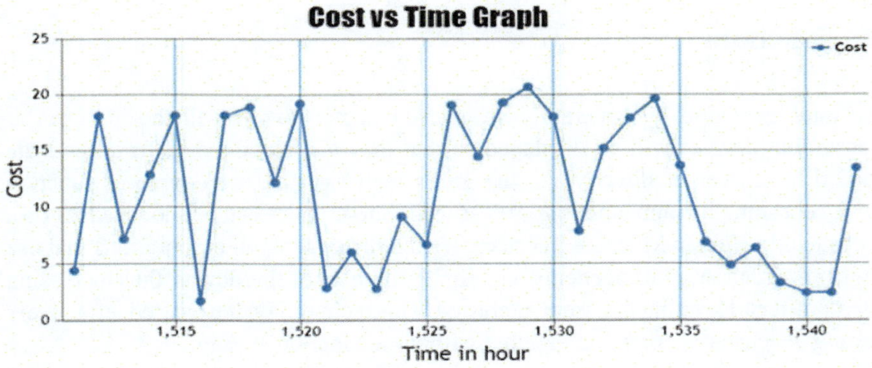

Fig. 5 Graph representing cost values against time

There is a website through which remote management and monitoring of the grid is made possible for the controller (Fig. 6).

Table 3 Voltage transferred by MCU and temp

MCU output V given to an AC (air conditioner) (V)	Temperature control of an air conditioner
0.000	Appliance off
0.825	Fixed at 32 °C
1.650	Between 32 and 25 °C
2.475	Between 25 and 20 °C
3.300	Any temp as per user

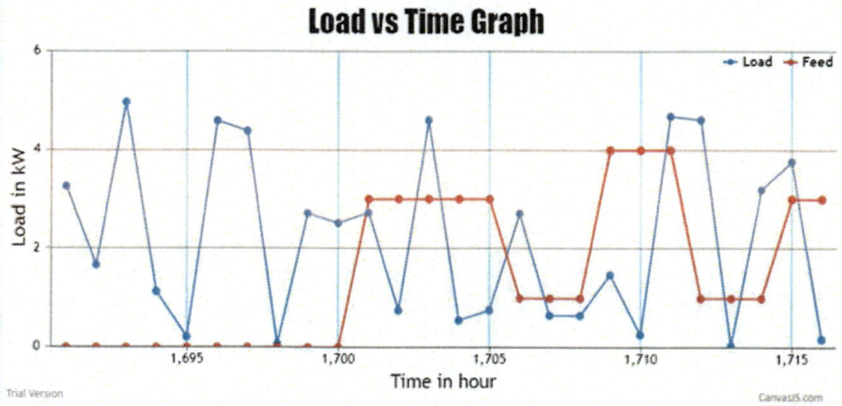

Fig. 6 Changing load and signal values shown in a graphical representation together

6 Conclusion

An intelligent grid by generating an adafruit io network for all the high power consuming devices will be displayed so that they can receive data from the state load dispatch center directly. As the grid's load variates, the degree of function heavy consumption applaince variates as well. Surveillance can be shown either on a website or a Human Machine Interface. Using this system, all corporations that use high power can save lots of energy and trim the number of shutdowns, thus increasing their profit and keeping the users interested. This technology, if executed, effectively revamp electricity generation together with dissemination system.

Acknowledgements This research did receive funding from Mitsubishi Electric India Pvt Lmt, NIT Jamshedpur Alumni base and TEQIP-3 research fund . The author would like to thank her own students Abhishek Dayal, Ayush Ranjan, Dhanesh Tolia, K Arun, Akshit Samadhiya for developing the IOT and PLC codes and creating a working template model of the above research. This work was an outcome of dedication and motivation provided by Dr. Kumari Namrata, Associate Professor(EED), NIT JAMSHEDPUR.

References

1. Hill K, Dong Z, Meng DJ, Zheng Y (2017) Coordinated dispatch of virtual energy storage systems in smart dstribution networks for loading management. IEEE Trans Syst Man Cybern Syst 49(4):776–786
2. Pau M, Collotta G (2017) An innovative approach for forcasting of energy requirements to improve a smart home management system based on BLE. IEEE Trans Green Commun Networking 1(1):112–120
3. Arun SL, Selvan MP (2018) Intelligent residential energy management system for dynamic demand response in smart buildings. IEEE Syst J 12(2):1329–1340
4. Minoli D, Sohraby K, Occhioggrosso B (2017) IoT considerations, requirements, and architectures for smart buildings—energy optimization and next-generation building management systems. IEEE Internet Things J 4(1):269–283
5. Battula S, Nunna HK, Doolla S, Srinivasan D (2018) Energy management in smart distribution systems with vehicle-to-grid integrated microgrids. IEEE Trans Smart Grid 9(2):4004–4016
6. Kumar N, Zeadally S, Misra SC (2016) Mobile cloud networking for efficient energy management in smart grid cyber-physical systems. IEEE Wireless Commun 23(5):100–108
7. Wang X, Zhang Y, Chen T, Giannakis GB (2016) Dynamic energy management for smart-grid-powered coordinated multipoint systems. IEEE J Sel Areas Commun 34(5):1348–1359
8. Venayagamoorthy GK, Sharma RK, Gautam PK, Ahmadi A (2016) Dynamic energy management system for a smart microgrid. IEEE Trans Neural Networks Learn Syst 27(8):1643–1656
9. Mondal A, Obaidat MS, Misra S (2017) Distributed home energy management system with storage in smart grid using game theory. IEEE Syst J 11(3):1857–1866
10. Krishnan R, Wang K, Ouyang Z, Shu L, He L (2015) A game theory-based energy management system using price elasticity for smart grids. IEEE Trans Ind Inform 11(6):1607–1616
11. Bracco S, Bracco M, Delfino F, Procopio R (2017) An energy management system for the savona campus smart polygeneration microgrid. IEEE Syst J 11(3):1799–1809
12. Park WK, Han J., Choi CS, Lee I, Kim SH (2014) Smart home energy management system including renewable energy based on ZigBee and PLC. IEEE Trans Consum Electron 60(2):198–202
13. Han J, Choi CS, Park WK, Lee I, Kim SH (2014) PLC-based photovoltaic system management for smart home energy management system. IEEE Trans Consum Electron 60(2):184–189
14. Tsai CY, Hsieh HI, Hsieh GC (2014) Photovoltaic burp charge system on energy-saving configuration by smart charge management. IEEE Trans Power Electron 29(4):1777–1790
15. Hu Q, Li F (2013) Hardware design of smart home energy management system with dynamic price response. IEEE Trans Smart Grid 4(4):1878–1887
16. Xu Z, Jia QS, Guan X, Shen J (2013) Smart management of multiple energy systems in automotive painting shop. IEEE Trans Autom Sci Eng 10(3):603–614
17. Joo SK, Jo HC, Kim S (2013) Smart heating and air conditioning scheduling method incorporating customer convenience for home energy management system. IEEE Trans Consum Electron 59(2):316–322

Modelling and Analysis of a Grid Connected Squirrel Cage Induction Generator

Amit Aniket and Atma Ram Gupta

Abstract At present major part of the electricity is generated by using non-renewable resources like coal. So there is need to shift from this conventional source of energy to a non-conventional one like Solar, Wind etc. In this paper the focus is on Wind Energy to meet the demands of electricity in our country. In the past decades there has been an increase in Wind energy power generation. Wind energy is directly fed to the power grids. But it is a challenge to connect wind turbines to the grids as power system stability becomes a major concern. There has been many improvements in the power system stability and the design of Wind turbines used in Wind Energy Conversion Systems (WECS). There are two main types–DFIG and SCIG. This paper presents the SCIG and two case studies to understand how it behaves when connected to the grid. A simulation of WECS is done in MATLAB Simulink to understand it better. Wind generators can be modelled in different reference frames like Space Vector, direct quadrature (dq), etc. In this paper modelling is done using Space Vector model and making some changes to it to obtain dq reference frame model and hence modelling is done in arbitrary reference frame.

Keywords Squirrel cage induction generator (SCIG) · Reference frame theory · Wind energy · Mathematical model

Abbreviations

WECS	Wind Energy Conversion Systems
IG	Induction Generator
SCIG	Squirrel Cage Induction Generator
DFIG	Doubly Fed Induction Generator

A. Aniket (✉) · A. R. Gupta
National Institute of Technology, Kurukshetra, Haryana, India
e-mail: connectamit29@gmail.com

A. R. Gupta
Shri Phanishwar Nath Renu Engineering College, Araria, India

K. Namrata et al. (eds.), *Smart Energy and Advancement in Power Technologies*, Lecture Notes in Electrical Engineering 926,
https://doi.org/10.1007/978-981-19-4971-5_8

WRSG	Wound Rotor Synchronous Generator
PMSG	Permanent Magnet Synchronous Generator
DGC	Direct Grid Connection
Dq	Direct Quadrature
MATLAB	MATrix LABoratory
Eqv	Equivalent
Ckt.	Circuit

1 Introduction

This paper gives an insight into the Wind Energy Conversion Systems. In WECS DFIG and SCIG are widely used. The constructional features of both these generators are same with some minor differences in the rotor structure. Laminations of thin silicon steel are used to make the stator which helps in reducing iron losses due to eddy currents. Three phase winding is done along the canal formed by the stacked laminations. A rotating magnetic field can be generated by connecting the stator winding to a 3-φ supply. This induces a 3-phase voltage which then gives the rotor current in the rotor bars. Eventually electromagnetic torque is generated when current and rotating field interacts with each other. Reference frame theory [1] is used to present different models of the induction generator. A simulation of WECS [2] is done in MATLAB Simulink to understand it better.

The stator construction of DFIG [3] is similar to that of SCIG. But the rotor part is different. The rotor is connected to the power converter through slip rings which controls the rotor current, rotor speed and grid current. It is not necessary to have the same rotor frequency and grid frequency. The active and reactive power can also be controlled by using this arrangement.

The induction generator can be modelled in two ways–space vector theory and dq-axis model [4]. The advantage of using a space vector model is that the equations used are very clear and can be easily expressed using an eqv. Ckt. But the disadvantage is that it contains complex variables which makes the analysis a bit tough since both real and imaginary parts are used. Now coming to the dq-axis model, the use of complex variable is eliminated but two eqv. Circuits have to be used. So, in this paper both the models [5] are discussed and some conversions are also used to convert the data from one reference frame to other.

2 Space Vector Model

- Assumptions
  - (1) the induction generator has a symmetrical structure.

- (2) All the 3-phases are balanced in the IG
- (3) core losses are negligible
- (4) Linear magnetic core for both stator and rotor

- The space vector model can be represented by using 3 equations that are: [2]

- voltage equations,
- flux linkage equations,
- motion equation

2.1 Voltage Equation

$$\begin{cases} \overrightarrow{v_s} = R_s \overrightarrow{i_s} + p\overrightarrow{\emptyset_s} + j\omega\overrightarrow{\emptyset_s} \\ \overrightarrow{v_r} = R_r \overrightarrow{i_r} + p\overrightarrow{\emptyset_r} + j(\omega - \omega_r)\overrightarrow{\emptyset_r} \end{cases} \tag{1}$$

where

$\overrightarrow{v_s}$, $\overrightarrow{v_r}$	voltage vectors (stator and rotor) (V)
$\overrightarrow{\emptyset_s}$, $\overrightarrow{\emptyset_r}$	flux linkage vectors (Wb)
$\overrightarrow{i_s}$, $\overrightarrow{i_r}$	current vectors (A)
R_s, R_r	winding resistances (Ω)
ω	arbitrary reference frame's rotating speed (rad/s)
ω_r	electrical angular speed of rotor (rad/s)
p	derivative operator $\left(p = \frac{\mathrm{d}}{\mathrm{d}t}\right)$

2.2 Stator and Rotor Flux Linkages Equation

$$\begin{cases} \overrightarrow{\emptyset_s} = (L_{\mathrm{ls}} + L_m)\overrightarrow{i_s} + L_m\overrightarrow{i_r} = L_s\overrightarrow{i_s} + L_m\overrightarrow{i_r} \\ \overrightarrow{\emptyset_r} = (L_{\mathrm{lr}} + L_m)\overrightarrow{i_r} + L_m\overrightarrow{i_s} = L_r\overrightarrow{i_r} + L_m\overrightarrow{i_s} \end{cases} \tag{2}$$

where

$L_s = L_{\mathrm{ls}} + L_m$	self-inductance (stator) (H)
$L_r = L_{\mathrm{lr}} + L_m$	self-inductance (rotor) (H)
L_{ls}, L_{lr}	leakage inductances (stator and rotor) (H)
L_m	magnetizing inductance (H)

All the rotor-side parameters including the variables in above eq. are written by referring to the stator side.

2.3 Motion Equation

$$\begin{cases} J\frac{d\omega_m}{dt} = T_{\text{em}} - T_{\text{mech}} \\ T_{\text{em}} = \frac{3N_P}{2} R_e\left(j\overrightarrow{\emptyset_s}\,\overrightarrow{i_s}*\right) = -\frac{3N_P}{2} R_e\left(j\overrightarrow{\emptyset_r}\,\overrightarrow{i_r}*\right) \end{cases} \tag{3}$$

where

J	moment of inertia (rotor) (kgm^2)
T_{mech}	mechanical torque (N m)
N_P	No. of pole pairs
T_{em}	electromagnetic torque (N m)
ω_m	mechanical speed of rotor, $\omega = \omega_r/N_P$ (rad/s)

3 dq Reference Frame Model

To obtain the dq-axis model of an IG from space vector model, the space vectors are decomposed to the corresponding d- and q- axis components.

$$\begin{cases} \overrightarrow{v_s} = v_{s(d)} + jv_{s(q)};\ \overrightarrow{i_s} = i_{s(d)} + ji_{s(q)};\ \overrightarrow{\emptyset_s} = \emptyset_{s(d)} + j\emptyset_{s(q)} \\ \overrightarrow{v_r} = v_{r(d)} + jv_{r(q)};\ \overrightarrow{i_r} = i_{r(d)} + ji_{r(q)};\ \overrightarrow{\emptyset_r} = \emptyset_{r(d)} + j\emptyset_{r(q)} \end{cases} \tag{4}$$

- Substituting Eq. (4) into Voltage Eq. (1) of space vector model (Sect. 2.1) and equating real and imaginary components of the equations, we get the *dq-axis* voltage equations for the IG.

$$\begin{cases} v_{s(d)} = R_s i_{s(d)} + p\emptyset_{s(d)} - \omega\emptyset_{s(q)} \\ v_{s(q)} = R_s i_{s(q)} + p\emptyset_{s(q)} + \omega\emptyset_{s(d)} \\ v_{r(d)} = R_r i_{r(d)} + p\emptyset_{r(d)} - (\omega - \omega_r)\emptyset_{r(q)} \\ v_{r(q)} = R_r i_{r(q)} + p\emptyset_{r(q)} + (\omega - \omega_r)\emptyset_{r(d)} \end{cases} \tag{5}$$

- Substituting Eq. (4) into Eq. (2) we get the *dq-axis* flux linkages

$$\begin{aligned} \emptyset_{s(d)} &= (L_{ls} + L_m)i_{s(d)} + L_m i_{r(d)} = L_s i_{s(d)} + L_m i_{r(d)} \\ \emptyset_{s(d)} &= (L_{ls} + L_m)i_{s(q)} + L_m i_{r(q)} = L_s i_{s(q)} + L_m i_{r(q)} \\ \emptyset_{r(d)} &= (L_{lr} + L_m)i_{r(d)} + L_m i_{s(d)} = L_s i_{r(d)} + L_m i_{s(d)} \\ \emptyset_{r(q)} &= (L_{lr} + L_m)i_{r(q)} + L_m i_{s(q)} = L_s i_{r(q)} + L_m i_{s(q)} \end{aligned} \tag{6}$$

- Expressions for the torque

$$T_{\text{em}} = \begin{cases} \frac{3N_P}{2}\left(i_{s(q)}\emptyset_{s(d)} - i_{s(d)}\emptyset_{s(q)}\right) \\ \frac{3N_P L_m}{2}\left(i_{s(q)}\emptyset_{r(d)} - i_{s(d)}\emptyset_{r(q)}\right) \\ \frac{3N_P L_m}{2L_r}\left(i_{s(q)}\emptyset_{r(d)} - i_{s(d)}\emptyset_{r(q)}\right) \end{cases} \tag{7}$$

4 Simulation Model

- Rearranging Eq. (5) we get

$$\begin{cases} \emptyset_{s(d)} = \frac{1}{S}\left(v_{s(d)} - R_s i_{s(d)} + \omega\emptyset_{s(q)}\right) \\ \emptyset_{s(q)} = \frac{1}{S}\left(v_{s(q)} - R_s i_{s(q)} - \omega\emptyset_{s(d)}\right) \\ \emptyset_{r(d)} = \frac{1}{S}\left(v_{r(d)} - R_r i_{r(d)} + (\omega - \omega_r)\emptyset_{r(q)}\right) \\ \emptyset_{r(q)} = \frac{1}{S}\left(v_{r(q)} - R_r i_{r(q)} - (\omega - \omega_r)\emptyset_{r(d)}\right) \end{cases} \tag{8}$$

P is the derivative operator in Eq. (5)
S represents Laplace operator which replaced p in the Eq. (5)
$1/S$ represents an integrator.

The dq-axis flux linkage Eqs. (8) are converted into a matrix form for simplicity:

$$\begin{bmatrix} \emptyset_{s(d)} \\ \emptyset_{s(q)} \\ \emptyset_{r(d)} \\ \emptyset_{r(q)} \end{bmatrix} = \begin{bmatrix} L_s & 0 & L_m & 0 \\ 0 & L_s & 0 & L_m \\ L_m & 0 & L_r & 0 \\ 0 & L_m & 0 & L_r \end{bmatrix} \cdot \begin{bmatrix} i_{s(d)} \\ i_{s(q)} \\ i_{r(d)} \\ i_{r(q)} \end{bmatrix} \tag{9}$$

- Multiplying inverse of L (matrix of inductance) on both sides of the Eq. (9):-

$$[\lambda] = [L][i] \rightarrow [L]^{-1}[\lambda] = [L]^{-1}[L][i] \rightarrow [i] = [L]^{-1}[\lambda] \tag{10}$$

$$\begin{bmatrix} i_{s(d)} \\ i_{s(q)} \\ i_{r(d)} \\ i_{r(q)} \end{bmatrix} = \frac{1}{k} \begin{bmatrix} L_r & 0 & -L_m & 0 \\ 0 & L_r & 0 & -L_m \\ -L_m & 0 & L_s & 0 \\ 0 & -L_m & 0 & L_s \end{bmatrix} \cdot \begin{bmatrix} \emptyset_{s(d)} \\ \emptyset_{s(q)} \\ \emptyset_{r(d)} \\ \emptyset_{r(q)} \end{bmatrix} \tag{11}$$

where $k = L_s L_r - L_m^2$.

- The torque and rotor speed equations for the model can be expressed as

$$\begin{cases} \omega_r = \frac{N_P}{JS}(T_{em} - T_{mech}) & (a) \\ T_{em} = \frac{3N_P}{2}\left(i_{s(q)}\emptyset_{s(d)} - i_{s(d)}\emptyset_{s(q)}\right) & (b) \end{cases} \tag{12}$$

Table 1 2.3 MW, 690 V, 50 Hz SCIG parameters

Sr. No.	Parameters	SI unit	p.u
1	Output power (Rated)	2.30 MW	–
2	Line-to-line voltage (Rated)	690 V (rms)	–
3	Phase voltage (Rated)	398.84 V (rms)	1.0
4	Stator frequency (Rated)	50 Hz	1.0
5	Power factor (Rated)	0.888	–
6	Rotor speed (Rated)	1512 rpm	1.0
7	Synchronous speed, ω_s	1500 rpm	–
8	Slip (Rated)	−0.008	–
9	Pole pairs, N_P	2	–
10	Mechanical torque (Rated), T_m	14.74 kN m	1.0
11	Rated stator flux linkage	1.2748 Wb (rms)	1.0053
12	Rated rotor flux linkage	1.2096 Wb (rms)	0.9539
13	Stator winding resistance, R_s	1.405 Ohm	0.006
14	Rotor winding resistance, R_r	1.395 Ohm	0.008
15	Stator leakage inductance, L_{ls}	0.178039 mH	0.111
16	Rotor leakage inductance, L_{lr}	0.178039 mH	0.111
17	Magnetizing inductance, L_m	0.1722 mH	3.6481
18	Moment of inertia, J	0.0131 kg.m2	–
19	Inertia time constant, H	5.8078 s	–
20	Base flux linkage	1.2681 Wb (rms)	1.0

- Based on Eqs. (8), (11) and (12), a simulation for Induction Generator can be developed as depicted in Fig. 3 where
  a. $\mathrm{k} = (L_s * L_r) - L_m^2$
- The other constants defined in this model are listed in Table 1.
- The dq/abc block used in Fig. 3 is a subsystem in MATLAB Simulink which is used to convert the stator currents from dq reference frame to abc reference

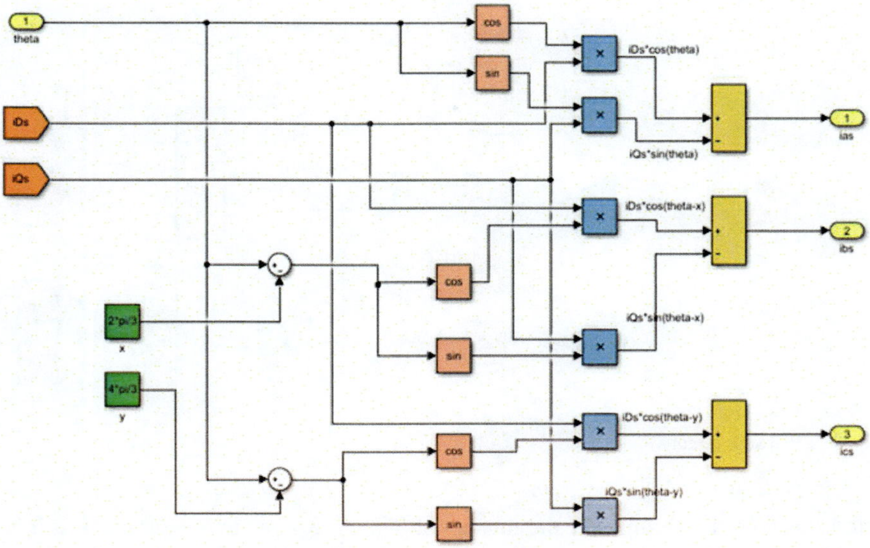

Fig. 1 dq/abc transformation of stator current

frame by performing the matrix calculation of Eq. (13). This matrix calculation is implemented or modelled as block diagram in Simulink which shown in Fig. 1.

$$\begin{bmatrix} x_a \\ x_b \\ x_c \end{bmatrix} = \begin{bmatrix} \cos\theta & -\sin\theta \\ \cos\left(\theta - \frac{2\pi}{3}\right) & -\sin\left(\theta - \frac{2\pi}{3}\right) \\ \cos\left(\theta - \frac{4\pi}{3}\right) & -\sin\left(\theta - \frac{4\pi}{3}\right) \end{bmatrix} \cdot \begin{bmatrix} x_d \\ x_q \end{bmatrix} \tag{13}$$

- The Load subsystem is shown in Fig. 2. It is used to provide a load to the system. Stator currents ($i_{s(d)}, i_{s(q)}$) in synchronous reference frame are converted into the reference frame where space vector is aligned with the rotor ($i_{\mathrm{Ds}}, i_{\mathrm{Qs}}$). Values of the variables used are as follows:

  b. $V_{\mathrm{Ds}} = 398.84393$ V
  c. $V_{\mathrm{Qs}} = 0$

5 Dynamic Performance of SCIG with DGC

Initially there is no motion of the blades and the turbine is in a parking mode. The blades are pitched out of the wind. The blades starts moving due to wind and the turbine and generator starts rotating. After some time the speed reaches to an operative level and the generator starts producing current. Slowly the generator reaches the

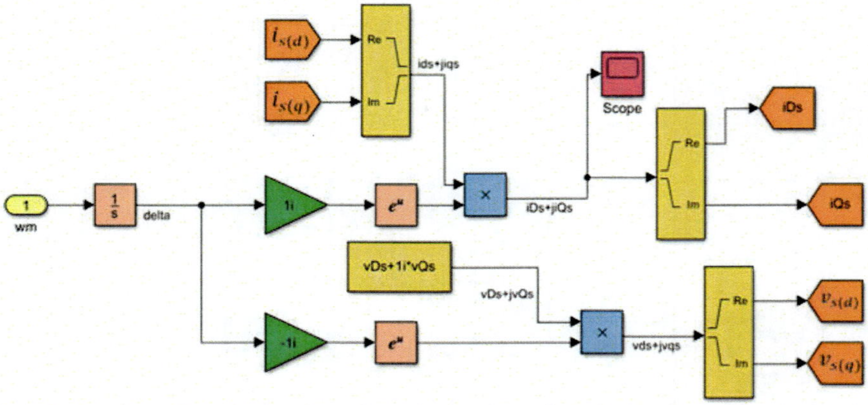

Fig. 2 Load subsystem

rated speed i.e. 1450 rpm. The parameters of SCIG are given in the Table 1 [6]. It is then connected to the grid by closing the circuit breaker. The plots for stator currents (i_{as}, i_{bs}, i_{cs}), T_e and w_r are shown in Sect. 5.1. Transients can be observed in the plot. By the magnetization of generator core due to stator current, electromagnetic torque T_e is produced. When the generator rotates below synchronous speed, positive torque is produced in motoring mode due to which the turbine accelerates. DC offset currents can be observed in the stator currents in Fig. 4. Now the operation becomes steady state with $T_e = 14.74$ N m (Fig. 5). The sum of these offset currents in i_{as}, i_{bs} and i_{cs} is zero due to 3-phase balanced system. At $t = 0.239$ s, the generator reaches the synchronous speed (1500 rpm) which can be observed in Fig. 6.

Now, if we keep the rotor speed constant, the motion equation is simplified to $d\omega_r/dt = 0$ and $w_r = w_s$. The transients [7] when the speed reaches to the synchronous speed (1500 rpm) are studied and shown in Figs. 4, 5 and 6. The measurements of Figs. 5 and 6 are shown in Tables 2 and 3. If we observe the dynamic response [8] after making this change and connecting it to the grid [9], we can see that there are oscillations in the generator torque T_e [10] due to very high inrush current where the peak value is greater than 150 A. There are huge oscillations in the generator torque T_e due to high amplitude of the stator currents. Since the rotor speed is kept constant in this case, the motion equation are eliminated and the transients are even more than the first case. So there are two problems that we face in SCIG that are–first, there is huge stator current and second there are oscillations in torque [11] which makes it unstable.

In the plot of Fig. 4 it is observed that there is excessive inrush current [12] during the starting of system because the generator is directly connected to the grid [13]. The peak values of the current reaches to 150 A and torque reaches to about 325 Nm. There are high oscillations during the starting of system [14]. These high oscillations have a very negative impact on the grid [9]. There is also a very high mechanical stress due to high torque oscillations. This shows that SCIG cannot be directly connected

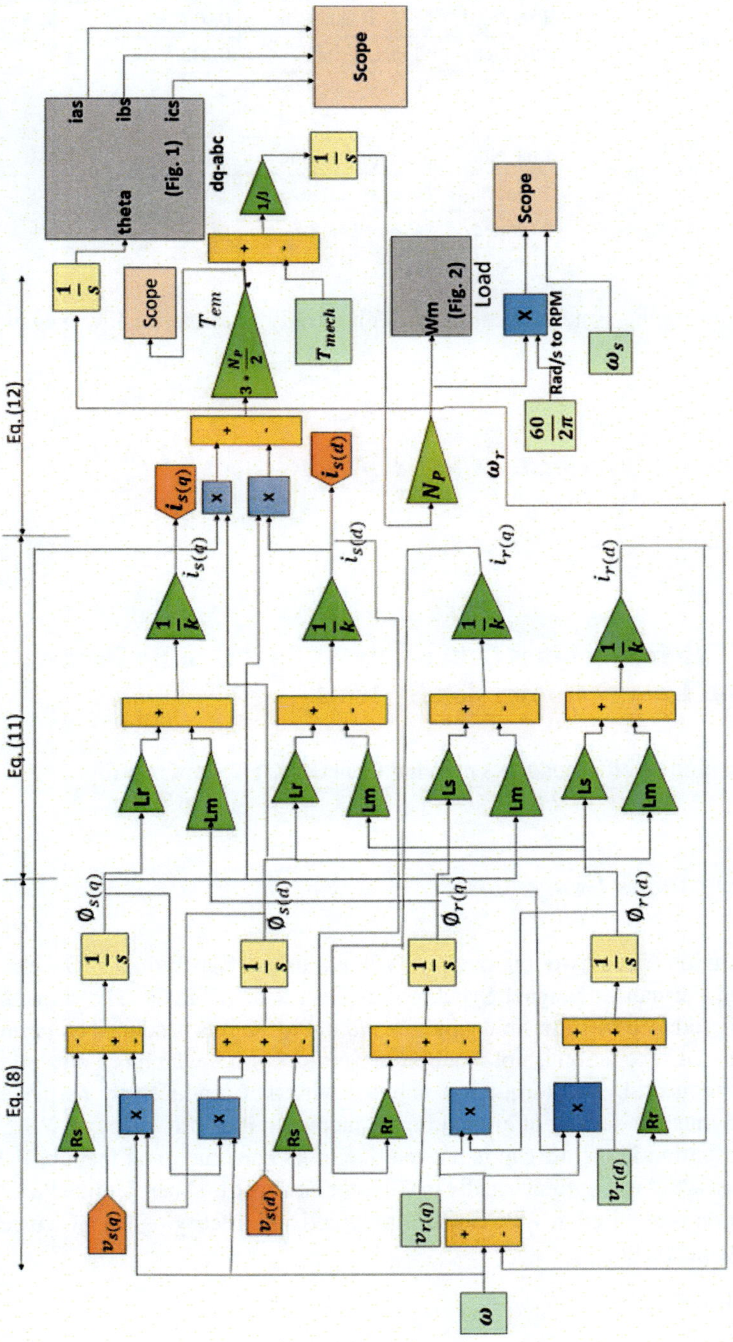

Fig. 3 Simulink model for dynamic performance of SCIG

Table 2 Measurements for Torque-Time of Fig. 5

Line No.	Type of value	Time (s)	T_e (Nm)
1 (bold)	Max. value	0.017	325.6
4 (dotted)	Steady state	0.232	14.75

Table 3 Measurements for Speed-Time curve of Fig. 6

Line No.	Type of value	Time (s)	RPM
1 (bold)	Max. value	0	2900
4 (dotted)	Steady state	0.239	1494

to the grid during the system startup [15]. This is only possible for low megawatt turbines.

5.1 *Dynamic Performance Curves of SCIG*

See Figs. 4 and 5.

6 Some Other Machines (Alternatives) That Can Be Used in Wind Energy Conversion Systems

The classification of different types of Wind Generators are shown in Fig. 7.

6.1 *Synchronous Generator*

These are divided into two main types–Wound Rotor Synchronous Generator (WRSG) and Permanent Magnet Synchronous Generator (PMSG). The rotor flux in WRSG is produced by rotor field winding and in PMSG it is produced by permanent magnets. DC excitation is not required in case of PMSG, which is a plus point. The MMF produced by the torque is combined with that produced by permanent magnets and thus the density of air gap flux is increased in PMSG. Now in WRSG the flux distribution in the air gap is uneven due to its constructional features. The field windings are places symmetrically in the rotor on the pole shoes. Unlike PMSG, a DC excitation is required in WRSG that is provided after rectifying the AC current generated by the exciter.

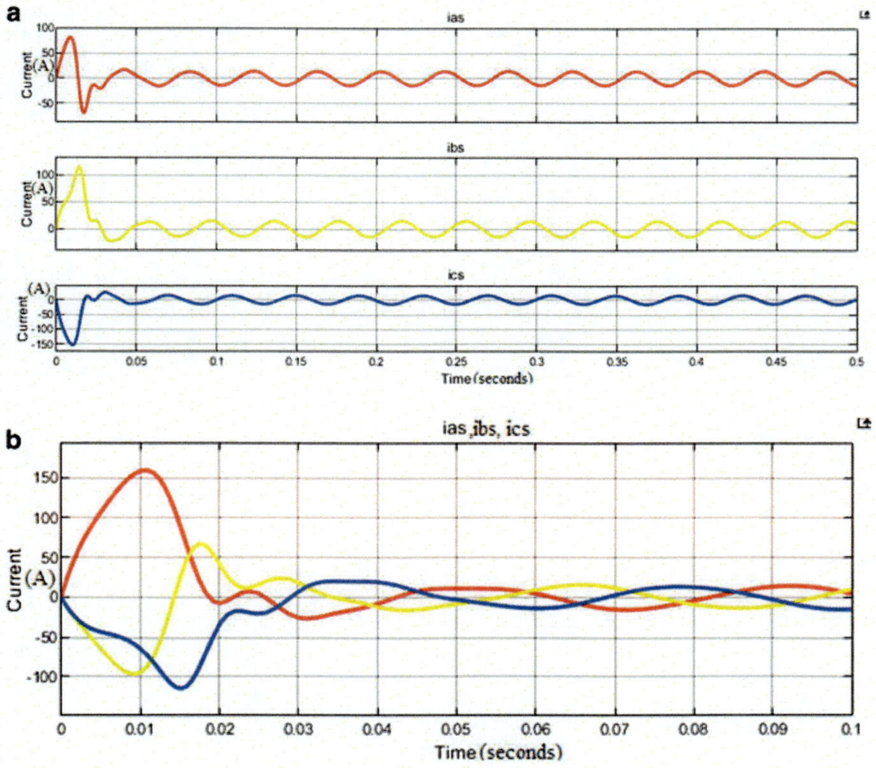

Fig. 4 **a** Separate plots for stator currents, **b** stator currents in a single plot

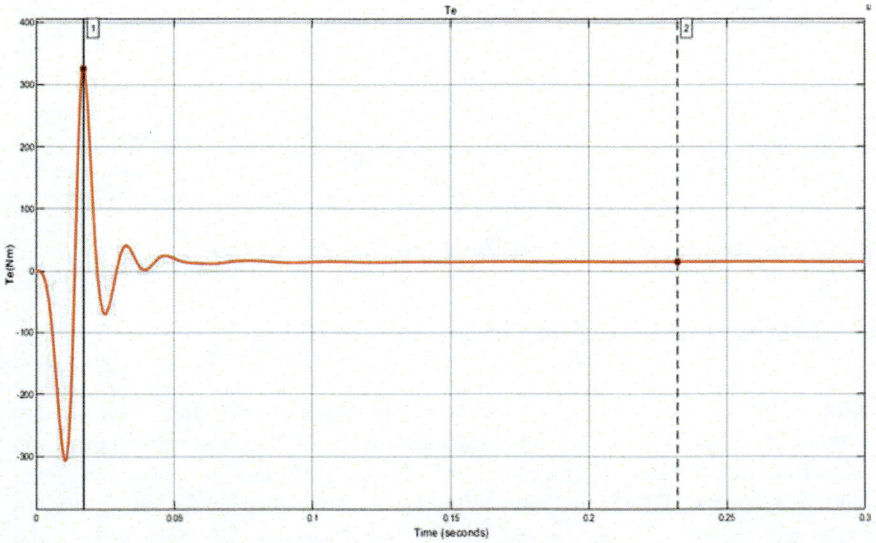

Fig. 5 Torque versus time plot

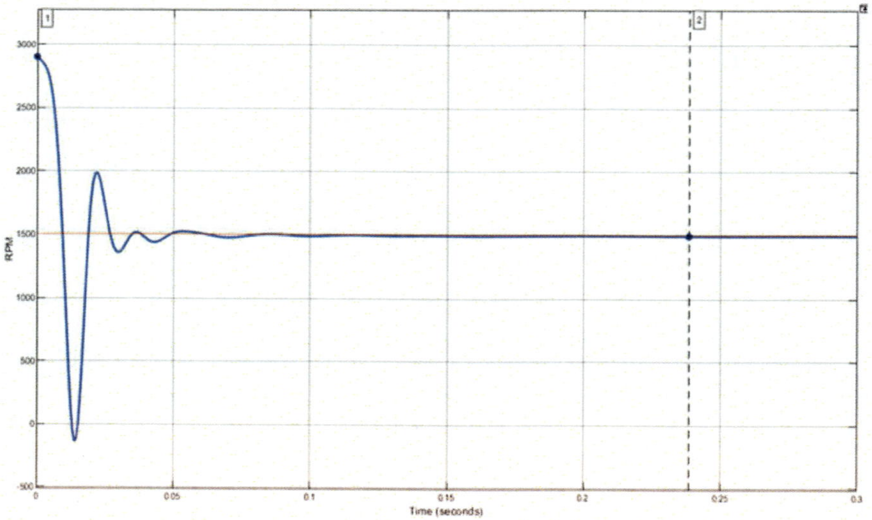

Fig. 6 Speed-time curve

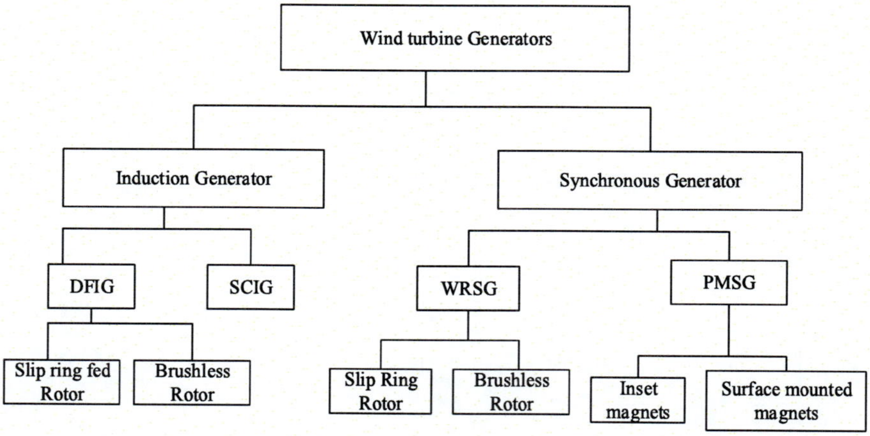

Fig. 7 Classification of electric generators used in wind turbines

6.2 *Doubly Fed Induction Generator (DFIG)*

The main characteristic of this generator is that the rotor currents can be controlled to get variable speeds by the use of some external devices. Power flows in one direction only in rotor but in stator, it is bidirectional. The stator of the generator is connected to the grid directly. DFIGs are more efficient in variable speed operation than in fixed speed operation. It can be operated at speed variation of about 30% of the synchronous speed. At present, it is the most used generator for Wind Turbines.

Table 4 Comparison between different types of electric generators used in wind turbines

Generator	Advantages	Disadvantages
SCIG	• Simple and rugged in construction • Inexpensive • Minimum maintenance required • Robust operation • Fixed-speed applications	• High mechanical stress • Energy yield low • Energy losses are high in gears • Less focus on efficiency and power flow control
DFIG	• Converts used are of low cost • Compact in size • Less mechanical stress • Converter losses are less • Robust power control	• Brush or slip ring required • Gear losses are high • Control schemes very complex
WRSG	• Mechanical stress is low • Brushes or Slip rings absent • Energy yield is high • Power control is robust	• Large size • Expensive converters used • Converter losses are high • Cost of Cu winding is high
PMSG	• Mechanical stress low • Brushes or slip rings absent • Less Cu losses • Energy yield is very high • Power control is robust	• Large space required • Construction is very complex • Converters are very expensive • Converter losses are high • Magnet used is very expensive • Issue with Demagnetization

A detailed analysis is done on the operation of SCIG in this paper but a comparative study is also done and their advantages and disadvantages are presented in tabular format in Table 4 so that different generators can be compared at a glance.

7 Conclusions

A Simulink model was created by using mathematical expressions for the generator. Using this model, case studies can be done on DGC of SCIG during System Startup. It is verified that SCIG cannot be directly connected to the grid during the system startup. This is only possible for low megawatt turbines. For high megawatt turbines, high oscillations have a very negative impact on the grid. There is also a very high mechanical stress due to high torque oscillations. In short, there are two problems that we face in SCIG that are–first, there is huge stator current and second there are oscillations in the torque which make it unstable. But scope is also there as new ways and technologies are developed to reduce the high torque oscillations and inrush current. Some other alternatives to SCIG are discussed in Sect. 6 with their advantages and disadvantages.

Acknowledgements With profound respect, I would like to thank Dr. Atma Ram Gupta, Associate Professor, Shri Phanishwar Nath Renu Engineering College Araria former Assistant Professor, Electrical Engineering Department, NIT Kurukshetra, and PhD research scholar Mr. Abhineet Prakash from IIT Patna for providing me insight knowledge and expertise that greatly helped me in writing this paper. I express my profound thanks to Dr. Sanjoy Kumar Parida, Associate Professor, EED, IIT, Patna, for giving me the opportunity and assistance to work on this paper. I will always be grateful to them for giving me their valuable time and support.

References

1. McCalley J Machine Transformations. http://home.eng.iastate.edu/~jdm/, Slide 1–20. http://home.engineering.iastate.edu/~jdm/wind/Transformations.ppt
2. Wu B (2011) Lang Y, Zargari N, Kouro S (2011) Power conversion and control of wind energy systems. In: Wind generators and modeling. Wiley-IEEE Press, USA, pp 49–85. https://doi.org/10.1002/9781118029008.ch3
3. Nam NH (2019) Modeling, algorithm control and simulation of variable-speed doubly-fed induction generator in grid connected operation. In: 2019 26th international workshop on electric drives: improvement in efficiency of electric drives (IWED), pp 1–5. https://doi.org/10.1109/IWED.2019.8664338
4. Rao J, Qu R, Li D, Gao Y (2016) An improved dq-axis coordinate system model for interior permanent magnet machines. IEEE Conf Electromagnet Field Comput (CEFC) 2016:1–1. https://doi.org/10.1109/CEFC.2016.7816057
5. Asmine M et al (2011) Model validation for wind turbine generator models. IEEE Trans Power Syst 26(3):1769–1782. https://doi.org/10.1109/TPWRS.2010.2092794
6. Wu B, Lang Y, Zargari N, Kouro S (2011) Power conversion and control of wind energy systems. In: Wind generators modelling. Wiley-IEEE Press, USA, pp 319–326. https://doi.org/10.1002/9781118029008.app2
7. Kushwaha SKS, Samuel P, Mohanty SR (2017) Transient stability analysis of SCIG based marine current farm and doubly fed induction generator based offshore wind farm using bridge type fault current limiter. In: 2017 4th IEEE Uttar Pradesh section international conference on electrical, computer and electronics (UPCON, pp 494–499. https://doi.org/10.1109/UPCON.2017.8251099.
8. Boudjemaa M, Rachid C (2013) Dynamic response of SCIG with direct grid connection. In: 4th international conference on power engineering, energy and electrical drives, pp 1322–1327. https://doi.org/10.1109/PowerEng.2013.6635805
9. Benchagra M, Maaroufi M, Ouassaid M (2011) Study and analysis on the control of SCIG and its responses to grid voltage unbalance. Int Conf Multimedia Comput Syst 2011:1–5. https://doi.org/10.1109/ICMCS.2011.5945737
10. Tian G, Wang S, Pang Y (2015) Mechanism analysis and control strategy of dynamic torque improvement for wind farm with squirrel cage induction generator under unbalanced voltage by STATCOM. In: International conference on renewable power generation (RPG 2015), pp 1–6. https://doi.org/10.1049/cp.2015.0466.
11. Ma P, Wang Q, Li Y, Jiang S, Zhao M (2021) Research on torque ripple suppression of the slotted limited angle torque motor. IEEE Trans Magn 57(2):1–6. Art no. 8200106. https://doi.org/10.1109/TMAG.2020.3006018
12. Shimizu H, Mutsuura K, Yokomizu Y, Matsumura T (2005) Inrush-current-limiting with high T/sub c/superconductor. IEEE Trans Appl Supercond 15(2):2071–2073. https://doi.org/10.1109/TASC.2005.849454
13. Gorski DA, Balkowiec T, Koczara W (2018) Grid connection of a converter controlled squirrel-cage induction generator. In: 2018 7th international conference on renewable energy research and applications (ICRERA), pp 348–353. https://doi.org/10.1109/ICRERA.2018.8566730
14. Thampatty KCS (2019) Design of MRAC based TCSC for damping sub-synchronous oscillations in SCIG based wind farm. In: TENCON 2019—2019 IEEE region 10 conference (TENCON), pp 1846–1852. https://doi.org/10.1109/TENCON.2019.8929682

15. Rezaee M, Harley RG, Heydari H (2015) Designing an efficient PI-based voltage control method for squirrel-cage induction generators in islanding/weak grid-connection conditions. In: 2015 Clemson University power systems conference (PSC), pp 1–7. https://doi.org/10.1109/PSC.2015.7101681

Application of PSAT for Voltage Stability Improvement Using FACTS Devices

Shweta Kumari, Lalit Kumar, Manoj Kumar Kar, and Sanjay Kumar

Abstract The utilization of maximum demand increases exponentially in the real life from the base load to peak load. The power demands may be industrial or residential. In the system network, the high-power demand of consumers can be achieved by the integration of flexible AC transmission system (FACTS) controllers. In the present life, FACTS device is latest technology to improve the transmission capability, and hence, more power is transmitted to the distribution side. In the system network, when the load on the power grid increases, some of the load bus voltages decrease due to which the voltage instability problem occurs in the system. So, it is necessary to identify the weakest bus of the system. By employing the FACTS devices, it enhances the power carrying capacity, reduces the transmission loss, and improves voltage profile of the system with the help of PSAT in MATLAB.

Keywords Continuous power flow (CPF) · Power flow analysis · Power system analysis toolbox (PSAT) · FACTS · Voltage profile

1 Introduction

In day-to-day life, voltage stability has a significant impact on the power system network due to the possibility of voltage collapse. This paper deals with the performance analysis for voltage enhancement, computation of voltage collapse, and improvement of power system stability using FACTS controller like static

S. Kumari (✉) · L. Kumar · M. K. Kar · S. Kumar
EE Department, NIT Jamshedpur, Jharkhand, India
e-mail: shwetatk17@gmail.com

M. K. Kar
e-mail: mkkar@aissmscoe.com

S. Kumar
e-mail: sanjay.ee@nitjsr.ac.in

M. K. Kar
Electrical Engineering Department, AISSMS College of Engineering, Pune, India

K. Namrata et al. (eds.), *Smart Energy and Advancement in Power Technologies*, Lecture Notes in Electrical Engineering 926,
https://doi.org/10.1007/978-981-19-4971-5_9

VAR compensator (SVC), static synchronous compensator (STATCOM), static synchronous series compensator (SSSC), etc. In the real scenario, the demand of power system changes very rapidly because of which voltage stability issues are the biggest concern in the power system. So, for providing the better power flow in the system, both voltage stability analysis and load flow analysis become very important. Due to rapid increase in the demand of the power system, a new line injection is very difficult in the network causing voltage instability problems. As a result, the system may suffer from blackout. In order to avoid the aforementioned problem, the power system designing and planning for maintaining voltage profile are necessary.

2 Background

The features, algorithms, and different case studies have been discussed in [1] to demonstrate the capabilities of PSAT. The consequences of different types of faults on the IEEE 9 bus test system using power world simulator is discussed in [2]. The improvement of power system dynamics can be achieved using series-connected FACTS devices, [3]. The loadability of power system is enhanced by using FACTS devices, and is presented in [4]. In [5, 6], PV curve has been plotted with SVC using PSAT to improve the voltage stability. The authors in [7] used FACTs devices for allocating the optimal position in order to improve the power transfer ability of the system. The methods and strategy of sensitivity for optimal placement of FACTS controller is presented in [8]. This paper justifies the necessity of optimal location of FACTs controllers to solve the power system security problem. Due to this, the improvement in power carrying capability of the system is justified in [9]. In this work, the voltage stability of steady state with has been improved using fuzzy logic based approach. The power quality improvement of an interconnected grid system using STATCOM is presented in [10]. The use of different metaheuristic methods for deciding the control variable setting of FACTS controllers in the IEEE test system is presented [11]. In [12], the authors show how to reduce active power loss and preserve voltage stability in the power system network in which UPFC devices provide reactive and active power to the system to achieve the performance. In [13], the efficacy of an artificial neural network to predict system voltage instability is presented. Because two crucial parameters, power loss minimization and voltage compensation, allow distribution systems to work smoothly in both normal and abnormal situations, the paper in [14] uses improved harmony search algorithm method on a balanced radial distribution system.

3 Modeling of FACTS Controller

i. STATCOM is a FACTS family electrical device which is used as a compensating device. In the power system, STATCOM is based on the power electronics converter like voltage source converter (VSC), and it can act like either sink or source of reactive power of power system. When this device is connected to a source of power system, it gives a real AC power. STATCOM is a part of FACTS group of devices. Generally, STATCOM is installed for the purpose of to the backup of electrical network that has a low power factor and voltage regulation. The use of static synchronous compensator is for voltage profile improvement in the power system network. Regarding the system, there are two objectives using this controller: First is to make the phase difference zero, and the other objective is to basically synchronize the output of the voltage source converter; these two objectives of the controller reduce the error.

ii. SSSC is a 2nd generation of FACTS controller: SSSC is a compensator that can provide the controllable compensating voltage. It can insert the voltage in series with the line and control the impedance. The basic elements of this device are DC voltage source, coupling transformer, line impedance, etc.

The basic construction of a static synchronous series compensator is shown in Fig. 1. There are two loads connected on both sides of the source end and load end. The transmission line is having its impedance. SSSC is the series FACTS device. In this case, it uses a VSC to inject the controllable voltage with line current of the power system network through a transformer connected in series. SSSC provides the *C* or *L* compensation which does not depend on the transmission lines. The application of static synchronous series compensator is low-frequency power oscillation network. The advantages of SSSC can provide the real power compensation in the network as well as enhance the power transfer capability of the system. The limitations of SSSC are:

(a) It is costly system; (b) due to heavy loading, shunt reactor is required; and (c) control system is also required for the VSC.

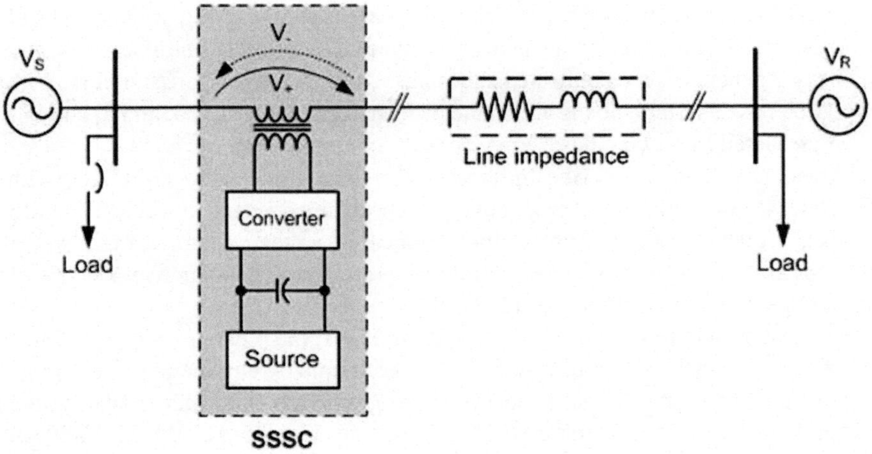

Fig. 1 Construction of SSSC

4 Power System Analysis Toolbox (PSAT)

PSAT is a MATLAB toolbox. Using this toolbox, it can analyze and control the electric power system which is available in online. With the help of PSAT power flow (PF), continuous power flow (CPF), optimal power flow (OPF), and simulation in time domain can be plotted easily. The advantages of power system analysis toolbox (PSAT) are: (a) It can provide the complete graphical interface and simulation network of the system; (b) it can easily access all the graphical user interface (GUI); and (c) it can provide the Simulink-based library which are user-friendly of toolkit in the design of the system.

STEPS TO FIND WEAK BUS OF THE SYSTEM:

Step 1: First make an individual line diagram of the test system.
Step 2: Read the data of each and every component of the test system.
Step 3: Then, keep the loads and data files.
Step 4: By using Newton Raphson, solve the power flow of the system.
Step 5: Then, simulate the PF results.
Step 6: Then, one by one eliminate each transmission parameters and simulate the PF and CPF for identifying the maximum loading parameters for each transmission lines.
Step 7: Set out the results and identify the most critical parameters in the transmission lines.

5 Results and Discussion

In this paper, the IEEE 9-bus system using PSAT is simulated and the results with or without FACTS devices are presented below.

The IEEE 9-bus test system with 100 MVA base is shown in Fig. 2. It consists of three two winding transformers which are connected between bus numbers 1, 2, and 3. The two generators are connected at bus nos. 2 and 3. Six transmission lines with 100 MVA base and three loads are connected between bus nos. 5, 6, and 8, and bus 1 is represented as a swing bus.

Tables 1, 2, and 3 describe the analysis of load flow without FACTS, with STATCOM, and with SSSC, respectively. The obtained result in Tables 2 and 3 is superior than Table 1 in terms of improvement of voltage, minimum real power loss, bus active, and reactive power. Figure 3 presents the comparative analysis of bus voltage with and without FACTS devices.

Here, after modeling the 9-bus system using PSAT, the most loading parameters are solved by using continuous power flow of each time by eliminating every transmission line step by step. Most severe line is recognized which includes to the system instability. So, with the integration of FACTS devices like STATCOM and SSSC on the IEEE 9-bus system, the voltage profiles are enhanced, and the comparison of these two FACTS devices SSSC gives the better results.

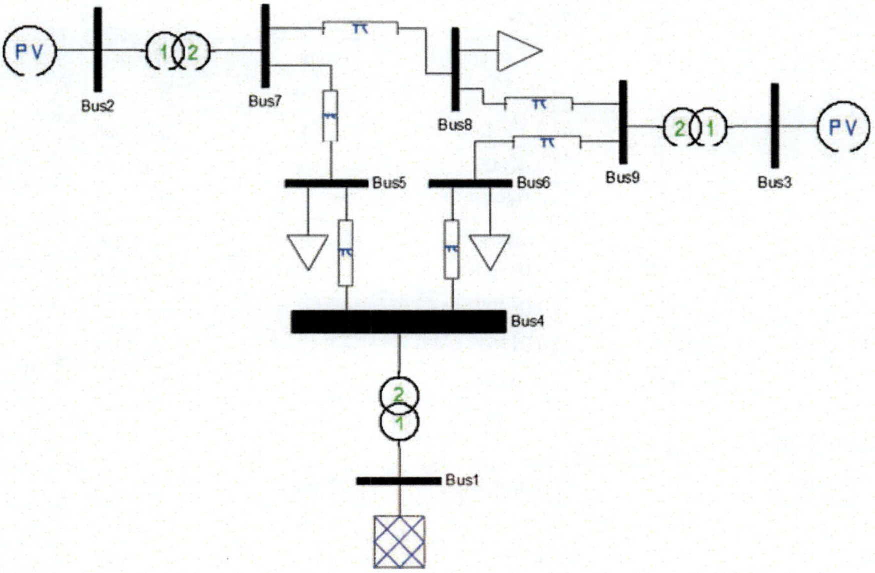

Fig. 2 IEEE 9-bus test system using PSAT

Table 1 PF from PSAT without FACTS device

Bus nos	V/p.u	Phase angle	P/p.u	Q/p.u
1	1.04	0	0.77429	0.82129
2	1.025	0.16844	1.63	0.50666
3	1.025	0.08408	0.85	0.40814
4	0.98769	−0.03543	0	0
5	0.9355	−0.07004	−1.25	−0.5
6	0.95738	−0.06224	−0.9	−0.3
7	0.98275	0.07219	0	0
8	0.96628	0.01613	−1	−0.35
9	0.99437	0.0392	0	0

6 Conclusion

In this paper, the system performance is analyzed using IEEE 9-bus system. The siting of FACTS controller is decided based upon the power flow analysis. The appropriate placement of FACTS devices with PSAT has been studied. The objectives like real power loss, voltage magnitude enhancement, and current carrying capacity of the transmission lines are evaluated. In this paper, with the help of PSAT, the power flow and line voltage stability are evaluated with and without FACTS devices. It is also

Table 2 PF from PSAT with STATCOM

Bus nos	V/p.u	Phase angle	P/p.u	Q/p.u
1	1.04	0	0.6231	0.21917
2	1.025	0.13646	1.3632	0.09298
3	1.025	0.07006	0.71084	0.18092
4	1.0224	−0.0317	0	0
5	1.026	−0.06355	−1.0454	0.39964
6	0.99143	−0.05162	−0.75266	−0.25089
7	1.0094	0.05492	0	0
8	0.99183	0.01151	−0.83629	−0.2927
9	1.0085	0.03151	0	0

Table 3 PF from PSAT with SSSC

Bus nos	V/p.u	Phase angle	P/p.u	Q/p.u
1	1.04	0	0.76609	0.8643
2	1.025	0.038	1.63	0.45895
3	1.025	0.02365	0.85	0.40271
4	1.0351	−0.00041	0	0
5	1.0456	0.03712	−1.25	−0.5
6	1.0313	0.00545	−0.9	−0.3
7	1.0297	0.03381	0	0
8	1.0282	0.02735	−1	−0.35
9	1.0274	0.02154	0	0

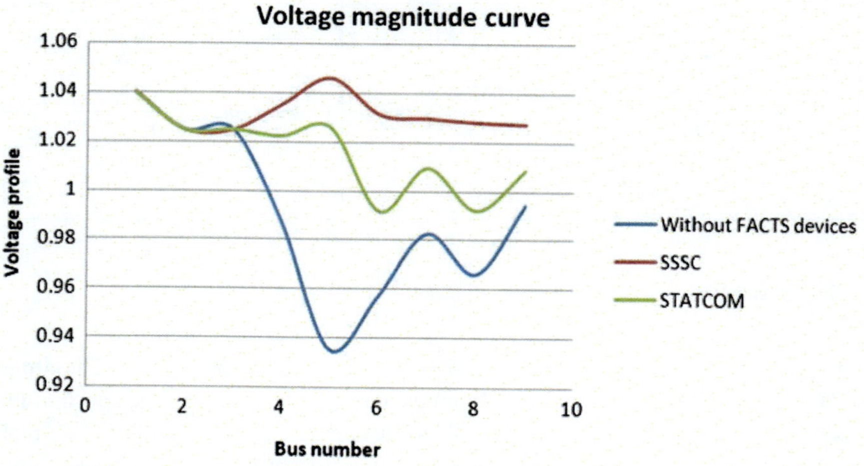

Fig. 3 Comparison of voltage curve with and without FACTS devices

included that the contribution of the FACTS devices enhances the voltage magnitude of the system. The result justifies the efficiency of incorporation of FACTS controllers in the system.

References

1. Milano F (2005) An open source power system analysis toolbox. IEEE Trans Power Syst 20(3):1199–1206
2. Rajesh Varma PV, Kar MK, Singh AK (2021) Transient analysis of a standard IEEE-9 bus oower system using power world simulator. In: Advances in smart grid automation and industry 4.0. Springer, Singapore, pp 233–243
3. Kar MK, Kumar S, Singh AK, Panigrahi S (2021) A modified sine cosine algorithm with ensemble search agent updating schemes for small signal stability analysis. Int Trans Elect Energy Sys 31(11):e13058
4. Kar MK, Kumar L, Kumar S, Singh AK (2020) Efficient operation of power system with FACTS controllers using evolutionary techniques. In: 7th international conference on signal processing and integrated networks (SPIN), IEEE, pp 962–965
5. Dwivedi AK, Vadhera S (2019) Reactive power sustainability and voltage stability with different FACTS devices using PSAT. In: 2019 6th international conference on signal processing and integrated networks (SPIN), pp 248–253
6. Kumar L, Raw BK, Gupta SK, Kumar S (2019) Voltage stability enhancement using shunt devices and identification of weak bus through voltage stability indices. In: 2019 4th international conference on recent trends on electronics, information, communication & technology (RTEICT), pp 247–251
7. Varma KS, Singh SN, Gupta HO (2001) FACTS devices location for enhancement of total transfer capability. In: Power engineering socity winter meeting, vol 2. IEEE, pp 522–527
8. Manikandam S, Arul P (2013) Optimal location of multiple FACTS devices using sensitivity methods. Int J Eng Trends Technol 4:10
9. Moger T, Dhadbanjan T (2017) Fuzzy logic approach for reactive power coordination in grid connected wind farms to improve steady state voltage stability. IET Renew Power Gener 11:351–361
10. Kar MK, Kumar S, Singh AK (2022) Power quality improvement of an interconnected grid system using PWM technique of D-STATCOM. In: Recent advances in power electronics and drives. Springer, Singapore, pp 31–41
11. Kar MK, Kumar S, Singh AK, Panigrahi S (2021) Reactive power management by using a modified differential evolution algorithm. Opt Control Appl Methods
12. Kumar L, Verma K, Kumar S (2021) UPFC modelling for augmentation of voltage stability and reduction of active power losses. In: Kumar J, Jena P (eds) Recent advances in power electronics and drives. Lecture notes in electrical engineering, vol 707. Springer, Singapore. https://doi.org/10.1007/978-981-15-8586-9_38
13. Singh P, Parida SK, Chauhan B, Choudhary N (2020) Online voltage stability assessment using artificial neural network considering voltage stability indices. In: 2020 21st national power systems conference (NPSC), pp 1–5
14. Varma PV, Kar MK, Singh AK (2021) Optimal sizing and location of DG for power loss reduction and voltage improvement of distribution system using IHSA Algorithm. In: IEEE 2nd international conference on applied electromagnetics, signal processing, & communication (AESPC), pp 1–5

An Economic Evaluation of Fuse Placement Strategy in Distribution Network for Reliability

Umesh Agarwal, Naveen Jain, and Manoj Kumawat

Abstract In context of restructuring of power system, it is utmost important to mention that protection devices are of prime importance among the several components of distribution network. The devices like fuse play an indispensable role in overhead distribution lines. In order to reduce the capital expenditure as well as the cost of outages, the protection devices should be placed optimally. In this situation, the economic allocation of the fuse cut-out considering various failure probabilities has not been studied earlier. Further, in the field of reliability and cost analysis, fuse has traditionally been regarded as 100% reliable. This paper extends the present reliability evaluation procedure to incorporate the economic allocation of fuse with their failure probability to assess the economic impact on the system reliability with assessment of priority level of fuse removal. In this paper, various conditions for the probability of fuse failure have considered and examined the economical placement strategy and impact of fuse failure on RBTS BUS-2 system.

Keywords Distribution system · Energy not served · Fuse failure probability · Reliability

1 Introduction

In the present scenario, reliability analysis of distribution system is getting more attention as 80% of customer interruption is due to failure in distribution system [1]. The main contribution of the protective device is to protect the system whenever any disturbance occurs and to isolate that faulty part from the system. These protective devices play an indispensable role in providing the reliable supply to the consumers.

U. Agarwal (✉) · N. Jain
Department of Electrical Engineering, College of Technology and Engineering, Maharana Pratap University of Agriculture and Technology, Udaipur, Rajasthan, India
e-mail: umeshbkb.agarwal@gmail.com

M. Kumawat
Department of Electrical and Electronics Engineering, National Institute of Technology Delhi, Delhi, India

K. Namrata et al. (eds.), *Smart Energy and Advancement in Power Technologies*, Lecture Notes in Electrical Engineering 926,
https://doi.org/10.1007/978-981-19-4971-5_10

In reality, however, protective devices may sometimes fail to respond when they are needed whereas they may respond without any need [2, 3]. This will lead to the unreliability in the system. In recent years, several efforts have been attempted to evaluate the impacts of different failure modes to know the effect on the power system reliability [4–7]. The Markov model to evaluate the routine tests effects on the reliability is described in [4]. This proposed model has been improved further in [5] by dividing the Markov model into two parts to expand the condition of the protection devices and power equipment. The effect of relay coordination on the reliability of an interconnected power system was studied in [6].

In majority of the research literature, it is considered that the power system is interconnected and relays are the prime device for its protection. However, this statement has less relevance in case of distribution network as it generally has radial configuration with fuse as a protective device. Modelling of fuse differs from the modelling of relay for reliability. The failure probability of relay is more as compared to fuse due to more numbers of elements with it. The fuse operates as a single element, and its own failure rate is considered, whereas the relay-circuit breaker protection system consists of several components such as transducers, relay, circuit breaker, etc. Hence, due to the individual failure rate of each element, the overall failure rate raises [7–15]. Therefore, there is an information gap in determining a suitable model for these devices.

This paper aims to highlight this issue with the two contributions. Firstly, a mathematical formulation is done to find the economic allocation of the fuse for distribution network by well-known reliability index "energy not served (ENS)". Finally, the consequence of fuse failure probability is considered with multiple fuse failure probabilities. The organization of the paper is as follows: Sect. 2 contains the problem formulation. Mathematical modelling considering fuse allocation and its malfunction probability is included in Sect. 3. In Sect. 4, network topology and case study are included with result analysis. Finally, Sect. 5 concludes the paper.

2 Problem Statement

In general, the protection devices are provided to protect the system and make the system more reliable. There are many cost-effective devices, which are easily available to develop protection scheme. Therefore, it is expected to carry out techno-economic impact analysis prior to implementation of a new protection scheme. Over-current relays (directional/bidirectional) with differential protection schemes have significant cost, and they are used in distribution system at starting end of the feeder in primary substation. Further, reclosers and sectionalizers are the supplementary category of protection devices, and they have considerably high investment cost. Therefore, they are installed at the beginning of distribution feeder. It is worth mentioning that the fuse is an inexpensive device and widely used for protection purpose in distribution network. In general, the fuse is connected at the primary side of distribution transformer, and it protects the main feeder and loads from failure.

In competitive market, cost control is the prime focus of all companies and industries. Hence, precise modelling and allocation of fuse cut-outs are quite important, which can lead to a cost saving approach. It is important to decrease in outage frequency and duration over a given time span with a particular set of customers to address the questions above. Since the fuse positioning at the distribution feeder end reduces the faulty zone, the advantage should be compared against the costs of the requisite amount of fuse cut-outs. In special circumstances, the failure probability of fuse leads to the unreliability. Therefore, an organized and in-depth examination should be done before considering the fuse failure probability. To achieve this goal, reliability modelling of the fuse cut-outs is proposed and investigated. In addition, efficient techno-economic approach of the fuse to make the reliable distribution network is proposed in this paper.

3 Mathematical Formulation

3.1 Modelling of Fuse Cut-out

There are lots of model to evaluate reliability with failure of relay available in literature [7]. However, the fuse modelling is different from relay, as fuse does not have any assessment state when current carrying element is up. Figure 1 represents the failure rate model of fuse cut-out. In general, fuse operated in normal state. Whenever any fault occurs in the system, then fuse has to operate by melting down to isolate the faulty section [5]. However, sometimes it may happen that fuse fails to operate, and then it will be the responsibility of backup protection to isolate the faulty section. Sometimes, it may happen that fuse trips unintentionally and de-energizes the system. In this situation, system moves to state-3 from state-1 directly.

In (1), failure rate is calculated for the system with fuse failure probability.

Fig. 1 Fuse cut-out model

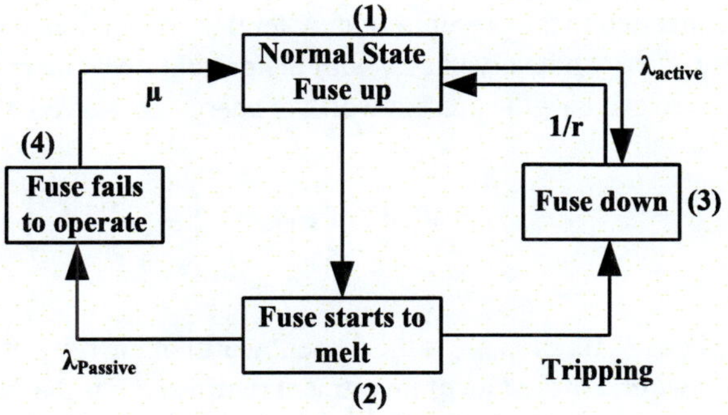

$$\lambda_{LP_i} = \left\{ \begin{array}{c} (\lambda_{\text{fuse opeartes}}) \times (P_{\text{fuse opeartes}}) \\ + (\lambda_{\text{fuse fails}}) \times (P_{\text{fuse failure}}) \end{array} \right\} \tag{1}$$

3.2 Formulation of Placement Strategy

If the annual failure/km of feeder is given as λ and average outage duration is given as *r*, then annual outage duration is obtained as [10];

$$U_i = (\lambda_i \times l_i) \times r_i \tag{2}$$

where l_i represents the feeder length.

The annual energy not served (ENS) (kWh/yr.) [10] is calculated as:

$$\text{ENS}_i = \sum_{i=1}^{\text{LP}} U_i \times \text{load}_i \tag{3}$$

This energy not served is to be converted into the worth of ENS. Therefore, a cost factor is considered for this conversion, and the worth of energy not served is calculated as [10]:

$$C_{\text{ENS}} = \sum_{i=1}^{\text{LP}} \text{ENS}_i \times C_i \tag{4}$$

where C_i is the cost of energy not served in \$/kWh and C_{ENS} is the annual cost of ENS.

Since C_{ENS} is a cost that happens every year during the network lifetime, it must be converted to the present value. After this conversion, it becomes possible to compare the energy not served cost with the present investment cost for fuse installation.

To evaluate the present worth of energy not served, the expression is as below [8]:

$$P.W.\text{ of } C_{\text{ENS}} = \left(\frac{1 - \frac{1}{(1+i)^n}}{i} \right) \times C_{\text{ENS}} \tag{5}$$

where $P.W.$ of C_{ENS} is the present worth of energy not served during the lifetime of network, *i* is the rate of investment return and *n* is the lifetime of the network in year. In fact, the proper allocation of the fuse can reduce the cost of energy not served and make some profit for the utility.

$$\text{DOC} = F_{\text{Capital}} - \left\{ \begin{array}{l} P.W.\text{ of } C_{\text{ENS}}(2) \\ -\ P.W.\text{ of } C_{\text{ENS}}(1) \end{array} \right\} \tag{6}$$

where DOC is the difference of cost and F_{Capital} is the capital cost for fuse placement. Equation (6) is used to evaluate the difference in the cost of fuse installation and cost of energy not served after and before the fuse removal. If the value is positive, the fuse installation at that particular location is not required. Otherwise, if it is negative, fuse at that location is required.

4 Case Study

4.1 Network Topology

The topology of the evaluated distribution network is shown in Fig. 2. The studied network is a part of RBTS (Bus-2) system [1]. This network consists of four circuit breakers (CB) connected at the starting point of each feeder. There are four feeders ($F1$, $F2$, $F3$ and $F4$) of 11 kV each. The network consists 20 transformers, 14 sectionalizing switches, 20 fuses and 22 load points. The total number of consumers at the network is 1908. The load data of various load points and consumers is shown in Table 1 and Table 2. It is observable from above-mentioned tables that $F1$ and $F4$ have the highest load as compared to other feeders, whereas feeder $F2$ has the minimum load (two consumers). Both loads of feeder $F2$ are directly connected to the feeder as these are large load points and don't require any transformation of voltage. The reliability data for the feeder and transformer are included in Table 3.

The technique for allocating fuse cut-outs has now been applied for the system shown in above Fig. 2. In (3), it is used to calculate the energy not served, and this ENS is converted to the cost of energy not served using (4). The present worth of fuse installation is considered 170$ in this paper [7]. Then, C_{ENS} converted in the present worth of cost of energy not served (PW of C_{ENS}) using (5). In this equation, the value of time is taken 20 years and rate of investment return [7] is considered 10%. The cost of energy not served is taken 0.1 $/kWh.

For base case, the PW of C_{ENS} is obtained as below:

$$P.W.\text{ of } C_{\text{ENS}}(1) = \left(\frac{1 - \frac{1}{(1+0.1)^{20}}}{0.1} \right) \times 13242 \times 0.1$$

$P.W.\text{ of } C_{\text{ENS}}(1) = 11,273.66\$$

Now, fuse is removed from branch 11 and the process of present worth calculation is repeated

$P.W.\text{ of } C_{\text{ENS}}(2) = 11,569.082\$$

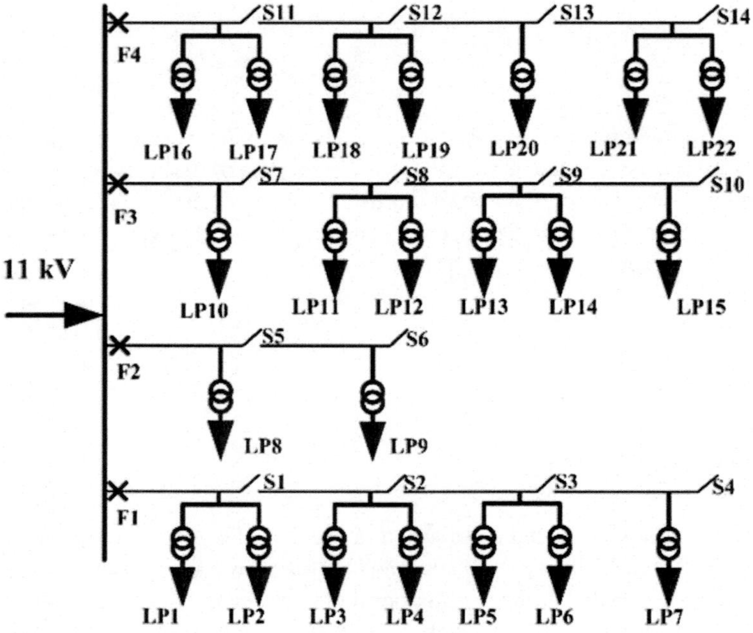

Fig. 2 RBTS bus-2 distribution network [1]

Table 1 Peak load in percent for each sector [1]

Customer type	Peak load (MW)	Sector peak (%)
Residential	7.25	36.25
Small users	3.50	17.50
Govt. & Inst	5.55	27.75
Commercial	3.70	18.50
Total	20	100

Table 2 Loading data of load points [1]

Load points	Load at various load points (MW)	
	Average	Peak
1–3, 10, 11	0.535	0.8668
12, 17–19	0.450	0.7291
8	1.00	1.6279
9	1.15	1.8721
4, 5, 13, 14, 20, 21	0.566	0.9167
6, 7, 15, 16, 22	0.454	0.7500
Total	12.291	20.00

Table 3 System reliability data [1]

Failure rate (failure/yr./km.)	Repair time (h)	Replacement time (h)	Switching time (h)
Feeder			
0.065	5	–	1
Transformer			
0.015	200	5	1

Here, difference of the cost is calculated, and cost is compared with the capital cost of fuse installation.

$$\text{DOC} = 170\$ - \{11,569.082 - 11,273.66\} = -125.42\$$$

The negative sign shows that removal of fuse from branch 11 is not justified; therefore, it should be preserved. Now again, the fuse cut-out is removed from branch 13 and calculation is done as below;

$$P.W.\text{ of } C_{\text{ENS}}(2) = \left(\frac{1 - \frac{1}{(1+0.1)^{20}}}{0.1}\right) \times 13339.03 \times 0.1$$

$$P.W.\text{ of } C_{\text{ENS}}(2) = 11,356.268\$$$

DOC is calculated from this value as $P.W.$ of $C_{\text{ENS}}(1)$ is considered same as the value of in previous calculation.

$$\text{DOC} = 170\$ - \{11,356.268 - 11,273.66\} = 87.392\$$$

As the DOC is positive, therefore, fuse can be removed from branch 13. The same process is applied on each branch, and $P.W.$ of C_{ENS} is calculated for fuse removal from all branches one by one. The priority level of the fuse cut-out removal is decided. Table 4 shows the results. From Table 4, it can be concluded that it is economical to remove the fuse cut-out from branch 15 and branch 13 with priority index of 1 and 2.

The analysis is carried further by removing the fuse cut-out from branch 15 (LP_9) and then from branch 13 (LP_8). The $P.W.$ of C_{ENS} is obtained as below:

The present worth of the cost of energy not served for branch 15 is taken as base now, and value is shown below:

$$P.W.\text{ of } C_{\text{ENS}}(1) = 11345.464\$$$

$$P.W.\text{ of } C_{\text{ENS}}(2) = \left(\frac{1 - \frac{1}{(1+0.1)^{20}}}{0.1}\right) \times 13423.69 \times 0.1$$

Table 4 Energy not served results and fuse removal significances

Branch no of removed fuse	ENS (kWh/year)	Present worth of ENS ($/year)	Priority of removal
No fuse removal	13,242	11,273.66	
2	13,711.07	11,673.007	11
3	13,813.48	11,760.194	21
5	13,826.85	11,771.577	22
6	13,708.32	11,670.666	10
8	13,778.83	11,730.694	18
9	13,806.35	11,754.124	19
11	13,589.0	11,569.082	6
13	13,339.03	11,356.268	2
15	13,326.34	11,345.464	1
17	13,471.55	11,469.09	3
19	13,691.47	11,656.320	9
20	13,763.23	11,717.413	16
22	13,735.42	11,693.74	13
23	13,764.03	11,718.095	17
25	13,477.97	11,474.555	4
27	13,743.79	11,700.86	14
28	13,670.22	11,638.23	8
30	13,665.63	11,634.32	7
31	13,758.12	11,713.063	15
33	13,550.15	11,536.006	5
35	13,723.82	11,683.86	12
36	13,807.0	11,754.677	20

$$P.W. \text{ of } C_{\text{ENS}}(2) = 11428.344\$$$

The difference of cost is calculated to get the economic justification for fuse placement.

$$\text{DOC} = 170\$ - \{11,428.344 - 11,345.464\} = 87.12\$$$

In this way, the whole system can be evaluated and fuse must be installed with proper economical and technical justification. The research is proceeded further including the effect of fuse failure probability. The results with various operational probability of fuse for various reliability indices are included in Table 5.

Table 5 Reliability indices for various fuse failure probabilities

Fuse operational probability (%)	95	90	80	70	60
SAIFI (Failure/Yr.)	0.267	0.286	0.320	0.357	0.393
SAIDI (Hours/yr.)	0.714	0.732	0.767	0.803	0.839
ECOST (k$/Yr.)	85.70	86.20	91.97	96.32	100.65
ENS (MWh/Yr.)	13.54	13.88	14.44	15.06	15.68

5 Conclusion

In this paper, the fuse cut-out placement strategy including the effect of hidden failure (fuse failure probability) has been proposed. The proposed methodology has been tested on RBTS BUS 2 system. The findings of the fuse cut-out allocation revealed that fuse cut-outs were installed improperly. Thus, the priority level of the fuse cut-out removal has been obtained for each feeder.

Further, this paper extended the system reliability evaluation procedure to incorporate the fuse failure probability. The analysis has been carried out with different operational probabilities. It is worth concluding that the fuse failure probability value can significantly alter the solution of cost/worth analysis. In the considered cases, the worst scenario is with 60% fuse operational probability, whereas the ECOST and ENS increased by 36% and 23.07%, respectively. This study underlines the importance of the fuse failure probability in reliability analysis.

References

1. Agarwal U, Jain N (2020) Reconfiguration of radial distribution network for reliability enhancement considering renewal energy sources. In: 2020 International conference on electrical and electronics engineering (ICE3)
2. Su S, Hu Y, He L, Yamashita K, Wang S (2019) An assessment procedure of distribution network reliability considering photovoltaic power integration. IEEE Access 7:60171–60185. https://doi.org/10.1109/ACCESS.2019.2911628
3. Li G, Bie Z, Xie H, Lin Y (2016) Customer satisfaction based reliability evaluation of active distribution networks. Appl Energy 162:1571–1578
4. Anderson PM, Chintaluri GM, Magbuhat SM, Ghajar RF (1997) An improved reliability model for redundant protective systems—Markov models. IEEE Trans Power Syst 12(2):573–578
5. Billinton R, Fotuhi-Firuzabad M, Sidhu TS (2002) Determination of the optimum routine test and self-checking intervals in protective relaying using a reliability model. IEEE Trans Power Syst 17(3):663–669
6. Yu X, Singh C (2004) A practical approach for integrated power system vulnerability analysis with protection failures. IEEE Trans Power Syst 19(4):1811–1820
7. Gilvanejad M, Abyaneh HA, Mazlumi K (2012) Fuse cut-out allocation in radial distribution system considering the effect of hidden failures. Int J Electr Power Energy Syst 42(1):575–582
8. Oskounejad MM (1996) Engineering economy or economic evaluation of industrial projects, 7th edn. Publication Center of Amirkabir University of Technology, Tehran
9. Allan RN, Billinton R, Sjarief I, Goel L, So KS (1991) A reliability test system for educational purposes-basic distribution system data and results. IEEE Trans Power Syst 6:813–820

10. Agarwal U, Jain N, Kumawat M, Maherchandani JK (2020) Weibull distribution based reliability analysis of radial distribution system with aging effect of transformer. In: 2020, 21st national power systems conference (NPSC), 2020, pp 1–6
11. Agarwal U, Jain N (2019) Distributed energy resources and supportive methodologies for their optimal planning under modern distribution network: a review. Technol Econ Smart Grids Sustain Energy 4:3
12. Agarwal U, Jain N, Kumawat M, Singh SN (2021) An assessment procedure for distribution network reliability considering load growth. In: 2nd electric power and renewable energy conference (EPREC-2021), NIT Jamshedpur (Accepted)
13. IEEE Guide for Electric Power Distribution Reliability Indices, IEEE Standard 1366, 2012
14. Wang J, Xiong X, Zhou N, Li Z, Weng S (2016) Time-varying failure rate simulation model of transmission lines and its application in power system risk assessment considering seasonal alternating meteorological disasters. IET Gener Transm Distrib 10:1582–1588
15. Zhong J, Li W, Wang C, Yu J, Xu R (2017) Determining optimal inspection intervals in maintenance considering equipment aging failures. IEEE Trans Power Syst 32(2):1474–1482

Improved Dynamic Performance in Grid Connected Wind Energy System Using Dynamic Voltage Restorer

Preeti Rani, Ved Parkash Arora, and Naveen Kumar Sharma

Abstract Wind system is one of the mostly utilized distribution generation resources due to its positive impacts compared to traditional generation sources. But, its intermittent nature cause power fluctuation and due to this behavior most of distribution generation sources are connected to grid to regulate the supply for utility appliances. Integrated wind energy system (WES) with grid to analyze the stability during various three-phase faults presented in this work. This proposed system consists of wind turbine, permanent magnet synchronous generator (PMSG), filter, and some switching devices. PMSG has many advantages compared with other available alternators to generate electricity such as its high efficiency and required less maintenance due to utilization of permanent magnet in the alternator. But, due to intermittent nature of WES, it leads to power quality (PQ) issues through various three-phase fault scenarios. To mitigate these issues, dynamic voltage restorer (DVR) a custom power device is used in series with grid network. To check the effectiveness of proposed DVR this paper presents proposed system in MATLAB/Simulink environments. Simulation results represent investigates the performance of DVR with specific controller for the stability of proposed system during various three-phase fault conditions.

Keywords DVR · PWM · WECS · PQ · PMSG

P. Rani (✉)
Electrical Engineering Department, GZSCCET, MRSPTU, Bathinda 151001, India
e-mail: pretty.singla15@gmail.com

V. P. Arora
Electrical Engineering Department, GZSCCET, MRSPTU, Bathinda 151001, India

N. K. Sharma
Electrical Engineering Department, IKG PTU, Jalandhar 144603, India

K. Namrata et al. (eds.), *Smart Energy and Advancement in Power Technologies*, Lecture Notes in Electrical Engineering 926,
https://doi.org/10.1007/978-981-19-4971-5_11

1 Introduction

Non-renewable energy sources such as natural gas and coal are replaced with distribution generation resources due to its ecofriendly nature and easy availability of these resources all over the world. Distribution generation resources can also be refer to renewable energy sources such as solar energy, wind energy, and hydro energy. In these days, utilization of WES is widely used to extract the power from wind especially in large-scale industries [1]. Due to high potential and number of positive impacts of WECS, in feature definitely it will become first choice of electricity market [2].

Basically in wind energy, electricity produced with help of converting the kinetic energy of wind into mechanical energy and that mechanical energy is fed to generator for producing electricity. Wind flows due to the uneven of atmosphere by variation of temperature, by rotation of earth so on [3]. Wind turbine is main part of wind farm that is used for to converting the kinetic energy of wind into mechanical energy. Wind turbine is two types, first is horizontal type wind turbine, and another is vertical wind turbine. The length of blades is main factor of wind farm which describes the amount electricity produces by wind turbine. Mostly horizontal type wind turbine prefers for installation of wind farm [4].

Due to intermittent nature of WES some PQ issues occur in network which may trip grid connected sensitive loads. Some of PQ issues are sag, swell, harmonics etc. Out of these sag is one of common PQ issue that generally occurs in the distribution network due to presence of high inductive load at the end users utility side [5]. PMSG may disconnect from the grid connected WES due to unbalance operation. Due to occurrence of tripping and disconnection consequently grid will lose its stability [6]. Thus to avoid this situation many of countries have designed their fault ride through capability codes to enhance the stability during PQ issues [7]. There are devices used at utility side to protect the appliances from PQ issues namely Custom Power Devices (CPD). This paper proposed DVR to mitigate PQ issues by injecting controlled voltage during fault period [8].

Most of industries utilized advanced switching converters and utility devices those are sensitive to unbalance operation [9]. The main reason for unbalance operation is changes in voltage magnitude and non-sinusoidal wave shape of measured signals. Self-operating system of commercial applications subjected to one of the most common PQ issue, namely, voltage sag because some of control devices well performed at the time of voltage sag [10]. As per European standard at each bus of power system network voltage sag limits to 4%. It has been found that distribution network exposed to unbalance operation of grid system due to power fluctuation in some particular frequency range 0.01 to 0.1 Hz [11].

Another most occurrence PQ issue is harmonics. In integrated wind energy system inverter and other switching converts are main reason to generate harmonics in measured signals because they are operated at high-frequency signals and penetrate high current harmonics with active power signals [12]. Harmonics affects the stability of grid system and cause overheating of utility equipments, tripping of switchgear devices of system.

In future, it is not acceptable that high integration of distribution generation sources will decrease the PQ without any penetration of PQ improvement topologies. It is also find that PQ issue caused by three phase fault at distribution network will make the difficult task to integrate WES with grid network. As per study it is found that there are number of power quality improvement topologies to mitigate the PQ issues cause by WES [13].

The main objective of this paper to present analysis of three-phase fault issues in distribution network with PMGH based WES and also presented the PQ improvement topology by investing the performance of DVR with SRF controller in order to enhance the reliability of proposed system. To verify the assessment of proposed DVR, the proposed system is modeled in MATLAB/Simulink environment. In Sect. 2, WES modeling is presented. Analysis on PQ presented in Sect. 3. DVR structure describe in Sect. 4. Control scheme of DVR topology is presented in Sect. 5. Section 6 represented simulation results and analysis. At the end, Sect. 7 represented conclusion.

2 WES Modeling

In WES, mechanical energy is extract from blowing air by employing wind turbine, then it is converted into electrical energy by using PMSG. Due to its stability and its less cost, this work proposed PMSG to deliver active power to grid.

Due to variant temperature intensity on earth, it creates variable pressure which is the reason to blow air from high pressure to low pressure. The speed of wind can be measured from the following equation [14, 15].

$$V_a = V_{10}\left(\frac{z}{10}\right)^{\alpha} \tag{1}$$

V_a Velocity of wind at some height (a meter) above ground (m/s).
V_{10} Velocity of wind 10 m above the ground (m/s) (Meteorological Center usually uses 10 m height fir reporting of wind speed).
α Power law exponent or index.

Wind input power can be calculated from following equation [16]

$$P_{wind} = \left(\frac{1}{2}\right)(\rho S V^3) \tag{2}$$

P_{wind} Power in wind (W).

ρ Density of air (kg/m^3).

S Swept area of turbine (m^2).
V Velocity of wind (m/s).

3 Power Quality

PQ can be defined as to sustain the voltage magnitude in specific range for the normal operation of utility appliances. There are number of reasons to generate PQ issues, but today main reason is advanced power electronic devices. Power electronic devices used as switching converters which is the main cause for nonsinusoidal wave shape and it is called as harmonics. There are some other PQ issues are sag, swell, interruptions, overvoltage, under voltage, etc. These issues generally occur due to utility appliance or due to three-phase faults in distribution network [17, 18].

As per literature survey, the percentage of occurrence of above mentioned various issues is different. Such that the percentage of occurrence of sag, swell, interruptions is 55% and harmonic occurrence is only 5% in European countries. In American countries occurrence of harmonic is 22% of total PQ issues. It means a common method cannot be applied to measure PQ issues for each country [19].

4 DVR Structure

DVR is a CPD employed in network in series through series transformer to inject control voltage in series with network. DVR with main components represented in Fig. 1. DVR consists of a battery as storage device to supply DC voltage to voltage source inverter (VSI) [20]. VSI is an insulated gate bipolar Transistor (IGBT) switch which is employed to convert DC supply in to ac but output is a controlled signal on the basis of voltage magnitude and time period to regulate the grid voltage. Controller is a main part of DVR to provide pulse width modulation (PWM) refers to gate pulse of VSI [21]. The utilization of DVR in grid connected WES is to inject three-phase ac voltage in series with network at DVR bus through series transformer. Three-phase voltage has controlled voltage magnitude and phase angle to regulate the grid voltage during abnormal condition of grid [22].

5 Control Scheme of DVR

There are number of control scheme to inject control voltage in series through transformer. This paper proposed synchronous reference frame (SRF) controller to prove the effectiveness of DVR during three-phase fault conditions. SRF controller operation depends on Clarks and Parks transformation as represented in Fig. 2. By

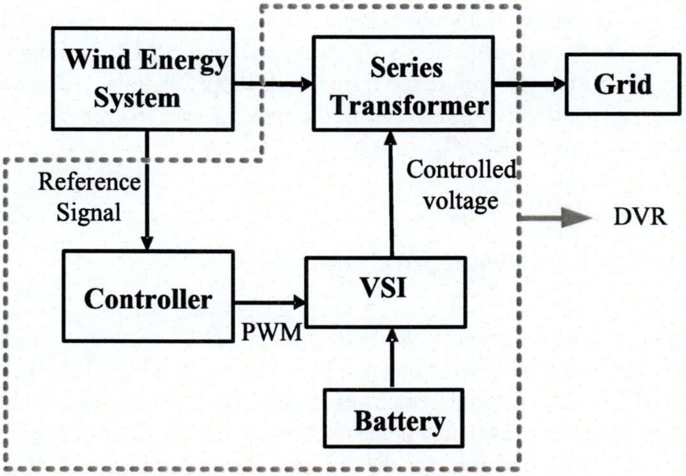

Fig. 1 Schematic diagram of DVR connected to grid connected WES

employing this transformation control scheme converts three-phase SRF (abc) to two phase rotating reference frame (dq0). In same manner, this transformation applied on reference voltage for getting reference voltage in two phase rotating reference frame (dq0) [23].

In this controller, a comparator is used to compare two phase rotating reference frame of grid connected WES supply voltage with reference voltage to provide error

Fig. 2 Flow chart of operation of SRF control of DVR

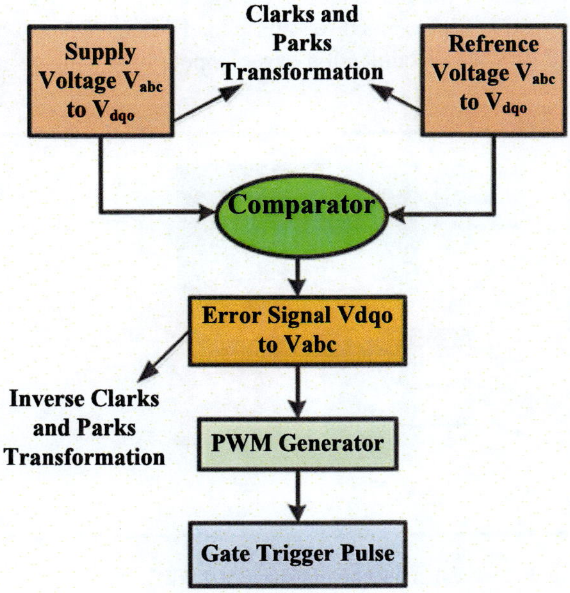

signal in the form of pulse width modulation. The reference signal consists of voltage amplitude, frequency and phase angle. Phase locked loop (PLL) is utilized to match phase angle of rotating dq0 reference frame with supply voltage phase angle. The output of controller provided to inverter in the form of gate trigger pulse to control the injected voltage of DVR [24].

6 Simulation and Result Analysis

To investigate the performance of DVR, the proposed system is modeled in MATLAB/Simulink environment shown in Fig. 3. In this work grid connected WES consists of PMSG to deliverer active power to grid. When three-phase fault occur in the grid connected system, system voltage regulated with SRF control with DVR. The proposed scheme utilized to enhance the stability of grid during various three-phase faults. The important parameters of proposed system depicted in Table 1.

6.1 *Case 1: Short-Circuit Fault: Three-Phase Line to Ground Fault (3L-G)*

In the first case, the performance of WES DVR with proposed control techniques is evaluated under three-phase line to ground fault condition. Real-time three-phase supply voltages are 1 p.u., and fundamental frequency is 50 Hz. In this DVR is required to maintain supply voltage at 1 p.u. across grid connected sensitive loads. Figure 4a represents three-phase load voltage in p.u. without DVR under sag period 0.2–0.3 s, DVR injection sag compensation voltage in volts by SRF controller is

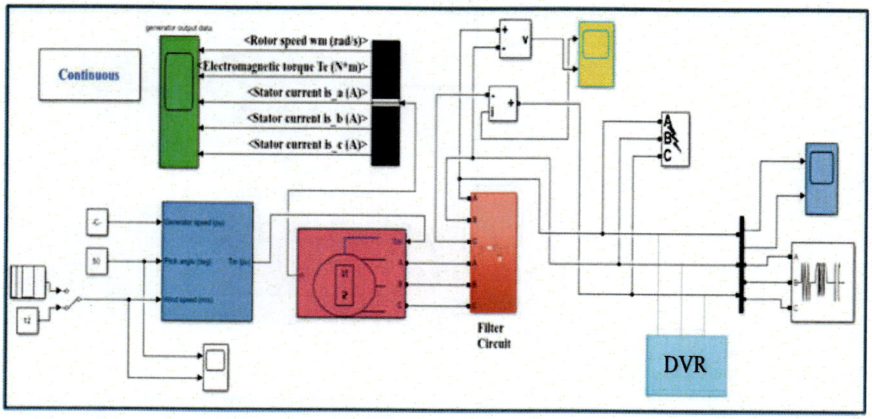

Fig. 3 Simulink model of grid connected WES

Table 1 Grid connected WES

Simulation parameters	Values
Base wind speed	8 m/s
Stator resistance	0.001 p.u
Flux linkage	0.09 p.u
Armature inductance	0.01 H
Rated voltage and frequency	562 V/50 Hz
Base rotational speed	1.0 p.u
Max. power at base wind speed	0.69 p.u
Series RLC branch	1 Ω, 0.001H, 0.0001C
Parallel RC branch	0.1 Ω, 0.0001C

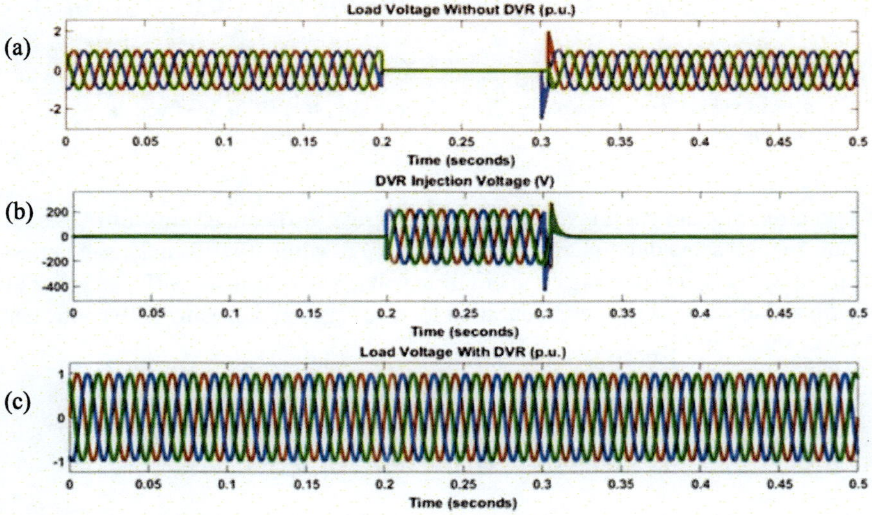

Fig. 4 **a** Load voltage with three-phase (3L-G) fault in p.u., **b** DVR injected voltage in volts, **c** Load voltage in p.u.

shown in Fig. 4b, three-phase sag compensation load voltage in p.u. utilizing SRF controller as depicted in Fig. 4c.

6.2 *Case 2: Unbalanced Sag: Two Phase Line to Ground Fault (2L-G)*

In the second case, the performance of WES DVR with proposed control techniques is evaluated under two phase line to ground fault condition. Real-time three-phase supply voltages are 1 p.u. and fundamental frequency is 50 Hz. In this, DVR is

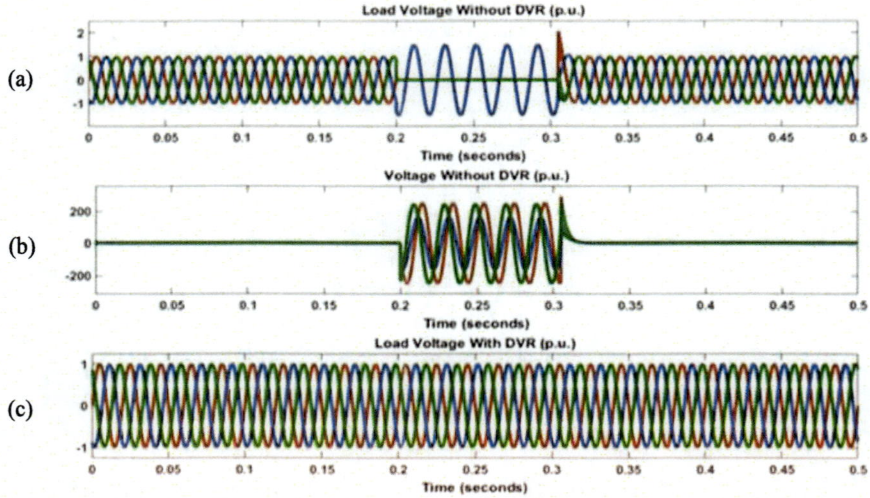

Fig. 5 **a** Load voltage with three-phase (2L-G) fault in p.u., **b** DVR injected voltage in volts, **c** Load voltage in p.u.

required to maintain supply voltage at 1 p.u. across grid connected sensitive load. Figure 5a represents three-phase load voltage in p.u. without DVR under sag period 0.2 to 0.3 s, DVR injection sag compensation voltage in volts using SRF controller in Fig. 5b and three-phase sag compensation load voltages in p.u. using SRF controller DVR as shown in Fig. 5c.

6.3 *Case 3: Total Harmonic Distortion Analysis*

By employing switching converters in proposed system with DVR cause distortion of voltage waveforms and in this work the Total Harmonic distortion (THD) in percentage measured at three buses with fast-Fourier transform (FFT) analysis in MATLAB software. Using SRF Control, the THD percent of DVR drops to 2.31%, compared to 7.65% without it. Figure 6 shows harmonic spectrum of proposed system with SRF control DVR.

7 Conclusion

A DVR-based control scheme to mitigate PQ issues in grid connected WES is presented. The execution of DVR with SRF controller to enhance the stability of grid connected WES with PMSG during various three-phase conditions, namely, short-circuit fault and unbalanced sag is also presented. The proposed control scheme

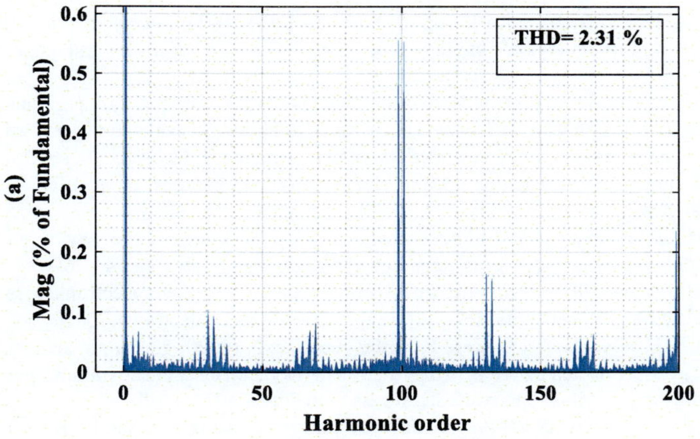

Fig. 6 Harmonic spectrum of grid connected WES with SRF control DVR

utilized for DVR is modeled in MATLAB/SIMULINK software to improve the dynamic of PMSG-based WES. Series compensation-based DVR prove its effective behavior with grid voltage regulation capability, THD control and reactive power control proficiency. Due to SRF controller simplicity, it utilized with DVR to inject controlled voltage in terms of magnitude sand time period through series transformer. The simulation is carried out to represent better functioning of DVR with SRF controller with compensation of various three-phase fault scenarios for improving dynamic performance of grid connected WES.

References

1. Bubshait S, Mortezaei A, Simões MG et al (2017) Power quality enhancement for a grid connected wind turbine energy system. IEEE Trans Ind Appl 53:2495–2505
2. Fekkak B, Menaa M, Loukriz A et al (2021) Control of grid-connected PMSG-based wind turbine system with back-to-back converters topology using a new PIL integration method. Int Trans Electr Energy Syst 3:e12882
3. Sivakumar TA, Linda MM (2020) Improving the dynamic performance of grid connected wind farms using modern UPFC. Microprocess Microsyst 4:103015
4. Kook KS, Liu Y, Atcitty S (2006) Mitigation of the wind generation integration related power quality issues by energy storage. Electr Power Qual Utilisation J 12:77–82
5. Naderi Y, Hosseini SH, Ghassemzadeh S et al (2020) Power quality issues of smart microgrids: applied techniques and decision making analysis. In: Decision making applications in modern power systems, pp 89–119
6. Boulouiha MH, Khodja M, Rahiel D, Allali A, Kaddour F, Denai M (2019) Power quality enhancement in electricity grids with wind energy using multicell converters and energy storage. J Renew Sustain Energy. 11:013302
7. Chan JY, Milanović JV (2015) Assessment of the economic value of voltage sag mitigation devices to sensitive industrial plants. IEEE Trans Power Delivery 30:2374–2382

8. Masetti C (2010) Revision of European Standard EN 50160 on power quality: reasons and solutions. In: Proceedings of 14th International conference on harmonics and quality of power, pp 1–7
9. Seme S, Lukač N, Štumberger B, Hadžiselimović M (2017) Power quality experimental analysis of grid-connected photovoltaic systems in urban distribution networks. Energy 139:1261–1266
10. Hafezi H, D'Antona G, Dedè A, Della Giustina D, Faranda R, Massa G (2016) Power quality conditioning in LV distribution networks: results by field demonstration. IEEE Trans Smart Grid 8:418–427
11. Bajaj M, Singh AK (2020) Grid integrated renewable DG systems: a review of power quality challenges and state-of-the-art mitigation techniques. Int J Energy Res 44:26–69
12. Kumar D, Zare F (2015) Harmonic analysis of grid connected power electronic systems in low voltage distribution networks. IEEE J Emerg Sel Top Power Electron 4:70–79
13. Khajeh KG, Solatialkaran D, Zare F, Mithulananthan N (2020) Harmonic analysis of multi-parallel grid-connected inverters in distribution networks: emission and immunity issues in the frequency range of 0–150 kHz. IEEE Access 8:56379–56402
14. Agalar S, Kaplan YA (2018) Power quality improvement using STS and DVR in wind energy system. Renew Energy 118:1031–1040
15. Ren G, Liu J, Wan J, Guo Y, Yu D (2017) Overview of wind power intermittency: Impacts, measurements, and mitigation solutions. Appl Energy 204:47–65
16. Molla EM, Liu CH, Kuo CC (2019) Power quality improvement using microsystem technology for wind power plant. Microsyst Technol 26:1799–1811
17. Sundarabalan CK, Selvi K (2013) Power quality enhancement in power distribution system using artificial intelligence based dynamic voltage restorer. Int J Electr Eng Inf 5:433
18. Mahela OP, Khan B, Alhelou HH, Tanwar S (2020) Assessment of power quality in the utility grid integrated with wind energy generation. IET Power Electron 13:2917–2925
19. Zejun D, Yongqiang Z, Yu X (2010) Economic loss evaluation and selective treatment of power quality. In: Proceeding in International conference on critical infrastructure, pp 1–4
20. Hassanein WS, Ahmed MM, Abed El-Raouf MO, Ashmawy MG, Mosaad MI (2020) Performance improvement of off-grid hybrid renewable energy system using dynamic voltage restorer. Alexandria Eng J 59:1567–1581
21. Rini AJA, Prabaharan N, Palanisamy K (2017) FRT capability in DFIG based wind turbines using DVR with combined feed-forward and feed-back control. Energy Procedia 138:1184–1189
22. Molla EM, Kuo CC (2020) Voltage sag enhancement of grid connected hybrid PV-wind power system using battery and SMES based dynamic voltage restorer. IEEE Access 8(2020):130003–130013
23. Pal R, Gupta S (2020) Topologies and control strategies implicated in dynamic voltage restorer (DVR) for power quality improvement. Iran J Sci Technol Trans Electr Eng 44:581–603

Power Generation from Thermoelectric Module for Nonlinear Loads

Kumari Namrata, Vrishank Tiwari, Priyanka Priyadarshini Sahoo, and Suman Kumari

Abstract Natural energy resources in the current era are fast depleting owing to the increase in the need for energy for the vast growth in industries and technologies. Means of finding new ways to provide renewable sources of energy are necessary to keep up with this growth. This increasing need for energy has led to the search for new alternative sources of energy. Some of these sources include geothermal energy that can be extracted from the earth or also the waste heat released from vehicles and machines (generally vehicles). A new method that has come up recently is the use of thermoelectric generators. In this case of thermal heat, thermoelectric generator modules can be used to extract the waste heat energy and convert it into useful electrical energy. This electrical energy output from the TEG module will be in DC form. Using an inverter, AC output can be obtained and then used for nonlinear loads. In this, a thermoelectric module is used which produces a voltage output due to two different (a hot and cold side) temperatures. For practical use, a neutral point clamped (NPC) three-level inverter is used along with an LUO converter. This paper aims to generate useful power from waste energy using TEG modules, an LUO converter, and an NPC multilevel inverter. This method can also reduce the bequest toward pollution and can be counted upon to prove an efficient contributor to green energy.

Keywords Renewable sources · Thermoelectric generators (TEGs) · Neutral point clamped inverter (NPC) · LUO converter

List of Symbols

AC	Alternating current.
CMLI	Cascaded H-bridge multilevel inverter.

K. Namrata · V. Tiwari · P. P. Sahoo (✉) · S. Kumari
NIT Jamshedpur, Jamshedpur, Jharkhand, India
e-mail: priyankasahooee@gmail.com

K. Namrata
e-mail: namratanitjsr04@gmail.com

K. Namrata et al. (eds.), *Smart Energy and Advancement in Power Technologies*, Lecture Notes in Electrical Engineering 926,
https://doi.org/10.1007/978-981-19-4971-5_12

CO	Carbon monoxide.
DC	Direct current.
I_{l1}	Current through input side inductor L_1.
I_{C1}	Current through converter capacitor C_1.
I_{C2}	Current through converter capacitor C_2.
MMCs	Modular multilevel connections.
NPC	Neutral point clamped.
PWM	Pulse width modulation.
PV	Photovoltaic.
TEG	Thermoelectric generators.
VSC	Voltage source converter.

1 Introduction

Energy is a major part of our life, is required in everyday fields such as industry, health sector, transportation, and home, and is obtained from resources such as fossil fuels, hydroelectric, geothermal, wind, and nuclear energy [1]. Although conventional resources are widely used, they can also run out in the near future. Also use of these energy resources is also a major source of air, water, and soil pollution. A large part of these resources when used to produce energy are generally lost as waste heat. The use of only non-conventional resources such as hydroelectric or geothermal may pose as eco-friendly, but they are not enough to match the energy needs of the growing rise in energy demand. A steadfast growth in such eco-friendly sources of energy or discoveries in the field of renewable energy sources will always be favorable to the environment. In this respect, the invention of thermoelectric generators came into existence. A new attribute of semiconductors is that when a temperature difference is given between two surfaces of semiconductors, electricity is produced [2]. This is called as Seebeck's effect [3]. The materials that are widely used in TEGs are generally lead telluride, bismuth telluride. The temperature at which a TEG is used is a maximum of 300 °C and a minimum of 30 °C. The efficiency of TEG is generally less than 15%; however, the fact that it saves energy from waste heat is an achievement. For example, this can be used in the application of extraction of waste heat from automobiles and can extract energy up to 0.9 kW. In today's world, we use PV as a very necessary technology. The efficiency that is produced by PV plants is comparatively less. The technology that is used continuously reduces its cost and requires technical advances and new research for efficiency increment. Therefore, research is done to reduce the losses that affect solar panels such as by sunlight, conditioning circuit, energy storage system.

A thermoelectric generator (TEG) is a device used likely in the place of the PV module because of the inefficiency and the involvement of various devices involved in the production of the power from PV modules. The TEG takes the input power from the temperature difference between the hot source and the sink [4], while on the other hand, PV module uses solar radiation energy to the usable power, hence involving various costly devices like MPPT unit and the battery, still have lower efficiency and requires lots of maintenance to get the output. It is found in the study that TEG supplies power for 24 h and gives the output four times more than that of the PV modules.

The paper "Thermoelectric Generator: A Review" written and published by Mratyunjay Singh, Ashutosh Mishra, and Sanjay Nirapure is referred to for the use of TEG as a clean source of energy. This is so because TEGs utilize the waste heat released from industrial and domestic sources and convert it into useful electrical energy. For example, in the case of automobiles, the waste heat given out from the exhaust is tapped by the TEG. This heat creates a temperature difference which will lead to the formation of a voltage potential based on the principle of the Seebeck effect [3].

The proposed paper consists of thermoelectric generator modules. They are connected to a heat source say, for example, suppose TEG is mechanically connected to an exhaust system of the vehicles. The hot carbon monoxide gas (CO) that produces during the start-up of the engine results in the occurrence of two parts, i.e., the hot part, created by the emission of hot CO, and the colder part of the vehicle, which will further be cooled down from the cool air from the surroundings [5]. Hence, the production of source and the sink happens automatically resulting in the production of voltage, due to Seebeck's effect. The produced voltage is not enough to run the load and also containing various ripples, now to reduce this, LUO converter comes in the picture, which is a bidirectional DIDO DC-DC converter, and the LUO converter will boost the output voltage and hence gives the output with higher voltage and low ripple [6]. Since it is a DIDO converter, means that it will take the power from the batter as well, a charge controller is being installed, which helps in keeping the SoC of the battery in the preferable percentage for increasing the battery life [7].

In the proposed paper, we have taken the voltage which is obtained from the TEG module is of very low magnitude. It is not enough to be put into considerable use. For this reason, a converter has to be used that will boost the voltage level. Here, LUO converter is used for the boosting of the voltage level. The paper "Comparative Performance Analysis of Boost converter and Luo converter for controlling of DC motor speed", published by Prof. Baria Himmatbhai Balvantbhai and Prof. Suryaprakash Singh, gave reason to choose an LUO converter over a DC-DC boost converter. The facts that an LUO converter reduces the unwanted ripples in the voltage output thereby rendering high-quality output voltage along with high transfer voltage gain gives reason enough to choose an LUO over a DC-DC boost converter.

The reference for the mathematical modeling of TEG in MATLAB/SIMULINK has been cited from the paper "Model Building and Simulation of Thermoelectric module using MATLAB/SIMULINK", published by Huan-Liang Tsai and Jium-Ming Lin. This paper affords the method of creating a TEG module in the software. This modeling is done in the paper to give an accurate measure of the voltage that is received from the TEG into the LUO converter and NPC inverter, after which the output will be obtained.

2 TEG Setup

2.1 Thermoelectric Generator (TEG)

Thermoelectric energy generation is a method of generating power that converts waste heat energy or any form of heat energy into electric voltage [4]. This is done by using two different temperatures; a hot temperature $t1$ and a cold temperature $t2$. In this method, no reactions or moving parts are required as the temperature difference is enough to generate a potential difference, based on the principle of the Seebeck effect [3].

Using this method, electricity can be produced from low heat sources such as the exhaust heat from automobiles and vehicles, heat released from industries as well as using solar and geothermal energy resources. A thermoelectric generator (TEG) is a device used likely in the place of the PV module because of the inefficiency and the involvement of various devices involved in the production of the power from PV modules. The TEG takes the input power from the temperature difference between the hot source and the sink [4], while on the other hand, PV module uses solar radiation energy to the usable power; hence, involving various costly devices like MPPT unit and the battery still has lower efficiency and requires lots of maintenance to get the output. It is found in the study that TEG supplies power for 24 h and gives the output four times more than that of the PV modules. A thermoelectric module generally comprises a thermocouple that has p-type and n-type semiconductors. These semiconductors are usually connected electrically in series and thermally in parallel.

Based on the figure of merit, materials that are typically used for making thermoelectric modules are bismuth telluride and lead telluride, both of which have a figure of merit of 1 at room temperature and 230 °C to 430 °C, respectively [8]. Other materials that are used are alloys of bismuth, antimony, selenium, and tellurium; alloys of lead; and alloys of silicon germanium. Various thermoelectric modules can be seen in Fig. 1.

2.2 LUO Converter

In this paper, an LUO DC-DC converter is used. A DC-to-DC converter is used in order to lower or higher output voltage required as per the load. In order to get the multiple output voltages according to the load, we will have two methods: First one will be to use the set of programmable batteries, which will change according to the load demand. Secondly, we will use the DC-DC converter to get the proposed output. I/p to this converter is given from the thermoelectric generator module [8]. The use of this converter reduces switching losses and minimizes switching stress. The gain and output variations are taken in simulation in MATLAB/SIMULINK. LUO converter is a DC-DC switching mode boost converter. It provides high gain and increases as

Fig. 1 Different types of thermoelectric modules

the stages of the converter increase in the arithmetic progression and conducts in the continuous conduction mode. The components of the LUO converter are inductor L1, two capacitors Cl and C2, two diodes, and an ideal switch. As the duty cycle of the converter increases, ripple current of the inductor increases. A typical block diagram of the LUO converter is shown in Fig. 2.

2.3 NPC-Level Inverter:

The multilevel electrical converter provides an acceptable answer for synthesizing the lower order harmonics for medium- and high-power systems. With the day-by-day increment in the cost of fuel and increase in the population, the demand for the renewable resources like solar, wind, biomass, etc. increases, and since they only give the DC output voltage, a need for a DC-AC converter is required [1]. The simulation results conferred during this paper verify the operation of the papered topology [9].

The modification on the feedback loop is done within the power circuit to get five levels of switching in the power circuit; for getting required output from the circuit, a diode along with electrical condenser (1nF) is used.

Disadvantages include:

- Reduction in overall power quality of the system.
- Increase in the level of the harmonics of the system.
- The overall efficiency becomes poor.

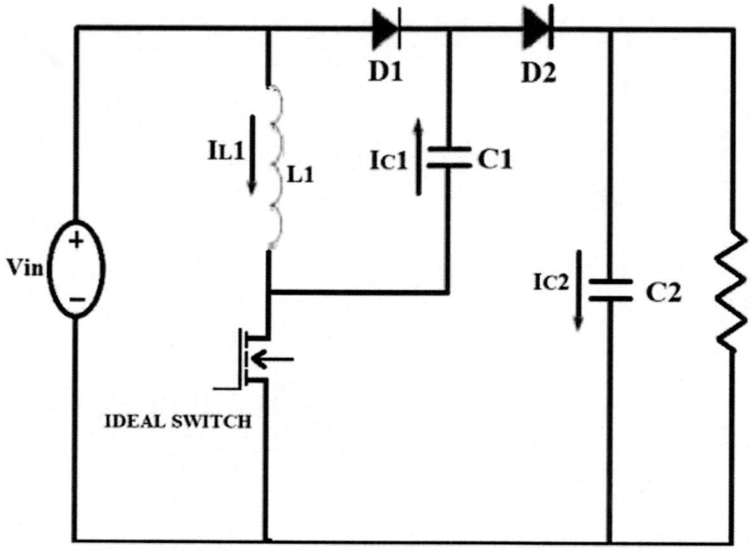

Fig. 2 LUO converter (block diagram)

In this paper, a digital logic control circuit is proposed for a solar-fed cascaded H-bridge multilevel inverter (CMLI) to achieve higher levels with a reduced number of switches without the requirement of bidirectional switches, filter components, detailed look-up tables, and output transformers. The techniques include "binary", "trinary", and modular multilevel connections (MMCs) to achieve five levels. The two-stage CMLI power circuit used to achieve levels by "binary" mode using digital switching technique comprises of counters and logic functions. The same power circuit is used to achieve five levels. In MMC, a five level is achieved with the single-stage inverter by the addition of i/p voltages using the embedded controller based on the proposed switching sequence [10].

The block diagram as shown in Fig. 3 can be explained as:

- I/p supply: AC.
- Driver circuit: It can be used to amplify the 5 V pulses to 12 V for using transistor technology and provided isolations for using an optocoupler. It has two functions:
  - Amplification.
  - Isolation.
- Inverter: It converts direct current (DC) to alternating current (AC).
- PWM generator: It is used to generate PWM pulses to make a switching signal.
- Load: Home appliances.

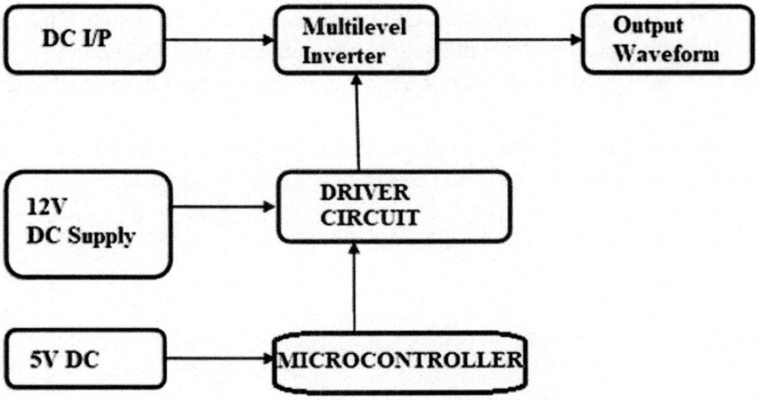

Fig. 3 NPC inverter (block diagram)

3 Simulations

The block diagram in Fig. 4 can be explained as:

- In the above block diagram, MMC is used which is a new type of VSC; it is been used for the high or medium voltage transmission and also used for the connection of the grid to the renewable energy to make the overall system a distributed energy system (DES).
- Reduction in the total harmonic distortions (THDs) is done by the use of MMC while increasing the overall voltage level, hence increasing the efficiency.
- The overall MMC module consists of a floating capacitor, a switch, and a diode.
- In order to get desired output power from the PV module, the solar panel is used (connection can be a series or parallel, depending upon the load of the system) [11].

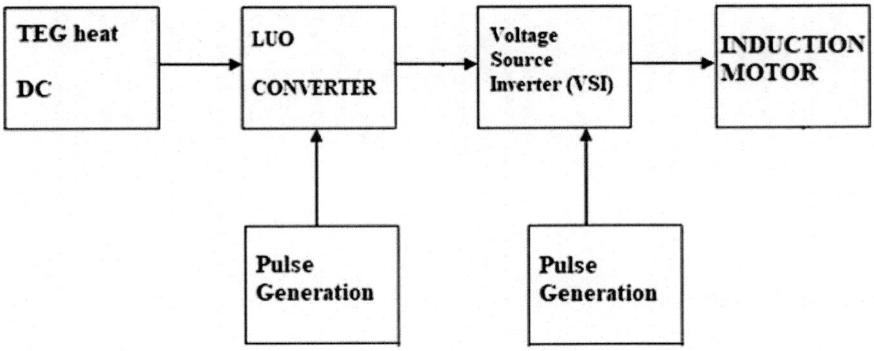

Fig. 4 Software block diagram

- The i/p of the PV is given to the LUO converter in order to increase the output DC voltage, the resultant DC voltage, i.e., the output of the LUO converter is then given to the MMC to convert it into AC voltage having low harmonics and high efficiency.
- For leveling the o/p according to the load, a pulse generator is being used which will in turn fire the switches of the inverters (Figs. 5, 6; Table 1).

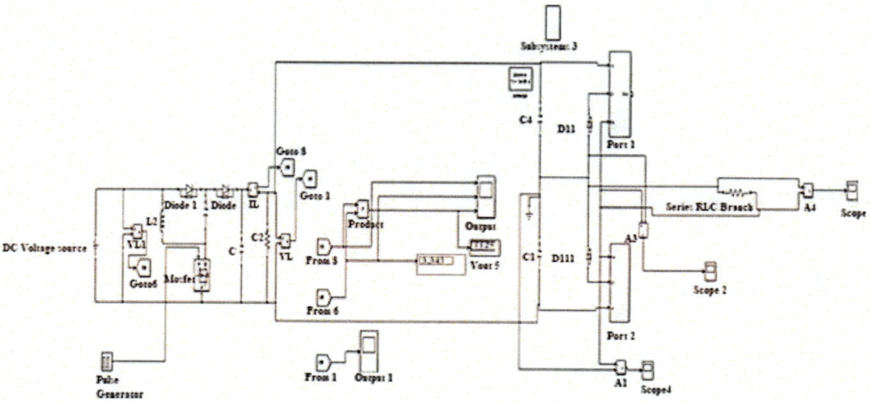

Fig. 5 Simulation with LUO converter

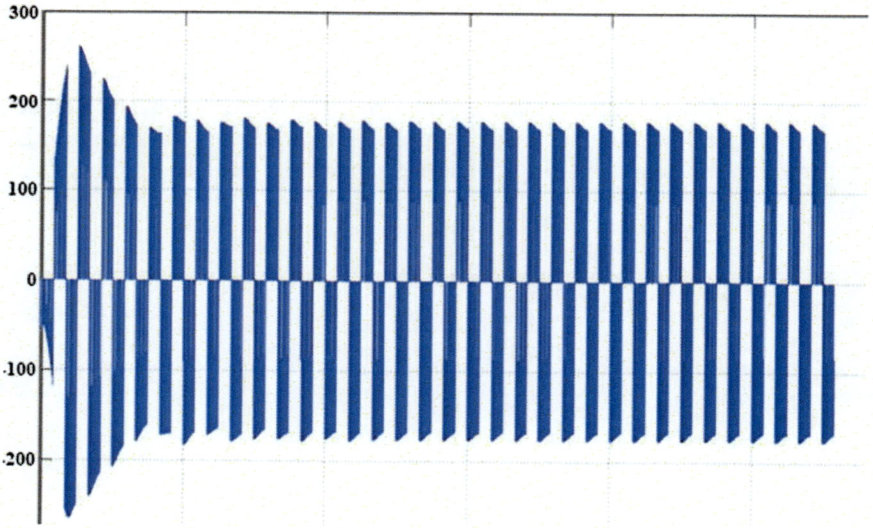

Fig. 6 LUO converter output

Table 1 Converter specifications

Input voltage	100 Vdc
Output voltage	335 Vdc
Output power	1384 W
Tapped inductor	L1 = 1.3 uH; L2 = 1.2 mH

4 TEG Modeling

See Fig. 7.

4.1 Parameters Used in Modeling and Their Equations [13, 14]

- Resistance, $R = \frac{V_m^2}{W_m}$
- Seebeck's Coefficient, $S = \frac{2Vm}{T_H - T_C}$
- Optimal resistance ratio, $m = \frac{\Delta T + \backslash\eta_{\text{th}}^{\max} T_C}{\backslash\eta_{\text{th}}^{\max} T_H}$
- The figure of merit, $Z = \frac{m_{\text{opt}}^2 - 1}{0.5(T_H - T_C)}$
- Current, $I = \frac{S(T_H - \Delta T) - S\left[(T_H - \Delta T)^2 - 2\Delta T Z\right]^{0.5}}{R}$
- Voltage, $V = -R(I - I_{\text{sc}})$
- Short circuit current, $I_{\text{SC}} = \frac{2W_m}{V_m}$
- Thermal Conductivity, $k = \frac{S^2}{RZ}$.

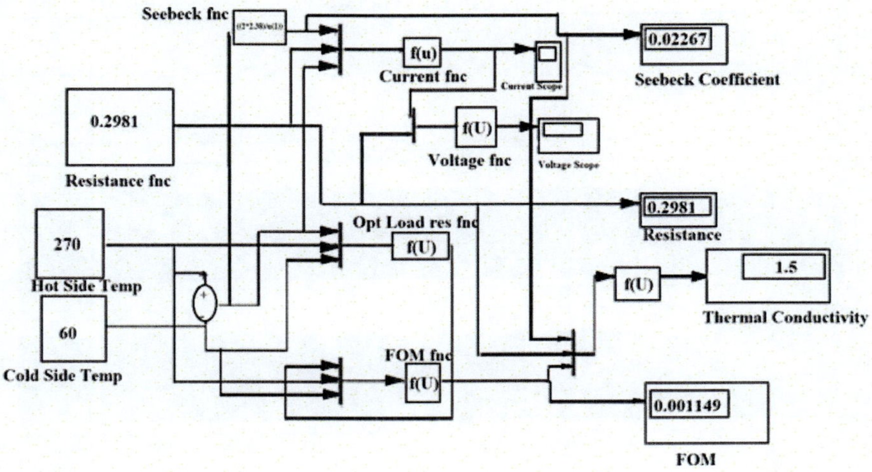

Fig. 7 Modeling of TEG [12]

The above equations were derived to design a TEG module in MATLAB/SIMULINK [15, 16]. The voltage output of this TEG is then used as the i/p to the LUO converter and the NPC inverter and finally given as AC output to nonlinear loads [17].

4.2 Comparison of LUO and Boost Converter

See Fig. 8.

Output with LUO converter is shown in Fig. 9.

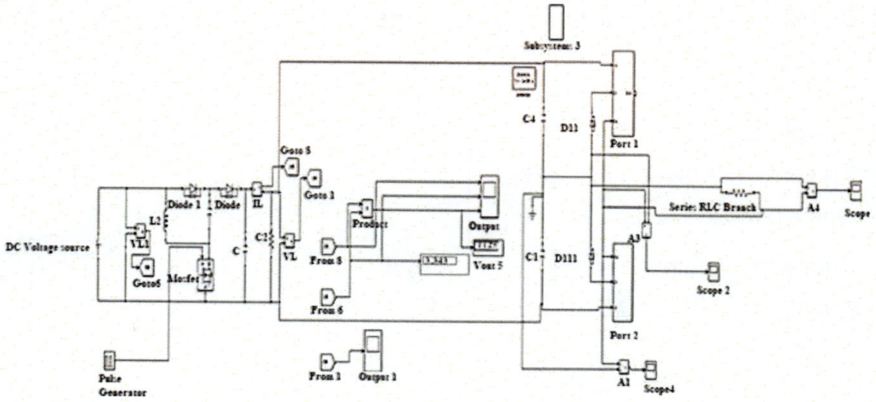

Fig. 8 LUO converter

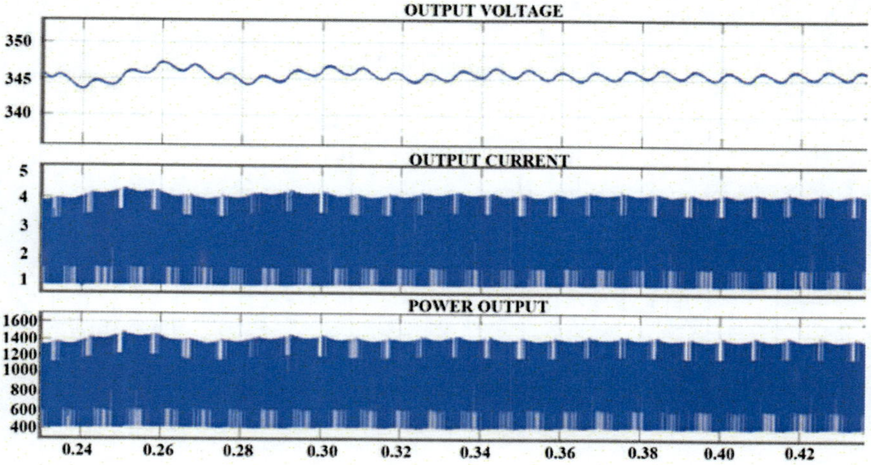

Fig. 9 Output voltage, output current, output power of LUO converter

4.3 Boost Converter

Boost converter simulation is shown in Fig. 10.

The output of the boost converter is shown in Fig. 11.

Table 2 shows the outputs from both the converters.

From Table 2, we can see that in both cases, the input voltage is set to 100 V. In the first case of the LUO converter, we can see that the output voltage is 345 V, and in the second case of the boost converter, we are getting an output voltage of 245 V.

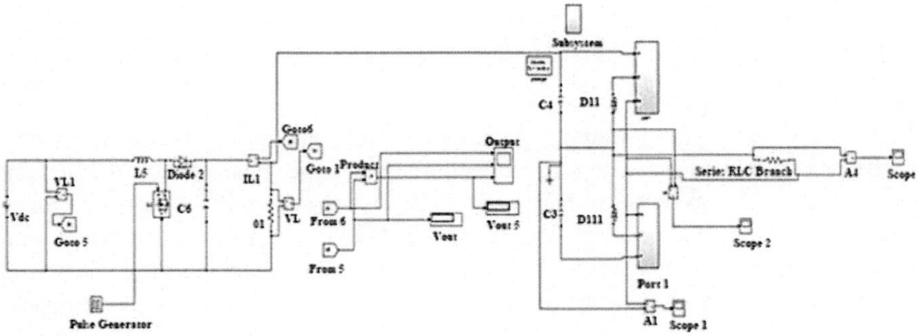

Fig. 10 Boost converter

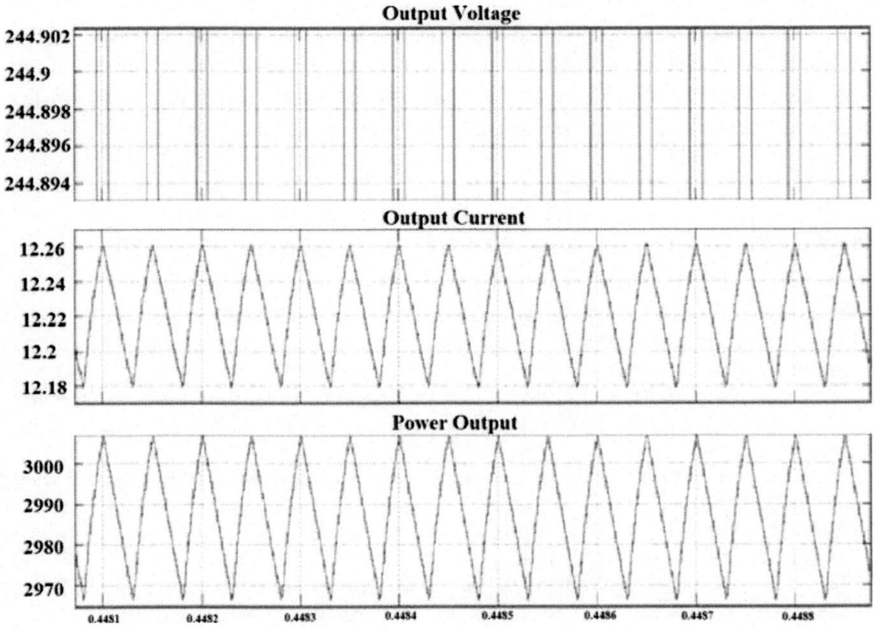

Fig. 11 Output of boost converter

Table 2 Comparison of two converters

With LUO converter	With boost converter
I/P voltage: 100 V	I/P voltage: 100 V
Output voltage: 345 V	Output voltage: 245 V

Plus, the output from the LUO converter is ripple-free. So, we can see that the LUO converter has much better output voltage quality; at the same time, the voltage value than that of the boost converter.

5 Conclusion

Thermoelectric energy is a strategy for creating power that converts waste heat energy or any type of heat energy into electric voltage. This paper revolves around the fact that the waste form of energy that is exiting from vehicles which are in the form of heat energy can be converted to electrical potential using Seebeck's effect, but the energy that we directly receive is not that much efficient and has other electrical disturbances as well, so with the use of power conditioning devices like converter, here the use of LUO converter and the neutral point clamped multilevel inverter is done which is novel. The use of these power electronics converters ensures that we get ripple-free and good power quality output. Then, this power can be fed to the nonlinear loads in domestic and industrial areas. This paper also proposes an idea of the clean and non-conventional form of energy which will boost in the near future, and due to the depletion of the existing form of energy, the present paper can prove to be one of the best alternative forms of energy extraction and also a boon for the co-generation process of electricity production organizations.

References

1. Loupis M, Papanikolaou N, Prousalidis J (2013) Fuel consumption reduction in marine power systems through thermoelectric energy recovery. In: Proceedings of the 2nd International MARINELIVE conference on all electric ship. Athens, Greece
2. Stecanella, Priscilla AJ et al (2015) Electricity generation using thermoelectric generator-TEG. In: 2015 IEEE 15th International conference on environment and electrical engineering (EEEIC). IEEE
3. Maran ALO et al (2016) Use of the Seebeck effect for energy harvesting. IEEE Latin Am Trans 14(9):4106–4114
4. Maharaj S, Govender P (2014) A prototype thermoelectric co-generation unit. In: Twenty-second domestic use of energy. IEEE
5. Ivanov K, Aleksandrov A (2019) Method of study of thermoelectric generators. In: 2019 16th Conference on electrical machines, drives and power systems (ELMA). IEEE
6. Luo FL (1997) Luo-converters, a series of new DC-DC step-up (boost) conversion circuits. In: Proceedings of Second international conference on power electronics and drive systems, vol 2. IEEE

7. Rocha RP et al (2009) An energy scavenging microsystem based on thermoelectricity for battery life extension in laptops. In: 2009 35th Annual conference of IEEE industrial electronics. IEEE
8. Lineykin S, Ben-Yaakov S (2007) Modeling and analysis of thermoelectric modules. IEEE Trans Ind Appl 43(2):505–512
9. du Toit Mouton H (2002) Natural balancing of three-level neutral-point-clamped PWM inverters. IEEE Trans Industr Electron 49(5):1017–1025
10. Sebaaly F, Kanaan HY, Moubayed N (2014) Three-level neutral-point-clamped inverters in transformerless PV systems—state of the art. In: MELECON 2014–2014 17th IEEE Mediterranean electro technical conference. IEEE
11. Lee HS (2010) Thermal design: heat sinks, thermoelectric heat pipes, compact heat exchangers, and solar cells. Wiley
12. Tsai H-L, Lin J-M (2010) Model building and simulation of thermoelectric module using Matlab/Simulink. J Electron Mater 39(9):2105
13. Prasad AR, Mini VP (2018) Design and modelling of thermal energy harvesting system for power transformer. In: 2018 4th International conference for convergence in technology (I2CT). IEEE
14. Deboever J, Grijalva S (2016) Modeling and optimal scheduling of integrated thermal and electrical energy microgrid. In: 2016 North American power symposium (NAPS). IEEE
15. Corry TM, Spira G (1962) Thermoelectric generator design, performance, and application. IRE Trans Mil Electron 1:34–40
16. Kanno T (2016) Transverse thermoelectric effect and its applications using synthetically or naturally anisotropic materials. In: 2016 IEEE International meeting for future of electron devices, Kansai (IMFEDK). IEEE, 2016.
17. Guizani M et al (2018) Performance analysis of Luo-converter for photovoltaic application. In: 2018 15th International multi-conference on systems, signals & devices (SSD). IEEE

Solar Energy-Based Battery Charger and Motion Sensing Alarm Using Arduino

Mukul Singh, Omveer Singh, M. A. Ansari, Vishwamitra Singh, and Simran Chaubey

Abstract This paper deals with the motion sensing with the help of motion detecting sensor powered by a 9 V rechargeable battery. The solar panels charge the 9 V battery using solar energy. Once the battery gets discharged, the solar panels are used to charge the battery again or to provide parallel supply of the energy to the circuit in the rarest case. The battery is charged during day time only when the device is not in use. This reduces the cost to huge levels over long period of time. This paper also shows a comparison of cost with respect to the existing motion detectors not using the solar energy as energy alternative. Also, it shows the impact of renewable source of energy in overall cost optimization of the device with this technology which has not been used ever before. The moving objects are detected by the PIR motion sensor which send the signal to Arduino which again sends the signal to piezoelectric buzzer. The piezoelectric buzzer starts playing the predefined alarming tune. Also, an LED is turned on as soon as the motion is detected. A one way rectifier is added in the circuitry of solar panel and the battery so that in the evening and night time the reverse charging of solar panels from battery does not take place.

Keywords Motion sensing alarm · Solar energy · PIR sensor · Solar panel · Arduino

1 Introduction

In today's world, everyone need safety and security at different places in our life like at our home, office, parking area, banks, etc. Outermost layer of protection for the

M. Singh (✉) · O. Singh · M. A. Ansari
Department of Electrical Engineering, Gautam Buddha University, Greater Noida, India
e-mail: gbubuddhams@gmail.com

V. Singh
Department of Electronics Engineering, Delhi Technological University, New Delhi, India

S. Chaubey
Department of Information Technology, Gautam Buddha University, Greater Noida, India

K. Namrata et al. (eds.), *Smart Energy and Advancement in Power Technologies*, Lecture Notes in Electrical Engineering 926,
https://doi.org/10.1007/978-981-19-4971-5_13

perimeter security can be guarded by motion sensing alarm. A motion sensor when attached with an Arduino can be used to warn us to act before the things get worse. For example, we can call the police before someone damages our property or steals something from our vicinity. We can use the motion detectors to stay alert of what is going on when we are not present somewhere. They can be placed at vulnerable points in houses, lockers, banks, and apartments and help us to be safe and sound [1, 2]. Also, they are resourceful and easily installable. Installing motion detectors and combining them with physical security and home automation will provide us not only a modern high-response physical security system, but we will also have highly energy efficient system [3, 4].

There are several methods to build a motion sensing alarm which includes different types of motion sensors like passive infrared sensor (using the person's own body heat, motion is detected) [5], microwave sensor, ultrasonic sensor, etc [6]. Hence, a device is needed to be built which can detect the motion of a moving object accurately without any false triggering in such a way that it does not affect the environment and make the future operating cost of motion sensor approximately nil. The device must be eco-friendly and uses renewable energy source so that the fuel cost is nil leading to a cost effective device.

2 Literature Survey

2.1 Solar Energy

Solar energy is received from sun in the form of radiant light and heat which is harnessed with the help of a many different technologies evolving from time to time like solar heating, photovoltaic, artificial photo synthesis, etc. Solar energy is a very precious source of renewable energy only on the basis of how we trap and distribute it or transform it into solar power. The use of photovoltaic systems, concentrated solar power are included in active solar techniques for harnessing energy which is oriented toward the Sun [7].

2.2 Motion Detection

The device that detects the motion of a moving object is called a motion detector. These devices are integrated as a system's component which automatically complete the task, i.e., alerts motion of someone in the restricted area [8]. Only by observing the movements of human or animal body, these passive motion sensing alarms based on infrared technology use heat detection of human. Radar technology is the basis of the working of active motion sensors which works by sending and receiving radio waves that comes back when they are interfered by a moving object. Combined motion detectors are not applicable to all areas even if they use several technologies. Different motion sensors work at different conditions, some work better outside than the others, based on the observation and performance. In active motion sensing alarms, radar technology is used [9].

Motion sensors are categorized as follows:

(a) Passive infrared sensors (PIR)
(b) Active infrared sensors
(c) Microwave alarm sensors
(d) Ultrasonic detectors
(e) Topographic motion detectors
(f) Gesture detectors.

Depending upon whether it is dark or light and integrated inside or outside, different motion sensing technologies are applied accordingly [10, 11].

2.3 *Working of PIR Sensor*

The input and output of sensors are dependent on many variables, so PIR sensors are more complicated. Figure 1 illustrates the working of a basic PIR sensor. When any heat source object moves in front of PIR sensor, variation in pulses is detected by the sensor, and the output signal is generated.

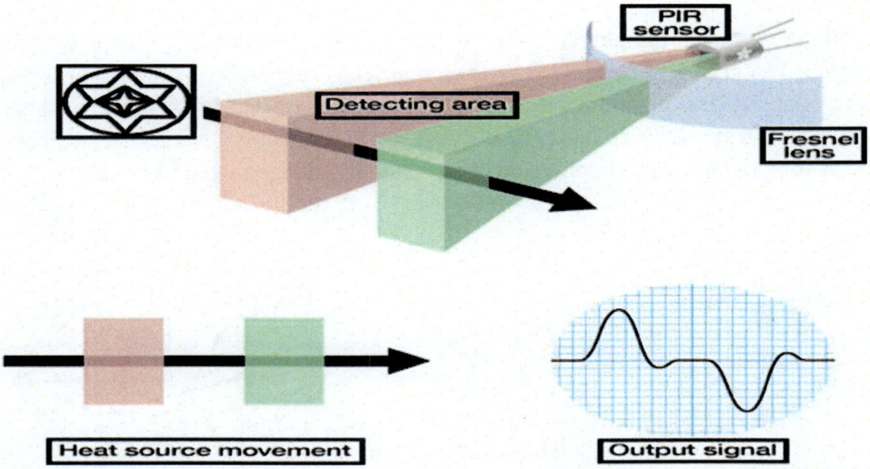

Fig. 1 Working of PIR sensor, producing output signal when heat source moves in front of it

2.4 *PIR Sensors and Lenses*

In a hermetically sealed metal can, the IR sensor is kept so that the immunity from noise/temperature/humidity can be improved [5]. The silicon is a transmissive material; a window is made for IR to pass through it. The coated silicon protects the sensing element. PIR sensors vary only in price and sensitivity as they are generic. Optics is very important in it. The PIR sensor and circuitry is fixed, so the breadth, range, and sensing pattern are changed by the lens easily.

The lens covers around two rectangles is the detection area but the desired detection area is needed to be much larger. In order to do that, a simple lens which condenses a large area into a small one just like those in camera is used. The PIR lenses are needed to be small, thin and moldable. So, the sensors are actually Fresnel lenses as shown above in Fig. 2.

3 Model Structure

Figure 3 shows the block diagram of for detection of motion. The 555 timer gives alarm through speaker as soon as the photoresistor detects the different phases. The pin2 is made high by photo transistor of 555 timer and will be set in mono-stable mode when there is no intrusion, and no output will be given in this configuration. The alarm will be made alert when the pin 2 of mono-stable timer gets low under the circumstances of an intrusion. The capacitor is responsible for alarm time duration [12–14].

Figure 4 shows the circuit design of PIR sensor. A beam of infrared light is produced by the IR sensor [5]. The laws which lay the basis of foundation of IR sensors are planks radiation law, Stephan Boltzmann law and Wien's displacement law. Any sort of photo radiation or IR radiation can possibly be detected by photo transistors. IR radiation is converted into current or voltage by photo transistor.

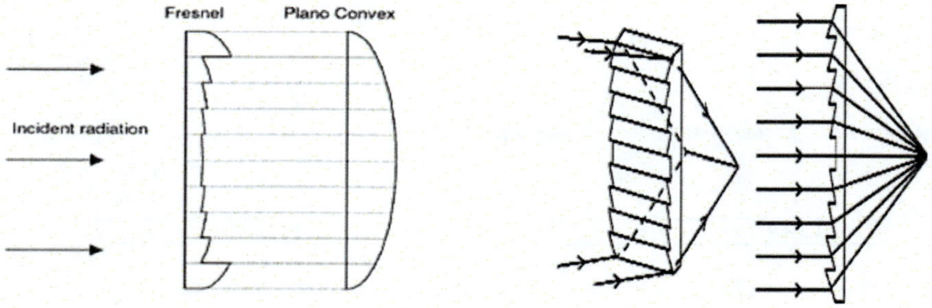

Fig. 2 Fresnel lens and large range of IR to the sensor because of condensing of light due to Fresnel lens

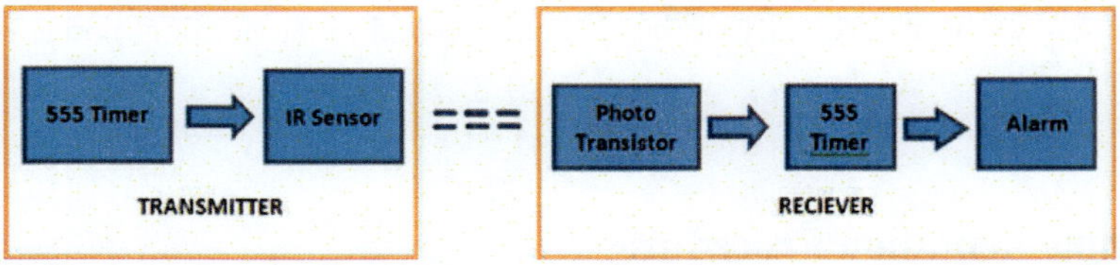

Fig. 3 Block diagram of motion detection

Fig. 4 Circuit design of PIR

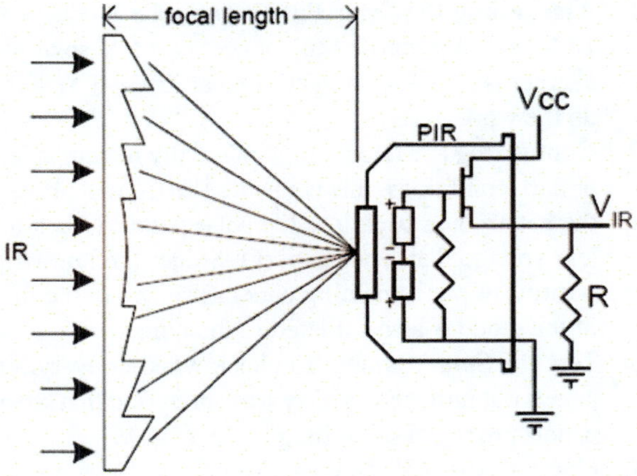

3.1 Solar Panels

Sunlight is used as source of energy which is absorbed by PV panels to produce electricity using connected assembly of typically 6 × 10 photovoltaic solar cell. [15, 16]. The area of a module is determined by the efficiency of that module, when provided with same output rating, i.e., 8% efficient 230 W module will have twice the area of a 16% efficient 230 W module. Only a limited amount of power is produced by a single solar module, so multiple modules installation is done to get the desired amount of power. An array of photovoltaic modules is generally used in a PV system for various purposes. The solar energy is available for free in abundant quantity. Hence, price of solar power is very less and kept falling in many countries. Solar power is very cheaper as compared to other means of energy generation. Solar energy should be used as in alternative in place of every non-renewable energy source [17, 18]. The voltage for the compatibility between solar panel and battery must be considered. How fast battery is charged by solar panels told by the amperes production of the solar panels. It is important to consider the voltage and the power requirements in most of the applications. Power of the products is usually indicated by the solar panel manufacturers (e.g., different kinds of models are of 20, 40, 65 W etc.). The relation is Watts = Volts × Amperes. The battery of the project work is at 9 V, and we have to do a series connection between two solar panels designed to work at 6 V each one. Velocity of the charge is related to the nominal power of the panel [19, 20].

4 Implementation Procedure

Steps included in recharging of cell using solar panel are [2]:

1. **Assembling the Solar Panel**: The solar modules are needed to produce around 12 V, so it was needed to connect two 6 V solar modules in parallel. This solar charger is a trickle charger rated at: 6 V—1 W—0.083 A—It is *slow*, but it will do the work.
2. **Connecting the Diode**: If accidentally battery is left plugged in during low light or dark conditions, this connected diode will prevent energy from flowing out of the battery and back into the solar panels (when the solar panels are producing less voltage than your battery). Solder the non-stripped side of the diode to the positive of our remaining panel as shown in the first picture. The remaining end of the diode is also soldered to the wire.
3. **Test the Solar Panels**: To check whether everything is working, take the solar panels out into the sun. Day was sunny and that when tested with the multi-meter, panels produced something close to 12 V.
4. **Connect the 9 V Battery Connector**: Take positive wire from the solar panel which is to be soldered and connect them to the positive terminal of the battery connector. Wires should be grounded.
5. **Prevent Accidental Short Circuits**: It is needed to be made sure that during any time there shall be no possibility of touching of positive and ground wires with each other (this creates a short circuit). Cover the exposed wires using electrical tape as shown in the picture. These possible short circuits may prove to be very dangerous as later on they can cause fires and even cause the explosion of the battery.
6. **Test the Charger One Last Time**: Test the charger one last time so as to be sure that nothing should get damage while the panels are mounted at a place and connected to the battery. Follow the same procedure as Step 6.
7. **Test the New Battery Charger**: After verifying that rechargeable battery is not charged till peak level. Carefully measure the voltage of battery with multi-meter. Make sure to write down the readings. Put the battery charger at a location where there is lot of sun rays falling on the battery charger [14]. After connecting the battery to the solar charger just for wait about 45–60 min because the charger will take time to charge as it is slow. Remove the battery from the connection after some time and check the voltage again of battery using multi-meter. As the value on the multi-meter as observed was increased, so it is concluded that project is successfully completed [15–20].

5 Analysis of the Model

The motion sensing alarm using solar energy was completed. Motion is sensed by the motion detecting PIR sensing. The different stages of the motion sensing are as shown in the demo.

Figure 5 shows that the motion sensing alarm is in switched off state, and no motion is being detected. During this stage, it must be confirmed that there is no motion in the nearby surroundings to ensure an efficient test run and practical working of the developed hardware. Multiple motion making subjects like humans or pet animal can also be placed at different distance and making them move at different times to check the range of the hardware. Figure 6 shows the alarm in switched 'On' state when all components are working properly and waiting for any motion to occur and to be detected. Until no motion has occurred, the green LED will not blink. The blinking of LED gives the signal of activation of the motion sensor and detection of the object. Figure 7 shows the motion detection stage when some motion has occurred in front of PIR sensor. LED glows at the same time the piezoelectric buzzer activates, and the alarm signal is triggered to produce sound from piezoelectric buzzer [5, 6]. Activation of both the signal such that the blinking of LED and the sound of buzzer confirms the presence of motion in the vicinity of the hardware model. The necessary condition for the sensor is that it should be kept stable at a fixed position by the user. Also, the solar panel will efficiently charge the battery under a clear sky. Solar panel should properly cleaned to avoid any losses due to non-absorption of energy by the panel from sun. The solar modules of higher voltage should be used, i.e., 12 V as that of the rated voltage of the battery is 9 V. It must be checked by the user to ensure that solar modules have a higher rating as compared to that of the battery used. As the rechargeable battery may not be fully charged due to any exceptional reason, so a 9 V battery can be used to avoid any irregularities in the motion detection as a backup power supply for just in case of any unwanted condition. We have also integrated a 9 V battery in our hardware model as a backup. Proper and complete working of the developed hardware model of solar energy-based motion sensing alarm is shown in Fig. 7. Various stages of battery charging using solar modules are observed during different time of the day, and critical discussion is done in the results section.

6 Results and Discussion

Table 1 shows the recorded observations of battery charging during day time. The project was tested on a sunny day about how fast (or slow) this charger is. The project was put on a rooftop of Tulsidas Boys Hostel built in Gautam Buddha University. The average battery gets discharged 165–170 h of continuous usage. However it was observed that charging is very slow, but it efficiently charged the battery and in one day 9 V battery can be charged near to full. Rest was charged in next two hours before sunlight goes out. If we use the solar panels for 10 years then we

Fig. 5 Switched OFF state

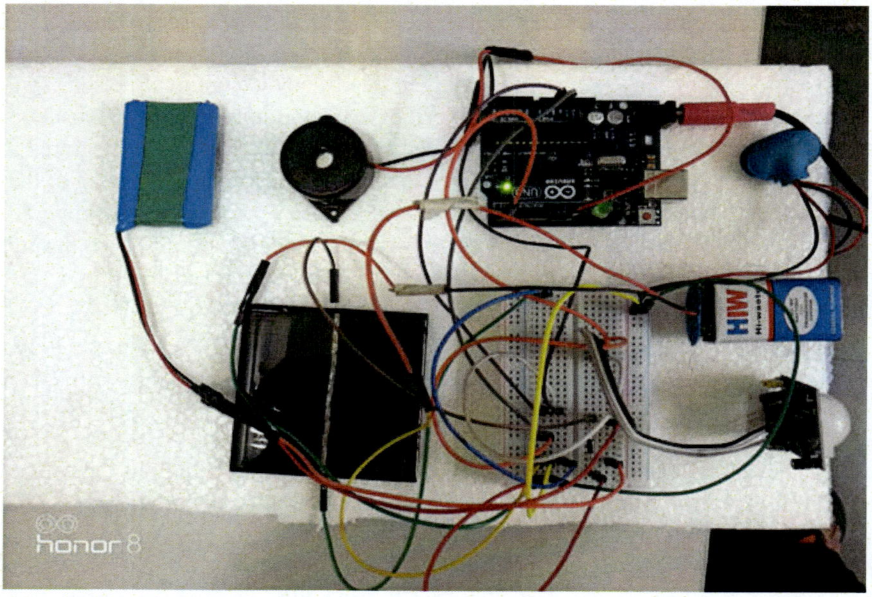

Fig. 6 Switched ON state

Fig. 7 Working state

will avoid new battery wastage. We have consider a total of 1 lakh people using this alarm among the current population of the world, i.e., 7,794,798,739 which is around 0.00001282906% of the case we have taken into consideration.

Normally without using the solar panels, the cost of the batteries would be around INR 12,48,00,00,000 for 1 lakh users in 10 years considering that battery gets worn out in 65 to 70 h and user single user has purchased a new battery every time worth INR 240. Whereas if the combination of solar panels and batteries is used then same amount spent would be around INR 3,90,00,000. The difference of the cost is around INR 12,09,00,00,000. Hence, if 0.00001282906% of the people uses the alarm, then INR 1,209 crores can be saved or earned simply by using solar energy. And if 1% people uses it, then total saving or profit would be around INR 94,23,91,16,75,451 approximately (which is around nine trillion four hundred twenty-three billion nine hundred eleven million six hundred seventy-five thousand four hundred fifty-one Indian rupees). Rather than saving or earning money, it is more about saving the nature, environment and future for upcoming generations by using clean energy.

Table 1 Observations of battery charging during day time

Time of day	Observations	Voltage of battery (V)
8:15 AM	Clear sunny sky with no clouds	8.25
9:15 AM	Clear sunny sky with no clouds	8.28
10:15 AM	Clear sunny sky with no clouds	8.30
11:15 AM	Clear sunny sky with no clouds	8.33
12:15 PM	Clear sunny sky with no clouds	8.39
1:15 PM	Clear sunny sky with no clouds	8.53
2:15 PM	Clear sunny sky with no clouds	8.71
3:15 PM	Clear sunny sky with no clouds	8.97

7 Conclusion

The motion was detected when there was any movement in front of the PIR sensor, and the alarming signal was produced by piezoelectric buzzer. The green LED was turned confirming that the motion was detected successfully. The solar panel charges the rechargeable lithium ion battery effectively providing the 9 V output. In the absence of the primary 9 V source, a supplementary 9 V battery was used so that motion detection process should not get obstructed when the 9 V battery discharges or any other problem with respect to supply source arises. Using the solar energy for recharging the battery and using in motion sensing alarm can lead to the eco-friendly, reliable, efficient device beneficial for upcoming generations.

References

1. Liu J, Tan JZ (2011) The analysis and application of solar energy PV power. In: 2011 International conference on advanced power system automation and protection
2. Liu K, Makaran J (2009) Design of a solar powered battery charger. In: 2009 IEEE electrical power & energy conference (EPEC)
3. Haque A, Zaheeruddin (2013) Research on solar photovoltaic (PV) energy conversion system: an overview. In: Third International conference on computational intelligence and information technology (CIIT 2013)
4. Suresh S, Bhavya J, Sakshi S, Varun K, Debarshi Y (2016) Home monitoring and security system. In: 2016 International conference on ICT in business industry & government (ICTBIG)
5. Chodon P, Adhikari DM, Nepal GC, Biswa R, Gyeltshen S, Chencho (2013) Passive infrared (PIR) sensor based security system. Int J Electr Electron Comput Syst 14(2)

6. Sathishkumar M, Rajini S (2015) Smart surveillance system using PIR sensor network and GSM. Int J Adv Res Comput Eng Technol (IJARCET) 4(1)
7. Nayyar A, Puri V (2016) A review of Arduino board's Lilypad's & Arduino shields. In: 2016 3rd International conference on computing for sustainable global development (ICCSGD)
8. Manchanda S, Sharma S (2016) Analysis of computer vision based techniques for motion detection. In: 2016 6th International conference—Cloud system and big data engineering (Confluence)
9. Chowdhury ZI, Imtiaz MH, Azam MM, Sumi MRA (2011) Design and implementation of pyroelectric infrared sensor based security system using microcontroller. In: Students' technology symposium (TechSym), 2011 IEEE, At Indian Institute of Technology (IIT), Kharagpur, India
10. Dhulekar PA, Gandhe ST, Shewale A (2019) Motion estimation for human activity surveillance. In: International conference on emerging trends & innovation in ICT
11. Tang K, Xu Y, Guo W (2019) An indoor human motion detection algorithm based on channel state information. In: 2017 International smart cities conference (ISC2)
12. Kushner D (2011) The making of Arduino. IEEE Spectrum
13. Nicholas B, Graham M, Gary TRR (2013) High temperature solar thermal central-receiver billboard design. In: International conference on solar energy
14. Louis L (2016) Working principle of Arduino and using it as a tool for study and research. Int J Control Autom Commun Syst (IJCACS) 1(2)
15. Singh M, Singh O, Kumar A (2019) Renewable energy sources integration in micro-grid including load patterns. In: 2019 3rd International conference on recent developments in control, automation & power engineering (RDCAPE), pp 88–93. https://doi.org/10.1109/RDCAPE47089.2019.8979036
16. Singh M, Rana V, Ansari MA, Dikshant SB, Singh P (2018) Power quality enhancement to sensitive loads with PV based microgrid system. In: 2018 International conference on sustainable energy, electronics, and computing systems (SEEMS). Greater Noida, India, pp 1–6. https://doi.org/10.1109/SEEMS.2018.8687334
17. Singh M, Singh O (2019) Phasor solution of a micro-grid to accelerate simulation speed. In: Proceedings of 2nd International conference on advanced computing and software engineering (ICACSE). Sultanpur, India. https://doi.org/10.2139/ssrn.3351025
18. Mahela OP, Khan B, Alhelou HH, Tanwar S (2020) Assessment of power quality in the utility grid integrated with wind energy generation. IET Power Electron 13:2917–2925
19. Islam MA (2020) Tailoring of the structural and optoelectronic properties of Zinc-Tin-Oxide thin films via oxygenation process for solar cell application. IEEE Access 8:193560–193568
20. Konyu M, Ketjoy N, Sirisamphanwong C (2020) Effect of dust on the solar spectrum and electricity generation of a photovoltaic module. IET Renew Power Gener 14(14):2759–2764

Convolution Neural Network and Continuous Wavelet Transform-Based Islanding Detection of Integrated DG with Phase Angle Between Voltage and Current

Ch. Rami Reddy, M. Kondalu, S. Ravindra, G. Srinivasa Rao, B. Srikanth Goud, and A. Narasimha Reddy

Abstract The growing electricity market is driving distributed generation forward (DG). Nearly the DG sources of nature are renewable. Islanding is one of the many complications of DG sources. Customers and their appliances can suffer from the insulation. The islanding will be detected in two seconds and the DG will have to be switched off in accordance with IEEE 1547 DG interconnection specifications. In this paper, an integrated island detection method for deeper learning is applied with Continuous Wavelet Transformations (CWT) and Convolution Neural Network (CNN). This technique transforms the details from the sequence of time into scalograms, later the images are used for training and testing of islanding and non-islanding incidents. The results are correlated with the reasoning approaches of the Artificial Neural Networks (ANN). The comparison reveals that the deep study method suggested effectively senses both islanding and non-islanding cases.

Keywords Distributed generation · Smart grid · Convolution neural network · Islanding detection · Scalogram

Ch. Rami Reddy (✉) · M. Kondalu
Electrical and Electronics Engineering, Malla Reddy Engineering College (A), Maisammaguda, Secunderabad, Telangana 500100, India
e-mail: crreddy229@gmail.com; crreddy229@mrec.ac.in

S. Ravindra
Electrical and Electronics Engineering, Vasi Reddy Venkatadri Institute of Technology, Nambur, Guntur, Andhra Pradesh 522508, India

G. Srinivasa Rao
Electrical and Electronics Engineering, CMR College of Engineering and Technology, Hyderabad, India

B. Srikanth Goud
Electrical and Electronics Engineering, Anurag University, Venkatapur, Ghatkesar, Medchal, Telangana 500088, India

A. Narasimha Reddy
Electronics and Communication Engineering, Malla Reddy Engineering College (A), Maisammaguda, Secunderabad, Telangana 500100, India

K. Namrata et al. (eds.), *Smart Energy and Advancement in Power Technologies*, Lecture Notes in Electrical Engineering 926,
https://doi.org/10.1007/978-981-19-4971-5_14

1 Introduction

DG systems are highly integrated and complicate the power system network. The islanding is one of the many problems due to such DG assimilation. DG feeds local loads until it is disconnected from the grid [1]. It may be deliberate or unexpected. The deliberate islanding occurs while utility is held. Due to a lack in the grid or uncertainties in the power system, unintended islanding can be caused [2]. Not only does it hurt consumers and personal equipment but it also cumbersome the grid [3]. Researchers suggest significant islanding identification approaches. They are explained here briefly. The methods of passive behavior are routinely checked by checking for passive parameters at the point of common coupling (PCC), compared to the predefined threshold value [4]. The passive components include voltage, current, frequency, impedance, phase angle, etc. The method affirms the islanding [5] if the parameter exceeds the value defined. Their thresholds were therefore subjected to a large non-detection zone (NDZ) and their difficulty [6, 7]. Active approaches are proposed in order to solve these demerits. The successful methods continuously inject a low-frequency harmonic signal and control PCC parameters [8]. The inserted signal would not impact the tracked parameters in a grid-related matter but in the islanding event, it results in a difference in the guidelines observed. The differences in the perticulator are used to locate the island [9, 10]. These proposals have no NDZ, but the level of power is degraded [11]. Hybrid approaches are proposed to eliminate the disadvantages of active methods. These are the combinations of active and passive methods [12]. The active solution supports this because the passive strategy suspects the insulated case [13]. The NDZ and impact efficiency of these approaches are not lower than the active process [14]. Data are collected from the utility and the DG [15] for the remote islanding approach. Researchers have been proposing various signal processing techniques to minimize NDZ and improve the efficiency of passive methods by removing the secret characteristics from the passive parameters [16–18]. Modules for the artificial intelligence learning distinguish islanding and non-islanding cases reliably without a threshold [19]. They have no NDZ, but big data are essential for model training [12, 20–25]. Due to the progress of smart grid technologies and potential complexities of the power system network, it is compulsory to produce an effective islanding detection technique. A new technique focused on profound learning is presented in which CWT and CNN are used for detection. First, the time series data from the PCC was translated into Scalogram CWT illustrations containing data from different islanding incidents. Subsequent to this Scalogram the proposed CNN model will be educated. This is the second attempt to adapt the strategies of image analysis for classifying islanding events. The rest of the document is structured in accordance with the following aspects. Section 2 explains how time series input to Scalogram is transformed. Section 3 outlines the method of testing and the planning of the data collection. Section 4 presents CNN architecture and preparation. Section 5 shows the findings and discussions. The conclusion is given in Sect. 6.

2 Time Series Data to Scalogram Image Conversion

The method of the time series transformation to Scalogram conversions is presented in this section. In preparation of the simple Scalogram image, the signal data from (1) is used [20]. The two frequencies, 10 and 200 Hz are consisting of amplitudes of 15 and 25, respectively. Randomly chosen amplitudes and frequencies for the clarification are shown. The transformation of a signal is done by the continuous wavelet transforms (CWT).

$$f(t) = 15\sin(2\pi \times 10 \times t) + 25\sin(2\pi \times 200 \times t) \tag{1}$$

The wavelet transform of any signal $f(t)$ can be specified as

$$X(u, s) = \int_{-\infty}^{+\infty} f(t)\frac{1}{\sqrt{s}}\psi^{*}(\frac{t-u}{s})dt \tag{2}$$

The time–frequency of a signal energy density is a Scalogram in a wavelet transformation. Simply put, a Scalogram is one observed incarnation of a wavelet that produces a time, frequency, and magnitude, respectively in color gradients at what end x, y, and z axis [21]. For Eq. (1), the time series effects of Scalogram are seen in Fig. 1. It is produced with Morse wavelets by applying the CWT. Figure 1 shows two 10 and 200 Hz frequencies, and two 15 and 25 amplitudes, respectively. Data may therefore be translated to Scalogram images at any time sequence. It is common knowledge that any monitored learning needs data set for network preparation and testing. In this article, Scalogram images from various time series events are used to create the data collection. The following segment outlines in depth the test system and planning for CNN training.

3 Test System and Data Set Preparation

For the testing of all controlled procedures, extensive training knowledge is needed. Standard data sets are required for issues of image recognition. For islanding detection methods certain basic data sets are not valid. Therefore the development of suitable data is suited to a typical test method. A PV source of a grid of 100 KW is seen in Fig. 2, such that a data set is generated. This paradigm was implemented such that the work presented could be satisfied. In the MATLAB/Simulink platform, the simulations are born. At $t = 0.4$ s, the islanding event is set up by opening the CB (Circuit Breaker). For six stages, a stage angle of 1000 samples per second is acquired between positive sequence part of voltage and the current at the PCC. To have these simulations, the PC has an i5 CPU, an 8 GB RAM, and a Windows 10 OS. Different islanding and not islanding activities are verified for the production of the

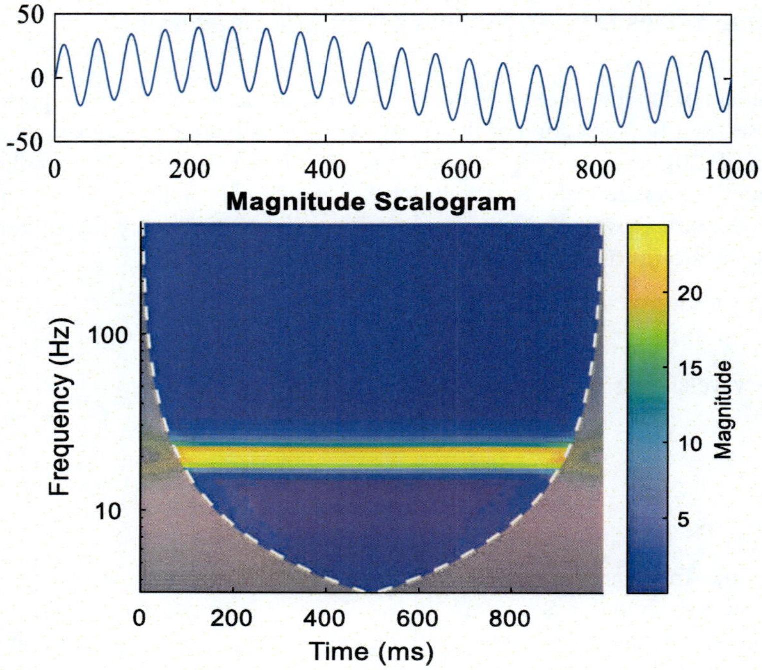

Fig. 1 Scalogram image of Eq. (1)

picture data and the effects are reported as time series drawings. For the generation of Scalogram images, CWT is added to each time series record. The phase angle Scalogram for integral and disconnected grid operations between positive sequence part of voltage and current at PCC is shown in Fig. 3. It is obvious that there is a good difference in signals of the island and the non-islanding incidents. The technique CNN for classifying images is used for detecting events in these images.

In the islanding condition, a zero or minimal power difference between DG and the load cannot be identified in most passive approaches. The condition is considered and the various islanding capsules are analyzed and used in the data settings in almost the worst power mismatches. The data collection often covers many islanding cases and non-islanding cases, including load switches, condenser banks, short-circuit defects, and motor switching accidents. For formation of datasets, there are a total of 300 islands and non-islands. This involves 150 islands and 150 activities non-islanding incidents. Table 1 list both cases.

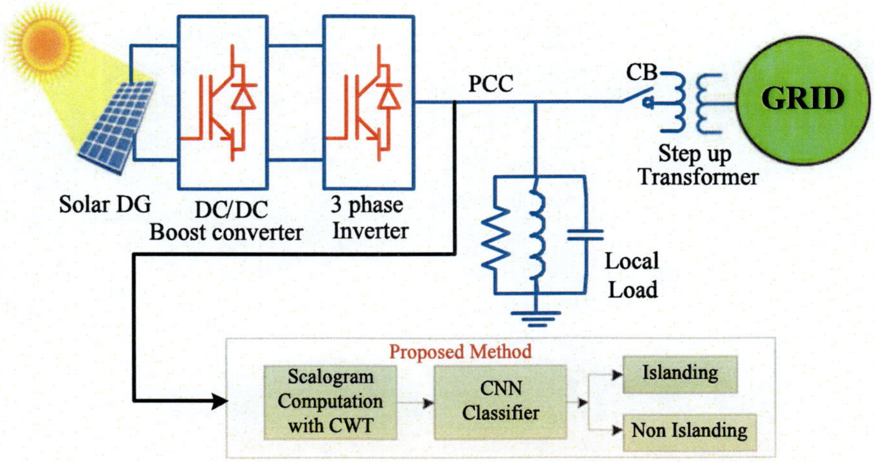

Fig. 2 Test system for implementation of the proposed method

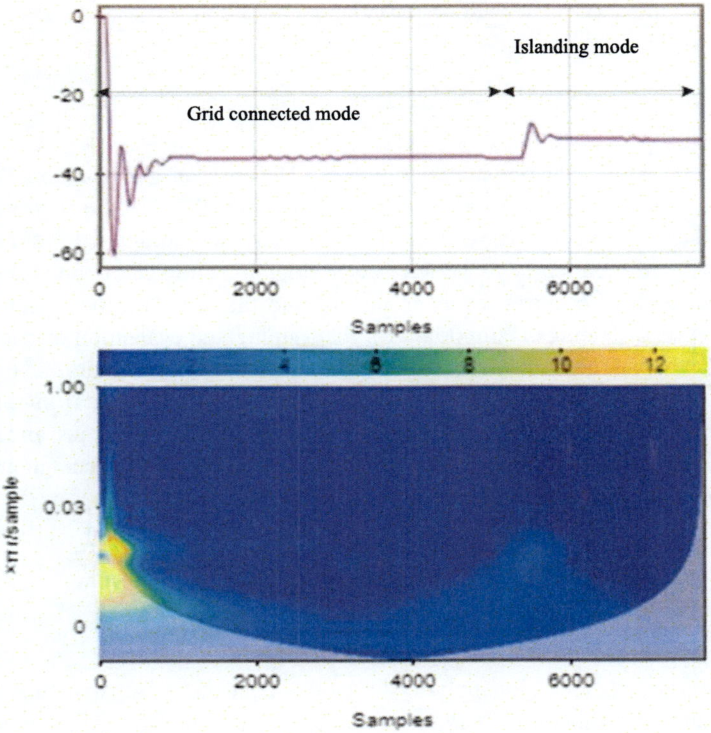

Fig. 3 Scalogram image variation for grid-connected and islanded data

Table 1 CNN design parameters

Training parameters	Design value
Optimizer	Stocastic gradient decent with momentum
Momentum	0.2
Learning rate	0.001
Maximum epochs	15
Mini batch size	10
Loss function	Cross entropy
Weight initialization	Random
Convolution layers	5
Kernals	$3 \times 3, 5 \times 5, 11 \times 11$
Drop out	0.5
Stride	2
ReLu	5
Max pooling layers	5
Fully connected layer	3

4 Methodology and CNN Design

The approach, design, and preparation specs for CNN are seen in this section. Figure 4 shows the stages in the planned method of islanding. PCC's time-server format acquires the phase angle between the positive sequence component voltage and current. This information is converted into Scalogram. The Scalogram is provided for classification of occurrences as an insight into the already experienced CNN. The extraction of features is important for the training and evaluation of controlled learning methods. These characteristics are the exactness of the solution. In the deep analysis of the CNN, these characteristics are automatically extracted from the input images. It consists of several layers, mostly for extracting characteristics, and is used to analyze only the final minority layers. Figure 5 shows CNN's general structure and many CNN slabs are listed here in short [22, 23].

4.1 Convolution Layer

Convolution is a procedure for applying the action of one on two separate functions in deep learning. Map functionality is pixels on the picture, while kernel functionality is an image. In addition, they are represented as a group of numbers. These arrays both yield the total. The picture's filter has been adjusted, based on the amount of time the camera was exposed to the first time. Until the whole image is covered. Output is generated from these calculations. With deep learning, instead of artificial neural

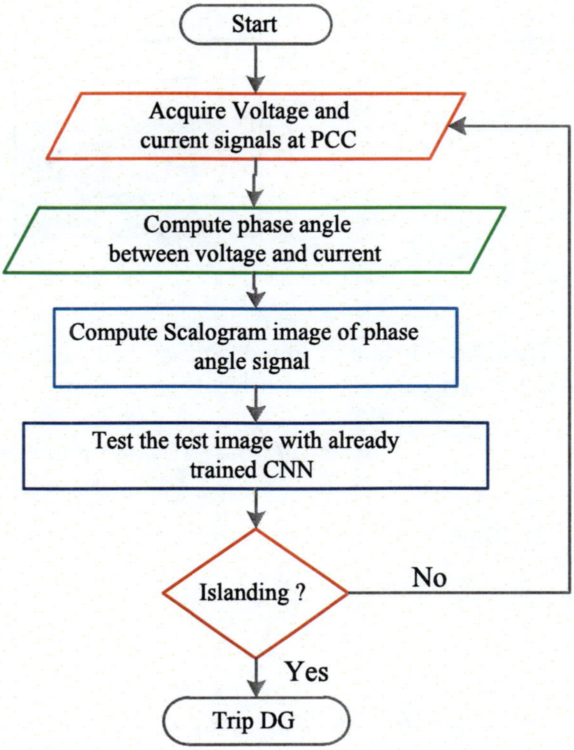

Fig. 4 Flow chart of the proposed method

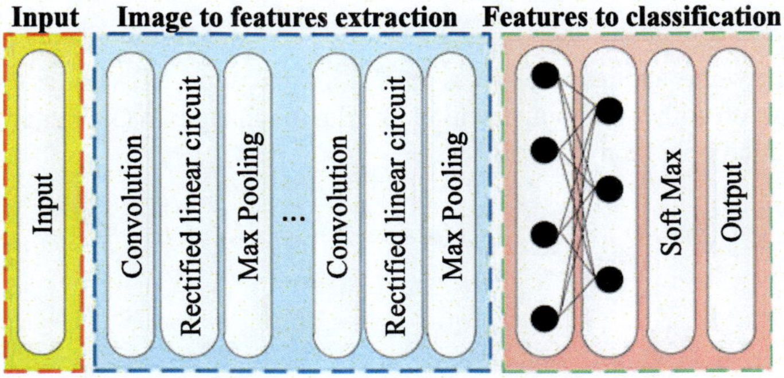

Fig. 5 Familiar design of CNN

networks, all input neurons are linked to all the output neurons. The convolution activity is represented by the $*$ operator. Output $f(x)$ is characterized when the input $I(x)$ is convoluted with the kernel $K(x)$ as (3)

$$f(x) = (I * K)(x) \tag{3}$$

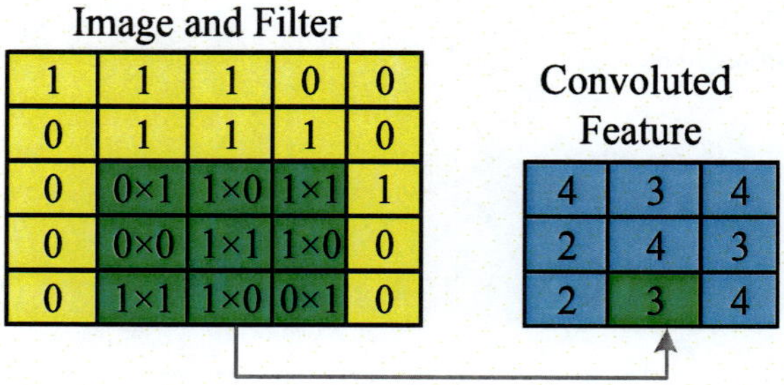

Fig. 6 Convolution operation in CNN

If x takes only integer attitudes, the discredited convolution can be defined as (4), which assumes the one-dimensional convolution

$$f(x) = \sum_a I(a) \cdot K(x - a) \tag{4}$$

The two-dimensional convolution with input $I(a, b)$ and filter $K(m, n)$ is illustrated as (5)

$$f(x) = \sum_m \sum_n I(m, n) \cdot K(a - m, b - n) \tag{5}$$

By commutative law, filter is flipped and Eq. (5) is corresponding to (6)

$$f(x) = \sum_m \sum_n I(a - m, b - n) \cdot K(m, n) \tag{6}$$

Neural networks appliance the cross-correlation operation, it is same as the convolution operation without flipping the filter, the Eq. (6) changes to (7). Figure 6 shows the convolution operation in detail.

$$f(x) = \sum_m \sum_n I(a + m, b + n) \cdot K(m, n) \tag{7}$$

4.2 *Rectified Liner Unit (ReLu) Layer*

To provide an interpolated response to changes in the pixel position, the gain in the convolution lamination smoothly increases or decreases as the pixel value changes. In activations, nonlinear transformations are usually using a tanh, Sigmoid, Rectified Linear Unit (ReLu). At the performance of the CNN, LR, ReLu activation is used. In this case, it is seen as Fig. 7.

Fig. 7 ReLu activation function

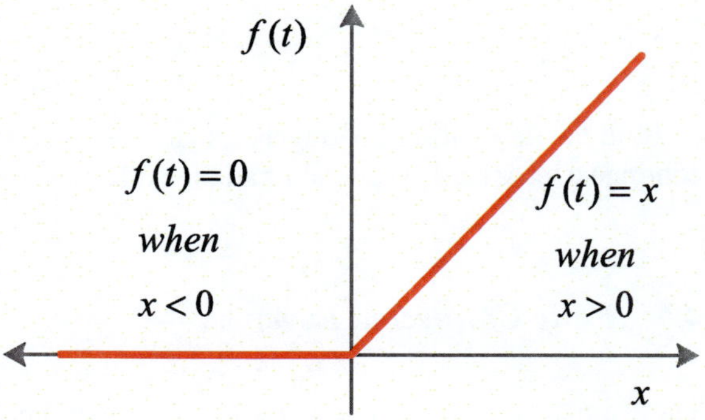

Fig. 8 Different pooling operations with 2 × 2 filter and stride 2

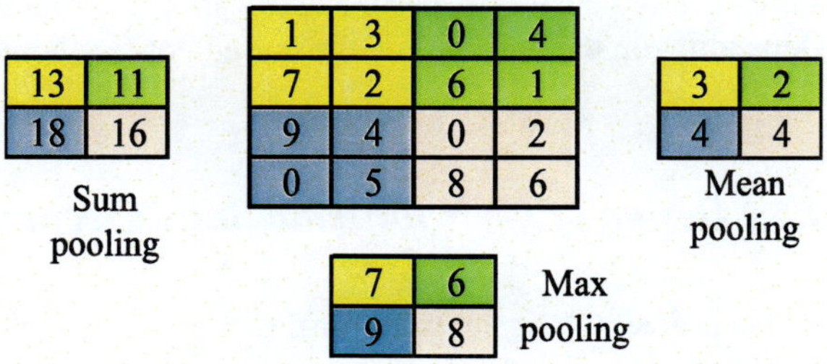

It is the nerve's input. Whether the input is negative, it gives zero output; if the input is positive, it produces the same result. By the complexity of the equations, this layer speeds up the creative process and solves the gradient issues.

4.3 Pooling Layer

If the resolutions of the extractions go down, so do the pools. To correct for compression, this layer actually accentuates the opposite phenomena. Max pooling, average pooling, and L2 normalization are three common types of automatic programming. Non-overlapping two-dimensional areas for data entry. In terms of production, maximum zone values would yield the maximum number of results for an ordinary pooling, the total of all values is used, and for all combined is the number. What is suggested is to solve the problem involves using the maximum pool (Fig. 8).

4.4 Softmax Layer

This layer offers real-valued softmax scores for each of all N-dimensional data. These are the classifications in mind It is mathematically stated as (8).

$$x_i = \frac{e^{z_i}}{\sum_{j=1}^{n} e^{z_j}} \tag{8}$$

If all the determined contingencies are in one dimension, so all the conceivable contingencies are nil. The total of all the odds must be one here.

4.5 Fully Connected Layer

These are the layers that output the CNN's data. This layer delivers the results. In a totally connected layer, the neurons are attached to every other layer. These features are used to calculate the basic classification contribution from the previous layers. Depending on the application, the ingredients are combined in various ways.

4.6 Design of CNN for Islanding Detection Approach

In this article, CNNs are used for the classification of different cases. A number of variables are used as the CNN is built. They are starting out with the previous two steps being well-supported. Thus, at the beginning of the training, all the hyperparameters are available. This will help to determine how many layers are needed to yield efficient printing. After the ideal hyper-parameter variance has been found, statistical slabs are used to guide CNN. It's starting with one sheet on the cutting mat initially. Three CNN operations can be found in any CNN layer: convolution, ReLu activation, and max pooling if the CNN has run on one layer, the procedure is replicated for the same procedures on the second. The amount of layers found shows that the high quality of eight architecture is better than that of seven layers. The CNN network configuration also provides eight layers for islanding and other grouping purposes. The next move is investigating filters. Comparing 3×3 kernel to the results from a 5×5 and $11 \times$, it is noticed that the 3×3 kernel has decent performance. It is successfully tested with the momentum system and stochastic descent method. The error-plus-plus-accuracy learning score of 0.001 serves our interests well. All of the details and specification parameters are presented in Tables 1 and 2.

5 Results and Discussion

The construction is in relation to 75% of data, and testing consists of 25% of data. 25% of the data remains entirely out of the network's view This data collection has different types of activities, such as turning off and on as well as rebooting the computer. Load transients, capacitor transients, fault transients, and ground bounce are all part of the non-islanding activities. Stochastic images are created from all these

Table 2 Different Scalogram simulated for data set preparation

Events	Number of cases
Islanding	110
Near zero power loading	40
Large and medium loading	70
Non-islanding	148
Capacitor switching (ON)	10
Capacitor switching (OFF)	10
Induction motor switching (ON)	10
Induction motor switching (OFF)	10
Load switching (ON)	10
Load switching (OFF)	10
Various types of fault switching	8
Grid-connected (Out of islanding area)	80

time series results. islanding and non-island light levels are seen in the horizontal or vertical linear variations in the scanner images In Figures 9, 10, 11, 12 and 13, a single test Scalogram is seen. A total of 65 instances (25 from islanding plus 40 non-islanding) were assessed. These are the only three samples that show some defect in the testing process.

6 Conclusion

This paper introduces a novel island detection approach that integrates CWT and CNN. Positive sequence voltage is captured as an angle to the present waveform. The data collection includes 258 examples of both normal and aberrant islanding. The CNN is trained on 75% of the data and kept for review. The percentage of research results that are non-islanding is 33%. This methodology has a consistency rating of 95%. CNNs have been shown to be effective in detecting islanding classifications.

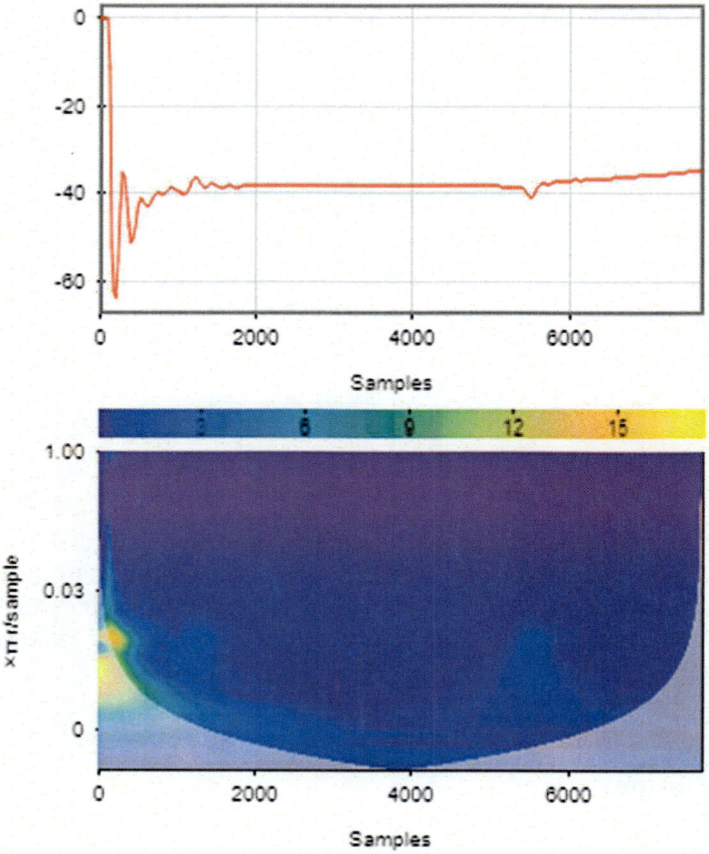

Fig. 9 Islanding case for 100% of load

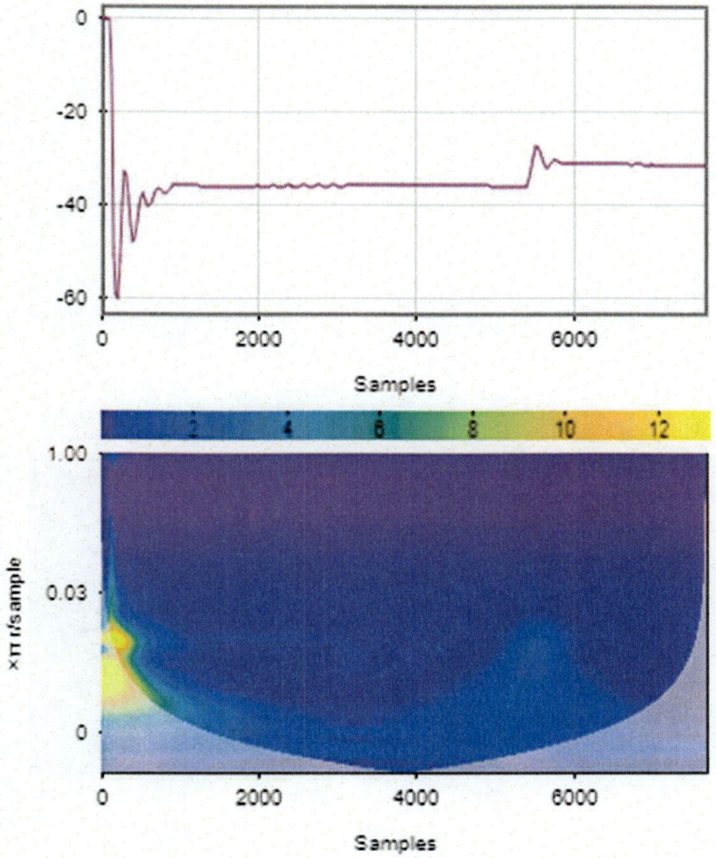

Fig. 10 Islanding case for 80% of load

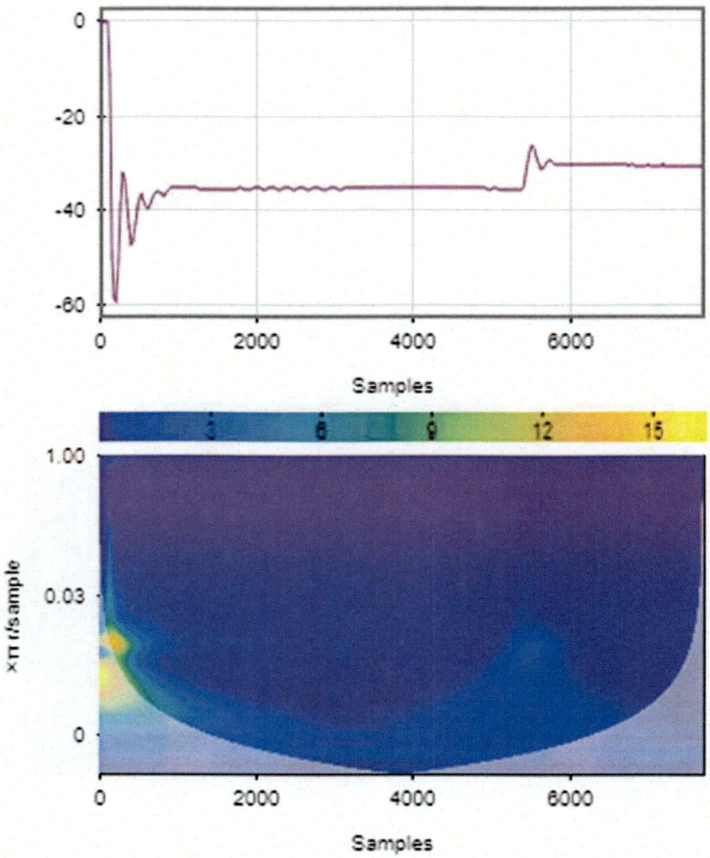

Fig. 11 Islanding case for 50% of load

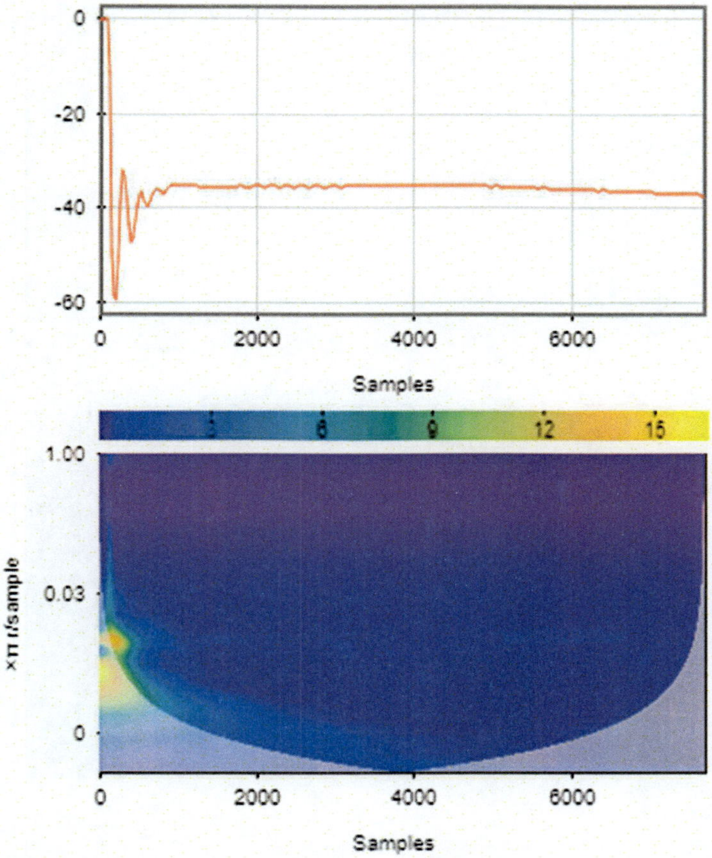

Fig. 12 Non-islanding case of capacitor switching

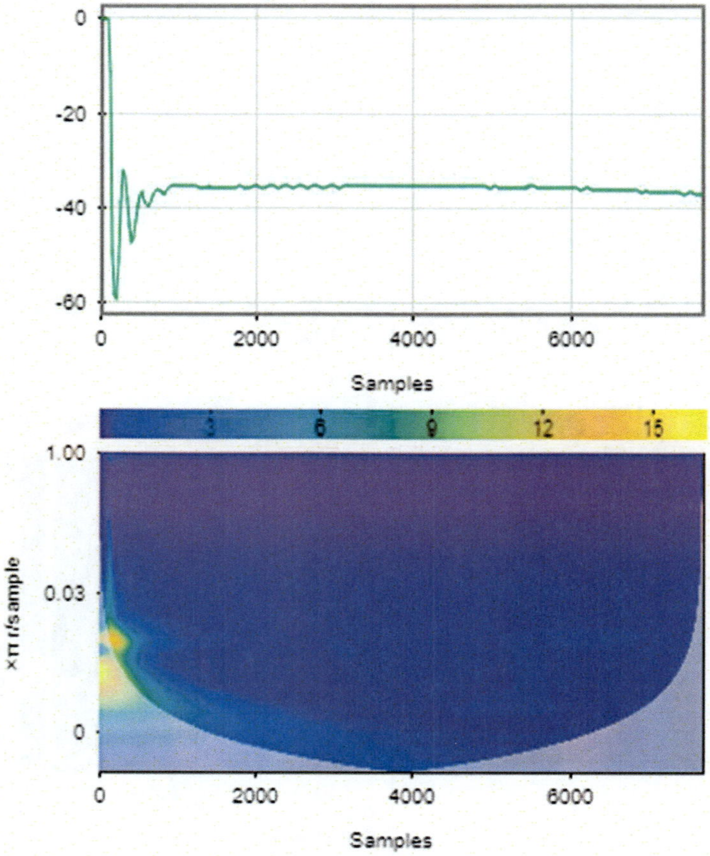

Fig. 13 Non-islanding case of induction motor switching

References

1. Bekhradian R, Davarpanah M, Sanaye-Pasand M (2019) Novel approach for secure islanding detection in synchronous generator based microgrids. IEEE Trans Power Delivery 34(2):457–466
2. Cui Q, El-Arroudi K, Joós G (2018) Islanding detection of hybrid distributed generation under reduced non-detection zone. IEEE Trans Smart Grid 9(5):5027–5037
3. Raza S, Mokhlis H, Arof H, Laghari JA, Mohamad H (2016) A sensitivity analysis of different power system parameters on islanding detection. IEEE Trans Sustain Energy 7(2):461–470
4. Reigosa DD, Briz F, Blanco Charro C, Guerrero JM (2017) Passive islanding detection using inverter nonlinear effects. IEEE Trans Power Electron 32(11):8434–8445
5. Marchesan G, Muraro MR, Cardoso G, Mariotto L, de Morais AP (2016) Passive method for distributed-generation island detection based on oscillation frequency. IEEE Trans Power Delivery 31(1):138–146
6. Salles D, Freitas W, Vieira JCM, Venkatesh B (2015) A Practical method for non detection zone estimation of passive anti-islanding schemes applied to synchronous distributed generators.

IEEE Trans Power Delivery 30(5):2066–2076
7. Saleh SA, Aljankawey AS, Meng R, Meng J, Chang L, Diduch CP (2016) Apparent power-based anti-islanding protection for distributed cogeneration systems. IEEE Trans Ind Appl 52(1):83–98
8. Wen B, Boroyevich D, Burgos R, Shen Z, Mattavelli P (2016) Impedance-based analysis of active frequency drift islanding detection for grid-tied inverter system. IEEE Trans Ind Appl 52(1):332–341
9. Murugesan S, Murali V (2020) Disturbance Injection Based Decentralized Identification of Accidental Islanding. IEEE Trans Industr Electron 67(5):3767–3775
10. Sivadas D, Vasudevan K (2020) An active islanding detection strategy with zero non detection zone for operation in single and multiple inverter mode using GPS synchronized pattern. IEEE Trans Ind Electron 67(7):5554–5564
11. Murugesan S, Murali V (2019) Hybrid analyzing technique based active islanding detection for multiple DGs. IEEE Trans Ind Inf 15(3):1311–1320
12. Kermany SD, Joorabian M, Deilami S, Masoum MAS (2017) Hybrid islanding detection in microgrid with multiple connection points to smart grids using fuzzy-neural network. IEEE Trans Power Syst 32(4):2640–2651
13. Khodaparastan M, Vahedi H, Khazaeli F, Oraee H (2017) A novel hybrid islanding detection method for inverter-based DGs using SFS and ROCOF. IEEE Trans Power Delivery 32(5):2162–2170
14. Chen X, Li Y, Crossley P (2019) A novel hybrid islanding detection method for grid-connected microgrids with multiple inverter-based distributed generators based on adaptive reactive power disturbance and passive criteria. IEEE Trans Power Electron 34(9):9342–9356
15. Xu W, Zhang G, Li C, Wang W, Wang G, Kliber J (2007) A power line signaling based technique for anti-islanding protection of distributed generators—part I: scheme and analysis. IEEE Trans Power Delivery 22(3):1758–1766
16. Mohanty SR, Kishor N, Ray PK, Catalo JPS (2015) Comparative study of advanced signal processing techniques for islanding detection in a hybrid distributed generation system. IEEE Trans Sustain Energy 6(1):122–131
17. Ray PK, Kishor N, Mohanty SR (2012) Islanding and power quality disturbance detection in grid-connected hybrid power system using wavelet and S-transform. IEEE Trans Smart Grid 3(3):1082–1094
18. Do HT, Zhang X, Nguyen NV, Li SS, Chu TT (2016) Passive-islanding detection method using the wavelet packet transform in grid-connected photovoltaic systems. IEEE Trans Power Electron 31(10):6955–6967
19. Khamis A, Xu Y, Dong ZY, Zhang R (2018) Faster detection of microgrid islanding events using an adaptive ensemble classifier. IEEE Trans Smart Grid 9(3):1889–1899
20. Manikonda SKG, Gaonkar DN (2019) IDM based on image classification with CNN. J Eng 2019(10):7256–7262
21. Sejdic E, Djurovic I, Stankovic L (2008) Quantitative performance analysis of scalogram as instantaneous frequency estimator. IEEE Trans Signal Process 56(8):3837–3845
22. Guo K et al (2018) Angel-eye: a complete design flow for mapping CNN onto embedded FPGA. IEEE Trans Comput Aided Des Integr Circuits Syst 37(1):35–47
23. Ker J, Wang L, Rao J, Lim T (2018) Deep learning applications in medical image analysis. IEEE Access 6:9375–9389
24. Govinda RS, Harinadha Reddy K, Reddy Ch (2021) Islanding detection parameters for integrated distributed generation. Recent Adv Electr Electron Eng (Formerly Recent Patents on Electrical & Electronic Engineering) 14(2):131–143
25. Reddy JR, Pandian A, Rami Reddy C (2020) An efficient learning based RFMFA technique for islanding detection scheme in distributed generation systems. Appl Soft Comput 96:106638

Analysis of Microgrid and Protection Schemes: A Review

Mukul Singh, Omveer Singh, and M. A. Ansari

Abstract The review paper presents a detailed analysis and review of microgrid and factors on which development of protection algorithms for microgrid-interfaced renewable energy sources depends. The review focuses on every aspect of the microgrid. It includes the factor affecting the protection of microgrid under different conditions. This is done after the investigation and literature review of various protection schemes currently in effect and are being implemented at various stages in a microgrid. To understand the microgrid protection schemes, understanding the recent protection challenges associated with AC microgrids with inverter-interfaced RES should be investigated thoroughly. The research paper includes microgrid classification, advantages of microgrid, characteristics of microgrid, microgrid protection schemes, limitation of microgrid protection schemes, and future scope of the paper. Major concern of the protection is specifically for the situation when microgrid switches from grid connected to island mode of operation. Exploration of the selected protection solutions and algorithms that would alter the regular protection methods are needed to be critically analyzed to achieve effective conclusions and implementable algorithms.

Keywords Microgrid · Solar PV · Wind farm · Renewable energy sources (RESs) · Protection schemes · Distributed energy resources (DERs)

1 Introduction

Many times, single power plant is not sufficient to give necessary value of electricity desired by load of consumers. This problem was solved by integrating the different power plant's energy supply with power grid. Yet, the problem of reaching the energy remained the same. Later on to resolve the issue, concept of islanded microgrid was introduced. Islanded microgrid is a self-sustained microgrid which is not connected

M. Singh (✉) · O. Singh · M. A. Ansari
Department of Electrical Engineering, Gautam Buddha University, Greater Noida, India
e-mail: gbubuddhams@gmail.com

K. Namrata et al. (eds.), *Smart Energy and Advancement in Power Technologies*,
Lecture Notes in Electrical Engineering 926,
https://doi.org/10.1007/978-981-19-4971-5_15

to main power grid. It has its own sources of energy/power. They are mostly renewable energy sources like wind and solar. Yet, another problem figured was the non-availability of sufficient amount of energy from limited amount of renewable energy plants. This problem was overcame by integrating a diesel generator along with renewable energy sources so that sufficient amount of energy is can be maintained in the microgrid [1]. Microgrid is independent from grid's power supply because of the availability of various alternate supply from different sources to MG.

Energy provided by the PV panels, wind and diesel generators are enough for fulfilling the desired necessity of user connected to microgrid [2]. Also, hybrid energy storage system (HESS) is applied instead of battery energy storage system to enhance the reliability of islanded microgrid by performing sensitivity analysis. The various types of models used are as follows: the maximum IMG load ability evaluation model, the Borgonovo method, the modified Kriging model, and Borgonovo indices are calculated based on the surrogate model [3]. Hence, the special behavior of microgrids has shown an urge to make the changes to regular protection strategies implemented to the distribution networks. Distributed energy resources are interfaced with the help of power electronic equipment usually produce bi-direction fault current. This leads to decrease in the capacity of the fault current leading to severe protection sensitivity failure and disruptions in the fault detection. It is highly preferential that the protection schemes have the feature to adapt itself to new conditions arising due to islanding and sectionalization after implementation of dynamic topologies.

2 Literature Review

Roa et al. [4] presented the work on simulation, designing, and implementation of the benchtop DC microgrid. He has shown the advantages of control system which is an alternative to future smart home applications. According to him, after comparison with AC systems, most of microgrids are AC as they are easier to integrate with the grid. Li et al. [1] presented the work aimed at building an integrated new energy mode service for modern agriculture by making a new energy work vehicle which will move in all seasons to avoid black tea withering. Opathella et al. [5] made a prototype of microgrid constituting three-phase converter bank which is AC–DC network connected with battery energy storage system, AC and DC loads. All network element details are included in his research. He concluded that power balance is can be managed by converters by comparing power flow analysis results to voltage measurements. Mirez et al. [2] performed various simulations for searching new methods of interconnection among microgrids by using novel strategies. He came to the conclusions that (a) It is important to have three mini-voltage scales even when microgrid bus contains a nominal voltage; (b) The new time scale of microgrid operation would be period; (c) Linear programming techniques should be used to optimize the cost of generation and storage; (d) Coordinated actions for tertiary control of microgrid should be represented to serve the purpose of interconnection of devices between microgrids. Johannes Hofer et al. [6] investigated the performance coupling of on-site PV generation of hybrid microgrid with residential AC and DC loads of building in simulation. Hybrid microgrid was designed and planned for implementation with PV system of residential building. Correia et al. [7] proposed the smart integration with DER's in microgrids through use of D-UPQC. The distributed unified power quality conditioners contain a photovoltaic and an ESS which can be operated in grid-connected mode (or islanded mode of operation if the grid fails). Simulation model is made where the results were obtained with distributed unified power quality conditioners (D-UPQCs) working as a link among photovoltaic (PV) systems, energy storage systems (ESS), and low-voltage (LV) grid. Their system provides high chance of improvement in quality of power in microgrid and a unitary power factor when connected to the low-voltage grid. Silvia Ma Lu et al. [3] reviewed simplified model of two-level voltage source converter (VSC) to join two DC sources with three-phase AC grid. The operation of whole microgrid was studied, and behavior was checked using MATLAB to analyze the results for proper performance. PV cell's working principle was simulated after modeling of an electric model and concluded that highest operation point of the PV system can be forced by MPPT control algorithm. He also tested VSC model based on six IGBT where SPWM was applied to achieve voltage modulation. Ali Hooshyar et al. [8] considered various generation conditions and various types of faults to analyze different types of relay for a microgrid. Problematic conditions and different scenarios were identified, and simulation was done in time domain where correct function of the relays was focused. Also, he laid emphasis on the ways for protection of DC microgrid and their requirement of equipment needed for protection. Myla Bharath Kumar et al. [9] presented the PV and wind microgrid's control and modeling after proper selection of appropriate wind turbine and AC/DC microgrid. System model for operation included solar irradiation level, uncertainty, ambient temperature, intermittent wind speed characteristics, and load. Without halting the utility grid's operation, many micro-sources are integrated in the microgrid. Luis I. Minchala et al. [10] presented the design and development of an EMS for a safe operation of islanded type of microgrid when load demand and generation capacity is imbalanced. Distributed controllers of integrated models of components of microgrid are tested in simulation platform. Simulation showed that frequency and voltage magnitude operated even when one of DC's was in fault condition but within a secured interval. Fouad et al. [11] presented a wind and solar-based microgrid system and addressed problems related to stability, operation, and control of the system. Technical issues which are involved in successful operation of microgrid are identified after modeling of system and analyzing the simulation results were identified related to generating units based on renewable power. All the loads are can be connected to microgrid as DG's has sufficient capacity to support all the loads. Katayoun Rahbar et al. [12] two cooperative microgrids are considered with each of them having renewable energy generator and ESS for energy management problem. Impacts of microgrids are studied on total energy cost using their ESS's and energy cooperation and proposed online algorithms having less complexity and valid arbitrary realization. Using clustering approach, another method is introduced for the more than two microgrids by applying online algorithms. Prof. Ing. Vincenzo Parenti

Castelli et al. [13] implemented a smart control system to handle peak shaving or load leveling (ESS application). Designing of an ESS prototype was done along with its development and integration with production system based in renewable energy to create a smart microgrid. Energy flow was managed in an efficient and intelligent way using power demand function. The energy produced can be given to the load, supplied to the grid, or can be used for battery storage. Rajdeep Chowdhury et al. [14] combined the models of micro-sources (photo voltaic array, PWM inverter, fuel cell stack) to make a microgrid simulated the developed model in Simulink. The designed system can be operated in grid connected and islanded, both mode of operations of a microgrid. They laid the groundwork for the development of a sophisticated model of microgrid which will allow to have more behavior understanding of microgrid. Sina Parhizi et al. [15] reviewed issues which concerns microgrids and provided research ideas related to microgrids. Author gave a review of microgrid introduction, benefits associated with them, their components and applications for grid performance enhancement which is continued by studies on various topics like control, operation, communications, economics of microgrid, protection. Author has given a broad view of integration of microgrid in power systems.

Manzar Ahmed et al. [16] presented a microgrid model having transient response to changing inputs and have steady state. Discussion on fuel cell current model, wind turbine, solar cells and micro turbines is made, and a complete model having various power sources, their power electronics, and load and mains model is built up. They selected 16 load points, and all of them were supported by microgrid. Harmonics and load efficiency were calculated, and conclusion came that load supporting points decrease due to which loss increases and limit reached out when gradual increase in efficiency of PV load was made. Daming Zhang et al. [17] studied the two modes of microgrid operation, i.e., grid-connected mode and islanded mode of operation when microgrid is integrated with solar panel-based distributed renewable energy generation and diesel-engine generator-based conventional energy generation. Various conditions are also tested when diesel generators are disconnected from the microgrids. The new equilibrium was achieved, and microgrid was still operational even after the disconnection of microgrid. Diesel engine operated harmoniously with microgrid in both modes of operation. Rui Huang et al. [18] proposed various solutions to the problems like standardization of information exchange, intermittency. Author performed and validated the approach on three microgrid test beds and concluded the fact that proper and gradual integration of distributed renewable energy generation into microgrid is possible. By using genetic algorithm and nonlinear programming, design for renewable distributed energy can be optimally sized and placed. V. Senthil Kumar et al. [19] carried out detailed simulation in MATLAB and ensured the stability of controller by Lyapunov method. Current control scheme along with nonlinear controller is proposed to control active and reactive power flow efficiently under various operating conditions as it is easier to implement when compared with different types of controllers. Also, proposed model has fast convergence and better stability of the tracking error. The "p-q" theory is used faster reference current estimation. Efficient use of renewable energy is facilitated by power control property in microgrid.

Ting Zhu et al. [20] proposed sharing of energy to balance harvested local energy and demand explicitly by using an alternative structure nearby homes. To determine particular homes who will share energy, novel approach of energy sharing is developed which minimized system losses and increased efficiency. Mismatch between consumed and harvested energy was addressed for individual homes to ensure energy sharing. Author concluded that energy loss up to 60% can be reduced from AC line, decreased battery size while achieving high energy efficiency of the system, and system is robust for cluster of homes and varied window size. Raffael Buhler et al. [21] discussed three concepts that integrate renewable energy sources which need ICT in distribution grid. Adequate ICT was developed to discuss new approach which is "fit and forget." Energy hub approach can be beneficial to both, microgrids and VPPs. This approach doesn't focus on a dedicated solution for a particular problem. Holistic solutions were searched after identifying conversion possibilities, energy inputs, flows and outputs. High reliability is demanded with high potentials for microgrid which are on islands and remote regions. Benjamin Kroposki et al. [22] provide alternative approaches for operation, design, and integration of DERs with microgrids. Author have covered key considerations for the successful operation, proper designing, and efficient planning of microgrid which include power quality, protection schemes and modifications, reserve margins, load shedding, demand response, frequency, voltage impact, and multiple PCC. Faisal Mohammed et al. [23] aimed at developing models which are suitable for designing and overall analysis. Individual sources of power of microgrid are modeled to analyze the transient and steady-state behavior. Models of fuel cell, photovoltaic cell, diesel engine, wind turbine, and micro-turbine were developed for investigation to understand sophisticated models of evolved microgrids.

PV array 1 and 5 were given constant surrounding temperature of 40 °C. PV array 2, 4, 6, 8 will be kept at 45 °C. PV array 3 and 7 were kept at 35 °C. Using the Perturb and Observe technique for the maximum power point tracking to maximize the output of the PV array and then analyzing the various outputs is done in the paper. Modeling of microgrid to accelerate simulation speed is also trending using phasor solutions. To damp out oscillations occurred due to disturbances in machines or generators, FACTS devices are used settle down at stable equilibrium point. One among the FACT devices is STATCOM. Single and double-pole tripping for distribution systems is used for protection during unbalanced short-circuits as fault-resilient, and smart microgrids do not trip the healthy phases sometimes. Selective phase tripping demands fault-type classification [24–43].

3 Classification of Microgrid

On the basis of above review, few sections are divided as follows [44–49]. Generally, they are classified into three types:

3.1 Residential Microgrid

This kind of microgrid is applied is residential colonies containing many houses, campuses of school, colleges, universities, and various communities. The major problem is caused by impulse to voltage of grid and it frequency at the time of transmitting the renewable energy generated back to main grid while building a residential microgrid. Other problems associated with residential microgrid are frequent transition between grid-connected mode and islanded mode, reliability, and stability of the system is also not ensured.

3.2 Remote Microgrid

This kind of microgrid is usually applied in rural areas, like islands, where renewable energy is present in abundant quantity as it is isolated from the main land and there nothing to obstruct the wind from flowing, and also, most of the time, sky is clear too thus providing abundant solar energy without much interference of human beings. But, it creates a disadvantage as a highly self-sustainable system is required for this kind of rural microgrid. Yet, advantages overcome the minor disadvantage which is the reduced carbon dioxide emission into the environment and reduced losses of transmission.

3.3 Mobile Microgrid

This kind of microgrid is applied where the moment of loads is their frequently at different time. Mostly, it is employed in military-based camps and naval systems as military camps shifts frequently and Navy's ships of cargo, maintenance, and defense sails to different locations throughout the sea. These types of microgrids are required because when required, they are highly flexible with fast movements from a place to another.

Throughout the world, a lot of progress is made on microgrid with lots of investigation going on, and various projects are adopted in U.S. CERT and NREL proposed the popular microgrid laboratories known today. Under the operation by American Electric Power at test beds near Columbus, Ohio the full-scale demonstration of microgrid concepts took place. Large-scale PV, fuel cell, wind turbine, and batteries were used to test another microgrid at a project of Santa Rita Jail organized by CERT. There is a lot of research going on for microgrids modeling. Development of microgrid helped in reduction of future expansion of transmission system and enhancement of non-conventional power penetration. CERTS was first architecture of microgrid which was shown by R.H. Lasseter. Barnes also proposed a microgrid system under "Microgrids European project." Various interfaces and connections

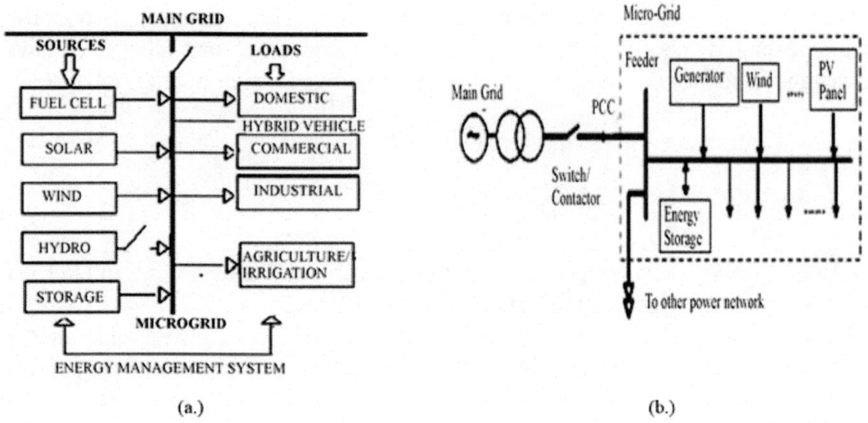

Fig. 1 **a** Microgrid architecture, **b** microgrid system

were made with network to connect with microgrid. This advancement in distributed connection with microgrids was remarkable. The microgrid needs sufficient flow of energy from various sources to attain system stability. The PV system alone is not capable to avail the sufficient amount of supply to microgrid. If there are two or more sources of supply, the requirement of power needed by load can be fulfilled at right time. Figure 1a shows the microgrid architecture, and Fig. 1b shows the microgrid systems.

4 Advantages of Microgrid

Fossil energy depletion, environmental pollution, and other problems can be resolved by converting these clean energies to obtain electrical energy which increase the diversified use of energy. Centralized power grid has many demerits like very huge cost and other operational problems associated due to which it is becoming difficult to fulfill the increasing user demand while maintain safety and reliability. Even when regular power system have various features like linking whole grid to form a much larger grid, distributed energy generation has provided strong support to power grid; increased reliability has reduced power grid's dependency, improved peak and valley performance. With the introduction of new technologies, idea of microgrid expanded. Microgrid is an entity that is a decentralized system. The microgrids have major application in power area. Microgrid can act as a power supply unit in a controlled manner; these are fast and take actions in few seconds for distribution to grid and to fulfill requirements of transmission. For users, microgrids can target and achieve particular needs. Energy exchange between microgrid and grid is done at PCC; the backup of each other is the two parts, thus improving reliability of power supply. It is shown that some of these forms of micro-power can be used to provide heat to users in the localized area by connecting them to thermal thus improving proper use of the generated energy. As in summary, it can be concluded that microgrids have advantages like they are reliable, resilient, and improved power quality [44–49].

5 Characteristics of Microgrid

1. **Uniqueness**—the microgrid is schedulable flexibly consisting of lots of load and micro-sources which can be called as small systems.
2. **Diversity**—the microgrid is composed of renewable and conventional energy sources which makes it very diverse. Also, the inclusion of various storage devices of energy is included in the microgrid system for stable operation. Other variety of loads are included in microgrid making it much more diverse.
3. **Controllability**—based on working conditions, microgrid can opt for various operation modes due to which there is an improved reliability and security of the system. This all is possible due to perfect control strategy.
4. **Interactivity**—Whenever it is required, microgrid avails the assistance to the main grid whenever required because microgrid has independent generation sources. Also, when microgrid needs power, it can take power from main grid.
5. **Independence**—till some extent, the demand of electricity by local loads can be fulfilled by microgrid; as under certain conditions, it can operate independently because of independent sources of energy associated with it.

6 Issues, Causes, and Remedies Related to Microgrid

Along with efficient working of microgrid, it involves some issues associated with it caused by various independent or collective reasons which are removed with the help of technological advancement in protection techniques, algorithms, and schemes. The various issues associated with microgrid are variation in intensity of current during fault, inappropriate working of (C.B.) circuit breaker during unwanted islanding of microgrid, poor standardization, improper relay inter-operability, unwanted fluctuations in frequency and voltage, overvoltage in the microgrid, non-identification of fault location, non-identification of fault current direction, non-isolation of area affected under sudden faults, low efficiency during high impedance faults, etc. The issues are mainly caused because of the increased complexity of microgrid and connection of multiple independent (DER's) distributed energy resources. To fulfill the requirement of the loads associated with microgrid, it imports the supply from main grid during grid-coupled mode. But, any fault on grid side, load side, DER side, or in the transmission system may trigger an unwanted islanding due to which microgrid parameters like voltage, active power, reactive power, frequency, and current vary very suddenly to high values and create disturbances and imbalances in the whole microgrid or a section of it. The loads such as houses and industries may experience a failure of supply of electricity for a duration dependent on the severity of the fault or issue associated [50–68].

After years of research for advancement in microgrid technology and its protection, various fault detection techniques based on signal processing such as wavelet (W.T.) [69], fast-Fourier (F.F.T.) [70], discrete Fourier (D.F.T), Hilbert-Huang (H.H.T.) [71], S-transforms [72], Park & Clark (P.C.T.) [73, 74], based on knowledge and memory such as fuzzy logic (F.L.) [75, 76], artificial neural network (A.N.N.) [77, 78], machine learning (M.L.) [79, 80], hybrid (A.N.N. and Fuzzy) [81], deep neural network (D.N.N.), adaptive neuro-fuzzy (A.N.F) [82] and based on model development such as multi-agent based (M.A.), impedance based (Z), traveling wave based (T.W.), sequence components based (S.C.), graph making based (G.M.), synchronized voltage and current based (S.V.C.) are developed and implemented for the microgrid protection to overcome the various issues faced by the microgrid during islanding as well as grid-connected conditions [83–85].

7 Microgrid Protection

Undertaking the several issues related to protection, there is an increased demand of new protection strategies. It led to the background study of the various existing microgrid protection schemes as investigated in Table 1 [86–91]. Table 1 shows the various schemes of protection implemented throughout the world in recent times for different working conditions which influence the working of protection schemes.

Table 1 Various protection schemes of microgrid

Protection schemes	Working of protection schemes
Protection using communication	Protection using digital relays and communication system, protection scheme employing a central computer and Ethernet ring communication
Voltage-based protection	Protection using disturbed voltage signal
Adaptive protection	Protection using low-voltage microgrid protection scheme
Differential protection	Protection using differential zone and differential current
Overcurrent protection	Protection using high-current microgrid protection scheme
Distance protection	Protection using Mho characteristics and DIgSILENT Power factory software on a MV microgrid
Multi-agent protection	Three layers of protection, i.e., equipment layer, substation layer, and system layer using Java Agent Development Framework

Table 2 Limitations and future scope of various protection schemes

Protection scheme	Limitations and future scope
Protection using communication	CT errors, mismatches, cost of relays and switching devices, cyberattacks
Voltage-based protection	Does not consider HIF, symmetrical faults, single-phase tripping
Adaptive protection	Large computational memory is required; upgradation of protective devices to digital counterparts is required
Differential protection	Mostly limited to power transformers and transmission networks. Ineffective under unbalanced loads
Overcurrent protection	Directional element is required for protection. This is redundant when fault current is unidirectional (Source and DG at same bus)
Distance protection	Improper selection of zones
Multi-agent protection	Lack of high-speed communication

8 Limitations and Future Scope

It is difficult to visualize a single scheme that can accurately protect an active distribution system due to the following mentioned problems. A hybrid scheme integrating traditional schemes (overcurrent, distance etc.) and modern schemes (employing PMUs, IEDs, and real-time communication and control) is a likely solution [92–95]. The choice of scheme will depend on the type of microgrid structure, type and control of DG sources, cost, required reliability, selectivity, sensitivity, security, and speed.

Table 2 shows the limitations and future scope of the schemes of protection associated with microgrid. Therefore, new algorithms are required to be developed for the protection of AC/DC microgrid. So, during the research work, development of protection algorithms should be focused for microgrid having inverter interface with renewable energy sources.

9 Conclusion

A combination of RES, i.e., solar and wind energy are used to generate the renewable energy which is transferred to the microgrid. The microgrid is connected to the loads like houses, small machines, shops, and industrial loads. Batteries are connected with the solar panels so that during night, those batteries are used as an optional source of energy. Also, a diesel engine generator is connected so that if microgrid doesn't have sufficient amount of energy to provide power to loads/consumers, then it can provide the necessary power for that duration of time when other primary sources of energy are unable to produce sufficient amount of energy. Model for such an integrated system is developed or can be made in future and simulation will be done. If required results are achieved, then practical implementation of the model will be tried to be achieved. Various types of protection strategies were analyzed to recommend the best protection strategy suited for the various developed models. This is an approach toward sustainable and clean energy integration for a better and safe future of mankind and every living species on earth. But, with this approach, many challenges arise as microgrids enable a multi-directional flow of power. Apart from the main grid to our homes, shops, etc., the flow of power from microgrids to both home and back to the main grid is the process. This multi-directional flow of power from microgrid and supply of power generated by multiple sources in a distributed system complicates the protection system and creates a lots of challenges for microgrid. Hence, development of a protection algorithm for microgrid is needed to be developed which can help in overcoming the recent issued and problems in the protection of microgrid interfaced with renewable energy sources via inverter.

References

1. Li H, Yuan Y, Zhang L, Huang T (2018) Research and application on microgrid integrated energy service technology for modern agriculture. In: 2nd IEEE conference on energy internet and energy system integration (EI2), Beijing, 20–22 Oct 2018
2. Mirez J (2017) A modelling and simulation of optimized interconnection between DC microgrids with novel strategies of voltage, power and control. In: IEEE Second international conference on DC microgrids (ICDCM). Nuremburg, 27–29 June 2017
3. Shafie-Khah M, Vahid-Ghavidel M, Somma MD, Graditi G, Siano P, Catalão JPS (2020) Management of renewable-based multi-energy microgrids in the presence of electric vehicles. IET Renew Power Gener 14(3):417–426
4. Roa DAA, Martinez JB, Feernandez XC (2019) A benchtop DC microgrid for renewable energy sources integration. In: IEEE 9th Annual computing and communication workshop and conference (CCWC). Las Vegas, 7–9 Jan 2019
5. Opathella C, Venkatesh B (2018) Prototype microgrid design and results. In: IEEE electrical power and energy conference (EPEC). Canada, 10–11 Oct 2018
6. Zhong W, Wang L, Liu Z, Hou S (2020) Reliability evaluation and improvement of islanded microgrid considering operation failures of power electronic equipment. J Modern Power Syst Clean Energy (JMPSCE) 8(1):111–123
7. Yahaya AA, Almuhaini M, Heydt GT (2020) Optimal design of hybrid DG systems for microgrid reliability enhancement. IET Gener Transm Distrib 14(5):816–823
8. Norouzi M, Aghaei J, Pirouzi S, Niknam T, Lehtonen M (2020) Flexible operation of grid-connected microgrid using ES. IET Gener Transm Distrib 14(2):254–264
9. Afzal M, Huang Q, Amin W, Umer K, Raza A, Naeem M (2020) Blockchain enabled distributed demand side management in community energy system with smart homes. IEEE Access 8:37428–37439
10. Pawitan GAH, Kim J (2020) MPC-based power management of renewable generation using multi-ESS guaranteeing SoC constraints and balancing. IEEE Access 8:12897–12906
11. Lu Z, Xu X, Yan Z, Wang H (2020) Density-based global sensitivity analysis of islanded microgrid loadability considering distributed energy resource integration. J Modern Power Syst Clean Energy 8(1):94–101
12. Hooshyar A, Iravani R (2017) Microgrid protection. Proc IEEE 105(7):1332–1353
13. Zhou Z, Chen M, Jiang J, Zhang D, Ye S, Liu C (2020) Analysis and design of a novel thyristor-based circuit breaker for DC microgrids. IEEE Trans Power Electron 35(3):2959–2968

14. Dabbaghjamanesh M, Fard AK, Mehraeen S, Zhang J, Dong ZY (2020) Sensitivity analysis of renewable energy integration on stochastic energy management of automated reconfigurable hybrid AC–DC microgrid considering DLR security constraint. IEEE Trans Ind Inf 16(1):120–131
15. Almuhaini M, Bizrah A, Heydt G, Khalid M (2019) Impact of wind speed modelling on the predictive reliability assessment of wind-based microgrids. IET Renew Power Gener 13(15):2947–2956
16. Francés AR, Uceda J (2019) Blackbox polytopic model with dynamic weighting functions for DC-DC converters. IEEE Access 7:160263–160273
17. Jayamaha DKJS, Lidula NWA, Rajapakse AD (2019) Wavelet-multi resolution analysis based ANN architecture for fault detection and localization in DC microgrids. IEEE Access 7:145371–145384
18. Siti MW, Tungadio DH, Sun Y, Mbungu NT, Tiako R (2019) Optimal frequency deviations control in microgrid interconnected systems. IET Renew Power Gener 13(13):2376–2382
19. Liu X, Liu Y, Liu L, Xiang Y, Yuan X (2019) Optimal planning of AC-DC hybrid transmission and distributed energy resource system: review and prospects. CSEE J Power Energy Syst 5(3):409–422
20. Jiayi H, Chuanwen J, Rong X (2018) A review on distributed energy resources and microgrid. Renew Sustain Energy Rev 12(9):2472–2483
21. Zoka Y, Sasaki H, Yorino N, Kawahara K, Liu CC (2014) An interaction problem of distributed generators installed in a MicroGrid. Proceedings of IEEE International conference of electric utility deregulation, restructuring and power technology. vol 2, pp 795–799
22. Bahramirad S, Khodaei A, Svachula J, Aguero JR (2015) Building resilient integrated grids: One neighborhood at a time. IEEE Electr Mag 3(1):48–55
23. Paquette AD, Divan DM (2014) Providing improved power quality in microgrids: difficulties in competing with existing power-quality solutions. IEEE Ind Appl Mag 20(5):34–43
24. Khodaei A (2014) Resiliency-oriented microgrid optimal scheduling. IEEE Trans Smart Grid 5(4):1584–1591
25. Mayhornetal E (2012) Optimal control of distributed energy resources using model predictive control. Proceedings of IEEE power energy society general meeting, pp 1–8
26. Nutkani IU, Loh PC, Blaabjerg F (2013) Cost-prioritized droop schemes for autonomous microgrids. In: Proceedings of IEEE energy conversion congress exposure, pp 1021–1025
27. Ross M, Hidalgo R, Abbey C, Joós G (2011) Energy storage system scheduling for an isolated microgrid. IET Renew Power Gener 5(2):117–123
28. Tasdighi M, Ghasemi H, Rahimi-Kian A (2014) Residential microgrid scheduling based on smart meters data and temperature dependent thermal load modeling. IEEE Trans Smart Grid 5(1):349–357
29. Krapels EN (2013) Microgrid development: good for society and utilities. IEEE Power Energy Mag 11(4):94–96
30. Bidram A, Davoudi A (2012) Hierarchical structure of microgrids control system. IEEE Trans Smart Grid 3(4):1963–1976
31. Guerrero JM, Vasquez JC, Matas J, deVicuña LG, Castilla M (2011) Hierarchical control of droop-controlled AC and DC microgrids—a general approach toward standardization. IEEE Trans Ind Electron 58(1):158–172
32. Pogaku N, Prodanovic M, Green TC (2007) Modeling, analysis and testing of autonomous operation of an inverter-based microgrid. IEEE Trans Power Electron 22(2):613–625
33. Vaccaro A, Popov M, Villacci D, Terzija V (2011) An integrated framework for smart microgrids modeling, monitoring, control, communication and verification. Proc IEEE 99(1):119–132
34. Kroposki B (2009) An integration facility to accelerate deployment of distributed energy resources in microgrids. In: Proceedings of IEEE power energy society general meeting, pp 1–4
35. Wang L (2012) Dynamic analysis of a microgrid system for supplying electrical loads in a sailing boat. In: Proceedings of IEEE power energy society general meeting, pp 1–7

36. Salehi V, Mazloomzadeh A, Mohammed O (2011) Real-time analysis for developed laboratory-based smart micro grid. In: Proceedings of IEEE power energy society general meeting, pp 1–8
37. Loh PC, Li D, Chai YK, Blaabjerg F (2013) Autonomous control of interlinking converter with energy storage in hybrid AC–DC microgrid. IEEE Trans Ind Appl 49(3):1374–1382
38. Dasgupta S, Sahoo SK, Panda SK, Amaratunga GAJ (2011) Single-phase inverter-control techniques for interfacing renewable energy sources with microgrid—Part II: Series-connected inverter topology to mitigate voltage-related problems along with active power flow control. IEEE Trans Power Electron 26(3):732–746
39. Bloemink JM, Iravani MR (2012) Control of a multiple source microgrid with built-in islanding detection and current limiting. IEEE Trans Power Delivery 27(4):2122–2132
40. Katiraei F, Iravani MR (2016) Power management strategies for a microgrid with multiple distributed generation units. IEEE Trans Power Syst 21(4):1821–1831
41. Hou C, Hu X, Hui D (2010) Hierarchical control techniques applied in micro-grid. Proceedings of international conference on power system technology, pp 1–5
42. Tsikalakis AG, Hatziargyriou ND (2018) Centralized control for optimizing microgrids operation. IEEE Trans Energy Convers 23(1):241–248
43. Parhizi S, Lotfi H, Khodaei A, Bahramirad S (2015) State of the art in research on microgrids: a review. IEEE Access 3
44. Singh M, Singh O, Kumar A (2019) Renewable energy sources integration in micro-grid including load patterns. In: 2019 3rd International conference on recent developments in control, automation & power engineering (RDCAPE). Noida, India, pp 88–93
45. Singh O, Singh M (2020) A comparative analysis on economic load dispatch problem using soft computing techniques. Int J Softw Sci Comput Intell (IJSSCI) 12(2). https://doi.org/10.4018/IJSSCI.2020040104
46. Singh M, Singh O (2019) Phasor solution of a micro-grid to accelerate simulation speed. In: Proceedings of 2nd International conference on advanced computing and software engineering (ICACSE). Sultanpur, India
47. Singh M, Rana V, Ansari MA, Dikshant SB, Singh P (2018) Power quality enhancement to sensitive loads with PV based microgrid system. In: 2018 International conference on sustainable energy, electronics, and computing systems (SEEMS). Greater Noida, India, pp 1–6
48. Singh M, Singh O, Singh V, Sharma T (2019) Modeling, simulation and comparison of DFIG based phasor model average model and detailed model of wind farm. In: Proceedings of 2nd International conference on advanced computing and software engineering (ICACSE)
49. Singh M, Ansari MA, Tripathi P, Wadhwani A (2018) VSC-HVDC transmission system and its dynamic stability analysis. In: 2018 International conference on computational and characterization techniques in engineering & sciences (CCTES). Lucknow, India, pp 177–182
50. Telukunta V, Pradhan J, Agrawal A, Singh M, Srivani SG (2017) Protection challenges under bulk penetration of renewable energy resources in power systems: a review. CSEE J Power Energy Syst 3:365–379
51. Beheshtaein S, Cuzner R, Savaghebi M, Guerrero JM (2019) Review on microgrids protection. IET Gener Transm Distrib 13:743–759
52. Mirsaeidi S, Dong X, Shi S, Wang B (2017) AC and DC microgrids: a review on protection issues and approaches. J Electr Eng Technol 12:2089–2098
53. Javed W, Chen D, Farrag ME, Xu Y (2019) System configuration, fault detection, location, isolation and restoration: a review on LVDC microgrid protections. Energies 12:1001
54. Beheshtaein S, Cuzner RM, Forouzesh M, Savaghebi M, Guerrero JM (2019) DC microgrid protection: a comprehensive review. IEEE J Emerg Sel Top Power Electron
55. Mirsaeidi S, Dong X, Shi S, Tzelepis D (2017) Challenges, advances and future directions in protection of hybrid AC/DC microgrids. IET Renew Power Gener 11:1495–1502
56. Mirsaeidi S, Dong X, Said DM (2018) Towards hybrid AC/DC microgrids: critical analysis and classification of protection strategies. Renew Sustain Energy Rev 90:97–103

57. Sarangi S, Sahu BK, Rout PK (2020) Distributed generation hybrid AC/DC microgrid protection: a critical review on issues, strategies, and future directions. Int J Energy Res 44:3347–3364
58. Barra P, Coury D, Fernandes R (2020) A survey on adaptive protection of microgrids and distribution systems with distributed generators. Renew Sustain Energy Rev 118:109524
59. Brearley BJ, Prabu RR (2017) A review on issues and approaches for microgrid protection. Renew Sustain Energy Rev 67:988–997
60. Mirsaeidi S, Said DM, Mustafa MW, Habibuddin MH, Ghaffari K (2014) An analytical literature review of the available techniques for the protection of micro-grids. Int J Electr Power Energy Syst 58:300–306
61. Manditereza PT, Bansal R (2016) Renewable distributed generation: the hidden challenges–a review from the protection perspective. Renew Sustain Energy Rev 58:1457–1465
62. Habib HF, Lashway CR, Mohammed OA (2017) A review of communication failure impacts on adaptive microgrid protection schemes and the use of energy storage as a contingency. IEEE Trans Ind Appl 54:1194–1207
63. Hosseini SA, Abyaneh HA, Sadeghi SHH, Razavi F, Nasiri A (2016) An overview of microgrid protection methods and the factors involved. Renew Sustain Energy Rev 64:174–186
64. Memon AA, Kauhaniemi K (2015) A critical review of AC microgrid protection issues and available solutions. Electr Power Syst Res 129:23–31
65. Gopalan SA, Sreeram V, Iu HH (2014) A review of coordination strategies and protection schemes for microgrids. Renew Sustain Energy Rev 32:222–228
66. Haron AR, Mohamed A, Shareef H (2012) A review on protection schemes and coordination techniques in microgrid system. J Appl Sci 12:101–112
67. Almutairy I (2016) A review of coordination strategies and techniques for overcoming challenges to microgrid protection. In: Proceedings of the 2016 Saudi Arabia smart grid (SASG). Jeddah, Saudi Arabia, pp 1–4
68. Basak P, Chowdhury S, Dey SH, Chowdhury S (2012) A literature review on integration of distributed energy resources in the perspective of control, protection and stability of microgrid. Renew Sustain Energy Rev 16:5545–5556
69. Pillay P, Bhattacharjee A (1996) Application of wavelets to model short-term power system disturbances. IEEE Trans Power Syst 11:2031–2037
70. Chaitanya BK, Soni AK, Yadav A (2018) Communication assisted fuzzy based adaptive protective relaying scheme for microgrid. J Power Technol 98:57–69
71. Thattai K, Sahoo A, Ravishankar J (2018) On-line and off-line fault detection techniques for inverter based islanded microgrid. In: Proceedings of the 2018 IEEE 12th International conference on compatibility, power electronics and power engineering (CPE-POWERENG 2018). Doha, Qatar, pp 1–6
72. Kar S, Samantaray SR (2014) Time-frequency transform-based differential scheme for microgrid protection. IET Gener Transm Distrib 8:310–320
73. Escudero R, Noel J, Elizondo J, Kirtley J (2017) Microgrid fault detection based on wavelet transformation and Park's vector approach. Electr Power Syst Res 152:401–410
74. Eslami R, Sadeghi SHH, Askarian-Abyaneh H, Nasiri A (2017) A novel method for fault detection in future renewable electric energy delivery and management microgrids, considering uncertainties in network topology. Electr Power Compon Syst 45:1118–1129
75. Bukhari SBA, Haider R, Zaman MSU, Oh Y, Cho G, Kim C (2018) An interval type-2 fuzzy logic based strategy for microgrid protection. Int J Electr Power Energy Syst 98:209–218
76. Wang H, Keerthipala W (1998) Fuzzy-neuro approach to fault classification for transmission line protection. IEEE Trans Power Delivery 13:1093–1104
77. Hong Y, Wei Y, Chang Y, Lee Y, Liu P (2014) Fault detection and location by static switches in microgrids using wavelet transform and adaptive network-based fuzzy inference system. Energies 7:2658–2675
78. Wang S, Chen H (2019) A novel deep learning method for the classification of power quality disturbances using deep convolutional neural network. Appl Energy 235:1126–1140

79. Baghaee HR, Mlakíc D, Nikolovski S, Dragičevíc T (2019) Support vector machine-based islanding and grid fault detection in active distribution networks. IEEE J Emerg Sel Top Power Electron 1–19
80. Samantaray S (2009) Decision tree-based fault zone identification and fault classification in flexible AC transmissions-based transmission line. IET Gener Transm Distrib 3:425–436
81. Lin H, Sun K, Tan Z, Liu C, Guerrero JM, Vasquez JC (2019) Adaptive protection combined with machine learning for microgrids. IET Gener Transm Distrib 13:770–779
82. Wang Y (2010) The tutorial: S transform. Graduate Institute of Communication Engineering, National Taiwan University, Taipei, Taiwan
83. Ding SX (2008) Model-based fault diagnosis techniques: design schemes, algorithms, and tools. Springer Science & Business Media, London, UK
84. Al Hassan HA, Reiman A, Reed GF, Mao Z, Grainger BM (2018) Model-based fault detection of inverter-based microgrids and a mathematical framework to analyze and avoid nuisance tripping and blinding scenarios. Energies 11:2152
85. Hare J, Shi X, Gupta S, Bazzi A (2016) Fault diagnostics in smart micro-grids: a survey. Renew Sustain Energy Rev 60:1114–1124
86. Singh M, Ansari MA, Pal NS, Kumawat S (2018) Comparison of control techniques for damping of oscillations in power system using STATCOM. In: 2018 International conference on computational and characterization techniques in engineering & sciences (CCTES). Lucknow, India, pp 188–191
87. Hooshyar A, Elsaadany EF, Pasand MS (2016) Fault type classification in microgrids including photovoltaic DGs. IEEE Trans Smart Grid 7(5):2218–2229
88. Hooshyar A, Iravani R (2018) A new directional element for microgrid protection. IEEE Trans Smart Grid 9(6):6862–6876
89. Rahmani S, Zare AR, Zare MR, Hooshyar A (2019) Voltage and frequency recovery in an islanded inverter-based microgrid considering load type and power factor. IEEE Trans Smart Grid 10(6):6237–6247
90. Chowdhury S, Chowdhury SP, Crossely P (2009) Microgrids and active distribution networks. The Institute of Engineering and Technology, London, UK
91. Hatziargyriou N (2014) Microgrids: architecture and control. IEEE Press, Wiley
92. Sharkh SM, Georgies MAA, Orfanoudakis I, Hussain B (2014) Power electronics converters for mirogrids. IEEE, Wiley
93. Yuksel E, Abdulkerim Eyvaz M (2018) Special topics in renewable energy system. Intechopen, London, United Kingdom
94. Carlos A, de Souza Z, Castilla M (2019) Microgrid design and implementation. Springer
95. Tabatabaei NM, Kabalci E, Bizo N (2020) Microgrid architectures, control, & protection methods. Springer

An Intelligent Approach for Defect Detection of Bearing

Papia Ray, Arpana Singh, and K. R. Satyajit

Abstract The problem of blending in empirical mode decomposition (EMD) method has led to the proposal of a new technique called ensemble empirical mode decomposition analysis (EEMD). As EEMD has a problem with reconstruction error, so a new technique has been proposed in this paper to check the ability to separate blending modes, and the method is complete ensemble empirical mode decomposition with adaptive noise (CEEMDAN). In this paper, Hilbert spectroscopic analysis was performed to verify the patterns extracted from different frequencies. Further, EEMD is implemented to correct the mode mixing problem in EMD. Moreover, due to noise in modes extracted from EEMD, CEEMDAN is implemented here. Other parameters such as mean value and entropy of patterns of all the discussed methods over here are calculated to find out the comparison between all three methods (EMD, EEMD and CEEMDAN). From the simulation results, it can be concluded that CEEMDAN is a better detector for defect in bearing.

Keywords Hilbert-Huang transform · Ensemble empirical mode decomposition · Empirical mode decomposition · Complete ensemble empirical mode decomposition and adaptive noise

1 Introduction

The fault information in the signal for each component of the bearing is in the vibration signal of the bearing [1]. The bearing has a signal which is vibrating which has wide frequency components, it comes with a huge noise, and the fault signals are easily drowned [2].

Developing signal processing techniques for noisy signals and non-stationary is a challenging task. Several techniques, like frequency for less-time conversion [3]

P. Ray (✉) · A. Singh · K. R. Satyajit
Department of Electrical Engineering, Veer Surendra Sai University of Technology, Burla, Sambalpur, Odisha, India
e-mail: papiaray_ee@vssut.ac.in

K. Namrata et al. (eds.), *Smart Energy and Advancement in Power Technologies*, Lecture Notes in Electrical Engineering 926,
https://doi.org/10.1007/978-981-19-4971-5_16

and wavelet transform [4], are proposed to eradicate the issues for such applications. These techniques require prior information regarding signals to be searched because it naturally absences its adaptation of the searched signals. The empirical mode decomposition (EMD) method is a self-adaptive method and is used to estimate nonlinear and non-stationary signals [5]. But EMD algorithm is used to deal with an erratic signal, and mode mixing is also an issue in it [6–8]. So, to overcome the problem of mode mingling, ensemble empirical analysis (EEMD) is introduced instead of EMD [9]. Some white noise is added to EMD with incomplete largeness of the searched signals. In addition, to remove added white noise residuals and reduce the time wasting of EEMD, complete ensemble empirical mode decomposition (CEEMDAN) was proposed to substitute the EEMD analysis as the customary for the EMD analysis [10]. Nevertheless, the CEEMDAN method moderately increases the calculation efficiency of the EEMD analysis so that the issue of EEMD analysis remains unsolved. Further betterments in case of EEMD can be achieved by CEEMDAN [11]. The method CEEMDAN overcomes the problems of both EMD and EEMD method.

1.1 Literature Review

The EMD is presented as adaptive method for analysing time–frequency data used in wide areas in the requests or applications to extract signals from generated facts in non-stationary and noisy nonlinear processes. But everything beneficial comes with a drawback and so does EMD [12]. A major drawback of the original EMD is the frequent appearance of mode mixing, which is defined as a single intrinsic mode function (IMF) either consisting of signals with widely varying scales, or a signal from a similar scale assessed in various components of the IMF [13].

A simple noise-assisted technique named ensemble EMD analysis (EEMD) is presented to alleviate mixing of the modes issue of EMD [14]. The EEMD obtains the true IMF components by doing the average of the complete experiments. Of course, each individual experiment may lead to very noisy results, as both the additive noise decomposition consists of the signal and the added white noise [15].

By applying CEEMDAN, it has been noticed here that it draws out the weaker features from noisy signal. So, for fault diagnosis of rolling element bearings, CEEMDAN is suggested. So, this analysis resolves the issue of calculating the concluding average and finds IMF which are less noisy.

The first part of the paper (Sect. 2) talks over the method of the decomposition of signals, i.e. empirical mode decomposition and ensemble empirical mode decomposition and complete ensemble empirical mode decomposition. The second part (Sect. 3) deals with the analysis and performance estimation of the decomposition methods of the simulated signal. The last part (Sect. 4) deals with the conclusion.

2 Methods of Defect Defection

Three methods have been discussed in this section, i.e. EMD, EEMD and CEEMDAN which are further examined in this paper for defect detection.

2.1 Empirical Mode Decomposition (EMD)

The IMFs can be reduced in components called intrinsic mode functions because of its self-changeable data decomposition method, i.e. EMD [16]. EMD is the most modifiable and systematic methods of decomposition, and it decomposes a simulated signal into several IMF signals that is written in steps below.

Step 1—Determine the local minima and maxima of the signal $f(t)$.
Step 2—Insert between maxima to obtain an upper envelope and minima to obtain a lower envelope.
Step 3—Compute the mean of the upper and lower envelope.
Step 4—Extract IMF.

2.2 Ensemble Empirical Mode Decomposition (EEMD)

The mode mixing problem can be eradicated by the EEMD by inducing white noise various times in it. In a background space that is filled with white noise, a decomposed signal is applied [17].

The stages for the calculation of EEMD are as mentioned below:

1. The addition of white noise with normal distribution is carried into the decomposed signal.
2. On the signal, EMD is done, and several white noises are added to acquire IMFs.
3. Repeat the above process by adding white noise in every step.
4. The average value of IMFs calculated is considered as the final output.

Figure 1 describes the proposed method for the decomposition methods. Firstly signal is added with a noise and is decomposed to produce intrinsic mode functions (IMFs) by using the EMD method. Further white noise is added multiple times to reduce the noise and mode mixing problem of EMD, and this is another method called EEMD.

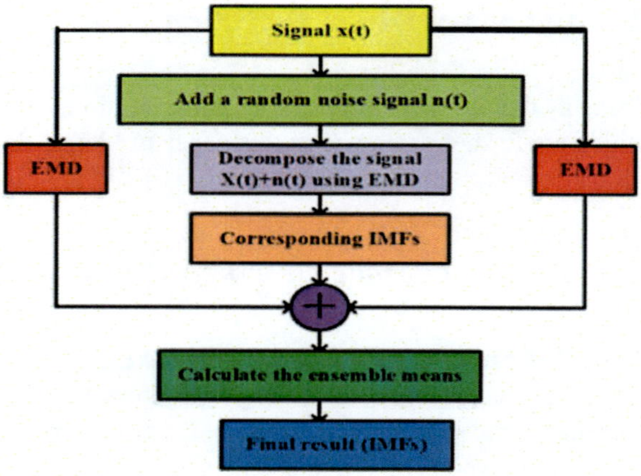

Fig. 1 Proposed method

2.3 *Complete Ensemble Empirical Mode Decomposition with Adaptive Noise (CEEMDAN)*

The CEEMDAN is a variation of the EEMD algorithm that provides an exact reconstruction of the original signal and a better spectral separation of the intrinsic mode functions (IMFs) [18]. The CEEMDAN is used to effectively analyse such kind of signals which produce such huge noise. So, this paper has a better method, i.e. CEEMDAN which removes the residual noise of the vibration signal of the bearing.

The foremost indication behind the projected CEEMDAN analysis is written below:

- Signal is made exactly like in EMD, and the first mode is extracted as like EMD by adding white noise to it. The average operation is performed on IMF1.
- By realising the noise, the first stage residue is calculated.
- By means of EMD method, second mode is computed by adding white noise to the residue of the first stage, and IMF2 is obtained.
- Similarly, the residual is found for the last residual component, and also, other IMFs are obtained.
- If the residual component is less than two extrema's, then decomposition is stopped.

3 Performance Evaluation of Decomposition Methods

In the rotating machines, the bearing due to exhaustive use creates noise producing vibration signals. These signals are classified in terms of Gaussian and Non-Gaussian noise [18]. Non-Gaussian is to be removed using decomposition methods.

The proposed signal is given in Eq. (1) as:

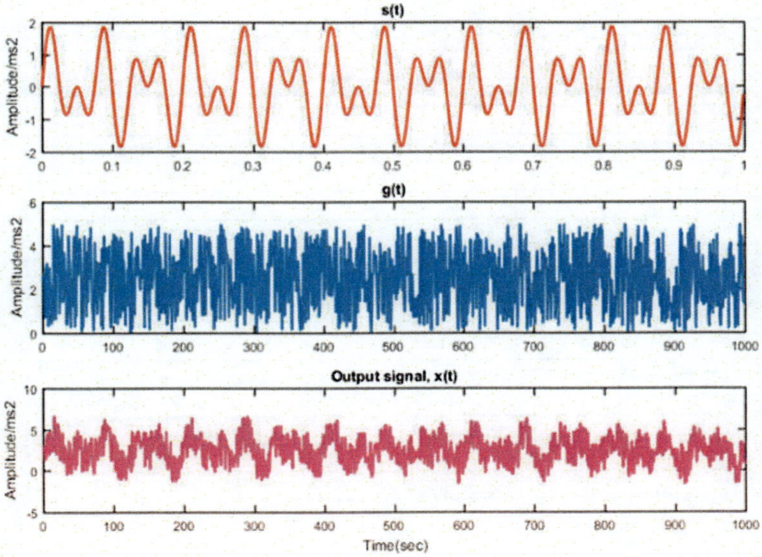

Fig. 2 Simulated signal

$$x(t) = s(t) + g(t) = \sin(30\pi t) + \sin(50\pi t) \tag{1}$$

$s(t)$: It has two sine wave signals frequencies of 15 and 25 Hz, $g(t)$: consists of random signal with Gaussian distribution. The sampling rates of $s(t)$ and $g(t)$ are 1000 Hz.

Figure 2 explains the three waveforms consisting of $s(t)$ having two frequencies, i.e. 15 and 25 Hz. $g(t)$ is the Gaussian distribution noise. The output which is the simulated signal is the summation of the Gaussian noise and the sinusoidal signals.

Moving forward we have the comparison done amongst the three decomposition methods, i.e. EMD, EEMD and CEEMDAN. These describe the difference between the true and false IMFs.

In Fig. 3, the IMFs of the simulated signal $s(t)$ are seen. The signal has true and false IMFs. It is seen from the extracted information of all IMFs as in Fig. 3 that IMF 1, 2, 3, 4 and 5 represents true IMFs whereas IMF 6 and 7 gives no information.

In Fig. 4, it is seen that IMFs have fewer frequency components than in the case of IMFs in EMD, and hence, noise is eliminated. Also, it has been noticed from Fig. 4 that we can extract information from the IMF 1, 2, 3, 4 and 5 but unable to extract it from the IMFs 6, 7 and 8 because they are false IMFs.

From Fig. 5, it has been depicted that the IMFs have more number of frequency components than in the case of IMFs in EMD and EEMD. Also, it can be seen from Fig. 5 that we can extract information from the IMF 1, 2, 3, 4 and 5 but unable to extract it from the IMFs 6, 7 and 8.

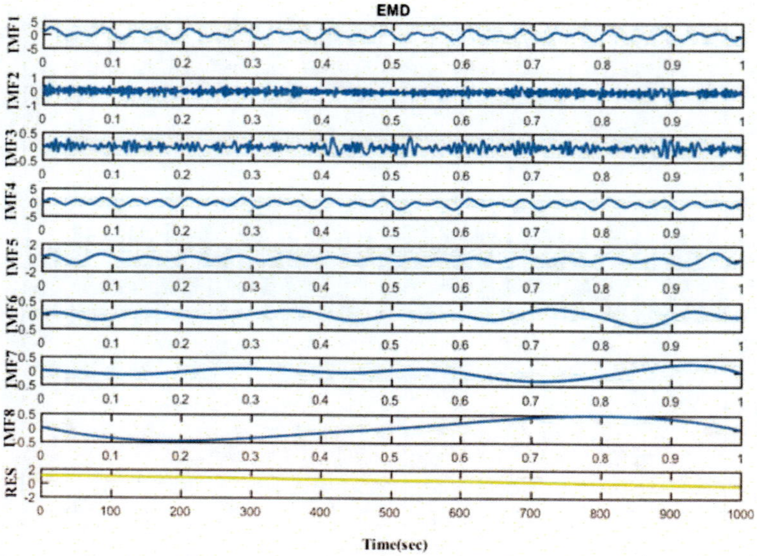

Fig. 3 Signal decomposition by EMD

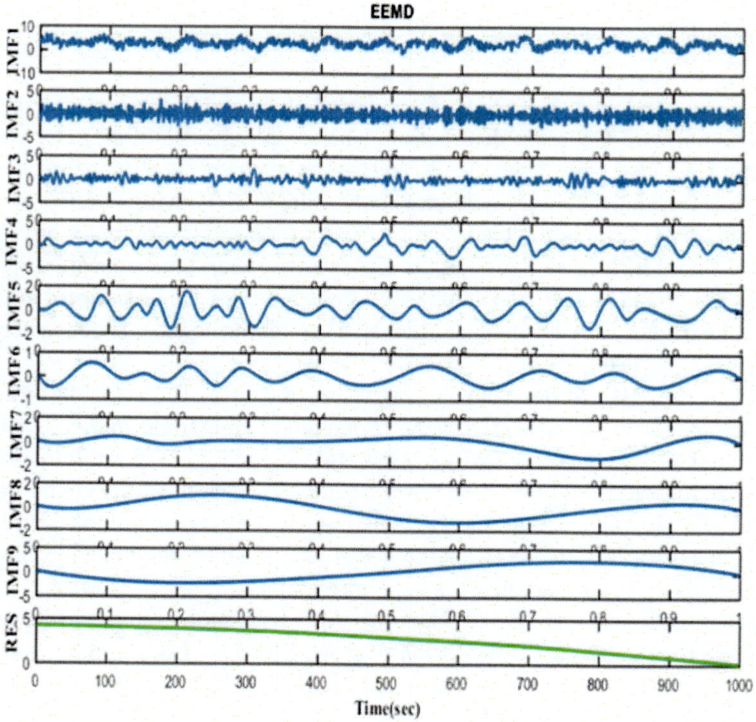

Fig. 4 Signal decomposition of EEMD

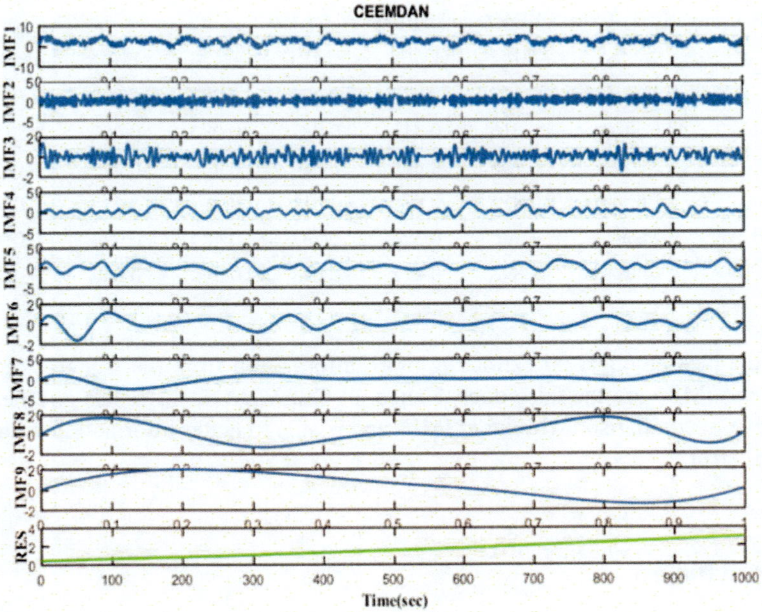

Fig. 5 Signal decomposition of CEEMDAN

As by the IMFs, it is concluded that some are true whereas some are false IMFs. So, with the parameters like mean value and entropy, it is easy to find out the qualitative value of the IMFs of the signal.

Below are the formulas to find the mean and entropy value for finding the qualitative IMFs

$$\text{Ex} = \frac{1}{n}\sum_{i=1}^{n} x_i \tag{2}$$

$$\text{En} = \sqrt{\frac{\pi}{2}} \times \frac{1}{n}\sum_{i=1}^{n} |x_i - \text{Ex}| \tag{3}$$

Ex is the mean value, and En is the entropy used to measure the qualitative features of the IMFs of the decomposed signals.

Equation (2) describes about the mean value, and Eq. (3) tells us about the entropy of the signal to be determined.

Table 1 shows the calculation done for the qualitative values of the IMF's produced by the decomposition method. IMF1 calculated for every method, i.e. EMD and EEMD extracts more value which means those are the true IMFs. Similarly, IMF2, IMF3 of the EMD extracts more value and hence are true ones, but IMF4, IMF5 and IMF6 are the false IMFs because of smaller values being extracted from it. And again in CEEMDAN, IMF1, IMF2, IMF3 and IMF4 are the true IMFs value with IMF1

Table 1 Comparison of EMD, EEMD and CEEMDAN

		IMF1	IMF2	IMF3	IMF4	IMF5	IMF6
EMD	Mean	0.312	0.217	0.287	0.087	0.029	0.012
	Entropy	0.516	0.412	0.482	0.080	0.057	0.018
EEMD	Mean	0.756	0.698	0.684	0.432	0.212	0.089
	Entropy	0.814	0.876	0.511	0.38	0.148	0.086
CEEMDAN	Mean	**0.945**	0.576	0.732	0.489	0.074	0.015
	Entropy	0.8**98**	0.477	0.698	0.358	0.056	0.010

being the best amongst all extracting more information from the signal. Whereas IMF5 and IMF6 are the false ones. Hence, in CEEMDAN, IMF1 has got the best ensemble in comparison with other IMFs and is able to better information with less residual error.

4 Conclusion

This paper proposes the decomposition methods to eradicate the non-Gaussian noise. This output of the simulated signal is therefore evaluated with the decomposition methods EMD, EEMD and CEEMDAN. With the EMD method, true IMFs information is extracted, but it comes with the mode mixing problem phenomenon. So, EEMD was performed, and here, it is seen that signal with fewer frequency components is present and mode mixing has been eradicated. Further to reduce the residual noise of the EEMD and to eliminate the noise successfully, CEEMDAN is carried out. So, we found the better method for noise assistance for true IMF signal. Calculation of the parameters like mean value and entropy is carried out for the modes to see the qualitative value of the IMF's. Hence, it concludes that CEEMDAN is a better method in assisting noise of the bearing of the machine.

References

1. Keshtan MN, Khajavi M (2016) Bearings fault diagnosis using vibrational signal analysis by EMD method. Res Nondestr Eval 27(3):155–174
2. Bin GF, Gao JJ, Li XJ et al (2012) Early fault diagnosis of rotating machinery based on wavelet packets—Empirical mode decomposition feature extraction and neural network. Mech Syst Signal Process 27(1):696–711
3. Satish L (2002) Short-time Fourier and wavelet transforms for fault detection in power transformers during impulse tests. Proc Inst Electr Eng Sci Meas Technol 145(2):77–84
4. Jiang X, Mahadevan S (2010) Wavelet spectrum analysis approach to model validation of dynamic systems. Mech Syst Signal Process 25(2):575–590

5. Huang NE, Shen Z, Long SR et al (1998) The empirical mode decomposition and the Hilbert spectrum for nonlinear and non-stationary time series analysis. Proc R Soc Lond A 454(1971):903–995
6. Ricci R, Pennacchi P (2011) Diagnostics of gear faults based on EMD and automatic selection of intrinsic mode functions. Mech Syst Signal Process 25(3):821–838
7. Cheng J, Yu D, Tang J, Yang Y (2008) Application of frequency family separation method based upon EMD and local Hilbert energy spectrum method to gear fault diagnosis. Mech Mach Theory 43(6):712–723
8. Lin L, Hongbing J (2009) Signal feature extraction based on an improved EMD method. Measurement 42(5):796–803
9. Wu ZH, Huang NE (2009) Ensemble empirical mode decomposition: a noise-assisted data analysis method. Adv Adapt Data Anal 1(1):1–41
10. Lei Y, He Z, Zi Y (2009) Application of the EEMD method to rotor fault diagnosis of rotating machinery. Mech Syst Signal Process 23(4):1327–1338
11. Torres ME, Colominas MA, Schlotthauer G, Flandrin P (2011) A complete ensemble empirical mode decomposition with adaptive noise. In: Proceedings of 36th 2011 IEEE international conference on acoustics, speech and signal processing (ICASSP). Prague, Czech Republic, pp 4144–4147
12. Huang NE, Shen Z, Long SR, Wu MC, Shih EH, Zheng Q, Tung CC, Liu HH (1998) The empirical mode decomposition method and the Hilbert spectrum for non-stationary time series analysis. Proc R Soc Lond 454A:903–995
13. Wu ZH, Huang NE (2009) Ensemble empirical mode decomposition: a noise-assisted data analysis method. Adv Adapt Data Anal 1:1–41. https://doi.org/10.1142/S1793536909000047
14. Galli AW, Heydt GT, Ribeiro PF (1996) Exploring the power of wavelet analysis. IEEE Comput Appl Power 9(4):37–44
15. Stockwell RG, Mansinha L, Lowe RP (1996) Localization of the complex spectrum: the S transform. IEEE Trans Signal Proc 44(4):998–1001
16. Li L, Dang R, Fan Y (2014) Modified EEMD de-noising method and its application in multiphase flow measurement. Chin J Sci Instrum 35(10):2365–2371
17. Jiang Y, Tang C, Zhang X, Jiao W, Li G, Huang T (2020) A novel rolling bearing defect detection method based on spectrum analysis and cloud model improved EEMD. IEEE Access 8(1):24323–24333
18. He S, Zi Y, Wan Z et al (2014) Application of adaptive multiwavelets via lifting scheme in bevel gear fault diagnosis. Chin J Sci Instrum 35(1):148–153

Comparison of Various Empirical Models to Estimate Monthly Mean Diffuse Solar Radiation for Humid-Subtropical Climate Region of India

Deepak Kumar Singh, Saibal Manna, and Ashok Kumar Akella

Abstract This research shows the performance analysis of different models in estimating monthly average diffuse solar radiation on the horizontal surface (R_d) of humid-subtropical climate region (HSCR) of India. The annual average diffuse, beam and global solar radiation (R_g) values have been obtained from Indian Metrological data as 7.82, 14.48, and 22.30 MJ/m2-day, respectively for Aligarh locations. The location has abundant potential to use solar energy under good conditions in the sky. The suitable estimation of R_d with unique input parameter (i.e., R_g) was investigated. A literature correlating diffuse fraction (DF) (R_d to R_g ratio) to the sky-clearness index (SCI) (R_g to extraterrestrial solar radiation ratio) was selected for 25 representative models. The aim is to get the best precise model among 25 models for the evaluation of R_d in the region under consideration. The result of 25 models had been compared with Indian Metrological data. In the context of well-known statistical indicators, distinctive statistical evaluation of the models was carried out. The outcomes of statistical test gave that model 10 provided smaller MBE (−0.0625), RMSE (0.9938), MPE (−0.0781), MAE (0.9125), and t-stat (0.2090) among all 25 models. This research contributes to the understanding of the importance of R_d, where the lack of essential devices, along with the significant operating and maintenance costs, is a notable barrier to assessment. The results of this paper are helpful for developing nations along with remote places having related weather conditions.

Keywords Diffuse solar radiation · Empirical models · Statistical error analysis · Aligarh India

1 Introduction

Solar power is the leading source of renewable energy, as of its infinite nature and global accessibility. A wide range of solar-powered applications can benefit from

D. K. Singh (✉) · S. Manna · A. K. Akella
Department of Electrical Engineering, National Institute of Technology Jamshedpur, Jamshedpur 831014, India
e-mail: dksingh1946@gmail.com

K. Namrata et al. (eds.), *Smart Energy and Advancement in Power Technologies*, Lecture Notes in Electrical Engineering 926,
https://doi.org/10.1007/978-981-19-4971-5_17

this sustainable energy source [1, 2]. It is required to determine the amount of solar radiation accessible in the area of concern to enhance the efficiency of solar system. For the purpose of building solar-powered appliances for the desired location, having precise data on solar radiation is crucial. This knowledge is also important to achieve longstanding solar energy systems meteorological predictions. Based on the data needed and the ultimate objective assigned for the system to complete, some methods for assessing solar radiation are more suited than others [3]. The functionality of devices in large-scale solar energy applications is linked to monthly mean of day-to-day incident solar power [4]. The usage of empirical models to measure solar radiation is a widely used methodology for evaluating or assessing solar power at a site of interest ahead of installing solar equipment [5].

Solar radiation can be calculated using satellite images or ground-based horizontal area observations employing instruments like pyranometers in weather monitoring stations [6]. Nonetheless, due to a series of roadblocks faced by a lack of essential equipment as well as expensive costs readily available. This is true in many impoverished countries and even remote places, where many of these equipment's operation and maintenance procedures do not include the collection of solar radiation data. However, in recent years, there has been significant work to establish weather stations of solar radiation but still, the data collected is insufficient and does not meet the requirement. In the case of R_d, this is especially true, as it has a broad range of properties based on the regional site's topography and environment. Thus, methods of estimation for quantifying R_d have been widely employed. R_g is a crucial and helpful parameter in the creation of empirical models for assessing R_d [7].

For Aligarh, India, they explored and created a new model for measuring diffused, direct, and global solar radiation with the correlation between SCI and DF [8]. Instead of devising new models, it is preferable to determine the applicability of existing models for estimating R_d for a certain location. From this perspective, a different statistical assessment is conducted utilizing several statistical indicators on 25 solar radiation models. The task is accomplished for the city of Aligarh (HSCR of India). This work encourages researchers to select the most relevant model for diffuse solar radiation from the available options and also to examine the region's solar radiation potential. The fundamental destination of this study is:

(i) To evaluate the monthly average daily diffuse (R_d), daily global (R_g), and extraterrestrial solar radiation (R_0) (MJ/m^2-day) for 25 models in Aligarh location.
(ii) To analyze every model with measured and estimated data employing statistical evaluation which consists of Mean percentage (MPE), Root mean square (RMSE), Mean absolute (MAE), Mean bias (MBE) error, and t-statistics (t-stats).

About Aligarh Location

Aligarh is situated on the north side of India in Uttar Pradesh state, 140 km southwest of India's capital city, New Delhi (28.61 °N, 77.20 °E). The city is located in the HSCR, which has a lot of monsoon and dry-winter weather. There is a great amount of sun radiation throughout the year in this region. In Aligarh, summer season varies from end of March to the beginning of June and the winter season from November to February. Moreover, temperature in winter varies from 6–12 °C (42.8–53.6 °F) and the lowest temperature varies 2–3 °C (35.6–37.4 °F). The region gets fair yearly rainfall (about 800 mm), beginning in June and running until the end of August. During the summer and winter seasons, it has mostly clear sunny days.

2 Solar Radiations on Horizontal Surface

The R_0 is expressed by given equation [9]

$$\overline{R}_0 = \frac{24}{\pi} R_{sc}\left(1 + 0.033\cos\left(\frac{360}{365}m\right)\right)\left(\cos\beta\cos\alpha\sin\lambda_s + \frac{\pi}{180}\lambda_s\sin\beta\sin\alpha\right) \tag{1}$$

where R_{sc} is solar constant (1.367kw/m^2) and α is latitude. The declination angle (β) is calculated from the below equation

$$\beta = 23.45\sin\frac{360}{365}(284 + m) \tag{2}$$

where, m is the total no. of days as appeared in Table 1.

The sunshine hour angle λ_s is determined by

$$S_{max} = \left(\frac{2}{15}\right)\lambda_s \qquad \lambda_s = \cos^{-1}(-\tan\beta\tan\alpha) \tag{3}$$

Table 1 Total number of days corresponding to month [9]

Month	Representative day
January	17
February	47
March	75
April	105
May	135
June	162
July	198
August	228
September	258
October	288
November	318
December	344

The following equation is used to find $\overline{R}_g$ [10]:

$$\frac{\overline{R}_g}{\overline{R}_0} = 0.3156 + 0.4520\left(\frac{\overline{S}}{\overline{S}_{\max}}\right) \tag{4}$$

where, S_{max} and S are monthly mean of the maximal feasible daily hours and daily hours of bright sunshine (hours).

3 Selections of Models

The models have been chosen based on their wide applications to a range of climates, geographies, and data sets, as evidenced by the literature. There are numerous functional forms Quadratic, Linear, Quartic, Cubic, and Exponential for models in the literature integrating DF with SCI. Furthermore, these models contain a variety of models available, ranging from historical to contemporary advances. Those that showed ludicrous estimated values have been excluded from the list of selected models amid first iteration. As a result, only models with plausible estimations are considered. Here, it should be noted that there are also various more similar models. According to a survey of the literature, R_g is the most frequently accepted measure for determining diffuse radiation.

Model-1

Solar radiation records were employed to estimate monthly average day-to-day short-wave radiation for latitudes 40 °N–40 °S on inclined and vertical surfaces [11].

$$\frac{\overline{R}_d}{\overline{R}_g} = 1.000 - 1.130\left(\frac{\overline{R}_g}{\overline{R}_o}\right) \tag{5}$$

Model-2

The interaction between total, diffuse and extraterrestrial solar radiation was discussed [12].

$$\frac{\overline{R}_d}{\overline{R}_g} = 0.84 - 0.62\left(\frac{\overline{R}_g}{\overline{R}_o}\right) \tag{6}$$

Model-3

Calculation of average insolation on titled surfaces on a monthly basis [13].

$$\frac{\overline{R}_{\mathrm{d}}}{\overline{R}_{\mathrm{g}}} = 1.390 - 4.027\left(\frac{\overline{R}_{\mathrm{g}}}{\overline{R}_{\mathrm{o}}}\right) + 5.531\left(\frac{\overline{R}_{\mathrm{g}}}{\overline{R}_{\mathrm{O}}}\right)^2 - 3.108\left(\frac{\overline{R}_{\mathrm{g}}}{\overline{R}_{\mathrm{o}}}\right)^3 \tag{7}$$

Model-4

They were able to retrace numerous models, by comparing total diffuse and horizontal insolation components as well as daily and hourly incident sunlight [14].

$$\frac{\overline{R}_{\mathrm{d}}}{\overline{R}_{\mathrm{g}}} = 1.3500 - 1.6075\left(\frac{\overline{R}_{\mathrm{g}}}{\overline{R}_{\mathrm{o}}}\right) \tag{8}$$

Model-5

It was the first to suggest such a combination of a sunshine duration and a SCI utilized in the assessment of South Africa's diffuse solar radiation models [15].

$$\frac{\overline{R}_{\mathrm{d}}}{\overline{R}_{\mathrm{g}}} = 1.403 - 1.672\left(\frac{\overline{R}_{\mathrm{g}}}{\overline{R}_{\mathrm{o}}}\right) \tag{9}$$

Model-6

It was observed that grouping the sunshine period and the SCI produces more accurate diffuse radiation estimation [16].

$$\frac{\overline{R}_{\mathrm{d}}}{\overline{R}_{\mathrm{g}}} = 0.924 - 0.894\left(\frac{\overline{R}_{\mathrm{g}}}{\overline{R}_{\mathrm{o}}}\right) \tag{10}$$

Model-7, 14, 16

Authors created other procedures in which DF was connected with sunshine period [17, 24, 26].

$$\frac{\overline{R}_{\mathrm{d}}}{\overline{R}_{\mathrm{g}}} = \frac{1}{1 + \exp\left(7.997\left(\left(\frac{\overline{R}_{\mathrm{g}}}{\overline{R}_{0}}\right) - 0.586\right)\right)} \tag{11}$$

$$\frac{\overline{R}_{\mathrm{d}}}{\overline{R}_{\mathrm{g}}} = 1.0212 - 1.1672\left(\frac{\overline{R}_{\mathrm{g}}}{\overline{R}_{\mathrm{o}}}\right) \tag{12}$$

$$\frac{\overline{R}_{\mathrm{d}}}{\overline{R}_{\mathrm{g}}} = \frac{1}{1 + \exp\left(-4.90 + 8.78\left(\frac{\overline{R}_{\mathrm{g}}}{\overline{R}_{\mathrm{o}}}\right)\right)} \tag{13}$$

Model-8

Evaluation of a diffuse sun irradiation model in the North Mediterranean Belt was discussed [18].

$$\frac{\overline{R}_{\rm d}}{\overline{R}_{\rm g}} = 0.724 + 2.738\left(\frac{\overline{R}_{\rm g}}{\overline{R}_{\rm o}}\right) - 8.32\left(\frac{\overline{R}_{\rm g}}{\overline{R}_{\rm O}}\right)^2 + 4.967\left(\frac{\overline{R}_{\rm g}}{\overline{R}_{\rm o}}\right)^3 \quad (14)$$

Model-9

Experiments on diffuse solar radiation for construction-related work were conducted in Athens, Greece [19].

$$\frac{\overline{R}_{\rm d}}{\overline{R}_{\rm g}} = 0.9995 - 0.05\left(\frac{\overline{R}_{\rm g}}{\overline{R}_{\rm o}}\right) - 2.4156\left(\frac{\overline{R}_{\rm g}}{\overline{R}_{\rm O}}\right)^2 + 1.4926\left(\frac{\overline{R}_{\rm g}}{\overline{R}_{\rm o}}\right)^3 \quad (15)$$

Model-10

For Izmir, Turkey, the DF of monthly and day-to-day global radiation was compared [20].

$$\frac{\overline{R}_{\rm d}}{\overline{R}_{\rm g}} = 1.7821 - 6.648\left(\frac{\overline{R}_{\rm g}}{\overline{R}_{\rm o}}\right) + 11.17\left(\frac{\overline{R}_{\rm g}}{\overline{R}_{\rm O}}\right)^2 - 6.5641\left(\frac{\overline{R}_{\rm g}}{\overline{R}_{\rm o}}\right)^3 \quad (16)$$

Model-11

The neural networking method was employed to examine the utilization of $R_{\rm g}$ and other meteorological data for $R_{\rm d}$ [21].

$$\frac{\overline{R}_{\rm d}}{\overline{R}_{\rm g}} = 0.90 + 1.1\left(\frac{\overline{R}_{\rm g}}{\overline{R}_{\rm o}}\right) - 4.5\left(\frac{\overline{R}_{\rm g}}{\overline{R}_{\rm O}}\right)^2 + 0.01\left(\frac{\overline{R}_{\rm g}}{\overline{R}_{\rm o}}\right)^3 + 3.14\left(\frac{\overline{R}_{\rm g}}{\overline{R}_{\rm o}}\right)^4 \quad (17)$$

Model-12

Estimation of $R_{\rm d}$ in China on day-to-day basis [22].

$$\frac{\overline{R}_{\rm d}}{\overline{R}_{\rm g}} = 1.0212 - 1.1672\left(\frac{\overline{R}_{\rm g}}{\overline{R}_{\rm o}}\right) \quad (18)$$

Model-13

Selection of models for $R_{\rm d}$ and $R_{\rm g}$ over Turkey's Central Black Sea regions [23].

$$\frac{\overline{R}_d}{\overline{R}_g} = 1.027 - 1.659\left(\frac{\overline{R}_g}{\overline{R}_o}\right) + 1.102\left(\frac{\overline{R}_g}{\overline{R}_O}\right)^2 - 0.402\left(\frac{\overline{R}_g}{\overline{R}_o}\right)^3 \quad (19)$$

Model-15

At Kampala, Uganda, empirical correlations were employed to forecast R_d levels [25].

$$\frac{\overline{R}_d}{\overline{R}_g} = 0.980 - 1.046\left(\frac{\overline{R}_g}{\overline{R}_o}\right) \quad (20)$$

Model-17

They found that grouping the sunshine period and SCI, rather than utilizing each parameter independently, was a more reliable technique for estimation in China [27].

$$\frac{\overline{R}_d}{\overline{R}_g} = 1.012 - 1.144\left(\frac{\overline{R}_g}{\overline{R}_o}\right) \quad (21)$$

Model-18

In Malaysia, they calculated diffuse and global solar radiation for five locations using fuzzy logic, nonlinear, linear, and ANN models and found that the ANN-based model best predicted R_d [28].

$$\frac{\overline{R}_d}{\overline{R}_g} = 0.9497 + 0.9270\left(\frac{\overline{R}_g}{\overline{R}_o}\right) - 4.8821\left(\frac{\overline{R}_g}{\overline{R}_O}\right)^2 + 3.2542\left(\frac{\overline{R}_g}{\overline{R}_o}\right)^3 \quad (22)$$

Model-19

Ekren created and tested models for calculating R_d in Turkey based on atmospheric parameters [29].

$$\frac{\overline{R}_d}{\overline{R}_g} = 0.8387 - 0.9229\left(\frac{\overline{R}_g}{\overline{R}_o}\right) \quad (23)$$

Model-20

They developed a model with multiple locations in Europe (based on data from 44 meteorological points in 11 nations) to find out the components of solar radiation by combining SCI and DF on a flat surface [30].

$$\frac{\overline{R}_d}{\overline{R}_g} = 0.9888 + 0.3950\left(\frac{\overline{R}_g}{\overline{R}_o}\right) - 3.7003\left(\frac{\overline{R}_g}{\overline{R}_O}\right)^2 + 2.2905\left(\frac{\overline{R}_g}{\overline{R}_o}\right)^3 \quad (24)$$

Model-21

They discussed about diffused solar fraction distribution in Taiwan [31].

$$\frac{\overline{R}_d}{\overline{R}_g} = 0.8066 + 1.9651\left(\frac{\overline{R}_g}{\overline{R}_o}\right) - 6.5435\left(\frac{\overline{R}_g}{\overline{R}_O}\right)^2 + 3.8590\left(\frac{\overline{R}_g}{\overline{R}_o}\right)^3 \quad (25)$$

Model-22

10 empirical models were created at six sites in Algeria to correlate the DF with the relative period of the sunshine. In addition, they created generalized correlations for estimating diffuse sun irradiation in different Algerian locations [32].

$$\frac{\overline{R}_d}{\overline{R}_g} = 1.832\exp\left(-2.090\frac{\overline{R}_g}{\overline{R}_o}\right) \quad (26)$$

Model-23

They established a diffuse radiation model in Tabass, Iran to recommend the appropriate tilt angle for solar surfaces [33].

$$\frac{\overline{R}_d}{\overline{R}_g} = 1.07884 - 1.22683\left(\frac{\overline{R}_g}{\overline{R}_o}\right) \quad (27)$$

Model-24

They established empirical models for R_d on slanted surfaces and analyze the optimal tilt angle for a place in Sindh, Pakistan, to evaluate solar power resources [34].

$$\frac{\overline{R}_d}{\overline{R}_g} = 0.966 - 1.044\left(\frac{\overline{R}_g}{\overline{R}_o}\right) \quad (28)$$

Model-25

Empirical modeling and observational characterization were demonstrated to calculate R_d, direct and R_g solar radiation in the Rio de Janeiro city [35].

$$\frac{\overline{R}_d}{\overline{R}_g} = 0.13 + 0.86\frac{1}{1 + \exp\left(-6.29 + 12.26\left(\frac{\overline{R}_g}{\overline{R}_o}\right)\right)} \quad (29)$$

4 Statistical Error Analysis

With the help of several statistical factors, the certainty and performance of the 25 preferred models have been examined. These are MBE, RMSE, MPE, MAE, and t-stats test indicators.

MBE—The mean bias error measures a model's overestimation or underestimation of R_d values. A negative MBE indicates less estimation, while a positive MBE indicates more estimation. The value of MBE should ideally be zero.

$$\text{MBE} = \frac{1}{r}\sum_{n=1}^{r}\left(\overline{R}_{n,e} - \overline{R}_{n,m}\right) \tag{30}$$

where, $R_{n,m}$, and $R_{n,e}$ are nth measured and estimated monthly average daily solar radiation.

RMSE—It is employed to judge model performance by comparing measured and estimated values. In comparison to models with high RMSE values, those with lesser values are recognized to have the best result. For perfect cases, it should be zero but it always has positive value. RMSE is expressed as

$$\text{RMSE} = \left[\frac{1}{r}\sum_{n=1}^{r}\left(\overline{R}_{n,e} - \overline{R}_{n,m}\right)^2\right]^{\frac{1}{2}} \tag{31}$$

MPE—It is defined as an evaluation of the magnitude of a value's error expressed as a percent of the measured or observed value.

$$\text{MPE} = \frac{1}{r}\sum_{n=1}^{r}\left(\frac{\overline{R}_{n,m} - \overline{R}_{n,e}}{\overline{R}_{n,m}}\right) \times 100 \tag{32}$$

MAE—This is the overall error value calculated as an absolute value by dividing the difference in estimated and measured values to the number of observations. MAE indicates the closeness between measured and estimated values.

$$\text{MAE} = \frac{1}{r}\sum_{n=1}^{r}\left|\overline{R}_{n,e} - \overline{R}_{n,m}\right| \tag{33}$$

t-stats—The models are additionally validated using t-statistics error analysis. The model has value around zero, and will have top performance in all models. The mathematical equation to calculate t-stats incorporating RMSE and MBE as:

$$t = \left[\frac{(r-1)\text{MBE}^2}{\text{RMSE}^2 - \text{MBE}^2}\right]^{\frac{1}{2}} \tag{34}$$

5 Results and Discussion

The calculated value of β, R_0, R_g is given in Table 2 and the observed value of R_d is also listed in the same table. The variation of R_0, R_g, and observed value of R_d on horizontal surface is illustrated in Fig. 1 and Fig. 2, respectively.

The calculated or estimated value of R_d (MJ/m^2-day) on horizontal surface at Aligarh for 25 different models is listed in Table 3. The bar plot of the 25 different diffuse radiation models is shown in Fig. 3. Statistical error analysis has been conducted on the basis of the statistical indicators stated in the Eqs. (30–34) of all 25 models and the results are compiled in Table 4. As it is known that, lower values for statistical error indicators imply a more precise estimate.

Based on analysis of statistical errors, Model 10 shows the best result with MBE = −0.0625, RMSE = 0.9938, MPE = −0.0781, MAE = 0.9125 and t-stat = 0.2090. After that, model 18 shows the better performance with MBE = −0.5042, RMSE = 1.3074, MPE = 4.9545, MAE = 1.1708 and t-stat = 1.3862. Afterwards, model 21 shows the good performance having MBE = −0.9875, RMSE = 1.7379, MPE = 10.5719, MAE = 1.4208 and t-stat = 2.2902. And then model 6 shows the good performance with MBE = −1.1208, RMSE = 1.4870, MPE = 13.1311, MAE = 1.1708 and t-stat = 3.8044. Model 19 has the highest values of the indicators MAE, MPE, RMSE, MBE, and t-stats. The different statistics error plots are given in Figs. 4, 5, 6, 7 and 8.

Table 2 Calculated value of β, R_0, R_g, and observed value of R_d

Month	β (°)	R_0 (MJ/m^2-day)	R_g (MJ/m^2-day)	R_d (observed) (MJ/m^2-day)
January	−20.92	22.5	17.0	6.50
February	−12.95	27	19.8	7.25
March	−2.42	32.3	22.8	8.0
April	9.41	37.1	25.2	8.1
May	18.80	40	26.4	9.0
June	23.08	40.9	26.6	9.3
July	21.18	40.3	26.4	10.1
August	13.45	38.1	25.6	8.1
September	2.22	34	23.6	7.0
October	−9.60	28.5	20.6	8.1
November	−18.91	23.5	17.6	6.5
December	−23.05	21.2	16.2	6.0

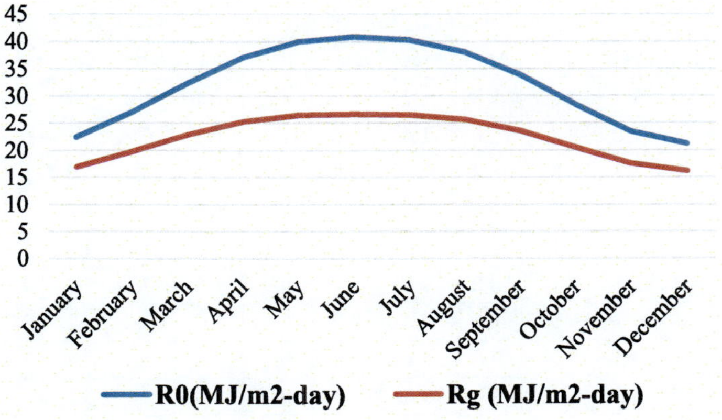

Fig. 1 Variation of R_0 and R_g on horizontal surface at Aligarh

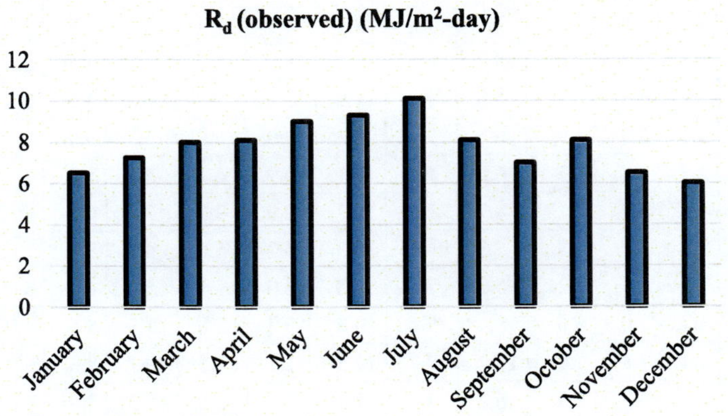

Fig. 2 Observed value of R_d on horizontal surface at Aligarh

6 Conclusion

The goal of this paper would be to analyze different empirical models for determining R_d that had been devised and listed in the literature. For all the months, diffuse component is found to be nearly always less than the beam components. During April–June and August–September, a high beam component is noted, indicating clear sky. For Aligarh, the annual mean diffuse, beam, and global solar radiation values have obtained as 7.80, 14.52, and 22.32 MJ/m^2-day, respectively. It is also observed that beam and R_g maximal values come to be 17.50 MJ/m^2-day in the month of August and 26.60 MJ/m^2-day in the month of June, respectively. Due to the existence of dense rainfall and clouds during the monsoon, R_d maximum value is recognized in July as 10.10 MJ/m^2-day. A comprehensive statistical analysis of 25 solar diffuse radiation models is carried out. Five statistical error tests have been

Table 3 Calculated or estimated value of R_d (MJ/m^2-day) for 25 different models

Models	Jan	Feb	Mar	Apr	May	Jun	July	Aug	Sept	Oct	Nov	Dec
Model 1	2.4	3.4	4.6	5.8	6.7	7.0	6.8	6.1	5.0	3.7	2.6	2.2
Model 2	6.3	7.6	9.2	10.6	11.4	11.6	11.4	10.8	9.7	8.1	6.6	5.9
Model 3	2.7	3.7	4.8	5.8	6.5	6.8	6.6	6.1	5.2	4.0	3.0	2.5
Model 4	2.3	3.4	4.9	6.5	7.6	8.1	7.8	6.9	5.5	3.9	2.5	1.9
Model 5	2.3	3.5	5.1	6.7	7.9	8.4	8.1	7.2	5.7	4.0	2.6	2.0
Model 6	4.2	5.3	6.7	8.0	8.8	9.1	8.9	8.3	7.1	5.7	4.5	3.9
Model 7	3.4	4.7	6.3	8.1	9.4	9.9	9.6	8.6	7.0	5.2	3.7	3.1
Model 8	3.1	4.3	5.9	7.6	8.8	9.3	9.1	8.1	6.5	4.8	3.4	2.8
Model 9	3.8	5.0	6.5	8.0	9.0	9.4	9.2	8.4	7.1	5.5	4.1	3.5
Model 10	5.2	6.4	7.9	9.1	9.8	10.0	9.9	9.4	8.3	6.9	5.5	4.8
Model 11	3.2	3.9	5.0	6.1	7.0	7.3	7.1	6.4	5.4	4.2	3.4	3.0
Model 12	2.3	3.3	4.5	5.7	6.6	6.9	6.8	6.1	5.0	3.6	2.6	2.1
Model 13	3.9	4.8	6.0	7.1	7.8	8.1	7.9	7.4	6.4	5.2	4.1	3.6
Model 14	2.3	3.3	4.5	5.7	6.6	6.9	6.8	6.1	5.0	3.6	2.6	2.1
Model 15	3.2	4.2	5.5	6.8	7.6	7.9	7.8	7.1	6.0	4.6	3.4	2.9
Model 16	2.5	3.5	4.9	6.5	7.6	8.1	7.9	6.9	5.5	3.9	2.7	2.3
Model 17	2.5	3.4	4.7	6.0	6.8	7.1	6.9	6.2	5.1	3.8	2.7	2.2
Model 18	4.5	5.7	7.2	8.7	9.8	10.2	9.9	9.1	7.7	6.1	4.8	4.2
Model 19	2.4	3.2	4.3	5.3	6.0	6.3	6.2	5.6	4.7	3.5	2.6	2.1
Model 20	2.7	3.8	5.2	6.7	7.8	8.2	8.0	7.1	5.8	4.2	3.0	2.4
Model 21	3.7	5.0	6.6	8.4	9.6	10.0	9.8	8.8	7.3	5.5	4.0	3.4
Model 22	6.4	7.8	9.6	11.2	12.2	12.5	12.3	11.5	10.1	8.3	6.7	6.0
Model 23	2.5	3.6	4.9	6.2	7.1	7.4	7.3	6.5	5.3	3.9	2.8	2.3
Model 24	3.0	4.0	5.2	6.5	7.3	7.6	7.4	6.8	5.7	4.3	3.2	2.7
Model 25	2.9	3.7	4.7	5.8	6.6	7.0	6.8	6.1	5.0	3.9	3.1	2.7

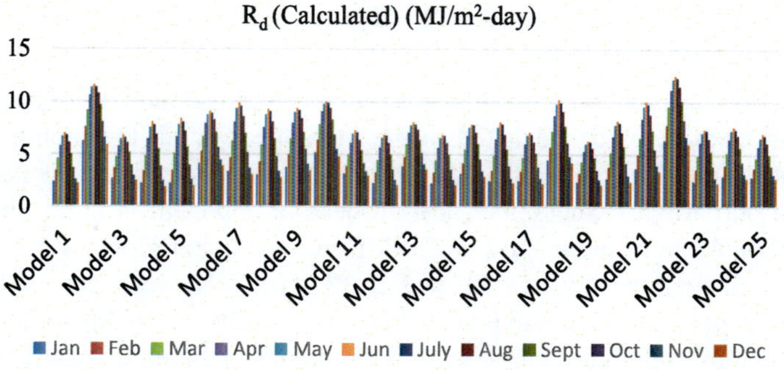

Fig. 3 Calculated value of R_d for 25 models on horizontal surface at Aligarh location

Table 4 Results of statistical error analysis utilizing several indicators for 25 selected models

Model	MBE	RMSE	MPE	MAE	t-stat
Model 1	−3.1375	3.2527	38.5175	3.1375	12.1270
Model 2	1.2708	1.7125	−17.016	1.3208	3.6718
Model 3	−3.0208	3.1074	37.4176	3.0208	13.7542
Model 4	−2.7208	2.9925	32.5365	2.7208	7.2429
Model 5	−2.5375	2.8667	30.0970	2.5375	6.3101
Model 6	−1.1208	1.4870	13.1311	1.1708	3.8044
Model 7	−1.2458	1.9296	13.7093	1.4958	2.8040
Model 8	−1.6875	2.1785	19.3805	1.6875	4.0621
Model 9	−1.2042	1.6907	13.7378	1.2875	3.3651
Model 10	−0.0625	0.9938	−0.0781	0.9125	0.2090
Model 11	−2.6625	2.7596	32.9014	2.6625	12.1722
Model 12	−3.2042	3.3208	39.3448	3.2042	12.1818
Model 13	−1.8042	1.9635	22.1032	1.8042	7.7238
Model 14	−3.2042	3.3208	39.3448	3.2042	12.1818
Model 15	−2.2458	2.4292	27.2654	2.2458	8.0448
Model 16	−2.6375	2.8858	31.6414	2.6375	7.4700
Model 17	−3.0458	3.1706	37.3372	3.0458	11.4705
Model 18	−0.5042	1.3074	4.9545	1.1708	1.3862
Model 19	−3.4792	3.5482	43.1879	3.4792	16.5730
Model 20	−2.4208	2.6790	28.9918	2.4208	7.0111
Model 21	−0.9875	1.7379	10.5719	1.4208	2.2902
Model 22	1.7208	2.2183	−23.1511	1.7375	4.0769
Model 23	−2.8458	3.0010	34.6671	2.8458	9.9092
Model 24	−2.5208	2.6752	30.7968	2.5208	9.3342
Model 25	−2.9708	3.0519	36.7637	2.9708	14.0972

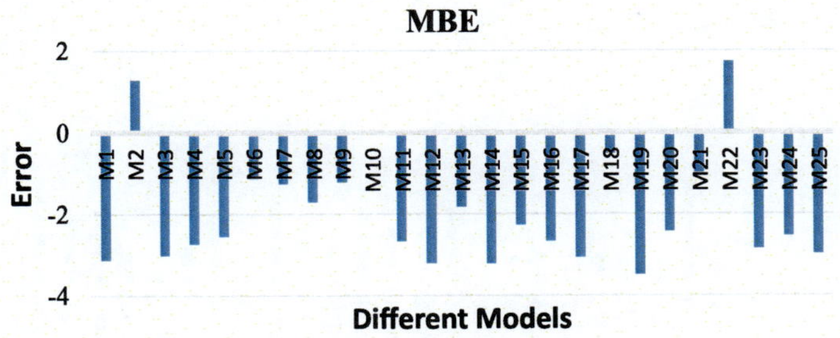

Fig. 4 MBE for 25 models

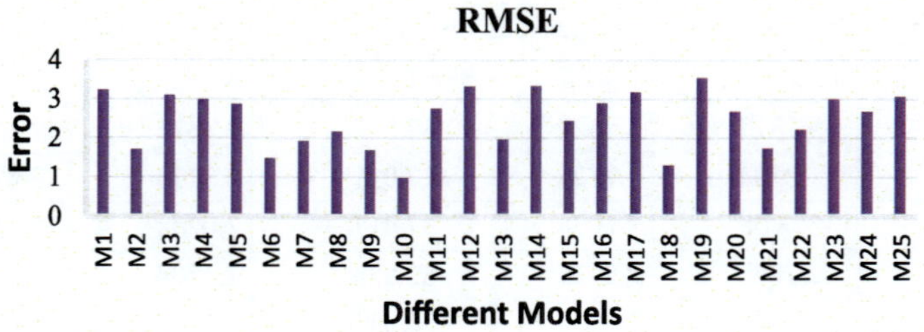

Fig. 5 RMSE for 25 models

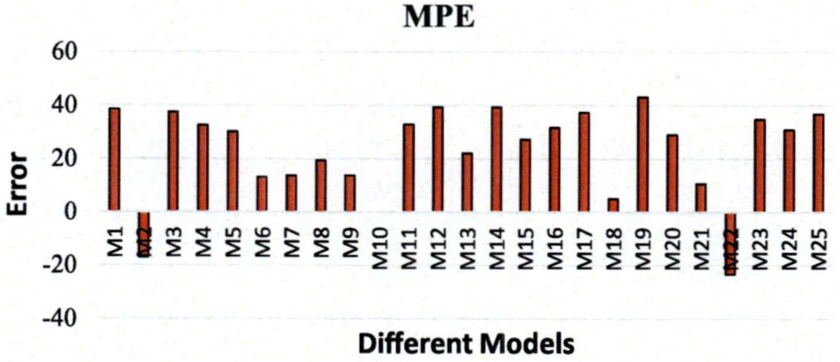

Fig. 6 MPE for 25 models

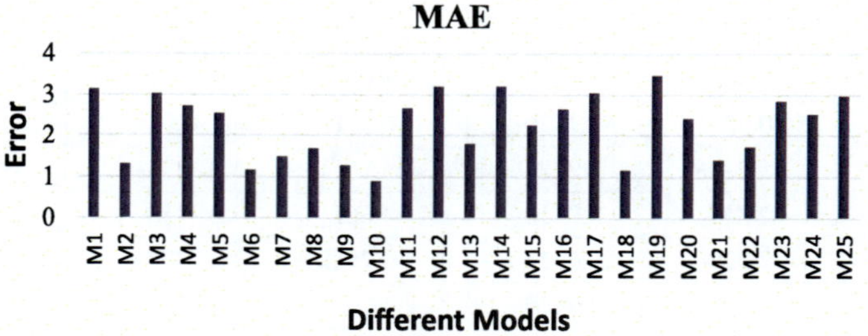

Fig. 7 MAE for 25 models

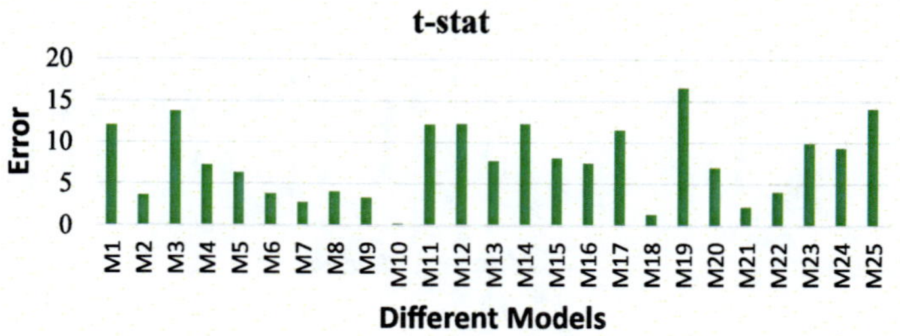

Fig. 8 t-stats for 25 models

employed to investigate the performance of the considered models. The performance of the 25 considered models have been tested using five statistical error tests, i.e., MBE, RMSE, MPE, MAE, and t-stats. Based on analysis of statistical errors, Model 10 has the best performance with MBE = −0.0625, RMSE = 0.9938, MPE = −0.0781, MAE = 0.9125 and t-stat = 0.2090. This comparison serves to identify the importance of the model for calculating diffuse solar radiation under similar climatological conditions around the world.

References

1. Mohammadi K, Shamshirband S, Tong CW, Alam KA, Petkovic D (2015) Potential of adaptive neuro-fuzzy system for prediction of daily global solar radiation by day of the year. Energy Convers Manag 93:406–413. https://doi.org/10.1016/j.enconman.2015.01.021
2. Benmouiza K, Cheknane A (2013) Forecasting hourly global solar radiation using hybrid k-means and nonlinear auto regressive neural network models. Energy Convers Manag 75:561–569. https://doi.org/10.1016/j.enconman.2013.07.003
3. Bakirci K (2017) Prediction of global solar radiation and comparison with satellite data. J Atmos Solar Terr Phys 152:41–49. https://doi.org/10.1016/j.jastp.2016.12.002
4. Makade R, Chakrabarti S, Jamil B, Sakhale C (2019) Estimation of global solar radiation for the tropical wet climatic region of India: a theory of experimentation approach. Renewable Energy 146:2044–2059. https://doi.org/10.1016/j.renene.2019.08.054
5. Mirzabe AH, Hajiahmad A, Keyhani A (2021) Assessment and categorization of empirical models for estimating monthly, daily, and hourly diffuse solar radiation: a case study of Iran. Sustain Energy Technol Assess 47:101330. ISSN 2213-1388. https://doi.org/10.1016/j.seta.2021.101330.
6. Chaiyapinunt S, Ruttanasupa P, Ariyapoonpong V, Duanmeesook K (2016) A shadowring device for measuring diffuse solar radiation on a vertical surface in a tropical zone. Sol Energy 136:629–638. https://doi.org/10.1016/j.solener.2016.07.044
7. Duffie JA, Beckman WA (2006) Solar engineering of thermal processes, 3rd edn. Wiley, New York
8. Jamil B, Akhtar N (2015) Statistical analysis of short-term solar radiation data over Aligarh (India) Chapter 61. In: Progress in clean energy, vol 2, pp 937–948. https://doi.org/10.1007/978-3-319-17031-2_61
9. Klein SA (1977) Calculation of monthly average insolation on tilted surfaces. Sol Energy 19:325–330. https://doi.org/10.1016/0038-092X(77)90001-9
10. Garg HP, Garg SN (1985) Correlation of monthly average daily global, diffuse and beam radiation with bright sunshine hours. Energy Convers Manag 25:409–417
11. Page JK (1964) The estimation of monthly mean values of daily total short wave radiation on vertical and inclined surfaces from sunshine records for latitudes 40°N-40°S. Proc. UN Conf. New Sources Energy 4:378–387
12. Tuller SE (1976) The relationship between diffuse, total and extraterrestrial solar radiation. Sol Energy 18:259–263. https://doi.org/10.1016/0038-092X(76)90025-6
13. Klein SA (1977) Calculation of monthly average insolation on titled surfaces. Sol Energy 19:325–329. https://doi.org/10.1016/0038-092X(77)90001-9
14. Hawas MM, Muneer T (1984) Study of diffuse and global radiation characteristic in India. Energy Convers Manag 24:143–149. https://doi.org/10.1016/0196-8904(84)90026-8
15. Gopinathan KK (1988) Empirical correlations for diffuse solar radiation. Sol Energy 40:369–370. https://doi.org/10.1016/0038-092X(88)90009-6
16. Trabea A (1999) A multiple linear correlation for diffuse radiation from global solar radiation and sunshine data over Egypt. Renewable Energy 17:411–420. https://doi.org/10.1016/S0960-1481(98)00124-4

17. Boland JW, Scott L, Luther M (2001) Modelling the diffuse fraction of global solar radiation on a horizontal surface. Environmetrics 12:103–116. https://doi.org/10.1002/1099-095X(200103)12:2%3c103::AID-ENV447%3e3.0.CO;2-2
18. De Miguel A, Bilbao J, Aguiar R, Kambezidis H, Negro E (2001) Diffuse solar irradiation model evaluation in the North Mediterranean Belt area. Sol Energy 70:143–153. https://doi.org/10.1016/S0038-092X(00)00135-3
19. Karatasou S, Santamouris M, Geros V (2003) Analysis of experimental data on diffuse solar radiation in Athens, Greece, for building applications. Int J Sustain Energ 23:1–11. https://doi.org/10.1080/0142591031000148597
20. Ulgen K, Hepbasli A (2003) Comparison of the diffuse fraction of daily and monthly global radiation for Izmir, Turkey. Energy Sources 25:637–649
21. Soares J, Oliveira AP, Boznar MZ, Mlakar P, Escobedo JF, Machado AJ (2004) Modeling hourly diffuse solar radiation in the city of Sao Paulo using neural network technique. Appl Energy 79:201–214. https://doi.org/10.1016/j.apenergy.2003.11.004
22. Jin Z, Yezheng W, Gang Y (2004) Estimation of daily diffuse solar radiation in China. Renewable Energy 29:1537–1548. https://doi.org/10.1016/j.renene.2004.01.014
23. Tehran S, Sari A (2005) Model selection for global and diffuse radiation over the Central Black Sea (CBS) region of Turkey. Energy Convers Manage 46:605–613. https://doi.org/10.1016/j.enconman.2004.04.004
24. Haydar A, Balli O, Hepbasli A (2006) Estimating the horizontal diffuse solar radiation over the Central Anatolia region of Turkey. Energy Convers Manag 47:2240–2249. https://doi.org/10.1016/j.enconman.2005.11.024
25. Mubiru J, Banda EJKB (2007) Performance of empirical correlations for predicting monthly mean daily diffuse solar radiation values at Kampala, Uganda. Theoret Appl Climatol 88:127–131. https://doi.org/10.1007/s00704-006-0249-1
26. Boland J, Ridley B, Brown B (2008) Models of diffuse solar radiation. Renewable Energy 33:575–584. https://doi.org/10.1016/j.renene.2007.04.012
27. Jiang Y (2009) Correlation for diffuse radiation from global solar radiation and sunshine data at Beijing, China. ASCE J Energy Eng C 107. https://doi.org/10.1061/(ASCE)07339402(2009)135:4(107)
28. Khatib T, Mohamed A, Mahmoud M, Sopian K (2011) Modeling of daily solar energy on a horizontal surface for five main sites in Malaysia. Int J Green Energy, 8(8):795–819. https://doi.org/10.1080/15435075.2011.602156
29. Ekren O (2012) Developing and evaluation of various correlations for diffuse solar radiation for Urla (Izmir, Turkey). J Solar Energy Eng 135. https://doi.org/10.1115/1.4007855
30. Bortolini M, Gamberi M, Graziani A, Manzini R, Mora C (2013) Multi-location model for the estimation of the horizontal daily diffuse fraction of solar radiation in Europe. Energy Convers Manag 67:208–216. https://doi.org/10.1016/j.enconman.2012.11.008
31. Kuo CW, Chang WC, Chang KC (2014) Distribution of solar diffuse fraction in Taiwan. Energy Procedia 57:1120–1129. https://doi.org/10.1016/j.egypro.2014.10.098
32. Boukelia TE, Mecibah MS, Meriche IE (2014) General models for estimation of the monthly mean daily diffuse solar radiation (Case study: Algeria). Energy Convers Manag 81:211–219. https://doi.org/10.1016/j.enconman.2014.02.035
33. Khorasanizadeh H, Mohammadi K, Mostafaeipour A (2014) Establishing a diffuse solar radiation model for determining the optimum tilt angle of solar surfaces in Tabass, Iran. Energy Convers Manage 78:805–814
34. Khahro S, Tabbassum K, Talpur S, Alvi M, Liao X, Dong L (2015) Evaluation of solar energy resources by establishing empirical models for diffuse solar radiation on tilted surface and analysis for optimum tilt angle for a prospective location in southern region of Sindh, Pakistan. Int J Electr Power Energy Syst 64:1073–1080. https://doi.org/10.1016/j.ijepes.2014.09.001
35. Filho EPM, Oliveira AP, Vita WA, Mesquita FLL, Codato G, Escobedo JF, Cassol M, Franca JRA (2016) Global, diffuse and direct solar radiation at the surface in the city of Rio de Janeiro: observational characterization and empirical modeling. Renewable Energy 91:64–74. https://doi.org/10.1016/j.renene.2016.01.040

Review on Data Compression Methods of Smart Grid Power System Using Wavelet Transform

Rakhi Jadhav and Anurag Mahajan

Abstract This paper starts with the existing electrical grid using conventional energy sources and very recently using renewable energy sources. Smarter technology is needed to accommodate those renewable energies for smooth functioning of power grid. The smart grid along with application of digital signal processing and digital two-way communications record, analyze and control the power grid from generation to usage. Power signal has to be monitored, and proper actions have to be taken quickly if any abnormality is noticed. The power signal data is huge, and hence, the system becomes slow and complex. The need of data compression, reduction of noise and complexity, regeneration of original signal is discussed as cost of data storage and the data transfer is reduced and the data analysis is easy. The detailed survey has been made on the digital signal processing techniques and algorithms developed by researchers to compress and denoise smart grid data. Comparative study of the digital signal processing techniques has been done and observed the importance of wavelet transform in terms of compression ratio, signal-to-noise ratio and reconstructional error for various types of simulated or real electrical signals. The results based on the literature review discussed the conclusion and the future scope for wavelet transform in smart grid to get the better results.

Keywords Discrete wavelet transform · Embedded zero-tree wavelet transform · Wavelet packet decomposition · Singular value decomposition · Daubechies

1 Introduction

Electrical grid generation is either thermal, hydro, nuclear and very recently renewable. The transmission is either of high-voltage AC or high-voltage DC; the distribution system mostly it is AC type, recently DC type too. The worldwide energy demand will be increased. The electricity consumption in India is very larger than

R. Jadhav (✉) · A. Mahajan
Electronics and Telecommunication Engineering, Symbiosis Institute of Technology, Symbiosis International (Deemed University), Pune, Maharashtra 412115, India
e-mail: rakhi.jadhav.phd2019@sitpune.edu.in

K. Namrata et al. (eds.), *Smart Energy and Advancement in Power Technologies*,
Lecture Notes in Electrical Engineering 926,
https://doi.org/10.1007/978-981-19-4971-5_18

generation capacity. Hence, energy needs to be efficiently used, and losses can be reduced in future. The renewable energy contribution in India is 17% to achieve the low carbon emission in future. So, smarter technology is needed to accommodate those renewables for smooth functioning of grid. It is very difficult to go for huge nuclear or hydro or thermal power plants taking 6 to 10 years time from now. Only one way is to put renewable energy sources as high as possible. Most of the thermal power plants today are old and need quick replacement which is very costly. At present, we have all manual reading and poor efficiency of conventional generation systems; most of the transmission and distribution lines are over loaded, reduced equipment life, increasing incidence of faults, operational constraints, voltage and frequency limits, uncertainty of renewable sources. It is aimed in next 5 years the energy installation has to grow [1].

1.1 Smart Grid

The digital two-way communication between the electric utility and user is making the grid smart. Customers will no longer have to wait for monthly statement of energy consumption and get timely picture through smart meters. So consumers can manage usage pattern based on the price of energy to make monthly electricity bill as minimum as possible. Smart grid helps to adjust your supply to save money and select suitable time to buy it and save more by producing rooftop solar energy. There are service providers and operations sectors. The participants join for sailing a trailing of the power in the market. The consumer domain is residential if the load is less than or equal to 20 kw and commercial if load is within 20–200 kw and industrial if it is greater than or equal to 200 kw. It has dedicated communication system for data flow from consumer to consumer or may be from one smart grid to the other. The energy service interface ESI is for the customer domain which communicates with the automatic meter infrastructure (AMI) which is connected to the internet. The consumer domain is mainly the automation of premises and home, industrial automation, microgeneration and their corresponding functions. Microgeneration includes all the type of distributed generations like solar, wind, fuel cells also batteries. In the market domain, the grid services are purchased and sold. The small participants sell their power in the market. The service provider will help in the billing, customer account management, enhanced management of use and home energy generation. For green energy, it is required to produce energy from wind, solar or fuel cell. The major part is the control, measurement and protection. So we have to measure voltage, current, frequency and power using supervisory control and data acquisition (SCADA). The control to manage the power flow and the reliability of the system is important otherwise power network will collapse. The generated power should be equal to demand otherwise frequency will be fluctuated. It sends power to consumers through various substations. The voltage is produced within 20–23 kV which is transformed to 400 kV, again the 400 kV is stepped down to 230 kV, further 132 kV, then 66 kV, then 11 kV, then 440 V at the customer end.

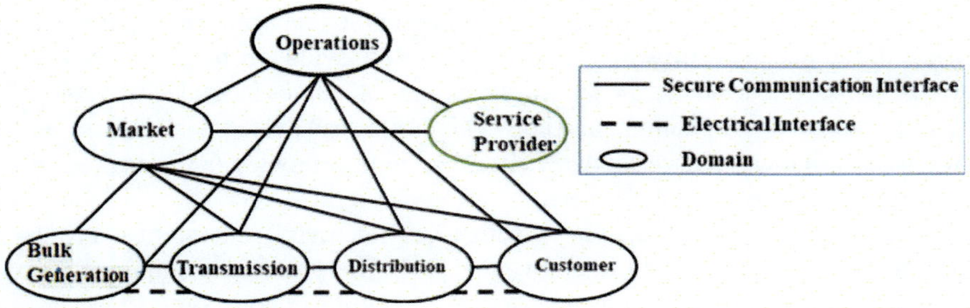

Fig. 1 Smart grid communication

The regional transmission operator (RTO) operates the transmission independently between the generation and the load with stable voltage and frequency [1–5].

A smart grid as shown in Fig. 1 is an electricity network that can intelligently integrate the actions of all users connected to generators, consumers for sustainable, economic and secure electricity supply.

There are following advantages of smart grid [1].

- It transmits energy effectively and restores quickly after disturbances. It reduces functional and administrative cost. Its peak demand is reduced and hence reduces cost of electricity. It has large integration of renewable energy and customer energy production.
- The existing grid has limited sensors and partially digital whereas smart grid is expected to be fully digital. The existing grid is of manual monitoring, and smart grid will be self-monitoring with sensors everywhere.
- The existing grid communicates one way and smart grid in two way. The existing grid generates centrally, and smart grid generation will be distributed.
- The existing grid suffers failures and blackouts with limited control, restores manually, but smart grid avoids them with full control mechanism and self-healing.
- The smart grid will have smart infrastructure with smart energy and information systems. It has smart communication, regulation, protection system and improved security.
- The smart energy system will have smart generation, transmission and distribution. The smart information system will have smart metering, sensors, phasor measurement units as well as information management, smart routers.

The smart grid power system is based on the information and communication technology which sense, analyze and control the grid from power production to its usage. As in [6], the smart grid meets power demand in reliable and economic manner with respect to local changes in usage. A smart grid has non-conventional energy sources and smart meters for measurement. Design and operation of smart grids can be using methods of digital signal processing [7–9]. There is burden on the system due to data transfer and collection. Data compression techniques reduce the data in a compact form, and data denoising will improve the signal quality. As in [10], the smart grid is an electrical system and has an automatic control over energy distribution, faults detection and its repair. It works according to the management of panel with respect to the decision software. A smart grid can provide electrical supply from various sources like wind energy, solar systems and sometimes from electric vehicles to end users via two-way communications. This system can overcome the issues very fast within it that can reduce the work. It has an aim to provide uninterrupted and power quality supply to all customers.

The smart grid is shown in Fig. 2 which has electricity generators, transmission and distribution substations as given in [11]. The phasor measurement units (PMU) for real time measurement are installed at multiple remote points. It has integrated communications like power line communications (PLC), supervisory control and data acquisition (SCADA), power flow controllers, smart power meters, collectors and control nodes at distribution and transmission systems [12, 13]. Substations provide voltage to multiple locations as in [14]. The substation equipment are transformers, circuit breakers, isolator switches, capacitors, inductors and protective relays. The smart grid collects the real-time data on large scale and monitors the system in an efficient way. The smart meters operate in the two ways between system and users by collecting the meter readings and providing it to the collectors for the billing. These collected readings are sent to the control center of the utility services. The objective as in [15] for data compression is to compress the power signal data at the sending end without loss of valuable contents in the data and then reconstruct the compressed data at receiving end. As the data is huge, it can slow down the system and the data are not transferred quickly. The data compression will enable the system to use less space and send the data to collection and control centers in compact size. The reconstructed signal will enable the system to use it reliably for the desired applications as the care will be taken to have very small reconstructional error. There is two-way flow of electricity and information with the widely distributed network. There is constant rise in the development of technology of the smart grid.

The smart meters measure electricity for billing which are foundations of smart grid [15, 16]. Automatic meter reading known as AMR which is automatically collecting detection, consumptions and status from energy meters and sending it to a central system for bills, solving problems and analyzing. Automatic meter infrastructure communicates two way with the meters. Therefore, nearly all of this information is available in real time [1]. It is possible to collect power quality disturbances after compression from all sources [17–20]. The phasor measurement units (PMU) mainly measure the power to decide healthy condition of the system. It provides loss of mains protection, monitor fault, locate disturbances, monitoring power quality, etc. [21]. Hence, the information management is being done from metering, sensing and monitoring and handles modeling of data, analysis, integration and improvement. Reliable and effective information exchange is a key to the success of the future smart grid to respond to any event in time; it must guarantee security and privacy. Smart communication is wireless, cellular communications and cognitive radio. The wired communication can be through fiber optic or power line. Wireless communication has low installation cost, also more suitable for remote applications. The cognitive radio is a secondary radio and suitable for backup. Fiber optic has electromagnetic and radio interference immunity for high-voltage operating environment. It has large

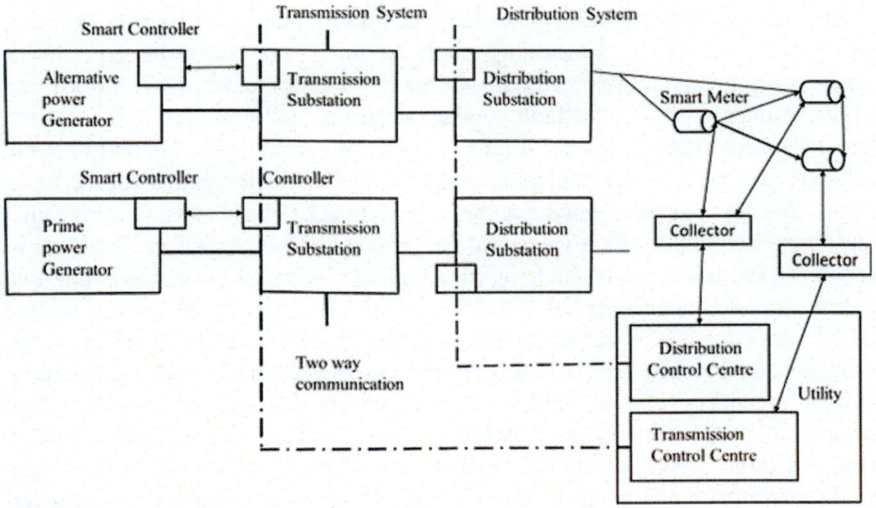

Fig. 2 Smart grid power system

bandwidth and immunity. The instalment cost of optical fibers may be high, used as a cost effective for high speed transmission [1]. Power line communication (PLC) carries data on a conductor also used for transmission [12, 22]. The narrowband PLC is used for smart metering infrastructure. Smart management increases reliability of system [1].

Parameters of Electric Signals. Steady-state and disturbance parameters affect power quality. Steady-state parameters have harmonics, deviation in frequency, flicker, voltage fluctuations and unbalance. Electric signal disturbances include supply outages, momentary or transient overvoltage or surges, momentary interruptions, voltage sag and swell. Frequency deviation is a variation of the fundamental frequency, e.g., 50 Hz in India, may be out of limit from steady state due to faults on the power system, disconnection of heavy load. Harmonics are the unwanted multiples of fundamental frequency due to nonlinear loads causing distortion in the waveform which leads to power quality problems. Transients are impulses or high-frequency oscillations caused by a sudden change of state may be internal or nearby. The transients are produced due to inductor and capacitor as they don't change their stored energy instantly just before the switching action and just after the switching action. Interharmonics are frequencies whose components are sinusoidal, have frequency non-integer of the fundamental value and have much shorter duration. They can be either in the form of discrete frequencies or in the wide-band spectrum. They are categorized into overloads, oscillations and distortion. Swells and sags are the rise or fall respectively 10% or more in the root mean square value of voltage between half cycle and 1 min. A voltage sag occurs due to rapid change of loads, e.g., short circuit. Voltage swells are less common than voltage sags (dips) and occur due to system fault conditions like single line to ground fault [10, 23].

Electrical fault is an abnormal state due to equipment failures such as transformers and rotating machines, etc. due to high fault currents. They interrupt electric flows, damage equipment and even cause death. Faulty voltages and currents differ from nominal values. Short-circuit fault conductors touch each other and large current flow. The open-circuit fault is the failure of one or two conductors. The short-circuit faults are symmetrical and unsymmetrical. The faults which occur in all the three phases are known as the symmetrical fault. System is balanced even after the fault. The symmetrical faults mainly occur at the terminal of the generators. The fault is unsymmetrical if current is changing in magnitude. After an unsymmetrical fault, system becomes unbalanced [24–27].

The transistors, other semiconductor devices, inductors and transformers are nonlinear components. They cause harmonics in an electrical system. They carry short fragment of current, and current is not proportional to voltage. The harmonic distortion has adverse effects on the equipment. The nonlinear devices occupy a large percentage of energy demand. Harmonic currents and voltages produced in a three-phase four-wire system are third-order, zero sequence harmonics (its odd multiples 3rd, 9th, 15th, 21st, etc., displaced by zero degrees). These third-order, zero sequence harmonics add up resulting excessive current in the neutral [10, 23].

Smart Grid Data. In smart grid, large data is developed by distribution transformers, current transformers, voltage transformers, smart meters, security systems. In [4, 15], it specifies that smart grid data is mainly composed of consumer details, connected load, maximum demand and daily usage pattern of power. Consumer details are population data, surrounding area details, etc. The consumer details comprised of consumer load, system of services and load demand. It has data of electricity generation, transmission, distribution, consumer services. Business data includes climatic, geographic information, building details and so on. The classification of customers is done by grouping them into a small number of classes for tariff formulation. The data characteristics are size, speed, variety and value. The real-time huge power production and its consumption in residential, commercial and industrial applications are measured and collected. An electric supply system puts pressure on the data processing to detect the fault, restore the power after the fault and decide future load demand. The smart grid power system includes large complex data of parameters like voltage, current, frequency, power, etc. It also has data of business and consumer details. The smart grid data helps to support power generation, its use and recovery of system failures. The analysis of power grid data helps to improve consumer services and also make changes in existing supply system. The data is multisource generated by sensing, monitoring and communicating devices. The data needs to be properly integrated for its effective use. Valuable information needs to be extracted from multiple sensing and metering devices and analyzed for improving the grid to function properly [20, 23, 27]. Hence, selection of dataset and its analysis is very important which is difficult to achieve as in [6]. The grid has to be protected to secure its privacy and operate in its best capacity as in [1]. Effective compression needs low computational complexity for data encryption. The low complexity leads to low computational time, and the compression algorithm is fast and efficient [28, 29]. The following are the evaluation parameters for the data compression.

Compression Ratio (CR): It is a ratio of original signal size to compressed signal size.

Signal-to-Noise Ratio (SNR or S/N): It is a ratio of original signal-to-noise signal that degrades it.

Mean Square Error (MSE): It measures the error of signal quality for compressed signal as compared to original signal [11].

2 Review of Literature

There are two broad types of data compression methods, viz. lossy and lossless compression as in [10, 30]. For lossless method, the signal is almost restored and measured parameter is compression ratio (CR). In Lossy method, the reconstructed signal may be different and the evaluation parameters are compression ratio and distortion between them.

2.1 Lossless Coding Methods

There are Lempel–Ziv (LZ) and high-order delta modulation with Huffman coding in lossless compression methods. As in [31], the proposed one-dimensional (1D) method is Huffman coding for high order delta modulation (HDM). It has compression ratio equal to 2:1. It has better compression ratio than traditional Huffman coding. Huffman recovers complete data, but it is very time consuming. HDM is less complex and has high compression. It is suitable for high sampling frequency power quality data. As in [10], the electric power signals have characteristics such as impulse, harmonics, flicker, sag, swell. LZ is simple to encode and decode 1D signal and has low compression ratio 5:1. High compression ratio is achieved by lossy compression methods.

2.2 Lossy Coding Methods

There are following lossy methods for electric signals, viz. (i) parametric, (ii) mixed parametric and transform, (iii) transform-based, (iv) singular value decomposition (SVD) as given in [10].

Parametric. The electric signals are mostly formed by sources, transmission wires and loads. There are resistors, inductors and capacitors (RLC) in the circuits whose transient behavior is modeled by damped sinusoids. As in [10], the parametric damped sinusoids modeling technique gives CR > 16:1, signal-to-noise ratio (SNR) of distortion value > 31 dB. The transform-based methods give better CR and SNR.

Mixed Parametric and Transform. The electric circuits are formed by sinusoids and transients as given in [10]. It states that wavelets are preferred for transients or short duration signals not for sinusoidal signals due to short bandwidth and arbitrary frequency value. Hence, the above hybrid coding method is used. Compression of voltage and current parameters for fundamental, harmonic and transient components gives CR 16:1, mean square error (MSE) –30 dB. The transform-based methods give better CR and MSE.

Transform Coding Methods. The high compression ratio can be obtained with lossy coding method with degradation of the signal as in [10]. It states that the distortion should be such that the compressed signals can be used for the analysis and diagnosis of power signals. The encoding process is composed of transformation, quantization and entropy coding. They use linear transform such as the lapped orthogonal, Hartley and the discrete cosine transform (DCT), but discrete wavelet and wavelet packets are widely used transforms. The discrete Fourier transform (DFT) and discrete cosine transform have their basic components as functions of sine and cosine. They analyze stationary and periodic signals due to localizing good frequency domain. DFT is unable to represent signals which are not periodic, have transients and variable frequency. Hence, wavelet-based transforms are widely used for the compression of electrical disturbances as they localize good in the domain of time and frequency. It can concentrate large signal energy in few coefficients in the transient and non-periodic pattern [24, 25, 32–39].

The discrete wavelet transform (DWT) is a computational algorithm. The wavelets are constructed from scaling functions. In pyramid algorithm, only approximation branches are used for the next level of resolution.

Figures 3 and 4 are as given in [40]. The 1-D DWT computation is done by a two-channel filtering unit having low-pass filter (LPF), high-pass filter (HPF) and a pair of down samplers. As in [41], the process is repeated until a suitable resolution is obtained. There are different types of filters used like Daubechies (db), Symlets, Coiflets where G and H are conjugates of g and h. Let x, z and x' be the original signal, compressed signal and the reconstructed signal, respectively. The data compression and regeneration is measured in terms of compression ratio (CR), normalized mean square error (NMSE), respectively as in [40].

$$\mathrm{CR} = \frac{x}{z} \tag{1}$$

whereas percentage CR is the reciprocal of compression ratio.

$$\mathrm{NMSE} = \frac{\left\|x - x'\right\|^2}{\|x\|^2} \tag{2}$$

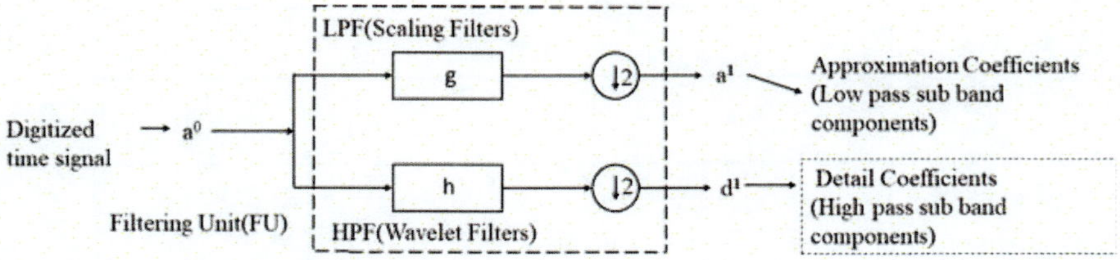

Fig. 3 Computation of one-level 1D DWT

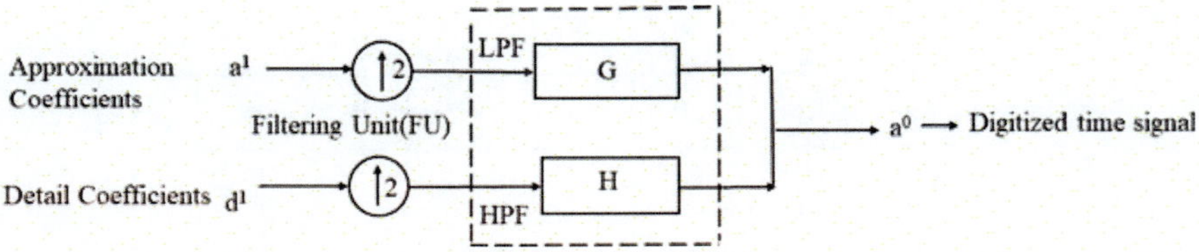

Fig. 4 One-level signal reconstruction of a^0 using pyramid algorithm in reverse

The wavelet packet transform (WPT) is a complete binary tree form of discrete wavelet transform (DWT) algorithm which decomposes approximation and detail signals to the next resolution level giving the best basis selection. Each subtree is a subspace in the form of scaling and wavelet function. The suitable representation of the signal is required to achieve the complete signal from WPT. The basis of best value or coefficients is calculated using non-normalized Shannon entropy after computing all approximation and detail coefficients up to a certain resolution as in [41].

In Fig. 5 as in [41], the minimum entropy, i.e., maximum information is given by the best basis for distribution of its coefficients. The orthogonal addition of two subspaces of children is equal to the parent subspace. The best basis finds entropy related to each subspace, then compare the entropy of parent and two children's subspaces. If the entropy of parent is not greater, then omit two children from the tree. If the entropy of parent is greater, then keep its two children. The process is continued until the signal is reconstructed same as at the top. The criteria of minimum description length (MDL) can suppress noise and compress signal, free from threshold which is useful for real data to calculate the noise. It gives the compromise between the reconstructed error and the retained coefficients.

Discrete Wavelet Transform. As in [40], wavelet transform is used to compress actual disturbance data. It is three-scale wavelet decomposition using Daubechies wavelets. The detail signals contain wavelet coefficients. The coefficients of disturbance events are retained discarding other coefficients. It is tried to achieve the smooth signal. It compresses the signal and the white noise. It can detect fault. The cost of storing data is reduced as CR is 3 to 6. The data transmission cost and time are reduced, and NMSE is 10^{-5} to 10^{-6}.

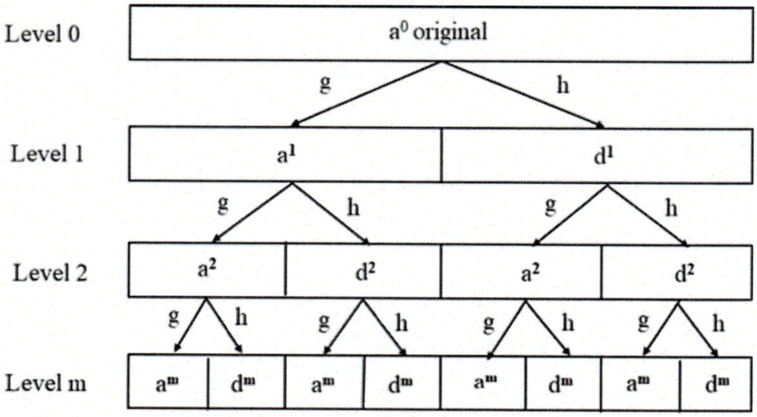

Fig. 5 Wavelet packet decomposition of a^0

As in [42], the slantlet transform (SLT) is an orthogonal discrete wavelet transform. Its time localization is improved and has zero moments of value two. It is used to compress different power quality events. It is a two-level filter bank keeping the views of complexity. For an impulse at CR of value 10, the retained energy for DWT is 91.13% and for SLT it is 94.01%. The MSE with slantlet is −16.98 dB and −13.56 dB with DWT. So SLT is superior than DWT in case of MSE. Same CR could be obtained with both SLT and DWT; only error needs to be improved in DWT.

As in [43], power quality transient data are compressed using B-splines up to three-scale and then a modified wavelet transform (S-transform) is applied to reconstruct and analyze the data. It has better time frequency localization property. The results for an impulsive transient signal are CR equal to 7.1429, NMSE 10^{-4} to 10^{-6} with SNR 30 dB.

As in [44], the voltage and current waveforms in electric arc furnaces are compressed where voltage flicker and current distortions due to harmonics are major concerns. The vector quantization can store more data; hence, it is used for the harmonics. The threshold coding is better in case of voltage flicker having different power cycles. The DWT with multiresolution analysis (MRA) is used to compress the data. The threshold and vector quantization coding are used to extract the characteristic data to eliminate the redundancy. The results for Daubechies-4 (db4) using 3-level MRA are CR equal to 6.12 for the voltage waveform with threshold coding (TH3) and CR equal to 4.08 for the current waveform for vector quantization (VQ3). The average CR is 5.1and NMSE 1.37E−03.

As in [45], it has introduced wavelet transform-based enhanced data compression method (EDCM). The notching phenomenon, low-frequency oscillatory transient signals and transient signals are verified with Daubechies 4 and decomposition at second to sixth level. It has applied 6-level decompositions to calculate the fundamental sinusoidal separating it from the transient ones and comparing with standard disturbance compression method (SDCM). The CR is 15.01:1 with SNR 35.56 dB for 6-level EDCM, and CR is 17.22:1 with SNR 40.78 dB for 6-level SDCM for fundamental to harmonic and interharmonic ratio (FHR) equal to 0 and threshold 6%. The CR is 16.45:1 with SNR 41.76 dB for 6-level EDCM and CR is 7.02:1 with SNR 41.69 dB for 6-level SDCM for fundamental to harmonic and interharmonic ratio (FHR) equal to infinity and threshold 6%. EDCM performance in terms of CR and SNR is worse than SDCM as FHR approaches 0 (maximum harmonics).

As in [11], the voltage sag signal is simulated and decomposed by using DWT-based MRA with Daubechies wavelets from db1 to db8 and decomposition scales from 1 to 4. The Daubechies wavelet 5 and scale 2 are selected. The proposed method improved CR, SNR and reduced MSE as compared to DCT, and the results are CR 2.875, MSE 14.05 and SNR 28.88 dB.

As in [21], wavelet transform is used to compress disturbances of voltage and frequency due to 3-phase faults using Daubechies 2 (db2) and Coiflet 1(coif1) up to 7th level of decomposition. The results suggest that voltage disturbance db2 compression at level 3 is better than coif1 with error below 2%. The Coif1 has significant distortions for both. Frequency disturbance with db2 and coif1 perform in same way. For voltage disturbance with db2, CR is 74.63 and error is 0.8067% at 7th level of decomposition. Coif1 distorts the signal more at level 3. For frequency disturbance with db2, CR is better at 7th level and Coif1 distorts the signal more at level 7.

As in [39], the order and scale of WT are required to be properly selected, better CR and SNR.

As in [46], the embedded zero-tree wavelet transform (EZWT) is used for the variable compression of the transient signal to match with variable bandwidth of transmission line. The high compression can be obtained by decomposing signal to its maximum extent as the high compression depends on zero-tree of wavelet scales. This work is compared with the previous work in [40]. For Daubechies 6 wavelet, the variable CR is 16:1 with NMSE 4.5×10^{-5} for normal capacitor switching disturbance signal which is higher with same NMSE than previous work [40].

As in [47], the encoded noisy signal generated by MATLAB is denoised on Field Programmable Gate Arrays (FPGA) technology by WPT better than WT.

As in [41], the single-phase to ground fault is compressed using minimum description length criterion for selecting suitable wavelet filter with DWT and WPT for resolution level four. The best basis is selected by minimum entropy. It uses 10 Daubechies wavelets, 5 Coiflets wavelets and 7 Symlets wavelets, so total 22 wavelets are used. The Symlet 7 filter gives CR equal to 36.36 for WPT which is slightly higher for DWT and gives CR equal to 84 for WPT combined with lossless coding LZH. It gives same MSE equal to 0.10 for both which is slightly less than DWT.

As in [48], compression of disturbance data is done with wavelet packet enhanced arithmetic coding by separating high- and low-frequency components with best basis. The data is decomposed with 10 different levels of Daubechies wavelet coefficients to find the best basis. The disturbance data can be reconstructed by using inverse wavelet packet transform. The results are CR equal to 6.95:1 and NMSE equal to 8.51×10^{-5} at 8th level of decomposition. It is compared with different compression methods combining coding methods.

As in [49], the three power quality disturbances biexponential impulse, oscillatory transient and voltage sag signals are compressed with wavelet decomposition and spline interpolation technique. The impulse signal is decomposed with WT into high CR 11.43 with NMSE $8.569e^{-006}$ using a Daubechies 4 at 4th level of decomposition and the threshold value 10. The second signal is decomposed with WT into high CR 8.1 using a Daubechies 5 at 4th level of decomposition and the threshold value 10 with increased NMSE $3.169e^{-005}$. The third signal is decomposed with WT into high CR 11.05 using a Daubechies 5 at 4th level with threshold value of 10 with NMSE $1.67e^{-005}$. The highest CR of the three signals is 63.99 for db4 and 4th level with the threshold value of 3, decimation factor 4. The NMSE error is from 10^{-2} to 10^{-4}. The CR obtained is 4 times greater, and NMSE error increases with CR in spline as compared to WT.

As in [50], the Embedded Zero-tree Wavelet Transform (EZWT) is used to compress noisy electrical signal and compared with wavelet transform (WT). Biorthogonal wavelet of support width 4, filter length 4 (bior4.4) at level 3 is used. The results are CR 1.45 with NRMSE 1.61×10^{-2} for EZWT and CR 1.45 with NRMSE 1.84×10^{-2} for WT. The results are CR 7.64 with NRMSE 4.51×10^{-2} for EZWT and CR 7.64 with NRMSE 4.55×10^{-2} for WT. Hence as CR increases, NRMSE also increases for both EZWT and WT, but NRMSE is less in EZWT than WT.

As in [51], the real and simulated PMU disturbance data are compressed and denoised with wavelet packet decomposition (WPD) and compared with wavelet decomposition (WD). The testing signals of 20 different types are simulated with Db 2 at 6th level. Its CR with WD is 12.5, NRMSE 2.5×10^{-4} and SNR after denoising is 40 dB. Its CR for WPD is 12.5 with NRMSE 3×10^{-5} and SNR after denoising is 50 dB for simulated PMU signal of sinusoidal noise. Hence, NRMSE and SNR are improved in WPD than WD.

As in [52], the Embedded Zero-tree Wavelet Transform (EZWT) gives variable data compression and denoising for images. The analysis, compression and denoising are done for the phasor measurement unit (PMU) and power system data and gives variable compression ratios and data denoising.

As in [53], the wavelet packet decomposition is suggested to compress, denoise and analyze smart grid system and results are compared with wavelet decomposition for phasor measurement unit (PMU), frequency disturbance recorder (FDR) and load voltage. Thirty candidate wavelets with decomposition level 2 to 11 are used. Thus, 20 different testing signals with Daubechies 2 at 6th level are selected. It gives CR equal to 50, NRMSE in the range of 10^{-5} and SNR 30 dB. The results are better than WD method, but it has high time complexity.

As in [54], wavelet packet transform (WPT) with MDL algorithm is applied for compression of PMU, FDR and load voltage data. The weighted Shannon entropy is used to calculate the best basis. The 15 wavelet functions db2 to db10, coif1 to coif5; and sym2 to sym10 are applied to these 19 test signals of three phase source, line and load signal voltages to decompose at levels 2 to 8. The Symlet 8 (sym8) and scale 4 are selected for the compression of the data. Entropy calculates the signal concentration. The best basis can be obtained by minimizing the entropy. The results are CR 1.37 with NRMSE 6.5×10^{-9} for FDR, CR 1.37 with NRMSE of 1.5×10^{-10} for PMU and CR 1.37 with NRMSE of 5.3×10^{-7} for load voltage. The results are CR 4.55 with NRMSE 3.03×10^{-7} for FDR, CR 4.55 with NRMSE of 1.5×10^{-8} for PMU and CR 4.55 with NRMSE of 1.5×10^{-4} for load voltage. It shows that as CR increases, NRMSE increases.

The high impedance faults will not provide sufficient fault current to operate fuses or protecting devices, hence need to detect such faults in low voltage distribution system. In this paper disturbances in transformer neutral current due to fault are detected by DWT and used multiresolution analysis (MRA). The plotting area below absolute value of detail coefficients is used for index of fault. The accuracy is checked by harmonic analysis. The results obtained on test system are valid for different loads and their operating conditions in low voltage system. The fault can be located anywhere in the feeder in a distribution system and can be detected in 0.25 s maximum time [26].

Singular Value Decomposition (SVD). It is a lossy compression method as discussed in [55]. It forms the data matrix X in a smart grid with m meters and time instant t in each row. The matrix X is comprised of three matrices. For r SVs (singular values) for a given compression ratio, form the submatrices. X can be approximated with only significant r SVs. XR is the approximated X. These submatrices store the total number of elements $(m + t + 1) \times r$. The loss of information is calculated. If the loss is not acceptable before sending the data, then data is compressed again by increasing r and repeating the same procedure. The data matrix is reconstructed on receiving side. Let x, $(x - x')$ and x' be the original signal, noise signal and reconstructed signal respectively for total N discrete samples in the original signal. The SNR is the ratio of original signal and the noise signal called signal-to-noise ratio [56].

$$\mathrm{SNR} = 10\log_{10}\left(\frac{x^2}{(x - x')^2}\right) \tag{3}$$

As in [11], the MSE is mean square error which calculates the error of signal quality with respect to the original signal after compression.

$$\mathrm{MSE} = \frac{\sum_{i=1}^{N}\left(x - x'\right)^2}{N} \tag{4}$$

In [55], SVD method is proposed to compress the real data from meters at many substations of an UK company. SVD is complex in computations. For high CRs, reconstruction errors are high in DWT and for small CRs, DWT give small errors. High CRs can be obtained with DWT; only reconstruction errors need to be reduced.

In [57], lossy compression by wavelet domain singular value decomposition (WDSVD) is given. The signal is split into number of cycles and is combined into 2D matrix. The SVD reduces the data for every submatrix using Bayesian rules to set adequate threshold. It has low computations and better CR-PRD as compared to

SVD and 2D DWT. It monitors flicker, harmonics, interharmonics and suitable for multiple power signal monitoring.

In [56], SVD a lossy compression method is given. The set of data is simulated and compressed in the IEEE fourteen bus system for harmonic signal, fault signal and voltage signal. Singular value decomposition (SVD) has complexity of computations. The results show that DWT has better SNR than SVD, but SVD gives better MSE than DWT. Hence, the error needs to be improved in DWT.

3 Results and Discussion

The comparison between different data compression methods for smart grid system using discrete wavelet transform is given in Table 1.

As per Table 1 from the best of existing design, the compression ratio obtained with DWT is 44.84 and the reconstruction error 0.5620% at the 6th level. The compression ratio obtained with wavelet packet decomposition (WPD) is 50, reconstruction error NRMSE is in the range of 10^{-5} and SNR is 30 dB at the 6th level. WPD gives better CR, less reconstruction error and better SNR as compared to DWT. The applications requiring high CR but can compromise with large error can use DWT, whereas applications requiring very small error can use WPD.

4 Conclusions

The different data compression methods are discussed in this paper. It is observed that discrete wavelet transform is better than other methods as lossless methods have low CR and time consuming. SVD is more complex due to increase in matrices with increasing numbers of meters. DWT has better CR and SNR than SVD, but SVD has low reconstruction error. The DWT has good localization in time & frequency, suitable for transient and non-periodic signals, better representation and features preservation. It can detect, analyze disturbances and fault and can reduce noise. EZWT is suitable for images, and it is asked to use it for compression of electric signals. The compression by wavelet packet decomposition gives better CR and SNR and small MSE, but it has huge tables of computations which make the method complex. The cryptographic functions depend on the computational complexity of the compression algorithm. If computational complexity is low to moderate, then smart meters need low power and processor usage, low memory hence cost less. It makes synchronization of compression algorithm fast to resume after temporary disconnection of communication link. Low complexity needs computational time low and compression algorithm is fast. DWT is used for the research on the power quality analysis. WPD is required to be used for any type of signal and its analysis. It is suggested to use WPD for fault detection and fault analysis. In the future, the compression ratio (CR), signal-to-noise ratio (SNR) can be increased and the

Table 1 Comparison of data compression methods

S. No.	Data compression method	Details of signal	Compression ratio (CR)	Reconstruction error	Signal-to-noise ratio (SNR)
1	DWT [40]	Actual disturbance data (db, scale 3)	3–6	NMSE 10^{-5} to 10^{-6}	NA
2	B-spline and S transform [43]	Impulsive transient data (scale 3)	7.1429	NMSE 10^{-4} to 10^{-6}	30 dB
3	SLT orthogonal discrete wavelet transform [42]	Impulse signal (scale 2)	10 for SLT and DWT	MSE −16.98 dB −13.56 dB	NA
4	DWT and MRA threshold and vector quantization [44]	Voltage flicker and current distortions due to harmonics (db 4, level 3)	5.1	NMSE 1.37E−03	NA
5	DWT with EDCM compared to SDCM [45]	Transient signals (db 4, scale 6)	15.01:1 with EDCM & FHR 0	MSE −30 dB	35.56 dB with EDCM
6	WT [21]	For voltage disturbance (db2, 6th level) (db2, 7th level)	CR 44.84 74.63	0.5620% 0.8067%	NA
7	EZWT [46]	Transient signal (db 6)	16:1	NMSE 4.5×10^{-5}	NA
8	EZWT [50]	Noisy electrical signal (bior4.4, 3rd level)	7.64 (same in WT)	NRMSE 4.51×10^{-2} less than WT	NA
9	WPT with MDL [41]	Single-phase to ground fault (Symlet 7, scale 4)	CR 36.36 (WPT) CR 84 (WPT and LZH)	MSE 0.10	NA
10	Wavelet packet Enhancement arithmetic coding [48]	Power disturbance data (8th level)	6.95:1	NMSE 8.51×10^{-5}	NA

(continued)

Table 1 (continued)

S. No.	Data compression method	Details of signal	Compression ratio (CR)	Reconstruction error	Signal-to-noise ratio (SNR)
11	WD and spline interpolation technique [49]	Impulse, oscillatory transient and voltage sag signals (Db4, 4th level)	63.99 (4 times higher than WT)	NMSE 10^{-2} to 10^{-4} higher than WT	NA
12	WPD [51]	Simulated PMU data for sinusoidal noise (Db2, 6th-level)	12.5	NRMSE 3×10^{-5} (less than WD)	50 dB
13	WPD [53]	Frequency disturbance PMU, load voltage (Db2, 6th level)	50	NRMSE range 10^{-5}	30 dB
14	WPT with MDL [54]	3 signals Frequency disturbance, PMU signal, Load voltage (sym8, scale 4)	4.55	NRMSE 3.03×10^{-7} 1.5×10^{-8} 1.5×10^{-4}	NA

complexity and reconstructional error can be decreased further using discrete wavelet transform and wavelet packet transform respectively. The database for the behavior of the smart grid equipment for fault signals can be established. The database for applications and conditions suitable for compression techniques for power signals can be established. The database can be prepared for the evaluation parameters for comparison and testing. The data compression has a scope in smart meters, energy theft calculations, load forecasting, demand side management, fault protection, power quality and fault analysis, automation of distribution and substation, management of load during feeder outage, etc.

References

1. Tomar A, Kandari R (2021) Advances in smart grid power system. Academic Press
2. Bose A (2010) Smart transmission grid applications and their supporting infrastructure. IEEE Trans Smart Grid 1(1):11–19
3. Li F, Qiao W, Sun H, Wan H, Wang J, Xia Y, Xu Z, Zhang P (2010) Smart transmission grid: vision and framework. IEEE Trans Smart Grid 1(2):168–177
4. Stragier J, Hauttekeete L, De Marez L (2010) Introducing smart grids in residential contexts: consumers' perception of smart household appliances. In: 2010 IEEE conference on innovation

technologies for an efficient and reliable electricity supply (CITRES 2010), pp 135–142
5. Asaad M, Ahmad F, Alam MS, Sarfraz M (2019) Smart grid and Indian experience: a review. Resour Policy
6. Uddin Z, Ahmad A, Qamar A, Altaf M (2018) Recent advances of the signal processing techniques in future smart grids. Human-Centric Comput Inf Sci 8(1)
7. Silva LRM, Duque CA, Ribeiro PF (2015) Smart signal processing for an evolving electric grid. EURASIP J Adv Signal Process 2015(1)
8. Silva LRM, Duque CA, Ribeiro PF (2015) Recent developments on signal processing for smart grids. In: IEEE power energy society general meeting
9. Carvalho TCO, Duque CA, Silveira PM, Ribeiro PF (2013) Considerations on signal processing for power systems in the context of smart grids. IEEE Power and Energy Society General Meeting, pp 1–5
10. Tcheou MP, Lovisolo L, Ribeiro MV, Da Silva EAB, Rodrigues MAM, Romano JMT, Diniz PSR (2014) The compression of electric signal waveforms for smart grids: state of the art and future trends. IEEE Trans Smart Grid 5(1):291–302
11. Karthika S, Rathika P (2018) Wavelet transform based compression of electric signal waveforms for smart grid applications, vol 7, no 4, pp 5419–5426
12. Berger LT, Schwager A, Escudero-Garzás JJ (2013) Power line communications for smart grid applications. J Electr Comput Eng 2013
13. Yan Y, Qian Y, Sharif H, Tipper D (2013) A survey on smart grid communication infrastructures: motivations, requirements and challenges. IEEE Commun Surv Tutorials 15(1):5–20
14. Jiang Z, Li F, Qiao W, Sun H, Wan H, Wang J, Xia Y, Xu Z, Zhang P (2009) A vision of smart transmission grids. In: 2009 IEEE Power and energy society general meeting (PES '09), pp 1–10
15. Wen L, Zhou K, Yang S, Li L (2018) Compression of smart meter big data: A survey. Renew Sustain Energy Rev 91:59–69
16. Wijaya TK, Eberle J, Aberer K (2013) Symbolic representation of smart meter data. ACM international conference proceeding series, pp 242–248
17. Mehta K, Russell BD (1989) Data compression for digital data from power systems disturbances: requirements and technique evaluation. IEEE Trans Power Delivery 4(3):1683–1688
18. Bollen MHJ, Gu IYH (2006) Signal processing of power quality disturbances. Wiley-IEEE Press, New York
19. Bollen MHJ, Das R, Djokic S, Ciufo P, Meyer J, Rönnberg SK, Zavoda F (2017) Power quality concerns in implementing smart distribution-grid applications. IEEE Trans Smart Grid 8(1):391–399
20. Ribeiro MV, Szczupak J, Iravani MR, Gu IYH, Dash PK, Mamishev AV (2007) Emerging signal processing techniques for power quality applications. EURASIP J Adv Signal Process 2007
21. Bhuiyan BA, Absar MW, Roy A (2018) Performance comparison of various wavelets in compression of PMU generated data in smart grid. In: 3rd International conference on electrical information and communication technology (EICT 2017), pp 1–6
22. Galli S, Scaglione A, Wang Z (2011) For the grid and through the grid: the role of power line communications in the smart grid. Proc IEEE
23. Albu MM, Neurohr R, Apetrei D, Silvas I, Federenciuc D (2011) Monitoring voltage and frequency in smart distribution grids. A case study on data compression and accessibility, pp 1–6
24. Dhend MH, Chile RH (2016) Efficient fault diagnosis in smart grid using non-conventional mother wavelet function. In: Asia-Pacific power energy engineering conference (APPEEC), pp 342–347
25. Dhend MH, Chile RH (2017) Fault diagnosis of smart grid distribution system by using smart sensors and Symlet wavelet function. J Electron Test Theory Appl 33(3):329–338
26. Vineeth N, Sreejaya P (2020) High impedance fault detection in low voltage distribution systems using wavelet and harmonic fault indices. In: 2020 IEEE International conference on power electronics, smart grid and renewable energy (PESGRE 2020), pp 1–6

27. Hlalele TS, Sun Y, Wang Z (2019) Faults classification and identification on smart grid: part-a status review. Procedia Manuf 35:601–606
28. Unterweger A, Engel D (2015) Resumable load data compression in smart grids. IEEE Trans Smart Grid 6(2):919–929
29. Singh, S (2019) Improving quality of service in the smart grid using data compression and encryption technique. Int J Adv Res Ideas Innov Technol 5(3):2172–2176
30. Jumar R, Maaß H, Hagenmeyer V (2018) Comparison of lossless compression schemes for high rate electrical grid time series for smart grid monitoring and analysis. Comput Electr Eng 71:465–476
31. Zhang D, Bi Y, Zhao J (2009) A new data compression algorithm for power quality online monitoring. In: 1st International conference on sustainable power generation and supply, SUPERGEN '09, no 50707016, pp 1–4
32. Dekhandji FZ (2017) Detection of power quality disturbances using discrete wavelet transform. In: 2017 5th International Conference on Electrical Engineering - Boumerdes (ICEE-B 2017), pp 1–5
33. Cheng L, Ji X, Zhang F, Huang H, Gao S (2018) Wavelet-based data compression for wide-area measurement data of oscillations. J Mod Power Syst Clean Energy 6(6):1128–1140
34. Andrade LCM, Oleskovicz M, Fernandes RAS (2016) Adaptive threshold based on wavelet transform applied to the segmentation of single and combined power quality disturbances. Appl Soft Comput J 38:967–977
35. De Andrade LCM, Oleskovicz M, Fernandes RAS (2014) Analysis of wavelet transform applied to the segmentation of disturbance signals with different sampling rates. In: IEEE power energy society general meeting
36. De Andrade LCM, Nanjundaswamy T, Oleskovicz M, Fernandes RAS, Rose K (2019) Advances in classification and compression of power quality signals. J Control Autom Electr Syst 30(3):402–412
37. Karimi M, Mokhtari H, Iravani MR (2001) Wavelet based on-line disturbance detection for power quality applications. In: Proceedings of IEEE power engineering society transmission and distribution conference (Winter Meeting), vol. 1, p 175
38. Santoso S, Grady WM, Powers EJ, Lainoree J, Bhatf SC (1999) Characterization of distribution power quality events with Fourier and wavelet transforms. IEEE Power Eng Rev 19(2):48
39. Littler TB, Morrow DJ (1999) Wavelets for the analysis and compression of power system disturbances. IEEE Trans Power Delivery 14(2):358–364
40. Santoso S, Powers EJ, Grady WM (1997) Power quality disturbance data compression using wavelet transform methods. IEEE Trans Power Delivery 12(3):1250–1257
41. Hamid EY, Kawasaki ZI (2002) Wavelet-based data compression of power system disturbances using the minimum description length criterion. IEEE Trans Power Deliv 17(2):460–466
42. Panda G, Dash PK, Pradhan AK, Meher SK (2002) Data compression of power quality events using the Slantlet transform. IEEE Trans Power Deliv 17(2):662–667
43. Dash PK, Panda G, Sahoo DK, Panigrahi BK (2002) Power quality disturbance data compression, detection, and classification using integrated spline wavelet and S-transform. IEEE Power Eng Rev 22(7):61
44. Wu CJ, Fu TH, Huang HP (2003) Data compression technique in recording electric arc furnace voltage and current waveforms for tracking power quality. Proceedings of IEEE power engineering society transmission and distribution conference, vol 1, pp 383–388
45. Ribeiro MV, Romano JMT, Duque CA (2004) An improved method for signal processing and compression in power quality evaluation. IEEE Trans Power Delivery 19(2):464–471
46. Chung J, Powers EJ, Grady WM, Bhatt SC (1999) Variable rate power disturbance signal compression using embedded zerotree wavelet transform coding. IEEE Eng Soc Winter Meet 2:1305–1309
47. Mahmoud MI, Dessouky MIMD, Deyab S, Elfouly FH (2007) Signal denoising by wavelet packet transform on FPGA technology. http://ubicc.org/files/pdf/183ubiccike_183.pdf. Last accessed 12 Aug 2021

48. Huang SJ, Jou MJ (2004) Application of arithmetic coding for electric power disturbance data compression with wavelet packet enhancement. IEEE Trans Power Syst 19(3):1334–1341
49. Găşpăresc G (2010) Data compression of power quality disturbances using wavelet transform and spline interpolation method. In: 2010 9th conference on environmental and electrical engineering (EEEIC 2010), pp 285–288
50. Khan J, Bhuiyan S, Murphy G, Arline M (2013) Embedded zerotree wavelet based data compression for smart grid. In: 2013 IEEE industry applications society annual meeting, pp 1–8
51. Khan J, Bhuiyan S, Murphy G, Williams J (2014) PMU data analysis in smart grid using WPD, pp 1–5
52. Khan J, Bhuiyan S, Murphy G, Arline M (2015) Embedded Zerotree wavelet based data denoising and compression for smart grid. IEEE Trans Ind Appl 9994(c):1–11
53. Khan J, Bhuiyan S, Murphy G, Williams J (2016) Data denoising and compression for smart grid communication. In: IEEE transactions on signal and information processing over networks, vol 2, no 2, pp 200–214
54. Bhuiyan S, Khan J, Murphy G (2018) Weighted entropy for data compression in smart grid. 2018 IEEE industrial and applications society annual meeting (IAS 2018), pp 1–6
55. De Souza JCS, Lessa Assis TM, Pal BC (2017) Data compression in smart distribution systems via singular value decomposition. IEEE Trans Smart Grid 8(1):275–284
56. Karthika S, Rathika P (2019) An efficient data compression algorithm for smart distribution systems using singular value decomposition. In: IEEE International conference on intelligent techniques in control, optimization and signal processing (INCOS 2019), pp 1–7
57. Li Q, Zhang M, He S, Li S (2018) Instructions data compression for smart grid monitoring using wavelet domain singular value decomposition. In: Proceedings of 30th Chinese control decision conference (CCDC 2018), pp 5171–5175

Scheduling of Demand-Side Resources for a Residential Air Conditioning Load

Kishor C. Muley and Sandeep Bhongade

Abstract The home energy management (HEM) system is a significant portion of the smart grid, which can conceivably empower demand response programs in dynamic pricing environment by utilizing thermostatically-controllable appliances (TCAs). The primary aim of this assessment is to bound the effective cost of residential heating, ventilation, and air conditioning (HVAC) loads for the entire residential building considering day ahead scheduling without violating the set of end-user comfort inclinations. This paper studies the possible benefits of implementing a simple load control scheme to meet the assumption of inelastic energy demand. This is achieved by utilizing the mathematical formulation of HVAC loads. The results of the proposed approach will be entrenched on the MATLAB platform using the concurrent electricity pricing of utility in addition to real-time outside temperature information which achieves remarkable diminishing in building energy utilization cost. The association between the planned load control and the commonly used ON/OFF load control method shows that noteworthy energy and cost-saving are achieved.

Keywords Home energy management · Smart grid · Demand response · HVAC

1 Introduction

Increase in energy demand, need to transform the traditional grid into a smart grid, that can be achieved by replacement of traditional control demand-side management (DSM) of electrical energy resources into smart electrical energy resources.

Optimal energy management is one of the effective solutions for eco-friendly use of energy at all levels of consumer (i.e., residential, commercial, and industrial). Demand response (DR) and load management are responsible for demand-side management as well as a financial regulating system for electric power supply. In

K. C. Muley · S. Bhongade (✉)
Department of Electrical Engineering, Shri G. S. Institute of Technology and Science, Indore, India
e-mail: bhongadesandeep@gmail.com

K. Namrata et al. (eds.), *Smart Energy and Advancement in Power Technologies*, Lecture Notes in Electrical Engineering 926,
https://doi.org/10.1007/978-981-19-4971-5_19

demand response, voluntary rationing is accomplished by price incentives offering lower net unit pricing in ex-change encouraging consumers for reduced power consumption in peak periods. DR is a beneficiary for both utility and consumer [1]. The load management initiated for shifting the peak load of the consumer to an off-peak period by encouraging the consumer to participate in the demand response program for minimization of cost of energy.

DSM is utilized for all categories of loads: manufacturing, commercial, and housing. Many authors suggested that residential and commercial buildings account for a significant portion of total electricity generation like in the United States, commercial buildings usage 54% of total electricity, while sub-urban houses consume about 40% [2, 3].

Residential loads account for a large share of electrical energy consumption up to 37% of total electrical energy consumption.

As per the international energy agency's report, annual electricity consumption is continuously increasing near about 3.4% worldwide [SOURCE]. Considering these rates, residential electricity consumption needs to be managed to improve energy efficiency as well as customer benefits. It can be achieved through an intelligent control system called a home energy management system (HEMS or HEM).

HEM is based on the smart home framework concept and has smart controllers/controllable devices and advanced communication technologies. The end user chooses to maintain energy efficiency and comfort. However, HEMS is optimized for comfort and energy. Various authors suggested that monitoring only real-time energy consumption can lead to a reduction in energy demand of about 25%, enabling end users to use energy responsibly. Moreover, HEM can reduce peak demand, save electricity bills, and meet DSM demand in smart grids [4]. To reduce power consumption without affecting user comfort, DR solution is the best option, and mainly, two types of DR programs are considered, i.e., incentive base DR and time-based DR.

Incentive-based DR program is based on ON and OFF management with deadline notifications; however, time-based DR requires consumers to participate to schedule their energy consumption time based on changes in price signals. DR is the most efficient technology that reduces electricity bills. The main aim to provide multiple DR and DSM methods is to reduce the load/burden of electricity resulting in energy saving [5]. This study mainly focused on the mathematical modeling of HVAC load and the energy management scheme.

2 Related Work

Home appliance scheduling algorithm based on predetermined preference and customer comfort preference under demand response is proposed in the study, the selective appliances as space cooling devices, water heaters, dryers, and electric vehicles to control on/off for home energy management [6]. The Dijkstra algorithm proposed in the previous study [7] shows the best performance, and the cost of peak hours can be reduced by considering autonomous home energy management of dispatchable devices. Pradhan et al. proposed a consumer benefit-based demand response model in a dynamic price environment, a peak load reduction algorithm, and ESS peak load reduction algorithm for heating, ventilation, and air conditioning (HVAC) systems are proposed, and the fundamental motive of this article was to examine and simulate an air conditioning system with a power management function by modifying the temperature set points inside the building structure [8]. HVAC devices debts for the biggest part of residential premises power consumption. The research work in [9] emphasises on lessen the running time of HVAC through converting the indoor temperature constraint on the forecasted peak time. The end result suggests that the energy cost is minimized through peak load deduction and energy utilization discount. The proposed set of rules is demonstrated through MATLAB simulations. The article [10] focuses on demand response modeling of split type air conditioner system on the basis of factors that affect the energy consumption of air conditioner. Yoon et al. [11] suggest heating, ventilating, and air conditioning (HVAC) control strategy for peak load reduction in a residential building in a real-time price environment. Mubbashir Ali et al. [12] build up a systematic equation to assess the adaptability of the HVAC load. This model is intended to amplify or limit power utilization during explicit hours of the day by modifying the HVAC load while upholding the user's thermal comfort. In [13], DR approach for residential heating, ventilation, and air conditioning (HVAC) installations that violate end-user comfort is proposed, the methodology designed to minimize the average discomfort of end users participating in the DR program. The purpose of [14] is to develop a consumer demand-side response (DSR) model to help end user consumers/aggregators and utility to minimize energy costs when peak electricity prices occur during peak seasons, the model allows consumers to independently and actively manage air conditioning loads through aggregators. The inspiration driving this article [15] is to present the improvement of a load model for conventional controllable loads, such as space cooling/warming, water heaters, clothes garments dryers, and electric vehicles. This paper proposes [16] an algorithm for demand-side resource scheduling of housing loads based on dynamic pricing, the particle swarm optimization algorithm was used to analyze the optimal load control operation under the real-time pricing scheme. The results show that compared with the control operation based on conventional ON / OFF, this scheme significantly reduces the energy consumption of the building. This article [17] encounters the hypothesis of rigid requirements by inspecting the potential advantages of implementing a simple load control system. This scheme permits consumers to change the energy consumption pattern from high prices during the day to low prices and explore the benefits to individual consumers by using California's price and demand data through examples applied to residential air conditioners.

The present study proposes a new demand-side resource scheduling algorithm for residential housing electrical energy management. The main focus of this paper is as follows:

- An algorithm is developed to limit the costs of energy consumption of residential buildings loads under the imperatives of consumer comfort.
- The algorithm is simple and can be applied to medium or large buildings.
- Scheduling considers dynamic pricing without fail to comply with consumer inclinations for comfort during the day.

Comparison of total costs minimization of building energy consumption with normal on/off method.

3 Mathematical Model of HVAC Load

The model and parameter presented in this paper are based on the control strategies of major load such as HVAC which is responsible for increase in energy consummation at residential premises [16, 17]. This will be model by Eq. (1) as follows;

$$\mathrm{Tin1}(i) = \in \mathrm{Tin} + (1 - \in)\left[\mathrm{Ti}_{\mathrm{out}}(t) + \mathrm{COP}\left(\frac{\mathrm{qi}(i)}{A}\right)\right] \tag{1}$$

Equation (1) represents the indoor temperature of the home as a function of energy consumed by the HVAC load for load control strategy.

Where

Tin1	Indoor temperature of next time step. (°F).
$\in$	System inertia (p.u.)
Tin	Default indoor temperature (°F).
Tiout	Real-time outdoor temperature (°F).
COP	Coefficient of performance.
qi	Energy consumed in each time slot (kW).
A	Thermal conductivity (kW/°F).

3.1 Real-Time Outdoor Temperature Data

Real-time local temperature data for consumer demand response is acquired from [18]. Figure 1 represents the outdoor temperature data of Delhi.

This data is used for the air conditioner load control strategy for the energy management scheme.

3.2 Real-Time Pricing Data

The electricity real-time retail pricing data for the present study is shown in Fig. 2.

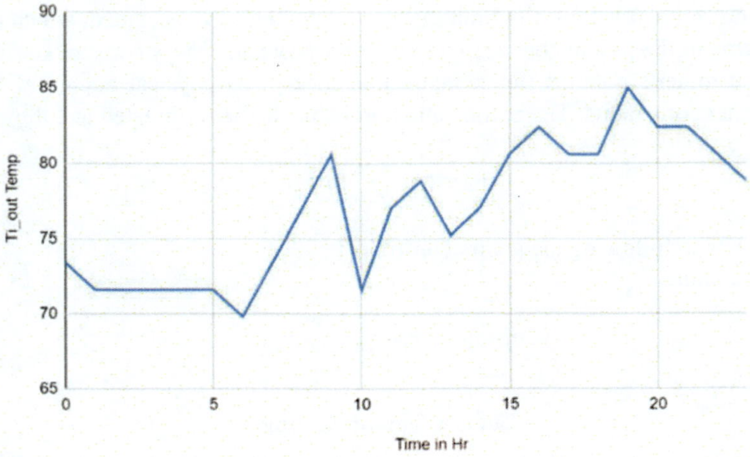

Fig. 1 Outdoor temperature versus time

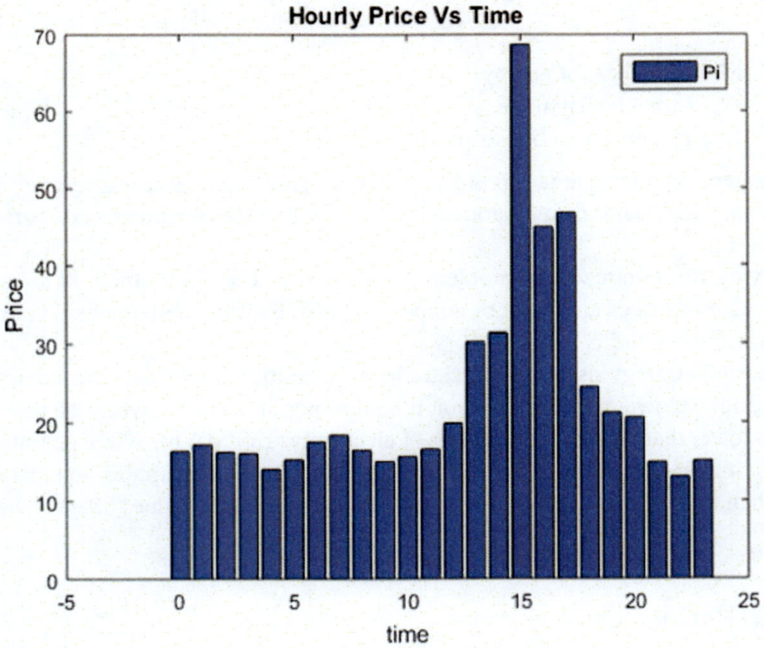

Fig. 2 Real-time price ($/MWh) versus time

The objective function is to limit the total cost of energy consumed, which is the summation of the product between the real-time cost and energy consumed, subject to the customer's comfort (temperature tolerances) and physical boundary of air conditioner constraints. The optimization problem can be formulated as follows:

$$\min C = \sum \mathrm{Pi} \cdot \mathrm{qi} \tag{2}$$

Subjects to following inequality constraints.
Temperature

$$\text{Tiout} > \text{Tin for Cooling} \tag{3}$$

$$\text{Tiout} < \text{Tin for Heating} \tag{4}$$

$$\text{qimin} < \text{qi} < \text{qi max} \tag{5}$$

where

min C Minimum cost of energy.
Pi Real-time hourly price.
qi(i) Energy consumed by AC in an hour i.

In an energy management problem, a four-degree thermostat temperature band ensures that the operating temperature is always within the consumer's comfort zone (75–79 °F).

Solving the minimization problem as mention in Eq. (2) requires knowing all hourly data of prices and outdoor temperatures for the time under consideration as shown in Figs. 1 and 2.

The methodology used will operate the air conditioner between two predefined limits, and the lower temperature limit is coolest possible (75 °F) when the real-time price is lower than the predefined price of electricity (13$/MWh). While operates the air conditioner, the hottest possible, i.e., near the upper-temperature boundary (79 °F), when the real-time prices are higher than the predefined value (68$/MWh).

3.3 HVAC Parameters

The HVAC parameters for the present study are given in Table 1.

3.4 Real-Time Air Conditioner Load Control Strategy

Pseudocode for real-time air conditioner load control strategy is shown as follows;

Table 1 Parameters of the HVAC

S. No	Parameter name	Value	Unit
1	T_0 = Desired temperature	77	°F
2	A = Thermal conductivity	0.14	kW/°C
3	COP = Coefficient of performance	2.5	p.u
4	ϵ = System inertia	0.93	p.u
5	Tin_min = Minimum temperature	75	°F
6	Tin_max = Maximum temperature	79	°F
7	d = Allowable deviation	2	°F
8	Pi_min = Lower price boundary	13.0	$/MWh
9	Pi_max = Upper price boundary	68.6	$/MWh
10	qi_min = Minimum power output of AC	1.5	kW
11	qi_max = Maximum power output of AC	3.5	kW

```
*******************************************************
Begin
Read the input data A, COP, Tin, E, Pi, qi, t
for i = 1:24 {termination criterion}
if Ti_out (i) > Tin
qi(i) = (Ti_out(i)*A-Tin*A)/COP
end
if Ti_out(i) < Tin
qi(i) = (Tin*A-Ti_out(i)*A)/COP
end
end
Limiting the Energy consumed for AC in an hour i for i = 1:24
qmin ≤ Qi ≤ qmax
end
Evaluate the room temperature using Eq. 1
Limiting the room temperature
for i = 1:24
Tmin ≤ Tin ≤ Tmax
End
Evaluate the price using Eq. 2 and power consumption
*******************************************************
```

4 Results and Discussion

In the present study, two cases are discussed for HVAC load:

1. Without load control
2. With load control.

These two cases are simulated on MATLAB environment for validation.

1. **Without Load Control**

For the without load control method, the simulated indoor temperature is shown in Fig. 3. The simulation results justify that despite real-time varying prices, the indoor temperature always remains constant at the desired temperature (77 °F) without load control of HVAC as presented in Fig. 4.

The variation of simulated indoor temperature with real-time outdoor temperature is shown in Fig. 5.

Figure 5 presents the variation of simulated indoor temperature with real-time outdoor temperature for without load control method. It is observed from here, that

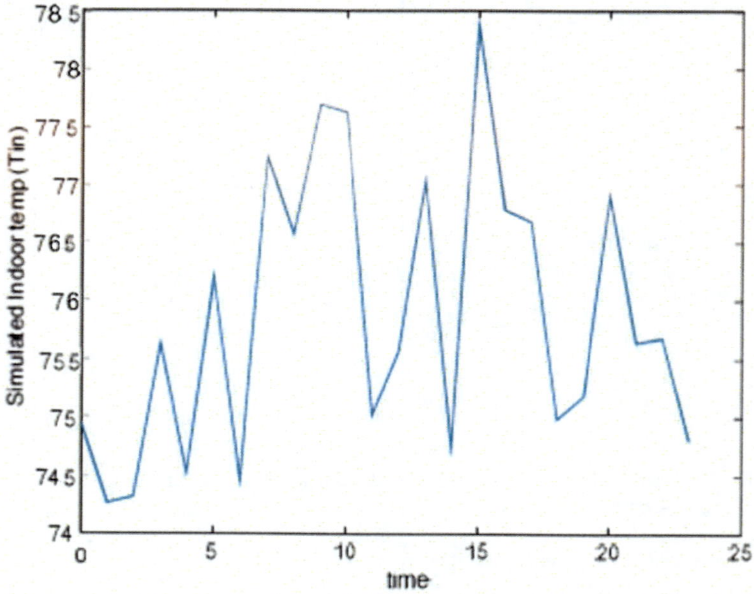

Fig. 3 Simulated indoor temperature (T versus Tin)

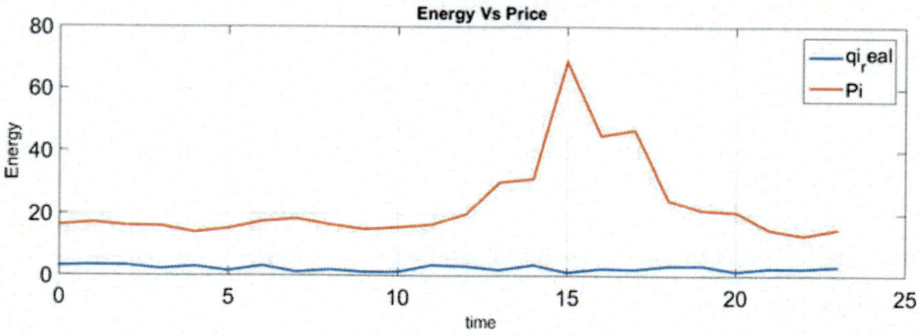

Fig. 4 Simulated indoor temperature and price versus time

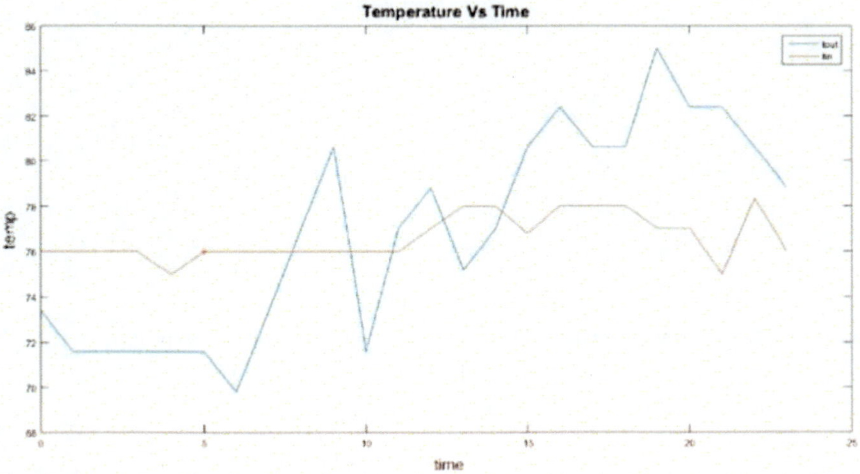

Fig. 5 Simulated indoor temperature (Tin) Vs Time (T) for without load control method

Table 2 Result of HVAC load (without load control)

Strategy	No-load control
Consumption (MWh)	58.228
Total cost ($/MWh)	44.37
Mean_qi	2.4262
Tin	77 °F

the consumer comfort desired temperature is fluctuating for a large change in outdoor temperature.

This increases total power consumption and means the energy consumed over one day, which ultimately increases the cost of energy consumption as represented in Table 2.

2. **With Load Control**

The main objective of the present study is achieved by applying a load control strategy. In this method, air conditioner is operated in between the defined band of temperature (i.e., 75–79 °F) and prices (Pi_min – Pi_max). The outcome obtained from simulation in MATLAB environment is summaries as:

(i) Thermostat temperature is controlled as per the comfort preferences of consumer as illustrated in Fig. 6.
(ii) Real-time hourly pricing is regulated using the load control method for operation of HVAC load in the user comfort zone of preferences as presented in Fig. 7, which results in a reduction in cost.
(iii) Simulation results in the reduction of total power consumption using the load control technique at real-time pricing as shown in Fig. 8.

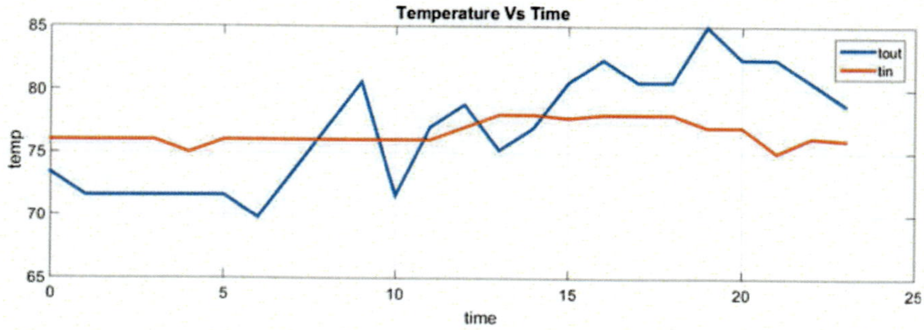

Fig. 6 Simulated indoor temperature (Tin) Vs Time (T) for with load control method

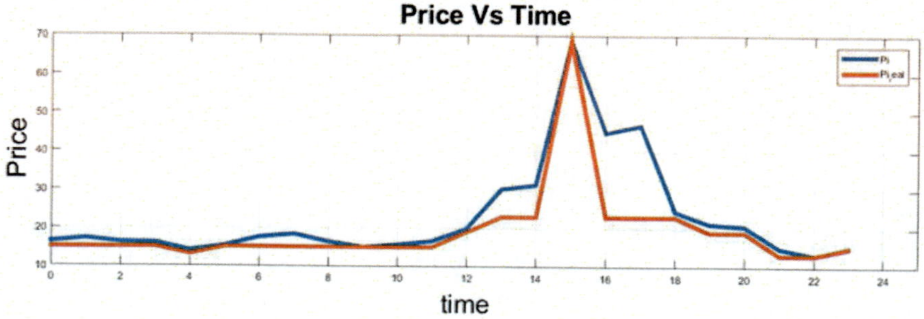

Fig. 7 Hourly pricing versus time

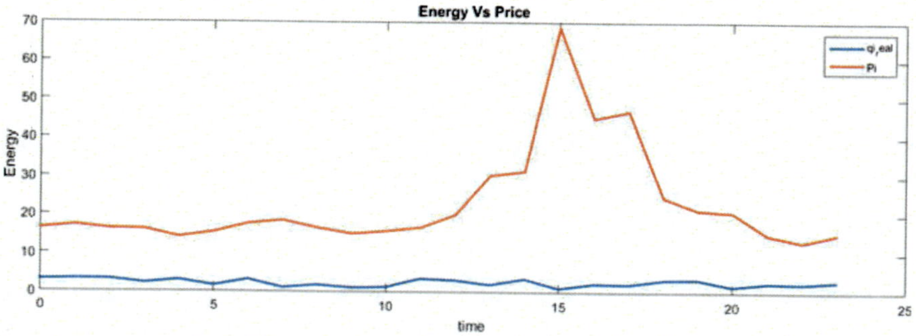

Fig. 8 Energy consumption and real-time price versus time

(iv) The regulated energy consumption.

In Fig. 6, it is showing the load control method using the proposed algorithm, thermostat temperature is controlled as per the comfort preferences of the consumer, which results in a reduction of the total consumption and total cost of energy consumption over a day.

Table 3 Result of HVAC load (with load control)

Strategy	With load control
Consumption (MWh)	36.6287
Total cost ($/MWh)	22.5
Mean_qi	1.5262
Tin	77 °F

It is observed from Table 3 that with load control, the mean energy consumption and cost of energy are minimized.

Comparison of simulated results obtained without load control and load control suggests that the proposed load control method results in financial as well as operational independence of heating, ventilation, and air conditioning (HVAC) system at the consumer end.

5 Conclusion

The scheduling of the home energy management system in the actual pricing plan is optimized for the user's comfort preferences. In some cases, operational preferences and constraints are randomized for investigation purposes. The comparison between the proposed and the normal ON/OFF load control operation shows that the former accomplishes improved outcomes in terms of price of energy consumption and energy saving. Therefore, the study can be further extended to incorporate all the loads used in residential homes and aggregate them into buildings.

References

1. Gabbar HA (2018) Energy conservation in residential, commercial and industrial facilities. Wiley, New Jersey, pp 225–246
2. Chua KJ, Chou SK, Yang WM, Yan (2013) Achieving better energy-efficient air conditioning: are view of technologies and strategies. Appl Energy 104:87–104
3. US Department of Energy (2010) Buildings energy data book, pp 1–271
4. Khan AA, Razzaq S, Khan A, Fatima Khursheed O (2015) HEMS s and enabled demand response in electricity market: an overview. Renew Sustain Energy Rev 42:773–785
5. Faisal HM, Javaid N, Ali Khan Z, Mussadaq F, Akhtar M, Abbasi RA (2020) Towards efficient energy management in a smart home using updated population, vol. 926. Springer Nature Switzerland, pp 39–52
6. Pipattanasomporn M, Kuzlu M, Rahman S (2012) An algorithm for intelligent home energy management and demand response analysis. IEEE Trans Smart Grid 3(4):2166–2173
7. Basit A, Sidhu GAS, Mahmood A, Gao F (2015) Efficient and autonomous energy management techniques for the future smart homes. IEEE Trans Smart Grid 99:1–10
8. Pradhan V, Balijepalli VSKM, Khaparde SA (2016) An effective model for demand response management systems of residential electricity consumers. IEEE Syst J 10(2):434–445

9. Kim N-K, Shim M-H, Won D (2018) Building energy management strategy using an HVAC system and energy storage system. Energy
10. Dejvisesa J, Tanthanuch N (2016) A simplified airconditioning systems model with energy management. Procedia Comput Sci 86:361–364. Published by Elsevier 2016 International Electrical Engineering Congress, iEECON2016, 2–4 March 2016, Chiang Mai, Thailand
11. Yoon JH, Baldick R, Novoselac A (2014) Dynamic demand response controller based on real-time retail price for residential buildings. IEEE Trans Smart Grid 5(1):121–129
12. Ali M, Safdarian A, Lehtonen M (2014) Demand response potential of residential HVAC loads considering users preferences. In: 2014 5th IEEE PES innovative smart grid technologies Europe (ISGT Europe), October 12–15 (2014). IEEE, Istanbul
13. Erdinc O, Tascikaraoglu A, Paterakis NG, Eren Y, Catalão JPS End-user comfort oriented day-ahead planning for responsive residential HVAC demand aggregation considering weather forecasts. IEEE Trans Smart Grid
14. Marwan S (2017) Optimise energy cost for air conditioning based on the market price under demand side response model. Int J Electr Comput Eng (IJECE) 7(3):1125–1132
15. Shao S, Pipattanasomporn M, Rahman S (2013) Development of physical-based demand response-enabled residential load models. IEEE Trans Power Syst 28(2)
16. Tiptipakorn S, Lee W-J (2007) A residential consumer-centered load control strategy in real-time electricity pricing environment. In: Power symposium, 2007. NAPS'07. 39th North American. NM IEEE, Las Cruces, pp 505–510
17. Ilic M, Black J, Watz J (2002) Potential benefits of implementing load control. In: IEEE power engineering society winter meeting, 2002, vol 1, no 27–31, pp 177–182
18. www.findlocalweather.com

Implementation of MPPT in Grid Interfaced Solar-PV Generating Unit

Narayan Prasad Gupta and Priyanka Paliwal

Abstract The concept of renewable energy leads to increase the electricity production worldwide. Renewable energy resources can only fulfil future demand without affecting the environment. Harnessing solar energy is challenging as solar insolation varies with time. For preserving solar output voltage at a constant value, maximum power point tracking (MPPT) is employed. This paper presents real time implementation of grid-connected (GC) solar-PV system with perturbs & observe and incremental conductance algorithm-based MPPT tracking. The GC-solar generating unit is analyzed for varying solar insolation and loading. The obtained experimental results demonstrate the competency and high robustness of the control scheme.

Keywords Perturb & observe · Incremental conductance · MPPT · Solar cell · Grid inverter

1 Introduction

In recent time, the emergency of power is confronted everywhere in the world due to the increment in domestic and modern interest. Renewable energy resources have the capacity to meet future energy demand with less impact on the environment. It is trying for analysts to investigate more supportable energy assets. Sustainable power sources are the best accessible alternatives to satisfy the interest of intensity [1]. The primary sustainable sources-based frameworks are PV, Wind, micro-hydro turbine, Fuel cell, etc. Solar-PV energy is believed as a dependable, capable and good source. It is advantageous in sense of contamination-free, extended life, low support, etc. As the efficiency of solar-PV is around 30–40% so researchers are working in this area to achieve maximum power through the PV framework [2]. The drawback of

N. P. Gupta (✉) · P. Paliwal
Electrical Engineering Department, Maulana Azad National Institute of Technology, Bhopal, India
e-mail: narayan.irig@gmail.com

P. Paliwal
e-mail: priyanka.manit@yahoo.com

K. Namrata et al. (eds.), *Smart Energy and Advancement in Power Technologies*,
Lecture Notes in Electrical Engineering 926,
https://doi.org/10.1007/978-981-19-4971-5_20

PV source in terms of night insolation and partial shading condition is the focus of research. The main hurdle is its high initial investment. Hence, an augmentation in the viability and productivity of maximum power point tracking assumes a significant job in expanding the transformation proficiency just as diminishing the cost of a solar-PV system [3].

The nonlinear characteristic of solar cell requires control techniques which overcome the drawback of changing weather condition, variation in temperature, partial shading and insolation [4]. The intensity of a PV framework is greatest at the lap point of the PV characteristics, termed as MPP [5]. As the MPP continues evolving as indicated by the fluctuating illumination levels, a MPPT technique is utilized to follow the MPP of the framework. Different sorts of MPPT techniques have been created and executed over a long time [6]. These MPPT techniques can be arranged based on numerous highlights including sun-powered effectiveness, dynamic reaction, sensors necessity, cost, etc. [7]. An assortment of MPPT techniques has been accounted for in literary works. MPPT described in the literature are: constant voltage control (CVC), perturb and observe, incremental conductance, hybrid MPPT, etc. MPPT techniques can be classified as conventional, AI-based and hybrid methods [8–12]. Further conventional MPPT techniques are broadly categorized as current-based, voltage-based, curve fitting-based, gradient decent, sliding mode, incremental resistance, constant voltage, P&O, IC, transient-based, power feedback-based, etc. [13–15]. Hybrid MPPT techniques include modified P&O, modified FLC and neuro-fuzzy-based, etc. The constant voltage control-based method is simple with ease of implementation yet lacks in accuracy and efficiency. P&O MPPT has considerable power loss and low efficiency. In general conventional methods have slow transient response with environment changing conditions. AI-based MPPT methods have high efficiency and accuracy in tracking. In hybrid MPPT methods online and offline methods are grouped so that to provide high accuracy in tracking and efficient in operation. In context to the above literature review, this paper presents an insight of P&O and IC with parameters variations, i.e., voltage, power and dynamic response, etc. Also, different solar radiations are considered to prove the efficacy system.

2 Solar-PV Modeling and MPPT

The photo-voltaic structure is categorized into two configurations namely, stand-alone structure and grid-connected structure. A simple P–N junction diode is utilized for converting solar irradiation into electricity. Figure 1 describes an equivalent network of a PV cell. Solar-PV-array is a combination of cells in series and parallel.

$$i_d = I_s\left[\mathrm{e}^{\alpha\left(v_{\mathrm{pv}}+R_s i_{\mathrm{pv}}\right)} - 1\right] \tag{1}$$

where $\alpha = q/\mathrm{AkTC}$, the PV-array output current, i_{pv} is obtained by

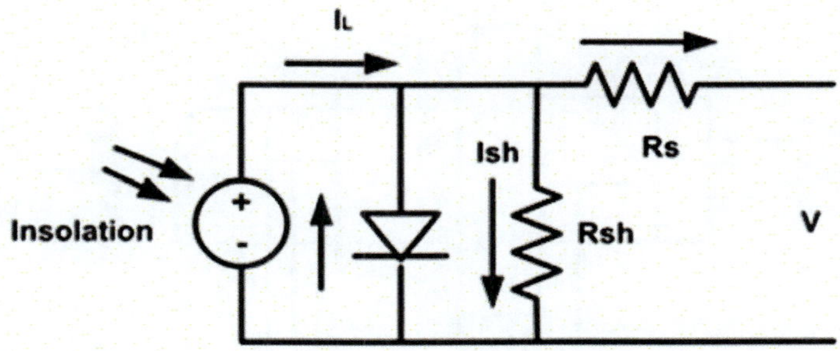

Fig. 1 Solar-PV Modeling

$$i_{\text{pv}} = N_p I_L - N_p I_s \left[e^{\alpha\left(\frac{v_{\text{pv}}}{N_s} + \frac{R_s i_{\text{pv}}}{N_p}\right)} - 1 \right] - \frac{N_p}{R_{\text{sh}}} \left(\frac{v_{\text{pv}}}{N_s} + \frac{R_s i_{\text{pv}}}{N_p} \right) \tag{2}$$

The current source output I_L is given by

$$i_L = \frac{G}{100} [i_{\text{sc}} + k_i \{T_c - T_{\text{ref}}\}] \tag{3}$$

The solar cells are joined in series–parallel combination to cater load demand in form of PV. The I–V nature of a solar array is nonlinear leading difficulty in measuring maximum value at which it operate. Figure 2a shows the characteristic I–V and photo-voltaic curve for a fixed level of solar irradiation and temperature. The output of each solar cell is chiefly subjected to insolation, temperature and partial criteria of shading shown in Fig. 2b and c.

MPP is tracked for exploiting maximum efficiency from a PV-array. Figure 3 shows a basic schematic of MPPT system. Various algorithms for MPPT already reported in literatures, among various described MPPT methods better choice in terms of cost and performance can be P&O and IC.

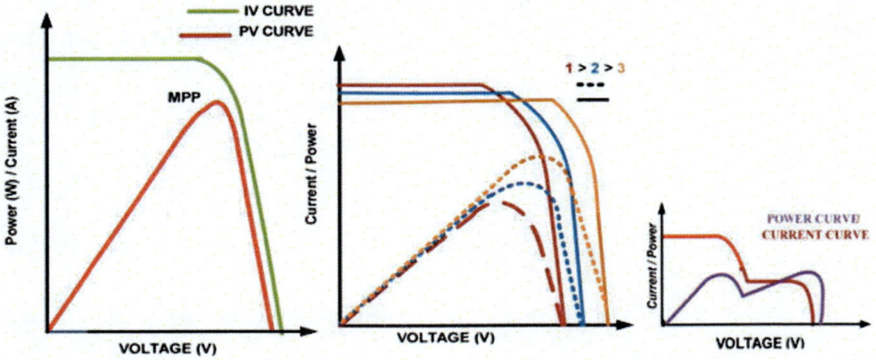

Fig. 2 Solar-PV under **a** I–V and PV curves under fixed insolation, **b** varying insolation and **c** partial shading

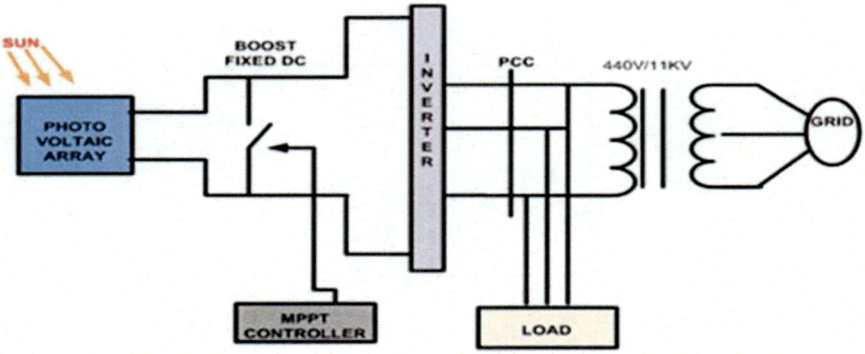

Fig. 3 Maximum power point tracker system

2.1 *Perturb and Observe Control*

This technique utilizes perturbations of reference voltage/current as depicted by the flowchart appeared in Fig. 4. The basic function of algorithm is to follow V_{mpp}. This is achieved by small and constant perturbation to PV voltage while the deviation in output power (dP) is estimated [13–15]. The P&O MPPT has two principle downsides. Difficulty in step size, large step size results in chance of algorithm to oscillate around maximum power point and small step size results in low speed of convergence. The solution to this problem is variable perturbation size. Table 1 presents positive and negative changes in reference.

2.2 *Incremental Conductance Control*

This method overcomes the limitations of P&O in terms of fast tracking. It stops perturbation when MPP is achieved. Figure 5 shows the flow chart [1–5]. The Increment depends on the gradient of PV-array which is 0 at MPP, negative at right and positive at left position.

$$0 \leq \mathrm{dp/dv} \quad \text{left} \tag{4}$$

$$0 \geq \mathrm{dp/dv} \quad \text{right} \tag{5}$$

$$\mathrm{dp/dv} = 0 \quad \text{MPP} \tag{6}$$

$$\mathrm{dp/dv} = \mathrm{d(iv)/dv} \quad = I + V\mathrm{di/dv} \tag{7}$$

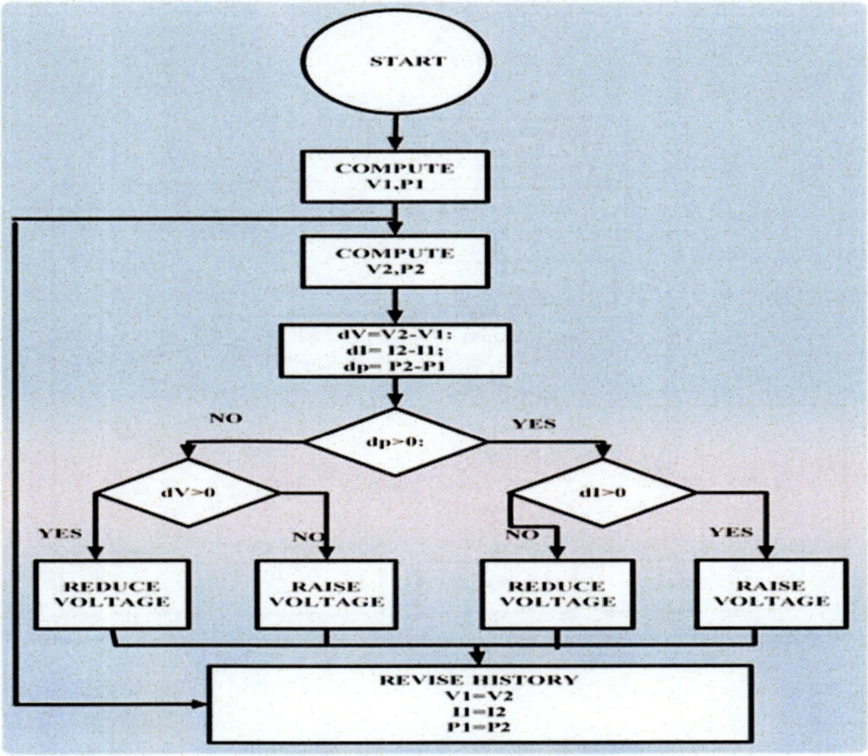

Fig. 4 Flow chart for P&O MPPT

Table 1 Perturb and observe

dV sign	dP sign	Next step track
+	+	+C
−	−	+C
−	+	−C
+	−	−C

$$-i/v \leq \text{di/dv} \quad \text{left} \tag{8}$$

$$-i/v \geq \text{di/dv} \quad \text{right} \tag{9}$$

$$\text{di/dv} = 0 \quad \text{MPP} \tag{10}$$

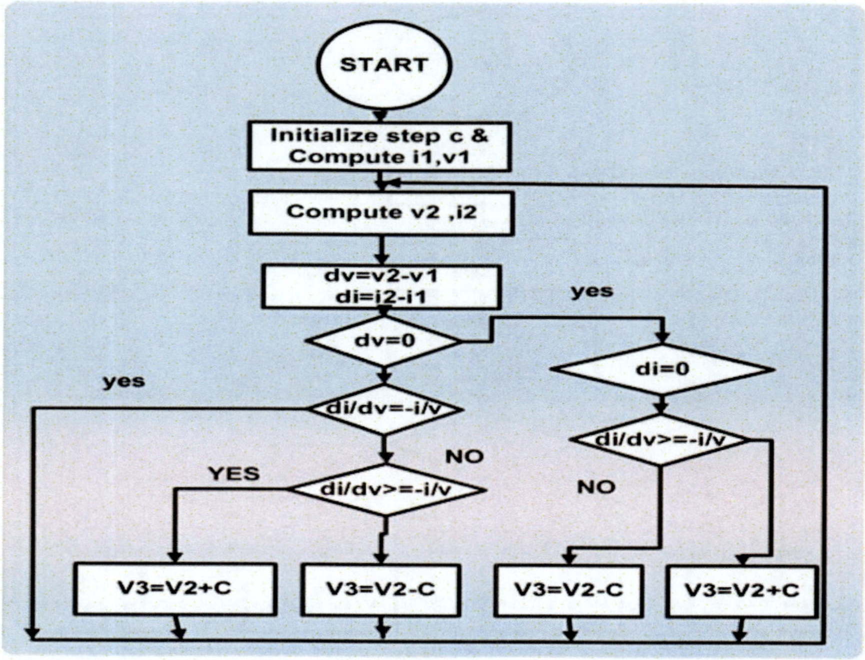

Fig. 5 Flow chart for IC-MPPT

3 Grid Inverter Control

Figure 6 represents control structure for grid inverter. The synchronous frame is utilized for generating firing pulses for grid inverter VSC. The PCC active and reactive power is measured and is matched with the set values so that error can be generated. This error signal is fed to PI controller to acquire I_{dref} and I_{qref}. These d and q-axis current reference are represented by

$$I_{\mathrm{dref}} = \left(k_p + \frac{k_i}{s}\right)(P_{\mathrm{ref}} - P_{\mathrm{meas}}) \tag{11}$$

$$I_{\mathrm{qref}} = \left(k_p + \frac{k_i}{s}\right)(Q_{\mathrm{ref}} - Q_{\mathrm{meas}}) \tag{12}$$

This I_{dref} and I_{qref} are now matched with measured i_d and i_q. This will generate error signals which are further processed with PI controller. The conditioned signal is now fed to generate firing pulses of 3-level grid inverter.

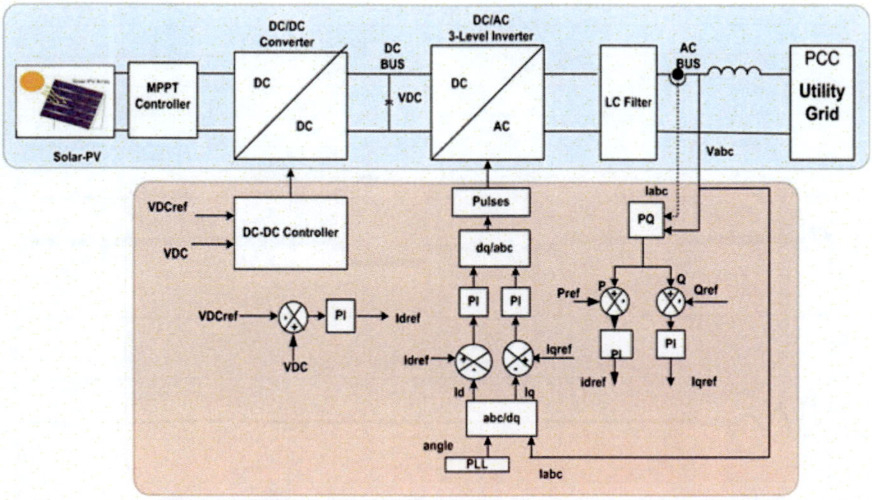

Fig. 6 Control structure of grid inverter

4 Results and Discussion

The grid tied solar-PV system is simulated in MATLAB. The considered variation in irradiance is 1000, 800, 600 W/m^2 with constant temperature as 25 °C. The output voltage of solar-PV cell is inverted. A 11 kV utility grid is interfaced to the solar-PV generating unit, with the aid of step down transformer. The dynamics in load is initially at 2 kW and after some time it will be 5 kW.

4.1 Performance of Grid Tied PV System with P&O Method

At irradiance of 1000 W per meter square, voltage of solar-PV cell = 272 V, as irradiance reduces to 800 W per meter square, in parallel voltage also goes down to 250 V. Furthermore when voltage is reduced again at 600 W per meter square, the voltage also reduces to 190 V. Moreover again if the irradiance value is increased to 1000 W per meter square the value of voltage before MPPT is also increased to 272 V. Figure 7 shows the variation in voltage with irradiance before and after MPPT.

The change in output power with variations in insolation is observed in Fig. 8. Three phases voltage is fed to the load of 2 kW, increased to 5 kW at $t = 2.2$ s. the load voltage is fairly constant while load current changes as per loading conditions. The P&O MPPT algorithm is simulated under different loading condition with variation in solar radiations. It oscillates close to MPP accurately and give maximum output power in every condition.

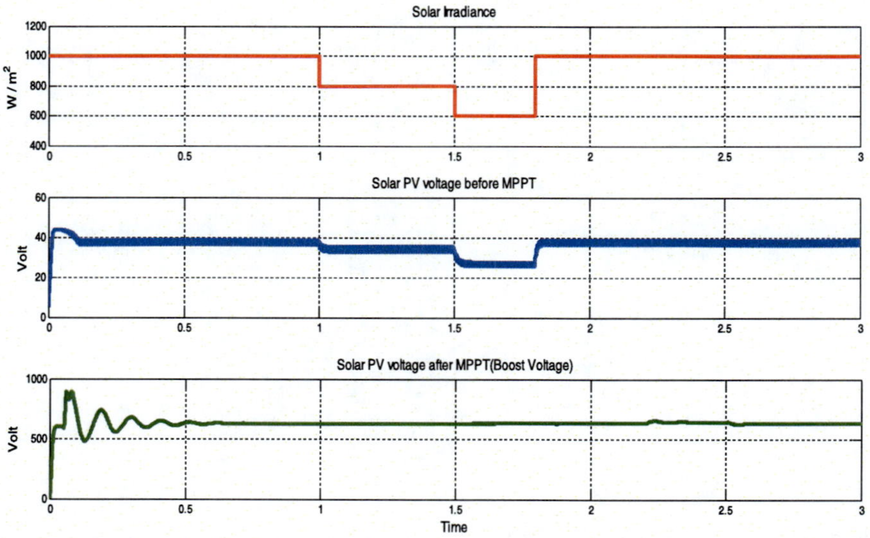

Fig. 7 Solar-PV voltage with P&O

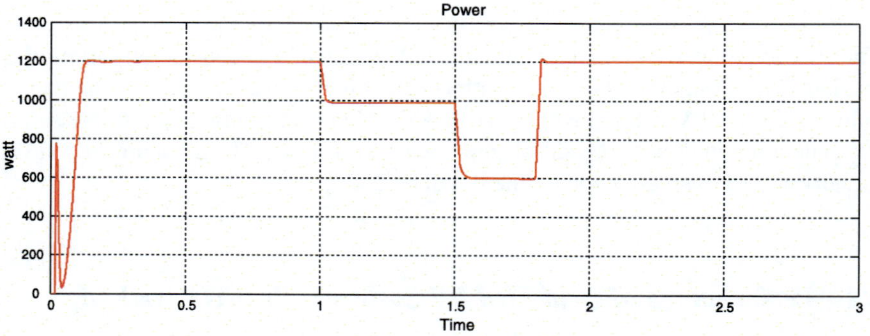

Fig. 8 Power obtained from PV cell with P&O MPPT

4.2 *Performance of Grid Tied PV System with IC Method*

The solar output voltage with variation in irradiance before after MPPT is shown in Fig. 9, which is slightly more than obtained with P&O method. The change in output power with variations in insolation is shown in Fig. 10, which is higher than that obtained with P&O method. Figure 11 shows 3-level inverter voltages waveform of 440 V at frequency of 50 Hz.

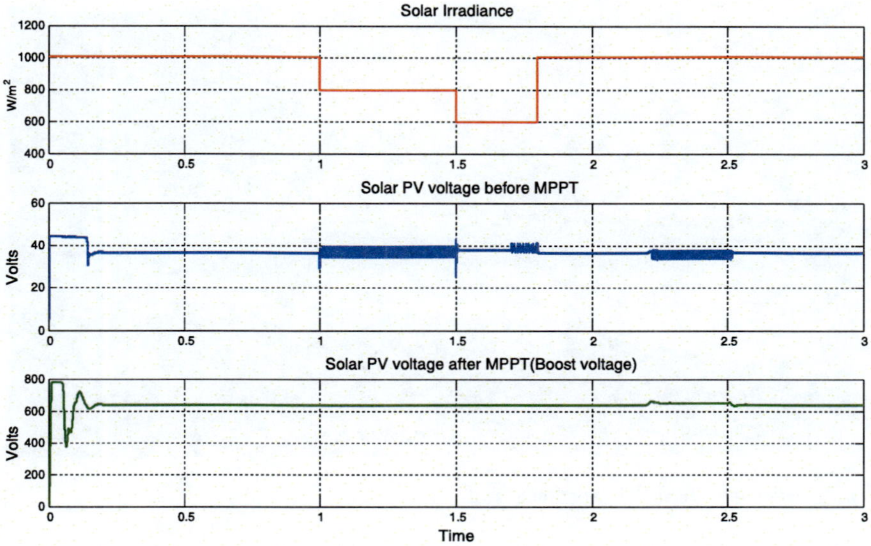

Fig. 9 Solar-PV voltage with IC

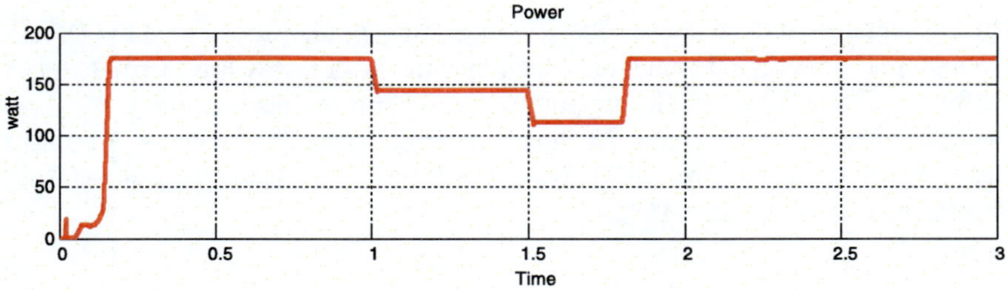

Fig. 10 Power obtained from PV cell with IC-MPPT

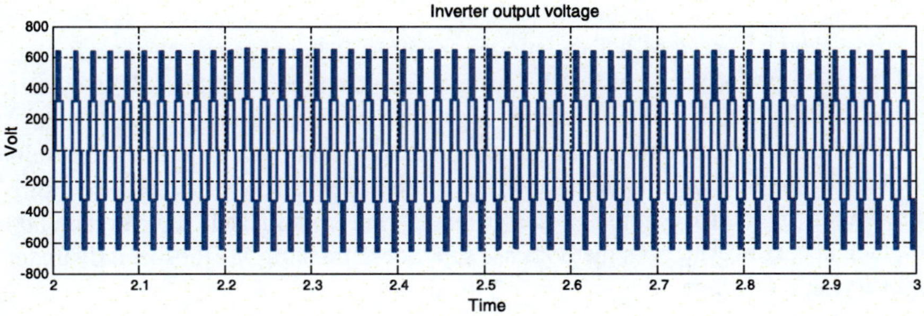

Fig. 11 Inverter output voltage

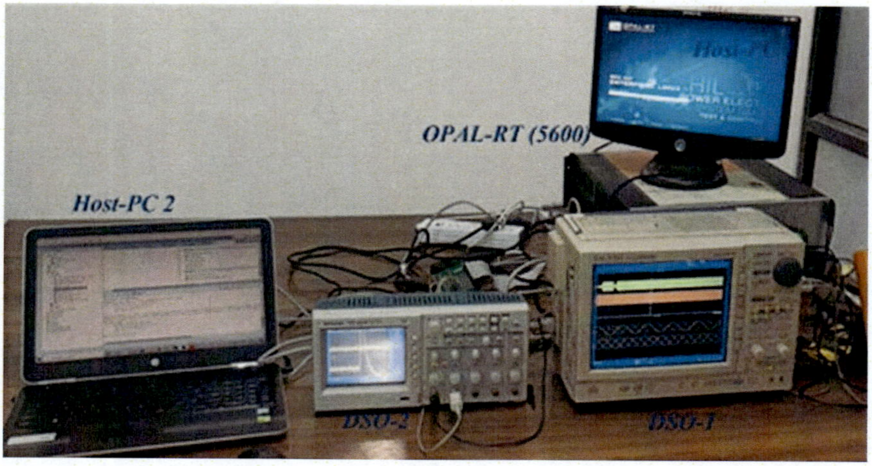

Fig. 12 Real time OPAL-RT experiment setup

4.3 Experimental Validation of Grid Tied PV System

The grid-connected three phase solar-PV generating unit is validated on the OPAL-RT-5600 real time-digital simulator which operates with fast Xilinx Artix-7 FPGA platforms. The obtained results are equivalent to hardware results. Figure 12 show real time opal-RT setup. Figure 13 shows the three phase PCC voltage. Figure 14 presents various waveforms with P&O-based MPPT and Fig. 15 presents various waveforms with IC-based MPPT.

5 Conclusion

The P&O and IC-based MPPT techniques are effectively improving the steady state and dynamic performance of the PV system. It is examined that the MPPT system tracks the MPP quickly and effectively regardless of fluctuations even with sudden and rapid change in the external environment. It is evident from experimental results that with IC method, in comparison to P&O, more power can be harnessed yet perturb observation method is simple in design and is not complicated like incremental conductance. However, both methods are superior in tracking MPP, especially under fast varying conditions.

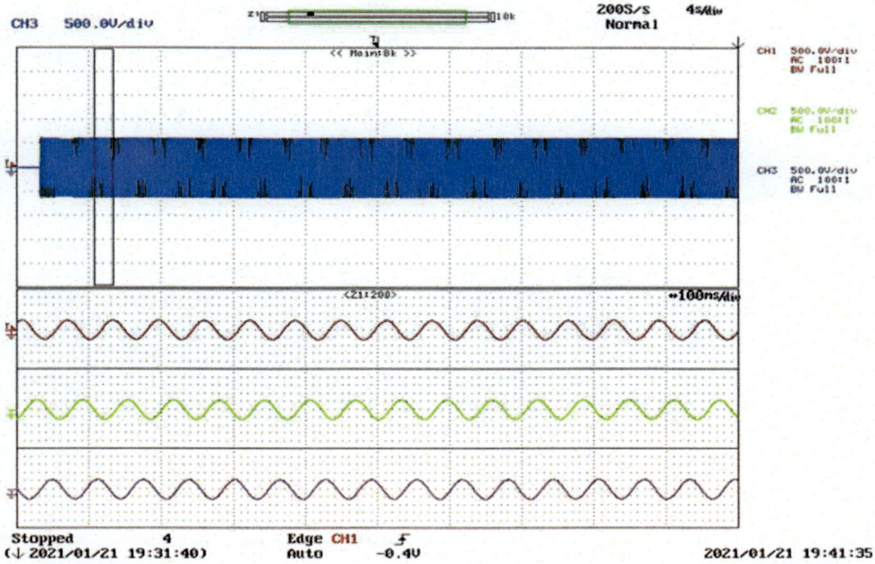

Fig. 13 Real time three phase PCC voltage waveform

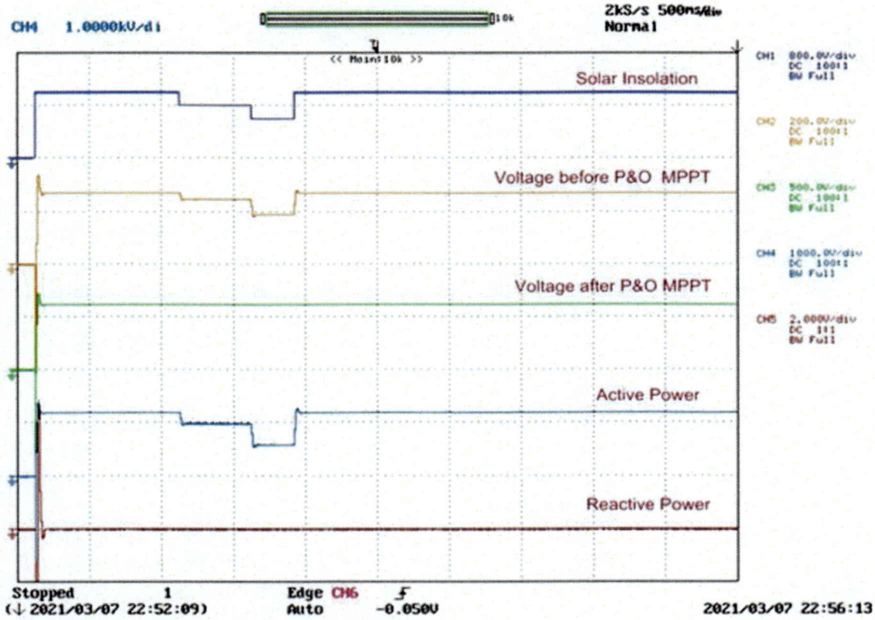

Fig. 14 Real time waveforms with P&O MPPT

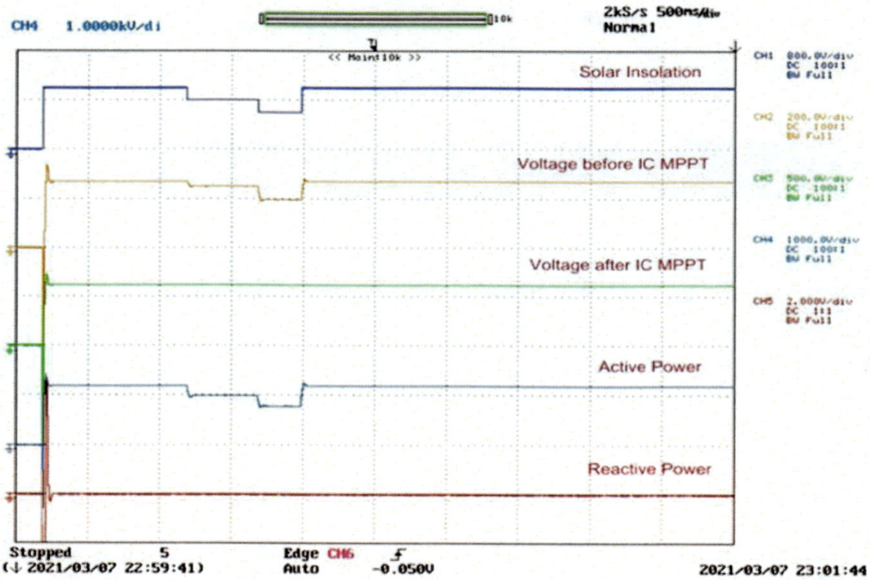

Fig. 15 Real time waveforms with IC MPPT

References

1. Verma D, Nema S, Shandhiya AM, Dash SK (2016) Maximum power point tracking techniques: recaptulation in solar photovoltaic systems. J Renew Energy Rev 54:1018–1034
2. Nema RK, Nema S, Agoihotri G (2009) Computer Simulation based study of photovoltaic cells/Modules and their experimental verification. Int J Recent Trends Eng 1(3)
3. Nema RK, Nema S, Agnihotri G (2010) Design, development and simulation of PC-based scheme for characterization of solar photo voltaic modules. Int J Power Electron 2(3)
4. Nema S, Nema RK, Agnihotri G (2011) Inverter topologies and control structure in photo voltaic applications: a review. J Renew Sustain Energy
5. Verma D, Nema S, Shandilya AM (2016) A different approach to design non-isolated. DC–DC converters for maximum power point tracking in solar photovoltaic systems. J Circ Syst Comput 25(8)
6. Nema S, Nema RK, Agnihotri G, Ramani, Krishnan A (2010) Matlab/Simulink based study of photo voltaic cells/modules/array and their experimental verification, vol 1, issue 3, pp 487–500
7. Nguyen XH, Nguyen MP (2015) Modulation and low mathematical modeling of photovoltaic cell/module/arrays with tags in Matlab/Simulink. Environ Syst Res
8. Anurag C, Saini RP (2014) A review on integrated renewable energy system based power generation for stand-alone applications: configurations, storage options sizing methodologies and control. Renew Sustain Energy Rev 38:120–199
9. Ray PK, Das SR, Mohanty A (2019) Fuzzy controller designed PV based custom power device for power quality enhancement. IEEE Trans Energy Convers 34(1):405–414
10. Ishaque K, Zainal S (2013) A deterministic particles warm optimization maximum power point tracker for photovoltaic system under partial shading condition. IEEE Trans Ind Electron 60(8):3195–3206
11. Kamarzaman NA, Tan CW (2014) A comprehensive review of maximum power point tracking algorithms for photovoltaic systems. Renew Sustain Energy Rev 37:585–598

12. Zhang W, Mao P, Chan X (2016) A review of maximum power point tracking methods for photovoltaic system. In: 2016 IEEE International conference on sustainable energy technologies (ICSET). Hanoi, pp 230–234. https://doi.org/10.1109/ICSET.2016.7811787
13. Gupta NP, Paliwal P (2021) Regulation of hybrid micro grid under transient operations. Int J Power Electron 13(1):83–111. https://doi.org/10.1504/IJPELEC.2021.10030583
14. Gupta NP, Paliwal P (2021) Novel droop integrated technique for regulation of islanded and grid connected hybrid microgrid. Int J Power Energy Convers 12(2):89–114. https://doi.org/10.1504/IJPEC.2021.114483
15. Faldu KS, Kulkarni PS (2020) Maximization of the output power from photovoltaic array under partial shading conditions. In: IEEE International students' conference on electrical, electronics and computer science (SCEECS), pp 1–6. https://doi.org/10.1109/SCEECS48394.2020.136

Islanding Detection of Integrated DG with Phase Angle Between Voltage and Current

B. Srikanth Goud, Ch. Rami Reddy, M. Kondalu, B. Nagi Reddy, G. Srinivasa Rao, and Ch. Naga Sai Kalyan

Abstract The integration of renewable energy systems is enhancing in daily life for supplying the global demand of electric energy. The concerned problem with integration of such distributed generation (DG) is islanding. The islanding may damage the consumers and equipment. As per the IEEE 1547 DG integration specifications, it must be identified in 2 s. In this article, a novel passive recognition approach occupying on the rate of change of phase angle between positive sequence voltage and current is (RCPABPSVAC) proposed. The existing passive methods failed to detect balanced and low-power mismatch islanding cases. The suggested approach can do it and strongly classifies the non-islanding situations with the islanding situations. The simulations are implemented on MATLAB/Simulink environment.

Keywords Distributed generation · Smart grid · Phase angle between voltage and current · Islanding detection · Non-islanding cases

B. Srikanth Goud (✉)
Electrical and Electronics Engineering, Anurag University, Venkatapur, Ghatkesar, Medchal, Telangana 500088, India
e-mail: Srikanth.b@anuraghyd.ac.in

Ch. Rami Reddy · M. Kondalu
Electrical and Electronics Engineering, Malla Reddy Engineering College, Maisammaguda, Secunderabad, Telangana 500100, India

B. Nagi Reddy
Electrical and Electronics Engineering, Vignana Bharathi Institute of Technology, Hyderabad, Ghatkesar, Telangana, India

G. Srinivasa Rao
Electrical and Electronics Engineering, CMR College of Engineering and Technology, Hyderabad, Telangana, India

Ch. Naga Sai Kalyan
Electrical and Electronics Engineering, Vasireddy Venkatadri Institute of Technology, Guntur, AP, India

K. Namrata et al. (eds.), *Smart Energy and Advancement in Power Technologies*, Lecture Notes in Electrical Engineering 926,
https://doi.org/10.1007/978-981-19-4971-5_21

1 Introduction

The regular use of non-renewable systems causes climate changes and is decreasing. Hence, the universe considers renewable energy sources (RES). The RES connected into the utility at the consumer stage is termed as DG. They produce islanding in the distribution network. Although a segment of a network is separated against the grid network and extended to production to regional load integrated is termed as electric islanding [1–5]. The theory of formation of islanding in the DG associated grid is depicted in Fig. 1. When the circuit breaker (CB) is isolated, the DG supplies to the load associated in regional area. In power network, the islanding is two categories, they are unintentional and intentional islanding [6–8]. It may cause due to breakdown in the system, intended disconnection based on CB for regular repair and an effect of description. From the DG linkage requirements, the islanding must be recognized in 2 s immediately islanding with its apparatus [9–15]. The islanding is hazardous to humans and apparatus associated to the power system, and it produces asynchronous association with utility voltage and frequency deviate their standards. The islanding recognition approaches are categorized as remote and local islanding recognition approaches [16–21]. The local approaches recommend the local signals at point of common coupling (PCC) for situation detection [15, 22–24]. The remote approaches are communication sharing techniques, after getting the information from pair of grid and user side they detect the islanding [25–27].

The regional islanding recognition approaches later categorized as passive, hybrid, and islanding recognition techniques. They are facing with high-non-detection zone (NDZ) related to hybrid and active approaches. The area of parameters where any recognition technique declines to recognize it is termed as NDZ. As a result of perturbance injections, the active strategies and hybrid strategies have consequences on reducing the quality in the power supplied [28–30]. In this article, a new passive recognition technique is suggested based on RCPABPSVAC. This method recognizes the islanding when the angle among positive sequence voltage and current is more than the threshold value.

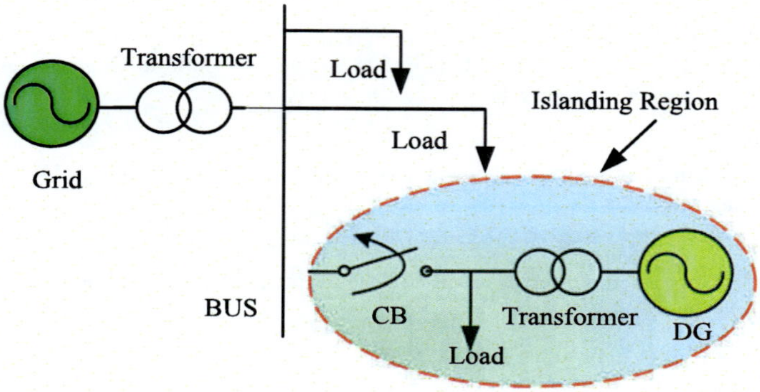

Fig. 1 Structure islanding in power network

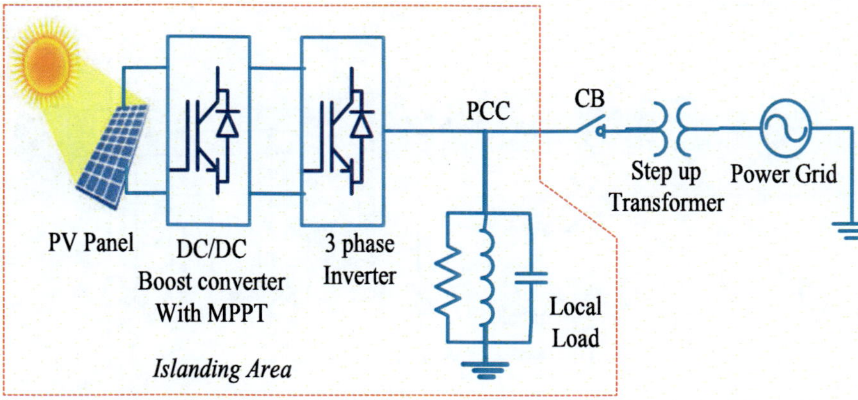

Fig. 2 Block diagram of the test system

2 Test System

A 100 KW solar system is integrated with a 120 kV main network through a DC-to-DC converter, voltage source inverter, filters, CB, transmission lines transformer as shown in Fig. 2. The solar system has 330 solar panels which has 66 strings of 5 series associated strings are placed in parallel to generate one kilo watt power (66 * 5 * 305.2 W = 100.7 KW). Each panel has an open circuit voltage of 64.2 V and a short circuit current of 5.96 A at 1000 W/m^2 solar irradiance and 25 °C ambient heat. The DC/DC structure increases the output voltage of solar panels by using MPPT techniques. This structure uses incremental conductance and integral regulator approach for the same. This boost converter steps the output voltage to 500 V and gives it as input to the inverter.

The inverter increases the 500 V DC voltage into 260 V AC voltage at the output of the inverter. After sending it through filters, the voltage and current harmonics are eliminated. The ripple free voltage is stepped up and integrated to the 25 kV feeder and 120 kV grid. The control circuit for integration is shown in Fig. 3.

3 Proposed Islanding Recognition Approach

Passive islanding recognition approach detects the islanding based on the recognition of changes in the passive signals [9]. The process of proposed approach is depicted in Fig. 4. It finds the islanding when change in passive signal RCPABPSVAC is above the setting value.

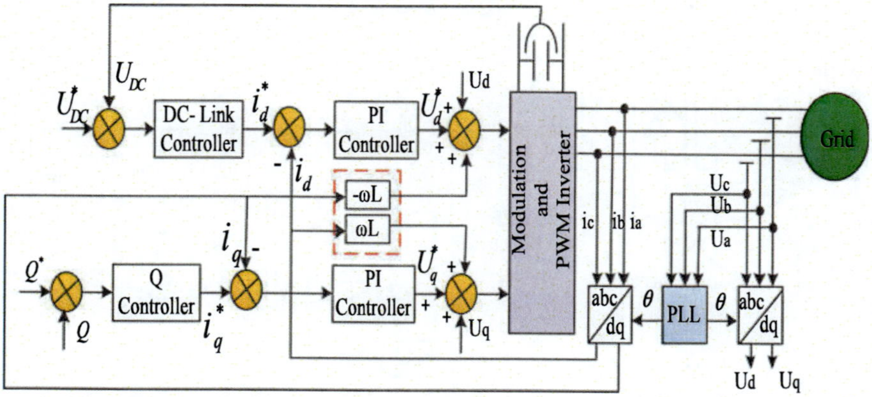

Fig. 3 Control circuit for grid integration of PV system

Fig. 4 Flowchart of proposed method

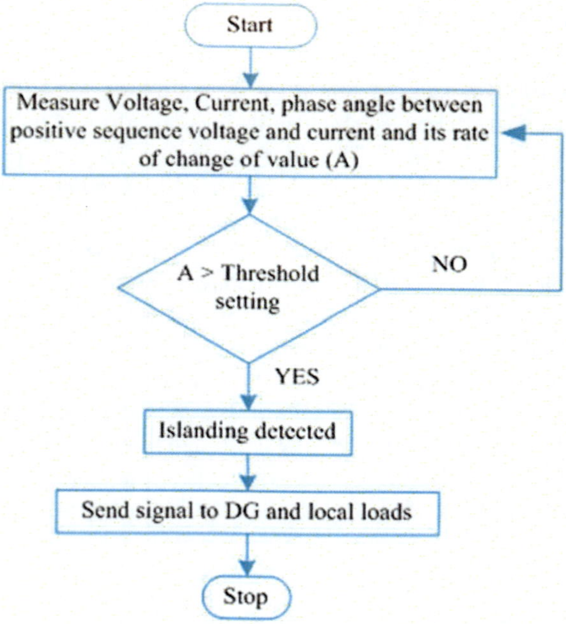

4 Simulation Results

The simulation results of the proposed method are implemented on MATLB/Simulink 2018b environment. The computer simulations are carried on the test system shown on Fig. 2, with proposed islanding approach for various power mismatch cases. Figures 5, 6, 7 and 8 indicates the simulation results for islanding cases when load is 100 kW, 70 kW, 50 kW, and 120 kW, respectively. At $t = 0.5$ s, the islanding is

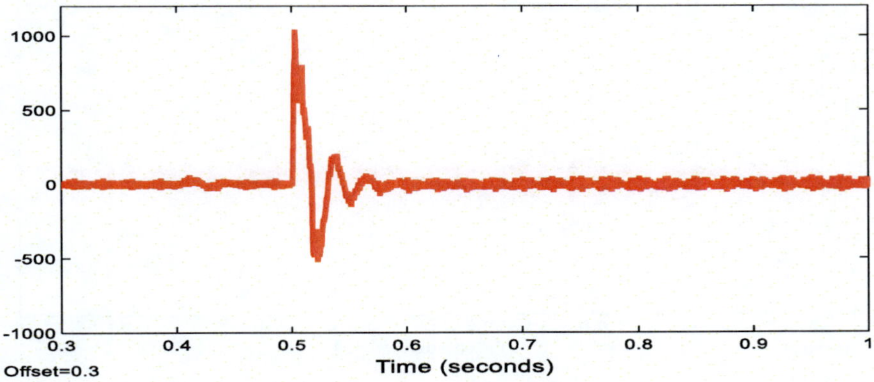

Fig. 5 RCPABPSVAC for 100 kW load

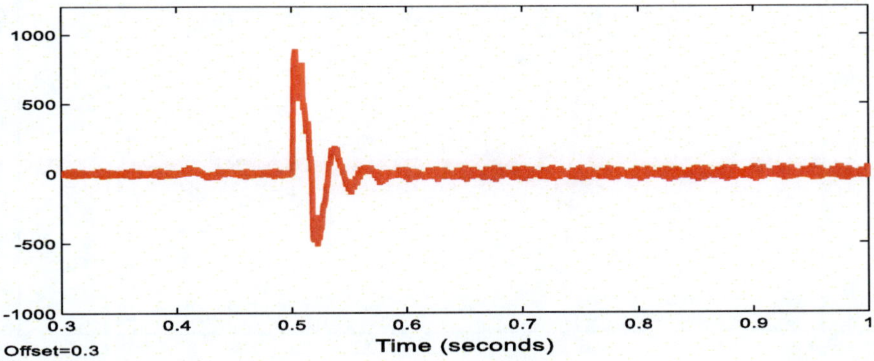

Fig. 6 RCPABPSVAC for 70 kW load

produced in the system. Figures 9 and 10 indicates the simulations for non-islanding cases. In these two cases, the switching is done at $t = 0.4$ s and 0.8 s, respectively. From islanding and non-islanding cases, it is clearly observed that the oscillations in the RCPABPSVAC are clearly discriminating. Hence by fixing a suitable level set value to the suggested technique, the islanding can be confirmed hardly in less than 20 ms.

5 Conclusion

This article projects a novel islanding recognition approach based on the changes in phase angle among positive sequence voltage and current signals. The suggested approach detects the islanding in lower than 20 ms which is very small compared to the IEEE 1547 DG integration standards. The suggested method recognized the

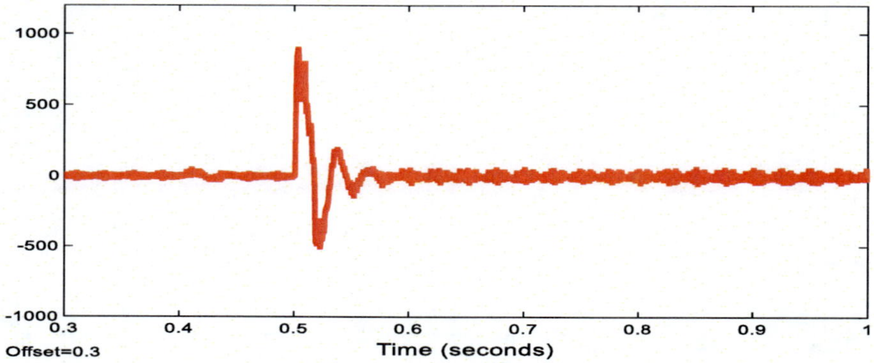

Fig. 7 RCPABPSVAC for 50 kW load

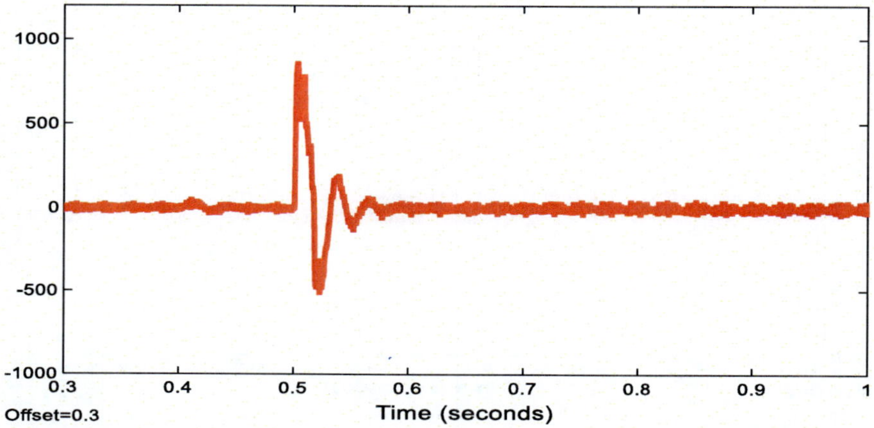

Fig. 8 RCPABPSVAC for 120 kW load

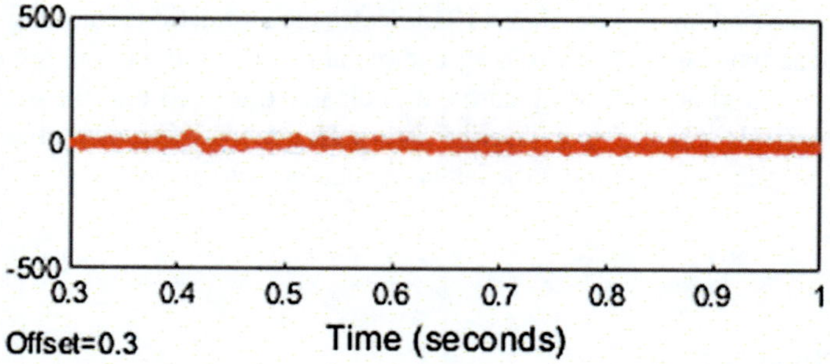

Fig. 9 RCPABPSVAC for capacitor switching

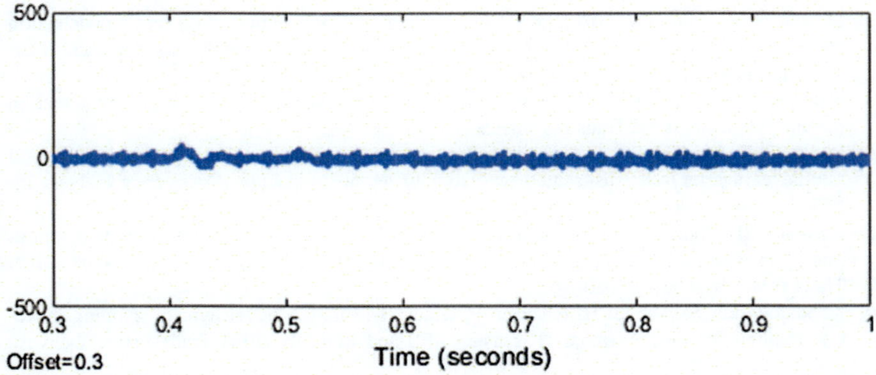

Fig. 10 RCPABPSVAC for 50HP induction motor switching

balanced islanding, near power mismatch islanding situations and non-islanding situations very exactly. Hence, the suggested technique is very speed and accurate in islanding detection process.

References

1. Goud BS, Reddy CR (2020) Essentials for grid integration of hybrid renewable energy systems: a brief review. Int J Renew Energy Res (IJRER) 10(2):813–830
2. Bekhradian R, Davarpanah M, Sanaye Pasand M (2019) Novel approach for secure islanding detection in synchronous generator based microgrids. IEEE Trans Power Deliv 34(2):457–466
3. Cui Q, El-Arroudi K, Joós G (2018) Islanding detection of hybrid distributed generation under reduced non-detection zone. IEEE Trans Smart Grid 9(5):5027–5037
4. Ch RR, Reddy KH (2019). Islanding detection techniques for grid integrated DG—A review. Int J Renew Energy Res (IJRER) 9(2):960–977
5. Raza S, Mokhlis H, Arof H, Laghari JA, Mohamad H (2016) A sensitivity analysis of different power system parameters on islanding detection. IEEE Trans Sustain Energy 7(2):461–470
6. Reigosa DD, Briz F, Blanco Charro C, Guerrero JM (2017) Passive islanding detection using inverter nonlinear effects. In: IEEE Trans Power Electron 32(11):8434–8445
7. Marchesan G, Muraro MR, Cardoso G, Mariotto L, de Morais AP (2016) Passive method for distributed-generation island detection based on oscillation frequency. IEEE Trans Power Delivery 31(1):138–146
8. Salles D, Freitas W, Vieira JCM, Venkatesh B (2015) A practical method for non detection zone estimation of passive anti-islanding schemes applied to synchronous distributed generators. IEEE Trans Power Delivery 30(5):2066–2076
9. Saleh SA, Aljankawey AS, Meng R, Meng J, Chang L, Diduch CP (2016) Apparent power-based anti-islanding protection for distributed cogeneration systems. IEEE Trans Ind Appl 52(1):83–98
10. Wen B, Boroyevich D, Burgos R, Shen Z, Mattavelli P (2016) Impedance-based analysis of active frequency drift islanding detection for grid-tied inverter system. IEEE Trans Ind Appl 52(1):332–341
11. Murugesan S, Murali V (2020) Disturbance injection based decentralized identification of accidental islanding. IEEE Trans Ind Electron 67(5):3767–3775

12. Sivadas D, Vasudevan K (2020) An active islanding detection strategy with zero non detection zone for operation in single and multiple inverter mode using GPS synchronized pattern. IEEE Trans Ind Electron 67(7):5554–5564
13. Murugesan S, Murali V (2019) Hybrid analyzing technique based active islanding detection for multiple DGs. IEEE Trans Ind Inf 15(3):1311–1320
14. Reddy C, Reddy KH (2018) A passive islanding detection method for neutral point clamped multilevel inverter based distributed generation using rate of change of frequency analysis. Int J Electr Comput Eng (2088-8708) 8(4)
15. Kermany SD, Joorabian M, Deilami S, Masoum MAS (2017) Hybrid islanding detection in microgrid with multiple connection points to smart grids using fuzzy-neural network. IEEE Trans Power Syst 32(4):2640–2651
16. Khodaparastan M, Vahedi H, Khazaeli F, Oraee H (2017) A novel hybrid islanding detection method for inverter-based DGs using SFS and ROCOF. IEEE Trans Power Delivery 32(5):2162–2170
17. Chen X, Li Y, Crossley P (2019) A novel hybrid islanding detection method for grid-connected microgrids with multiple inverter-based distributed generators based on adaptive reactive power disturbance and passive criteria. IEEE Trans Power Electron 34(9):9342–9356
18. Xu W, Zhang G, Li C, Wang W, Wang G, Kliber J (2007) A power line signaling based technique for anti-islanding protection of distributed generators—Part I: scheme and analysis. IEEE Trans Power Delivery 22(3):1758–1766
19. Mohanty SR, Kishor N, Ray PK, Catalo JPS (2015) Comparative study of advanced signal processing techniques for islanding detection in a hybrid distributed generation system. IEEE Trans Sustain Energy 6(1):122–131
20. Ray PK, Kishor N, Mohanty SR (2012) Islanding and Power quality disturbance detection in grid-connected hybrid power system using wavelet and S-transform. IEEE Trans Smart Grid 3(3):1082–1094
21. Do HT, Zhang X, Nguyen NV, Li SS, Chu TT (2016) Passive-islanding detection method using the wavelet packet transform in grid-connected photovoltaic systems. IEEE Trans Power Electron 31(10):6955–6967
22. Khamis A, Xu Y, Dong ZY, Zhang R (2018) Faster Detection of microgrid islanding events using an adaptive ensemble classifier. IEEE Transactions on Smart Grid 9(3):1889–1899
23. Manikonda SKG, Gaonkar DN (2019) IDM based on image classification with CNN. J Eng 2019(10):7256–7262
24. Ch RR, Reddy KH (2018) An efficient passive islanding detection method for integrated DG system with zero NDZ. Int J Renew Energy Res (IJRER) 8(4):1994–2002
25. Reddy CR, Reddy KH (2018) Islanding detection for inverter based distributed generation with Low frequency current harmonic injection through Q controller and ROCOF analysis. J Electr Syst 14(2):179–191
26. Reddy CR, Reddy KH (2019) Passive islanding detection technique for integrated distributed generation at zero power balanced islanding. Int J Integr Eng 11(6):126–137
27. Reddy CR, Reddy KH (2019) A new passive islanding detection technique for integrated distributed generation system using rate of change of regulator voltage over reactive power at balanced islanding. J Electr Eng Technol 14(2):527–534
28. Reddy JR, Pandian A, Reddy CR (2020) An efficient learning based RFMFA technique for islanding detection scheme in distributed generation systems. Appl Soft Comput 106638
29. Goud BS, Reddy BN, Pratyusha M, Kumar CV, Rekha R (2020) Review of islanding detection parameters in smart grids. In: 2020 8th International conference on smart grid (icSmartGrid). IEEE, pp 78–89
30. Reddy CR, Reddy KH. NDZ analysis of various passive islanding detection methods for integrated dg system over balanced islanding

An ANN Model for Load Performance Evaluation of a IEEE 9-Bus Radial Microgrid Distribution Feeder

Yuvraj Praveen Soni and E. Fernandez

Abstract ANN models offer convenient direct solutions to distribution feeder load flow performance studies. They have an advantage in providing direct results, thereby avoiding problems of convergence that may be encountered using other iterative load flow techniques. This paper develops an ANN model for a radial distribution feeder and demonstrates the model's suitability in total system loss and maximum voltage regulation computation. The trained network is utilized to study the effects on voltage regulation and total system loss corresponding to individual changes in active power loading, reactive power loading, line resistance, and reactance as measured from the generation bus. The network's inputs are total active load, reactive load, line resistance, and reactance. These parameters are varied from 50 to 150% of its base value, and output represents maximum voltage regulation and total system loss corresponding to each variation. The obtained ANN is proven to be a suitable tool for performance evaluation. Also, it has been seen that active power loading and resistance have a significantly higher effect on voltage regulation and system loss as compared to reactive power loading and line reactance.

Keywords Radial distribution system · Microgrid · Performance · Voltage regulation · System losses · ANN

1 Introduction

The microgrid can be seen as a breakdown of large and complex power system structure into a locally installed system that contains distributed generators locally to supply power to its connected loads. Planning a distribution system is tedious, requiring rigorous analysis of extensive data incorporating complex mathematical equations and iterative techniques that sometimes fail to converge. This brings out the need for an efficient tool that can be utilized to overcome the complex analysis

Y. P. Soni (✉) · E. Fernandez
Indian Institute of Technology Roorkee, Roorkee, Uttarakhand, India
e-mail: yuvraj.soni21092@gmail.com

K. Namrata et al. (eds.), *Smart Energy and Advancement in Power Technologies*, Lecture Notes in Electrical Engineering 926,
https://doi.org/10.1007/978-981-19-4971-5_22

providing a reliable solution with efficient time and without convergence failure. Numerous literature are available, which discusses several techniques to evaluate the radial distribution system's performance parameter.

Generally, the load flow technique prioritizes any planner to evaluate the voltage magnitude at every bus and the total system loss occurring in the distribution system. Bompard et al. [1] and González-Morán et al. [2] discusses the backward and forward sweep (BFS) method, which is the most commonly used load flow technique suitable for the distribution system. However, the BFS method needs to be developed for its computational time to solve complex mathematical equations. Augugliaro et al. [3] and Kongjeen et al. [4] suggested a modified BFS algorithm that improves the computational time to compute bus voltage magnitude of the radial distribution system. Even the modified BFS techniques failed to converge due to non-positive sequence impedance, to which Ju et al. [5] discusses its feasible solution.

Looking onto computational time and convergence failure in traditional techniques, research is directed toward artificial intelligence implementation for solving the problem related to the distribution system. Neural network is a well-developed technique applied for broader applications such as proposed by Cho et al. [6], Heo et al. [7] and Himmelblau et al. [8] for fault detection. Kashem et al. [9] and Oh et al. [10] proposed network reconfiguration for the loss minimization. Muller et al. [11] proposed an artificial neural network (ANN) for solving load flow in IEEE 30-bus, 57-bus, and 11-bus system. Du et al. [12], Bai et al. [13] and Singh and Srivastava [14] discuss the potential of neural networks in evaluating the network contingency.

ANN has proved its potential in the distribution system for voltage computation which is generally computed through traditional techniques. Further, maximum voltage regulation and total system loss are calculated by determining the worst bus (corresponding to lowest voltage magnitude) and the sum of all losses occurring in the system. In this paper, ANN is developed for IEEE 9-bus radial distribution system for performance evaluation considering maximum voltage regulation and total system loss computation. To train the model, we need practical load data variations. However, because of the unavailability of this, we make use of synthetic data that has been generated using the forward–backward sweep method by varying the input parameters. System performance is further investigated for changing active, reactive loads, line resistance, and reactance from 50 to 200% of its base value seen from its generation bus. The paper is organized as follows—Section 2 discusses the performance parameter traditional computation. Section 3 discusses the development of the ANN network, followed by results and discussion in Sect. 4. The paper conclusion is discussed in Sect. 5.

2 Voltage Regulation and System Loss Computation

The two most crucial parameters required to be monitored are voltage at every bus and total system loss. These parameters are generally calculated through iterative load flow techniques [1–3]. Through load flow, voltage and current at every node and branch are known to be further utilized for voltage regulation and branch power dissipation calculation.

2.1 *Maximum Voltage Regulation*

Voltage regulation is a measure of change in voltage magnitude (compared to generating bus voltage magnitude) at the specified bus with any parameter variation in the system. The bus voltages are computed through BFS load flow and Eq. (1) to determine voltage regulation for every node.

$$\mathrm{VR}_i = \frac{V_1 - V_i}{V_i} \tag{1}$$

for $i = 1, 2, \ldots, N$
where V_1 corresponds to voltage magnitude of bus to which power source is connected which is usually considered as 1 pu. N is total number of buses in the distribution system.

For any system, identifying the stable or unstable operation, targeting the worst bus, indicates that the remaining other buses are comparatively more stable than the worst bus, giving the instant idea about the system performance. A conventional method requires additional steps to determine the worst bus, and then its voltage regulation is calculated through Eq. (2). The maximum voltage regulation indicates the worst voltage bus whose magnitude is greatly suffered (voltage dip) due to variation in the parameter in the distribution system.

$$\mathrm{VR}_{\max} = \max(\mathrm{VR}_i) \tag{2}$$

2.2 *Total System Loss*

The total system loss adds all losses or power that are not billed through the generation company. These are dissipated in line resistance during power supply which depends upon the current magnitude flowing through the specified branch. The conventional way of loss calculation is given by Eq. (3), through which total system loss can be computed as shown in Eq. (4).

$$\mathrm{Loss}_i = I_i^2 * r_i \tag{3}$$

$$\mathrm{Loss}_{\mathrm{system}} = \sum_{i=1}^{M} \mathrm{Loss}_i \tag{4}$$

where I_i is ith branch current and r represents line resistance. M is the total number of branches in the distribution system.

3 Development of ANN

ANN is referred from the human brain made up of millions of neurons to achieve the targeted outcome. The ANN neurons are connected with synaptic weights, which carry the information from input to output layer, giving approximate results for the input applied. An ANN network is developed in this work to determine voltage regulation corresponding to worst bus and total system loss for the IEEE 9-bus radial distribution system. The line and load data for IEEE 9-bus radial distribution system is referred from Ref. [15]. To train the model, we need practical load data variations. However, because of the unavailability of this, we make use of synthetic data that has been generated using the forward–backward sweep method by varying the input parameters.

The developed ANN network, as shown in Fig. 1, consists of three layers: (1) input layer, (2) hidden layer, and (3) output layer. The input and output layers have the neurons equal to the number of input and output. The developed model consists of 36 input neurons that involve nine active power loading measurements, nine reactive power loading measurements, and nine line resistance and reactance parameters. Two output neurons at the output layer give the maximum voltage regulation and total system loss measurement. The hidden layer can have many layers; however, determining its layer size is a considerable task done by observing error obtained

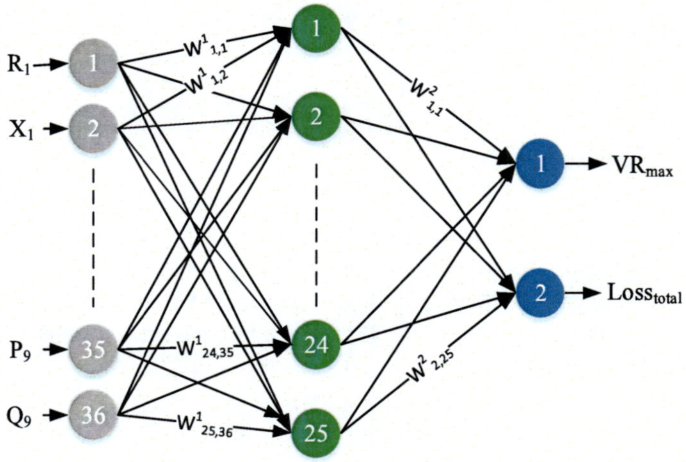

Fig. 1 Schematic representation of developed ANN

between output and targeted value with different layer numbers. Finally, the best structure is chosen.

In the developed ANN architecture, one layer of a hidden layer with 25 neurons is considered for the study as it produces the nearest approximate value to its targeted output. The training is compiled through Levenberg–Marquardt (LM) algorithm, and the mean squared error is used to decide the layer criterion. Each neuron from the first layer is connected with neurons in the second, and second layers are connected with the third layer with synaptic weights. $W^{\alpha}_{\beta,\gamma}$ shows weights between two neurons where α represent layer number, β represents neuron position at $(i+1)$th layer, and γ represents neuron position at ith layer.

Figures 2 and 3 represent performance and error histogram of developed neural network. The designed network can evaluate the maximum voltage regulation and total system losses with less error, as shown in Fig. 3. The mean squared error of 14.1692 is obtained at epoch 24, which is considered the best-obtained result.

4 Results and Discussions

The trained neural network is found to be suitable for evaluating system performance. The network requires loading data and line data to calculate the desired result. Effect

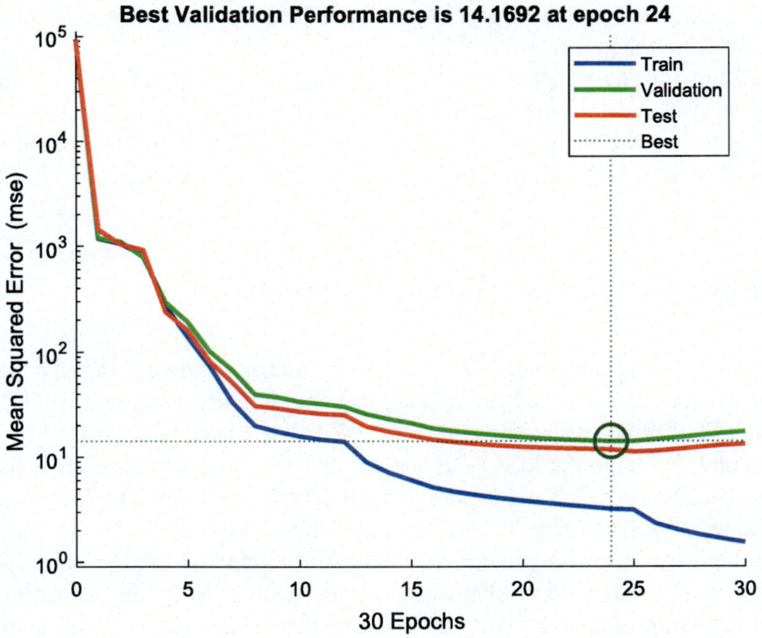

Fig. 2 Performance of developed ANN

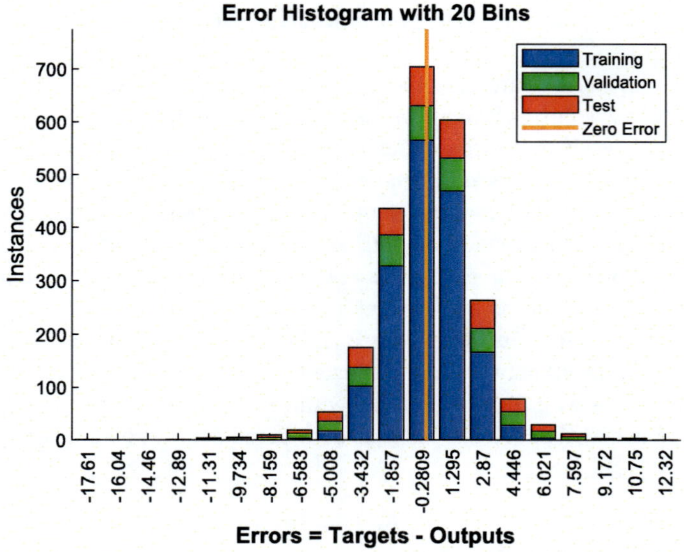

Fig. 3 Error histogram of developed ANN

on maximum voltage regulation and total system loss is observed with varying loading and line impedance values from 50 to 200% of its base value. All the inputs are measured from the generation bus, so any changes concern the total quantity calculated from the generation end.

The following study analyzes maximum voltage regulation and total system loss on varying system parameters, namely line resistance, reactance, and load active and reactive power. All these parameters are changed individually to observe the effect on given system performance through individual parameter variation.

4.1 Effect on Maximum Voltage Regulation

Figure 4 represents the changes in maximum voltage regulation through system parameter variation. At base loading and line condition, the maximum voltage regulation of IEEE 9-bus with generation at 23 kV at bus 1 is equal to 18.67%.

The total active power loading is varied, and the effect on maximum voltage regulation is observed. It is worth noting that the last bus (9th) of the system is found to be the worst bus, and thus, the voltage regulation is measured at the 9th bus. The relation between voltage regulation and active power loading is quite linear, which increases with an increase in total active power loading. At 200% of loading, the voltage regulation is approximately 50% means there is about a 50% dip in voltage magnitude. Similarly, an increase in total reactive power loading increases voltage regulation from 17.6 at 50% loading to about 23.02% at 200% loading.

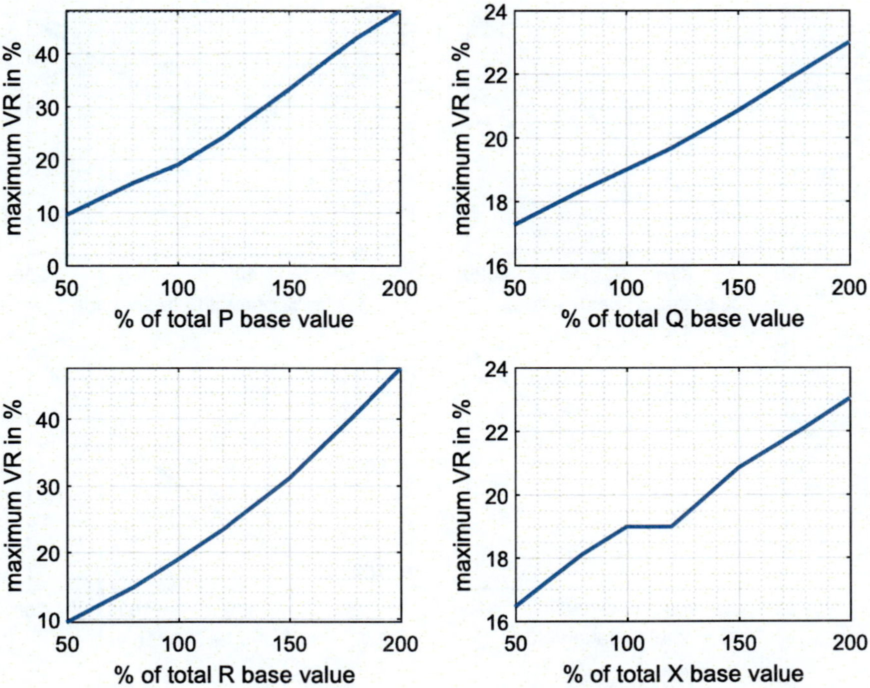

Fig. 4 Effect on maximum voltage regulation on IEEE 9-bus distribution system

The line resistance and reactance also affect voltage regulation, as evident from Fig. 4. Relatively active power and resistance have more effects as compared to reactive loading and line reactance. The voltage regulation value is also relatively closer for active load and line resistance variation and similarly for reactive loading and line reactance.

4.2 *Effect on Total System Loss*

Figure 5 represents the changes in total system loss through system parameter variation. At base loading and line condition, the total loss of IEEE 9-bus with generation at 23 kV at bus 1 is equal to 770 kW. Now the total active power loading is reduced to 50% of total load (at the base condition), and total system loss is reduced to 215.29 kW. When the active power loading increases, total system loss also increases, which is evident due to the current demand rise. Similarly, reactive power loading also linearly affects the entire system loss.

Effect on system loss on changing line resistance is a somewhat exponential curve that explains that above 170% of its base value, any increase in resistance

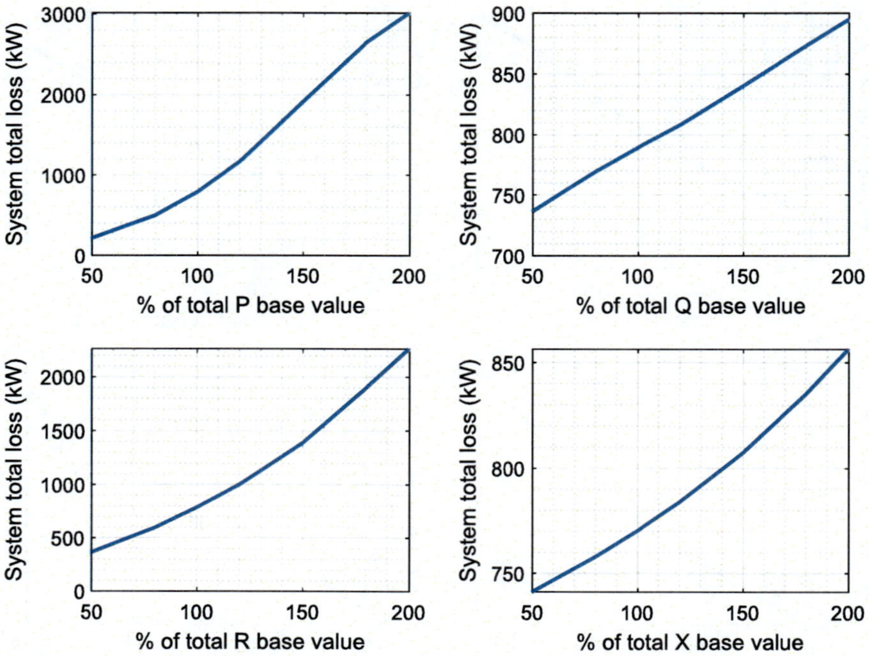

Fig. 5 Effect on total system loss on IEEE 9-bus distribution system

will significantly increase total system loss similar to line reactance. Identical to the earlier case, active power loading and line resistance have more effect than reactive power loading and line reactance.

5 Conclusion

The decentralized power structure is seen as an upcoming way of electrification installed in the distribution sector to supply power to consumers. In this paper, ANN is developed, which is proven to be a decent tool to be applied in the distribution system for performance evaluation. It eliminates the chances of convergence failure and slow computation, unlike conventional techniques. The obtained network is trained and found to produce results with a mean squared error of 14.1692, which is significantly low compared to the total load being served in the IEEE 9-bus radial distribution system. Further, the developed neural network is utilized to investigate maximum voltage regulation and total system loss with varying line and loading parameters individually from 50 to 200% of its base condition. It is observed that active power loading and resistance have more effects on voltage regulation and total system loss than reactive power loading and reactance. The developed ANN tool can evaluate the IEEE 9-bus radial distribution system, which gives almost the nearest values corresponding to any line or loading change in the system.

References

1. Bompard E, Carpaneto E, Chicco G, Napoli R (2000) Convergence of the backward/forward sweep method for the load-flow analysis of radial distribution systems. Int J Electr Power Energy Syst 22(7):521–530
2. González-Morán C, Arboleya P, Mohamed B (2017) Matrix backward forward sweep for unbalanced power flow in $\alpha\beta 0$ frame. Electric Power Syst Res 148:273–281
3. Augugliaro A, Dusonchet L, Favuzza S, Ippolito MG, Sanseverino ER (2008) A new backward/forward method for solving radial distribution networks with PV nodes. Electric Power Syst Res 78(3):330–336
4. Kongjeen Y, Bhumkittipich K, Mithulananthan N, Amiri IS, Yupapin P (2019) A modified backward and forward sweep method for microgrid load flow analysis under different electric vehicle load mathematical models. Electric Power Syst Res 168:46–54
5. Ju Y, Wu W, Zhang B (2015) Convergence problem in forward/backward sweep power flow method caused by non-positive-sequence impedance of distributed generators and its solution. Int J Electr Power Energy 65:463–466
6. Cho S, Choi M, Gao Z, Moan T (2021) Fault detection and diagnosis of a blade pitch system in a floating wind turbine based on Kalman filters and artificial neural networks. Renew Energy 169:1–13
7. Heo S, Lee JH (2018) Fault detection and classification using artificial neural networks. IFAC-PapersOnLine. 51(18):470–475
8. Himmelblau DM, Barker RW, Suewatanakul W (1992) Fault classification with the aid of artificial neural networks. In: IFAC symposia series, fault detection, supervision and safety for technical processes 1991, Pergamon, pp 541–545
9. Kashem MA, Jasmon GB, Mohamed A, Moghavvemi M (1998) Artificial neural network approach to network reconfiguration for loss minimization in distribution networks. Int J Electr Power Energy Syst 20(4):247–258
10. Oh SH, Yoon YT, Kim SW (2020) Online reconfiguration scheme of self-sufficient distribution network based on a reinforcement learning approach. Applied Energy 280:115900
11. Müller HH, Rider MJ, Castro CA (2010) Artificial neural networks for load flow and external equivalents studies. Electric Power Syst Res 80(9):1033–1041
12. Du Y, Li F, Li J, Zheng T (2019) Achieving 100× acceleration for N-1 contingency screening with uncertain scenarios using deep convolutional neural network. IEEE Trans Power Syst 34(4):3303–3305
13. Bai X, Tan J (2020) Contingency-based voltage stability monitoring via neural network with multi-level feature fusion. IFAC-PapersOnLine 53(2):13483–13488
14. Singh R, Srivastava L (2007) Line flow contingency selection and ranking using cascade neural network. Neurocomputing 70(16–18):2645–2650
15. Soni YP, Fernandez E (2020) Regression modelling of voltage profile in an IEEE 9-Bus radial microgrid feeder. In: 2020 IEEE students conference on engineering & systems (SCES). IEEE, Prayagraj, India, pp 1–5

Controlling of Solar Powered LED Street Lights Using Auto-intensity Control Mechanism

Nishant Kumar and Munna Kumar

Abstract The use of electrical energy plays an important role in day-to-day life. From household to industrial applications, it plays a severe role in operation. While the focus has been given in utilization of power from renewable sources of energy like solar, wind, etc. Thus, production of energy from these sources is quite limited, so there should be intension in saving of energy while it is not in use, i.e. efficient utilization of energy. On evolvement of microgrid, where the use of DGs is taking place, so set-up of small generation units is required, but it's also important to utilize the power in an efficient manner. The conservation of electrical energy is quite simple than giving more effort in producing more amount of energy as per need of the system. One of the major wastes of power can be controlled in proper operation of fan/lights in various places as per requirement. In this paper, a novel idea of controlling of LED-based street lights by monitoring the intensity of the light is being described. By identifying the intensity of the light proper switching of light can be done and thus saves wastage of electrical energy. First the modelling of solar cell is done followed by proper description of LDR, i.e. light detecting resistor, and controlled circuit using Arduino UNO is sensing the intensity and performing the switching accordingly.

Keywords Street lights · Auto-intensity control · LDR · Arduino UNO

1 Introduction

The use of renewable sources of energy is to a great extent nowadays. Every country is trying to meet their load demand by producing power from renewable sources of energy. The use of coal, petroleum, etc. can be easily done, but while seeing their production progress one can analyse that it will take lots of year to get renewed by

N. Kumar (✉) · M. Kumar
Department of Electrical Engineering, NIT Jamshedpur, Jamshedpur, Jharkhand, India
e-mail: krnishant125@gmail.com

M. Kumar
e-mail: 2019rsee004@nitjsr.ac.in

K. Namrata et al. (eds.), *Smart Energy and Advancement in Power Technologies*,
Lecture Notes in Electrical Engineering 926,
https://doi.org/10.1007/978-981-19-4971-5_23

nature [1, 2]. Thus, for next two decades there will be shortage of instantaneous production of power. Therefore, the best way is to meet the demand by generating power from R.E, i.e. renewable source of energy with an efficient working mechanism [3]. The evolvement of microgrid is common nowadays due to maintaining constant power supply to operate the critical loads. So, the use of DGs or the distributed generation units is quite important. The power obtained from these DGs is quite limited so it should be utilized efficiently. It is also observed that out of total power produced from R.E sources of energy, more than 80% gets produced from solar and wind. The installation of wind farm is quite difficult of place suitability as the places where wind speed is less or the place where wind speed is too high, and is difficult to place wind farm, but easily solar plant can be established due to falling of rays of sun on various places which is easily available.

In this paper, the efficient utilization of solar energy is presented. One of the applications of solar energy is for street lighting [4]. But it is important to switch on the LEDs when the intensity is less or falls below its critical value but if it is above the critical level, there is no need of switching it on. If this process is not managed efficiently, it will resemble loss in power or wastage of energy. This work is done with an aim of energy conservation with the help of auto-intensity control of LED based street lights, and it can easily replace the normal lighting strategy of the present scenario [5]. If the light is sufficient, i.e. in case of days, no lightning is required and the automatic sensing of light to put itself off conserves the energy, whereas in case of nights when there is dark, then sensing of street light and putting it into on state will make the system beneficial for humankind [6–8].

The sequential description of the system is done. First a brief description of solar PV cell is represented where the output current can be found out by using KCL or Kirchhoff's current law followed by the Simulink modelling of solar PV cell. After that, the small description of the lighting system is presented and how its intensity is used to change the switching state, i.e. ON and OFF states, performed is stated. All the parameters used are discussed briefly, and lastly the operation of how the entire process works to run the system automatically is being discussed. Thus, this lighting system can replace the current pattern of lighting system by saving the losses of energy.

2 Methodology

For sensing the intensity of the light, the work has been divided into certain modules for clear understanding. The main idea is to conserve the power of the LED in street lighting by operating it in an efficient manner. All these are analysed with respect to a photovoltaic system. The two main factors those are involved in determining the power from solar PV cell are temperature and solar irradiance. The use of sensor, i.e. light detecting resistor (LDR) to sense the light and accordingly perform the switching, is done so that the power is saved and making the system to work in an efficient manner.

2.1 Photovoltaic System

The photovoltaic system works on absorption of solar rays emitted by the sun. The light rays or the photons come and hit the solar photovoltaic panels and are being absorbed by the materials. Here, the semiconductor material used is Silicon that is basically doped. Two layers of Silicon are present. One with doped with a greater number of electrons, and other is doped with a smaller number of electrons known as n and p layers. Under normal condition, the barrier gets formed between the layers and the electrons cannot flow while the load is connected. The rays of the sun or the photon particles excite the electrons. And this helps to cross the junction, and the flow of current will take place in opposite to the flow of electrons. By this mechanism, the solar energy is converted to electrical form of energy. The mathematical expression of the load or output current can be determined from the equivalent circuit of solar PV cell [9]. As single diode is used in this case; this representation is known as dingle diode modelling of solar cell. In ideal circuit, just the drop of series and parallel resistances is neglected. The equivalent circuit of PV solar cell is represented in Fig. 1.

The output current (I) from the above figure of a single PV cell can be expressed as follows:

$$I = I_{\mathrm{L}} - I_{\mathrm{o}}\left[\exp\left(\frac{V + IR_{\mathrm{s}}}{\eta V_{\mathrm{t}}}\right) - 1\right] - \left[\frac{V + IR_{\mathrm{s}}}{R_{\mathrm{sh}}}\right] \tag{1}$$

where I_{L} represents the light-generated current in the cell, I_{o} is the diode saturation current, V is the output cell voltage, V_{t} is the thermal voltage, and R_{s} is the series resistance. R_{sh} is the shunt resistance, η is the diode ideality factor ranging between 1 and 2. Though the power of a single solar cell is not sufficient to meet the load demand so, a greater number of cells are connected in series–parallel fashion to supply as per the demand of the customer.

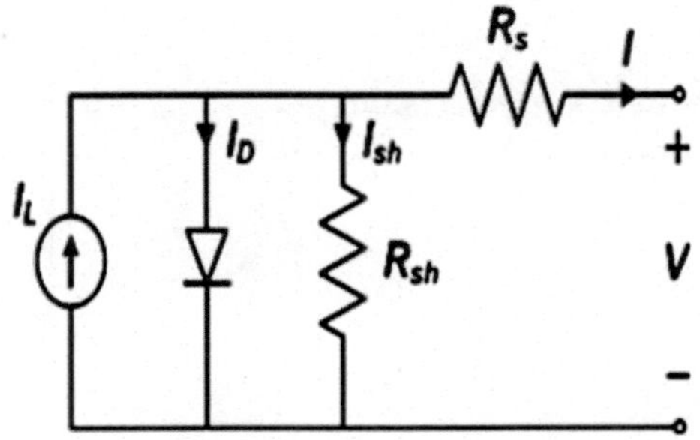

Fig. 1 Equivalent circuit of photovoltaic cell

2.2 Simulink Model

From the mathematical model, the Simulink model is formulated so that the behaviour of the solar cell can be analysed. The inputs of the model are basically temperature, solar irradiance and the supply voltage. The output is basically power, obtained by measuring the current and voltage value at the load terminal. With the increase in solar irradiance, it is observed that the power production increases when keeping temperature constant. And if the same is repeated with keeping irradiance constant and increase the temperature, then it is observed there is slight decrease in power. The model has been created in MATLAB, and solar PV cell is represented by Figs. 2 and 3, respectively [10].

In the above model, the voltage is not varied and the variable parameters are temperature and solar irradiance. These two parameters are varied, and the effect of power is seen as the ultimate aim is to see the variation or change in power. In an entire year, the solar radiation is less in amount for month December, whereas it reaches to the peak value in the month of May. So, it is observed that the power production gets varied throughout the year. It is highest in month of May, whereas lowest in month of December. In Fig. 2, it is seen that a "PV cell" is seen in the

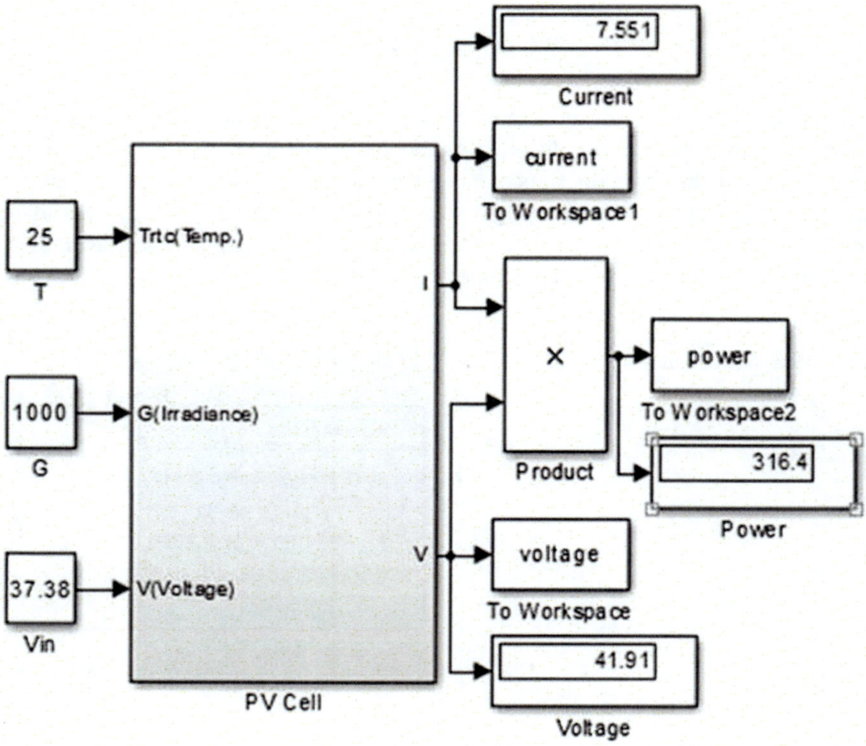

Fig. 2 Simulink modelling of solar PV cell

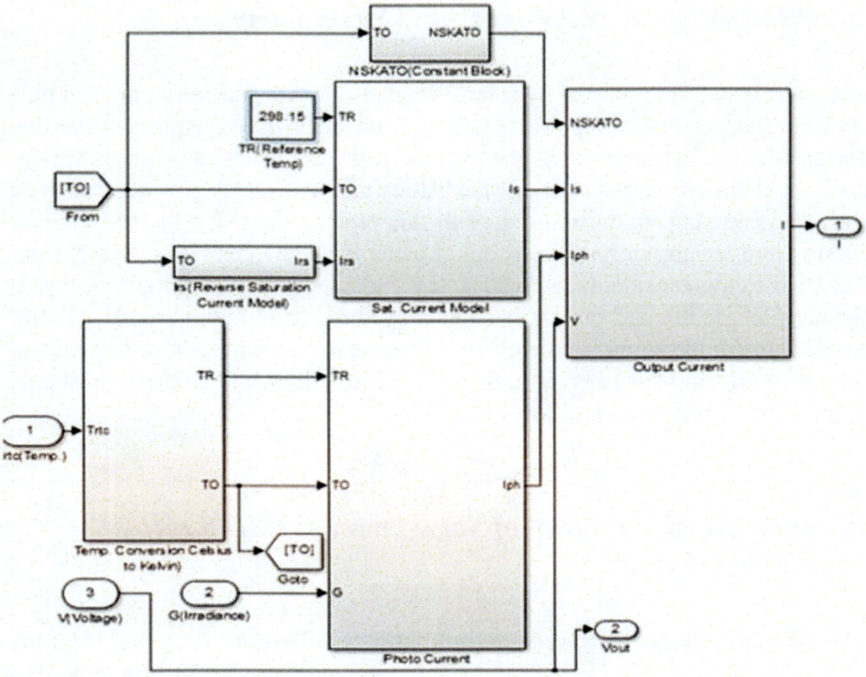

Fig. 3 Inner structure of PV cell block

simulation model. It is basically a subsystem, and the inner representation is shown in Fig. 3.

In Fig. 4, the variation of power is just represented to identify how the modelled cell behaved with respect to the ideal solar cell.

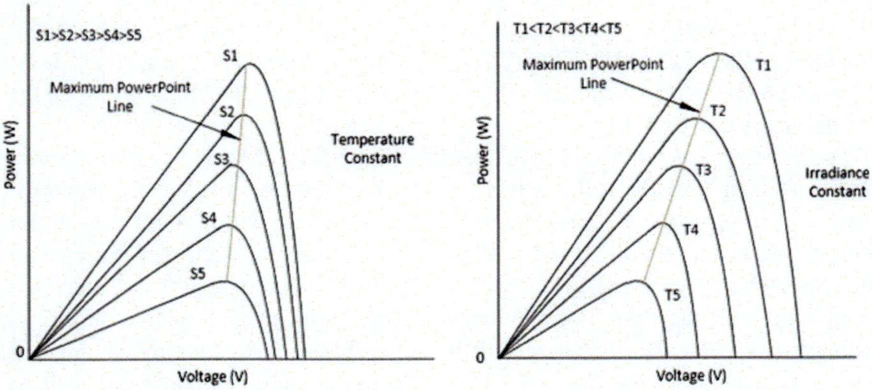

Fig. 4 Effect of change of power keeping temperature and solar irradiance constant respectively

2.3 Principle of Solar Powered LED Street Lamp

Solar street lamps are powered by solar PV panels and are generally mounted on a pole-like structure. The solar powered LED lights work on the principle of converting solar energy that is absorbed by photovoltaic cells into electrical form of energy. This form of energy is used to charge the battery that will supply power to the street lights. The components those are used in this process are PV cells, LED lamps, pole structure, controller box consisting of battery and charger. Nowadays, the use of LEDs is to a great extent due to advantages like low power requirement, higher efficiency, longer life and can easily be employed for street light purpose [11, 12]. The operation of the lamps should be like when the intensity of light in surrounding is less than critical value then only the LED lamp will turn ON otherwise should remain in OFF position to save loss of energy.

2.4 Auto-intensity Control of Solar Powered LED Street Lighting System

The basic block diagram of auto-intensity control of solar street light [13–15] using Arduino is represented in Fig. 5. The LDR or the light detecting resistor is used as a sensor that will sense the intensity of the light up to a certain level and give input signal to the Arduino. The internal working of Arduino helps to connect the external circuit consisting of street light to attach at the output port for proper completion of circuit. The programming of the Arduino is done with respect to the ambient light. The working of the sensor plays a vital role in determining or sensing the light that is desired only. The operation can be seen from the flow chart where it is noted that depending up on the intensity of the light, the sensor sends the signal to operate the pins. If the light is less than desired amount, then the sensor sends the signal to increase the light by switching the pins ON in which LEDs are connected. Thus, the light will get enhanced. And if it is observed that there is presence of adequate light, then the signals are sent to OFF in which LEDs are connected to the pins so that none is ON as adequate light is present. Thus, by adopting proper switching sequence the loss of energy can be done.

There are various blocks combined together to perform the task accurately. From Fig. 5, it is seen that solar cells are there for using solar energy into electrical energy. There is also presence of charge controller circuit which will monitor the current flowing within the circuit. If the voltage is 12 V, then the controller circuit will allow flow of current to charge the battery and when the voltage rises to 13.5 V, the controller circuit stops the flow of current to avoid high current flow into the circuit and destroy it. The use of rechargeable battery is basically to store the charge and getting discharged by supplying the load. The use of voltage divider circuit is to change or convert a large input voltage to a small output voltage by taking some fraction of the input voltage. By taking two series resistors, this conversion can be

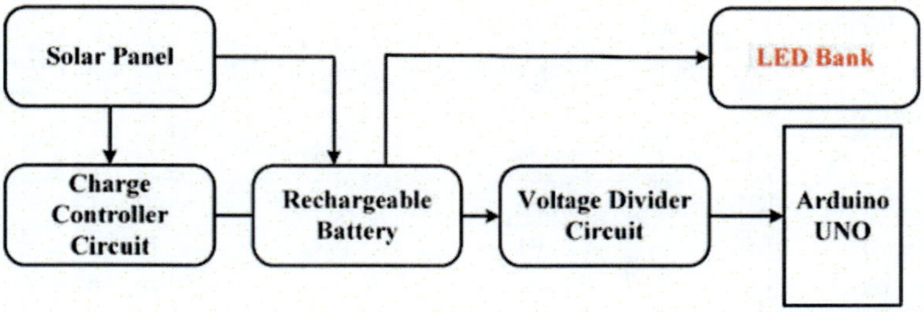

Fig. 5 Basic block diagram model of auto-intensity control of LED-based street lights

easily done so that the output or load will have a smaller voltage. The Arduino UNO is basically a microcontroller circuit that will help to control the light intensity at different period of time throughout the day. It will generate PWM waves, at necessary time using RTC or real-time clock, and the I/O pins are interfaced to the street light circuit.

3 Results and Discussion

The generated power from the solar PV array varies continuously with tilt angle, sunshine hours, latitude and longitude of location. With the change of these parameters, the generation of power gets affected. For the purpose of analysis, Jamshedpur is taken as site under consideration and required data is calculated for 7 October 2020. The latitude and longitude of the site being 22.8046°N and 86.2029°E, respectively. The sunrise and sunset times were observed to be 5.40 am and 5.24 pm respectively, obtained from the meteorological data available for site. With the help of this data, the power from the PV cell was calculated [13, 14]. Here, Jamshedpur is taken arbitrarily only so, this same process can be analysed for other places also. The values of power with adjustable array and non-adjustable array, i.e. static and dynamic power, are stated in Table 1. The average available power in static position was calculated to be 148.88 W, and average available power in dynamic position was calculated to be 186.31 W. Static power is basically the fixed amount of generated power, and it is found by taking the worst-case analysis so this power is always produced for a period of time. Dynamic power can get varied as it is calculated by taking the radiation parameters for best-case analysis. But in normal weather condition, there might be abnormality into the system thus the dynamic power falls to zero. The hourly variation of output power is as shown in Fig. 7.

The power available from the PV cell was utilized to supply the LED street lights. The LDR which is used as sensing device gives signal to the relay module. The Arduino UNO is an open-source microcontroller based on ATmega328P controller, consisting of programmable 14 digital and 6 analogy pins. It is powered by external dc supply of 9 V. The Arduino is so programmed via a type B USB cable such that

Table 1 Calculation of static and dynamic power

Time period	Military time	Sun angle (ϕ)	Pstatic (W)	Pdynamic (W)
09:00–10:00	9.50	42.95	182.12	133.30
10:00–11:00	10.50	33.71	189.25	157.43
11:00–12:00	11.50	27.95	192.69	170.21
12:00–01:00	12.50	26.01	193.69	174.07
01:00–02:00	13.50	28.01	192.66	170.09
02:00–03:00	14.50	33.84	189.16	157.12
03:00–04:00	15.50	43.13	181.96	132.80
04:00–05:00	16.50	55.32	168.87	96.08

the intensity of the street light can be controlled depending on the available luminous flux. The flow chart of Arduino programming is as shown in Fig. 6. With the help of Arduino, it is possible to control the intensity of the street light automatically, thus resulting in optimum power consumption. This technique is quite new as in most of the cases it can be seen that proper lighting coordination is not present. In case of days, there are lights turned on and it will waste power significantly. Thus, if there is presence of minimum intensity of light, then the LEDs should be kept off. There is presence of two types of power, i.e. the static one and the dynamic one. The static sources are minimum or the base values and the dynamics can be produced or also not. If the power requirement is below static level, then there is no requirement to switch on the LEDs, as sufficient power is already present for lighting. And if the requirement increases above the static level, then turn ON of LEDs are required so proper signal is sent to make it ON.

Thus, a proper coordinating system can help to analyse or perform the work in an efficient manner. It is the work of the sensor to function suitably as per the requirement of light and gives proper signal to switch on the LEDs those are connected to the ports respectively. If sufficient lights are present, the ports are closed and the LEDs would behave in off state as no connection or supply is fed to the respective ports. Thus, saving of energy is done as the production of power from renewable energy of sources are quite limited and to meet the increasing demand of the customer, the system should operate efficiently to minimize loss in the system.

In Fig. 6, the working of the entire process is explained. When the luminous flux is less than 15, then pin 7 is put ON and pin 6 is put in OFF state as only pin 7 is adequate for lighting the system. When the intensity is more than 15 and less than 100 that is higher than the previous stage then pin 7 is put OFF and pin 6 have to be put ON to maintain the desired light intensity. And when it is greater than 100 then both the pins are put OFF. Thus, by switching ON and OFF the pins, the switching of LEDs is done to achieve the desired light intensity. In Fig. 7, with the variation of sun angle at different interval of time throughout the day the static and dynamic power are plotted and accordingly the switching is done to achieve the desired light intensity over an area.

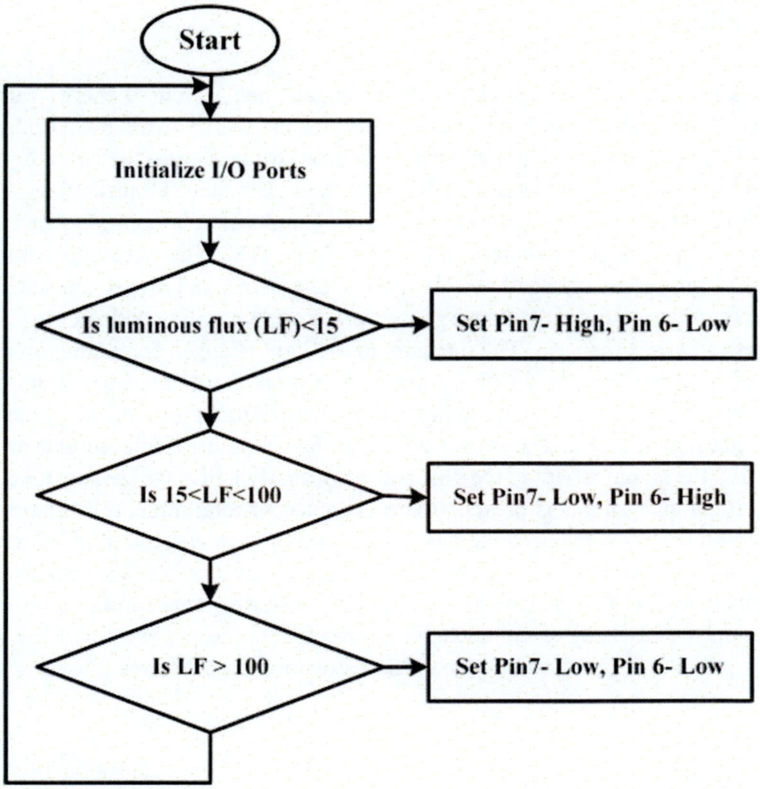

Fig. 6 Flow chart of Arduino programming

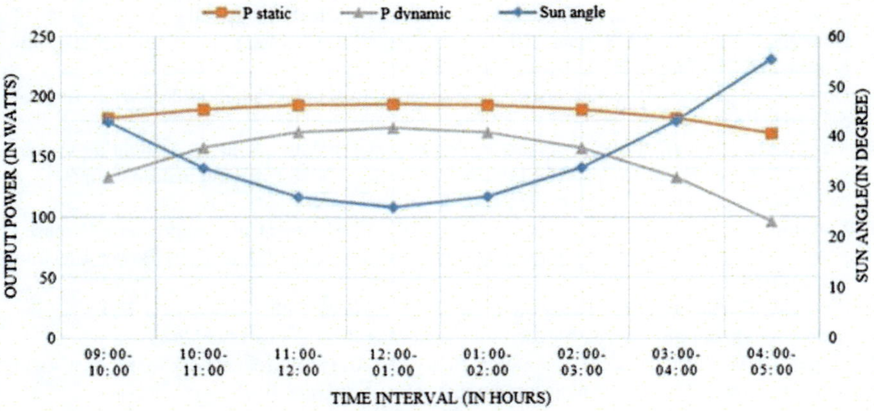

Fig. 7 Hourly variation of output PV cell

4 Conclusion

This work aims at providing an approach to mitigate the problem of energy wastage in street light by auto-intensity control [16, 17]. Solar energy is the most dominant source of renewable energy and thus has been used for powering the street lighting mechanism. From the solar radiation data calculated for the city Jamshedpur, it can be concluded that considered site receives large chunks of solar energy, which thus boosts the idea of solar powered LED street lights [18]. The conventional street lighting system can be replaced by solar powered systems, and use of auto-intensity control technique will result in energy conservation, which is needed considering recent energy crisis scenario. The intensity of the light changes depending on availability of luminous flux, lesser the luminous flux more is the intensity of the light. The working of the LDR is quite important in proper functioning of the system for proper light management. The sensor will send the signal if sufficient lights are not present and the signal is sent to the port and accordingly LEDs will get switched on. Further, if adequate lights is present, there is no proper requirement of lighting the LEDs are no supply is given to the ports. Thus, a proper management of light takes place to make the area having the minimum lighting. Further, there is also efficient management of energy as when it is not required the sensor senses and automatically switches off the LEDs by disconnecting the supply to the respective port. This work will thus prove to be an alternative to current street lighting systems.

References

1. Mahajan B, Namrata K, Varshney A (2018) Estimation of solar radiation flux using MATLAB for Amravati, Maharashtra, India. In: Carbon—Science and technology, vol 1, issue 1, pp 34–38
2. Namrata K, Sharma SP, Saxena SBL (2012) Prediction of solar power plant capacity for Jamshedpur, Jharkhand, India. IJSAT 2:138–142
3. Gopinathan KK (1988) A new model for estimating total solar radiation. Sol Wind Technol 5(1):107–109
4. Vieira JAB, Mota AB (2010) Implementation of a stand-alone photovoltaic lighting system with MPPT battery charging and LED current control. In: 2010 IEEE international conference on control applications (CCA), pp 185–190, 8–10 Sept 2010
5. Koutroulis E, Kalaitzakis K, Voulgaris NC (2001) Development of a microcontrollerbased photovoltaic maximum power point tracking control system. IEEE Trans Power Electron 16(1):46–54
6. Devi DA, Kumar A (2012) Design and implementation of CPLD based solar power saving system for street lights and automatic traffic controller. Int J Sci Res Publ 2(11)
7. Denardin GW, Barriquello CH, Campos A, Pinto RA (2011) Control network for modern street lighting systems. In: IEEE symposium on industrial electronics (ISIE), pp 1282–1289
8. Mohamaddoust R, Haghighat AT, Motahari Sharif MJ, Capanni N (2011) A novel design of an automatic lighting control system for a wireless sensor network with increased sensor lifetime and reduced sensor numbers. Sensors 11:8933–8952
9. Das S, Samadhiya A, Namrata K (2020) Mathematical modelling based solar PV module and its Simulation in comparison with datasheet of JAPG-71320/4BB solar module. In: National

conference on research and developments in material processing, modelling and characterization (RDMPMC-2020). Next generation materials and processing technologies, conference date-26–27th Aug 2020. Springer. https://doi.org/10.1007/978-981-16-0182-8
10. Das S, Namrata K (2020) Improving the performance analysis of MPPT controller unit of a PV generation system using optimization technique based on SMO. In: national conference on research and developments in material processing, modelling and characterization (RDMPMC-2020). Next generation materials and processing technologies, conference date-26–27th Aug 2020. Springer.https://doi.org/10.1007/978-981-16-0182-8
11. Costa MAD, Costa GH, Dos Santos AS, Schuch L, Pinheiro JR (2009) A high efficiency autonomous street lighting system based on solar energy and LEDs. In: Proceedings of Power Electronics Conference, pp 265–273, 1 Oct 2009
12. Mullner R, Riener A (2011) An energy efficient pedestrian aware smart street lighting system. Int J Pervasive Comput Commun 7(2):147–161
13. Rajput KY, Khatav G, Pujari M, Yadav P (2013) Intelligent street lighting system using GSM. Int J Eng Sci Invent 2(3):60–69
14. Saleem AL, Raja SR, Sachin DNS, Sachin HS, Usha et al (2015) Street light monitoring and control system. Int J Eng Tech 1(2):68–71
15. Tong X, Hu C, Rui T, Shen W. Economical optimization for multi-microgrid based on Stackelberg game. In: 2019 14th IEEE conference on industrial electronics and applications (ICIEA), pp 1760–1764
16. Akmal M, Al-Naemi F, Iqbal N, Al-Tarabsheh A, Meegahapola L. Impact of distribution PV generation on relay coordination and power quality. In: 2019 IEEE Milan PowerTech, pp 1–6
17. Shakirov VA. Estimation of global solar radiation and performance of photovoltaic modules using long-term meterological data. In: 2018 international multi-conference on industrial engineering and modern technologies (FarEastCon), pp 1–4
18. Talbi K, Harrouni S. Evaluating semi-emperical models for global solar radiation on the inclined surfaces in south of Algeria. In: 2018 International conference on electrical sciences and technologies in Maghreb (CISTEM), pp 1–6

Combined Economic Emission Dispatch of a Hybrid Energy System Using Quantum-Inspired Tidal Firefly Algorithm

Kapil Deo Bodha, Vinod Kumar Yadav, and Vivekananda Mukherjee

Abstract Population-based optimization algorithms are very efficient in handling the optimization problem, their integration with quantum mechanics principle greatly enhances its capability. In this manuscript, quantum-inspired tidal firefly algorithm (QITFA) is applied to resolve the combined economic emission dispatch (CEED) problem of a hybrid power system. The QITFA is applied to solve the CEED of an IEEE-30 bus system considering transmission line losses. Price penalty factor (PPF) approach is utilized to transform the many objective functions into singly objective functions. The results are collated with three other optimization techniques to prove the superiority of the quantum-inspired algorithm. The effect of increased renewable penetration on the operating cost and emission of power systems is also studied. From the analysis, it is seen that for every 10% increase in the renewable penetration in power system, the fuel cost reduces by 12% and emission is reduced by up to 25%.

Keywords Renewable energy · Power system optimization · Wind energy · Solar PV · Quantum mechanics

1 Introduction

The CEED aims to minimize the fuel cost and emission at the simultaneously without violating any constraint [1]. The CEED is a complex optimization problem and difficult to solve by conventional optimization techniques. Population-based technique has been proved to efficiently solve the CEED problem [2]. Many population-based algorithms were proposed in due course of time, some of the noteworthy algorithms applied to solve the CEED problem are particle swarm optimization (PSO) technique [3], Gravitational Search algorithm (GSA) [4] and firefly algorithm (FA) [5].

K. D. Bodha (✉) · V. Mukherjee
IIT(ISM) Dhanbad, Dhanbad, Jharkhand 826004, India
e-mail: bodhas@gmail.com

V. K. Yadav
Delhi Technological University, Delhi, India

K. Namrata et al. (eds.), *Smart Energy and Advancement in Power Technologies*, Lecture Notes in Electrical Engineering 926,
https://doi.org/10.1007/978-981-19-4971-5_24

Though these algorithms efficiently handle the optimization problem but they have a tendency to get trapped in local optima. Combination of metaheuristics can help reduce these drawbacks, like in [6], GSA and PSO techniques are combined. PSO enhances the capability of GSA by providing it with memory element. Other recent trend in optimization algorithms is to combine the classical algorithm with quantum mechanics principle. Inclusion of quantum mechanics principle greatly enhances the capability as in quantum space, the particle can move freely in entire feasible search space to find the optimized solution [7]. Many quantum mechanics-based algorithms were utilized to tackle the CEED problem of power systems, some of the noted work includes [8–10].

In this paper a newly developed, Quantum-inspired tidal firefly algorithm (QITFA) is utilized to solve the CEED of a hybrid solar thermal wind power system. Economic dispatch (ED) and CEED test cases are also considered to verify the effectiveness of the algorithm. PPF is utilized to transform the multi-objective function into a singly objective function [11]. The algorithm is applied to solve the CEED problem of an IEEE-30 bus, -6 unit system for a load demand of 1200 MW. The effect of increased renewable penetration on the operating cost and emission of power system is also analyzed.

2 Problem Statement

CEED comprises two objectives that have to be minimized at the same time, these objective functions can be formulated as shown.

2.1 Formulation of the Multi-Objective Problem

CEED comprises two objectives that have to be minimized at the same time, these objective functions can be formulated as shown in (1)

$$\text{Minimize}(\text{K}) = \sum_{i=1}^{n} \text{FuelCost}_i(P_i) + \text{EmissionCost}_i(P_i) \tag{1}$$

where, K represents the objective function to be optimized. $\text{FuelCost}_i(P_i)$ is the fuel cost ($/h) and $\text{EmissionCost}_i(P_i)$ is the emission in kg/h of the ith generator, respectively. n represents the total number of generators. Following constraint has to be considered during dispatching.

2.2 Power Balance Constraint

The total power generated should be equal to the sum of power demand and transmission line losses as expressed in (2)

$$\sum_{i=1}^{n} P_i - (P_{\text{linelosses}} + P_{\text{load}}) = 0 \tag{2}$$

where P_i is the power generated by the ith generating unit of the power system. Transmission losses is given by $P_{\text{linelosses}}$ and total demand of the system is given by P_{load}.

2.3 Inequality Constraint

The generation limits of the generators should not be violated during scheduling. It can be expressed as (3)

$$P_{i\_\text{minimum}} \leq P_i \leq P_{i\_\text{maximum}} \tag{3}$$

where P_{i_minimum} and P_{i_maximum} are the lowest and highest generating limit of the ith generator, respectively. $\text{FuelCost}_i(P_i)$, $\text{EmissionCost}_i(P_i)$ in Eq. (1) and $P_{\text{linelosses}}$ in (2) are formulated as shown [12]

$$\text{FuelCost}_i(P_i) = \alpha_i P_i^2 + \beta_i P_i + \gamma_i \tag{4}$$

where α_i,β_i and γ_i are the coefficient of fuel cost of the ith generating unit.

The emission objective function can be expressed as in (5)

$$\text{EmissionCost}_i(P_i) = D_i P_i^2 + E_i P_i + F_i + G_i * \exp(H_i * P_i) \tag{5}$$

where D_i,E_i,F_i,G_i and H_i represents the emission coefficients of the ith generator. Using the kron loss formula, transmission line losses can be calculated as in (6)

$$P_{\text{linelosses}} = \sum_{i=1}^{n} \sum_{j=1}^{n} P_i B_{ij} P_j \tag{6}$$

where B is the loss coefficient matrix.

The price penalty factor (PPF) is utilized to convert Eq. (1) into a single objective function [13]. The updated equation to be optimized can be expressed as (7)

$$\text{Min}\, C = \sum_{i=1}^{n} \left[\left(\alpha_i P_i^2 + \beta_i P_i + \gamma_i \right) + PF_i \left(D_i P_i^2 + E_i P_i + F_i + G_i * \exp(H_i * P_i) \right) \right] \tag{7}$$

where PF_i is the max/max price penalty factor and can be expressed as (8).

$$PF_i = \frac{\alpha_i P_{i\_\text{maximum}}^2 + \beta_i P_{i\_\text{maximum}} + \gamma_i}{D_i P_{i\_\text{maximum}}^2 + E_i P_{i\_\text{maximum}} + F_i + G_i * \exp\left(H_i * P_{i\_\text{maximum}} \right)} \tag{8}$$

2.4 *Renewable Sources Considered*

Solar energy and wind energy penetration are considered. It can be calculated as.

2.4.1 Solar Power

The power produced by a PV panel can be expressed as (9)

$$P_{\text{solar}} = k_1 R_s [1 + k_2 \left(t_j - t_{\text{reference}} \right) \tag{9}$$

where P_{solar} is the total solar power generated by the PV panel. k_1 represents the characteristic dispersion of panels, it's value is taken as 0.095, R_s is the solar radiation. t_j is the cell junction temperature (°C) and and $t_{\text{reference}}$ is the reference temperature of the solar panels at 25 °C. k_2 is the drift in panels temperature taken as 0.47%/°C. Equation (9) can be modified by adding another parameter k_3 in the equation to enhance its accuracy. The updated equation is expressed as (10) [14]

$$P_{\text{solar}} = k_1 \left[1 + k_2 \left(t_j - t_{\text{reference}} \right) \right] (k_3 + R_s) \tag{10}$$

2.4.2 Wind Power

Equation (11) gives the power generated by the wind turbine.

$$P_{\text{wind}} = \frac{1}{2} . \rho . E_{\text{fa}} . A' . S^3 . 10^{-3} \tag{11}$$

where P_{wind} denotes the power generated by the windmill (MW), ρ (1.225 kg/m^3) is the density of air, S is the wind speed (m/s), A' is the area covered by wind (m^2) and E_{fa} represents the efficiency factor, its value is dependent on speed of the wind [15].

2.5 Constraint Handling Process

Constraint handling in a CEED is a vital process that ensures that none of the constraints are violated at any step during the scheduling process. Generator 1 is chosen as slack generator. The power balance equation in (2) can be rewritten as (12) [6]

$$P_{\text{slackbus}} + \sum_{\substack{i=1 \\ i \neq s}}^{n} P_i + P_{\text{wind}} + P_{\text{solar}} = (P_{\text{linelosses}} + P_{\text{load}}) \tag{12}$$

Considering transmission losses, Eq. (12) can be written as in (13).

$$\begin{aligned} P_{\text{slackbus}} &+ \sum_{\substack{i=1 \\ i \neq s}}^{n} P_i + P_{\text{wind}} + P_{\text{solar}} \\ &= \sum_{\substack{i=1 \\ i \neq s}}^{n} \sum_{\substack{j=1 \\ J \neq s}}^{n} P_i B_{ij} P_j + \sum_{\substack{j=1 \\ j \neq s}}^{n} P_j (B_{js+} B_{sj}) P_{\text{slackbus}} + B_{ss} P_{\text{slackbus}}^2 \\ &+ \sum_{\substack{i=1 \\ i \neq s}}^{n} B_{i0} P_i + B_{S0} P_{\text{slackbus}} + B_{00} + P_{\text{load}} \end{aligned} \tag{13}$$

Equation (13) can be expressed as in (14)

$$x_1 P_{\text{slackbus}}^2 + y_1 P_{\text{slackbus}} + z_1 = 0 \tag{14}$$

where

$$x_1 = B_{ss} \tag{15}$$

$$y_1 = \sum_{\substack{j=1 \\ j \neq s}}^{n} P_j (B_{js+} B_{sj}) + B_{S0} - 1 \tag{16}$$

$$z_1 = P_{\text{demand}} + \sum_{\substack{i=1 \\ i \neq s}}^{n} \sum_{\substack{j=1 \\ j \neq s}}^{n} P_i B_{ij} P_j + \sum_{\substack{i=1 \\ i \neq S}}^{n} B_{i0} P_i - \sum_{\substack{i=1 \\ i \neq s}}^{n} P_{gi} - P_{\text{wind}} - P_{\text{solar}} \tag{17}$$

The positive roots of this equation can be calculated by (18).

$$P_{\text{slackbus}} = \frac{-y_1 \pm \sqrt{y_1^2 - 4x_1 z_1}}{2x_1} \quad \text{where } (y_1^2 - 4x_1 z_1) \geq 0 \tag{18}$$

Positive roots of (18) are only considered. The slack generator power P_{slackbus} is set to its lowest values if the roots are out of bound. If the roots are in inbound then the root with the lower value is considered [16]

The slack generation gap ($P_{\text{slackgengap}}$) is calculated using (19).

$$P_{\text{slackgengap}} = \sum_{i=1}^{n} P_i + (P_{\text{wind}} + P_{\text{solar}}) - (P_{\text{linelosses}} + P_{\text{load}}) \tag{19}$$

Repeat this procedure until the $P_{\text{slackgengap}}$ given by (19) reduces to zero.

3 QITFA

3.1 Tidal Firefly Algorithm

FA is a population-based algorithm that takes clue from the mating pattern of fireflies. The fireflies are attracted towards brighter fireflies and move towards them with a certain velocity. Some assumptions made for this algorithm are.

1. All fireflies are unisex.
2. Fireflies get attracted toward their brighter counterpart.
3. The brightness of the firefly diminishes with the increase in gap between them.

The movement equation of the firefly moving towards brighter firefly is given by (20)

$$y_{i,m}^{t+\Delta t} = y_{i,m}^{t} + \eta(d_{ij}) * (y_{j,m}^{t} - y_{i,m}^{t}) + \gamma * \left(\text{rand}\left(-\frac{1}{2}\right)\right) \tag{20}$$

where $y_{i,m}^{t+\Delta t}$ is the modified position of the ith firefly in the mth dimension, γ is the damping agent that lies between [0, 1], rand is random variable between [0, 1] [17].$\eta(d_{ij})$ is the attraction factor and it is given by (21)

$$\eta\left(d_{ij}\right) = \eta_0 \times \mathrm{e}^{-\varepsilon d_{ij}^2} \quad (m \geq 1) \tag{21}$$

where η_0 and ε are random numbers lying in [0, 1] and d_{ij} is the distance between two fireflies, it can be expressed as (22)

$$d_{ij} = \sqrt{\sum_{m=1}^{L}\left(y_{i,m} - y_{j,m}\right)^2} \tag{22}$$

where $y_{i,m}$ and $y_{j,m}$ gives the position of the fireflies, L is the total number of dimensions.

The attraction factor is replaced by a tidal force factor. The new updated equation of tidal firefly algorithm becomes

$$y_{i,m}^{t+\Delta t} = y_{i,m}^{t} + \Delta h_{ij} * \left(y_{j,m}^{t} - y_{i,m}^{t}\right) + \gamma * \left(\mathrm{rand}\left(-\frac{1}{2}\right)\right) \tag{23}$$

Here, Δh_{ij} is the difference of near ($h_{near}^{y_i^t}$) to far ($h_{far}^{y_j^t}$) tidal force. Mathematically it is:

$$\Delta h_{ij} = h_{near}^{y_i^t} - h_{far}^{y_j^t} \tag{24}$$

where

$$h_{\mathrm{near}}^{y_i^t} = \frac{G * \mathrm{mass}_1 * \mathrm{mass}_2}{\left(y_{i,m}^t\right) * \left(y_{i,m}^t\right)} \tag{25}$$

$$h_{\mathrm{far}}^{y_j^t} = \frac{G * \mathrm{mass}_1 * \mathrm{mass}_2}{\left(y_{j,m}^t\right) * \left(y_{j,m}^t\right)} \tag{26}$$

where G is the gravitational constant, mass_1 and mass_2 are the mass of particles 1 and 2, respectively.

3.2 *Quantum-inspired Metaheuristic*

In quantum space the particles follow the basic notion of quantum mechanics, they can move freely in the entire search space. The particles moving in the quantum space does not have a defined trajectory, their velocity and position cannot be calculated simultaneously. The particle position update equation can be expressed using Monte Carlo stochastic simulation as (27).

$$y_{i,d}^{t+\Delta t} = y_{\mathrm{bestm}}^{t} \pm q \cos^{-1}\left[\pm\sqrt{u}\right] \tag{27}$$

where u lies between [0, 1],

where y^t_{bestm} is the location of the brightest firefly in mth dimension, q can be expressed as (28)

$$q = \frac{2}{g}\left|y^t_{i,m} - y^t_{bestm}\right| \tag{28}$$

Here, g is calculated as in (29)

$$g = \frac{it_{\max}}{it_{\max} \times (e + 0.5) - e * it} \tag{29}$$

The updated equation of quantum-inspired tidal firefly algorithm can be expressed as (30)

$$y^{t+\Delta t}_{i,m} = y^t_{bestm} \pm \frac{1}{0.5 * g}\cos^{-1}\left[\pm\sqrt{u}\right]\left|y^t_{i,m} - y^t_{bestm}\right| \tag{30}$$

QITFA flowchart is shown in Fig. 1.

4 Results and Discussion

The QITFA is applied to resolve the CEED of a power system. Two test cases are considered. In case-1, CEED is done without considering renewable energy penetration. In case-2, CEED is done for different levels of renewable energy penetration. IEEE-30 bus, 6 unit test system is considered for the analysis. The fuel cost coefficient, emission coefficient and loss coefficients are taken from [14]. The results are collated with three techniques viz. Quasi Oppositional Teaching Learning based Optimization (QOTLBO), Differential Evolution (DE) and Teaching Learning based Optimization (TLBO).

4.1 Case-1

CEED is done for a load of 1200 MW. line losses (MW) are also considered. The results are displayed in Table 1. It is evident from the results that QITFA gives the lowest fuel cost compared to all other techniques referred to. It gives a fuel cost saving of up to 800$/h. Figure 2 displays the convergence graph of QITFA. It can be observed that QITFA exhibits excellent convergence characteristics and converge within 100 iterations.

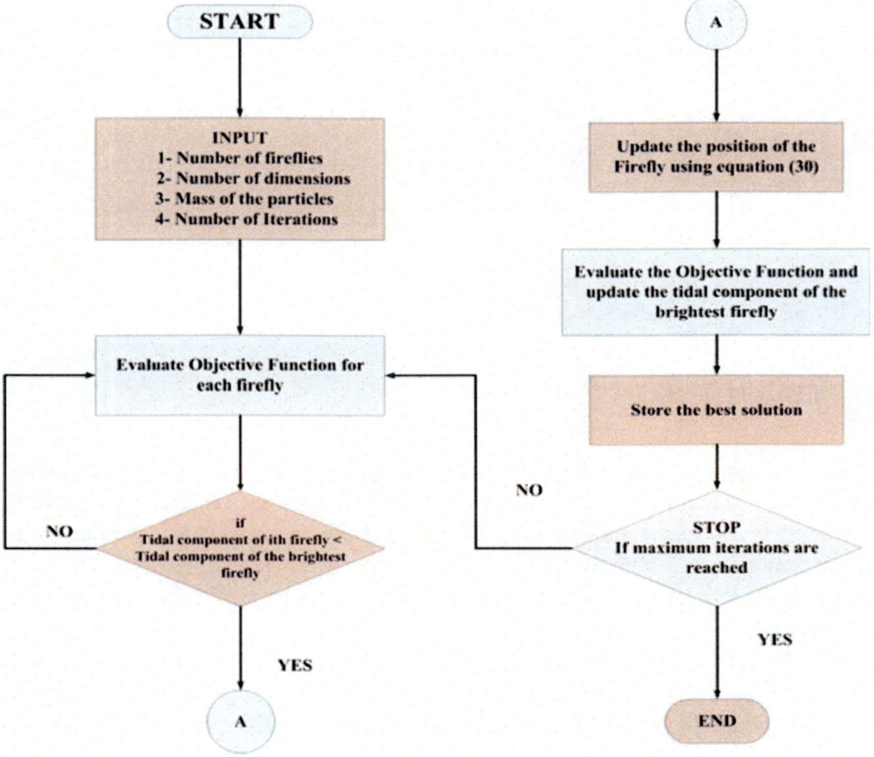

Fig. 1 Flowchart of QITFA

Table 1 Comparison of CEED results for different techniques

Generator	QOTLBO	DE	TLBO	QITFA
POG-1 (MW)	107.3101	108.6284	107.8651	89.0587
POG-2 (MW)	121.4970	115.9456	121.5676	98.0708
POG-3 (MW)	206.5010	206.7969	206.1771	218.2383
POG-4 (MW)	206.5826	210.0000	205.1879	208.4506
POG-5 (MW)	304.9838	301.8884	306.5555	323.8522
POG-6 (MW)	304.6036	308.4127	304.1423	315.0000
Fuel cost ($/h)	64,912	64,843	64,922	64,199
Emission (lb/h)	1281	1286	1281	1332
Line losses (MW)	51.4781	51.700	51.4955	52.67

POG- Power of Generator

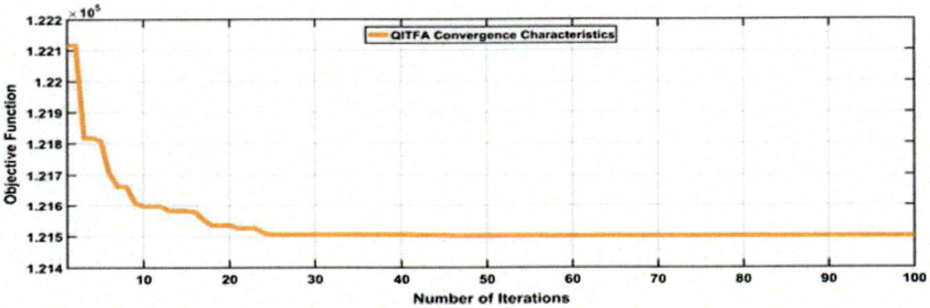

Fig. 2 Convergence characteristic of QITFA

4.2 Case II

In this case, renewable energy penetration is considered. Four levels of penetrations, 10%, 20%, 30% and 40% are considered. Penetration of above 30% of total load is avoided as it poses stability issues [7], but in many countries, there is renewable energy penetration as high as 70%.

The results are shown in Table 2. CEED for this system is formulated using QITFA technique.

From Table 2 it can be concluded that for every 10% increase in renewable energy penetration, the fuel cost reduces by 13% and emission reduces by 25%. There is also a significant decrease in transmission line losses as evident from the table.

Table 2 CEED with various levels of renewable energy penetration

Renewable energy penetration (%)	10	20	30	40
POG-1 (MW)	77.69	66.88	56.19	45.59
POG-2 (MW)	82.03	66.87	51.94	37.23
POG-3 (MW)	194.66	172.29	150.19	128.35
POG-4 (MW)	187.19	166.99	146.99	127.19
POG-5 (MW)	292.58	262.79	233.18	203.74
POG-6 (MW)	288.46	257.80	227.28	196.89
Fuel cost ($/h)	5724.3	5058.1	4418.99	3806.03
Emission (lb/h)	1070.9	842.14	646.977	484.63
Line losses (MW)	51.8586	33.64	25.7872	19.02

POG- Power of Generator

5 Conclusion

In this article, a newly developed algorithm based on quantum mechanics principle is applied to resolve the CEED of a power system. The results are collated with QOTLBO, TLBO and DE techniques. QITFA optimizes the problem efficiently and converges within 100 iterations. It is shown that QITFA provides a saving of up to 800 $/h. It is also applied to solve the CEED of a hybrid energy system with different levels of penetration. From the result analysis of this hybrid system it can be concluded that for every 10% increase in renewable energy penetration in system, the fuel cost and emission reduce by 13% and 25%, respectively.

References

1. Der Chen S, Chen JF (2003) A direct Newton-Raphson economic emission dispatch. Int J Electr Power Energy Syst. https://doi.org/10.1016/S0142-0615(02)00075-3
2. Jeyakumar DN, Jayabarathi T, Raghunathan T (2006) Particle swarm optimization for various types of economic dispatch problems. Int J Electr Power Energy Syst. https://doi.org/10.1016/j.ijepes.2005.09.004
3. Abido MA (2009) Multiobjective particle swarm optimization for environmental/economic dispatch problem. Electr Power Syst Res. https://doi.org/10.1016/j.epsr.2009.02.005
4. Shaw B, Mukherjee V, Ghoshal SP (2012) A novel opposition-based gravitational search algorithm for combined economic and emission dispatch problems of power systems. Int J Electr Power Energy Syst. https://doi.org/10.1016/j.ijepes.2011.08.012
5. Apostolopoulos T, Vlachos A (2011) Application of the Firefly algorithm for solving the economic emissions load dispatch problem. Int J Comb. https://doi.org/10.1155/2011/523806
6. Bodha KD, Yadav VK, Mukherjee V (2021) A novel quantum inspired hybrid metaheuristic for dispatch of power system including solar photovoltaic generation. Energy Sources Part B Econ Plan Policy 00(00):1–26. https://doi.org/10.1080/15567249.2021.1933265
7. Bodha KD, Yadav VK, Mukherjee V (2020) Formulation and application of quantum-inspired tidal firefly technique for multiple-objective mixed cost-effective emission dispatch. Neural Comput Appl 32(13). https://doi.org/10.1007/s00521-019-04433-0
8. Mahdi FP, Vasant P, Abdullah-Al-Wadud M, Kallimani V, Watada J (2019) Quantum-behaved bat algorithm for many-objective combined economic emission dispatch problem using cubic criterion function. Neural Comput Appl. https://doi.org/10.1007/s00521-018-3399-z
9. Mahdi FP, Vasant P, Abdullah-Al-Wadud M, Watada J, Kallimani V (2018) A quantum-inspired particle swarm optimization approach for environmental/economic power dispatch problem using cubic criterion function. Int Trans Electr Energy Syst. https://doi.org/10.1002/etep.2497
10. Xin-gang Z, Ze-qi Z, Yi-min X, Jin M (2020) Economic-environmental dispatch of microgrid based on improved quantum particle swarm optimization, vol 195. Elsevier Ltd.
11. Benasla L, Belmadani A, Rahli M (2014) Spiral optimization algorithm for solving combined economic and emission dispatch. Int J Electr Power Energy Syst. https://doi.org/10.1016/j.ijepes.2014.04.037
12. Deobodha K, Mukherjee V, Kumaryadav V, Saurabh K, Anium S (2018) A levy flight based voltage particle swarm optimization for multiple-objective mixed cost-effective emission dispatch. https://doi.org/10.1109/CONFLUENCE.2018.8442919
13. Ryu H-S, Kim M-K (2020) Combined economic emission dispatch with environment-based demand response using WU-ABC algorithm. Energies 13(23):6450. https://doi.org/10.3390/en13236450

14. Kherfane N, Kherfane RL, Younes M, Khodja F (2014) Economic and emission dispatch with renewable energy using HSA. Energy Procedia 50:970–979. https://doi.org/10.1016/j.egypro.2014.06.116
15. Younes M, Khodja F, Kherfane RL (2014) Multi-objective economic emission dispatch solution using hybrid FFA (firefly algorithm) and considering wind power penetration. Energy 67:595–606. https://doi.org/10.1016/j.energy.2013.12.043
16. Yadav VK, Padhy NP, Gupta HO (2014) The evaluation of the efficacy of an ongoing reform initiative of an Indian electric utility. Energy Sources Part B Econ Plan Policy 9(3):291–300. https://doi.org/10.1080/15567249.2010.493918
17. Yadav VK, Padhy NP, Gupta HO (2015) A holistic approach model for realistic goal-setting for efficiency enhancement with application to the Indian power sector. Energy Sources Part B Econ Plan Policy 10(2):120–131. https://doi.org/10.1080/15567249.2010.512903

Management of Energy and Coordinated Control of PV/HESS in Islanded DC Microgrid

Mubassir Ahmad, Danish Ali, Ward Ul Hijaz Paul, Md Safdar Ali, and Haroon Ashfaq

Abstract A sharp increment in the demand of the electrical energy has opened the door for the utilization of renewable energy. But because of its intermittent nature it has some limitations and requires some kind of Energy Storage System (ESS). In this paper, an islanded DC microgrid of small-scale is considered which incorporates solar photovoltaic energy generators, power converters, local DC loads and a hybrid energy storage system of battery and super capacitor. Various cases of solar irradiance are simulated to authenticate the presented power management and control strategy to regulate the DC bus voltage. This work adopts bidirectional DC/DC converters to perform the transfer of power in both directions. The charging or discharging of the battery and super capacitor are controlled by the circuit according to the need and maintain the magnitude within limits. The SIMULINK is used for modeling of DC microgrid system with HESS.

Keywords Battery · Hybrid energy storage system · Microgrid · Dual DC/DC converter · Supercapacitor

1 Introduction

It is predicted by the International Energy Agency that a rise in carbon-dioxide emission up to 130% and an increase in consumption of oil up to 70% can go by 2050, which can probably increase the average temperature of earth by 6 °C [1]. To tackle the situation and deal with these problem renewable energy sources (RESs) are the best option that we are left with. Solar energy and wind energy are the best easily available renewable sources that are now a days in demand for electric power generation worldwide. In today's era most of the household load such as laptops, oven, air conditioning, bulbs, mobile phones, is running on DC powers running these equipment from AC causes a loss of countable amount of energy in converting of AC power to DC which can be around 10–25% [2]. In

M. Ahmad · D. Ali · W. U. H. Paul (✉) · M. S. Ali · H. Ashfaq
Jamia Millia Islamia, New Delhi 110025, India
e-mail: wardulhijazpaul@gmail.com; ward1909916@st.jmi.ac.in

K. Namrata et al. (eds.), *Smart Energy and Advancement in Power Technologies*, Lecture Notes in Electrical Engineering 926,
https://doi.org/10.1007/978-981-19-4971-5_25

most of the industries, use of DC electric furnaces consumes much less energy as compared to AC furnaces [3]. Microgrid (MG) can be defined as a network that contains distributed energy generators (DGs), loads and energy storage systems. An ESS can help in assimilation of renewable sources into the microgrid by straightening the fluctuations, improvement in quality of power, and also contribute to frequency and other auxiliary services [4, 5]. If we use ESSs in the DC microgrid, reliability of the system increases. However, using single Energy Storage device in the microgrid has also some challenges because of limited capabilities of single ESS.

Two most important attribute of any energy storage technique is its power density and energy density. In literatures and research, various types of ESSs have been discussed for adoption in DC microgrid. Various types of Storage devices such as batteries [6], Super capacitors [7], flywheel systems [8] and Superconducting Magnetic Energy Storage [9]. One of the possible combinations for Hybrid Energy Storage System is suggested by theory of Ragone plot [10, 11] which consists of battery as high energy storage unit and super capacitor (SC) as high power storage unit. Hybrid systems fulfill the need of both high power and high energy density requirement simultaneously and also helps in suppressing electrical stress of the battery and extends lifetime. An ESS technology like battery has low power density but it has high energy density. Technologies like SCs have high power density and low energy density so it supplies the demand of high power which is the main cause of decrease in life of ESSs.

2 Analysis of the System and Components

2.1 Topology Consideration

A Hybrid Energy Storage System can be linked to a microgrid in various ways. These different ways of connecting HESS to the microgrid is called topology. There are three mainly three types of topologies, passive, semi-active and active. Passive topology is a type of topology in which two different ESSs with same terminal voltage are connected together to DC bus in simple manner. This topology has some advantages such as it is efficient and simple, and cost-effective [10]. In semi-active topology one storage device is linked to the DC bus using a bidirectional DC/DC power converter as an interface while another device is linked directly to the DC [10]. This topology needs more space for its installation which increase cost of overall system. This topology is more reliable and controllable than passive topology. In Active topology each storage device utilizes a separate dual converter when linked to a DC bus. As a results complexity of the system, its cost and losses increase. One of the most important is that all power of the ESS can be regulated actively. In this work a parallel active topology is incorporated. Figure 1 shows parallel active hybrid ESS topology.

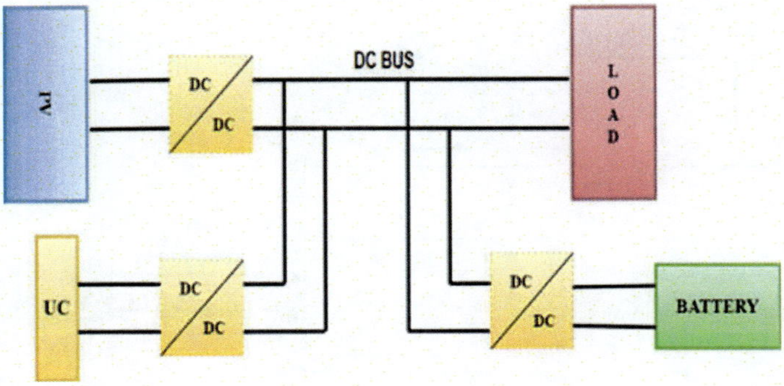

Fig.1 Parallel active topology

Energy management methodology is directly influenced by the type of topology used. A convenient topology should be chosen keeping in mind different factors such as efficiency, flexibility, controllability, complexity and cost [10].

2.2 *Modeling of the System*

A low voltage, low load and small-scale DC microgrid based on SPV is considered in this work as shown in Fig. 2. Solar PV is used as distributed energy recourse in this paper. Any type of local loads (resistive) has been considered as lumped. To regulate the output voltage and current of the solar PV, a boost converter is purposely connected at the output terminals of PV panel and other components such as battery, SC, boost converter, battery and SC with a BDC converter, and all these different components are linked to a common DC bus.

2.3 *PV Modeling*

Generally modeling of solar photovoltaic systems is nonlinear. Most of the parameters of PV array depend on solar irradiance and temperature. A well-known cell model is the Single diode circuit model and has been used by most of the researchers to study the behavior of energy generation of a PV cell [12]. Figure 3 shows the electrical circuit of solar cell model. The currents in the circuit can be represented by Eqs. (1) and (2).

$$I_{PV} = I_{SC} - I_{sh} - I_{D} \quad (1)$$

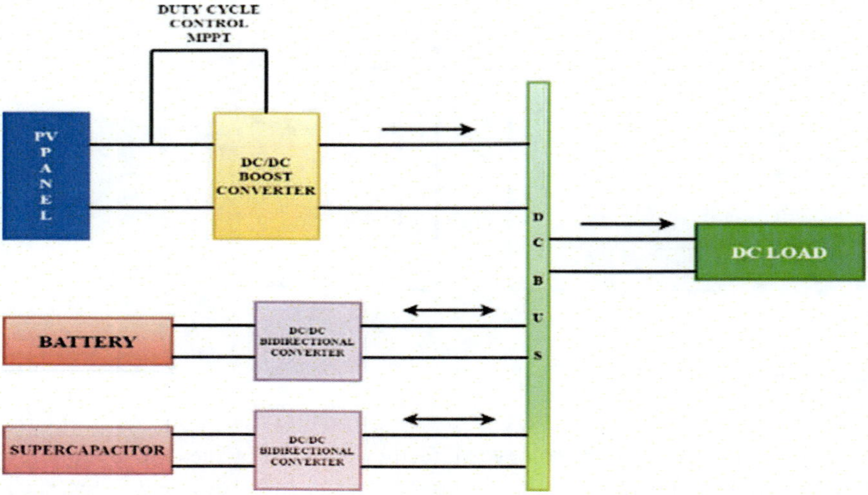

Fig. 2 Schematic of DC microgrid with HESS

Fig. 3 Single diode with series and shunt resistance-based PV cell model

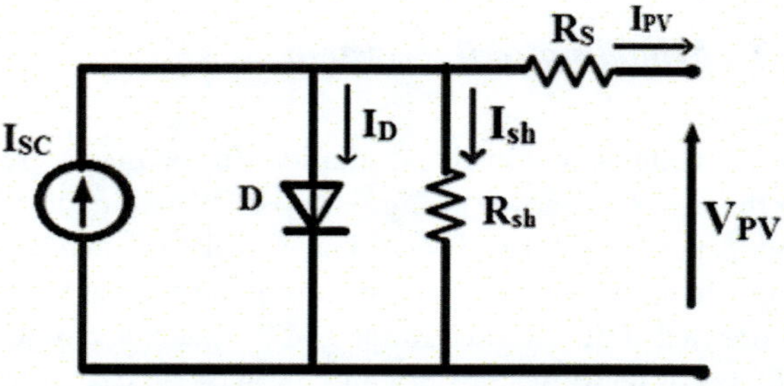

$$I_{PV} = N_P I_{SC} - N_P I_O \left\{ \exp\left[\frac{V_{PV} + I_{PV} R_S}{N_S . a} - 1 \right] - \frac{V_{PV} + I_{PV} R_S}{R_{sh}} \right\} \tag{2}$$

where,

I_{PV}, V_{PV}	PV panel current and voltage, respectively,
I_D	Diode current,
$R_{sh,}$ R_S	Parallel and Series resistance, respectively,
I_{SC}	Light generated current,
I_O, I_{sh}	Diode saturation current, Current through parallel resistor.
N_P, N_S	Number of parallel cells and number of series PV cells, respectively,
a	Ideality factor of diode after modification given by equation,

$$a = \frac{nkT}{q} \tag{3}$$

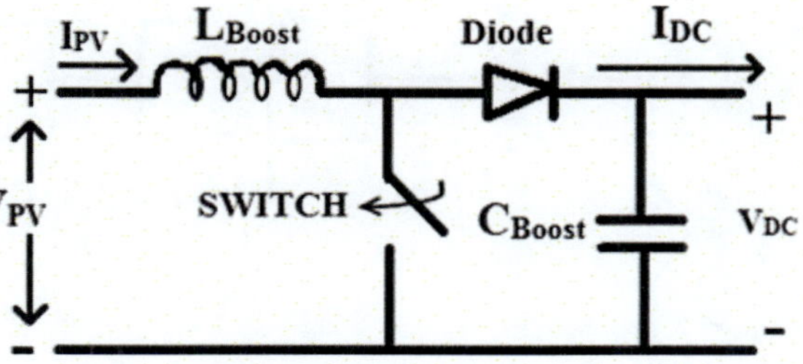

Fig. 4 Circuit diagram for DC boost converter

n dimensionless ideality factor (value for real diode is between 1 and 2)
k Boltzmann's constant
T, q Temperature of cell and electronic charge.

2.4 Modeling of Boost Converter

The boost converter is connected as an interface between Solar PV and DC bus. The most important job of this converter is to maintain the output voltage of PV array in stability limits. This is done by implementing a MPPT control algorithm which provides a reference voltage. Figure 4 shows the mathematical model of the boost converter.

To obtain the maximum power a maximum power point tracking (MPPT) algorithm is used which evaluates the output PV power at each instant by manipulating the operating points. To track the reference voltage generated from MPPT technique, boost converter controller generates a duty cycle [12]. P&O algorithm is utilized in this paper as MPPT.

2.5 Battery Modeling

A Lithium-ion battery is considered in this work as they have better response time and more cycle life. Generally, models are categorized into three main types, which are Thevenin model, impedance model and runtime-based models. The simplest is Thevenin model which uses a series resistor along with a parallel RC network is unable to predict battery runtime but it is one of the derivative model is good in predicting runtime of the battery [13, 14]. Impedance model is also unable to predict runtime as it can run only in a particular range of State of Charge and temperature [15]. Runtime-based model also have some cons and pros such as for a constant discharge current it can help in simulating DC voltage response and battery runtime, but it fails to do so for varying load current. [16].

A nonlinear model of battery is shown in Fig. 5. In this model, V_{bat} is output voltage of battery, I_{bat} is output current of battery and V_{OC} is open circuit voltage of battery, respectively. R_s is nothing, but instantaneous battery voltage drops. The

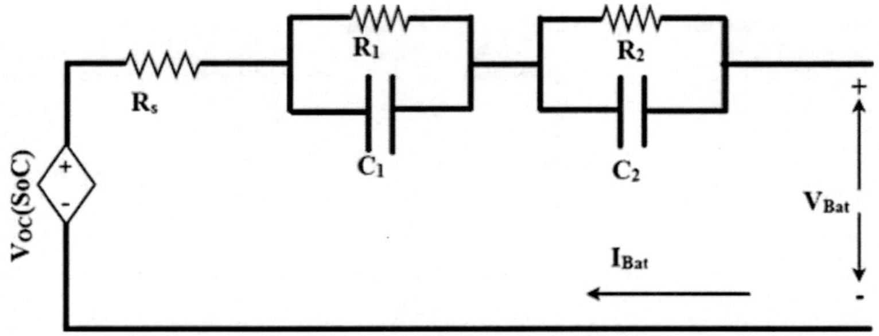

Fig. 5 Three state equivalent circuit for battery

parallel RC network (R1, C1) is short term time constant and RC network (R2, C2) represents long-term time constant of lithium ion battery step response [17].

2.6 *Bidirectional DC-DC Converter*

Each storage component of Hybrid Energy Storage system (battery and SC) is linked to a common DC bus using a dual DC-DC converter as an interface. Figure 6 Shows the schematic of dual bidirectional converter. It is called a dual converter because of its capability as it can work as a buck converter and as well as boost converter. Solar PV System with battery utilizes bidirectional converter to transfer power from one DC source to another DC source bi-directionally that from either source to another source. While working in buck mode it charges the ESS and while working in boost mode it transfers energy to the DC bus via dc link capacitor.

In this converter IGBTs are incorporated as a switch. A dual converter contains two IGBT switches. The working of the BDC depends on the difference of the DC bus voltage and ESS terminal voltage and whichever is higher it works accordingly as a buck converter or boost converter. The most essential purpose of BDC is to maintain the DC bus voltage at a specified value in every mode that is either charging or

Fig. 6 Buck boost converter for energy storage system

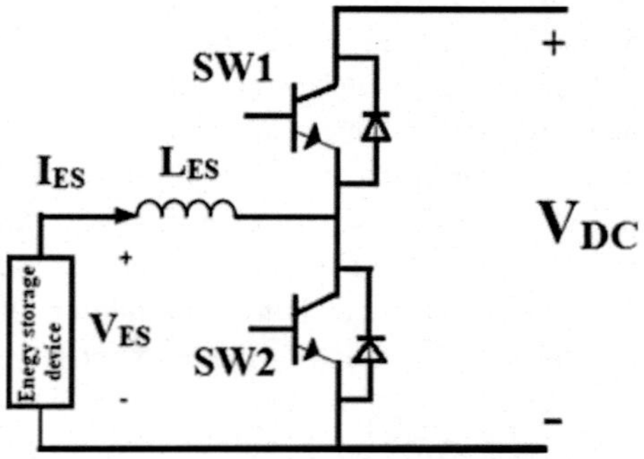

discharging [18]. The converter when operates as a boost converter it works in such a way that it increases the voltage of battery so that it matches the DC bus voltage.

Analysis of Charge/Discharge Control Circuit

The control circuit of HESS is shown in Fig. 7. As this is parallel active topology so each storage system utilizes bidirectional converter. When boost switch is closed, and buck switch is open then BDC works as a Boost converter. If HESS and DC bus are operating at different voltages then there is a flow of power from one direction to another depending upon which side is at higher voltage. Duty cycle of the switch which is needed to be modulated depends on the difference between the terminal voltage of HESS and bus voltage. When boost switch is closed then sources store the energy in the inductor by increasing current through it and then this inductor pushes this energy to HESS and helps in charging them by passing current through HESS. Now when boost switch is opened then to maintain steady current through inductor voltage through inductor changes its polarity and increases. As a result of this output voltage of boost converter is boosted and power flow from HESS to DC bus [19].

When bidirectional DC/DC converter works as a step-down converter then power flow from DC bus to HESS hence charging of HESS takes place. In this case buck switch is closed and boost switch is open.

There are four IGBTs in both BDC converters in the control circuit and all four IGBTs controlling is independent of each other. When the IGBT2 or IGBT4 works in PWM mode, then converter works as a step-down converter and flow of power from

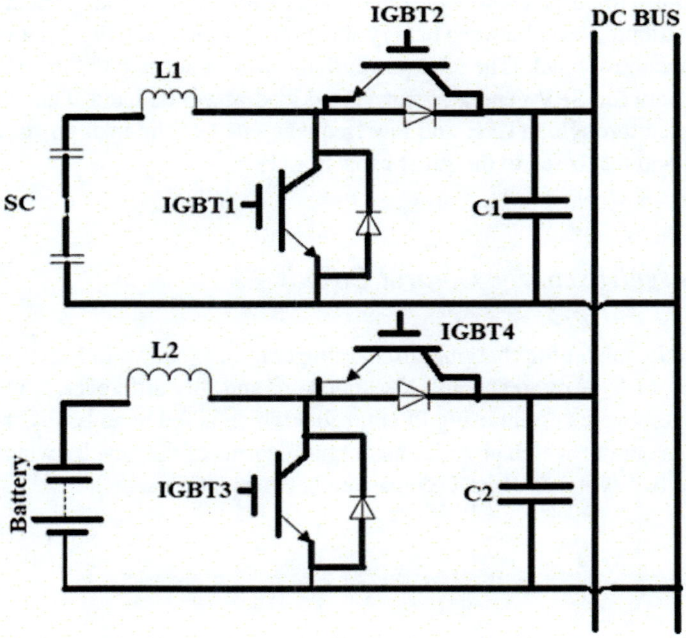

Fig. 7 Control circuit of hybrid energy storage system

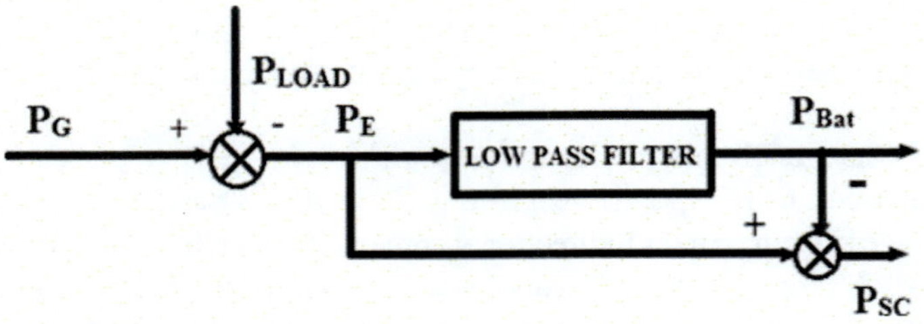

Fig. 8 Power distribution block

bus to ESSs and capacitor takes place via IGBT2 or IGT4 and inductor, hence in this case charging of HESS takes place. Whereas IGBT1 or IGBT3 does not operate in this mode that is they are open. Now when IGBT1 or IGBT3 operates in PWM mode, then BDC converter works as a step-up converter and flow of power from HESS to DC bus takes place.

3 Control System Methodology

3.1 Power Allocation

This is the most important part of the control scheme, so HESS management strategy needs to distribute power between battery and ultra-capacitor, and Fig. 8 shows power distribution control block. The total power that is to be allocated to the HESS is the difference between SPV generated power and load power demand. This difference is then passed through an LPF, and low frequency part of the total power is fed to the battery and rest is fed to the super capacitor [19].

3.2 Charge/Discharge Control Circuit

A SIMULINK model for charging discharging control is shown in Fig. 9. In which a difference of V_{DC} reference and V_{DC} is taken, and this difference is utilized to generate a reference current using PI controller which is fed to an LPF. Error is fed to a PWM generator which provides an output in terms of duty cycles which can be provided to two switches of BDC connected between DC bus.

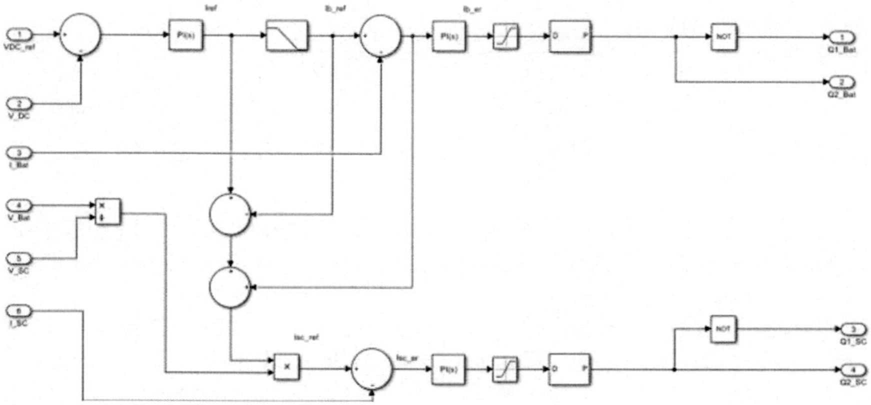

Fig. 9 SIMULINK model for control circuit of HESS

4 Experimental Validations

The performance of presented control method the model of the islanded microgrid with battery and SC model is implemented using SIMULINK platform which simulates a varying irradiance and constant load sourced by HESS elements and PV panels. Table 1 shows the parameter of simulations.

4.1 Operating Strategy

The system is studied with constant load conditions while the PV output is varied in two modes with different cases of PV power at standard conditions. Simulations are done in these conditions in two cases that is using only battery and with a SC and battery together and compare the benefits of using HESS. Bus voltage deviations

Table 1 System parameters

Parameter	value
PV rating	1000 W
Battery rating	50 Ah
Battery nominal voltage	24 V
Battery initial SOC	45%
DC link capacitance	0.175 mF
Load power	500 W
Capacitance rating	50 F
SC voltage	32 V
Bus voltage	50 V

Table 2 Operation modes of islanded microgrid

Modes	Cases	PV Power	ESS
Mode 1	Case 1	PPV = 0	Discharging
Mode 1	Case 2	$P_{PV} < P_{Load}$	Discharging
Mode 2	Case 1	$P_{PV} > P_{Load}$	Charging

are also discussed to authenticate the power management methodology. Load is kept constant at 500 watts. Standalone mode of operation with various modes and cases have been considered as listed in Table 2.

Load Power is constant (500 W) throughout the observation. This system is observed for a time range of 12 h daytime load profile and load is shown in Fig. 10. Figure 12 shows the DC bus voltage without SC connected and a stretched waveform of same is shown in Fig. 14. A rise of 9 V and dip of 11 V is seen from reference voltage (50 V) while operating without SC. Figure 13 shows bus voltage when SC and a stretched waveform of same is shown in Fig. 15. Very small fluctuations of around 0.5 V are seen in bus voltage. Highest value of fluctuation is 2.1 V. Hence it cleared that using capacitor with battery suppress the high voltage fluctuations and maintain the voltage within the stability limits. Using SC along with battery fluctuations in the battery power can be reduced up to an extent which improves the performance and life of the battery. Super capacitor absorbs or delivers high frequency component of the power. When there is sudden switching of load SC provides this peaky power at once as shown in Fig. 21. When fluctuation occurs and once again SC helps in suppressing these transients (Figs. 11, 16, 17, 18, 19 and 20).

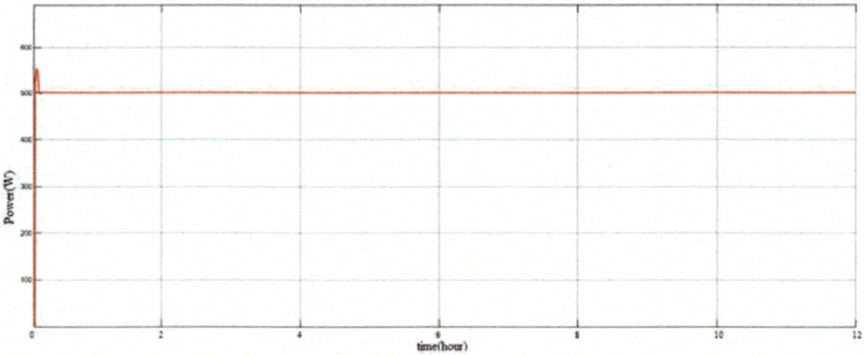

Fig.10 Constant DC load power

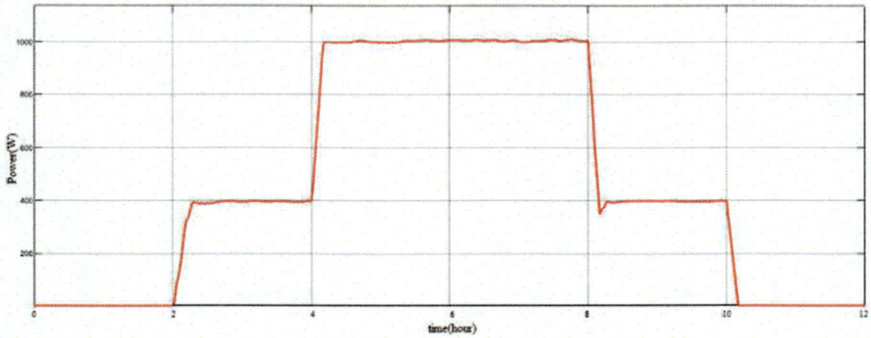

Fig. 11 PV power

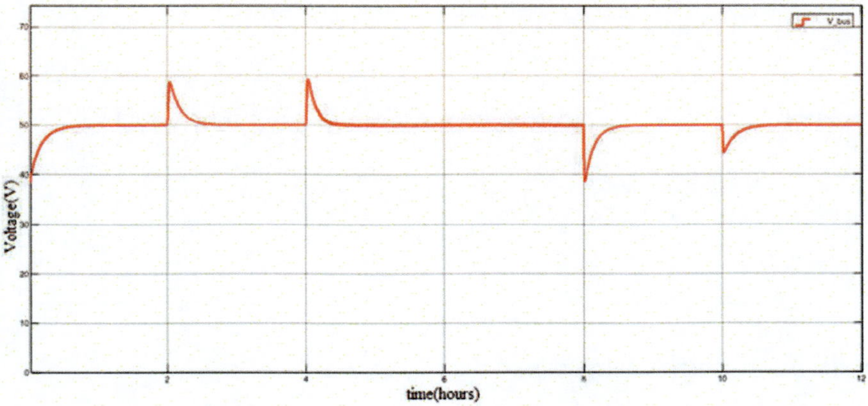

Fig.12 DC bus voltage without SC

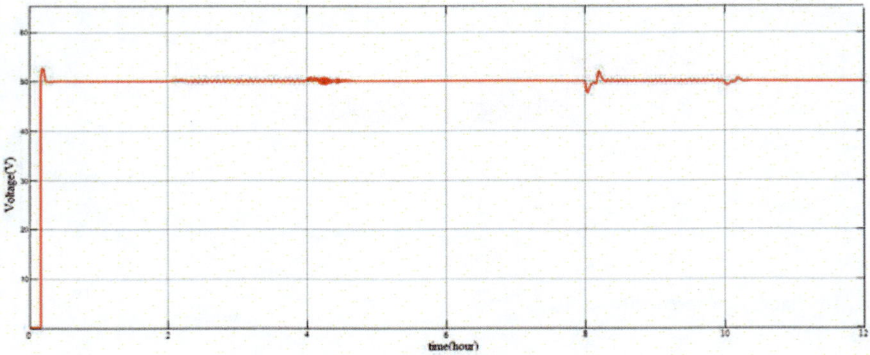

Fig. 13 DC bus voltage with SC

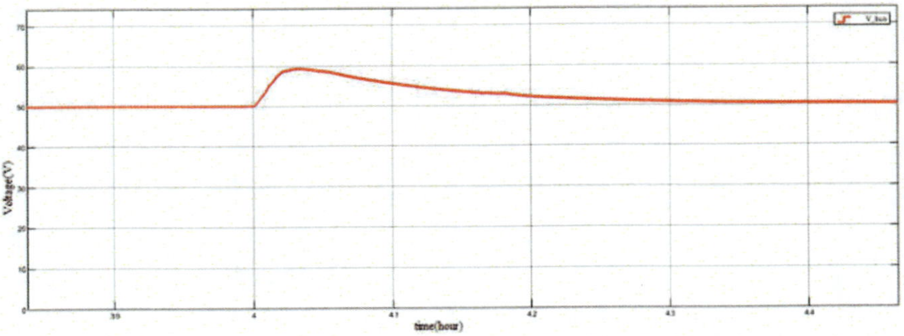

Fig. 14 Stretched DC bus voltage without SC

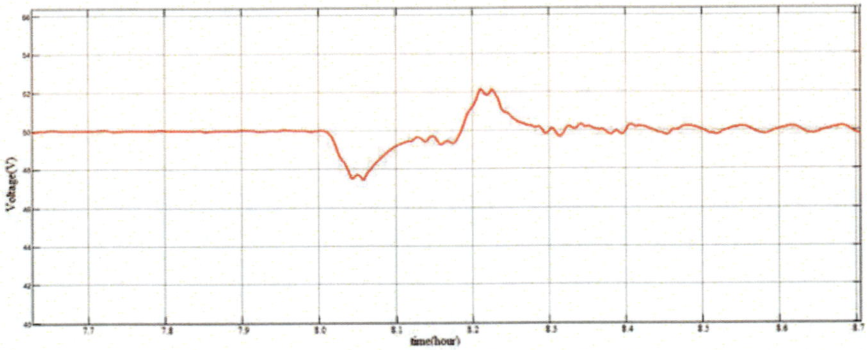

Fig. 15 Stretched DC bus voltage with SC

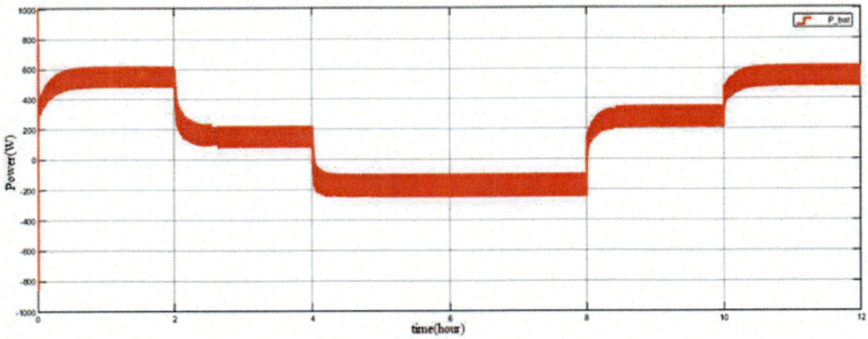

Fig. 16 Battery powered without SC

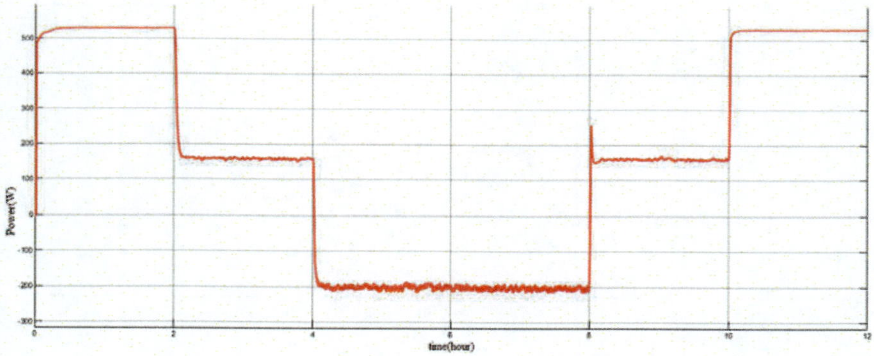

Fig. 17 Battery powered with SC

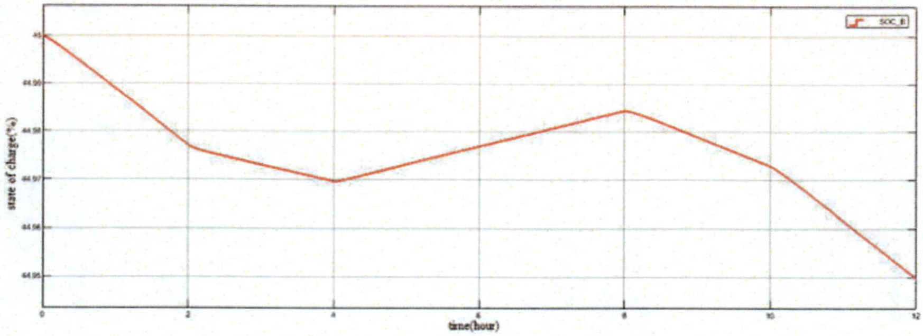

Fig. 18 Battery SOC without SC

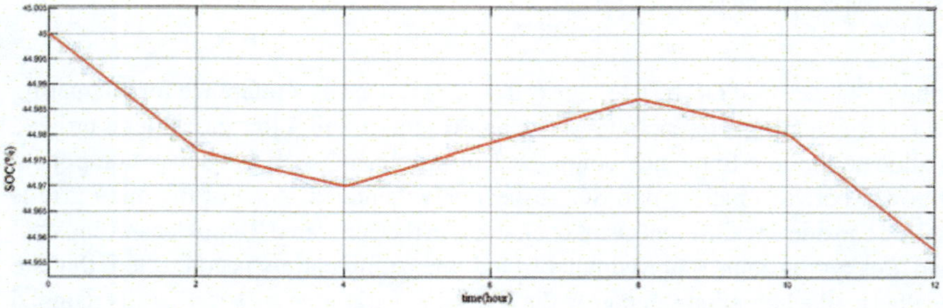

Fig. 19 Battery SOC with SC

5 Conclusion

A low level DC MG of low voltage with solar photovoltaic as renewable energy source and HESS as storage system is developed. A control methodology based on reference current generation is proposed to control the charge/discharge of HESS

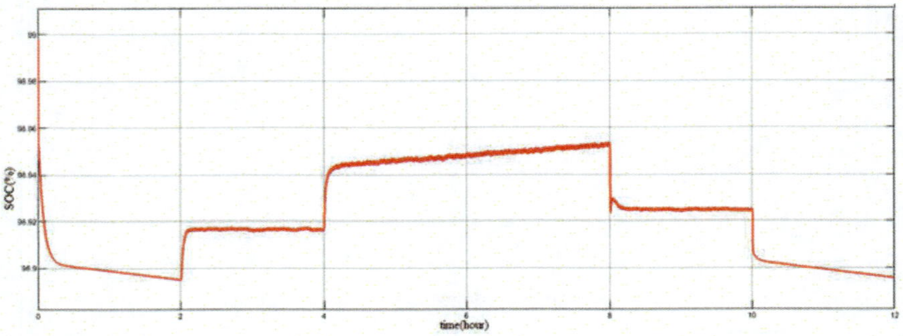

Fig. 20 Supercapacitor SOC

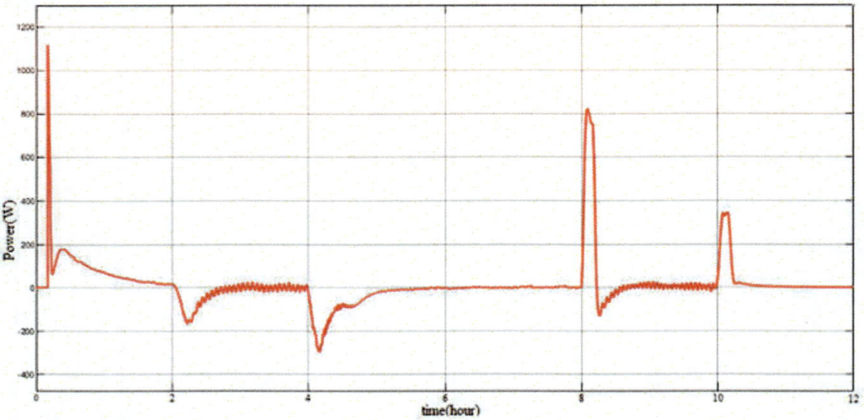

Fig. 21 Supercapacitor power

using BDC. Desired DC bus voltage (50 V DC) is used as a reference for comparing DC bus for creating reference current using PID controller. A low pass filter is utilized to allocate the low frequency component of power and high frequency component of power between battery and SC, respectively. Islanded mode of the microgrid is studied considering different cases of changes in the generated PV power and constant load. Peaky transients (approximately 11 V when only battery is connected) that are resulted by the immediate change in the PV power variations are successfully damped by the introduced control strategy energy with very small fluctuations (only around 2.1 V) left in voltage. The represented control methodology and microgrid system modeling is done using MATLAB and simulated using SIMULINK environment. The results that are obtained by simulations validate the effectiveness of the presented control methodology.

References

1. Aneke M, Wang M (2016) Energy storage technologies and real life applications—A state of the art review. Appl Energy 179:350–377
2. Patterson BT (2012) DC, come home: DC microgrids and the birth of the Enernet. IEEE Power Energy Magazine 10:60–69
3. Lazaroiu GC, Zaninelli D, Lazaroiu GC (2010) A control system for dc arc furnaces for power quality. Electric Power Syst Res 80(12):1498–1505
4. Chong LW, Wong YW, Rajkumar RK, Rajkumar RK, Isa D (2016) Hybrid energy storage systems and control strategies for stand-alone renewable energy power systems. Renew Sustain Energy Rev 66:174–189
5. Akhtar I, Paul WUH, Kirmani S, Asim M (2021) Cost analysis of 18 kW solar photovoltaic system for smart cities growth in India. In: Iqbal A, Malik H, Riyaz A, Abdellah K, Bayhan S (eds) Renewable power for sustainable growth. Lecture notes in electrical engineering, vol 723. Springer, Singapore. https://doi.org/10.1007/978-981-33-4080-0_63
6. Eghtedarpour N, Farjah E (2014) Distributed charge/discharge control of energy storages in a renewable-energy-based DC microgrid. Renew Power IET 8(1):45–57
7. Dou X, Quan X, Wu Z, Hu M, Sun J, Yang K, Xu M (2014) Improved control strategy for microgrid ultracapacitor energy storage systems. Energies 7(12):8095–8115
8. Park J-D (2010) Simple flywheel energy storage using squirrel-cage induction machine for DC bus microgrid. In: IEEE industrial electronics society conference, pp 3040–3045
9. Karaipoom T, Ngamroo I (2013) Enhancement of LVRT performance and alleviation of power fluctuation of DFIG wind turbine in DC microgrid by NI. In: SEMSIEEE international conference on applied superconductivity and electromagnetic devices, pp 207–208
10. Hajiaghasi S, Salemnia A, Hamzeh M (2019) Hybrid energy storage system for microgrid application: a review. 21:543–570
11. Ali D, Paul WUH, Ali MS, Ahmad M, Ashfaq H (2021) Optimal placement of distribution generation sources in hybrid generation network. Smart Grid and Renew Energy 12:65–80. https://doi.org/10.4236/sgre.2021.125005
12. Iqbal MM, Islam K (2017) Design and simulation of a PV system with battery storage using bidirectional DC-DC converter using MATLAB simulink. Int J Sci Technol Res 6(07). ISSN 2277-8616
13. Salameh ZM, Casacca MA, Lynch WA (1992) A mathematical model for lead-acid batteries. IEEE Trans Energy Convers 7(1):93–98
14. Paul WUH, Bhat MB, Kirmani S, Nahvi SA (2020) Data based controller design for PMDC motor setup using system identification. Stud Ind Place Names 40(10):11
15. Buller S, Thele M, Doncker RWAAD, Karden E (2005) Impedance-based simulation models of supercapacitors and Li–Ion batteries for power electronic applications. IEEE Trans Ind Appl 41(3):742–747
16. Benini L, Castelli G, Macii A, Macii E, Poncino M, Scarsi R (2001) Discrete-time battery models for system-level low-power design. IEEE Trans Very Large Scale Integr Syst 9(5):630–640
17. Min C, Rincon-Mora GA (2006) Accurate electrical battery model capable of predicting runtime and I-V performance. IEEE Trans Energy Convers 21(2):504–511
18. Hernandez GD (2016) Removal of direct current link harmonic ripple in single-phase voltage source inverter systems using supercapacitors. M.S. thesis. ECE Department, CA, USA
19. Hhuijuan W, Jiancheng Z (2016) Research on charging/discharging control strategy of battery-supercapacitor hybrid energy storage system in photovoltaic system. In: IEEE international power electronics and motion control conference. 978-5090-1210-7/16

Statistical Analysis of Household Power Supply Quality

Gaikwad Sachin Ramnath and R. Harikrishnan

Abstract The electricity is the fundamental need for the socioeconomic development of the nation. The access to electricity and its infrastructure development constitutes as the fundamental stage. It is crucial to focus on the quality of the power which is used in every household. This paper focuses on examining electricity power quality with voltage fluctuation and power outage. The purpose of this analysis is to focus on understanding the supply power quality in a rural region and its impact on social life. For achieving this purpose, a domestic category dataset is used, which belongs to a Kondhanur village, Pune district of Maharashtra state, India. This dataset consists of data about the voltage used in a household for the duration of one year from 1st January 2018 to 31st December 2018. The dataset uses time series data having the voltage recorded for every minute of the day for a year. In findings observed that, a minimum up to 116 V is recorded at morning time period (7th hour), and a maximum up to 262 V is recorded at the night time period (23rd hour). In one year, 45% times the available power supply is in required voltage range (221–240 V). The 3.84% times occurred the power failure/outage (0 V). Moreover, maximum 50.63% times occurred low and high voltage problems.

Keywords Descriptive statistics · Quality of electricity supply · Power failure · Voltage fluctuation · Time series data

1 Introduction

There are many methods and tools that have been developed to analyze the collected electricity data. The outcome of electricity data analysis is related to load analysis, load prediction, load management, power quality analysis, electrification analysis,

G. S. Ramnath · R. Harikrishnan (✉)
Symbiosis Institute of Technology (SIT), Symbiosis International Deemed University (SIDU), Pune, India
e-mail: dr.rhareish@gmail.com

G. S. Ramnath
e-mail: sachin.r.gaikwad@outlook.com

K. Namrata et al. (eds.), *Smart Energy and Advancement in Power Technologies*,
Lecture Notes in Electrical Engineering 926,
https://doi.org/10.1007/978-981-19-4971-5_26

voltage fluctuation analysis, and power outage analysis [1, 2]. At present, along with electricity connection, its power quality (PQ) with reliability and voltage fluctuation, stability parameters are equally important and statistically significant for socioeconomic development. Voltage fluctuation, power outage, and power blackout play a major role in affecting the safety and health of people. According to the previous studies, there are three main costs of power outages, namely direct financial costs, in-direct costs, and social costs. The financial cost includes loss of productivity, resume costs, appliance damage, and raw material spoilage. The in-direct cost includes loss related to income being postponed and economic loss in terms of investment and share. The social cost is about leisure time and risk to healthiness and safety. Among these power outage costs, author [3] focused on social cost and their impact on end consumers. In social cost analysis, the main challenge is to get sufficient datasets, it is due to the dynamic, complex problem which depends on various factors like, power outage duration, and its effects on different characteristics of individual consumers [4, 5]. So, author [3] filled the gap of lagging dataset for social cost by exploring the new source of data using Swedish Styrel system.

The authors [6, 7] build the outage prediction model (OPM) to predict the influence of storms on the power outage by applying the various tree-based ML regression algorithms, namely decision tree (DT), random forest (RF), boosted gradient tree (BT), and ensemble decision tree (ENS). The main challenge in OPM accuracy is the number of sample sizes to train the model. This work determined the optimal sample size for building the efficient OPM. This work also helps utilities to build their OPM by selecting optimal sample size. This OPM will act as a decision-making tool on storm events preparedness. Moreover, this work also proposed a new method to resolve the under and over-estimation biases exhibited by earlier studies in high severity and low-severity events, respectively [6].

Furthermore, smart grid (SG) allows the real-time bi-directional flow of electricity and information through a smart meter [4, 5]. The smart meter is a device which is capable to monitor and assess the real-time online performance of PQ parameters like notches, transients, momentary disorder, voltage swell and sag, voltage fluctuations, harmonic, and power factors. The author [8] performed the PQ assessment of voltage fluctuation by smart meter data through discrete wavelet transform (DWT) technique. The study proposed semi-supervised machine learning algorithm method which automatically detect disturbed voltage waveforms using a one-class support vector machine (OCSVM) with 93% accuracy. Moreover, for the classification of PQ issue, the multi-class SVM was applied with 90% accuracy.

In literature, different research aspects on PQ disturbance detection by using smart meter data were noted. The PQ problems are resolved into two main steps: First step is to find the more insights by feature extraction methods from voltage waveforms; second step is to train the model based on obtained feature sets and labeled data. Furthermore, the efficient feature extraction is one of the research direction in PQ disturbance detection. There are various efficient methods for feature extraction like Fourier transform (FT), S-transform (ST), wavelet transform (WT), fast Fourier transform (FFT), fast dyadic Fourier transform (FDFT), and root mean square (RMS). One more research direction in PQ is about the determination of optimal sets of features from smart meter signal data. These research directions are vital to perform the better assessment of PQ parameters in terms of voltage fluctuation and segregation through diverse artificial intelligence (AI) and machine learning (ML) techniques like fuzzy logic, neural network, SVM, decision tree, expert system, and hidden Markov model. Few researchers applied different research aspect for PQ disturbance detection, classification, with computational efficiency and processing speed [8, 9].

The main challenge about PQ research is to get an accurate and sufficient amount of voltage data. This arises due to the problem which have involvement of multi-dimensional variables, namely outage duration, voltage stability, and behavioral approach of individual consumers [3]. Furthermore, according to the United Nations (UNs) Sustainable Development Goals, one of its goals is to "ensure access to affordable, reliable, sustainable, and modern energy for all." Moreover, to achieve this goal, the World Bank and UNs have launched a Global Tracking Framework (GTF) that measures global development in energy access over time. GTF works beyond household electrification through the multi-tiered scale with multi-dimensional variables. On the same line, GTF studied the rate of electricity access across the globe including rural and urban locations database. This study helps GTF to check world-wide progress in the electrification of domestic category consumers.

Moreover, the electrification of households depends on various parameter, one among them being the per capita income of household. The additional key drivers to increase household electrification includes population density and urbanization [10]. Furthermore, there is an urgent need to move toward quality electricity supply for all. The quality of electricity supply directly influences the socioeconomic development of an individual. Moreover, the power supply data include multi-dimensional variables from which electricity access and individual's behavioral intervention can be analyzed. Despite, there is need to attract the attention of policy-makers and take initiatives on energy poverty and PQ issues in rural region [11].

For statistical analysis purpose, voltage fluctuation and power outage parameters are chosen, along with time-based data analysis approach. Moreover, the quality of power supply depends on the parameters which are considered for the purpose of analysis. For analysis, two significant parameters, namely time and season, are considered. Power demand and PQ are sensitive and vary with respect to time and seasons. So, the complete statistical data analysis is based on sample hour and minutes based. The selection of hour of the day depends on the on-peak time duration and off-peak-time duration. The considered hours are 7.00, 13.00, 19.00, and 23.00 from a day. Moreover, from each selected hour, 59th minute of the sample hour is considered.

The paper includes the following section: Sect. 2 discusses about data analysis methodology in materials and methods; Sect. 3 discusses the result analysis, and Sect. 4 is about the conclusion with future directions.

2 Materials and Methods

For data analysis consider the domestic category of Kondhanur village, Pune district of Maharashtra state India. The minute-wise voltage data of the 2018 year are taken for quality power supply analysis. The time series dataset is taken from Harvard dataverse Web site, which was collected by Prayas (Energy Group), Pune, India under the Electricity Supply Monitoring Initiative (ESMI). The structure of the voltage dataset includes location, date, hour, and minute. The purpose of ESMI is to enhance transparency and provides evidence-based feedback related to the quality of electrical power supply [12]. The data collection under ESMI was conducted by an electronic device. This device is like handheld radio with plug and play type. When the device is plugged into a plug point, then the device starts measuring voltage by the accuracy of ±4 V. Moreover, it records each minute's voltage value and then sends the same to the central server through a GPRS network [13].

Figure 1 shows the different time-based data used for the purpose of analysis. Figure 2 shows the three-step decision support system for power quality analysis in the proposed methodology. The original dataset includes various geographical location data under various categories like domestic, non-domestic, and agriculture, along with the voltage data for every hour and the minute of the day. After preprocessing the data, the dataset was modified for a yearly-duration from 1st January 2018 to 31st December 2018, which is used to achieve the pre-defined aim of the analysis.

3 Results and Discussion

Nowadays, every day and everywhere, time series data are generated. These time series data are having many applications not only in households but also in organizations and factories. The authors [14, 15] studied the prediction of multivariate time series data using a real-words dataset for various applications, namely traffic (road occupancy), electricity consumption, an exchange rate (daily exchange rates of foreign countries), solar power production, and polyphonic piano pieces. As per Fig. 1, different time series-based data analysis is performed, and results are shown in tables and graphs format.

3.1 One-Year Data Analysis with and Without Power Outage

The original dataset is having each hour and minute-wise voltage data of Kondhanpur location Pune India for the year 2018. This section deals with one-year data analysis with and without power outage, and following analysis has been carried out.

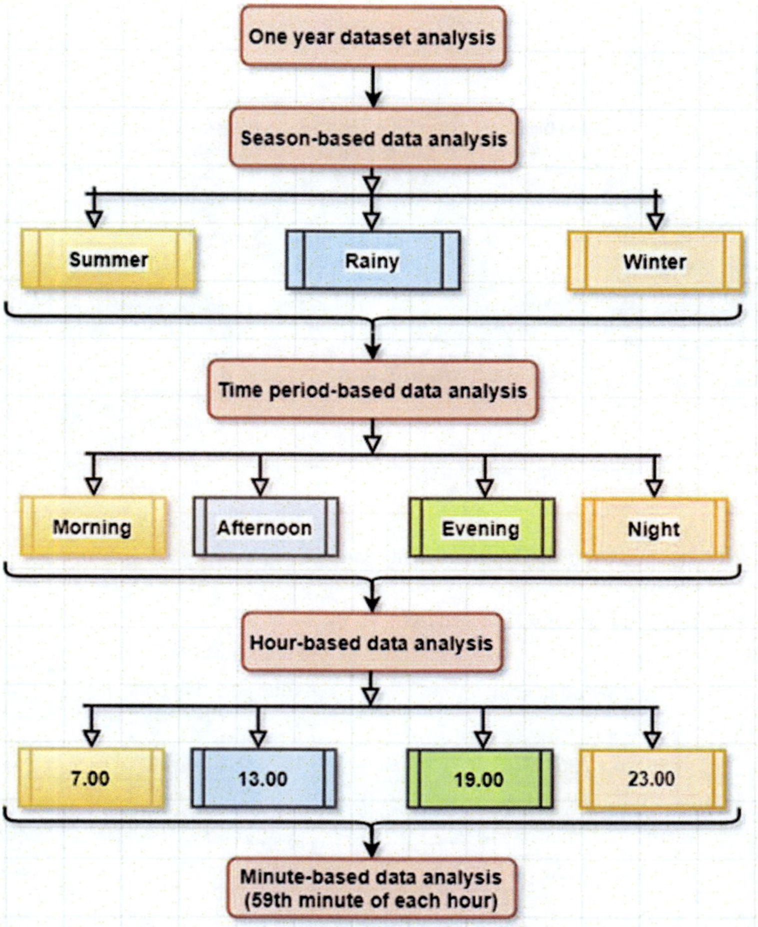

Fig. 1 Time series-based data analysis

Descriptive Statistics Analysis Tables 1 and 2 provide information about available voltage value on 59th minute of different hours' time period of based on seasons. It is observed that 23rd hour of the day is recorded maximum voltage up to 262 V in the year 2018 dataset. Table 1 shows that, maximum mean voltage is 230 V occurred at night time (23rd hour), and minimum mean voltage is 199 V occurred at morning time (7th hour). This variation in mean voltage is reported due to the variation in electricity demand.

Distribution Plot Analysis Figures 3, 4, 5, 6, 7, 8, 9, 10, 11, 12, 13, 14, 15, 16, 17 and 18 show the voltage fluctuation and density in a graphical format. Figure 3 shows the highest voltage variance (130 V) at morning time as compare to other time periods. Moreover, Fig. 4 shows that, maximum power outage issues are occurred at

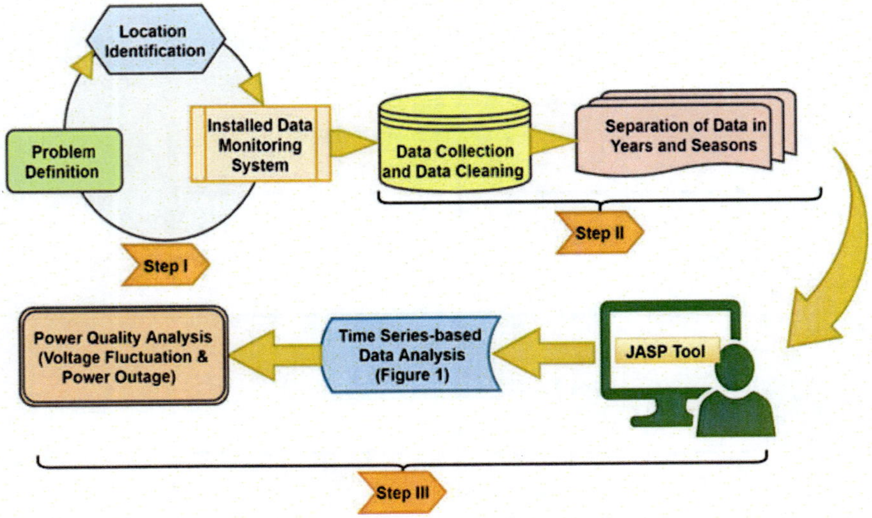

Fig. 2 Decision support system for power quality analysis

Table 1 One year and summer season analysis

Parameter ⇓	One year				Summer season			
	Time period in hour				Time period in hour			
	7	13	19	23	7	13	19	23
Valid	364	363	364	364	119	118	119	119
Missing	0	0	0	0	0	0	0	0
Mean	199	205	215	230	200	192	202	219
Median	209	226	221	234	206	218	208	221
Minimum	0	0	0	0	0	0	0	0
Maximum	246	256	257	262	236	239	239	240

Table 2 Rainy season and winter season analysis

Parameter ⇓	Rainy season				Winter season			
	Time period in hour				Time period in hour			
	7	13	19	23	7	13	19	23
Valid	119	118	119	119	123	123	123	123
Missing	0	0	0	0	0	0	0	0
Mean	200	192	202	219	200	211	227	238
Median	206	218	208	221	209	232	229	237
Minimum	0	0	0	0	0	0	192	222
Maximum	236	239	239	240	246	256	257	257

afternoon time (13th hour). Figures 7 and 11 show low voltage problems at mooring time (7th hour) in summer season and rainy season.

Box Plot Analysis Figures 19, 20, 21 and 22 show the voltage fluctuation on 59th minute as sample hour for different seasons. Comparing the four box plot analysis, the winter season has recorded no power outage except the 7th and 13th hour as shown in Fig. 22. Moreover, at morning, 7th hour shows more voltage variations compare to other hours in a year. The maximum power outage is occurred at afternoon time (13th hour) as shown in Figs. 19, 20, 21 and 22.

Fig. 3 One-year voltage at 7th hour

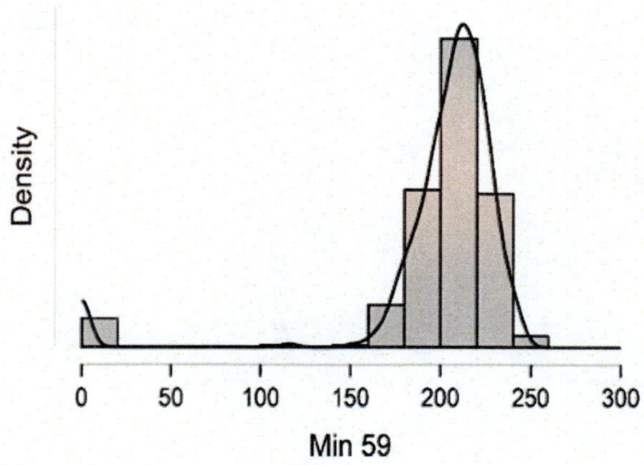

Fig. 4 One-year voltage at 13th hour

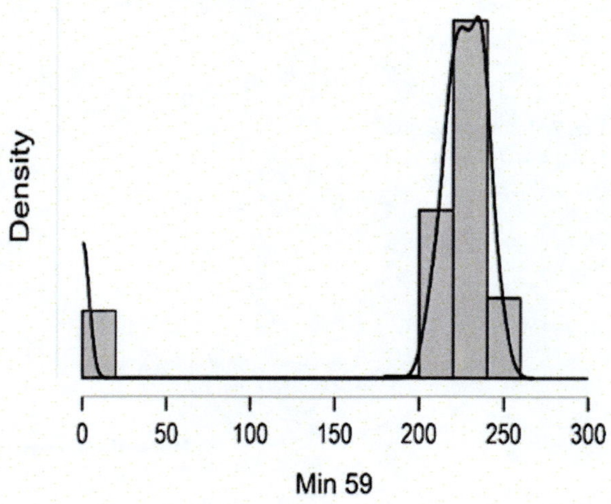

Fig. 5 One-year voltage at 19th hour

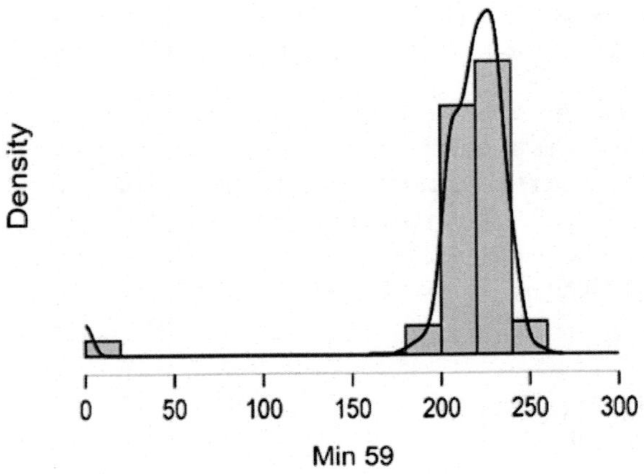

Fig. 6 One-year voltage at 23rd hour

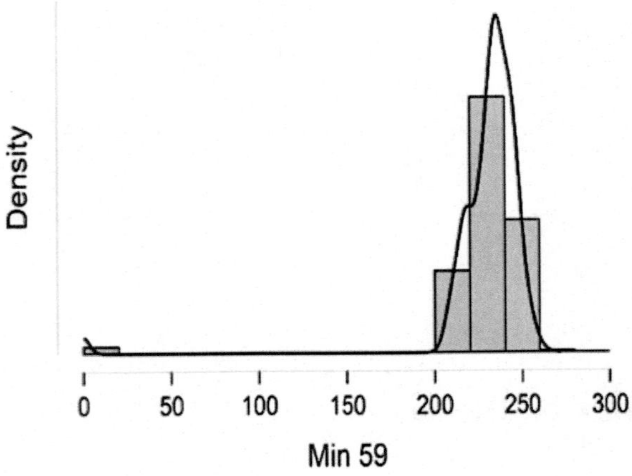

Fig. 7 Summer season voltage at 7th hour

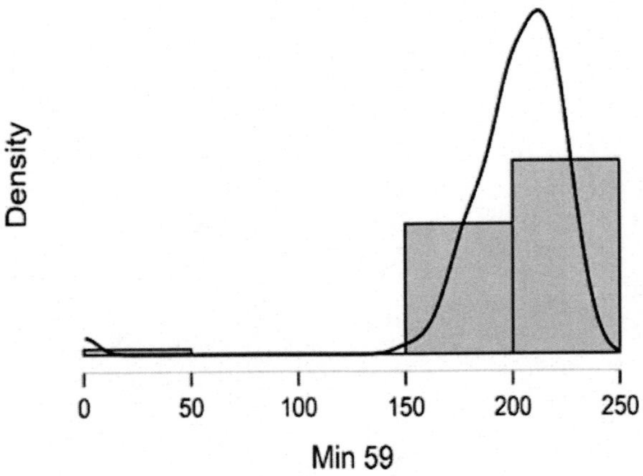

Fig. 8 Summer season voltage at 13th hour

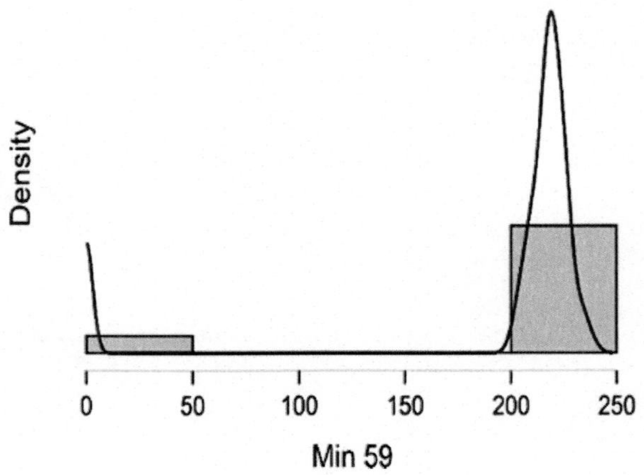

Fig. 9 Summer voltage at 19th hou

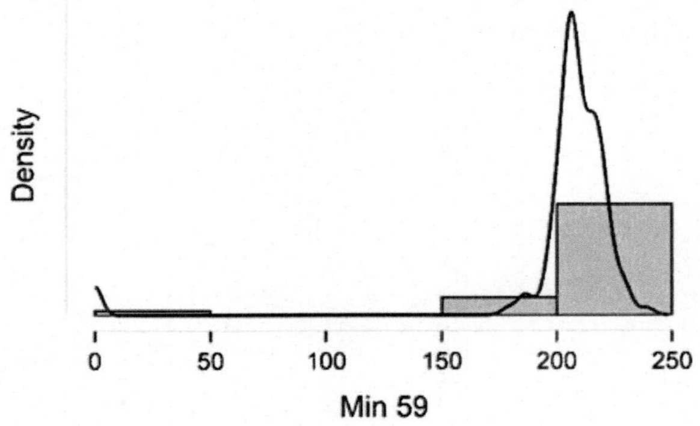

Fig. 10 Summer voltage at 23rd hour

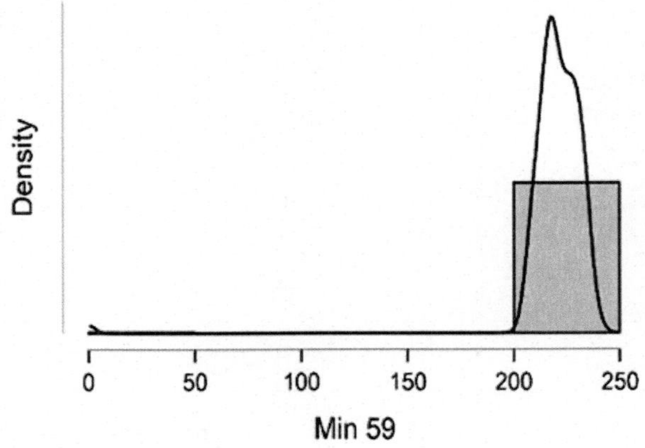

Fig. 11 Rainy season voltage at 7th hour

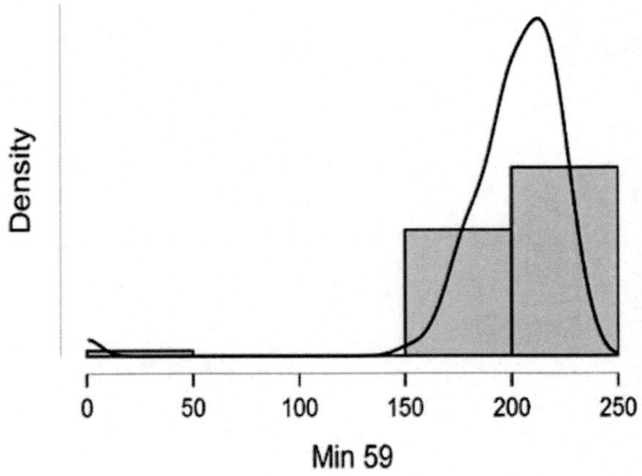

Fig. 12 Rainy season voltage at 13th hour

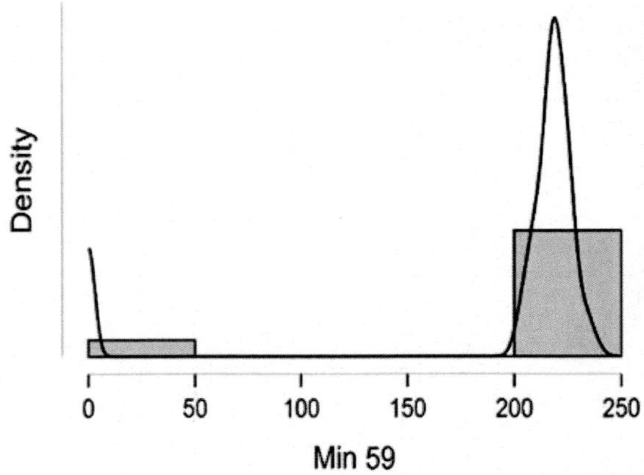

Fig. 13 Rainy season voltage at 19th hour

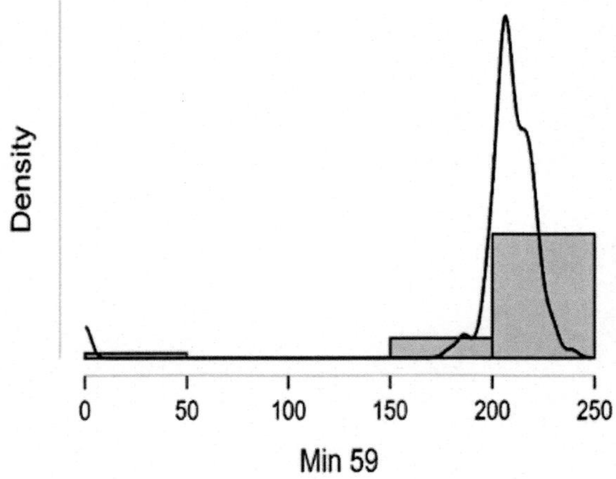

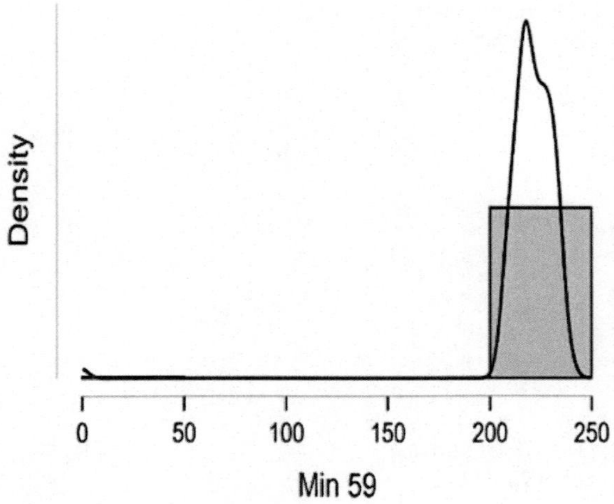

Fig. 14 Rainy season voltage at 23rd hour

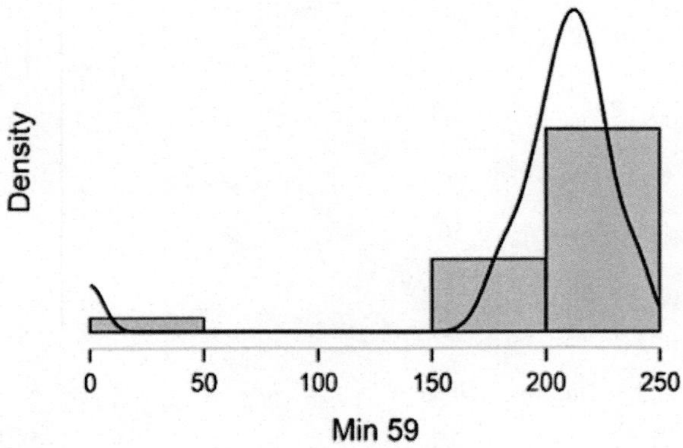

Fig. 15 Winter season voltage at 7th hour

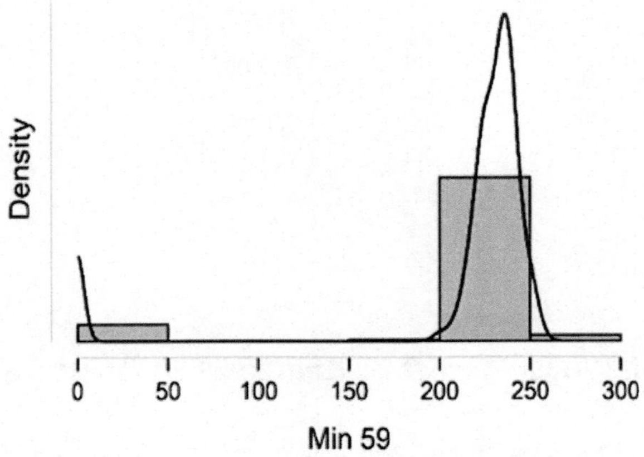

Fig. 16 Winter season voltage at 13th hour

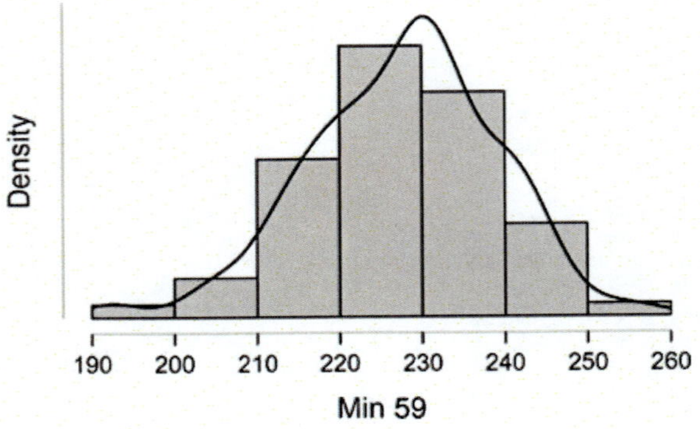

Fig. 17 Winter season voltage at 19th hour

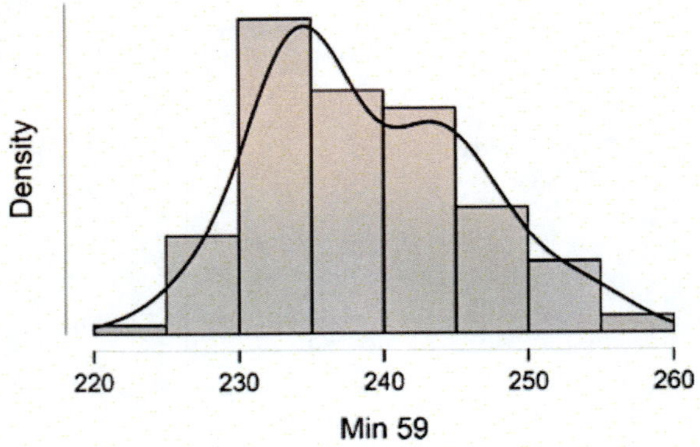

Fig. 18 Winter season voltage at 23rd hour

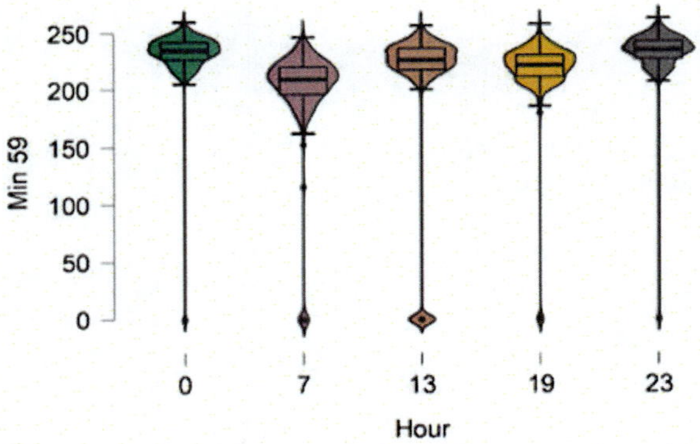

Fig. 19 One-year voltage analysis

3.2 One-Year Data Analysis with Power Outage

In this section, one-year data analysis with power outage is carried out. The data analysis of power outage of a year is shown in Table 3, and its box plot analysis shows in Fig. 23. The analysis shows that a total of forty times of a power outage is

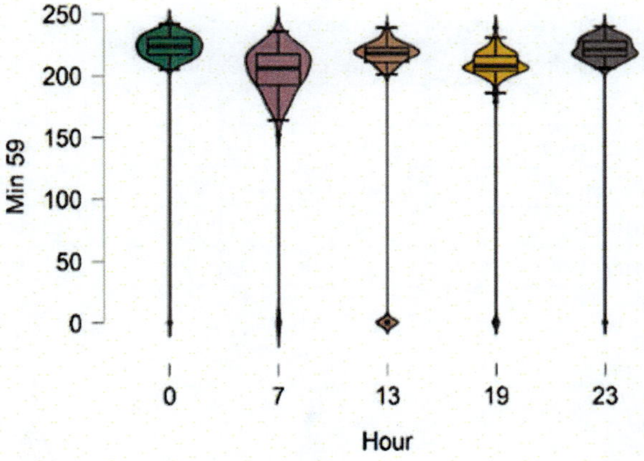

Fig. 20 Summer season voltage analysis

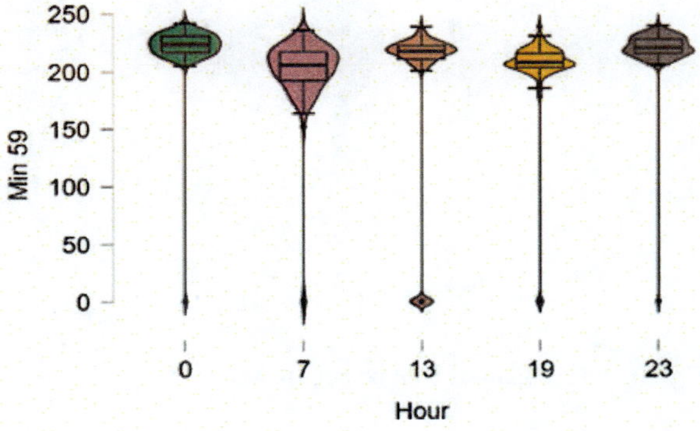

Fig. 21 Rainy season voltage analysis

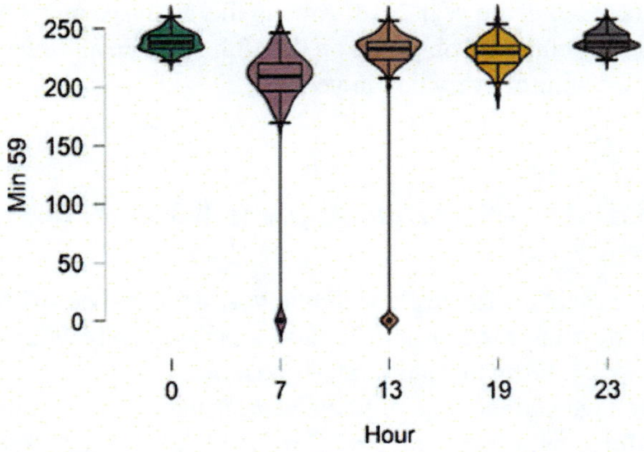

Fig. 22 Winter season voltage analysis

Table 3 1-year power outage analysis

Parameter	Time period in hour			
⇓	7	13	19	23
Frequency	8	24	5	3
Percentage	20	60	12.5	7.5
Min. volt	0	0	0	0
Max. volt	0	0	0	0

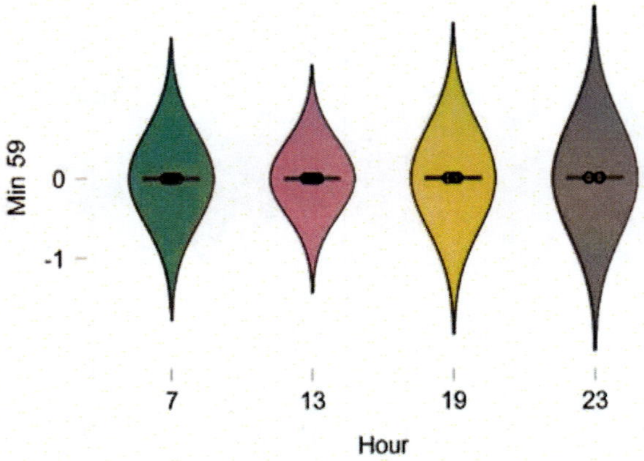

Fig. 23 1-year analysis with no hour power

recorded in the year 2018. Among this analysis, maximum twenty-four (60%) times power outage is observed on the 13th hour, and 3 times power outages are occurred 23rd hour of every day in year.

3.3 *One-Year Data Analysis Without Power Outage*

This sub-section deals with one-year data analysis with voltages. The data analysis with voltage is shown in Table 4, and its box plot analysis shows graphical format in Fig. 24. From the data analysis observed that, the 7th hour of the day provides larger voltage variations with 17 standard deviations as compare to other hours of the day. Moreover, means voltage is low at 7th hour around 208 V. The minimum observed voltage is 116 V at 7th hour, and maximum voltage is 262 V at 23rd hour.

Table 4 1-year without power

Parameter	Time period in hour			
⇓	7	13	19	23
Frequency	346	308	342	359
Mean	208	227	220	233
Std. deviat	17	11	12	11
Min. voltage	116	200	179	206
Max. voltage	246	256	257	262

Fig. 24 1-year analysis without power outage

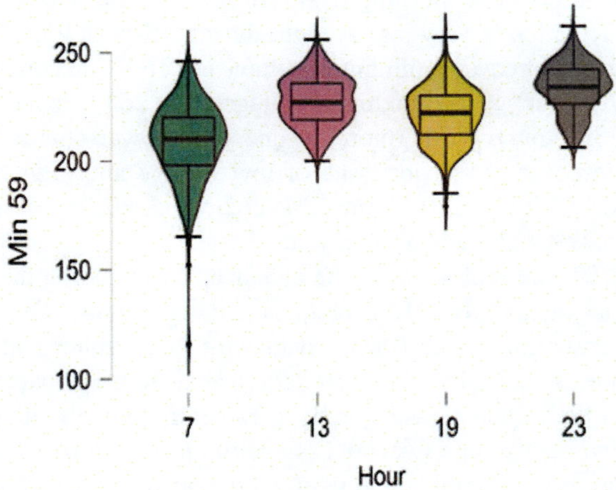

3.4 One-Year Data Power Quality Analysis

Table 5 shows one-year PQ in terms of voltage fluctuations and power failure at 59th minutes of sample hour 7th, 13th, 19th, and 23rd. The required voltage range is between 221 and 240 V which occurred only 45% times in one-year. In a year around 3.84% household consumers face power failures and around 50.63% household consumers are facing either low voltage or high voltage fluctuation problem or both.

Table 5 1-year power quality

S. No	Volt. status	Volt. range (V)	Count	% of count
1	Power outages	0	70	03.84
2	Low voltage	116–220	630	34.63
3	High voltage	241–262	286	15.73
4	Operating volts	221–240	833	45.79
		Total	1819	100%

4 Conclusion and Future Directions

In this paper, various micro-statistical time-based data analysis starting from yearly based up to minute-based method is applied to Kondhanpur village dataset, which is located in the Pune district of Maharashtra state in India. For the data analysis, dataset of the year 2018 is considered. Descriptive-graphical statistical operations are performed. Beyond electrification, this data analysis focuses on sustainable and quality electricity for all. From the above statistical data analysis, it confirms that voltage fluctuation is the issue and needs to be addressed at a particular time period of the day. According to time series voltage data analysis, maximum voltage fluctuation is observed at morning peak 7th hour period and 59th minute time, and voltage fluctuation is very less at night during off-peak 23rd hour period and 59th minute time. Moreover, a minimum voltage of 116 V is recorded in the morning time period (7th hour), and a maximum voltage up to 262 V is recorded at the nighttime period (23rd hour). Furthermore, in one year, power outage is recorded of 3.84% (0 V). Moreover, the major issue of low voltage and high voltage is observed which is 50.63% (116–220 V and 241–262 V). And the operating voltage is available of 45.79% (221–240 V) in a year.

The current work can be enhanced by including the following for power quality analysis. Apply ML techniques and algorithms, use smart meter data, use phasor measurement unit (PMU) data with large volume and more durations, consider questionnaire-based survey data to analyze the impact as social and economic cost. Further, other power quality parameters, namely, notches, transients, momentary disorder, voltage swell and sag, harmonic and power factors, can be considered. The future scope could be to work on the objective of sustainable electricity for all. The main challenge is to record the accurate measured data of multi-dimensional parameters of electricity supply. This would lead to get more insights on affordability, capacity, voltage stability, reliability, legality, health, and safety aspect.

Acknowledgements Authors are thankful to Symbiosis International Deemed University (SIDU) for allowing to carry out our research and to use resources to accomplish the objectives.

References

1. Wang Y, Chen Q, Hong T, Kang C (2019) Review of smart meter data analytics: applications, methodologies, and challenges. IEEE Trans Smart Grid 10(3):3125–3148
2. Oh S, Haberl JS, Baltazar JC (2020) Analysis methods for characterizing energy saving opportunities from home automation devices using smart meter data. Energy Build 216:109955
3. Landegren F, Johansson J, Samuelsson O (2019) Quality of supply regulations versus societal priorities regarding electricity outage consequences: case study in a Swedish context. Int J Crit Infrastruct Prot 26:100307
4. Gaikwad SR, Harikrishnan R (2021) Households electricity consumption analysis: a bibliometric approach. Library Philos Pract 1–21

5. Gaikwad SR, Harikrishnan R (2021) Social welfare maximization in smart grid: Review. In: IOP conference series: materials science and engineering. International conference on applied scientific computational intelligence using data science, ASCI-2020, pp 1–10
6. Yang F, Wanik DW, Cerrai D, Ehsan Bhuiyan MA, Anagnostou EN (2020) Quantifying uncertainty in machine learning-based power outage prediction model training: a tool for sustainable storm restoration. Sustainability (Switzerland) 12(4)
7. Cerrai D, Wanik DW, Bhuiyan MAE, Zhang X, Yang J, Frediani MEB, Anagnostou EN (2019) Predicting storm outages through new representations of weather and vegetation. IEEE Access 7(c):29639–29654
8. Parvez I, Aghili M, Sarwat AI, Rahman S, Alam F (2019) Online power quality disturbance detection by support vector machine in smart meter. J Mod Power Syst Clean Energy 7(5):1328–1339
9. Borges FAS, Fernandes RAS, Silva IN, Silva CBS (2016) Feature extraction and power quality disturbances classification using smart meters signals. IEEE Trans Ind Inf 12(2):824–833
10. Aklin M, Harish SP, Urpelainen J (2018) A global analysis of progress in household electrification. Energy Policy 122(November 2017):421–428
11. Aklin M, Cheng CY, Urpelainen J, Ganesan K, Jain A (2016) Factors affecting household satisfaction with electricity supply in rural India. Nat Energy 1(11):1–6
12. P. Prayas Energy Group (2018) Electricity supply monitoring initiative (ESMI)
13. Harvard-Dataverse (2018) Harvard Dataverse. https://dataverse.harvard.edu/dataverse/esmi. Last accessed 20 Apr 2021
14. Shih SY, Sun FK, Lee HY (2019) Temporal pattern attention for multivariate time series forecasting. Mach Learn 108(8–9):1421–1441
15. Lai G, Yang Y, Chang WC, Liu H (2018) Modeling long- and short-term temporal patterns with deep neural networks. SIGIR'18, 8–12 July 8–12 2018, Ann Arbor, MI, USA, pp 95–104

A Combined Sequence Approach Based on Cumulative Sum for Detection of Fault Amidst Power Swing for Line with Series Compensation

Alpana Mehta, Gyanu Gautam, Piyush Kumar, and Jitendra Kumar

Abstract Distance relays are subjected to unintended tripping at the time of power swing. Power Swing Blocking (PSB) function installed with distance relays prevents unwanted tripping. However, if a fault occurs amidst power swing, it must be detected and cleared. In a line with series compensation, fault detection at the time of power swing is challenging. The available techniques fail to detect fault in a line with series compensation. In this paper, a cumulative sum of zero-sequence and negative-sequence current is used to detect fault amidst power swing in a line with series compensation. The cumulative sum of zero-sequence component and negative sequence component helps in determining nature of fault along with the fault detection. The proposed technique has been tested for different fault conditions and compared with available techniques.

Keywords Power swing · Distance relay · MOV (Metal oxide varistor) · Cumulative sum

1 Introduction

Under steady-state conditions, the generated and consumed reactive and active power remains balanced. On the occurrence of system faults or due to line switching, generator disconnection, addition or removal of large loads causes a sudden change in electrical power. The mechanical input remains the same while the electrical power changes. Rotor angle oscillates among the generators, resulting in power swing. Severity of the disturbance decides whether the power swing is stable or unstable. Due to variation in three-phase power and oscillation of the rotor angle, the apparent impedance seen by a distance relay lies within the operational zone of the relay. It

A. Mehta (✉) · G. Gautam · P. Kumar · J. Kumar
Department of Electrical Engineering, National Institute of Technology, Jamshedpur, India
e-mail: alpanamehta077@gmail.com

J. Kumar
e-mail: jitendra.ee@nitjsr.ac.in

K. Namrata et al. (eds.), *Smart Energy and Advancement in Power Technologies*, Lecture Notes in Electrical Engineering 926,
https://doi.org/10.1007/978-981-19-4971-5_27

is misinterpreted by the relay as a fault, leading the relay to trip. A PSB is available in relays to prevent tripping of the relays amidst power swing [1]. The relay should identify the malfunction and operate rapidly to ensure stability.

With the recent development and growing population, to meet the growing electricity demand, the transmission line performance is being enhanced. For this, series compensation is being included in long transmission lines. For the lines with series compensation voltage drop, Ferranti effect and reactive impedance reduce. Due to compensation, fault detection in a line with series compensation becomes challenging. Frequency components of signal during fault depend on its location, level of compensation, type of fault, and MOV operation [2]. The apparent impedance oscillates, causing difficulty to differentiate between power swing and fault. The detection of fault amidst power swing in a line with series compensation finds limitations due to the use of series capacitors in transmission lines, resulting in voltage inversion, current inversion, subharmonic oscillations, and transients during the fault period. Several methods for detection amidst power swings have been proposed for regular transmission lines [3–5].

The conventional method [6–12] for fault detection in an uncompensated line includes measuring the rate of change of resistance at the relay location. Under steady-state conditions, the apparent resistance varies, while during the fault period it remains constant [13].

The magnitude and rate of change of Swing Center Voltage (SCV) is another technique. SCV is defined as the voltage at the location of two sources equivalent system where the voltage is zero and the angle between the sources is 180°. The SCV changes continuously under normal condition but remains fixed during fault [6].

A fault detector is also being used to detect high impedance ground fault amidst power swing using superimposed current components [11]. Prony method [11] based on the decaying dc in the current waveform is being employed in the fault detector for differentiating between power swing and fault. Even the voltage phase angles are employed for detecting faults amidst power swing.

Recent techniques involve Artificial Neural Network (ANN) [12] based adaptive systems and wavelet-based signal processing [14]. Wavelet transform creates a high-frequency signal at the inception of fault for detecting fault amidst power swing. The suitability of any detection technique depends on the location and fault condition in a line with series compensation.

Due to non-linear functioning of the series capacitor combination, the available techniques fail. The SCV magnitude and the rate of its change get affected as during fault period, current and voltage phase angle varies because of non-linear impedance across the MOV and series compensation [2, 15, 16]. The problem of voltage inversion in a line with series compensation makes it impossible for monitoring voltage phase angle at different locations making this method unsuitable. The superimposed current technique fails for high resistance and power angle tending to 180°. The fault current being low does not bypass the capacitor.

2 Challenges in Detection of Fault Amidst Power Swing for a Line with Series Compensation

In a line with series capacitor, MOV, and air gap poses protection problems depending upon the location and type of fault. Owing to its non-linear impedance, the conventional methods of fault detection find limitations in a line with series compensation.

Fault detection in a line with series compensation is difficult if a fault occurs at far ends. For faults occurring at far ends, or at a power angle close to 180°, the magnitude of fault current is low. This low current does not bypass the capacitor, resulting in sub-synchronous oscillations. These oscillations are a hindrance in the impedance estimation. Thus, making the fault detection during power swing a difficult task. For fault occurring closer to the relay, the fault current is large enough to bypass the capacitor.

The performance of the proposed technique has been demonstrated by a 400 kV, 50 Hz, 9 bus system. Three generators (600, 465, and 310 MVA) are connected between A and B buses [17]. Figure 1 [17] shows a test system that demonstrates the challenges with fault detection in a line with series compensation during power swing. Lines 1 and 2 are adjusted by 40%, and at the relay's end, capacitors are installed. MOV is used for the protection of series compensated lines. During fault period in impedance variation, it is not practically possible to use conventional positive sequence distance relay. For protection of a line with series compensation, MOV is used [17]. The figure shown below is single-line diagram of overall system. Let δ be power angle difference between A and B bus voltages. R is the distance relay used for the test purpose which is present at bus A. Suppose at 0.6 s. a 3-phase fault is created in line 2 which is then cleared after 0.1 s by the action of circuit breakers R3 and R4. Hence, power swing arises in the circuit. Since, it is a series of compensated lines, during power swing it is very hard to remove faults from the far endpoints. Suppose at 200 km far from the line a three-phase fault at 2.6 s is created, the corresponding Fig. 2 shows required current and voltage waveforms of the circuit at the instance.

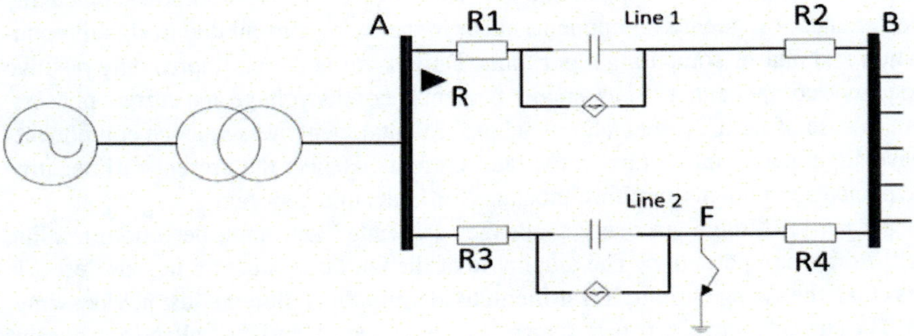

Fig. 1 The test system

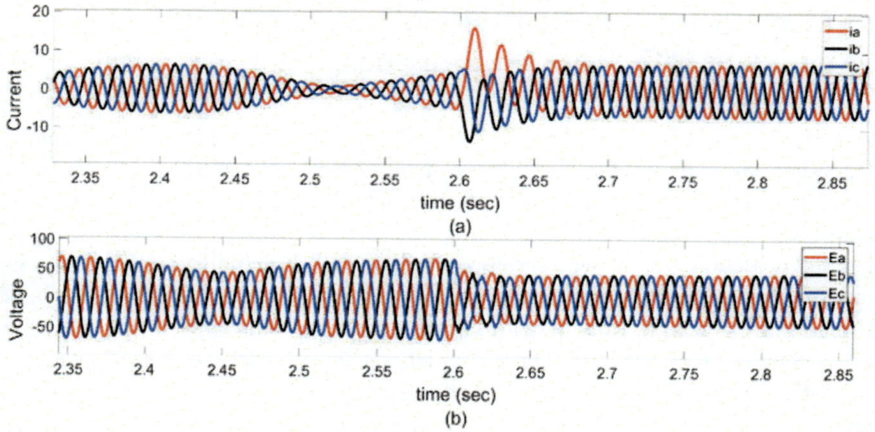

Fig. 2 Voltage and current waveforms when a fault is created at 2.6 s 200 km from the line. (all values in p.u.)

The main challenge in a line with series compensation is detecting far-end faults at the time of power swing. A three-phase symmetrical fault is created at a distance of 200 km from the relay location at 2.6 s and resulting waveforms of current and voltage are shown in Fig. 2. The figure shows how current and voltage are modulated with the swing frequency amidst power swing. Consequently, the fault identification cannot be done using sample-to-sample or cycle-to-cycle comparisons [18]. Current is lower than swing current under fault conditions. This does not bypass the capacitor resulting in sub-synchronous oscillation. These issues make the detection even more complex.

3 Proposed Technique

Although the power swing phenomenon is balanced in nature, it is noticed that a tiny percentage of sequence components occur in the swing current due to signal modulation and phasor computing algorithms. During pre-fault conditions, only positive sequence components exist as a major percentage of the voltage and current phasors, but in case of faults, (especially unbalanced faults), negative sequence components have significant contribution in the fault currents. Hence, the presence of negative sequence components signifies unbalance or faults in the system.

As per [17], there are many challenges associated with fault detection in a line with series compensation. The amplitude of the fault current for a far-bus defect is less than the swing current, it is difficult to identify this failure amidst power swing.

The current and voltage data for a relay site are saved, and the voltage and current phasors for each phase are approximated using the Discrete Fourier Transform (DFT) hence, the fundamental components are obtained. The Fortescue Theorem is then used to obtain the sequence components of the phase currents. The Cumulative Sum (CUSUM) technique is used to discover faults utilizing the sequence component gathered thus far. Using this technique, the CUSUM of both negative and zero-sequence components is produced, and the indices for both negative and zero sequences are then combined in an "OR" operation to generate the final trip signal.

An ag-fault and a three-phase fault at 2.6 s (with a slip frequency of 4.3 Hz) are made at the far end of the bus of the system depicted in Fig. 1 to investigate the fluctuation in $|\overline{I}_2|$ and $|\overline{I}_0|$ during swing and fault, and the results are presented in Figs. 3 and 4. For close-in faults, the fault current exceeds the swing current and MOV operates to bypass the series capacitor. The fault current in an ag-fault is reported to have a considerable magnitude of negative and zero-sequence components, and it oscillates due to components of modulating frequency in fault signals. In three-phase faults, due to initial transients, $|\overline{I}_2|$ and $|\overline{I}_0|$, vary vigorously at the point of occurrence of fault, but they have very small value because of the modulation of signal by the swing. Also, for ag-fault, the MOV operates only for a-phase, as a-phase current becomes higher than swing current, but for three-phase fault, MOV of all the three phases conducts and bypasses the series capacitor if fault current exceeds the swing current. The change in magnitude of sequence component-based technique serves the goal due to the tiny amount of sequence components left during the swing condition. A suitable threshold is selected and cumulative sum of $\Delta|\overline{I}_2|$ and $\Delta|\overline{I}_0|$ based technique has been used in this paper for better fault detection at the time of power swing. Being so versatile, cumulative sum (CUSUM) is used to detect sudden changes. The CUSUM-based approach largely has application in fault detection in a transmission line, based on sampled values of current signals [19].

When changes in the zero and negative sequence components of currents are provided as input signals, CUSUM is employed to produce a decent index for defect detection during power swing. The index calculation method is as follows,

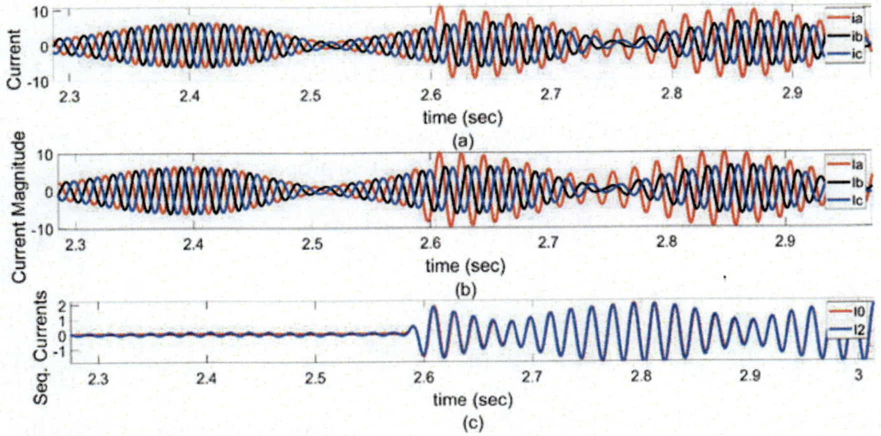

Fig. 3 Waveforms during ag-fault (all values in p.u.)

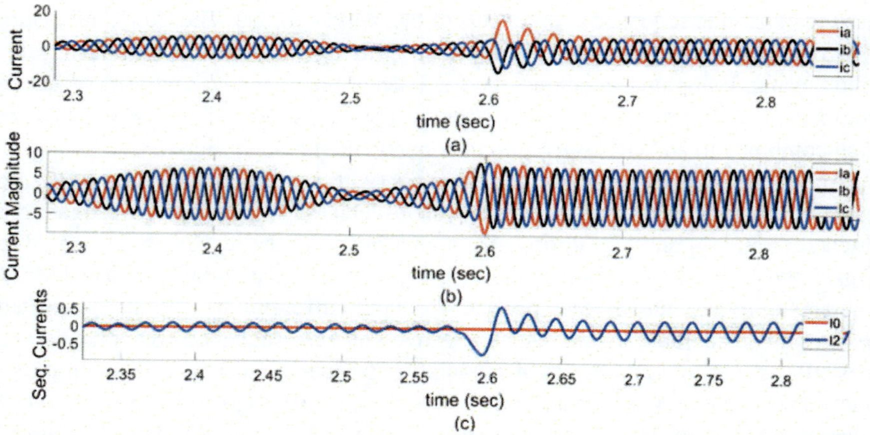

Fig. 4 Waveforms during three-phase fault. (all values in p.u.)

$$\overline{I}_2 = \frac{\left(\overline{I}_a + \alpha^2\overline{I}_b + \alpha\overline{I}_c\right)}{3} \tag{1}$$

$$\overline{I}_0 = \frac{\left(\overline{I}_a + \overline{I}_b + \overline{I}_c\right)}{3} \tag{2}$$

where, $\overline{I}_2$ represents negative sequence component, $\overline{I}_0$ represents zero-sequence component, and $\alpha = e^{j\frac{2\pi}{3}}$. $\overline{I}_a$, $\overline{I}_b$ and $\overline{I}_c$ are phase currents.

The derived signal for CUSUM is obtained as, (change in zero and negative sequence components),

$$d_0^k = \Delta\left|\overline{I}_0\right|^k = \left|\overline{I}_0\right|^k - \left|\overline{I}_0\right|^{k-1} \tag{3}$$

$$d_2^k = \Delta\left|\overline{I}_2\right|^k = \left|\overline{I}_2\right|^k - \left|\overline{I}_2\right|^{k-1} \tag{4}$$

Here, k represents the kth iteration.

For $d_0^k > \varepsilon_1$ and $d_2^k > \varepsilon_2$ the CUSUM test can be proposed as,

$$h_0^k = \max\left(h_0^{k-1} + d_0^k - \varepsilon_1\right) \tag{5}$$

And similarly,

$$h_2^k = \max\left(h_2^{k-1} + d_2^k - \varepsilon_2\right) \tag{6}$$

where the indices h_0^k and h_2^k are the test statistics and ε_1 and ε_2 represent the drift parameters. A fault is registered if,

$$h_0^k > h_1 \text{ or } h_2^k > h_2 \tag{7}$$

i.e., if any one of the above two conditions is true.

Where h_1 and h_2 are constants and should ideally be zero. To improve the performance of the detector, ε_1 and ε_2 provide effects equivalent to a low pass filter. When $d_0^k > \varepsilon_1$ and $d_2^k > \varepsilon_2$, the h_0^k and h_2^k values increase by a factor of $d_0^k - \varepsilon_1$ and $d_0^k - \varepsilon_2$, respectively. With the use of additional samples of current, the CUSUM approach gives a simple way to determine a fault condition (7). Following the evaluation of each pair of fault detection indices. In case of power swing condition only, the values of d_0^k and d_2^k will be less than ε_1 and ε_2, and hence h_0^k and h_2^k will be zero. The performance of the algorithm depends on the selection of values for ε_1, ε_2, h_1 and h_2. As stated earlier, power swing is balanced phenomenon, but a slight amount of negative sequence is obtained in the process of phasor expansion which gradually rises with the rise in slip frequency of swing cycle. For the CUSUM technique discussed for zero and negative sequence components, the values of ε_1 and ε_2 are selected such that d_0^k and d_2^k are zero amidst power swing (for both unbalanced and balanced power swing conditions), and finally helps to maintain the fault detection index h_0^k and h_2^k equal to zero.

In this paper the values of ε_1 and ε_2, set to 0.05 suits the purpose for a power swing with a slip frequency of 4.3 Hz. A typical power system's slip frequency is usually around 7 Hz. [14]. The numbers used for h_1 and h_2 allow the CUSUM algorithm to maintain the relaying scheme's balance of dependability vs. security and speed vs. accuracy criteria. The value of h_1 is selected as 50 and value of h_2 is selected as 20. [18]. The algorithm is tested for different cases of faults with delta near to 180°, and for faults at the far end of the transmission line.

4 Results

The proposed CUSUM technique [17] has been checked for symmetrical, asymmetrical, and high resistance faults, among other things. In a line with series compensation, the phenomenon of sub-synchronous resonance modulates voltage and current signals.

The basic component is estimated using the DFT. Considering phase—an as reference, sequence components are estimated. The reference value of ε_1 and ε_2 were set at 0.05 each and $h_1 = 50$ and $h_2 = 20$. The voltage and current waveforms reach their highest and minimum values when the power swing is kept close to 180°. At this instant, change in current and voltage will be insignificant. Thus, detecting fault during power swing becomes difficult when δ is close to 180°. The algorithm was tested at critical conditions, fault was created at 2.6 s when δ equals 175° with a slip frequency of 4.3 Hz. Faults that occur at far ends have a very small value of current, preventing MOV conduction and may result in current inversion. Faults are not identified under such conditions. Therefore, the testing of the algorithm is done

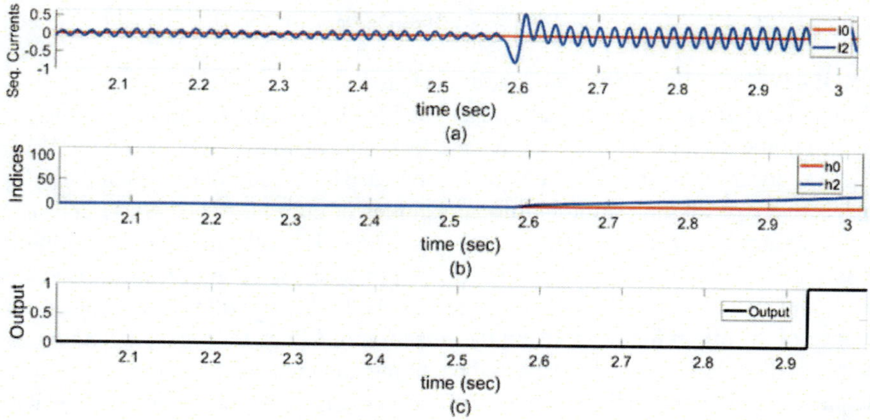

Fig. 5 Sequence currents, Index, and trip signal for three-phase fault amidst power swing (all values in p.u.)

at 150 km for phase faults, 240 km for line faults, and 200 km for symmetrical faults. In this paper, output 1 implies fault, and 0 implies no fault.

4.1 *Fault in the Line with Series Compensation with Three Phases*

In nature, both three-phase faults and power swing are balanced. The presence of negative sequence at the initial and subsequent periods helps in detecting fault at the time of power swing. A fault was induced at 2.6 s, 175° in the course of power swing at 200 km away from the relay position in line 1 for testing the algorithm. Unbalance is noted throughout the transient, and the onset of a 3-phase fault raises the CUSUM index. As seen in Fig. 5, the defect is identified shortly after it occurs, and a trip signal is generated instantly.

4.2 *Line-to-Ground Fault in a Line with Series Compensation with Low Resistance*

Negative sequence current is significant during power swing due to modulating frequency component in fault signals. Zero-sequence component is also present that makes fault detection possible even for asymmetrical power swings. The method was tested with a fault resistance of 5 Ω at 5 km for line-to-ground fault. Fault is initiated at 2.6 s and δ equals 175°. The CUSUM index is zero before the fault inception as

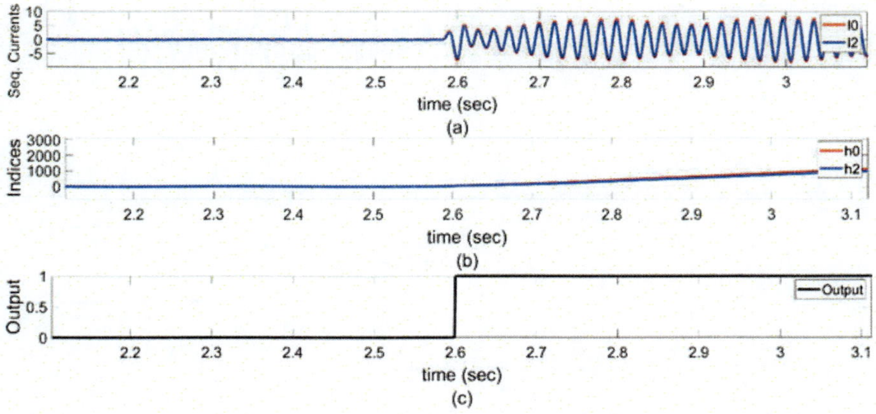

Fig. 6 Sequence currents, Index, and trip signal for ag-fault at the time of power swing (all values in p.u.)

in Fig. 6 and grows once the fault is detected. After fault detection, a trip signal is immediately generated.

4.3 High Fault Resistance in Line-To-Ground Fault

The fault detection with a high fault resistance implies a high value of fault current also δ equals 180°. An ag-fault with fault resistance 200 Ω was induced at 2.6 s at 175° amidst power swing at 150 km away from the relay position in line 1 for testing the algorithm. Figure 7 indicates that the fault is recognized and cleared after 2.6 s.

4.4 Line-to-Line Fault

Negative sequence component being significant because of the modulating frequency helps in fault detection even though the zero-sequence component is zero. At 240 km away from the relay location, a line-to-line fault with a fault resistance of 200 Ω was induced at 2.6 s at 175°. Figure 8 output 1 shows that the issue has been discovered and cleared after 2.6 s.

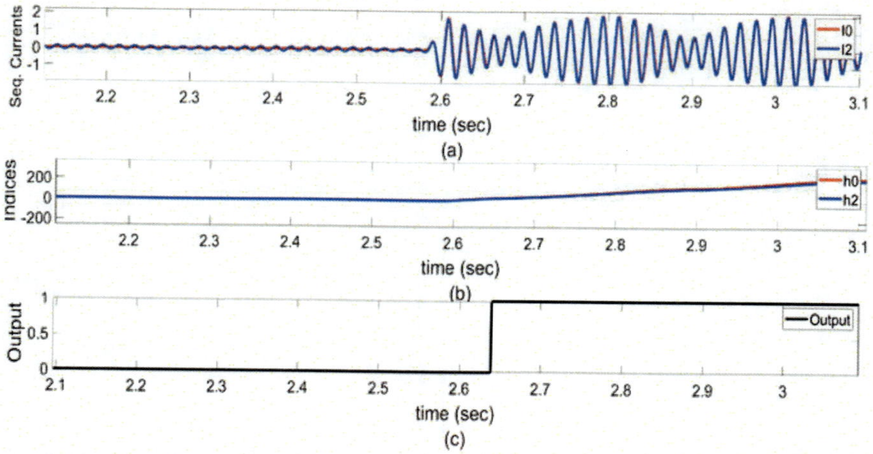

Fig. 7 ag-fault with high resistance

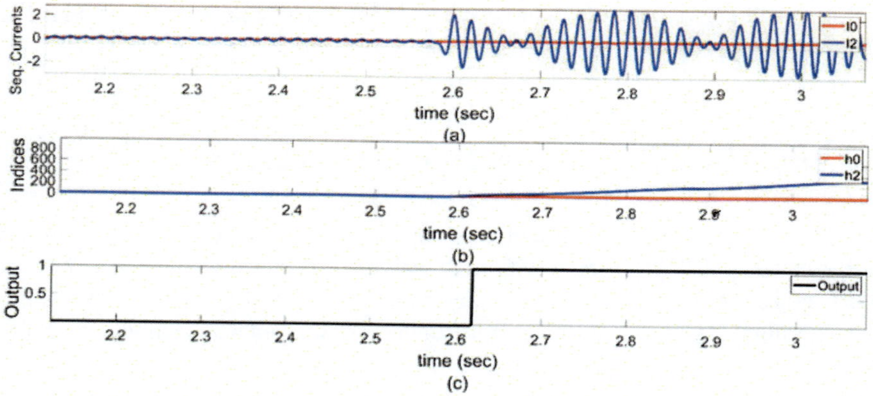

Fig. 8 Sequence currents, Index, and trip signal for line-to-line fault amidst power swing (all values in p.u.)

4.5 Double Line-to-Ground Fault in Series Compensated Line

In a LLG fault, both negative and zero-sequence components are significant to raise the CUSUM index and generate the trip signal. At 50 km from the relay station, an LLG fault with fault resistance 200 Ω was induced at 175°. After fault initiation, the problem is recognized and removed after 0.4 s using the aforementioned process, as shown in Fig. 9.

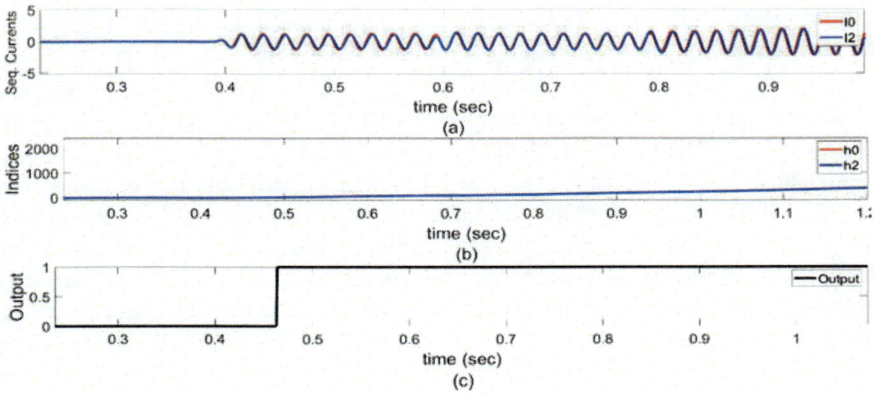

Fig. 9 Double line-to-ground fault result (all values in p.u.)

5 Comparative Study of the Algorithm

The conventional techniques of the rate of change of swing center voltage, superimposed current components, and rate of change of resistance is used to detect faults in an uncompensated line amidst power swing. The performance of the transmission line for the above techniques is compared with the change in R and SCV for the same system. It is clearly observed that for different types of faults, *R* and SCV should remain constant with very small magnitude amidst fault period. But in a line with series compensation, *R* and SCV are fluctuating amidst fault period. Also, the value of *R* when the SCV is high at the start of a fault, the fault detection process is noticeably delayed. As a result, *R* and SCV change in a line with series compensation, hence they cannot be used to detect a malfunction during a power swing.

Different simulations were run for various fault types at various fault sites and fault inception times. The frequency of the slide is kept constant at 4.3 Hz. For the 50 Hz system, the results clearly indicate that asymmetrical and symmetrical faults are detected within half a cycle for the 50 Hz system.

6 Conclusion

A defect detection technique for a line with series compensation is presented in this research amidst power swing. The problem is detected using cumulative sum of the changes in zero-sequence current and negative sequence current. The proposed technique's performance has been evaluated for various sorts of faults. When compared to existing techniques, the proposed method detects faults in half the time and is an improvement.

Zero-sequence components are present in a system if it involves grounding. During fault, presence of zero-sequence current $\overline{I}_0$, would mean that the resultant fault is some type of ground fault. If $\overline{I}_0$, is present in a significant amount, then its CUSUM will exceed the constant h_1, as is the result, shown in Figs. 6, 7, 9, and 10, signifying that a type of ground fault has occurred. If $\overline{I}_0$, is bare minimum such that h_0^k remains less than h_1, then the index will remain zero as shown in Figs. 5 and 8, meaning the fault which occurred is not a ground fault (Fig. 11).

Hence, by including $\overline{I}_0$, in the CUSUM technique for fault detection during power swing in a line with series compensation, the nature of the fault, whether it's a ground fault or not, can be determined, alongside with the detection of the fault, using the $\overline{I}_2$ CUSUM technique. The knowledge of type of fault also helps in deciding the protection approach of lines, as well as the severity of the fault and damage it can cause.

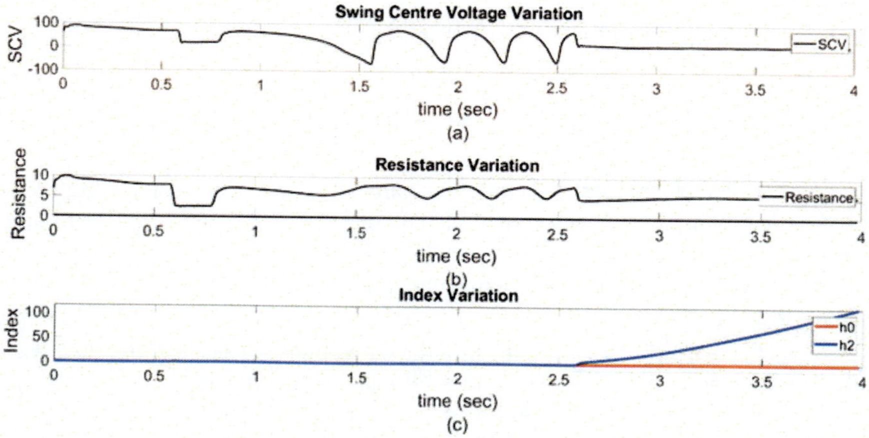

Fig. 10 Comparison of SCV, resistance variation, and index variation for ag-fault

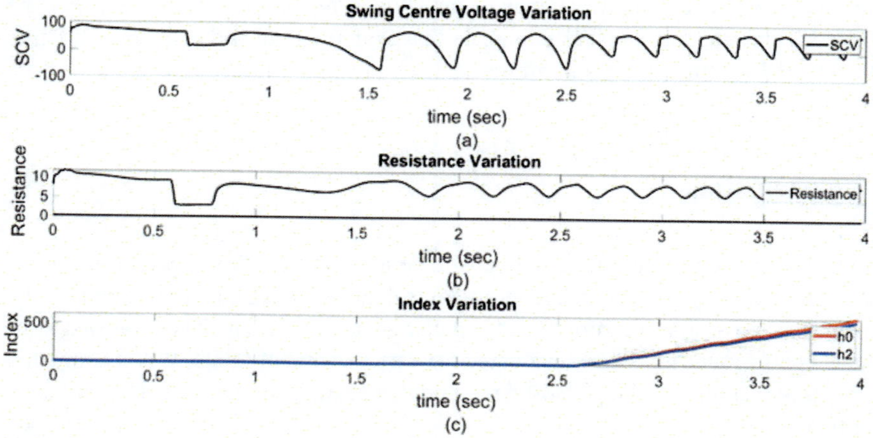

Fig. 11 Comparison of SCV, resistance variation, and index variation for three-phase fault

Since CUSUM of negative and zero sequences current component is used, the detection method is immune to Capacitor Control Voltage Transformer (CCVT) transients. CUSUM of zero and negative sequence component detects both symmetrical and asymmetrical faults at the time of symmetrical or asymmetrical power swing.

References

1. IEEE Power System Relaying Committee of the IEEE Power Engineering Society (2005) Power swing and out-of-step considerations on transmission line. Rep. PSRC WG D6, Jul 2005. Available http://www.pes-psrc.org
2. Abdelaziz AY, Ibrahim AM, Mansour MM, Talaat HE (2005) Modern approaches for protection of series compensated transmission lines. Elect Power Syst Res 75:85–98
3. Mechraoui A, Thomas DWP (1997) A new principle for high resistance earth fault detection during fast power swings for distance protection. IEEE Trans Power Deliv 12(4):1452–1457
4. Apostolov AP, Tholomier D, Richards SH (2004) Superimposed components based sub-cycle protection of transmission lines. In: Power systems conference and exposition, IEEE PES, vol 1, pp 592– 597
5. Lotfifard S, Faiz J, Kezunivic M (2010) Detection of symmetrical faults by distance relays during power swings. IEEE Trans Power Delivery 25(1):81–87
6. Benmouyal G, Hou D, Tziouvaras D (2004) Zero-setting power-swing blocking protection. In: Presented at the 31st annual western protective relay conference, Spokane, WA
7. Mechraoui A, Thomas DWP (1997) A new principle for high resistance earth fault detection during fast power swings for distance protection. IEEE Trans Power Del 12(4):1452–1457
8. Apostolov AP, Tholomier D, Richards SH (2004) Superimposed components based sub-cycle protection of transmission lines. In: Proceedings of IEEE power engineering society conference and exposition, vol 1, pp 592–597
9. Gao ZD Wang GB (1991) A new power swing block in distance protection based on a microcomputer-principle and performance analysis. In: Proceedings of international conference on advances in power system control, operation and management, Hong Kong, China, vol 2, pp 843–847
10. Lin X, Gao Y, Liu P (2008) A novel scheme to identify symmetrical faults occurring during power swings. IEEE Trans Power Del 23(1):73–78
11. Lotfifard S, Faiz J, Kezunovic M (2010) Detection of symmetrical faults by distance relays during power swings. IEEE Trans Power Del 25(1):81–87
12. Zadeh HK, Li Z (2008) A novel power swing blocking scheme using adaptive neuro-fuzzy inference system. Elect Power Syst Res 78:1138–1146
13. Gao ZD, Wang GB (1991) A new power swing block in distance protection based on a microcomputer-principle and performance analysis. In: International conference on advances in power system control, operation and management, Hongkong, vol 2, pp 843–847
14. Brahma SM (2007) Distance relay with out-of-step blocking function using wavelet transform. IEEE Trans Power Delivery 22(3):1360–1366
15. Alture HJ, Mooney JB, Alexander GE (2008) Advances in series compensated line protection. Available www.selinc.com/20081022, TP6340-01
16. Novosel D, Phadke AG, Saha MM, Lindahl S (1997) Problems and solutions for microprocessor protection of series compensated lines. In: Proceedings of conference developments in power system protection, pp 18–23
17. Nayak PK, Pradhan AK, Bajpai P (2013) Fault detection technique for series-compensated line. IEEE Trans Power Deliv 28(2)
18. Gustafsson F (2000) Adaptive filtering and change detection. Wiley, New York
19. Mohanty SR, Pradhan AK, Routray A (2008) A cumulative sum-based fault detector for power system relaying application. IEEE Trans Power Del 23(1):79–86

Simulation and Analysis of Hybrid Micro-grid Integrated with EV Load

Gopendra Kumar, Mukul Singh, M. A. Ansari, Omveer Singh, and Vimlesh Kumar Ray

Abstract A hybrid micro-grid behaviour under various connected EVs load is presented in this research paper. A novel concept in which a large number of electric vehicle loads and sources are combined into a single controllable unit system that provide heat and power to a localized region. The components of the micro-grid are defined using a variety of existing source models. In MATLAB/Simulink, a complete model is installed that includes a micro-grid, power sources such as solar, wind, and diesel generator models, as well as EVs load. The effect of combining different EVs load renewable energy sources with micro-grid is being investigated using a new combination of these models. The research paper examines various electrical parameters, i.e. state of charge, charge data and regulation data for wind farms, diesel generators, solar PV farms, and varying EVs loads. PV farm size and productivity, as well as wind farm speed, are amongst the various factors that influence energy output.

Keywords Micro-grid · EV load · Renewable energy sources (RES) · Distributed energy generation (DEG)

1 Introduction

With rising concerns about the greenhouse effect and emissions, clean and sustainable energy use is becoming increasingly common. Electric vehicles (EVs) have grown in popularity in recent years as a result of their environmentally friendly design and government incentive policies, posing significant grid challenges [1, 2]. As compared to conventional power loads, charging EVs has unique characteristics. Creation of the

G. Kumar · M. Singh (✉) · M. A. Ansari · O. Singh
Department of Electrical Engineering, Gautam Buddha University, Gr. Noida, India
e-mail: gbubuddhams@gmail.com

V. K. Ray
Department of Electronics Engineering, Gautam Buddha University, Gr. Noida, India

K. Namrata et al. (eds.), *Smart Energy and Advancement in Power Technologies*, Lecture Notes in Electrical Engineering 926,
https://doi.org/10.1007/978-981-19-4971-5_28

modern power grid, focus on practical error-less modelling of EVs power consumption, management and study of EV charging influence, is necessarily required. Several attempts are made to increase electricity output by harnessing energy from renewable sources such as solar, wind, hydro, geothermal, nuclear, and others. At different stages, these sources are used separately to produce electricity. The most difficult task is to transfer the energy that is produced, to everyone. Frequently, a single power plant is unable to supply the sufficient amount of electricity to meet the load of consumers. The problem was solved by combining the energy supply of various power plants with the power grid. Nonetheless, the issue of obtaining energy remained the same.

This issue was solved by combining a diesel generator with renewable energy sources, allowing the micro-grid to sustain a sufficient amount of energy. Intermittency, standardization for information sharing, site and size dependence and ambiguity are DER's main proposed solutions to the above-mentioned problem, which are based on current scenario studies of various existing models. The PV panels, wind turbines, and diesel generators provide enough power to the micro-grid to meet the needs of the micro-customer grid's at all times. Micro-grids can help to realize smart grid since they are a subset of the grid that consists of different units of energy storage and low voltage delivery on the demand side for better network management [3–6].

2 Literature Survey

Electric vehicle deployments on a wide scale are catching the attention of the power industry. For the electric power industry, electric vehicles present both opportunities and threats. They need a power source to recharge, but their battery can also provide power. EV fleets and buses provide opportunities that are distinct from those offered by individual automobiles. When a micro-grid is combined with solar panel-based distributed renewable energy generation and diesel-engine generator-based conventional energy generation, Zhang et al. [7] investigated the two modes of micro-grid operation. The effect of electric vehicles on the power grid during peak hours: EV arrive at home in the evening and plug into a power outlet to charge. When charging, an electric vehicle consumes a significant amount of electricity. As a result, a cluster of electric cars in and around a transformer adds to its load. The load factor of the grid decreases as the load increases during peak hours. A high load factor indicates relatively constant power consumption, whilst a low load factor indicates a device that experiences high loads on occasion. The utility must have enough power to meet peak demand whilst remaining idle rest of the time. Electric Vehicles that use the grid for charging during peak hours exacerbate the issue. The power utility struggles as a result of the increased demand.

The current in the power line rises during high demand periods. Increased losses in the transformer and transmission line are the result. As a consequence, the transmission line's voltage drop increases. At peak hours, the total network voltage decreases. It lowers the power factor and lowers the power efficiency. Finally, it leads to a strong active power demand in power plants. The ability of power utility grids to satisfy high power demand that occurs only during peak hours is hampered. During peak hours, charging electric vehicles adds a significant amount of load to the power grid. It is a significant burden on utilities to leave their power generating capacity idle during periods of low demand and switch it on during periods of high demand. An electric vehicle is used to balance the load.

The effect of electric vehicles on the power grid can be both positive and negative. Electric vehicles are massive loads that are plugged into the grid. Electric loads are not necessarily evenly distributed. The load factor can be minimized if an electric vehicle can supply electricity to the grid instead of drawing energy during peak load times. Thus, peak load losses, voltage drops, high generating station tension, and other disadvantages can be effectively minimized. Utility grids can encourage scheduled charging of electric vehicles because of the benefits of peak load shaving and load levelling. Zhu et al. [8] suggested specifically sharing resources to balance harvested local energy and demand by using an alternative structure near homes to balance harvested local energy and demand.

3 EV Load Components and Modelling Method

Energy storage device's charging characteristics have a significant impact on power consumption. The EV charger is a device that connects the battery to the grid. Electric Vehicles chargers are made to fulfil a variety of needs, and their control units adhere to industry specifications in order to control battery charging process. The Electric Vehicles charger has the greatest impact on charging performance and thus the load for charging when it comes to load modelling. The EV charger is a source of grid-dependent EV load characteristics that should be considered during the modelling process (Fig. 1).

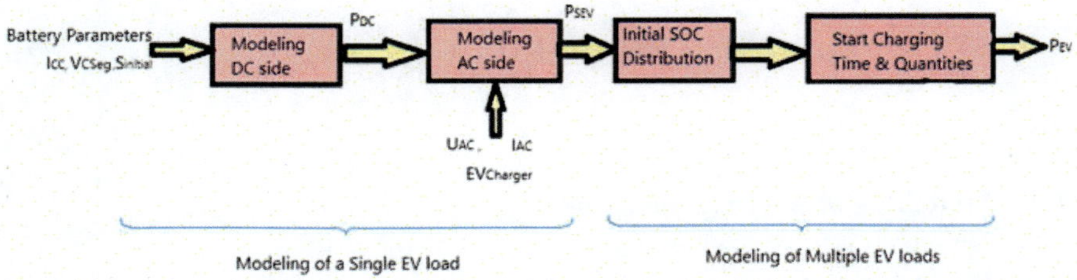

Fig. 1 A modelling of single and multiple EV loads

3.1 Modelling of a Single EV Load

The charging energy is divided into two type by the EV charger: AC and DC. To combine different battery models and EV chargers, the charging capacity of the two sides can be modelled separately [9, 10]. The battery charging power represents load model across dc side, whilst the ac side load model is represented by a single electric vehicle (EV load model). The lithium–ion battery model was charged with a single phase EV charger [11]. The characteristics of the dynamic energy storage device can be easily extracted from the battery datasheets of the manufacturers [12]. In the 20–90% SOC range, the constant current charging method applied. The charging process is then continued by a constant voltage charging technique till the battery is completely charged.

3.1.1 Modelling of DC Side Loads

In the charging phase, the dc side charging power is as follows:

$$P_{DC} = I_C \cdot V_C \tag{1}$$

Here in Eq. 1, I_C = charging current and V_C = terminal voltage of battery.

At this time, the lithium-ion battery's terminal voltage varies depending on the SOC frequency and charging current. The charging current decreases as the charging time increases, during the charging phase of the CV. At this time, the lithium–ion battery's terminal voltage varies depending on the charging current and SOC frequency. During the constant voltage charging cycle, the charging current decreases as the charging time increases. SOC Satisfy during the CC Charging Cycle.

Here in Eq. 2, I_{cc} = constant charging current, C = Battery rated capacity, and t = charging time of length (h).

$$S = S_{\text{initial}} + (I_{cc}/C)t \tag{2}$$

3.1.2 Modelling of the AC Side Load

The power factor of a single phase EV charger can be maintained at unity. The charging performance is unaffected by variations in power frequency or battery voltage. The charging current is maintained at a steady during the CC period, the grid voltage has the greatest impact on charging efficiency. The charge efficiency-to-grid voltage relationship can be expressed as a quadratic equation, which looks like this:

$$\eta_c = f_5(U_{AC}) = cU_{AC}^2 + c_1 U_{AC} + C_0 \tag{3}$$

Here η_c: Charge efficiency; $U_{AC:}$ grid ac voltage; c_0, c_1, c_2: constant coefficients.

3.2 *Modelling of Aggregate EV Loads*

3.2.1 SOC Distribution Model at the Start

Users' driving patterns, energy storage device capacities, and range anxiety all have an effect on the initial SOC distribution. When EVs are commonly used, it is rational to treat, as per the law of large numbers and the central limit theorem, the initial SOC distribution is a trimmed normal distribution with zero probability of all negative initial SOCs.

Here in Eqs. 4 and 5 we have, S_{start}: obtained initial charging SOC; N: received initial SOC; s_{start} and σ^2: Sample mean value and sample variance $N(\mu, \sigma^2)$: real distribution of the initial SOC,

$$S_{start} = 1/N = \left[\sum_{i=1}^{N} S_{start,i} \right] \tag{4}$$

$$\overline{\sigma^2} = 1/N = \left[\sum_{i=1}^{N} S_{start,i} - S_{start} \right] \tag{5}$$

3.2.2 Time to Start Charging Model

With the deployment of smart charging facilities in the near future, charging rules can drive EV customers' charging habits, and the wildness of the start charging time can be reduced. The quick charging habits of electric vehicle owners who are affected by unpredictable travel.

4 Model of the Proposed Work

See Fig. 2.

4.1 *Simulation Model of the System*

See Fig. 3.

4.2 System Detail

Figure 2 shows the micro-grid system connected with EV loads and Fig. 3 shows the developed Simulink model of micro-grid integrated with EV load. Table 1 displays the system details of the blocks used in Simulink model of Micro-grid.

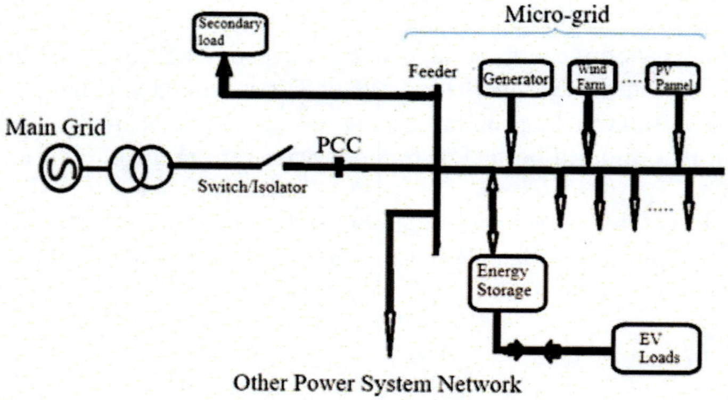

Fig. 2 Proposed micro-grid system connected with EV loads

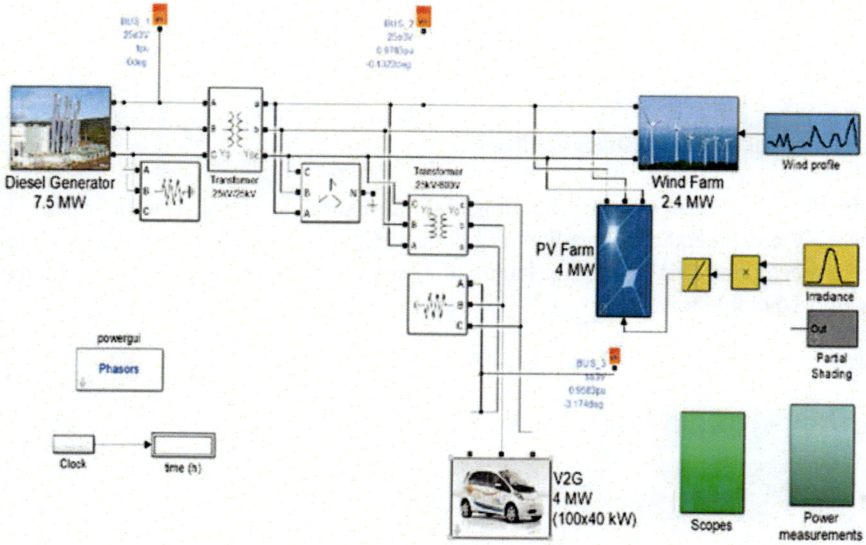

Fig. 3 Simulink model of the micro-grid connected proposed system

Table 1 System details of the developed micro-grid connected with EV

Serial number	Components name	Parameter rating
1	Diesel Generator	7.6 MW
2	Transformer1	25/25 kV
3	Transformer2	25 kV/600 V
4	Bus 1	25e3 V, 1 p.u., 0°
5	Bus 2	25e3 V, 0.9782 p.u., −0.1321°
6	PV farm	4 MW
7	Wind farm	2.5 MW
8	EV load	4 MW
9	Frequency	60 Hz
10	Grid supply	3 phase AC

5 Results and Discussion

Results of various electric vehicle loads connected with the micro-grid are shown below. The generated profile belongs to three types of EV with different type consumption profile based on the activities of the EV users. Load 1, 2, and 3 signifies the EV load 1, 2, and 3. All three loads behave in different manners and have different impacts on power regulation when charging from the grid. The profile of load 1 is when the EV is charged by the owners when they are at office and working. The profile of load 2 is when the EV is charged by the owners when they are at office and working but have to drive far and come back to office next day in one single charge. The profile of load 3 is when the EV is not charged by the owners when they are at office but is charged at home at any time of the day during 24 h. The variation in the profiles is dependent on duration of charge, discharge of the battery due to ride, possibility to ride back for a longer distance and during no obligations on charging the vehicles when at home [13–16].

5.1 EV Load 1 Connected to Micro-grid

Figure 4, 5, 6 shows regulation data, charging data, and state of charge of the battery of connected EV load 1 during 24 h of simulation based on the user profile of the EV load 1.

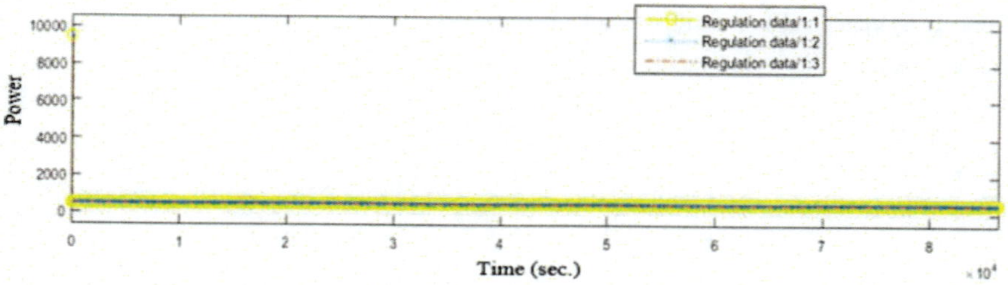

Fig. 4 Regulation data when load 1 connected

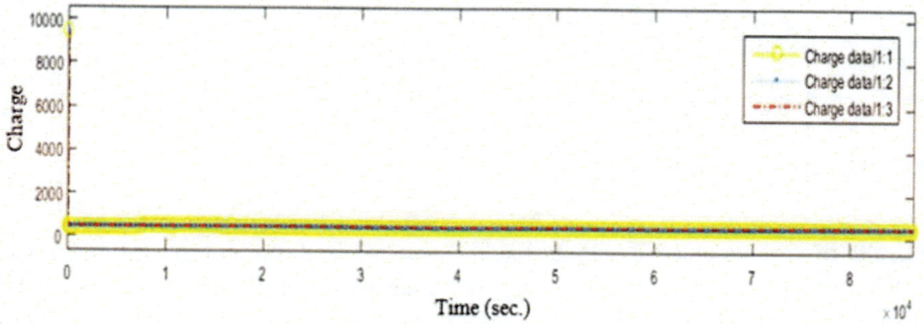

Fig. 5 Charging data when load 1 connected

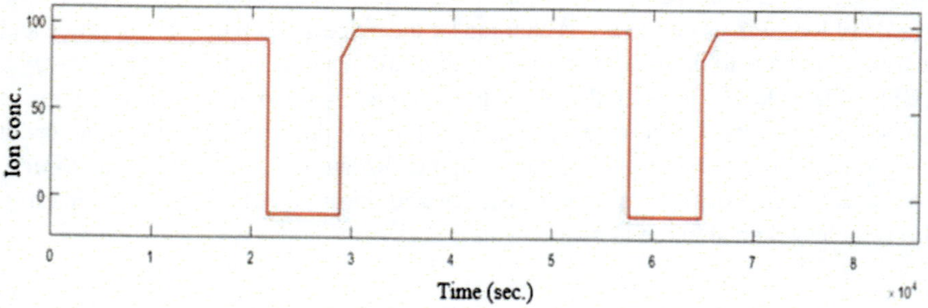

Fig. 6 State of charge when load 1 connected

5.2 *EV Load 2 Connected to Micro-grid*

Figures 7, 8 and 9 shows regulation data, charging data and state of charge of the battery of connected EV load 2 during 24 h of simulation based on the user profile of the EV load 2.

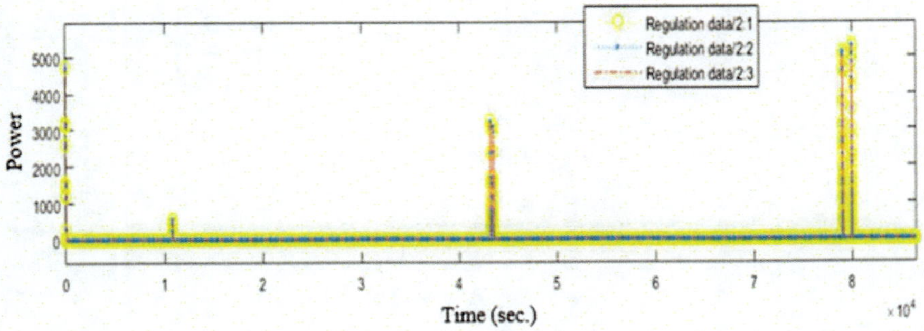

Fig. 7 Regulation data when load 2 connected

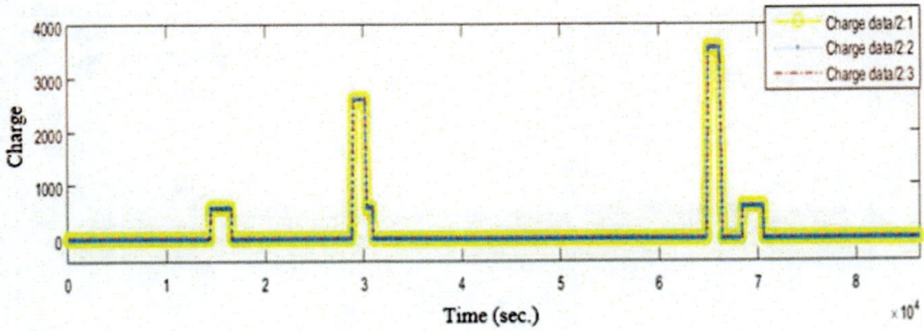

Fig. 8 Charging data when load 2 connected

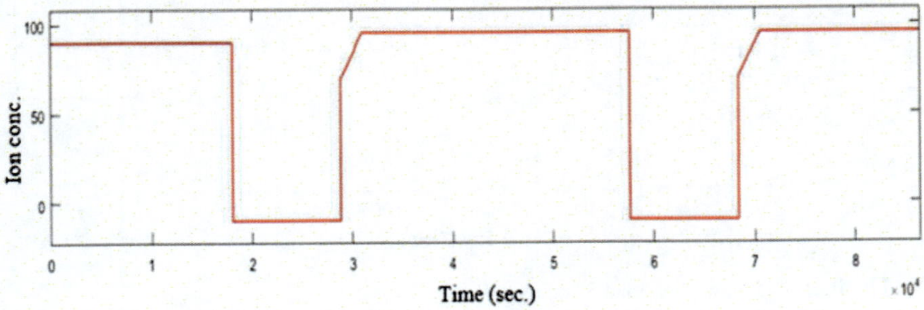

Fig. 9 State of charge when load 2 connected

5.3 EV Load 3 Connected to Micro-grid

Figures 10, 11 and 12 shows regulation data, charging data, and state of charge of the battery of connected EV load 3 during 24 h of simulation based on the user profile of the EV load 3.

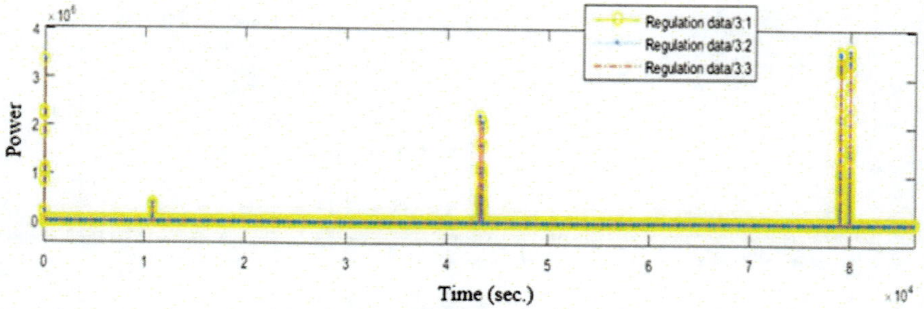

Fig. 10 Regulation data when load 3 connected

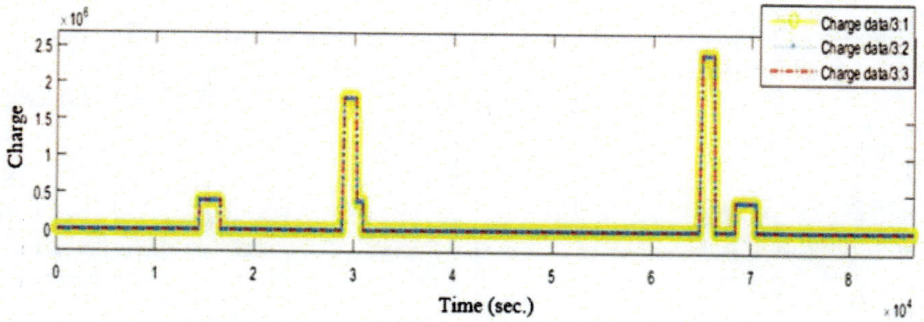

Fig. 11 Charging data when load 3 connected

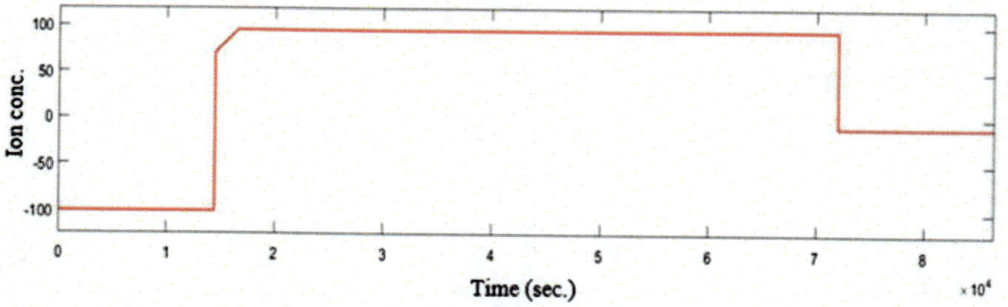

Fig. 12 State of charge when load 3 connected

6 Discussion and Conclusion

In this research paper simulation are run for twenty-four hour. Three different EV loads are connected to hybrid micro-grid. Individual such EV load connected acts once at a time in twenty hour time frame. Due to load variation grid parameter, State of charge (SOC) of vehicles battery changes are observed; Micro-grid performs better when Electric Vehicles (EV) load connected at off-Peak hours. Micro-grid performance worst when EV load connected at peak hours. Performance of the micro-grid observed average when EV load are connected other than peak and off-peak hour. At the point of charging, electric cars have an effect on the power utility grid. Uncontrolled charging of electric cars has a significant impact on the grid's efficiency (few of the parameter like power factor and voltage of the linked grid). The negative effect of EV on the electrical grid can be reduced along well-managed bidirectional charging. It is becoming increasingly important in a grid as EV grows. Impact of EV on the power grid will help to stabilize the system. Electric car charging that is timed to serve the grid as a micro-generator during peak hours and as load during off-peak hours is extremely beneficial to the power grid or micro-grid.

References

1. Diego A, Roa A, Martinez JB, Fernandez XC (2019) A benchtop DC microgrid for renewable energy sources integration. In: IEEE 9th annual computing and communication workshop and conference (CCWC), Las Vegas
2. Li H, Yuan Y, Zhang L, Huang T (2018) Research and application on microgrid integrated energy service technology for modern agriculture. In: 2nd IEEE conference on energy internet and energy system integration (EI2)
3. Azimian M, Amira V, Haddadipourb S (2020) Microgrid energy scheduling with demand response. Department of Electrical and Computer Engineering, Kashan Branch, Islamic Azad University, Kashan (Research gate article)
4. Luo C, Huang Y, Gupta V (2018) Stochastic dynamic pricing for EV charging stations with renewables integration and energy storage. IEEE Trans Smart Grid
5. Clairand JM, Javier RG, Bel CA (2018) Smart charging for electric vehicle aggregators considering users preferences. IEEE Access
6. Qian K, Zhou C, Allan M (2011) Modeling of load demand due to EV battery charging in distribution systems. IEEE Trans Power Syst 26(2):802–810
7. Liu P, Yu J, Mohammed E (2018) Decentralized PEV charging coordination to absorb surplus wind energy via stochastically staggered dual-tariffs schemes considering feeder-level regulations. IET Gener Trans Distrib
8. Shaaban MF, Atwa YM, El-Saadany EF (2013) PEVs modeling and impacts mitigation in distribution networks. IEEE Trans Power Syst 28(2):1122–1131
9. Ashtari A, Bibeau E, Shahidinejad S (2012) PEV Charging profile prediction and analysis based on vehicle usage data. IEEE Trans Smart Grid 3(1):341–350
10. Zhang H, Tang W, Hu Z (2014) A method for forecasting the spatial and temporal distribution of PEV charging load. In: IEEE PES general meeting conference & exposition, National Harbor, MD, pp 1–5
11. Ban M, Yu J (2015) Procedural simulation method for aggregating charging load model of private electric vehicle cluster. J Mod Power Syst Clean Energy 3(2):170–179
12. Dharmakeerthi CK, Mithulananthan N, Saha TK (2012) Modeling and planning of EV fast charging station in power grid. In: IEEE power & energy society general meeting, San Diego, CA, pp 1–8
13. Singh M, Singh O, Kumar A (2019) Renewable energy sources integration in microgrid including load patterns. In: 3rd IEEE international conference on recent developments in control, automation & power engineering
14. Singh M, Singh O, Singh V, Sharma T (2019) Modeling simulation and comparison of DFIG based phasor model average model and detailed model of wind farm. In: Proceedings of 2nd international conference on advanced computing and software engineering (ICACSE)

15. Singh M, Singh O (2019) Phasor solution of a micro-grid to accelerate simulation speed. In: Proceedings of 2nd international conference on advanced computing and software engineering (ICACSE)
16. Singh M, Rana V, Ansari MA, Saini B, Singh P (2018) Power quality enhancement to sensitive loads with PV based micro grid system. In: IEEE international conference on sustainable energy, electronics computing systems

Optimal Allocation of Wind-Based Distributed Generators in Power Distribution Systems Using Probabilistic Approach

Md. Safdar Ali, Abrar Ahmad, Ward Ul Hijaz Paul, Danish Ali, and Mubassir Ahmad

Abstract In recent times, the attention of distribution utilizes has been drawn toward the clean and renewable generation technologies like photovoltaics, wind energy, microturbine, fuel cells, etc., as these renewables have seen huge developments recently. And, these utilities are introducing these renewable in the distribution networks. The primary purpose of smart grids is to promote the introduction of renewable energy sources and to achieve greater reliability and efficiency of the network. In this paper, we try to analyze and model different kinds of uncertainties throughout the wind power output, and this has been modeled through the Monte Carlo simulation model. In order to optimize the location and efficiency of distributed generation, a chance-constrained stochastic optimization model is built while preserving the degree of wind power utilization. Later, we propose a hybrid algorithm to solve the simulation in which the differential evolution algorithm embedded in the Monte–Carlo simulation. The optimization techniques are analyzed on a 33 bus system and the efficacy of the proposed models is shown.

Keywords Distributed generators · Hybrid energy storage system · Microgrid · BESS · Wind systems

1 Introduction

Power systems analysis is a huge and significant portion of electrical engineering studies. It is primarily concerned with the generation and distribution of electrical power from the sending end to the receiving end as per customer requirements, resulting in a minimum amount of losses. The power at the end of the user frequently varies as a result of load fluctuations or disruptions within the transmission line width. The term power system stability is therefore of paramount importance in this field.

It is used to describe the system's ability to restore its activity to a stable state condition within a reasonable period of time after any transience or disruption in

Md. S. Ali · A. Ahmad · W. U. H. Paul (✉) · D. Ali · M. Ahmad
Jamia Millia Islamia, New Delhi 110025, India
e-mail: wardulhijazpaul@gmail.com; ward1909916@st.jmi.ac.in

K. Namrata et al. (eds.), *Smart Energy and Advancement in Power Technologies*, Lecture Notes in Electrical Engineering 926,
https://doi.org/10.1007/978-981-19-4971-5_29

the line has been encountered. In modern society, the electrical grid is important. Without it, it would struggle and break down a lot of society. Hence, ensuring the reliability of the electrical grid in all circumstances is important.

Wind power is a clean and green types of renewable energy from the point of view of total fuel utilization. When wind power can be distributed at affordable cost to customers without compromising the transmission system's consistency and safety.

System planning for distribution is the most extensive challenges faced by system planers, mainly in wind-power distributed generation units are placed in the network. In modern power systems, the level of penetration of wind power has grown rapidly. Because of stochastic character and inherent intermittent, wind energy addition may lead to imbalance in supply and demand and suggestive challenges to the safe and steady operation of power system network [1]. The use of energy storage systems (ESSs) is one favorable solution to improve renewable energy intermittencies. ESS technologies may escalate the efficiency and steadiness of the power system while providing other profits, including enhanced load control, congestion reduction of transmission, deferral of T&D improvements, reduction of carbon emissions, etc. The BESS has been characterized by pervasive concerns among different types of ESS due to its diverse benefits such as swift reaction and tranquil deployment [2]. While there has been significant progress in encouraging the use of BESS, effective implementation relies heavily on preparation. BESS optimum setting, sizing and operating rules can help regulate power flow, increase renewable permeation, and decrease energy loss.

As we know, we cannot upgrade the transmission network, but we can upgrade distribution network. Distribution network can be upgraded by wind and solar, etc. In this paper, we have used wind power with battery energy storage system (BESS). Throughout recent years, BESS planning throughout the delivery process has been widely studied [3]. The authors anticipated a two-stage stochastic model strategy to optimize the battery ESS considering the associations between the degree of wind energy permeation and the average load summary. Battery ESS power optimization model in a MG was suggested taking into account system steadfastness index, battery ESS guesstimate cost, and system operating cost [4]. The work in this paper is based on the DC model. Some important aspects such as shift of wattful and wattless power flow, steadiness of voltage, etc., are ignored in this work. BESS capacity is determined by the analytical method's historically recorded maximum value and the Monte-Carlo simulation sequential approach is used to find optimal ESS allocation and model system status [5]. An experimental approach was proposed for DG capability investment forecasting as part of a competitive auction on the electricity market. The finest sharing of distributed generator units was obtained from a distribution company's perspective through a cost-benefit study [6]. In the method of distribution using voltage stability sensitivities about changes in the active and reactive energy inserted at PQ buses. In the radial distribution system, an experimental approach is proposed to optimally assign distributed generator units to minimize energy losses [7]. The anticipated technique well thought-out various load profiles with load and DGs output varying in time. BESS integration, therefore, is worth further study to lower the curtailment in wind power. The increase in the use of wind power after deployment of ESS has been investigated. The analysis thus demonstrates that substantial work has been done to assign various DGs in the distribution system; nevertheless, most of the presented work on assumptions that the DGs production can be interpreted and controlled. The methods available are unable to find the exact solution due to the intermittent nature of the power output [8]. To obtain the optimum capacity and position the DGs in such a manner such that annual energy losses of distribution system are minimized this paper uses a probabilistic modeling methodology. The presented work is based on the direct current modeling and pays no attention to certain critical aspects such as change in wattful/wattless power flow, steadiness of voltage. With regard to uncertainty in wind power, and untimely energy demand, the newest study demonstrates a clear shift from conventional deterministic tactics to progressive probabilistic methods in BESS planning, it is normal for us to follow a probabilistic approach [9]. In addition, the effects of BESS incorporation on distributed generation operations are overlooked in the works described above. As a result, this paper also takes into account the operational costs and the power failure costs of the DG systems.

As we know, we cannot upgrade the transmission network, but we can upgrade distribution network. Distribution network can be upgraded by wind and solar etc. [10]. In this paper we have used wind power with battery energy storage system (BESS). Throughout recent years, BESS planning throughout the delivery process has been widely studied [11]. The authors anticipated a two-stage stochastic model methodology to optimize the BESS taking into consideration the associations between the degree of wind energy permeation and the average load summary. BESS power optimizing model in a MG was suggested taking into account system steadfastness index, Battery ESS guesstimate cost, and system running cost. BESS capacity is determined by the analytical method's historically recorded maximum value and the Monte Carlo simulation sequential approach has been used to find optimal ESS allocation and model system status [12]. An experimental approach was proposed for DG capability investment forecasting as part of a competitive auction on the electricity market. The finest sharing of distributed generator units was obtained from a distribution company's perspective through a cost-benefit analysis [13, 14]. In the radial distribution system, an experimental approach is proposed to optimally assign distributed generator units to minimize energy losses [15]. The increase in the use of wind power after deployment of ESS has been investigated. The analysis thus demonstrates that substantial work has been done to assign distribution generators in the distribution system. The methods available are unable to find the exact solution due to the intermittent nature of the power output. In this research, we will try to minimize the power losses at different busses.

Wind power generators, battery ESSs, demand of load, micro-turbines are included in the distribution network. The unit is associated to the grid via a connecting line with restricted capacity of generation. The distribution network's abstract architecture is shown in Fig. 1. Monte Carlo simulation approach produces a variety of scenarios. Errors in load forecasts, wind speed, turbine operating states data are integrated in all scenarios. Each situation is defined by a discrete distribution function and its probabilities.

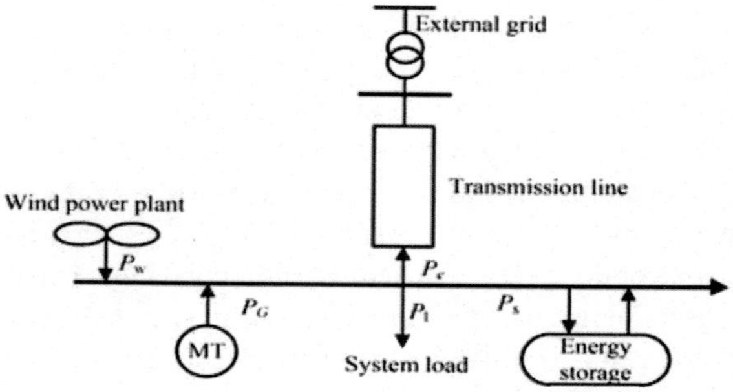

Fig. 1 Schematic of distribution network

2 System Modeling

2.1 *Wind Power Model*

Wind speed variations is based on the distribution of Weibull. The function of probability density is given as:

$$f(v) = \left(\frac{2v}{c^2}\right)\exp\left[-\left(\frac{v}{c}\right)^2\right] \tag{1}$$

$$V_{\mathrm{m}} = \int_0^\infty vf(v)\mathrm{d}v = \int_0^\infty \left(\frac{2v}{c^2}\right)\exp\left[-\left(\frac{v}{c}\right)^2\right]\mathrm{d}v = \frac{\sqrt{3.14}}{2}c \tag{2}$$

$$c = 1.128V_{\mathrm{m}} \tag{3}$$

$$P = \begin{cases} \dfrac{v - v_{\mathrm{in}}}{v - v_{\mathrm{in}}} P_{\mathrm{in}} v_{\mathrm{in}} v_{\mathrm{out}} \\ v_{\mathrm{i}} v_{\mathrm{r}} \\ P_{\mathrm{rated}} v_{\mathrm{r}} v_{\mathrm{out}} \end{cases} \tag{4}$$

where *c*, and *v* are scale factor, and wind speed, respectively.

If a location knows the value of the wind speed, the scaling index *c* can be determined as in (2)–(4).

For the Rayleigh distribution, small number of state has a good accuracy while large number of state has complexity problem.

2.2 Load Modeling

The machine peak load is required to meet the load form of the IEEE-RTS in order to go on with an appropriate scheduling results. According to this statement, a bunching method will divide the load into ten levels using the centroid categorization process which verifies that the option of 10 equivalent load rates (state) with different possibilities provide a sensible balance among correctness and rapid mathematical estimation.

3 Methodology

3.1 Problem Formulation

Converting the chance limit to its specified counterpart to the level of trust granted would be difficult. Since the variables of the decision are multi-dimensional and the constriction is not a function of analytical distribution. In this work, therefore, the Monte Carlo model is implemented to check whether the limit on chance holds. Monte Carlo simulation's application to the chance-bound probabilistic model has been comprehensively studied and its convergence exploration has been austerely verified. The techniques in detail are as follows:

1. Set $N = 0$.
2. Create S situations according to the random variables for a 24 h interval.
3. Test this situations for minimum chance.
4. Set $N = N + 1$ if this scenario has a chance limit. If not, then go for next step.
5. Continue steps again for S times from (1) to (5).
6. Find N/S. If N/S is greater or equal to the object value, then loop will continue; otherwise, Monte Carlo's simulation will end.

3.2 DE Algorithm Technique

To date, in order to solve the problem of BESS preparation various methods has been proposed, including nonlinear NIP programming, Bender's decomposition, genetic algorithms and PSO. To measure charge/discharge energy in each BESS and wind power, etc., the optimal power stream should be completely implemented in different scenarios in each time period. Taking into account the difficulty of the problem of optimization, we select the exploratory algorithm for solving the model. In this analysis, the Monte Carlo embedded Differential Evolution algorithm is used to find the location of Wind DG.

The DE algorithm is done by the modified IEEE 11 kV, 33-bus radial system. The Monte–Carlo embedded in this paper the setting of DE variables as crossover probability $P_{cr} = 0.7$, total no. of iterations = 100, and scaling factor $\mu = 0.8$. The confidence limit level of chance is set as $\alpha_i = 0.2$, $\beta_i = 0.2$. Two types of battery technology are being included in this review. The L/A battery is type I and the Zn/Br battery is type II. Because the power rating of BESS is coupled with the energy capacity, rating is set to be 20% of its total energy for BESS once a good BESS capacity solution is calculated. Matlab R 2018a conducted all the tests.

3.3 BESS Technology

First, we run our algorithm (without BESS) and further involve two different BESS technologies. The two cases showing different wind power consumption rates: not less than 81 and 86% of total wind power generation must be used in a day, although trust must be retained at least $1 - \beta_i = 81\%$ chance. The higher the α value, the higher the BESS capability must be mounted. BESS' increased capacity means that to put up the excess wind energy, larger BESS must be built. We can say that lead-acid batteries are a better choice because it is less costly. We also relate the variations in hourly wind output prior and after the deployment of the lead-acid battery. In high availability hour wind cycles (hour 2–6, hour 20–24), we find that wind power loss is reduced. The average rate of wind use rises from 66.7% without BES to 83.97% installed lead acid. It suggests that the ESS shows an essential role in increasing the rate of use of wind power.

4 Results and Summary

4.1 Without Optimization

From the Table 1, the power losses without optimization are 0.20268 MW, thus we have to further reduce this loss with the help of optimization techniques.

Table 1 Wind D.G location

Bus	Wind probability	Wind speed	Wind power	Power loss	Before	After
Bus 11	0.22518	4.5	0.25	Power loss (MW)	0.20268	0.17611

4.2 *Manual Method*

In manual method, the wind DG is randomly selected. In this case, we selected Bus 11 which reduced the loss up to 0.17611 MW. This is evident from Table 2. From the Fig. 2, it is clearly seen that after applying manual method (randomly selected), say bus 11, the average power loss has been reduced to 0.17611 MW. Also, from Fig. 3, it is observed that curve of voltage (pu) touches to unity which is desirable.

4.3 *Monte Carlo Embedded DE Solution*

From the Table 3, it is observed that after applying differential evolution algorithm the power loss has been reduced from 0.20268 to 0.17241 MW. Thus, there is a significant reduction in power loss. But, we would like to reduce it further using the hybrid optimizations.

4.4 *Hybrid Optimization*

In hybrid optimization, we would use voltage method, i.e., the optimization is done at that bus where the bus voltage is minimum. Using this optimization, the power losses reduce from 0.20268 MW to 0.14259 MW, which is very much significant. This is evident in Table 4 and Fig. 4. Thus, the hybrid optimization plays a crucial role in reducing the losses compared to other optimization techniques.

4.5 *Comparison of Results*

Results of different method have been summarized in the Table 5. It can be clearly seen that the objective value has been achieved using the hybrid optimization technique.

5 Conclusion

To assess the optimal allocation of distribution generation units based on wind energy to mitigate the loss of device power, the information for the three scenarios described in this paper is evaluated. In addition, the real annual power losses of the process were estimated for each case to examine the strength of the proposed modeling

Table 2 Branch information

Branch	From bus	To bus	From bus injection		To bus injection		Loss ($I^2 * Z$)	
			P (MW)	*Q* (MVAr)	*P* (MW)	*Q* (MVAr)	*P* (MW)	*Q* (MVAr)
1	1	2	3.921	2.44	−3.911	−2.43	0.0121	0.011
2	2	3	3.44	2.21	−3.39	−2.18	0.052	0.03
3	3	4	2.36	1.68	−2.34	−1.67	0.020	0.01
4	4	5	2.22	1.59	−2.20	−1.58	0.019	0.01
5	5	6	2.14	1.55	−2.11	−1.52	0.038	0.03
6	6	7	1.10	0.53	−1.09	−0.52	0.002	0.01
7	7	8	0.89	0.42	−0.89	−0.42	0.005	0.00
8	8	9	0.69	0.32	−0.68	−0.32	0.004	0.00
9	9	10	0.62	0.30	−0.62	−0.29	0.004	0.00
10	10	11	0.56	0.27	−0.56	−0.27	0.001	0.00
11	11	12	0.52	0.24	−0.51	−0.24	0.001	0.00
12	12	13	0.45	0.21	−0.45	−0.21	0.003	0.00
14	14	15	0.27	0.09	−0.27	−0.09	0.000	0.00
15	15	16	0.21	00.08	−0.21	−0.08	0.000	0.00
16	16	17	0.15	0.06	−0.15	−0.06	0.000	0.00
17	17	18	0.09	0.04	−0.09	−0.04	0.000	0.00
18	2	19	0.36	0.16	−0.36	−0.16	0.000	0.00
19	19	20	0.27	0.12	−0.27	−0.12	0.001	0.00
20	20	21	0.18	0.08	−0.18	−0.08	0.000	0.00
21	21	22	0.09	0.04	−0.09	−0.04	0.000	0.00
22	3	23	0.94	0.46	−0.94	−0.45	0.003	0.00
23	23	24	0.85	0.40	−0.84	−0.40	0.005	0.00
24	24	25	0.42	0.20	−0.42	−0.20	0.001	0.00
25	6	26	0.95	0.97	−0.95	−0.97	0.002	0.00
26	26	27	0.89	0.95	−0.88	−0.94	0.003	0.00
27	27	28	0.82	0.92	−0.81	−0.91	0.010	0.00
28	28	29	0.75	0.89	−0.75	−0.88	0.007	0.00
29	29	30	0.63	0.81	−0.62	−0.81	0.000	0.00
30	30	31	0.42	0.21	−0.42	−0.21	0.001	0.00
31	31	32	0.27	0.14	−0.27	−0.14	0.000	0.00
32	32	33	0.06	0.04	−0.06	−0.04	0.000	0.00
33	21	8	0.00	0.00	0.00	0.00	0.000	0.00
34	9	15	0.00	0.0	0.00	0.00	0.000	0.000
35	12	22	0.00	0.00	0.00	0.00	0.000	0.000
36	18	33	0.00	0.00	0.00	0.00	0.000	0.000
37	25	29	0.00	0.00	0.00	0.00	0.000	0.00

(continued)

Table 2 (continued)

Branch	From bus	To bus	From bus injection		To bus injection		Loss ($I^2 * Z$)	
			P (MW)	*Q* (MVAr)	*P* (MW)	*Q* (MVAr)	*P* (MW)	*Q* (MVAr)
			Total				0.20268	0.16

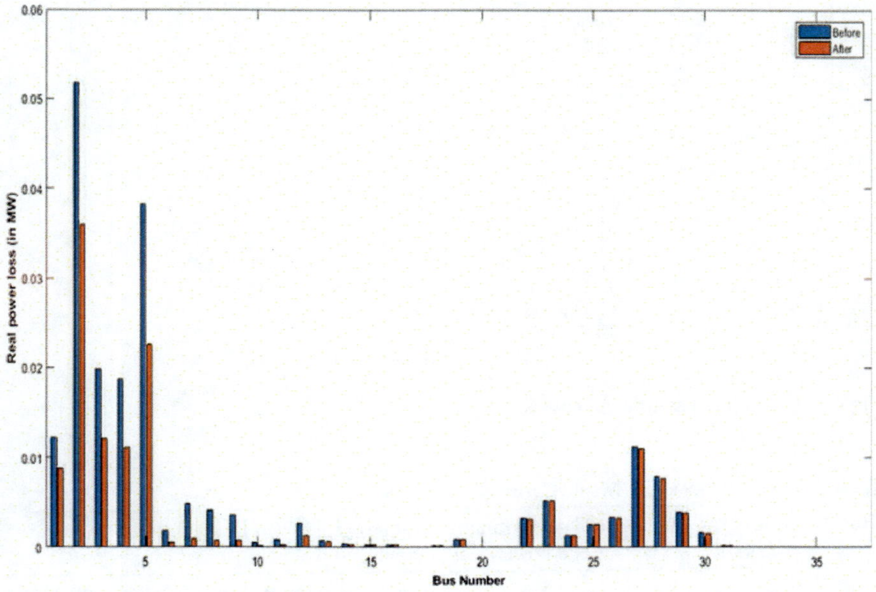

Fig. 2 Real-time power loss (MW) at different buses

algorithm. Optimization is done by Monte Carlo-based DE algorithm. Hybrid optimization is also done in which optimization is done by power, voltage, and manual method. The results reveal that the annual energy losses are significantly reduced regardless of the method used to allot the distributed generation units based on wind energy in the system. While comparing the annual total loss of energy determined in Monte Carlo embedded DE solution and the hybrid optimization method to the actual ones, the loss reduction in the hybrid optimization scenario was found to be similar to the actual loss calculation. These outcomes could be credited to the point that the proposed generation-load model considers every permissible combination of wind output energy and the load in the optimization problem in hybrid optimization, showing that this scenario is a decent illustration of the actual situation.

A stochastic method of optimization was anticipated in this paper to find the optimum position of wind DG in the DN with the goal of optimizing the use of wind energy. Uncertainties like the fluctuation of wind energy, variability in load demand, are taken into account. The cost of the operation of is determined depending upon the analysis of the scenario. To ensure the rate of use of wind power, the chance limit is implemented, and the Monte Carlo embedded DE solution is employed to

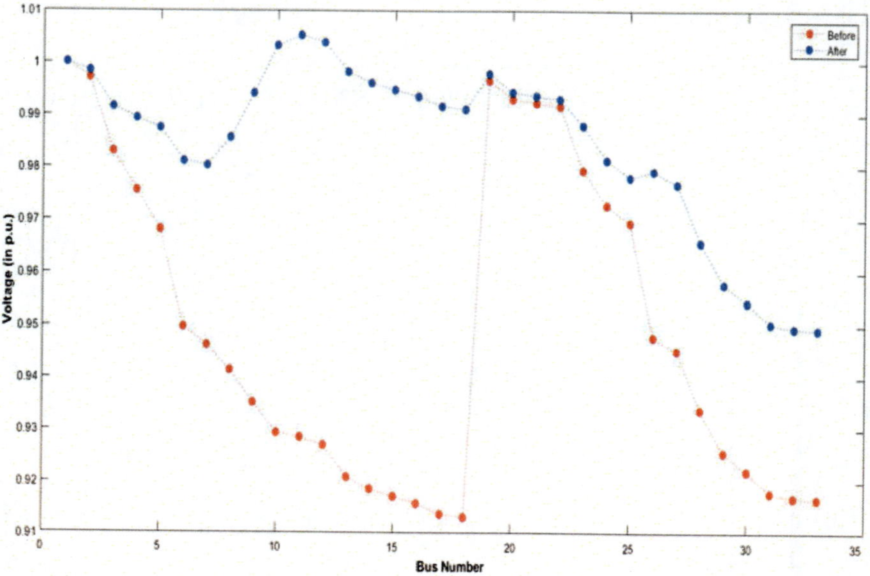

Fig. 3 Voltage (p.u.) at different busses

Table 3 Wind DG location (Monte Carlo DE algorithm)

Bus	Wind probability	Wind speed	Maximum power	Best power output	Power loss	Before	After
Bus 2	0.95013	19.5	5	3.8516	Power loss (MW)	0.20268	0.17241
Bus 8	0.23114	4.5	0.25	0.25			

Table 4 Wind DG location (hybrid voltage method)

Bus	Wind probability	Wind speed	Maximum power	Best power output	Power loss	Before	After
Bus 18	0.95013	19.5	5	0.61536	Power loss (MW)	0.20268	0.14259
Bus 17	0.23114	4.5	0.25	0.25			

find the solution of the anticipated problem. On the updated IEEE RTS 33-bus test platform, a series of simulation studies were carried out. The results of the simulation show the quality of the proposed model and system. The results showed that, based on capacity factor measurements, the expected approach could approximate the real loss estimates resulting in more accurate results.

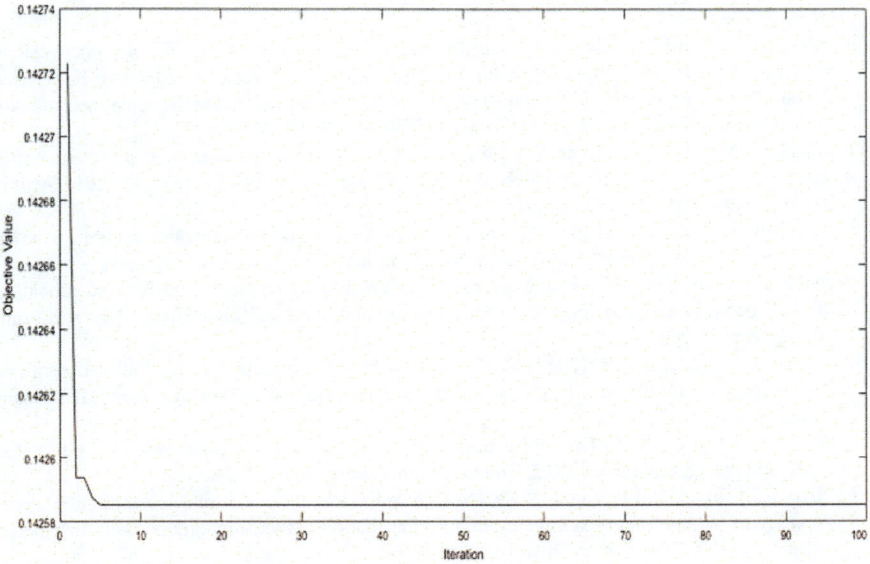

Fig. 4 Objective value at different iterations (voltage method)

Table 5 Comparison of results

Parameters		Methods				
		Manual	Monte–Carlo embedded DE solution		Hybrid optimization (voltage method)	
Bus		11	2	8	17	18
Wind probability		0.22518	0.95013	0.23114	0.23114	0.95013
Wind speed (m/s)		4.5	19.5	4.5	4.5	19.5
Wind DG output (MW)		0.35	3.8309	0.35	0.35	0.61536
Power loss (MW)	Before	0.20268	0.20268		0.20268	
	After	0.17611	0.17241		0.14259	

References

1. Zhu Y, Liu C, Sun K, Shi D, Wang Z (2019) Optimization of battery energy storage to improve power system oscillation damping. IEEE Trans Sustain Energy 10(3):1015–1024. https://doi.org/10.1109/TSTE.2018.2858262
2. Ehsan A, Yang Q (2018) Optimal integration and planning of renewable distributed generation in the power distribution networks: a review of analytical techniques. Appl Energy 210:44–59
3. Morshed MJ, Hmida JB, Fekih A (2018) A probabilistic multi-objective approach for power optimization in hybrid wind-PV-PEV systems. Appl Energy 211:1136–1149
4. CIGRE (2018) Impact of increasing contribution of dispersed generation on the power system. In: Working group, vol 37, no 23
5. EL-Khattam W, Salama MMA Distributed system planning using distributed generation. In: Proceeding of IEEE Canadian conference on electrical and computer engineering, Canada, vol

1, pp 579–82
6. Akhtar I, Paul WUH, Kirmani S, Asim M (2021) Cost analysis of 18 kW solar photovoltaic system for smart cities growth in India. In: Iqbal A, Malik H, Riyaz A, Abdellah K, Bayhan S (eds) Renewable power for sustainable growth. Lecture notes in electrical engineering, vol 723. Springer, Singapore. https://doi.org/10.1007/978-981-33-4080-0_63
7. Alzaidi KMS, Bayat O, Uçan ON (2019) Multiple DGs for reducing total power losses in radial distribution systems using hybrid WOA-SSA algorithm. Int J Photoenergy 2019(Article ID 2426538):20 p. https://doi.org/10.1155/2019/2426538
8. Paul WUH, Bhat MB, Kirmani S, Nahvi SA (2020) Data based controller design for pmdc motor setup using system identification. Stud Ind Place Names 40(10):11
9. Mahmoud PHA, Huy PD, Ramachandaramurthy VK (2017) A review of the optimal allocation of distributed generation: objectives, constraints, methods, and algorithms. Renew Sustain Energy Rev 75:293–312
10. Prasad S, Mallesham VKD (2017) Multi-objective hybrid estimation of distribution algorithm-interior point method-based meter placement for active distribution state estimation. IET Gener Trans Distrib 12(3):767–779
11. Xiong P, Singh C (2016) Optimal planning of storage capacity in power systems integrated with pind power generation. IEEE Trans Sustain Energy 7(1):232–240
12. Xiao J, Zhang Z, Bai L, Liang H (2016) Determination of the optimal installation site and capacity of battery energy storage system in distribution network integrated with distributed generation. IET Gener Trans Distrib 10(3):601–607
13. Zhan H, Wang C, Wang Y, Yang X, Zhang X, Wu C, Chen Y (2016) Relay protection coordination integrated optimal placement and sizing of distributed generation sources in distribution networks. In: Proceedings of IEEE power & energy society general meeting, July 2016, p 1
14. Prakash P, Khatod DK (2016) Optimal sizing and siting techniques for distributed generation in distribution systems: a review. Renew Sustain Energy Rev 57:111–130
15. Ali D, Paul W, Ali M, Ahmad M, Ashfaq H (2021) Optimal placement of distribution generation sources in hybrid generation network. Smart Grid Renew Energy 12:65–80. https://doi.org/10.4236/sgre.2021.125005

An Intelligent Mechanism for Utility and Active Customers in Demand Response Using Single and Double Q Learning Approach

Akhilesh Chandrakar and Priyanka Paliwal

Abstract The energy profiles of users in the traditional grid are non-compliant and intractable. However, with the evolution of smart grid, this need is fulfilled by customer-oriented programs known as demand side management (DSM) and demand response (DR). Q learning algorithm employed here so that enhanced benefits of DR made available to customers and retailers or utility. In this paper, various price versus demand functions along with their combinations have been used to tackle varying load profiles and comfort levels of different categories of customers. To represent customer response to incentive-based and price-based DR programs, composite demand functions (CDF) and dynamic price elasticity are put forward to examine user susceptibility to changing hourly price. This will help a retailer or any utility that will work as an agent to learn the customer environment and offer the most suitable price. Their learning capability is determined by principles of single and double q learning to demonstrate the comprehensive demand response (CDR) model that will yield best-suited benefit to users and utility. A study has been conducted by analyzing previous hourly demand and market prices of an area for the day.

Keywords Demand response · Q learning · Composite demand functions · Comprehensive demand response

1 Introduction

The residential sector constitutes about 25% of total energy consumption in the electricity sector in India. The household electricity consumption stands about 210 kWh per capita, and this is further expected to increase in the future [1]. Hence, there is a requirement for increased generation and transmission infrastructure and better communication between different levels in the industry. In recent years, the concept of the smart grid has emerged which is capable of tackling above challenges. It ensures better utilization of existing capacity, more reliability, legitimates tariff for

A. Chandrakar (✉) · P. Paliwal
Electrical Engineering Department, MANIT Bhopal, Bhopal, India
e-mail: akhil79022@gmail.com

K. Namrata et al. (eds.), *Smart Energy and Advancement in Power Technologies*, Lecture Notes in Electrical Engineering 926,
https://doi.org/10.1007/978-981-19-4971-5_30

customers, lower transmission and distribution losses. In addition, the system will have transparent policies so that customers and retailers will be able to contribute to the reliability and efficiency of the grid and also get benefitted in return [2, 3]. The users will be able to adjust their loads based on price or incentives offered from retailers or utility depending on the market conditions in smart grid [2, 4]. By shifting load to off-peak hours, users will be able to considerably lower their electricity costs [5]. A considerable effort has been done by the researchers to maximize the benefit of utility and customers by developing the best suitable strategies for prices in real-time situations. A variety of meta-heuristics have been studied for DSM and DR to improve the customer profile [6]. Genetic algorithm (GA) has been used for reducing loads at the customer level [7], binary particle swarm optimizations (BPSO) and GA has been applied for reducing electricity cost and user discomfort [2], ant colony optimization (ACO) has been utilized for reducing peak to average ratio par [8]. The bio-based algorithm such as whale optimization algorithm (WOA) has also been used to encourage productive electricity consumption [9]. These meta-heuristics are applied for DSM technique.

It is believed that DR is a subdivision of DSM. DR offered to the user can be subdivided as incentive based or price based. In incentive-based scheme, incentives and penalties are offered to the user to improve their energy profile. In price based, user profile improves as desired in terms of tariffs charged. The price-based schemes are further classified as real-time pricing (RTP), time of use (TOU) and critical peak pricing (CPP) [10]. In RTP, customers are charged on hourly basis. In TOU pricing, tariffs are set at the start of the day, and all conditions are specified based on demand levels. The CPP is quite like TOU with only difference that when peak demand arises, high prices are charged to the user.

In comparison with other meta-heuristics, q learning-based approach offers high quality solution due to its practical implications. Q learning is a categorization of reinforcement learning and represents a way by which it acquires the value of an action in particular state. It is a model-free learning algorithm and can handle varying environments and return rewards without any adaptations [11].

In this paper, a q learning-based framework for facilitating demand response program has been developed and analyzed. A retailer or utility is behaving as an agent whose benefits is dependent on its learning capability which is actually extracted by q learning [12]. Based on learning, real-time prices are determined on hourly basis. The study has been conducted on day-ahead wholesale hourly market prices to understand customer behavior and benefits are calculated if certain actions are followed by user in particular state [13]. A Q value is defined for each reward that has been assigned to particular action in each state. Based on Q value, optimum benefit is calculated.

To represent users hourly consumption, different demand versus price function as composite demand functions has been used such as linear, exponential, logarithmic, potential and hyperbolic [14]. The linear and exponential customers are willing to modify their profiles based on priced offered whereas logarithmic, potential and hyperbolic are reluctant. Figure 1 shows the variation of demand versus price methods.

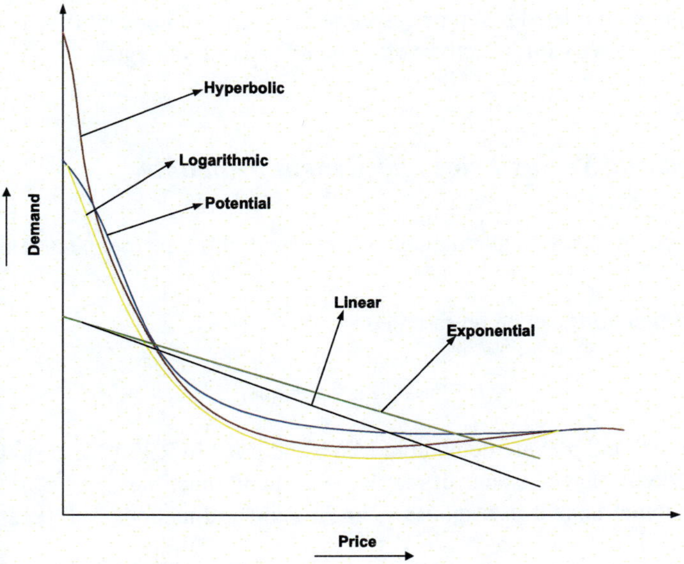

Fig. 1 Different demand versus price functions

Different demand versus price functions represent customer profiles which are combined and used as CDR model in which hourly prices are set based on hourly demand of user [15]. The mathematical representations are based on q learning which is calculated and depends on prices offered, dynamic price elasticity and change in customer demand. Dynamic price elasticity shows how the quantity demanded by consumers will respond to change in price [16].

CDR model had been developed using weighted coefficients of various demand versus price functions according to results from regression curve fittings. The CDF used to employ all the variety of users because of having buffer of adjusting weighting coefficients based on old retrieved data. Dynamic elasticity is been used in place of predetermined elasticity based on differentiating demand versus price functions. Using CDF and dynamic price elasticity, CDR is prepared and analyzed for calculating benefits for customers and utility.

Rest of paper is structured as Sect. 2 is devoted to modeling of customers based on CDF and dynamic price elasticity. Section 3 proposes agent-based q learning and its variants use on q learning limitations. Section 4 presents a case study and results based on discussion, and Sect. 5 finally concludes the paper.

2 Modeling User Energy Profile

This part involves modeling of user participation in DR programs. Whether incentive based and price based depending on utility perspective. Their participation is mathematically modeled based on response they offer and benefit they get for their consumption along with incentives offered and penalties charged.

2.1 Demand Versus Price and Benefit Function

This section involves mathematical modeling of different demand versus price functions.

1. *Linear functions* can be represented as:

$$d_{\text{lin}}(t) = x_{lin} + y_{\text{lin}}p(t) \tag{1}$$

where $d(t)$ and $p(t)$ are demand in MWh and price in RS/MWh at tth hour, and x and y represent constants of respective demand functions.

Customer benefit in response to linear demand functions can be represented as:

$$\begin{aligned} b_{\text{lin}}^{\text{cus}}(t) &= \text{bo}_{\text{lin}}^{\text{cus}}(t) + \text{po}(t)(d_{\text{lin}}(t) - \text{do}_{\text{lin}}(t)) \\ &\times \left(1 + \frac{(d_{\text{lin}}(t) - \text{do}_{\text{lin}}(t))}{2\text{pe}(t)\text{do}_{\text{lin}}(t)}\right) \end{aligned} \tag{2}$$

where $b^{\text{cus}}(t)$ is benefit of customer at tth hour, $b^{\text{cus}}(t)$, $\text{po}(t)$, $\text{do}(t)$ are the initial benefit, initial price and initial demand, respectively, and $\text{pe}(t)$ is dynamic price elasticity at tth hour.

2. The *logarithmic function* is represented as:

$$d_{\log}(t) = x_{\log} + b_{\log}\ln(p(t)) \tag{3}$$

Benefit equation corresponding to this function is:

$$\begin{aligned} b_{\log}^{\text{cus}}(t) &= \text{bo}_{\log}^{\text{cus}}(t) + \text{po}(t)\text{do}_{\log}(t)\text{pe}(t) \\ &\times \left\{\exp\left(\left(\frac{(d_{\log}(t) - \text{do}_{\log}(t))}{2\text{pe}(t)\text{do}(t)}\right) - 1\right)\right\} \end{aligned} \tag{4}$$

3. The *potential group* of customers can be expressed as:

$$d_{\text{pot}}(t) = x_{\text{pot}}(p(t)^{y_{\text{pot}}}) \tag{5}$$

Benefit function is as follows:

$$b_{\text{pot}}^{\text{cus}}(t) = \text{bo}_{\text{pot}}^{\text{cus}}(t) + \frac{\text{po}(t)\text{do}_{\text{pot}}(t)}{1 + \text{pe}^{-1}(t)}\left(\frac{d_{\text{pot}}(t)}{\text{do}_{\text{pot}}(t)}\right)^{\text{pe}^{-1}(t)} \tag{6}$$

4. Widely used model for *exponential function* can be expressed as:

$$d_{\exp}(t) = x_{\exp} \exp(y_{\exp} p(t)) \tag{7}$$

Benefit function for exponential function is as follows:

$$\begin{aligned} b_{\exp}^{\text{cus}}(t) &= \text{bo}_{\exp}^{\text{cus}}(t) + \text{po}(t) d_{\exp}(t) \\ &\times \left\{ \left(1 + \frac{1}{\text{pe}(t)} \ln\left(\frac{d_{\exp}(t)}{\text{do}_{\exp}(t)} \right) - 1 \right) \right\} \end{aligned} \tag{8}$$

5. *Hyperbolic demand function* is as follows:

$$d_{\text{hyp}}(t) = \frac{x_{\text{hyp}}}{y_{\text{hyp}} p(t)} \tag{9}$$

And their benefit functions is expressed as:

$$\begin{aligned} b_{\text{hyp}}^{\text{cus}}(t) &= \text{bo}_{\text{hyp}}^{\text{cus}}(t) + \text{po}(t) \text{do}_{\text{hyp}}(t) \text{pe}(t) \\ &\times \{ \ln[(d_{\text{hyp}}(t) - \text{do}_{\text{hyp}}(t))(1 - \text{pe}(t))] \} \end{aligned} \tag{10}$$

When customers participate in demand response programs, this benefit function will include tariffs, incentives and payments. Hence, customer total benefits when participating in demand response program can be expressed as follows:

$$\begin{aligned} \text{Tb}^{\text{cus}}(t) &= b^{\text{cus}}(t) + p(t) \times \text{do}(t) + \text{inc}(t) \\ &\times (d(t) - \text{do}(t)) - \text{pen}(t) \times (\text{re}(t) - (d(t) - \text{do}(t)) \end{aligned} \tag{11}$$

where $\text{Tb}^{\text{cus}}(t)$ is the total benefit when customer participated in demand response program, $\text{inc}(t)$ and pen (t) are incentives and penalties, and $\text{re}(t)$ is amount of load reduction done in DR programs.

Maximum total benefit can be found by equating $\frac{\partial (Tb^{cus}(t))}{\partial d(t)} = 0$, i.e.,

$$\frac{\partial (b^{\text{cus}}(t))}{\partial d(t)} = p(t) + \text{inc}(t) + \text{pen}(t) \tag{12}$$

By applying Eq. (12) on linear function and differentiating Eq. (2), we get,

$$\text{po}(t) \left[1 + \frac{(d_{\text{lin}}(t) - \text{do}_{\text{lin}}(t))}{\text{pe}(t) \text{do}_{\text{lin}}(t)} \right] = p(t) + \text{inc}(t) + \text{pen}(t) \tag{13}$$

The value of new demand can be expressed as:

$$d_{\text{lin}}(t) = \text{do}_{\text{lin}}(t) \left[1 + \text{pe}(t) \frac{p(t) + \text{inc}(t) + \text{pen}(t) - \text{po}(t)}{\text{po}(t)} \right] \tag{14}$$

Equation (14) demonstrates changes in user response in demand in an hourly basis according to price-based or incentive-based programs in which user participate.

Similarly, customer demand response for logarithmic, potential, exponential and hyperbolic is as follows:

$$d_{\log}(t) = \mathrm{do}_{\log}(t)\left[1 + \mathrm{pe}(t)\ln\left(\frac{p(t) + \mathrm{inc}(t) + \mathrm{pen}(t)}{\mathrm{po}(t)}\right)\right] \tag{15}$$

$$d_{\mathrm{pot}}(t) = \mathrm{do}_{\mathrm{pot}}(t)\left(\frac{p(t) + \mathrm{inc}(t) + \mathrm{pen}(t)}{\mathrm{po}(t)}\right)^{\mathrm{pe}(t)} \tag{16}$$

$$d_{\exp}(t) = \mathrm{do}_{\exp}(t)\exp\left(\mathrm{pe}(t)\frac{p(t) + \mathrm{inc}(t) + \mathrm{pen}(t) - \mathrm{po}(t)}{\mathrm{po}(t)}\right) \tag{17}$$

$$d_{\mathrm{hyp}}(t) = \mathrm{do}_{\mathrm{hyp}}(t)\left(1 + \frac{\mathrm{po}(t)\mathrm{do}(t)\mathrm{pe}(t)}{(p(t) + \mathrm{inc}(t) + \mathrm{pen}(t))(1 + \mathrm{pe}(t))}\right) \tag{18}$$

In case of incentive-based program, $p(t) = \mathrm{po}(t)$ and in case of price-based program, $\mathrm{inc}(t) = \mathrm{pen}(t) = 0$. Therefore, depending on DR programs customer demand response is examined, and optimum benefit is calculated.

2.2 *Dynamic Price Elasticity*

Fixed price elasticity is independent from demand functions and does not account the behavior of customers clearly [17]. Here, dynamic price elasticity been taken into account that will properly inspect users behavior. Mathematical modeling of dynamic price elasticity in this model as:

$$\mathrm{pe}(t) = \frac{\mathrm{po}(t)\partial d(t)}{\mathrm{do}(t)\partial p(t)} \tag{19}$$

Hourly elasticity has been used for calculating the respective benefits of demand versus price functions. For linear dynamic price, elasticity is expressed as:

$$\mathrm{pe}_{\mathrm{lin}}(t) = y_{\mathrm{lin}}\frac{\mathrm{po}(t)}{x_{\mathrm{lin}} + y_{\mathrm{lin}}\mathrm{po}(t)} \tag{20}$$

Similarly, for logarithmic, potential and exponential and hyperbolic, elasticity is respectively defined as follows:

$$\mathrm{pe}_{\log}(t) = \frac{y_{\log}}{x_{\log} + y_{\log}\ln\,\mathrm{po}(t)} \tag{21}$$

$$\mathrm{pe}_{\mathrm{pot}}(t) = y_{\mathrm{pot}} \tag{22}$$

$$\mathrm{pe}_{\mathrm{exp}}(t) = y_{\mathrm{exp}}\mathrm{po}(t) \tag{23}$$

$$\mathrm{pe}_{\mathrm{hyp}}(t) = y_{\mathrm{hyp}} \tag{24}$$

These values can be substituted in Eq. (2) to get respective benefits.

2.3 *Comprehensive Demand Response (CDR) Model*

Customers who follow linear demand function can cut short the load above certain limits of price. While potential group of customers continue to consume loads even if prices are high, the hyperbolic customers are sensitive to change their loads when prices are low but not during high prices. Due to this diversity, different group of customers are weighted depending on their accord and obtain the comprehensive response. Hence, divergent load profiles of customers are represented in comprehensive response as:

$$\begin{aligned} d(t) &= w_{\mathrm{lin}} d_{\mathrm{lin}}(t) + w_{\mathrm{log}} d_{\mathrm{log}}(t) + w_{\mathrm{pot}} d_{\mathrm{pot}}(t) \\ &\quad + w_{\mathrm{hyp}} d_{\mathrm{hyp}}(t) + w_{\mathrm{exp}} d_{\mathrm{exp}}(t) \end{aligned} \tag{25}$$

where w is weight associated with respective demand functions.

Comprehensive customer demand response is thus expressed as:

$$\begin{aligned} D(t) &= w_{\mathrm{lin}} \mathrm{do}_{\mathrm{lin}}(t)\left[1 + \mathrm{pe}(t)\frac{p(t) + \mathrm{inc}(t) + \mathrm{pen}(t) - \mathrm{po}(t)}{\mathrm{po}(t)}\right] \\ &\quad + w_{\mathrm{pot}} \mathrm{do}_{\mathrm{pot}}(t)\left(\frac{p(t) + \mathrm{inc}(t) + \mathrm{pen}(t)}{\mathrm{po}(t)}\right)^{\mathrm{pe}(t)} \\ &\quad + w_{\mathrm{log}} \mathrm{do}_{\mathrm{log}}(t)\left[1 + \mathrm{pe}(t)\ln\left(\frac{p(t) + \mathrm{inc}(t) + \mathrm{pen}(t)}{\mathrm{po}(t)}\right)\right] \\ &\quad + w_{\mathrm{exp}} \mathrm{do}_{\mathrm{exp}}(t)\exp\left(\mathrm{pe}(t)\frac{p(t) + \mathrm{inc}(t) + \mathrm{pen}(t) - \mathrm{po}(t)}{\mathrm{po}(t)}\right) \\ &\quad + w_{\mathrm{hyp}} \mathrm{do}_{\mathrm{hyp}}(t)\left(1 + \frac{\mathrm{po}(t)\mathrm{do}(t)\mathrm{pe}(t)}{(p(t) + \mathrm{inc}(t) + \mathrm{pen}(t))(1 + \mathrm{pe}(t))}\right) \end{aligned} \tag{26}$$

Equation (26) is the comprehensive demand response model ($D(t)$) which is used for evaluating customer performance in demand response programs.

2.4 Measurement of Weighting Coefficients

In order to evaluate customer demand response, initial x, y parameters of individual demand functions and then the weighting coefficients assigned to them is determined. For this, regression fitting curve method is used for analysis of divergent load profiles [18]. The relationship between historical data and demand response is developed for evaluating the coefficients [19]. After fitting curve, least square method (LSM) is applied to find the best weighting coefficients. A flow chart depicting different steps discussed above is presented in Fig. 2.

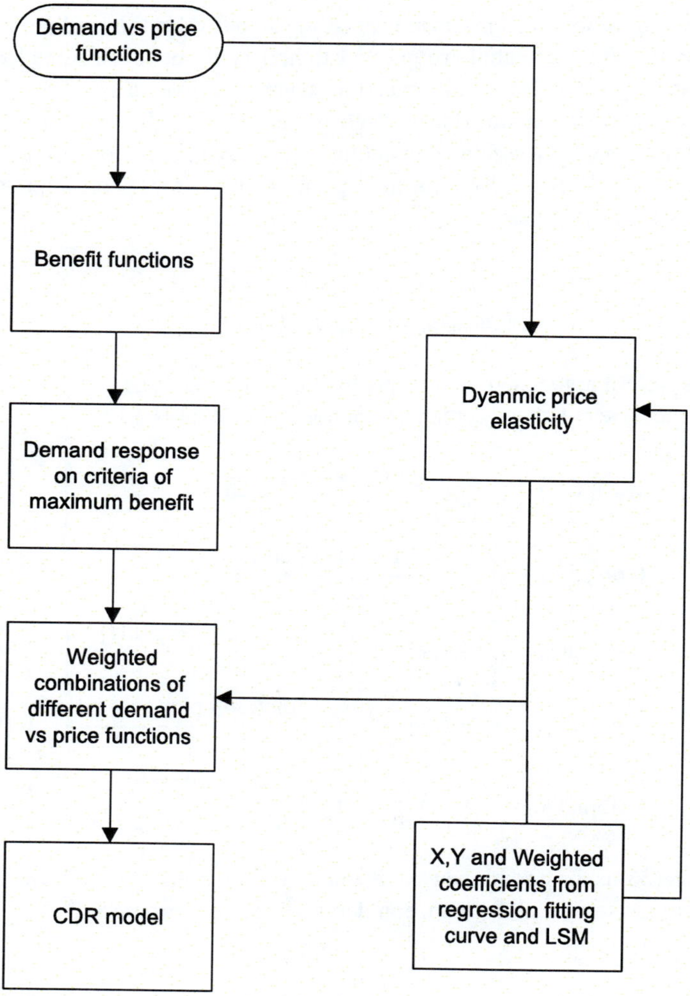

Fig. 2 Flowchart for evaluating CDR model

3 Application of Single and Double Q Learning Approach

In smart grid, customers receive real-time information regarding tariff and will adjust their loads accordingly. Equation 1 shows the CDR model for customer participation in DR programs in which they participate. It is believed that each retailer or utility has assigned agents that will represent section customers. It depends on intelligence of each agent how well they learn the user environment and acquire maximum benefit. This objective is fulfilled by q learning. In the following subsections, single q learning and its modified version double q learning has been discussed.

3.1 *Q Learning*

Customer's agent procures power in day-ahead market, adopts q learning for calculation of real-time hourly price and offers hourly prices to users based on its procurement cost, its adopted strategy, user behavior and its predicted benefit. The agents also learn from other competitor agents on how they are deciding prices for their customers. This raises competition among them and will lead to benefit of customers. The entire framework depends on each agent's learning capability and that is actually extracted by q learning [20]. It is believed that customer agents are unaware of customer environment and their behavior. They do not know what price to select that will maximize their benefits. So, they actually learn from past customer experiences which are indeed examined and extracted by q learning. Q learning portrays possibilities into actions without modeling and guide agents about determining price through rewards in each state [21]. Customer's agent will maximize its benefits while satisfying users through this q learning approach.

Let $s = \{s_1, s_2, s_3, \ldots s_n\}$ be the sets of feasible states and $a = \{a_1, a_2, a_3, \ldots a_n\}$ represent the sets of possible actions. At each learning stage, the agent at particular states selects a possible action. All possible actions in real-time offering can be defined as:

$a = \{\ a_1, a_2, a_3\}$

$$
\begin{aligned}
p_i(t) &\underset{a_1}{\longrightarrow} p_{i+1}(t) = p_i(t) + \Delta p \\
p_i(t) &\underset{a_2}{\longrightarrow} p_{i+1}(t) = p_i(t) - \Delta p \\
p_i(t) &\underset{a_3}{\longrightarrow} p_{i+1}(t) = p_i(t)
\end{aligned}
\tag{27}
$$

A policy of soft max and greedy policy is been made for deciding prices for users [22]. Based on these policies, a particular action is selected, and this policy actually depends on maximum probabilities and Boltzmann distributions.

This new price on each action updates the benefit value and takes an agent to a new state which is defined as: $s = \{s_1, s_2, s_3\}$.

$$s_i = \begin{cases} s_1 \text{ if } b(p_i(t)) > b(p_{i-1}(t)) \\ s_2 \text{ if } b(p_i(t)) < b(p_{i-1}(t)) \\ s_3 \text{ if } b(p_i(t)) = b(p_{i-1}(t)) \end{cases} \quad (28)$$

After setting state and action, the agent receives an award which is proportional to its benefits, and this rewards values update the Q value for particular state and its particular actions as follows:

$$R_i = s_{i+1} \times (100 + s_{i+1} + 1)1e - 3 \quad (29)$$

where R_i is the reward value at ith stage.

$$Q(s_i, a_i) = Q(s_i, a_i) + \alpha(R_i + \gamma \max Q(s_i, a_i) - Q(s_i, a_i)) \quad (30)$$

where $Q(s_i, a_i)$ is the qualitative value at particular state and action, α is learning rate and γ is discount factor.

These values are updated for limited number of learning stages L for each hour, and final stages are set as ultimate price which is used for calculating the benefit. Figure 3 presents a flow chart for q learning process in establishing real-time prices.

3.2 Double Q Learning

Single q learning involves a lot of over estimations and does not hold good for the stochastic process where there is large number of variations in demand between hours. In order to enhance the performance, different modifications have been proposed in q learning approach [23]. Double q learning is an improved version of single q learning which involves learning for 48 h and returns two Q values; each Q value is dependent on other [24]. The overall Q value is the average of two, and price is also the average of prices of hours of two days. For e.g., for first hour, price is average of price of first day first hour and second day first hour. It returns two Q values as follows:

$$\begin{aligned} Qa(s_i, a_i) &= Qa(s_i, a_i) + \alpha(R_i + \gamma \max Qb(s_i, a_i) - Qa(s_i, a_i)) \\ Qb(s_i, a_i) &= Qb(s_i, a_i) + \alpha(R_i + \gamma \max Qa(s_i, a_i) - Qb(s_i, a_i)) \end{aligned} \quad (31)$$

Remaining procedures remains same as single q learning.

3.3 Influence of Variables Used in Q Learning

1. *Learning rate* (*alpha*) (used in Eqs. 14 and 16) determines to what stretch newly gained information overwrites previous information. Its value ranges from 0 to

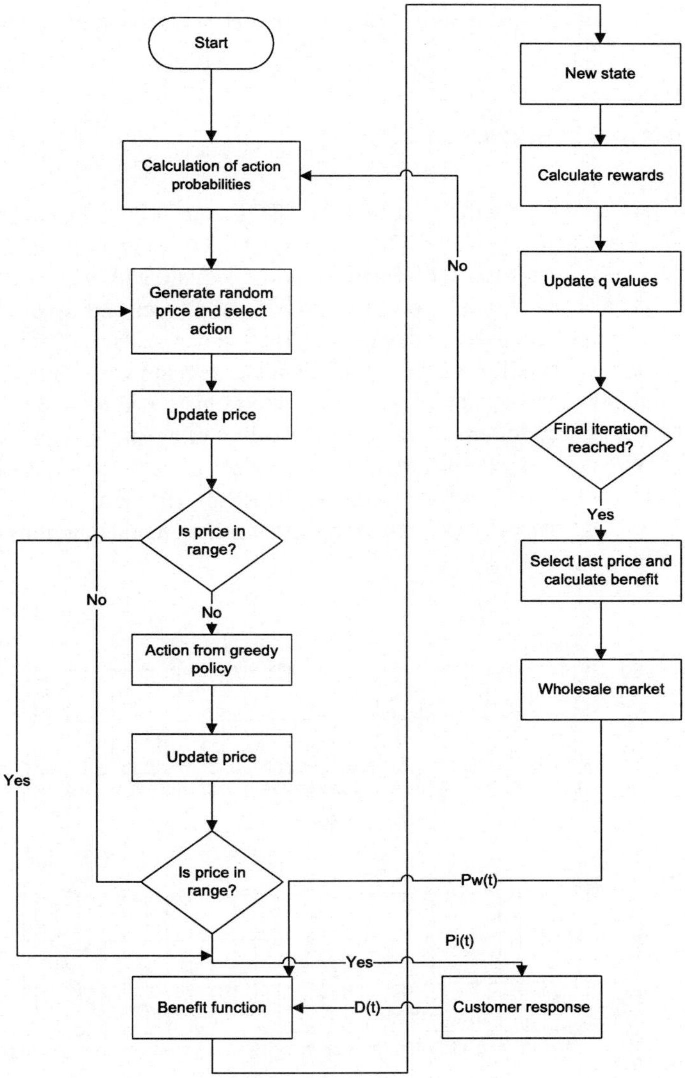

Fig. 3 Flowchart to determine customer response and benefit function using q learning

1. A zero value shows nothing is learned and value of 1 means agent learns only recent information. In deterministic terms, it is maintained constant. However, in stochastic environment, its value is decreased with increasing learning stage.
2. *Discount factor (gamma)* examines the importance of future rewards. A zero value shows an agent is short sighted which considers only near rewards, while value near 1 shows the agent values long-term high rewards. If value exceeds one, the process diverges. In this paper, high constant value of gamma close to one is taken for better results.

High initial value of q is assumed to ensure better exploration of environment.

4 Numerical Studies and Results

In this analysis, all three pricing programs RTP, TOU and CPP compared and respective benefits have been calculated. The results obtained using single and double q learning have been compared. The data for hourly demand and respective price for previous days has been obtained from IEX website of an area and used for analysis purpose [25]. The price and demand of $i-1$ day at tth hour is taken as initial price and demand of ith day. The CDR model with individual demand versus price functions is fitted to this previous day data. Figure 4a, b shows historical data of demand and price for two days used for enhancing learning ability of agent through q learning.

The coefficients x and y for different demand versus price functions are presented in Table 1 The coefficients vary according to historical load profiles.

Table 2 shows the respective weights assigned to each demand functions presented in CDR models.

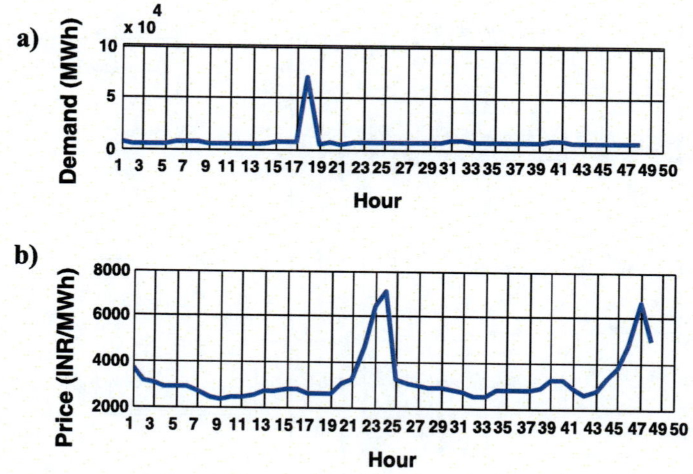

Fig. 4 **a** Original demand for 48 h, **b** original pricing for 48 h

Table 1 Coefficient of different demand versus price functions

Functions	X	Y
Linear	209.43	−0.441
Logarithmic	208.77	−23.556
Hyperbolic	209.45	−0.112
Exponential	209.565	−0.003
Potential	209.005	−0.215

Table 2 Weighting coefficients for CDR models

Functions	Weighting coefficients
Linear	0.455
Logarithmic	0.109
Hyperbolic	0.04
Exponential	0.396
Potential	0

The analysis has been done for the two cases viz. constant and variable learning rate. The obtained results are compared for utility and customer benefits. The performance of single q learning and double q learning is also compared for constant and variable learning rate. Different pricing schemes are also shown and compared with historical pricing for variable and constant learning rate. Maximum price set as $p(t) = 1.5\text{pw}(t)$ and minimum price as $\text{p}(t) = \text{pw}(t)$. The results have been discussed in detail in following subsections.

4.1 Pricing Schemes

Pricing schemes such as RTP and TOU are used and compared for single and double q learning at constant and variable learning rate 'alpha.' It is observed in Fig. 5a, b and Fig. 6a, b that variable alpha pricing for all three schemes is higher than constant alpha pricing. Real-time pricing is changing on hourly basis whereas TOU and CPP pricing is virtually constant, only changing at single hour. At constant alpha, double q pricing has offered lower value.

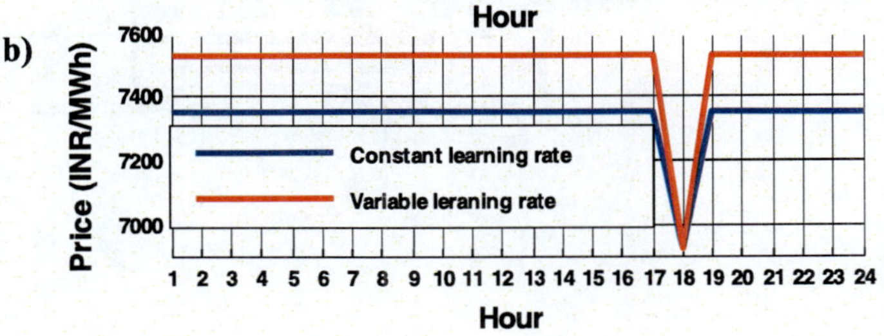

Fig.5 **a** RTP pricing scheme by single q learning, **b** TOU pricing scheme by single q learning

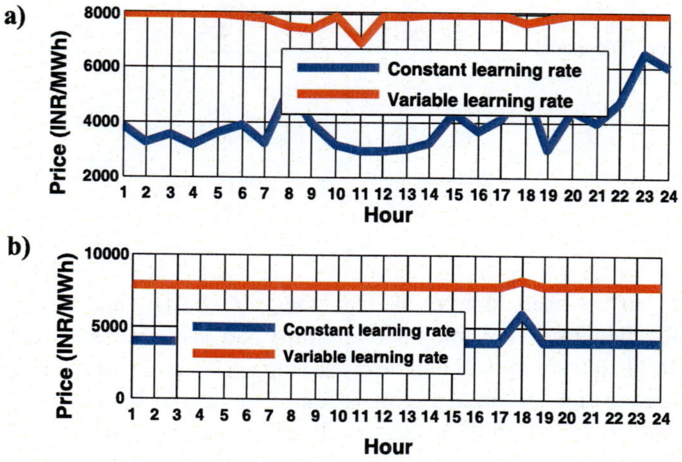

Fig. 6 **a** RTP pricing by double Q learning, **b** TOU pricing scheme by double q learning

4.2 Utility Benefits

Figure 7a, b presents a comparison of utility benefits obtained through constant and variable learning rate using single and double q learning. It has been observed that utility benefit increases for alpha variable both for single q learning and double q. It shows considerable increase in benefits in double q learning from alpha constant to alpha variable.

Fig. 7 **a** Utility benefit by single Q learning, **b** utility benefit by double q learning

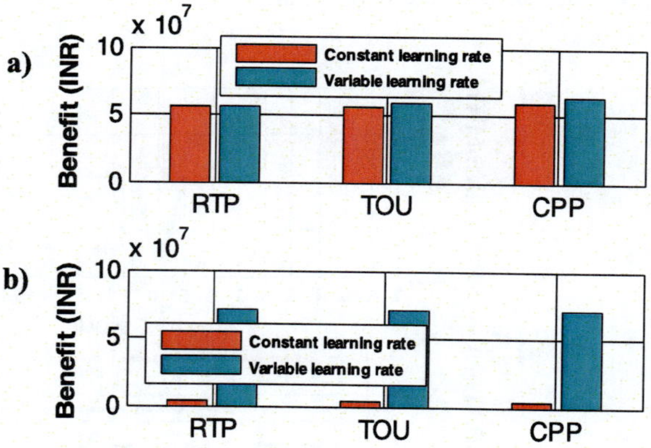

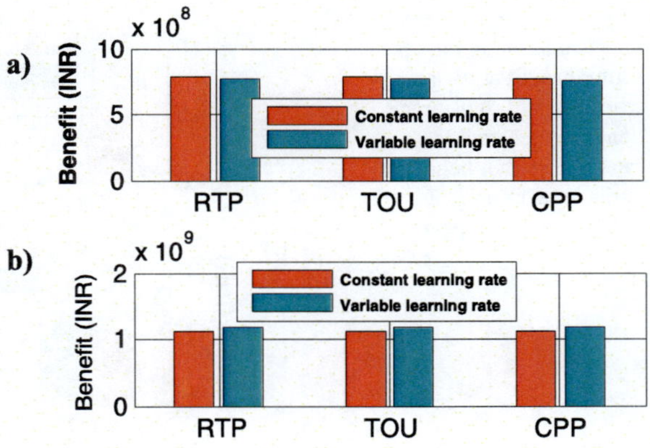

Fig. 8 **a** Customer benefit by single Q learning, **b** customer benefit by double q learning

4.3 Customer Benefits

Figure 8a, b shows customer benefits for alpha constant and alpha variable. The benefit decreases by small amount for variable alpha in single q learning.

4.4 Single Q Learning Versus Double Q Learning

Figure 9a–d shows a comparison of benefits offered through single and double q learning. It can be observed from Fig. 9 that at only in utility benefit at alpha constant, more benefit is within single q learning. For all other cases, benefits are more with double q learning.

5 Conclusion

The smart grid facilitates customers' potential to contribute in efficiency and reliability of grid. Adequately formulated DR programs can provide increased benefits.

In this paper, various demand versus price functions such as logarithmic, linear, hyperbolic, potential and exponential have been presented as CDF to represent divergent energy profiles. This helps to account the user behavior to facilitate utility or retailer to set proper DR programs for particular group of customers. This CDF is further stretched as weighted combination of these mathematical functions. Designing CDF resulted in CDR that account hourly changes in user demand and incorporates dynamic price elasticity. The CDR is then modeled so that user and retailer will be able to optimizes their benefit as it leads to better representation of old user energy profiles and lead to improved prediction of their demand in future.

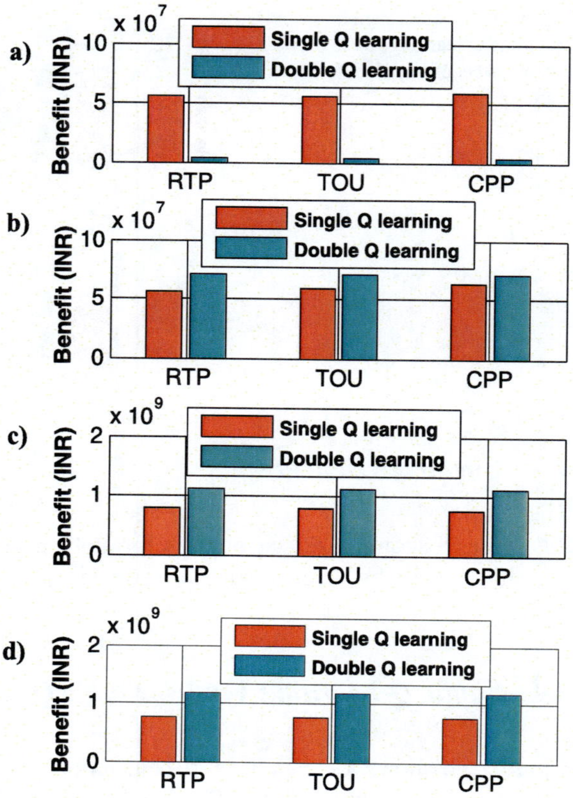

Fig. 9 **a** Utility benefit at constant learning rate, **b** utility benefit at variable learning rate, **c** customer benefit at constant learning rate, **d** customer benefit at variable learning rate

The CDR presents efficient hourly prices to user according to their load profiles and included DR programs.

An agent-based market set up by utility to represent groups of customers has been analyzed using q learning and its variant double q learning. Q learning makes agent intelligent to counter any changes in demand by effectively charging customers through tariffs, incentives and penalties to accomplish certain goals.

References

1. The World Bank (2008) Residential consumption of electricity in India background paper India: strategies for low carbon growth draft. World Bank, pp 1–73. http://www.moef.nic.in/downloads/public-information/Residentialpowerconsumption.pdf
2. Javaid N et al (2017) Towards cost and comfort based hybrid optimization for residential load scheduling in a smart grid. Energies 10(10):1–27. https://doi.org/10.3390/en10101546
3. Greening LA (2010) Demand response resources: Who is responsible for implementation in a deregulated market? Energy 35(4):1518–1525. https://doi.org/10.1016/j.energy.2009.12.013

4. Mahmood A, Javaid N (2016) Simulation study for optimized demand side management in smart grid simulation study for optimized demand side management in smart grid by CIIT/SP11-PEE-002/ISB in Electrical Engineering COMSATS Institute of Information Technology, Islamabad, Pakistan. https://doi.org/10.13140/RG.2.2.24746.52169
5. Ullah I, Rasheed MB, Alquthami T, Tayyaba S (2020) A residential load scheduling with the integration of on-site PV and energy storage systems in micro-grid. Sustain 12(1). https://doi.org/10.3390/su12010184
6. Chandrakar A, Paliwal P (2021) A technical survey on energy management and demand management mechanism of end users
7. Ogunjuyigbe ASO, Ayodele TR, Akinola OA (2017) User satisfaction-induced demand side load management in residential buildings with user budget constraint. Appl Energy 187:352–366. https://doi.org/10.1016/j.apenergy.2016.11.071
8. Bharathi C, Rekha D, Vijayakumar V (2017) Genetic algorithm based demand side management for smart grid. Wirel Pers Commun 93(2):481–502. https://doi.org/10.1007/s11277-017-3959-z
9. Sharma AK, Saxena A (2019) A demand side management control strategy using Whale optimization algorithm. SN Appl. Sci. 1(8):1–15. https://doi.org/10.1007/s42452-019-0899-0
10. Barbato A, Capone A (2014) Optimization models and methods for demand-side management of residential users: a survey. Energies 7(9):5787–5824. https://doi.org/10.3390/en7095787
11. Jang B, Kim M, Harerimana G, Kim JW (2019) Q-learning algorithms : a comprehensive classification and applications. IEEE Access 1. https://doi.org/10.1109/ACCESS.2019.2941229
12. Rudek R (2005) Introduction to multi-agent modified Q-learning routing for computer networks. Wrocław University of Technology
13. Kaliappan AT (2013) Flexible power consumption management using Q learning techniques in a smart home, pp 342–347
14. Yusta JM, Khodr HM, Urdaneta AJ (2007) Optimal pricing of default customers in electrical distribution systems : Effect behavior performance of demand response models, vol 77, pp 548–558. https://doi.org/10.1016/j.epsr.2006.05.001
15. Withagen C (1990) Spot pricing of electricity. Eur J Polit Econ 6(4):591–593. https://doi.org/10.1016/0176-2680(90)90014-a
16. He YX, Yang LF, He HY, Luo T, Wang YJ (2011) Electricity demand price elasticity in China based on computable general equilibrium model analysis, vol 36, pp 1115–1123. https://doi.org/10.1016/j.energy.2010.11.038
17. Aalami HA, Moghaddam MP, Yousefi GR (2010) Modeling and prioritizing demand response programs in power markets. Electr Power Syst Res 80(4):426–435. https://doi.org/10.1016/j.epsr.2009.10.007
18. Song K, Baek Y, Hong DH, Jang G (2005) Short-term load forecasting for the holidays using, vol 20, no 1, pp 96–101
19. Youse S, Moghaddam MP, Majd VJ (2011) Optimal real time pricing in an agent-based retail market using a comprehensive demand response model, vol 36, pp 5716–5727. https://doi.org/10.1016/j.energy.2011.06.045
20. Srihari SN. Q-learning topics in Q-learning
21. Manju MS (2011) An analysis of Q-learning algorithms with strategies of reward function. Int J Comput Sci Eng 3(2):814–820
22. Syafiie S, Tadeo F, Martinez E (2004) Softmax and -greedy policies applied to process control. IFAC Proc 37(12):729–734. https://doi.org/10.1016/S1474-6670(17)31556-2
23. Wang Y, Pavel L (2014) A modified Q-learning algorithm for potential games, vol 19, no 3. IFAC
24. Van Hasselt H, Group AC, Wiskunde C. Double Q-learning, pp 1–9
25. https://www.iexindia.com/marketdata/areaprice.aspx, x

Role of IoT in Smart City: A Review

Harpreet Kaur Channi

Abstract Taking advantage of IoT technologies for smart cities and embedded devices today leads to economic growth, infrastructure and the climate strengthened, transit networks strengthened, and public asset management costs reduced. Nowadays, cities are based on creating clever towns to deal with the increasing population and urbanization and globalization. Smart city is a philosophy of using technology and linked data sensors to develop and become efficient in terms of infrastructural and municipal operations which covers the maintenance of public lands, transit networks, residents, electrical sources, water sources, communication systems, local authorities, and other municipal facilities. Smart cities have a crucial role to play in keeping communities secure through linked devices and IoT apps. The introduction of an IoT-based intelligent city and its embedded technologies aims to enhance the efficiency, output and interactiveness of urban services, maximize capital, and cut costs. However, it is not an easy call for governments to analyze the challenges in terms of economic returns, value to citizens, implications on organizational structure, operational needs and how the investments fit within the realm of political and governmental strategy. This chapter will address the key role of IoT in smart cities and position of worldwide cities in the use of IoT.

Keywords Infrastructure · Globalization · Transportation · Sensors · IoT

1 Introduction

It is possible to describe a smart city as a city or municipality that uses smart technology to improve the quality of life for its residents. Through the use of information and communication technology, a smart city aims to improve the operational efficiency and quality of municipal services in order to decrease waste (ICT). Electricity supply, public transit systems, sanitation, solid waste management, etc., are all areas where smart technology for smart cities may be applied. IoT may be regarded as a

H. K. Channi (✉)
Department of Electrical Engineering, Chandigarh University, Mohali, India
e-mail: harpreetchanni@yahoo.in

K. Namrata et al. (eds.), *Smart Energy and Advancement in Power Technologies*, Lecture Notes in Electrical Engineering 926,
https://doi.org/10.1007/978-981-19-4971-5_31

vast network made up by billions of devices and sensors that are capable of communicating with the Internet in order to gather and exchange information. Anything that can be linked to the Internet can become IoT enabled. As an example, if a light bulb can be turned on and off using an app on your phone, it has become an IoT-enabled item. From the idea that city infrastructures such as street lighting and transit systems may be packed with sensors to better meet human needs, IoT and smart cities have a link. Worldwide cities are becoming cleverer. Cleaner air and water, greater accessibility, more effective infrastructure are steps being developed in smart cities worldwide to foster greener more cleaner urban environment [1]. These programs are enabled by technology such as the Internet of Things (IoT), big data analytics, and ICT, which provide the technological basis for introducing smart city projects. Intelligent communities use technical technologies to boost municipal amenities and the lives of people in order to obtain important information such as pollution, electricity consumption and air quality municipalities consumption IoT sensors, networks, and applications [2]. This data will then be used with technical innovations to enhance community facilities, including infrastructure, transport, and public safety, as seen in Fig. 1. Through which we will have smart city projects, IoT is an important technology. The "stuff" of the IoT goods, sensors, and software captures the data that make productive solutions in technology. Clever water meters track the quality and usage of water, warn water providers to spills or future leakage, for example. Big data processing has to be applied in intelligent city projects in order to incorporate intelligent city facilities, IoT produces vast datasets which need to be analyzed and analyzed. The IoT data collected is sorted, analyzed, and processed by large-scale data systems, part of an ICT system in the field.

Fig. 1 Smart city

2 The Intelligent City Components and Its Effect

2.1 Intelligent Networks

Smart cities have advanced connected sidewalks, intelligent car parks, smart lighting, and other transportation technologies in the global market for intelligent urban services [3].

- *Smart Lighting*: In order to provide customized lighting and demand-based lighting in different areas, local authorities can control smart lighting in real time, as Fig. 2 shows. Intelligent lighting also leads to optimizing daylight by dimming out occupancy-free sectors, for example, parking lots can be dimmed during working hours such that, if a car approaches, it can be felt to illuminate the required fields, so others can sit in a diffuse region.
- *Connected Streets*: Smart wired roads will gather data and relay information and resources to and from millions of devices, including traffic, road blockages, and bridges. This allows the effective use of capital and people to enhance public transport and the urban climate.
- *Management of smart parking*: Smart parking control technology can be used to identify the vacancy of a car in different public areas. The in-ground vehicle detection sensors from smart parking are fundamental technologies which play a major role in smart parking, which revolutionizes how cars can locate parking space in malls which town centers. Wireless sensors are installed into parking fields, transferring time, and space period data by local signal processors to a central parking control program. Smart parking eliminates pollution, pollution controls in cars, compliance costs, and driver fatigue reduction. Any system needs

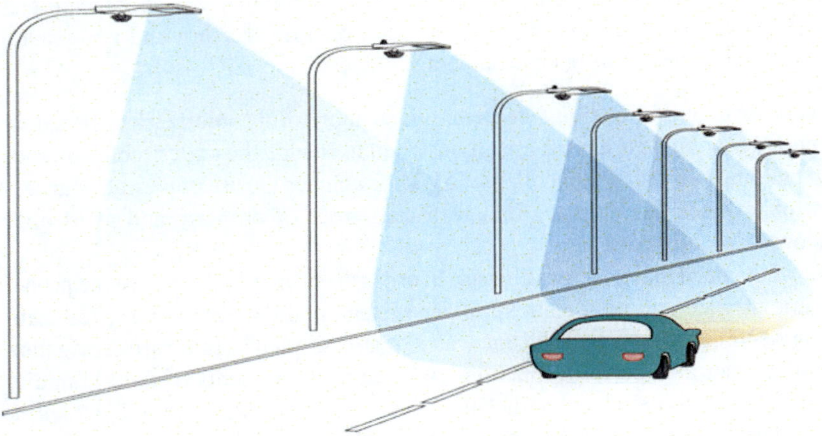

Fig. 2 Smart street lightning

Fig. 3 Smart charging station

to have a stable cloud platform connectivity to successfully implement intelligent parking technologies [4].

- *Associated Charging Stations*: Fig. 3 also indicates that intelligent technology includes charging points in parking structures, car buses, shopping centers and homes, airports and bus stations in the area. IoT incorporates electronic charging systems for cars to streamline the process of electricity charging and change the impacts of the energy grid.

2.2 Smart Buildings and Properties

Smart buildings employ different technologies as seen in Fig. 4 to ensure the protection and security of structures, the conservation of infrastructure and the general well-being of the neighboring environment.

- *Structures of protection and defense*: Video monitoring, biometrics, IP safety cameras, and wireless alarms are all designed to reduce the undesirable entrance into buildings and danger of collapse [5]. This includes the use of perimeter access control to block the entrance into restricted areas of property and to identify persons in prohibited areas.
- *Shrewd Orchard and Irrigator System*: In order to ensure plants receive the proper volume of water, an intelligent sprinkler system, coordinated with related technology and cloud, can be used for water treatment plants. Innovative irrigation machines can also perform functions, such as measurements of field humidity and fertilizers, to help local authorities save on water charges (intelligent sprinkler machines use weather forecasts to change their schedules to prevent the grass from rising peacefully).

Fig. 4 Smart building

- *Smart Space heating and Aeration*: Intelligent ventilation and heating systems control a variety of parameters, including temperature, sound, noise, building humidity, film theaters, and historical monuments [6]. The installation of the wireless sensor network is necessary for adequate heating and ventilation. They also collect data to automate HVAC systems and to improve their building effectiveness and performance.

2.3 Smart Industrial Environment

Industrial environments offer important possibilities for creating Internet of Things and digital technology applications that can be used in the following areas:

- Forest Fire Detection: Helps to identify warning areas when tracking combustion gases and preemptive fire situations.
- Air/Noise Pollution: Helps in the control of factory CO_2 emissions, car-emitted pollution, and farm-generated toxic gases.
- Snow Level Monitoring: Helps to assess the ski slope condition in real time, allowing defense firms to prevent avalanche.
- Landslide and Avalanche Avoidance: Helps to monitor soil moisture, field density, and to track harmful soil patterns.
- Earthquake Early Detection: Helps to predict tremor chances by using centralized sensors at different tremor locations.
- Fluid presence: Aids with the monitoring of moisture with data centers, structures, storage facilities to reduce deterioration and falls. Liquid presence

- Radiation Levels: Helps to produce contamination warnings for remote radiation exposure levels at nuclear power plants.
- Toxic and dangerous materials: Assistance to identify gas levels and leakage in chemical and manufacturing fields, and indoor mining environments [7].

2.4 Smart City Services

The smart city IoT solution includes public security and emergency services. The following are the key areas in which IoT can help:

- *Smart Kiosk*: In order to provide public services such as Wi-Fi, 24 IP monitoring cameras and analytics, digital signaling for publicity and public announcement, smart kiosks play an important role. Free video calls and free mobile charging station and the incorporation of the environmental sensor can also be introduced in some situations. Restaurants, shopping stores, and events are available in the nearby smart kiosks. It also helps tourists to monitor and connect with smartphones to provide additional information on request [8].
- *Monitoring of Risky Areas*: Sensors (cameras, street lighting) or actuators for real-time tracking may be used in dangerous environments or places vulnerable to injuries. Such sensors will alert people to the immediate avoidance of these areas when they detect crimes or malfunctions.
- *Public Security*: In public organizations and houses, IoT sensors can be installed to safeguard citizens and provide fire and police departments with real-time information on detecting robbery.
- *Fire/Explosion Management*: Intelligent fire sensors are able to identify and take instantly drastic measures, such as identification of falsified warnings, fire warning, information for firemen and the ambulance, blocking surrounding streets/buildings, assisting residents to relinquish, and coordinating emergency drones and robots.
- *Automatic Healthcare Dispatch*: Smart medical devices can be deployed in public places to provide patients, such as medication and medicines for patients, with 24/7 health care. These apps may also be used to call an ambulance for emergency patients.

2.5 Smart Energy Management

In integrated energy services, IoT systems may be used in smart cities, as shown in Fig. 5.

- *Smart Grid*: They are remotely controlled, integrated energy networks that deliver power or gas from generational sources. Manufacturing, industrial and in transmission and delivery projects, intelligent grid systems may be used. Specific IoT

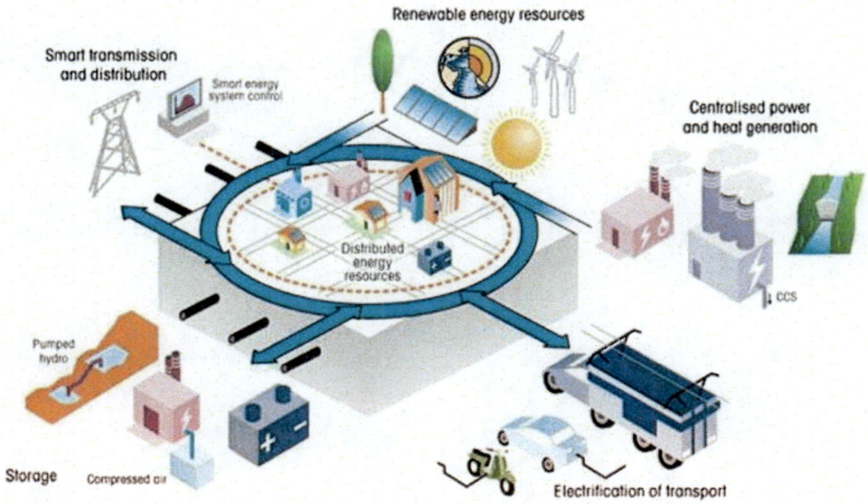

Fig. 5 Smart energy management

solutions such as gateways can be used both at transmission level and for energy conservation [9, 10]. For example, gateways may provide businesses with high accessibility and real-time monitoring a wider view of the dynamics of energy delivery. It also develops an energy distribution optimization demand-response mechanism for utilities based on utilization patterns.

- *Smart Meters*: Smart meters can be used in domestic and industrial electrical and gas meters for determining real-time energy usage information. Customers and utilities in smart meters will track electricity consumption [11]. In reality, energy tracking, alerts, and public dashboards now can be accessed from the Web through the embedded mobile apps.

2.6 Smart Water Management

Smart water management as shown in Fig. 6 is possible with IoT and connected devices:

- Management of potable water: track the quality of drinking water in urban areas.
- Chemical leakage: detects spills and pollution from river plants.
- Remote swimming pool measurement: automatically measures the temperature of the water.
- Water contamination: regulation of leaks and contaminants in the water. Marine emissions.

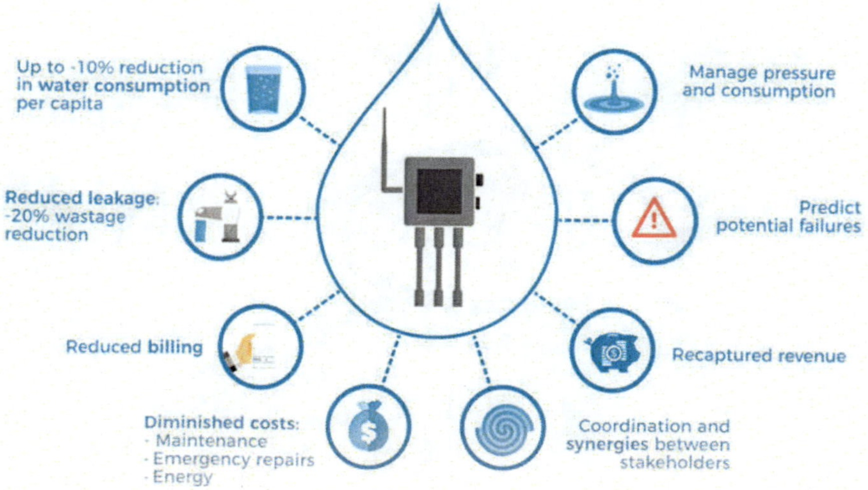

Fig. 6 Smart water management

- Water Outflows: senses the presence of liquid outside tanks and changes in pressure within pipelines.
- River Floods: tracks fluctuations in channel, dam, and reservoir water levels.

2.7 *Smart Waste Management*

Intelligent waste management systems assist communities and waste service administrators in minimizing waste, reducing maintenance costs, and solving environmental challenges related to inadequate waste disposal, as Fig. 7 demonstrates.

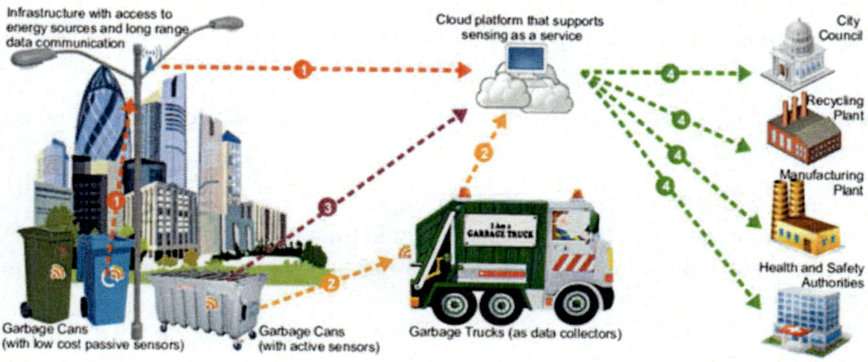

Fig. 7 Smart waste management

Incorporating an intelligent city has high odds of changing human lives and developing public services and operations in general. The main approaches to clever city deployment [12, 13] are smart sensor networks, the Internet of Things (IoT), and mobile devices.

3 Foundation Layers for Smart City Operations

Layer 1: Technological base includes—*Information and Communication Technology (ICT)*: People and administrators of the network use to connect. Governments can use ICTs to evaluate citizens' data and feedback and make city services changes [14].

Internet of Things (IoT): Internet linked tools and field instruments that can contact and transmit data to the control center. IoT is used by IoT systems in autonomous communities to collect intelligence and constantly solve problems.

Sensors: The data are obtained through city-wide sensors. Sensors can capture different information types including temperature, light, sound, people, or vehicle numbers.

Layer 2: Designated applications—the provision of city authorities and/or residents' information networks to boost municipal operations.

Layer 3: Application usage—achieving a sufficient mass of usage in the city or by known consumers.

4 Benefits of Smart City Solutions

In order to create an intelligent environment, it will use technologies and data with the goal of delivering improved living conditions, according to this study from the Mckinsey Global Institute [15]. Many town dwellers believe that changes in the following areas will provide a decent quality of life, as shown in Table 1.

Public safety: Real-time crime detection systems, for example, use predictive modeling in order to track patterns of crime and locate trouble zones, forecasts crime events. The security in these areas can be improved by police powers.

Faster commutes: The improvement of everyday travel time is essential for the good quality of life for most urban residents. Cities with intelligent mobility technologies will slash switching times on average by as little as 20%. Multimodal approaches for mobility allow passengers to select from all available modes of transport. As residents prefer the transport mode which best suits them at the moment, this leads to less automobile use.

Economic prosperity: Intelligent cities tend to attract companies and talent in technology. In addition, this raises risk capital in cities via their intelligent policies cities like London and New York have been getting an influx of money.

Table 1 Benefits of smart city solution

Smart energy	Smart buildings	Smart transportation
– Smart metering – Management of customer response and demand side – Automated delivery – Generation spread – Renewable energy fusion and open energy – Management and regulation of the network	– Regulation of light – Regulation of heating – Quality of electricity – Production of local electricity – security, regulation of occupancy – Synergies of energy conservation, luxury, and protection – Construction as network: multiple integration (HVAC, lighting, plug loads, fire, protection, sustainable mobility, storage, etc.) – Software: performance, automation and reporting, analytics and management of big data	– A transport to everything – Control of driver behavior – Open-source Web apps Regulation of traffic and fleet – Services focused on actual or near real-time information for drivers and passengers – Incorporated public transit
Smart water	*Smart waste*	*Smart physical safety/security*
– Regulation of stresses – Remote control and management predictive – Integrated water control systems – Intelligent estimation Power management and efficiency	– Regulation of waste – Therapy to excess water Cleaning of the area – Waste sorting – Collection of waste	– Video and reporting tracking – Seamless natural-manmade catastrophe coordination
Smart health care	*Smart education*	
– Sanitary adequacy – Controlling and reducing diseases – Intelligent hospitals – Entertainment in fitness, including study – Home and remote health systems like surveillance – Management of electronic documents	– Interactive work system open work – Internet access to multimedia content of the world class leveraging shared technology – Massive open online course (MOOC)	

Greener environment: Each smart city policy is focused on environmental issues. Intelligent city management and urban planning [16] was planned to reduce waste and emissions. The result is fewer cars in the streets and a reduced emissions when people go and use multimodal mobility. The intelligent management of your resources is also characterized by smart cities.

5 Top 10 Smart Cities in the World

Many cities worldwide have already adopted technologies linked to travel, such as the digital mobility network, to solve their problems. The following cities have been recognized for steps that have been taking to make smart cities actually motive [17]. These are city "operational networks" and that gather data from around the city, enabling intelligent urban governance and contact between cities and people.

5.1 Columbus Smart City

Columbus Ohio won a competition initiated by the United States Transportation Department in 2016. They have spent 40 million dollars in their award money in executing their innovative comprehensive mobility program in their region. The Columbus Plan focused on involving the private sector in the transformation of the city and collecting data in order to ensure that planned improvements are worthwhile for residents and visitors.

- The smart Columbus operating system (or SCOS) is the heart of smart city operations.
- SCOS allows decision-making in many areas in the city in real time, ranging from public transport and traffic management to health, water, and waste management. SCOS provides citizens with access to information, such as public transportation and traffic conditions.
- Machines for data collection for linking vehicles and services.
- Self-sufficient electric movement in different environments.
- Special centers for transportation knowledge exposure. Such junctions are linked physically to certain transport platforms.

5.2 Singapore Smart City

The Global Smart City Efficiency Index was developed in 2017 and assessed the accessibility, wellness, competitiveness, and protection of 20 smart cities. In all four regions, Singapore ranked number one. Singapore uses the digital Singapore network to tackle connectivity, safety, competitiveness, and security concerns. Digital Singapore is a detailed 3D model of the island-built area. This allows city planners to predict the effect of various structures, such as the effects of a new construction on road traffic or on the growth of parking regions.

The Vehicle to All (V2X, formalized in IEEE 802.11p) is another ambitious initiative. This project would build a network linking all the accessibility elements of the region. Various tools can connect, build an accurate and detailed image of the roads, promote protection, and reduce congestion.

5.3 Copenhagen Smart City

Denmark is a leading global player in environmental initiatives and has some of the advanced climate policies in the world. The government supports intelligent city projects and access to public information sources, including essential data regarding citizens, businesses, and real estates, unavailable in most other countries, and makes them a political priority in the country. Individuals, firms, and nonprofits exploit this information and share it via open platforms to digitize services and drive smart city innovation.

Copenhagen is taking a range of intelligent community measures in line with the ecological aspirations of the region. Copenhagen linking is the core goal. In order to manage traffic, the initiative entails monitoring connected vehicles, including personal devices such as cellular phones, reducing congestion and thus minimize air emissions. In addition, tracking devices provide fast access to traffic, parking, and transport information. In order to deter fraud and track patterns of motion, Copenhagen uses RFID tag to monitor items such as cars, bikes, and even home equipment. The smart grid sensor platform has also been used to leverage cheap wireless devices that collect millions of data points about the present urban conditions.

5.4 Barcelona Smart City

Barcelona is rapidly becoming a smart capital. The Sentilo Open-Source Project has been developed to integrate sensor data and make this accessible to city-wide information systems. Some of the steps taken to make the city greener and more effective. A number of intelligent design technologies are used by Sentilo:

- Street light operations to meet traffic patterns in real time.
- Parking area control.
- Intelligent transport systems make city trips really convenient.
- Reduce noise and traffic waste generation by the putting of special bags in dwellings for disposal at designated underground sites.

5.5 Dubai Smart City

Dubai launched its Smart City initiative in 2013 in the United Arab Emirates. The goal of the project is to turn Dubai into one of the world's smartest towns, directly sponsored by Dubai's ruler, Mohammed bin Rashid Al Maktoum. The initiative focuses on public–private partnership and the application of ICT and the Internet of Things in both physical and wireless urban infrastructures.

The following developments are, among others, introduced by Smart Dubai:

- Intelligent Dubai Index—a Web site that updates the status of targets, as Dubai becomes a clever city.
- Dubai blockchain—a project to digitize state details and data. This will improve and increase administration.
- Travel from Abu-Dhabi to Dubai with a hyperloop, which will make your journey even quicker.
- Volocopters—taxis and police autonomous.

5.6 *Helsinki Smart City*

Helsinki utilizes BioTope, an EU sponsored initiative, as an open-source framework. Innovative and collaborative broad data processing is encouraged by this initiative. In addition to encouraging local start-ups and projects, Helsinki authorities are now collaborating with other European smart cities. Finland made carriages legalistic and the Smart Region of Helsinki introduced the SOHjOA project, which adds driverless mini busses to the transport network of Helsinki. Since 2016, these robo busses are checked by actual riders on public roads. They are cost-effective vehicles that can carry up to 9 persons and work in tandem with regular traffic and travel.

5.7 *Tokyo Smart City*

Tokyo is as green as possible and aims to reduce CO_2 emissions by 75% by 2020. The municipal authority has established legislation such as construction limits and a duty on businesses to reduce or otherwise pay indirect carbon tax emissions to this purpose. A proposed framework will also include a formal climate plan that provides a success review. Tokyo's decentralized open-network version is known as Cpass.io, which connects government data on the Internet of Things (IoT) devices and other big data outlets to develop smart applications around Tokyo. CPass entitles to share information, information, such as flow of people or vehicles, accidents, dangers, and sudden changes in climate, with the public and businesses in Tokyo's city government or other government agencies. Developers from both the government and private sectors will use the data to create public-use applications.

5.8 *Oslo Smart City*

The capital of Norway is consistently listed on the global list of intelligent cities. This is part of his attempts to combat climate change. Buildings make up about 40% of the world's energy usage, and Oslo uses sensors to monitor illumination, heating, and refrigeration over a wide range. The tower's target to reduce the emissions of electric cars, an intelligent infrastructure and power storage technologies by 36% by 2020 and as many as 95% by 2030. More than two thousand charging points for electric cars, which are allowed free parking, storage, and transportation on ferries have no sales tax due to users. Norway announced plans for the construction, on 260 acres, of a sustainable smart city near the Oslo airport, to develop communities driven by technology. It is only supplied with clean electricity, and surplus is re-sold to the grid. Automatic street and building lighting, waste management, and security systems will be operated by sensor-based systems. Only, electric cars are required, but planners want to drive themselves at the end of the day.

5.9 Copenhagen Smart City

The Danish capital is pushing toward intelligent growth with radical environmental policies of its own. The Copenhagen Solutions Lab Incubator was awarded the 2017 Copenhagen Solutions Lab Grant, which tracks traffic, air pollution, garbage, energy usage, and other products and compares their activities in real time. It integrates in real-time electric vehicle parking, traffic lights, building and smart meters and charging systems and optimizes power, traffic, and the environment energy usage. The opportunity to evaluate, calculate, and compare all such data is meant to increase service delivery performance. Cyclists and other people in Copenhagen use an app that is built with all this knowledge and that takes you around the urban streets and shows you how easily you have to pedal to reach the next green light. Another path aims to develop reviews by leveraging customer input. This also tests cycled time and expended calories. Another community forum warns of bike-related police actions and riders may use details to avoid fines.

5.10 Boston Smart City

The area was one of the first innovative projects to innovate. In order to become a regional hub for creativity, Boston opened its creativity district in its port, and it is credited for helping to create over 200 start-ups. The master plan focuses on community participation or “participatory social planning” by using a collection of apps to collect information about parking space, comment on service issues, or to connect with others. You can report potholes, vandalism, or monitor a school child’s bus from anywhere in the area. Chinatown communitarian is a simulation of a computer game that allows the neighborhood to prepare and grow. The traffic snarls from Boston are established, and mobility microhubs are the foundation for a mobility program. It will provide a digital kiosk with information on bus, train and bicycle shares, parts of cars and other services in real time. In known congested districts, smart sensors link the microhubs with networked traffic signals [18].

6 Position of Indian Cities in the Use of IoT

Industry 4.0, also known as Industry 4.0, is the result of the advent of the Internet of Things (IoT) in India. In addition to the new "Digital India" program established by the government, IoT plays a prominent position in an expanding IoT business and technological setting. IoT investments in India reached USD 5 billion in 2019 according to a recent research by Zinnov in June 2020 and are anticipated to reach USD 15 billion by 2021. Over 60% of India's GDP and 80% of its tax income come from cities. New urbanization and efforts such as the 100 smart cities project are seen as a way for the Indian government's economy to break out of its sub-8% growth. India now has 68 cities with a population of above 1 million people. There are just 35 of these cities in the entire European continent. Estimates predict that the demand for critical public services such as water supply, sewage treatment, solid waste disposal, and transportation will grow by 4.5–8 times the current demand. Twenty-five percent of India's urban population lives in shantytowns. More than 40% of the population of Mumbai, Vishakhapatnam, Meerut, Vijaywada, and Jabalpur live in poverty.

Information received from sensors and other IoT devices in Kohima, India, is being utilized to regulate street lighting, guarantee public safety, and provide a wide range of citizen services to residents. For example, the Vande Bharat Express, India's first semihigh-speed train, utilizes a collision-avoidance system that includes sensors and other Internet of things devices to prevent accidents caused by human mistake or equipment malfunction.

A company in India that sells health and luxury teas, Tea Tantrum, uses Internet of Things (IoT) technology to monitor moisture content and maintain component proportions in certain of its products. Indian IoT installations are anticipated to exceed $17 bn by 2021, according to Constellation Research founder and lead analyst Ray Wang.

Nearly, 31% of India's present population lives in cities, which generate 63% of the country's GDP (Census 2011). 40% of the Indian population and 75% of India's GDP will be living in cities by 2030. Physical, institutional, social, and economic infrastructure must be developed comprehensively. To improve the quality of life and to attract people and investment, all of these factors must be considered in order to create an upward spiral of growth and development. The creation of smart cities is a step in the right direction in this regard.

They focus on the most pressing needs and the biggest potential to enhance people's lives. Using digital and information technology, urban planning best practices, public–private partnerships, and policy reform, they strive to make a difference in their communities. They always put the needs of others above their own.

To promote cities that offer basic infrastructure and reasonable quality of life for their inhabitants as well as create a clean and sustainable environment, the use of smart solutions is the mission's aim. Sustainability and inclusion are the primary goals, with a view toward creating a repeatable model that would serve as a beacon for other communities. As a result of the Smart Cities Mission, comparable smart cities can be created in different locations and portions of the country.

7 Conclusion

There are billions of digital gadgets that can connect and interact with each other via the network/Internet and can be remotely monitored and controlled under the Internet of Things (IoT). The Internet of Things (IoT) consists only of smart sensors and other connected devices. As an example, IoT collects weather data. Using IoT, cities can better control traffic, reduce pollution, optimize infrastructure, and keep residents safe and clean. IoT offers networking and various network and facilities management for cities. There are a wide variety of applications, from smart mobility networks to quality of life upgrades, intelligent waste management, recycling of energy, intelligent health facilities, etc. All of them seek to accomplish the same purpose, however, that of sustainable growth, by using approaches that minimize energy costs, make usage of natural resources, and contribute to an atmosphere that is cleaner and healthier. In reality, intelligent cities would take time to realize their full potential. But, it is clear the possibility of their existence.

With its immense potential, the Internet of Things is helping to turn towns into "smart cities" throughout the world. Improved traffic management, greater safety, lower pollution levels, and lower energy usage are some of the beneficial outcomes, along with a better quality of life for residents.

With the aid of the Internet of Things, communities can enhance energy distribution, expedite garbage collection, reduce traffic congestion, and even improve air quality. Using data from sensors and automobiles, connected traffic signals, for example, alter their timing and frequency in response to real-time traffic, decreasing road congestion.

References

1. Su K, Li J, Fu H (2011) Smart city and the applications. In: 2011 International conference on electronics, communications and control (ICECC). IEEE, pp 1028–1031
2. Washburn D, Sindhu U, Balaouras S, Dines RA, Hayes N, Nelson LE (2009) Helping CIOs understand "smart city" initiatives. Growth 17(2):1–17
3. Nam T, Pardo TA (2011) Conceptualizing smart city with dimensions of technology, people, and institutions. In: Proceedings of the 12th annual international digital government research conference: digital government innovation in challenging times, pp 282–291
4. Vanolo A (2014) Smart mentality: the smart city as disciplinary strategy. Urban Stud 51(5):883–898
5. Lombardi P, Giordano S, Farouh H, Yousef W (2012) Modelling the smart city performance. Innov: Eur J Soc Sci Res 25(2):137–149
6. Murtuza A, Fatima M, Kumar S, Anand R (2017) Design and implementation of solar based smart street lightning system. In: 2017 International conference on smart technologies for smart nation (SmartTechCon). IEEE, pp 283–287
7. Badgelwar SS, Pande HM (2017) Survey on energy efficient smart street light system. In: 2017 International conference on I-SMAC (IoT in social, mobile, analytics and cloud) (I-SMAC). IEEE, pp 866–869

8. Meena B, Ram M, Sambasivanayak R (2017) Smart street lightning system for effective power utilisation with accident avoidance. Int J Emerg Technol Comput Sci Electron (IJETCSE) 24(11)
9. Hall RE, Bowerman B, Braverman J, Taylor J, Todosow H, Von Wimmersperg U (2000) The vision of a smart city (No. BNL-67902; 04042). Brookhaven National Lab., Upton, NY (US)
10. Angelidou M (2014) Smart city policies: a spatial approach. Cities 41:S3–S11
11. Kamienski C, Soininen JP, Taumberger M, Dantas R, Toscano A, Salmon Cinotti T, Torre Neto A (2019) Smart water management platform: Iot-based precision irrigation for agriculture. Sensors 19(2):276
12. Robles T, Alcarria R, de Andrés DM, de la Cruz MN, Calero R, Iglesias S, Lopez M (2015) An IoT based reference architecture for smart water management processes. J Wirel Mob Netw Ubiquitous Comput Dependable Appl 6(1):4–23
13. Lee SW, Sarp S, Jeon DJ, Kim JH (2015) Smart water grid: the future water management platform. Desalin Water Treat 55(2):339–346
14. Pellicer S, Santa G, Bleda AL, Maestre R, Jara AJ, Skarmeta AG (2013) A global perspective of smart cities: a survey. In: 2013 Seventh international conference on innovative mobile and Internet services in ubiquitous computing. IEEE, pp 439–444
15. Okai E, Feng X, Sant P (2018) Smart cities survey. In: 2018 IEEE 20th International conference on high performance computing and communications; IEEE 16th international conference on smart city; IEEE 4th international conference on data science and systems (HPCC/SmartCity/DSS). IEEE, pp 1726–1730
16. Coletta C, Evans L, Heaphy L, Kitchin R (eds) (2018) Creating smart cities. Routledge
17. Cerrudo C (2015) An emerging US (and world) threat: cities wide open to cyber attacks. Secur Smart Cities 17:137–151
18. https://www.einfochips.com/blog/understanding-the-role-of-smart-city-and-its-components-in-the-iot-era/ (accessed on 10/7/2020)

Estimation and Comparison of Monthly Global Solar Radiation Between Empirical Models and ANN Method at Visakhapatnam, India

Kumaresh Pal, A. K. Akella, K. Namrata, S. Lakshmi Prasanna, and Anshuman Bhuyan

Abstract The sun's global radiation is a crucial aspect for evaluating the radiation of sun, as it gives the entire solar availability at a given place and can be calculated by the equipment. However, it is unfeasible to measure solar radiation in many areas due to maintenance and high price of the equipment used for measurement. Unfortunately, every place can't spare the price for the equipment attributable to the above factors. Hence, empirical models were established as a substitute to roughly calculate the data. This work is to compare the empirical models and find the suitable one to approximate the global monthly sun's radiation on parallel plane surfaces in the city, Visakhapatnam. The value of measured global solar radiance data facilitates the approximation of global radiation. And the execution of the empirical models is assessed using statistical error tests and in the end following the observations, it's declared that ANN model was reliable and accurate.

Keywords Sun's global radiation · Statistical error · Empirical models · ANN model

K. Pal (✉) · A. K. Akella · K. Namrata
Department of EE, NIT Jamshedpur, Jamshedpur, India
e-mail: kumaresh.pal@rediffmail.com

A. K. Akella
e-mail: akakella@rediffmail.com

K. Namrata
e-mail: namrata.ee@nitjsr.ac.in

S. Lakshmi Prasanna
Department of EE, Arka Jain University, Jamshedpur, India
e-mail: shivratrilakshmiprasanna@gmail.com

A. Bhuyan
Department of EE, ITER, Siksha 'O' Anusandhan University, Bhubaneswar, India
e-mail: bhuyananshuman@gmail.com

K. Namrata et al. (eds.), *Smart Energy and Advancement in Power Technologies*, Lecture Notes in Electrical Engineering 926,
https://doi.org/10.1007/978-981-19-4971-5_32

1 Introduction

India is blessed with rich resources of solar energy. The mean intensity of the sun's radiation received is 200 MV/km^2 but utilizing this energy efficiently is a crucial challenge. Sun's global radiation data is vital for optimal design of the prediction of system performance and measured sun's radiation data is the best source for accurate values of sun's global radiation. The precise way to measure this is by using pyranometers but they are expensive and require high maintenance. The second best method was to develop empirical models to approximate the radiation for locations where measurement was difficult. This work focuses on estimating the sun's radiation data on a parallel plane surface for city Visakhapatnam located in India. The data for irradiance was taken from national solar radiation database. The measured monthly global radiation was taken from national solar radiation database and the regression constants for b and a are taken as 0.47 and 0.28, respectively (Sukhatme and Modi) [1]. Measured data is compared between the following models: (i) Rietveld [2], (ii) Ogleman [3], (iii) Akinoglu [4], (iv) Glover [5], (v) Gopinathan [6], (vi) Linear Regression [7], (vii) ANN (Artificial Neural Network) model, to find out which is the uttermost efficient for modelling of sun's global radiation on the parallel plane surface. There are several models to approximate the monthly and daily mean sun's global radiation in various regions by various combinations of measured parameters. And the accessibility of a solar radiation model, in a specific region, proves to be worthy in predicting the aggregate value of power that would be generated from a particular solar energy system. The area used for our study is Visakhapatnam, located in India, which is the most populated city of Andhra Pradesh. Next to Chennai, Visakhapatnam is the largest city in the eastern coast of South India. It is located at 17.68 N, 83.21 E and is 28 m above sea level. Since it's a coastal area it has higher temperatures than the other parts of the state. It has a wet and dry climate and the mean temperature varies between 24.7 and 30.6 °C. The weather conditions of this study area is tropically savanna type and its temperature remains unchanged (avg. 30 °C) due to proximity of the ocean Bay of Bengal. Basic ideology: Solar radiation extends out to the earth's surface in the manner of direct solar radiation, diffuse solar radiation and reflected radiation. Sun's radiation received over the external surface of the earth's aerosphere is called as extraterrestrial radiation. The average solar irradiance value is 1361 W/m^2. The angle formed between the plane, the perpendicular to a line between earth's axis, sun and the earth is acknowledged as the sun's angle of decline. On a clear sunny day at noon, around 25% of sun's radiation is dispersed and absorbed as it progresses through into the atmosphere. Hence, only 1000 W/m^2 of incident radiation arrives at the earth's surface. The source of the radiation occurs straight from one of the biggest stars, i.e. the sun and so it's known as direct irradiance (or beam irradiance).

Figure 1 shows the outline of the measures followed to estimate the almost suitable model for estimating the global radiation.

Figure 2 shows the variance in generation of solar energy for the state of Andhra Pradesh, India for all months.

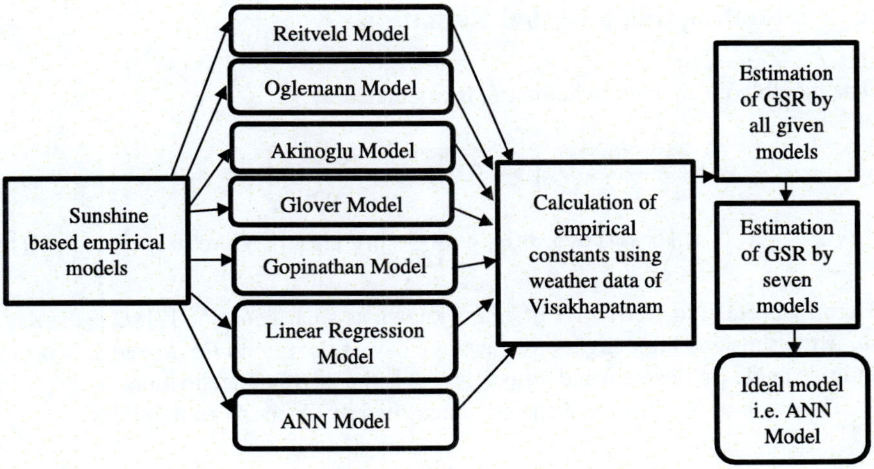

Fig. 1 Outline of procedure followed to evaluate best suitable model

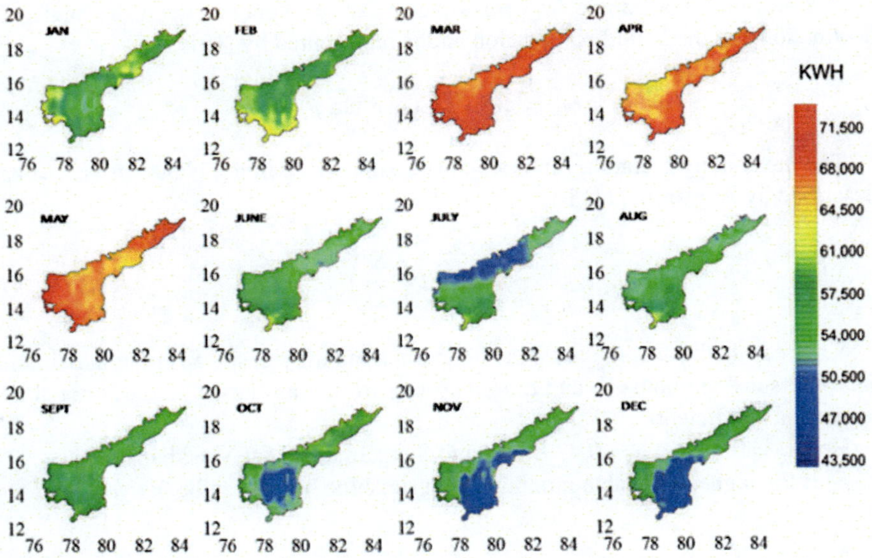

Fig. 2 Solar energy generation per month for Andhra Pradesh state, India, showing the variation in different months. *Source* Solar GIS

2 Estimating Sun's Global Radiation

The monthly mean daily radiation, I_0 is calculated by

$$I_0 = \frac{24 \times 3600}{\pi} I_{sc}\left[1 + 0.033\cos\left(\frac{360D}{365}\right)\right] \times \left(\cos\phi\cos\delta\sin\omega_s + \frac{\pi\omega_s}{180}\sin\phi\sin\delta\right)\text{kWh/m}^2\,\text{day}^{-1} \quad (1)$$

Here I_{sc} is equal to 1.367 kW/m^2 and is known as solar constant, [8], D is the day, ω_s is the sunshine hour angle of an average day of the month (in degrees), ϕ is the latitude angle (in degrees) and δ (in degrees) is the angle of declination.

The angle of declination, δ can be represented by Cooper's equation [9]

$$\delta = 23.34\sin\frac{360}{365}(284 + D) \quad (2)$$

The sunshine hour angle (ω_s) for a particular site depends upon the angle of declination and the latitudinal location and is represented by [10]

$$\omega_s = \cos^{-1}(-\tan\delta\,\tan\phi) \quad (3)$$

The mean daily radiation (on a monthly basis) on a parallel plane surface I_d, as described by Angstrom [11] is:

$$\frac{I_d}{I_0} = a + b\left(\frac{S}{S_o}\right) \quad (4)$$

S represents the average sunshine hours on a monthly basis, S_0 is the maximum probable sunshine hours, b and a are two-angstrom constants, familiarly noted as regression coefficients.

Here $a = 0.28$ and $b = 0.47$ (Modi and Sukhatme) [1] for Visakhapatnam.

S_0 is the highest probable monthly mean sunshine hours, can be derived from

$$S_0 = \left(\frac{2}{15}\right)\omega_s \quad (5)$$

S is the mean daily sunshine hours, S_0 is the possible sunshine duration (Fig. 3).

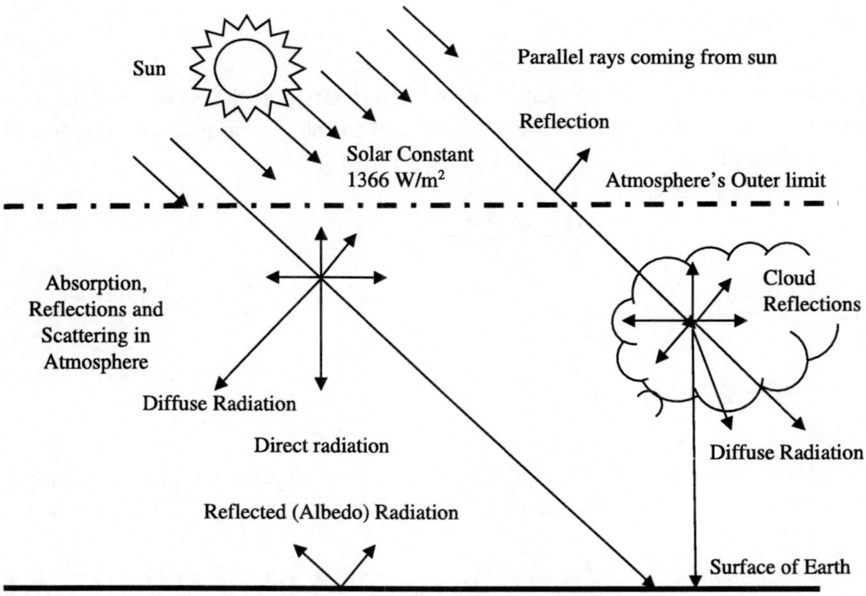

Fig. 3 Total Sun's global radiation. *Source* Solar Radiation Handbook [12]

3 Various Models' Descriptions for Predicting Sun's Global Radiation

The following seven models are being used for the study to predict the monthly mean sun's global radiations on a parallel plane surface for the city Visakhapatnam.

3.1 Reitveld Model

Reitveld [2] observed that a relationship exists amongst the regression coefficients (*b*, *a*) and the mean relative sunshine duration. He stated that his model gave an estimation that is twice as accurate for $\frac{S}{S_0}$, with the condition: $\frac{S}{S_0} < 0.4000$.

$$\frac{I_d}{I_0} = \left(\frac{S}{S_0}\right)b + a \tag{6}$$

3.2 Ogleman Model [3]

The model described by Ogleman observed that standard deviation of sunshine duration for better model parameter estimation and formulated a quadratic relation for solar radiation:

$$\frac{I_{\mathrm{d}}}{I_0} = \left(\frac{S}{S_0}\right)b + a - \left(\frac{S}{S_0}\right)^2 c \tag{7}$$

3.3 Akinoglu Model

Ecevit and Akinoglu [4] formulated a correlation in quadratic equation form to predict the values of sun's radiation and took same values for a and b. The equation is as:

$$\frac{I_{\mathrm{d}}}{I_0} = \left(\frac{S}{S_0}\right)b + a - \left(\frac{S}{S_0}\right)^{-2} c \tag{8}$$

3.4 Glover Model

Glover and McCulloch [5] presented the variation by latitude of the regression coefficients, b and a in the conventional equation of Angstrom type:

$$\frac{I_{\mathrm{d}}}{I_0} = \left(\frac{s}{s_0}\right)b + a \tag{9}$$

$$\frac{I_{\mathrm{d}}}{I_0} = 0.52\left(\frac{s}{s_0}\right) + 0.29\cos\phi \tag{10}$$

3.5 Gopinathan Model

Gopinathan [6] presented the variation as latitudinal location, air temperature, the site's elevation and the average relative humidification.

$$\frac{I_{\mathrm{d}}}{I_o} = \left(\frac{S}{S_0}\right)b + a \tag{11}$$

3.6 Linear Regression Model

Linear Regression model [7] is developed from Angstrom [11] and Prescott [7], Angstrom [11] equation modifies as

$$\frac{I_{\mathrm{d}}}{I_0} = a + b\left(\frac{S}{S_0}\right) \tag{12}$$

3.7 Artificial Neural Network (ANN) Model

ANN imitates the exact nature of the biological neural networks functioning in the brain. Neurons are the basic blocks of any Artificial Neural Network and in well-defined terms are known as the interconnected identical processing units which are countless in number. Global radiation was estimated by Emad using Artificial Neural Network (ANN) method based on the location coordinates, number of days and sunshine hours. A basic neural network has an input layer, a hidden layer and lastly an output layer. Other than these they also have transfer function, weight and a neuron. ANN models using Artificial neural networks are applied for prediction, simulation, forecasting bankruptcy situations, diagnosis of tumour issues, non-linear mapping, pattern recognition, classification and non-linear mapping. Input parameters are the neurons that form the input layer. The mean sun's daily global radiation on a monthly basis depicts a single neuron in the output layer (Fig. 4).

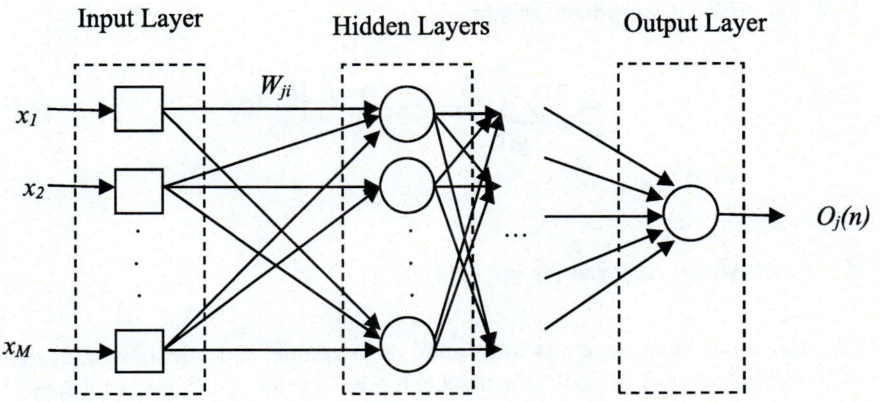

Fig. 4 Basic architecture of ANN model [13]

4 Validation of Models with Statistical Errors

For assessing, comparing and analyzing all the average global radiation approximation models, there are various parameters [14].

4.1 *Mean Bias Error (MBE)*

MBE test is applied for finding out the error between the calculated and measured data and provides information on long-lived performance. A positive value shows overrated result and a negative value shows underrated result and ideally, it should always be zero. One negativity of the MBE test is that underestimation in one observation gets cancelled by the overestimation in another observation.

$$\mathrm{MBE} = \frac{1}{S}\sum_{1}^{S} I_{i,\mathrm{calc}} - I_{i,\mathrm{meas}} \tag{13}$$

4.2 *Mean Percentage Error (MPE)*

The difference in the mean daily sun's global radiation data, being approximated by the models taken from the measured values is known as mean percentage error. Deviation is in percentage form and is also used in forecast errors. An MPE *r* between −10% and +10% is considered acceptable.

$$\mathrm{MPE} = \frac{1}{S}\sum_{1}^{S}\left[\frac{\left(I_{i,\mathrm{calc}} - I_{i,\mathrm{meas}}\right)}{I_{i,\mathrm{meas}}}\right] \times 100 \tag{14}$$

4.3 *Root Mean Square Error (RMSE)*

RMSE measures the average mismatch amid the measured and calculated data and produces results on short-lived performance. It gives a value of the level of disperse produced by the model of regression and explains the repeatability and readability of the model. The RMSE value should be closer to zero and is always positive [15].

$$\mathrm{RMSE} = \left[\frac{1}{S}\sum_{1}^{S}\left(I_{i,\mathrm{calc}} - I_{i,\mathrm{meas}}\right)^{2}\right]^{1/2} \tag{15}$$

4.4 The Nash–Sutcliffe Equation (NSE)

Any model is highly functioning when the value of NSE is closer to 1. For enhanced results and reformed comparison, this parameter has also opted as an evaluation criterion [16].

$$\mathrm{NSE} = 1 - \frac{\sum_1^S (I_{i,\mathrm{calc}} - I_{i,\mathrm{meas}})^2}{\sum_1^S \left(I'_{\mathrm{meas}} - I_{i,\mathrm{meas}}\right)^2} \tag{16}$$

4.5 Mean Absolute Percentage Error (MAPE)

To overcome the drawback of MBE test, MAPE is conducted. The MAPE avoids the error cancellation problem which is frequently observed in MBE. The accuracy is been shown in percentage form. It represents the absolute mean deviation in percentage form between the calculated and measured values [17].

$$\mathrm{MAPE} = \frac{1}{S}\sum_1^S \left| \frac{I_{i,\mathrm{meas}} - I_{i,\mathrm{calc}}}{I_{i,\mathrm{meas}}} \right| \tag{17}$$

4.6 t-Statistics Test

The t-test was introduced aby Stone (1993) and allows the comparison of models at the same time as MBE and RMSE may not be sufficient for assessing the outcome of the model [18]. It determines the statistical importance of the approximated value on a more accurate level. The analytical t value is to be estimated from standard statistical tables. The approximated t value must be less than the analytical value as the model will perform better for smaller values [19].

$$t = \left[\frac{(S-1)\mathrm{MBE}^2}{\mathrm{RMSE}^2 - \mathrm{MBE}^2} \right]^{1/2} \tag{18}$$

Table 1 Comparison between measured and approximated mean Sun's global radiation data (MJ m^{-2} day^{-1}) on a monthly basis for city Visakhapatnam, India

Month	Rietveld model	Ogleman model	Akinoglu model	Glover model	Gopinathan model	Linear regression model	ANN model	Measured data
Jan	18.871	18.022	17.949	18.996	18.479	16.758	17.432	17.422
Feb	21.599	20.747	20.7	21.819	21.237	19.092	19.77	20.01
Mar	22.293	21.623	21.647	22.756	23.198	20.13	21.72	21.82
Apr	24.831	24.038	24.045	25.332	24.578	22.73	22.81	22.99
May	24.421	23.991	24.073	25.215	24.678	21.92	22.14	22.18
Jun	17.064	17.713	17.458	19.062	19.685	16.25	17.28	17.49
Jul	15.561	17.203	16.802	18.562	19.472	16.777	16.01	16.02
Aug	16.764	17.464	17.07	18.865	19.705	16.534	16.53	16.35
Sept	17.568	18.02	17.921	19.106	19.547	16.728	16.94	17.06
Oct	18.153	18.132	18.481	19.011	18.988	16.931	17.65	17.62
Nov	17.771	15.668	17.121	18.039	17.629	15.781	16.31	16.4
Dec	17.851	15.588	16.963	17.964	17.485	15.841	16.26	16.32

5 Result Analysis

Table 1 given represents a comparison between the measured and approximate data of mean sun's global radiation per month of each individual model.

Figure 5 shows the graphic presentation of the measured and approximated value of mean global solar radiation per month of each individual model under consideration in study area.

The below-given Table 2 shows the calculated data of the statistical error parameters (MBE, MPE, NSE, MAPE, RMSE and *t*-stat) for each individual model under consideration at the mentioned area of study, i.e. Visakhapatnam [22–25].

Figure 6 shows the graphical representation of the statistical error parameters (MBE, MPE, NSE, MAPE, RMSE and *t*-stat) for each individual model under consideration at study area [26–28, 29].

6 Conclusion

Accessing global data of sun's solar radiation has become a necessity for affordable design and efficient implementation of solar energy. However, in many locations, measuring instruments are unavailable, so atmospheric parameters are being used to estimate the global radiation in that particular region. Seven empirical models in addition to ANN are considered in the paper and validated to approximate the mean global radiation received per month on a parallel plane surface in Visakhapatnam,

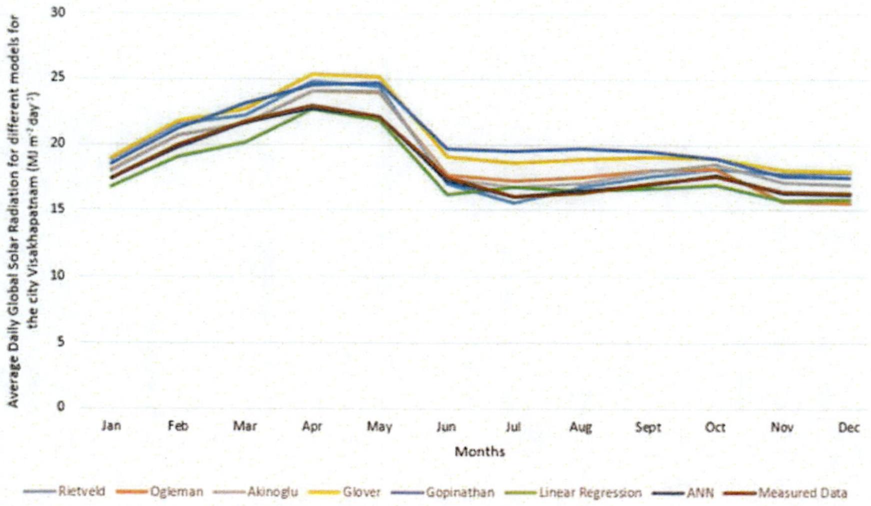

Fig. 5 The approximated and measured mean Sun's global radiation on a monthly basis for city Visakhapatnam, India [20, 21]

Table 2 The calculated data of the statistical parameters for all models at Visakhapatnam, India

Statistical errors	Rietveld model	Ogleman model	Akinoglu model	Glover model	Gopinathan model	Linear regression model	ANN model
MBE	0.9221	0.5439	0.7123	1.9204	1.9166	−0.5180	−0.0690
MPE	0.0479	0.0283	0.0385	0.1054	0.1069	−0.0270	−0.0030
RMSE	1.2446	0.9249	0.8673	2.0019	2.0856	0.7967	0.1308
NSE	0.7417	0.8573	0.8746	0.3317	0.2746	0.8942	0.9971
MAPE	0.0568	0.0448	0.0401	0.1054	0.1069	0.0366	0.0057
t-stat	3.6583	2.4115	4.7746	11.2680	7.7282	2.8334	2.0664

situated in Andhra. And most appropriate model is selected for the above-given location.

On the basis of available climatic parameters of sun-light hours, lowest and highest temperature and relative humidification these models were studied. As per the statistical evaluation of the empirical models, ANN was determined to be far more accurate and advanced in comparison with the other models; hence it could be utilized in approximation of global solar radiation in Visakhapatnam and also for the places having similar weather conditions.

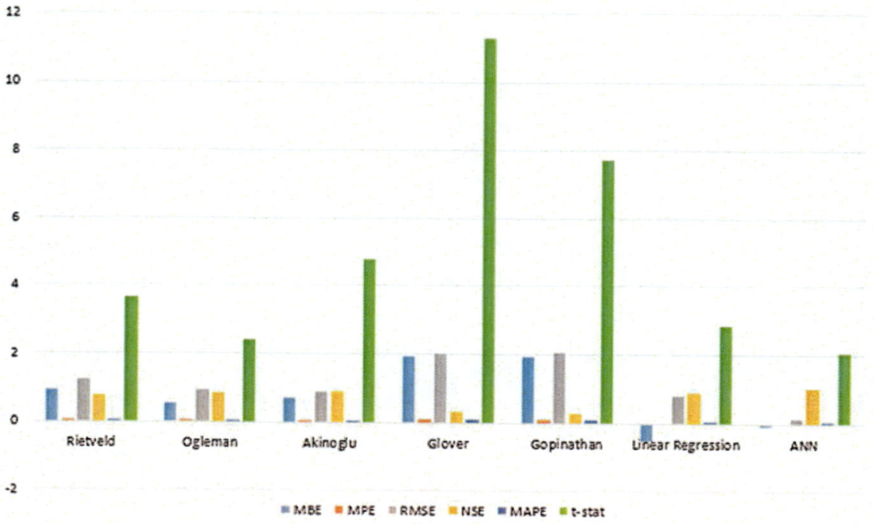

Fig. 6 Estimation of MBE, MPE, NSE, MAPE, RMSE and *t*-stat for all prescribed models [30–34]

References

1. Sukhatme SP (1997) Solar energy principles of thermal collection and storage. Tata McGraw-Hill
2. Rietveld MR (1978) A new method for estimating the regression coefficients in the formula relating solar radiation to sunshine. Agric Meteorol 19(3):243–252
3. Ecevit A, Ogelman H, Tasdemiroglu E (1984) A new method for estimating solar radiation from bright sunshine data. Sol Energy 33(6):619–625
4. Ecevit A, Akinoglu BG (1990) Construction of a quadratic model using modified Angstrom coefficients to estimate global solar radiation. Sol Energy 45(2):85–92
5. McGulloch JDG, Glover J (1958) The empirical relation between solar radiation and hours of sunshine. Q J R Meteorol Soc 84:172–175
6. Gopinathan KK (1988) A general formula for computing the coefficients of the correlation connecting global solar radiation to sunshine duration. Sol Energy 41(6):499–502
7. Prescott JA (1940) Evaporation from a water surface in relation to solar radiation. Trans Roy Soc Sci, Aust 64:25–114
8. London J, Frohlich C (1986) Revised instruction manual on radiation instruments and measurements. World Climate Research Program, Publication Series, No. 7
9. Cooper PI (1969) The absorption of solar radiation in solar stills. Sol Energy 12:3
10. Solanki CS, Sangani CS (2008) Estimation of monthly average direct normal solar radiation using elevation angle for any location. Sol Energy Mater Sol Cells 92:38–44
11. Angstrom A (1924) Solar and terrestrial radiation. Q J T Met Soc 50:121–126
12. Solar Radiation Handbook (2008) Solar Energy Centre, MNRE
13. Fagbenle RO (1990) Estimation of total solar radiation in Nigeria using meteorological data. Niger J Renew Energy 1:1–10
14. Namrata K, Sharma SP, Seksena SBL (2015) Comparision of different models for estimation of diffuse solar radiation in Jharkhand (India) region. Appl Solar Energy 51(3):65–70
15. Emad A, Ahmed and Adam, M. El-Nouby (2013) Estimate of global solar radiation by using artificial neural network in Qena, Upper Egypt. J Clean Energy Technol 1(2)
16. Khatib T, Mohamed A, Mahmoud M, Sopian K (2012) Estimating global solar energy using multilayer perception artificial neural network. Int J Energy 6(1)

17. Jiang Y (2009) Computation of monthly mean daily global solar radiation in China using artificial neural networks and comparison with other empirical models. Energy 34(9)
18. Sidel JL, Stone H (1993) The role of sensory evaluation in the food industry. Food Qual Prefer 4:65–73. Academic Press, San Diego
19. Abdul Azeez MA (2011) Artificial neural network estimation of global solar radiation using meteorological parameters in Gusau, Nigeria. Arch Appl Sci Res 3(2):586–595
20. Mubiru J (2011) Using artificial neural networks to predict direct solar irradiation. Adv Artif Neural Syst, Article ID 142054
21. Shukla KN, Sudhakar K, Rangnekar S (2015) Comparative study of isotropic and anisotropic sky models to estimate solar radiation incident on tilted surface: a case study for Bhopal, India. Energy Rep 1:96–103
22. Garg HP and Garg SN (1985) Correlation of monthly average daily global, diffuse and beam radiation with bright sunshine hours. Energy Convers Manag 25:409–417
23. Ahmad MJ, Tiwari GN (2011) Solar radiation models –a review. Int J Energy Res 35:271–290
24. Pham DT, Liu X (1995) Neural networks for identification, predication and control. Springer, London
25. Khatib T, Sopian K, Mohamed A, Mahmoud M (2012) Estimating global solar energy using multilayer perception artificial neural network. Int J Energy 6(1)
26. Beckman WA, Duffie JA (2006) Solar engineering of thermal processes, 3rd edn. Wiley, New York
27. Namrata K, Sharma SP, Saksena SBL (2012) Comparison of estimated daily global solar radiation using different empirical models. Int J Sci Adv Technol 2(4) (ISSN 2221-8386)
28. Jordan RC, Liu BYH (1960) The inter-relationship and characteristic distribution of direct, diffuse and total solar radiation. Sol Energy 4(3):1–9
29. Cankaya S (2009) A comparative study of some estimation methods for parameters and effects of outliers in simple regression model for research on small ruminants. Trop Anim Health Prod 41:35–41
30. Kumar R, Aggarwal RK, Sharma JD (2013) New regression model to estimate global solar radiation using ANN. Adv Energy Eng (AEE) 1:66–73
31. Kumar R, Aggarwal RK, Sharma JD (2015) Comparison of regression and artificial neural network models for estimation of global solar radiations. Ren Sust Energy Rev 52:1294–1299
32. Yang K, Koike T (2005) A general model to estimate hourly and daily solar radiation for hydrological studies. Water Resour Res 41:1–13
33. Lam JC, Li DHW (1996) Correlation between global solar radiation and its direct and diffuse components. Build Environ 31:527–535
34. SolarGIS. http://solargis.info/doc/postermaps

Absolute Point Positioning Algorithm for Navigation Applications

P. Sirish Kumar and V. B. S. Srilatha Indira Dutt

Abstract In this paper, one static positioning algorithm (i.e. lease square estimator) is implemented and one kinematic positioning algorithm designated as correntropy extended Kalman filter (CEKF) is proposed for precise GNSS/GPS applications and GAGAN-based aircraft landings. The proposed algorithm uses correntropy criterion (CC) as an optimal criterion, a local similarity measure, unlike minimum mean square error (MMSE). Also, it uses an iterative approach called fixed point for renovating the rearward estimates. The simulation results show that the proposed algorithm outperforms the static positioning algorithm in two-dimensional and three-dimensional surface.

Keywords Least square estimator · Correntropy extended Kalman filter · Correntropy · Minimum mean square error · GPS

1 Introduction

Global positioning system (GPS) allows each receiving unit/object (like ship, aircraft, and handheld navigator) to find its position anywhere globally. With the advent of GPS, navigation has been transformable, particularly in the area of civil aviation. Though the satellite-based systems of navigation often higher accuracies compared to traditional navigational aids. The positional accuracy readings are influenced by many factors of error like ionosphere or troposphere related delay, the bias of satellite clock, a proper receiving unit, and multiple paths of propagation as well as receiver/object geometry concerning satellites. Due to these discrepancies, standalone GPS does not satisfy the aircraft landing stipulations, such as CAT-I. Precision approaches in which the desirable horizontal and vertical accuracies are 16 m and

P. Sirish Kumar (✉)
Department of ECE, Aditya Institute of Technology and Management, Tekkali 532201, India
e-mail: sirishdg@gmail.com

V. B. S. Srilatha Indira Dutt
Department of ECE, GITAM (Deemed to Be University), Visakhapatnam 530048, India

K. Namrata et al. (eds.), *Smart Energy and Advancement in Power Technologies*, Lecture Notes in Electrical Engineering 926,
https://doi.org/10.1007/978-981-19-4971-5_33

4.5–7 m, respectively. There is an imperative need for improvement of the performance of GPS; this is done by augmentation where the general name is space-based augmentation system (SBAS). All the SBAS requires to satisfy the enroute navigation accuracy (accuracy of 2.2 nmi, the integrity of 2 min) and vertical guidance (horizontal accuracy of 220 m, vertical accuracy of 20 m, and integrity of 10 s) [1].

GPS is the only full-fledged global satellite constellation system in GNSS; it has 32 satellites more significant than the nominal figure of 24 satellites. All over this globe, a minimum of 14–18 satellites signal systems are available and operating fully for various sources since 1995 in both civilian and military fields [2]. For continuous worldwide coverage, the arrangement of GPS satellites is such that four satellites are arranged in each of 6 orbits. It is worth noting that after the breakup of Soviet Russia, there were many setbacks in the GLONASS operation affecting the full-scale deployment (23 operational) in orbit. Even the GALILEO system has only 12 satellites, which unable to have continuous coverage. GPS is a satellite-centric, all-weather, 24×7, worldwide navigational system, including the time factor [3]. It is specifically developed to provide exact position, speed, and time-related data using a shared worldwide grid meant for infinite appropriate users.

Currently used primary navigation systems for air traffic control will be replaced by GNSS/GPS-based navigation systems by the year 2017 as per the ICAO. The work presented in this paper is aimed at strengthening the GNSS/GPS-based navigation system. The investigated algorithms (i.e. LSE, and CEKF) used in this article are useful for navigation and surveying in urban canyons and dense forest areas where the satellite visibility is low [4]. The iterative algorithms implemented for receiver position estimation can also solve various other optimization problems encountered in the control and automation industry. They are also useful for geographic information system (GIS) and accurate aerial mapping applications.

2 Literature Study

Precise positioning and robust surveillance systems are crucial to secure countries such as India, with a long geographical frontier and an expansive coastline. This has led to the development and advancement of many of these systems. Aviation, maritime, mining, military, medicine, and agriculture are some of these system's real-world applications. The improvement of a single frequency real-time point positioning system (PPS) with the error mitigation mechanism for obtaining the best accuracy. Reference [4, 5] has described the Indian plan for a satellite-based navigating framework for civil aviation in which the augmentation systems like GBAS, ABAS, and SBAS involving ground, aircraft, and space, respectively. Reference [6] have worked on the theoretical framework of error analysis in cooperative positioning depending on local position information providing a group of mobile entities relative positions. For urban canyons, [7] have proposed a novel technique for integrating WIFI positioning technology and GPS in the availability of two GPS satellites. Four satellites need to be observed by the GPS to provide a 3-D solution.

In the proposed method, the pseudo-range observations are used to generate time difference of arrival (TDOA) measurement, which gives a hyperboloid surface for obtaining the possible location. The integration of Wi-Fi fingerprint technology leads to accuracy improvement of over 50%. Typically, positioning accuracy is reduced by the satellite geometry, receiver dimensions, and error related to atmosphere and multi-path propagation; dilution of precision (DOP) is a parameter of the satellite receiver's geometrical dimension [8]. The proposed method utilizes DOP's relationship with a combined constellation of GPS and IRNSS to have greater positioning accuracy required for CAT-I aircraft landing, survey work, and disaster management.

The common causes of inaccuracy are incorrect object tracking and disconnection of the GPS signal while using an application. Reference [9] have proposed a new closed-form expression for the pseudo-range GPS equation being suitable for a system of two linear equations, which gave a difference in pseudo-range. The user's position becomes the domain line of intersecting planes or hyperbola of revolution such that expression in 3-D has every degenerate of unique geometry. This approach offers an in-depth observation of the solution's characterization geometry, resolving the pseudo-ranges extended and unique properties. Reference [10] have proposed a scheme for evaluating a direct algorithm for computing the clock discrepancy of a single GPS receiver with an exact solution achieved with four satellites. When more than four satellites are available, an algorithm is combining all available GPS pseudo-ranges measurements.

From the above study, it can be summarizing that the basic premises and system details of positioning systems (i.e. source localization and global positioning system), measurement techniques, GPS signal structure, and GPS observables, iterative and recursive navigational techniques covering linear algebra are essential. Thorough knowledge of the concepts of GPS, positioning solution, estimation algorithms, and GPS applications are required from the literature to work with real-time GPS data. Several authors deal with GPS architecture, trilateration principle, GPS signals, signal characteristics (L1 and L2), mathematical description for positioning solution, and pseudorandom codes sent by GPS satellites have been discussed by many researchers.

3 Least Square Estimator

This section describes the pseudo-range estimation procedure with the TOA and satellite transmission times, mathematical formulation of the navigation problem, and linearization of the navigation equations corresponding to the different satellite visibility conditions. The linearized navigation equations are then used in the navigation solution (least square estimator) [11, 12].

The three-point GPS navigation problem with an over-determined case is a non-linear least-squares problem and required to solve for the minimum value concerning the unknown variables represented in Eq. (1).

$$\arg\min_{\vec{U}}\left(\text{obj}\,(\Pr_i, \vec{U})\right) \quad (1)$$

The residual vector can be given as

$$r_i = \Pr_i - f\left(\vec{U}\right) \quad (2)$$

The minimum value of the objective function, 'obj' in Eq. (1), is determined when its gradient is made equal to zero.

$$\nabla\text{obj} = \frac{\partial\,\text{obj}\left(\Pr_i, \vec{U}\right)}{\partial\vec{U}} = 0 \quad (3)$$

Since the model contains three parameters in Eq. (3), i.e. $\vec{U} = [x_u, y_u, z_u]^T$, three gradient equations are formed. The gradient for Eq. (1) is represented as shown in Eq. (4)

$$\frac{\partial\,\text{obj}\left(\Pr_i, \vec{U}\right)}{\partial\vec{U}} = \sum_{i=1}^{m} r_i \frac{\partial r_i}{\partial\vec{U}} = 0 \quad (4)$$

For nonlinearity, the gradient equation does not have a closed-form solution. So, initial values are chosen, and refinement is done iteratively to be changed in successive approximation as shown in Eq. (5).

$$\vec{U}_{K+1} = \vec{U}_K + \delta\vec{U} \quad (5)$$

The derivative in Eq. (4) is represented as

$$\left.\frac{\partial\,\text{obj}\left(\Pr_i, \vec{U}\right)}{\partial\vec{U}}\right|_{\vec{U}=\vec{U}_K} = \left(-\sum_{i=1}^{m} J_{ji}^T \times r_i\right)_{\vec{U}=\vec{U}_K} = 0 \quad (6)$$

Here, 'j' is the number of unknown parameters or size of $\vec{U}$, at every iteration, the function $f\left(\vec{U}\right)$ is linearized concerning $\vec{U} = \vec{U}_K$, and the change in measurement or error in measurement is computed as follows

$$\delta\Pr_i = \Pr_i - f\left(\vec{U}_K\right) \quad (7)$$

where the residual r_i is given as

$$r_i = \delta \Pr_i + f\left(\vec{U}_K\right) - f\left(\vec{U}\right) = \delta \Pr_i - \sum_{i=1}^{m} J_{ij} \times \delta \vec{U} \tag{8}$$

Finally, the gradient is determined by putting Eq. (8) in Eq. (6)

$$-\sum_{i=1}^{m} J_{ji}^T \times \left(\delta \Pr_i - \sum_{i=1}^{m} J_{ij} \times \delta \vec{U}\right) = 0$$

$$\sum_{i=1}^{m} J_{ji}^T \times \delta \Pr_i = \sum_{i=1}^{m} J_{ji}^T J_{ij} \times \delta \vec{U} \Rightarrow J^T \times \delta \vec{P} r = J^T J \times \delta \vec{U} \tag{9}$$

Furthermore, the unknown parameter $\delta \vec{U}$ is given as in Eq. (10), which is used to update the receiver position vector $\vec{U}_K$, successively.

$$\delta \vec{U} = \left(J^T J\right)^{-1} J^T \delta \vec{P} r \tag{10}$$

Here, in Eq. (10), the term $J^T J$ gives the Hessian matrix with approximation 'Hess.'

3.1 Steps in Computing Navigation Solution with Least Square Estimator

Step 1. Collect the pseudo-range measurements $\Pr_1, \Pr_{2,\ldots,} \Pr_m$ from 'm' satellites.

Step 2. Initialize the receiver position vector $\vec{U}_k = [x_{uK},\ y_{uK},\ z_{uK}]^T$ (i.e. $K = 0$).

Step 3. Compute the range measurements $\Pr_{1K}\ \Pr_{2K} \ldots \Pr_{mK}$ using the receiver position in step 2.

Step 4. Calculate the Jacobian matrix '$J_{m\times 3}$'.

Step 5. Calculate the error in the measurement vector $\delta \vec{P} r = [\delta \Pr_1 \quad \delta \Pr_2 \ldots \Pr_m]^T$ from steps 1 and 3.

Step 6. Determine the updated value or change in position $\delta \vec{U} = [\delta x_u \quad \delta y_u \quad \delta z_u]^T$

Step 7. Update the receiver position vector in Step 2, with the change in the position vector obtained in Step 6. (i.e. $\vec{U}_{K+1} = \vec{U}_K + \delta \vec{U}$)

Step 8. Repeat Steps 3–7 for a fixed number of times or until a threshold criterion is satisfied (like $\sqrt{(\delta x_u)^2 + (\delta y_u)^2 + (\delta z_u)^2} \leq$ threshold).

Step 9. $\vec{U}_{K+1} = \left[x_{uK+1}, y_{uK+1}, z_{uK+1}\right]^T$ is the final estimate of the receiver position.

Hence, the minimum requirement of four satellites has drawn attention to the development of other navigation algorithms. In addition to the reason mentioned above, i.e. the nonexistence of inverse in an over-determined case (number of unknowns < number of measurements), many other factors contribute to the same problem with three satellites as well. A traditional algorithm is implemented to identify a robust and reliable algorithm that efficiently deals with the above issues. So, the conventional algorithm implemented is least square estimator (LSE) that provides precise and robust position estimates.

4 Correntropy Extended Kalman Filter

The EKF endeavours honourably under Gaussian commotions. In any case, its yield can basically breakdown under non-Gaussian clamours, especially, when imprudent commotions upset the essential framework. EKF's fundamental intention is to be executed by the MMSE rule, which gathers just second-order error signal insights and is defenceless against huge disparities. In this article, another KF is performed, which can be best enforced in non-Gaussian boisterous frameworks, with CC is tailored to the error calculations for second and higher orders [13, 14].

4.1 *Mathematical Model of Algorithm*

The previous segment includes the linear equations

$$\begin{bmatrix} \hat{\mathbf{s}}(n|n-1) \\ \mathbf{m}(n) \end{bmatrix} = \begin{bmatrix} \mathbf{I} \\ \mathbf{B}(n) \end{bmatrix} \mathbf{s}(n) + u(n) \tag{11}$$

And

$$u(n) = \begin{bmatrix} -\big(\mathbf{s}(n) - \hat{\mathbf{s}}(n|n-1)\big) \\ \mathbf{q}(n) \end{bmatrix}, = \mathbf{D}(n)\mathbf{D}^T(n) \tag{12}$$

where $\mathbf{D}(n)$ is matrix of cholesky decomposition.

By $\mathbf{D}^{-1}(n)$ multiplication for Eq. (11), they are seeing to both sides

$$\mathbf{F}(n) = \mathbf{V}(n)\mathbf{s}(n) + \mathbf{e}(n) \tag{13}$$

Here,

$$\mathbf{F}(n) = \mathbf{D}^{-1}(n) \begin{bmatrix} \hat{\mathbf{s}}(n|n-1) \\ \mathbf{m}(n) \end{bmatrix}, \ \mathbf{V}(n) = \mathbf{D}^{-1}(n) \begin{bmatrix} \mathbf{I} \\ \mathbf{B}(n) \end{bmatrix},$$

$\mathbf{e}(n) = \mathbf{D}^{-1}(n)u(n)$, since $E\big[\mathbf{e}(n)\mathbf{e}^T(n)\big] = \mathbf{I}$,

It proposes the following cost function for correntropy:

$$Z_L(\mathbf{s}(n)) = \frac{1}{L}\sum_{i=1}^{L} J_\sigma(f_i(n) - \mathbf{v}_i(n)\mathbf{s}(n)) \tag{14}$$

The optimum $\mathbf{s}(n)$ approximation is given according to the criterion of correntropy

$$\hat{\mathbf{s}}(n) = \text{argmax}_{\mathbf{s}(n)} Z_L(\mathbf{s}(n)) = \text{argmax}_{\mathbf{s}(n)} \sum_{i=1}^{L} J_\sigma(e_i(n)) \tag{15}$$

$$e_i(n) = f_i(n) - \mathbf{v}_i(n)\mathbf{s}(n) \tag{16}$$

By solving, it allows the desired solution

$$\frac{\partial Z_L(\mathbf{s}(n))}{\partial \mathbf{s}(n)} = \sum_{i=1}^{L} [J_\sigma\left(e_i(n)\mathbf{v}_i^T(n)\right)(f_i(n) - \mathbf{v}_i(n)\mathbf{s}(n))] = 0 \tag{17}$$

This proceeds quickly

$$\mathbf{s}(n) = \left(\sum_{i=1}^{L}[J_\sigma\left(e_i(n)\mathbf{v}_i^T(n)\mathbf{v}_i(n)\right]\right)^{-1} \times \left(\sum_{i=1}^{L}[J_\sigma\left(e_i(n)\mathbf{v}_i^T(n)f_i(n)\right]\right) \tag{18}$$

Since $e_i(n) = f_i(n) - \mathbf{v}_i(n)\mathbf{s}(n)$.
Equation (18) is the fixed-point equation [11] of $\mathbf{s}(n)$ and can be rephrased as

$$\mathbf{s}(n) = g(\mathbf{s}(n)) \tag{19}$$

With

$$g(\mathbf{s}(n)) = \left(\sum_{\mathrm{i}=1}^{\mathrm{L}}[J_\sigma\left(f_i(n) - \mathbf{v}_i(n)\mathbf{s}(n)\mathbf{v}_i^T(n)\mathbf{v}_i(n)\right]\right)^{-1} \times \left(\sum_{\mathrm{i}=1}^{\mathrm{L}}[J_\sigma\left(f_i(n) - \mathbf{v}_i(n)\mathbf{s}(n)\right)\mathbf{v}_i^T(n)f_i(n)]\right)$$

Clearly get the approach of fixed point given under expression

$$\hat{\mathbf{s}}(n)_{t+1} = g\left(\hat{\mathbf{s}}(n)_t\right) \tag{20}$$

Can be rewritten as, at fixed-point iteration 't' the desired solution is indicated by $\hat{\mathbf{s}}(n)_t$.

Equation (18) is a fixed point expression and can also be interpreted as

$$\mathbf{s}(n) = \left(\mathbf{V}^T(n)\mathbf{H}(n)\mathbf{V}(n)\right)^{-1}\mathbf{V}^T(n)\mathbf{H}(n)\mathbf{F}(n) \quad (21)$$

where $\mathbf{H}(n) = \begin{bmatrix} \mathbf{H}_x(n) & 0 \\ 0 & \mathbf{H}_y(n) \end{bmatrix}$, $With \mathbf{H}_x(n) = \text{diag}(J_\sigma(e_1(n)), \ldots J_\sigma(e_a(n)))$, $\mathbf{H}_y(n) = \text{diag}(J_\sigma(e_{a+1}(n)), \ldots J_\sigma(e_{a+b}(n)))$.

Equation (21) may also be represented in some other way as

$$\mathbf{s}(n) = \hat{\mathbf{s}}(n|n-1) + \tilde{\mathbf{N}}(n)\left(\mathbf{m}(n) - (n)\hat{\mathbf{s}}(n|n-1)\right) \quad (22)$$

Here,

$$\left.\begin{aligned} \overline{\mathbf{N}}(n) = \overline{\mathbf{C}}(n|n-1)\mathbf{B}^T(n)\left(\mathbf{B}(n)\overline{\mathbf{C}}(n|n-1)\mathbf{B}^T(n) + \overline{\mathbf{Q}}(n)\right)^{-1} \\ \overline{\mathbf{C}}(n|n-1) = \mathbf{D}_c(n|n-1)\mathbf{H}_x^{-1}(n)\mathbf{D}_c^T(n|n-1) \\ \overline{\mathbf{Q}}(n) = \mathbf{D}_q(n)\mathbf{H}_y^{-1}(n)\mathbf{D}_q^T(n) \end{aligned}\right\} \quad (23)$$

It can be proved that correntropy approaches correlation for the kernel's sizes, which is larger than recommended. The kernel size can regulate the usage of higher-order moment information for correlating; this is a beautiful and unique concept with proper selection; the outcomes for correntropy are equivalent to robust statistical approaches. Furthermore, due to the smooth dependency of correntropy in kernel size, the kernel size output's sensitivity must be much smaller than the choice of thresholds. However, kernel size estimation methods must be deployed appropriately for practical applications.

5 Results and Discussion

The data used for analysing the two algorithms, i.e. least square estimator (LSE), and correntropy extended Kalman filter (CEKF) performance concerning IISc, Bangalore GPS receiver, is collected for over 22 h at a sampling rate of 30 s and resulted in 2640 epochs of data (i.e. 22 h × 120 epochs/Hour). The data used to evaluate the navigation algorithms performance concerning the IISc, Bangalore GPS receiver is collected from the Scripps Orbit and Permanent Array Centre (SOPAC) database on 1 January 2019. The obtained estimated receiver positions are compared with the surveyed position to measure the position error using the two algorithms, i.e. LSE, and CEKF. And these results are presented in Tables 1 and 2, respectively, and the given values are the estimated mean values of 22 h of data.

The following key points are observed from the overall analysis of the results. They are

Table 1 Estimated mean position comparison among two algorithms

True receiver position: X = 1,337,936.309 m; Y = 6,070,317.116 m; Z = 1,427,876.908 m

GPS time (hours)	Estimated position in X-direction		Estimated position in Y-direction		Estimated position in Z-direction	
	LSE	CEKF	LSE	CEKF	LSE	CEKF
00–01	1,337,903.12	1,337,913.12	6,070,350.23	6,070,335.23	1,427,889.27	1,427,886.27
01–02	1,337,899.93	1,337,910.31	6,070,339.69	6,070,328.71	1,427,883.12	1,427,883.02
02–03	1,337,891.54	1,337,900.67	6,070,327.83	6,070,312.89	1,427,881.78	1,427,881.60
03–04	1,337,890.93	1,337,900.96	6,070,339.67	6,070,328.78	1,427,885.15	1,427,881.93
04–05	1,337,893.99	1,337,904.01	6,070,350.44	6,070,335.42	1,427,884.55	1,427,880.96
05–06	1,337,896.30	1,337,905.63	6,070,343.93	6,070,329.33	1,427,883.87	1,427,882.04
06–07	1,337,901.57	1,337,911.80	6,070,341.76	6,070,327.86	1,427,883.53	1,427,880.20
07–08	1,337,903.29	1,337,914.85	6,070,347.08	6,070,335.16	1,427,882.49	1,427,879.75
08–09	1,337,906.33	1,337,919.72	6,070,349.25	6,070,334.42	1,427,885.71	1,427,884.14
09–10	1,337,910.22	1,337,920.95	6,070,348.58	6,070,335.11	1,427,889.48	1,427,885.14
10–11	1,337,911.86	1,337,923.50	6,070,356.41	6,070,341.24	1,427,884.32	1,427,880.68
11–12	1,337,909.82	1,337,919.83	6,070,359.91	6,070,346.41	1,427,883.29	1,427,878.42
12–13	1,337,903.74	1,337,913.20	6,070,364.76	6,070,351.20	1,427,882.36	1,427,877.51
13–14	1,337,901.95	1,337,911.35	6,070,360.82	6,070,343.99	1,427,881.82	1,427,878.04
14–15	1,337,896.45	1,337,905.24	6,070,346.28	6,070,330.80	1,427,882.28	1,427,878.67
15–16	1,337,894.50	1,337,903.29	6,070,344.96	6,070,329.23	1,427,877.98	1,427,874.84
16–17	1,337,903.15	1,337,913.97	6,070,333.43	6,070,327.84	1,427,881.53	1,427,880.79
17–18	1,337,906.92	1,337,916.38	6,070,333.17	6,070,327.91	1,427,882.97	1,427,880.30
18–19	1,337,905.62	1,337,916.31	6,070,345.23	6,070,331.19	1,427,881.25	1,427,879.19
19–20	1,337,908.06	1,337,918.24	6,070,341.68	6,070,326.48	1,427,878.51	1,427,875.62
20–21	1,337,904.13	1,337,915.31	6,070,337.12	6,070,322.43	1,427,879.72	1,427,875.36
21–22	1,337,901.35	1,337,911.90	6,070,343.15	6,070,328.12	1,427,886.40	1,427,883.54
Mean	1,337,902.04	1,337,912.30	6,070,345.70	6,070,332.26	1,427,883.24	1,427,880.36
Standard deviation	5.98	6.59	9.23	7.51	2.88	2.43
Variance	35.75	43.49	85.21	56.37	8.30	5.92

i. The proposed CEKF algorithm provides significant accuracy improvement in position estimation with a mean position error of 24.01 m, 15.52 m, and 3.90 m which are lesser by 10.26 m, 13.06 m, and 2.43 m concerning X-, Y-, Z-directions due to LSE, respectively.

ii. Vertical position error (i.e. error in Z-direction) of the proposed CEKF algorithm (i.e. 3.90 m) is found to be less than the required vertical position accuracy of CAT-I (4.5–7 m) aircraft landings, i.e. [3.90 m] < [4.5–7 m]. This is accomplished by neglecting other correctable error sources.

Table 2 Mean position error comparison among two algorithms

GPS time (hours)	Mean position error (in meters)					
	X-direction		*Y*-direction		*Z*-direction	
	LSE	CEKF	LSE	CEKF	LSE	CEKF
00–01	33.19	23.19	33.11	18.11	12.36	9.36
01–02	36.38	26.00	22.57	11.59	6.21	6.11
02–03	44.77	35.64	10.72	4.23	4.87	4.69
03–04	45.38	35.35	22.56	11.66	8.24	5.02
04–05	42.32	32.30	33.33	18.30	7.64	4.05
05–06	40.01	30.68	26.82	12.21	6.96	5.13
06–07	34.74	24.51	24.65	10.74	6.62	3.29
07–08	33.02	21.46	29.96	18.04	5.58	2.84
08–09	29.98	16.59	32.14	17.30	8.80	7.23
09–10	26.09	15.36	31.47	17.99	12.57	8.23
10–11	24.46	12.81	39.30	24.12	7.41	3.77
11–12	26.49	16.48	42.79	29.29	6.38	1.51
12–13	32.57	23.11	47.64	34.08	5.45	0.60
13–14	34.36	24.96	43.70	26.87	4.91	1.13
14–15	39.87	31.07	29.17	13.68	5.37	1.76
15–16	41.81	33.02	27.85	12.11	1.07	2.07
16–17	33.16	22.34	16.31	10.72	4.62	3.88
17–18	29.39	19.93	16.06	10.79	6.06	3.39
18–19	30.69	20.00	28.11	14.07	4.34	2.28
19–20	28.26	18.07	24.57	9.36	1.60	1.29
20–21	32.18	21.00	20.01	5.31	2.81	1.55
21–22	34.96	24.41	26.03	11.00	9.49	6.63
Mean	34.27	24.01	28.58	15.52	6.33	3.90
Standard deviation	5.98	6.59	9.23	7.51	2.88	2.43
Variance	35.75	43.49	85.21	56.37	8.30	5.92

iii. It is observed that from the 3-D surface diagram shown in Fig. 2, the receiver's estimated mean positions over an hour due to the CEKF algorithm seem close to the true receiver position (surveyed position) with less variance in contrast to the LSE algorithm.
iv. Figure 5 shows that the final estimated receiver mean position due to the CEKF algorithm is more proximal to the true receiver position (surveyed position) than LSE algorithm.

Table 1 provides the comparison of the estimated mean position over an hour due to the two algorithms (i.e. LSE & CEKF). Further, the observed minimum and maximum

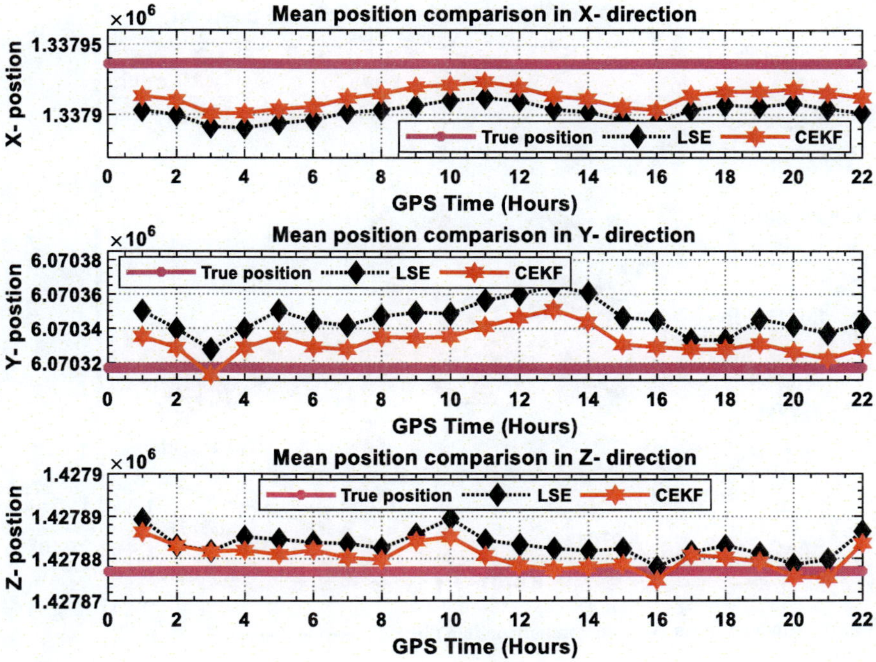

Fig.1 Estimated mean position comparison in a two-dimensional surface

position estimates corresponding to the two algorithms are shown in Tables 1 and 2 with bold numbers.

Figure 1 provides a graphical representation of the estimated receiver mean position over an hour compared with the true receiver position (magenta colour solid line) in *X*-, *Y*-, and *Z*-directions, respectively, in a two-dimensional surface.

Similarly, Fig. 2 shows that the estimated receiver mean positions over an hour reported due to the two algorithms are compared with the true receiver position in a three-dimensional surface.

It is observed that from Fig. 2 that the estimated receiver mean positions over an hour due to the CEKF algorithm are accurate, precise and seems close to the true receiver position with minimal variance in contrast to the LSE algorithm.

Table 2 given below compares the error in estimated mean position over an hour due to the LSE and CEKF algorithms (in *X*-, *Y*-, and *Z*-directions). The observed minimum and maximum error in position estimate corresponding to the two algorithms is shown in the following table with bold numbers simultaneously with the mean, standard deviation, and variance.

In addition to the above plots, mean position error over an hour is compared in receiver *X*-, *Y*-, and *Z*-directions reported with the two algorithms (LSE and CEKF) in a two-dimensional surface and is shown in Fig. 3.

Figure 4 represents the mean position error comparison over an hour among the two algorithms in a three-dimensional surface in addition to the above plots.

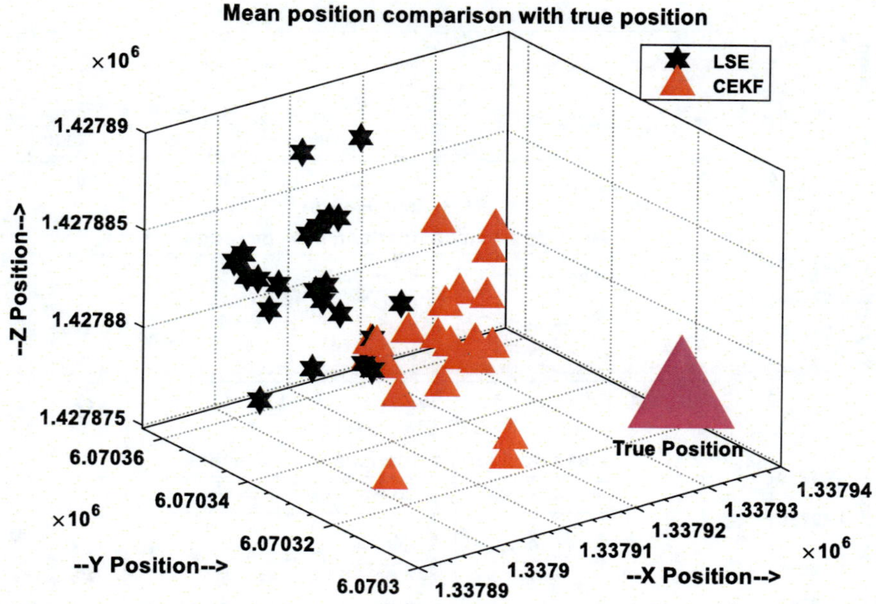

Fig.2 Estimated mean position comparison in a three-dimensional surface

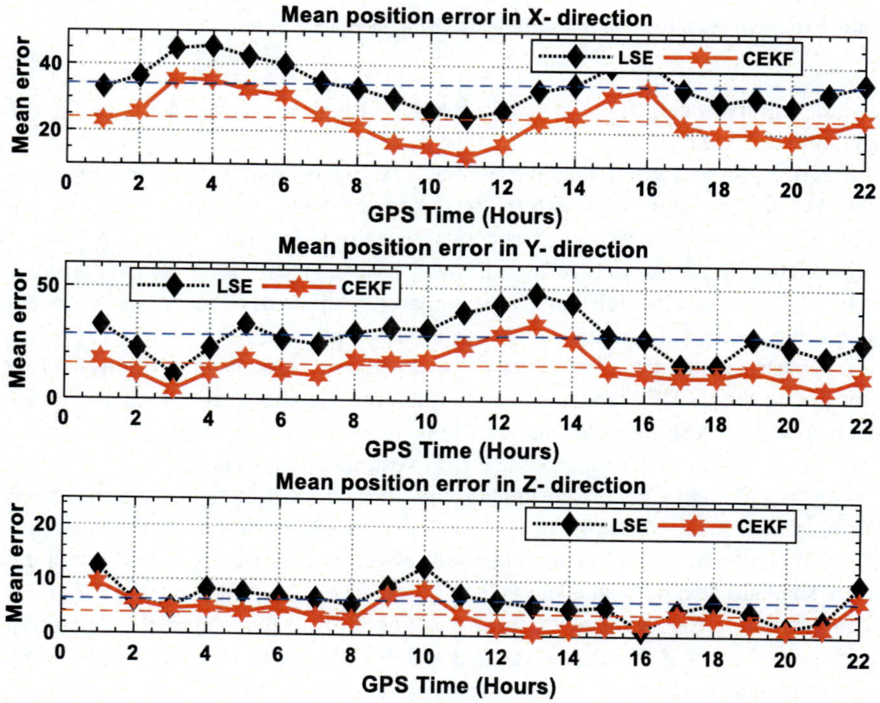

Fig. 3 Mean position error comparison in a two-dimensional surface

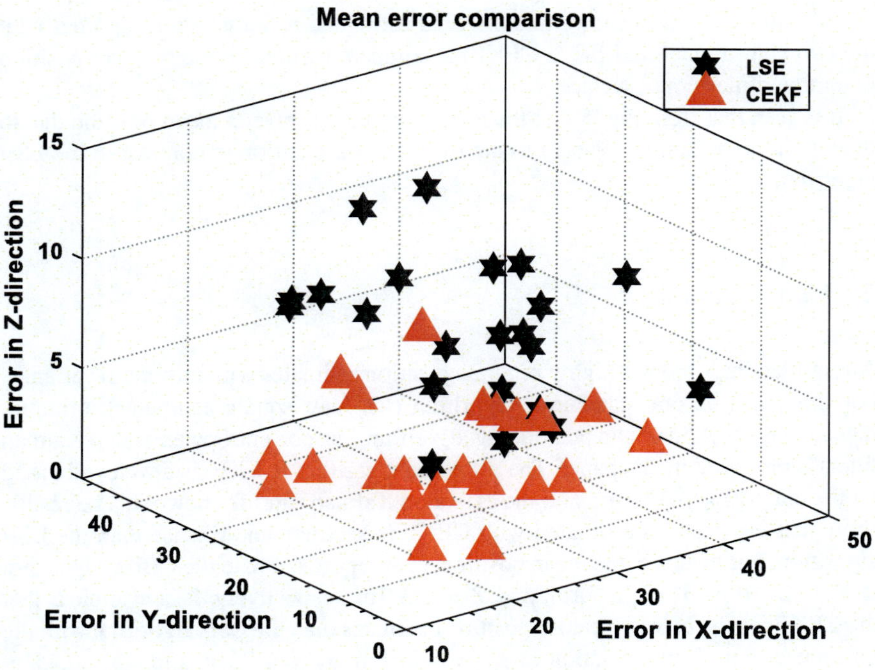

Fig. 4 Mean position error comparison in a three-dimensional surface

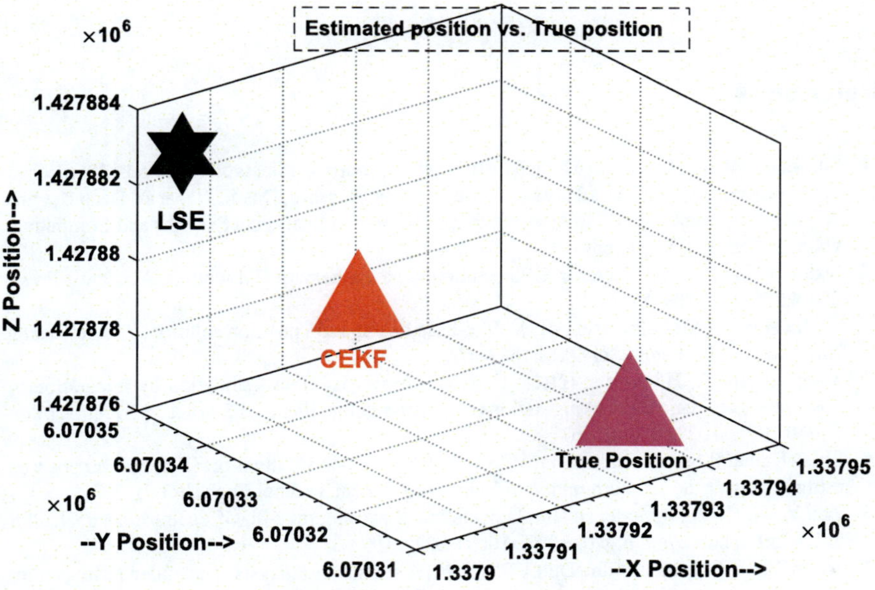

Fig.5 Estimated final mean position comparison among two algorithms in a 3-D surface

Additionally, Fig. 5 shows that the final estimated mean position reported with the two algorithms (i.e. LSE & CEKF) is compared with the true receiver position in a three-dimensional surface.

It is seen that from Fig. 5 that the final estimated receiver's mean position due to the CEKF algorithm is adjacent to the true receiver position in contrast to the LSE algorithm.

6 Conclusion

Aircraft landings require highly accurate position estimates with minimum variance. For this purpose, one standalone algorithm (i.e. least square estimator) is implemented, and one kinematic positioning algorithm (i.e. correntropy extended Kalman filter) is proposed in this paper. The proposed algorithm (CEKF) is developed based on the correntropy criterion and the extended Kalman filter framework. The results show that the position estimated with CEKF is accurate and precise than the LSE algorithm. For instance, the mean position error reported due to the CEKF algorithm is $X = 24.01$ m, $Y = 15.52$ m, and $Z = 3.90$ m, respectively. It is also clear that the CEKF results in less variance position estimates than the LSE algorithm with the difference in position deviation of $\sigma_x = -0.61$ m, $\sigma_y = 1.72$ m, and $\sigma_z = 0.45$ m. Hence, the proposed CEKF algorithm can be used in critical GPS applications like CAT-I aircraft landing and GAGAN user's navigation.

References

1. Sasibhushana Rao G, Lavanya B et al (2019) Satellite horizon effects on temporal GPS receiver position accuracy over coastal area of South India. In: Bansal J, Das K, Nagar A, Deep K, Ojha A (eds) Soft computing for problem solving. Advances in Intelligent Systems and Computing, vol 817. Springer, Singapore
2. Oxley A (2017) Uncertainties in GPS positioning: a mathematical discourse. Academic Press. ISBN:978-0-12-809787-8
3. Santra A et al (2019) Precision of satellite-based navigation position solution: a review using NavIC data. J Inf Optim Sci 40(8):1683–1691
4. Yang Y, Yang Y, Hu X et al (2021) BeiDou-3 broadcast clock estimation by integration of observations of regional tracking stations and inter-satellite links. GPS Solut 25:57. https://doi.org/10.1007/s10291-020-01067-x
5. Sirish Kumar P, Srilatha Indira Dutt VBS (2019) Navigation solutions for GPS receiver position estimation over the southern region of India. Int J Recent Technol Eng (IJRTE), 7(6)
6. Kou Y, Wu H (2021) Model and implementation of pseudorange-bias-free linear channel. GPS Solut 25:53. https://doi.org/10.1007/s10291-021-01093-3
7. Sirish Kumar P, Srilatha Indira Dutt VBS (2020) A sensitivity analysis of extended Kalman filter for GPS position estimation with and without clock offset. Mater Today: Proc 33:3626–3629
8. Wang X, Liang M (2018) GPS positioning method based on Kalman Filtering. In: 2018 International conference on robots & intelligent system (ICRIS), Changsha, China, pp 77–80. https://doi.org/10.1109/ICRIS.2018.00028

9. Zhao S, Chen B, Príncipe JC (2011) Kernel adaptive filtering with maximum correntropy criterion. In: The 2011 International joint conference on neural networks, San Jose, CA, USA, pp 2012–2017. https://doi.org/10.1109/IJCNN.2011.6033473
10. Liu T, Chen H, Chen Q et al (2021) Characteristics of phase bias from CNES and its application in multi-frequency and multi-GNSS precise point positioning with ambiguity resolution. GPS Solut 25:58. https://doi.org/10.1007/s10291-021-01100-7
11. Chen et al (2015) Convergence of a fixed-point algorithm under maximum correntropy criterion. IEEE Signal Process Lett 22(10):1723–1727
12. Anjasmara IM, Pratomo DG, Ristanto W (2019) Accuracy analysis of GNSS (GPS, GLONASS and BEIDOU) observation for positioning. In: E3S Web of conferences, vol 94. EDP Sciences
13. Ashok Kumar N et al (2018) Extended Kalman filter for GPS receiver position estimation. In: Intelligent engineering informatics. Springer, Singapore, pp 481–488
14. Sirish Kumar P et al (2020) Implementation of new navigation algorithm based on cross-correntropy for precise positioning in low latitude regions of South India. Int J Speech Technol 23(4):747–756

Islanding Detection Techniques in Distribution System

Ayushman Priyadarshi, Ratneswar Sahoo, Vrishank Tiwari, and Jitendra Kumar

Abstract The depletion of non-renewable energy sources together with advancement in the new technologies has fueled the growth of various unconventional sources of energy like wind energy, solar energy, etc., so that customer's energy demand is met. This has popularized the distributed generations (DG) concept whose adoption has expanded as a result of market deregulation and environmental concerns. But connecting DG to grid may pose a serious problem known as islanding. When a segment of a distributed energy network becomes separated from the utility but is still energized by DGs, this is known as islanding. Thus, the capability to detect islanding is a critical necessity for connecting DG to the power distributed system. Failure to trip the DGs during islanding can cause a slew of issues for various equipment and the loads they are linked to. In order to protect itself, consumer appliances, and line personnel, a DG should disconnect within 100–300 ms after the main supply is lost. To accomplish these objectives, each DG mandatorily provided with some equipment that can detect the islanding operation of DGs. Some equipment which can achieve such results are vector surge relay, ROCOF, etc. An efficient island detection technique (IDT) should have a negligible non-detection zone (NDZ), yet the techniques discussed above have inherent non-detection zone. In this paper, two signal processing approach for islanding detection is used, namely wavelet transform and S-transform. These two approaches are explored in the paper in depth, along with their uses and implication on a DG.

Keywords Distributed generators (DG) · Islanding · Wavelet transform · S-transform · Artificial neural network (ANN)

A. Priyadarshi (✉) · R. Sahoo · V. Tiwari · J. Kumar
NIT Jamshedpur, Jamshedpur, Jharkhand, India
e-mail: ayushmansrs@gmail.com

J. Kumar
e-mail: jitendra.ee@nitjsr.ac.in

K. Namrata et al. (eds.), *Smart Energy and Advancement in Power Technologies*, Lecture Notes in Electrical Engineering 926,
https://doi.org/10.1007/978-981-19-4971-5_34

1 Introduction

With rising power demand, the need for customer reliability as well as satisfactory power quality demands and the evolving popularity of alternative power generation sources such as wind, water turbines, fuel cells and photovoltaic systems, and distributed networks involving distributed generations (DG) is becoming increasingly significant for most power companies. As a result of market deregulation and environmental awareness, distributed generation (DG) has recently acquired traction and appeal [1, 2]. However, various issues, such as islanding detection, limit the utility of DG. When a segment of the distribution system becomes electrically isolated from the rest of the power system but is still energized by the distribution system, it leads to islanding. As a result, it is critical that the power distributed system, while employing DG, can differentiate between islanding situations and non-islanding ones.

The various issues with islanding include:

- If DG sources continue to supply the system even after main sources are opened and have been tagged out, then it can jeopardize the safety of line personnel.
- Various parameters like voltage, frequency, phase angle, energy, and so on may not be maintained within a defined permitted range, posing a major risk. The DG interconnection may not be effectively grounding the islanded system.
- Instantaneous reclosing may result in DG reclosing that is out of phase which, when happens during a voltage peak, will result in a significant switching transient of capacitive nature, which can be dangerous to system equipment.
- This poses a number of concerns like the damage of various electric components as a result of fluctuations in voltage and frequency.

For detection of islanding, a negative sequence voltage signal is used [3]. The energy content highlights the preference of the S-transform over the wavelet transform in terms of localizing and identifying islanding occurrences. The standard deviation (SD) of the S-transform contour as well as the wavelet transform signal is investigated by measuring its energy and is used to distinguish islanding from non-islanding conditions [3]. Wavelet transform is first utilized to analyze grid-connected voltage signal [4], according to which this technique is employed for islanding identification. As a result, wavelet denoising and non-stationary signal identification are achieved [4]. The study of non-stationary data can be used to detect islanding [4]. The wavelet reconstruction algorithm is used to extract the waveform of non-stationary signals, which can then be analyzed to differentiate islanding and disruptions. The conventional techniques that are being used for the detection of the islanding of the DG include.

(a) **Remote Signaling**—Transfer trip scheme and power line signaling-based scheme.
(b) **Passive Techniques**—Voltage unbalances [5], harmonic distortion [5], rate of change of frequency [6], rate of change of output power [7], and rate of change of frequency over power [8].

(c) **Active Techniques**—Reactive power export error detection [9] and phase or frequency shift method [10].
(d) **Hybrid Technique**—Positive feedback and voltage imbalances [11], voltage re-active power shift [12].

This paper tries to address the issue that develops during conventional techniques by incorporating the use of negative sequence components (such as current and voltage) that are detected at the desired DG position. When compared to non-islanding circumstances, a good islanding detection is assured by implying time variation method of the negative sequence impedance. Furthermore, the data of SD of the negative sequence impedance for one cycle accurately distinguishes islanding from non-islanding condition. As a result, the proposed wavelet transform and S-transform methods, which are based on negative sequence signals, provide excellent islanding detection techniques that outperform certain earlier techniques.

The main objectives of the proposed model are

1. By application of wavelet transform or S-transform, the negative sequence voltage signal is pre-processed. Here, negative sequence quantities are considered as they quantify the amount of unbalance in a power system.
2. Using a feature extraction method, the computational overhead of processing a large feature set is minimized, using energy as the extracted feature.
3. For each of the islanding scenarios, the total harmonic distortion (THD) is noted, and the result is graphically examined using histogram.
4. The method which is proposed in this paper is developed to be resistant to parameter changes.

2 Artificial Intelligence-Based Method

This topic discusses how to detect islanding in a distribution line quickly and accurately. It will be concluded that methods based on neural network combined with signal processing techniques have the merits of greater accuracy and swift response.

A methodology of the proposed method based on hybrid AI to detect islanding in a distribution line is discussed. Signal processing techniques such as wavelet transform aggregated with artificial neural networks (ANNs) and the S-transform methodology are employed in this study. Further, the wavelet transform is used to compress the overall data set's volume and convert it into a set of features. Finally, an ANN is employed to detect and differentiate between islanding and non-islanding scenarios.

(A) Wavelet transform

Wavelet transform is a type of time–frequency analysis that has multi-resolution feature and can be used to express the signal's characteristics in a time–frequency field [12]. The shape of a time–frequency window varies while the size remains constant, i.e., the time window and frequency window are both variable. In the low-frequency section, it has higher frequency resolution and lower time resolutions, whereas in the high frequency segment, it has higher time resolution and lower frequency resolution. It is ideal for detecting non-stationary signal disturbances. Time-scale information of a wavelet allows for the sorting of time-varying features [13, 14]. Thus, wavelets are an excellent tool for evaluating transient non-stationary signal because of this feature. As a result, wavelet would be more effective in supervising transient time-varying signals, encouraging the use of such a technology to improve detection capabilities. The elimination of unwanted high-frequency content and impulse components superimposed on the signal is the primary aim of wavelet transform analysis. Wavelet analysis' efficiency is due to its rapid pyramid method. There are two faces of the algorithm. The wavelet transform (DWT) is computed using the forward algorithm (decomposition phase), which decomposes the signal into component wavelets, whereas the inverse wavelet transform (IDWT) is computed using the backward approach (reconstruction phase) which reconstructs the original signal from the component wavelets. To deconstruct the signal into low- and high-frequency components, it uses linear filters, low- and high-pass analog devices, and combines these filters with down sampling. Wavelet transform of the sampled waveform, by implementing DWT, can be obtained by the following equation:

$$\mathrm{DWT}(p, q) = 2^{p/2} \sum_{p} \sum_{q} x(q) * \emptyset * \left(\frac{t1 - q2^{p}}{2^{p}} \right) \tag{1}$$

where the discretized mother wavelet is given by

$$\phi_{p,q}(\mathrm{t}) = \frac{1 * \omega}{\sqrt{a_0 p}} * \left(\frac{t - qb_0 a_0^p}{a_0^p} \right) \tag{2}$$

a_0 and b_0 are the fixed constants with $a_0 > 1$, $b_0 > 1$, and $p, q \in Z$, where Z is set of integers. The base function is defined discretely by selecting $a = a_0^p$ and $b = qb_0a_0^p$ in order to prevent redundant information.

The decomposition of the generated signal and discrete wavelet transform (DWT)-based method's development is depicted in Fig. 1, from which we assess fault situations by generating detailed and approximation coefficients using artificial neural network (ANN).

In the figure as shown in Fig. 1, the 10-level wavelet decomposition of the sampling frequency, i.e., 20 kHz. The system frequency is 50 Hz. So, initially, we are having 400 samples/cycle. After 10th level decomposition, we are left with three samples.

(B) **S-transform**

The S-transform combines the elements of the short time transform and the wavelet transform to create a time–frequency spectral localization method. It is a phase-corrected extension of the continuous wavelet transform that employs an analysis window whose width decreases with frequency to provide dependent resolution. The S-transform has the unique property of being entirely convertible from time domain to frequency domain both forward and inversely. The S-transform provides

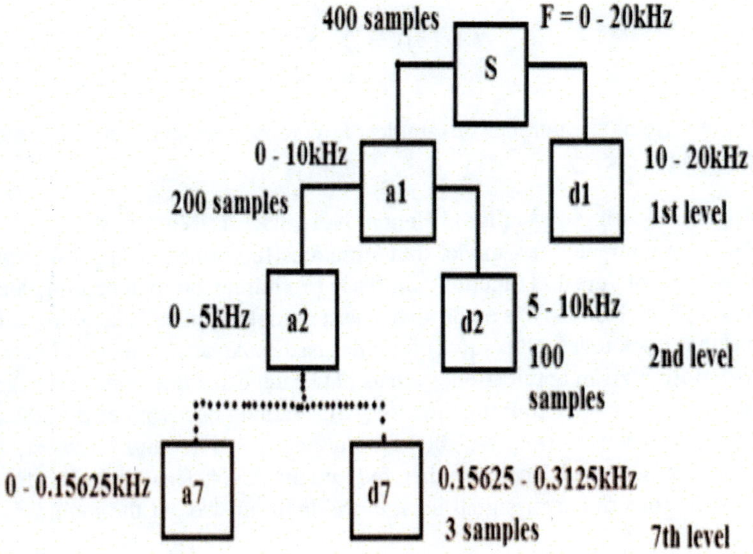

Fig. 1 Decomposition in discrete wavelet transform, DWT

good frequency domain clarity at low frequencies and good time domain clarity at high frequencies. The S-transform expression of a function x (t) is given by:

$$S(T,f) = \int_{-\infty}^{\infty} x(t)\frac{|f|}{\sqrt{2\pi}}e^{-\frac{(T-t)^2 f^2}{2}}e^{-i\omega t}\,\mathrm{d}t \tag{3}$$

where t represents time, f represents frequency, and T is a control parameter for the Gaussian window's position on time axis. In signal $x(t)$, percentage phase (Ø) and frequency (f) are given by

$$Ø(T,f) = a\tan\left\{\frac{\text{imag}(S(T,f))}{\text{real}(S(T,f))}\right\} \tag{4}$$

$$F(T,f) = \frac{1}{2\pi}\frac{d}{dt}\{2\pi f T + \varphi(T,f)\} \tag{5}$$

Feature Extraction

Feature extraction is a technique for reducing a huge data collection into a set of features in order to minimize its dimension. The statistics collected from a decomposed signal at each sub-band are listed below [15].

Energy of a signal is given by the sum of the square of the value of the signal.

$$E(x(S)) = \sum_{j=1}^{S} x^2(j) \quad (6)$$

where S represents the number of samples, $x(S)$ represents the energy function and E is the signal's energy.

Training and Testing of Artificial Neural Network (ANN)
An ANN is a computerized model that simulates the biological neural networks' functions and structure and functions similarly to a human brain in making decisions and reaching conclusions. It is made up of a vast number of linked clusters of artificial neutrons that work together to solve particular issues. An artificial neural network is customized for a given application, such as data categorization, pattern recognition, and regression, through a training process [16]. Training necessitates the adjustment of synaptic connections (i.e., weights) among neurons. First and foremost, a large variety of data is created. These are then fed into the ANN. This works in a feedback manner and sends the error signal back to the input all we get the required output with less error.

3 Islanding Detection in a Distribution Network

Besides the utility grid which supplies power to various loads, distributed generators are also present in a distribution network. So, early islanding detection is quite necessary. To investigate the performance of the technique for islanding detection proposed in this paper, a simulation model is given in Table 1.

A 20 km, 25 kV feeder line connects the two 9 MW wind farms to the 120 kV utility. Each wind farm is also connected to 500 kw resistive load. The dual feed

Table 1 Shows rating of various equipment of grid which has been considered

S. No.	Equipment	Rating
1	Alternator	Rated power = 2500 MVA, frequency = 50 Hz, rated voltage = V_{base} = 120 kV
2	Distributed generations	Two wind farms each having six 1.5 MW DFIG turbines are considered. The wind farm is connected to the distribution system using 25 kV, 20 km feeder line
3	Transformer T1	Rated power = 47 MVA, frequency = 50 Hz, rated voltage = 120/25 kV, V_{base} = 25 kV
4	Transformer T2	Rated power = 1.75 * 6 MVA, frequency = 50 Hz, rated voltage = 25/0.4 kV, V_{base} = 0.4 kV
5	Transformer T3	Rated power = 1.75*6 MVA, frequency = 50 Hz, rated voltage = 25/0.4 kV, V_{base} = 0.4 kV
6	Distribution lines	Π-configuration, 20 km each, rated voltage = V_{base} = 25 kV, rated power = 20 MVA

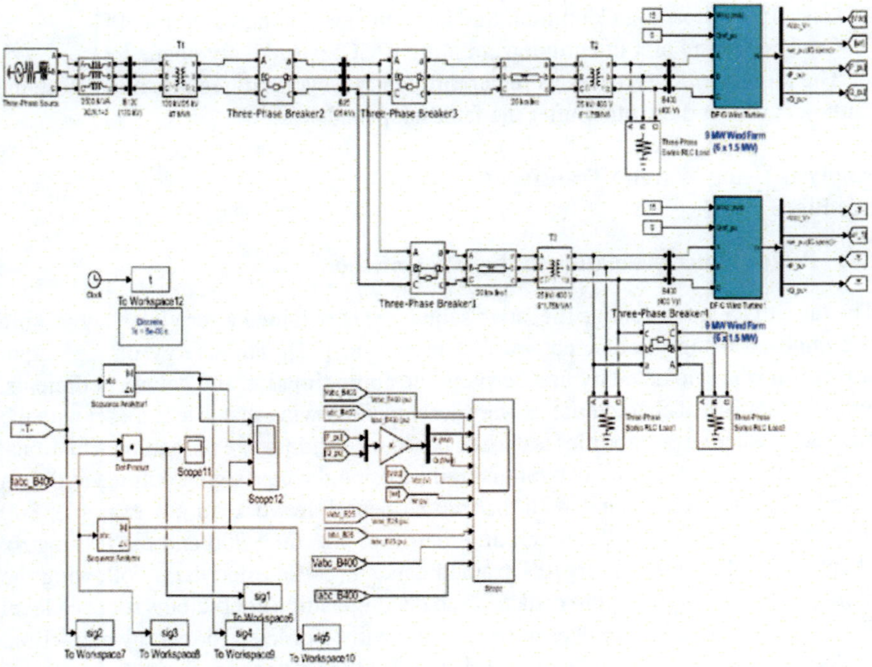

Fig. 2 Simulink model

induction generator is a kind of wind turbine (DFIG). A wound rotor induction generator and an AC/DC/AC IGBT-based PWM converter make up a DFIG. The stator of the DFIG is directly connected to the 50 Hz utility, while the input to the rotor of the DFIG is a variable frequency provided via the AC/DC/AC converter. Negative sequence voltage is collected from the DG end for analyzing purpose.

Figure 2 shows the detailed model of a distributed system with two doubly-fed induction generators working in parallel.

Proposed Hybrid AI-Based Method

Here, two methods are described, one using wavelet transform-based approach and another one is with S-transform based approach. The algorithm is described in Sect. 3.2.1. This paper investigates five different cases.

Case 1: Normal Operation: All the circuit breakers (three-phase) CB1, CB2, CB3 are in closed conditions.

Case 2: Islanding Condition: Initially, the contacts of CB1 are closed but later opened within a time of 0.01–0.015. During this period, the circuit breaker 2 and circuit breaker 3 are in closed operation. When CB1 is turned off, the whole distribution system is disconnected from the utility grid.

Case 3: DG trip: CB1 is always online. With a transition period of 0.01–0.015, CB3 is initially closed and then opened. In the line where CB2 is linked, there is another DG. This is the DG trip scenario.

Case 4: Overloading Condition: Suddenly the load is increased by 50%.

Case 5: When point of common coupling (PCC) is suddenly opened.

The aforementioned cases were put through two cycles of testing, one before the fault occurred and the other after the fault happened.

Analysis Using Wavelet Transform
See Fig. 3.

(A) Performance evaluation of the DWT method

The target DG site's voltage (negative sequence) is obtained after which signals are decomposed using the discrete wavelet transform [17]. The energy of the output waveform was calculated by first deriving the approximation and detail coefficients, and then, statistical characteristics were derived. Following that, a feature set with the greatest predictive possibilities is chosen based on a wide range of loading situations. Further, an artificial neural network is trained using the best signal outputs and then tested for each case of fault with the trained neural network. In our example, 300 samples are obtained, with the training matrix of size 60 * 300 and the test matrix of size 30 * 300, resulting in 70% trained data and 30% tested data. Following the training and testing, the approximation and decomposition coefficients for each level db1, db2, db3, and db4 are determined, which will be utilized for energy extraction.

As shown in Fig. 4, the ANN network used for the proposed technique is a type of feed forward backpropagation network, with training function Levenberg–Marquardt [18]. The hidden layer has a size of 10. MSE performance function determines the performance of the network [19]. The size of the input layer is determined by the total data set being considered, which in our case is 300. The output layer has a size of 1.

Analysis with S-transform
See Fig. 5.

(B) Performance evaluation of the S-transform method

For islanding detection in distributed generations, the suggested approach employs a time–frequency transform known as the S-transform. The S-transform is a time–frequency spectral localizing method that yields the time–frequency contours of voltage signals extracted at the target DG location. This technique is based on the spectral energy content of the voltage signal's negative sequence component. The approach has been put to the test in terms of detecting islanding. At the target DG position, negative sequence voltage is obtained. The obtained signal's S-transform was found. The S contours' energy is calculated. Then, the energy's cumulative sum (CUMSUM) is calculated. If CUMSUM $\geq$ Threshold value, then islanding case. If CUMSUM $\leq$ Threshold value, then non-islanding case.

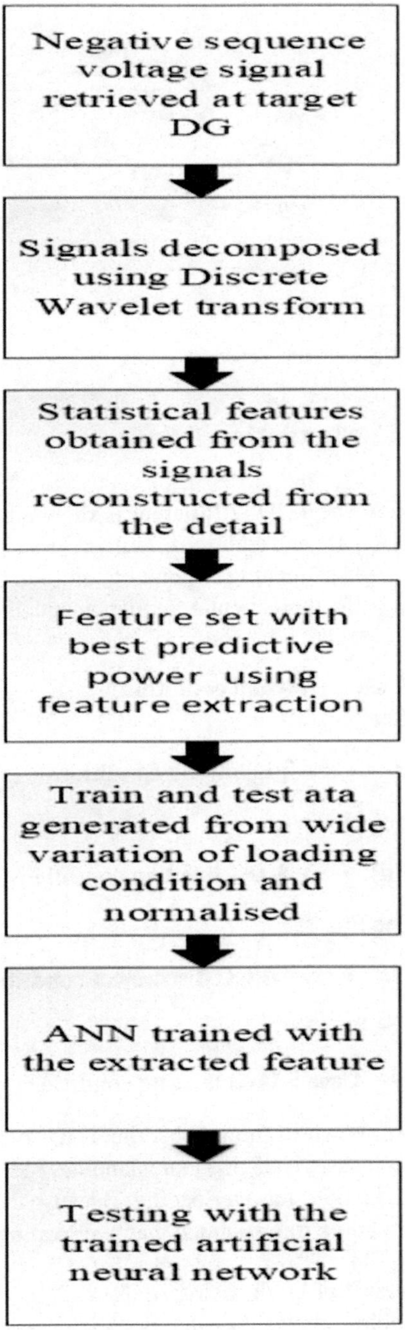

Fig. 3 Pre-processing technique using wavelet transform

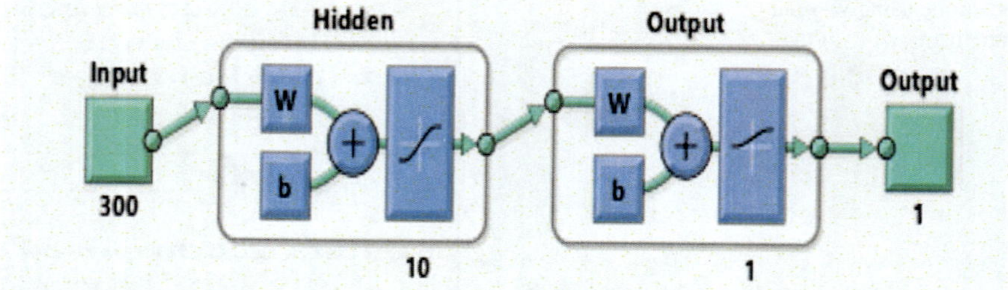

Fig. 4 Neural network training

4 Simulation Results

(i) The THD [20] output is shown in the form of histogram charts for the normal case and additional fault cases such as islanding condition, distributed generator trip, opening of point of common coupling, and DG overloading condition, with the fundamental frequency equal to 50 Hz.

Case 1 (Normal condition)
See Fig. 6.

(ii) **Case 2-Islanding condition**

See Fig. 7.

(iii) **Case 3-DG line trip condition**

See Fig. 8.

(iv) **Case 4-PCC disconnect condition**

See Fig. 9.

(v) **Case 5-Overloading condition**

As shown in the graphs, the THD for the normal case is 5.03%, but the THD for fault cases is 85.39% for islanding, 85.35% for DG trip, 85.29% for PCC disconnect, and 5.05% for overloading. Because there is not much variation for the overloading situation, it is assumed that overloading is not a case of islanding.

Result of DWT Using ANN
Table 2 shows the alterations in energy for various scenarios. **ΔE** is the difference in energy during normal operation and during islanding operation.

Table 2 shows that the energy change in islanding situation is 0.167, compared to 0.0042, 0.038, 0.0256, and 0.0063 for the non-islanding conditions. As a result, it

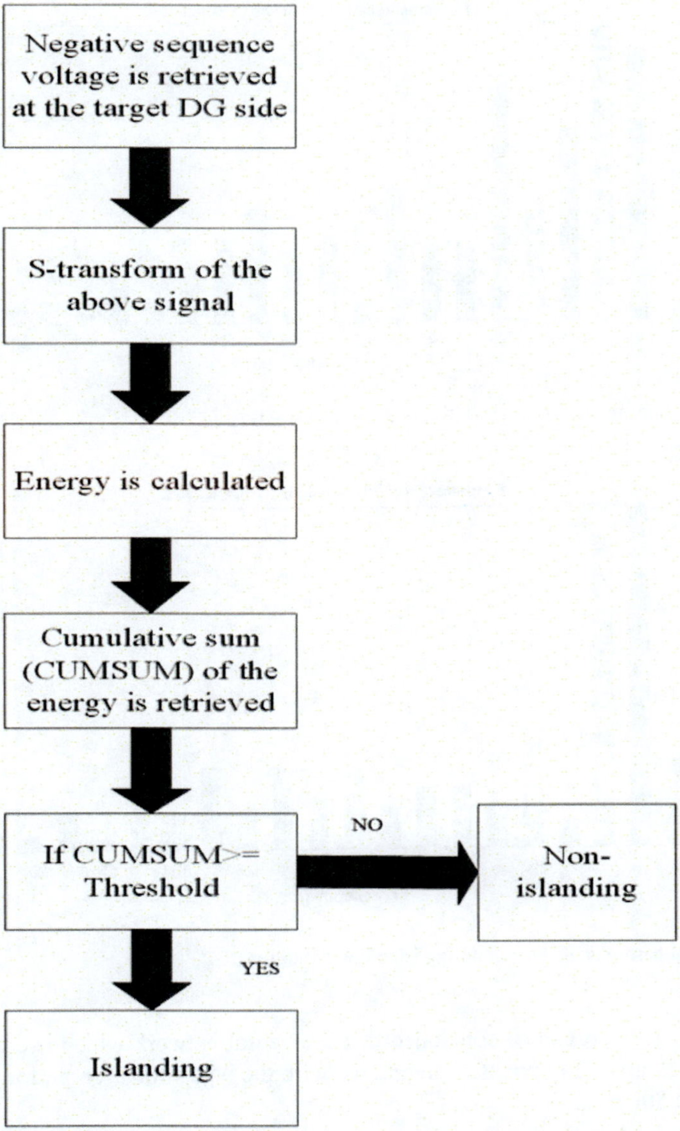

Fig. 5 Detection technique using S-transform

can be concluded that any islanding state has a considerably greater energy change than the usual condition and therefore can be easily distinguished from non-islanding cases.

After the conclusion of training process of ANN which converged in 14 iterations, the MSE is obtained which can be seen in the training performance plot (TPP) given in Fig. 10.

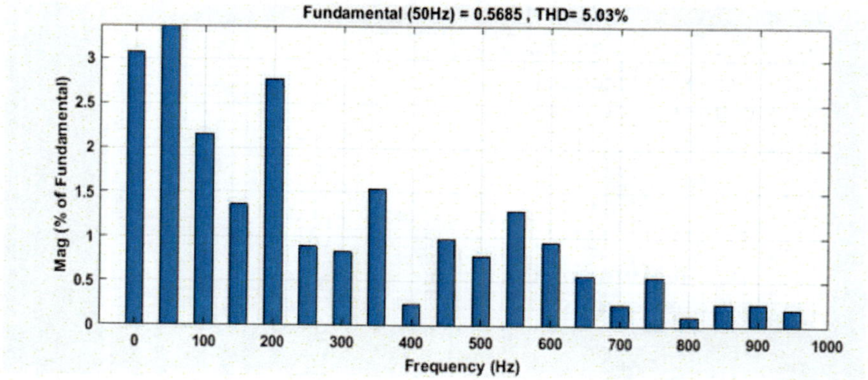

Fig. 6 Total harmonic distortion during normal condition

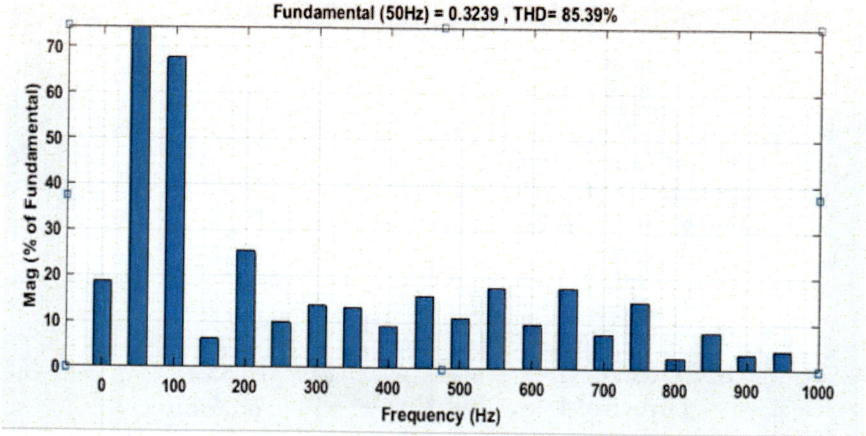

Fig. 7 Total harmonic distortion during islanding condition

From Fig. 11, it can be concluded that the neural network which is created has been trained up to 14 epochs or iterations, with the best validation performance of 0.10731 at 8th eproach.

Without and with islanding, input samples for retrieved energy of negative sequence voltage were taken. The following training and testing plots are generated using 0 as the target value for non-islanding case and 1 as the target value for islanding case, as illustrated in Fig. 12.

After the neural network had been trained, the best linear regression that relates the target to the output was plotted to evaluate its performance. Figure 3.16 depicts a regression plot of the trained neural network's output (y) versus the actual target value (T) for training (91%), testing (86%), validation (98%), and all regressions

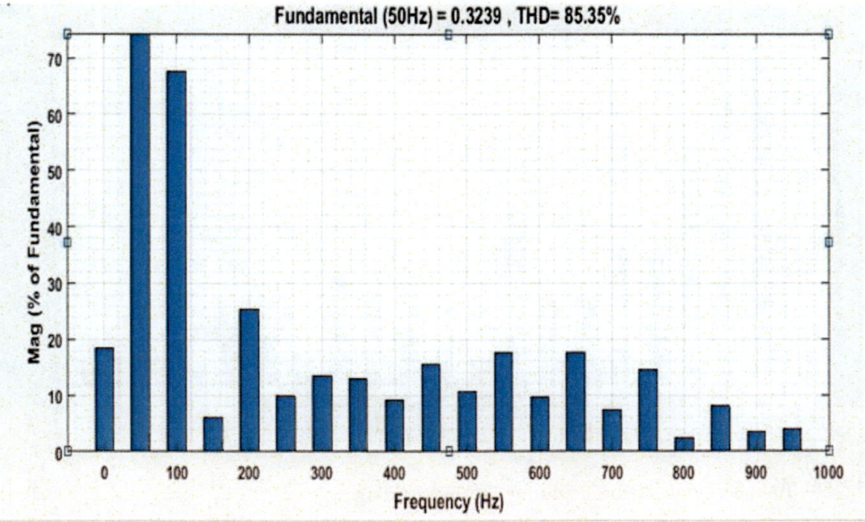

Fig. 8 Total harmonic distortion during distributed generator trip

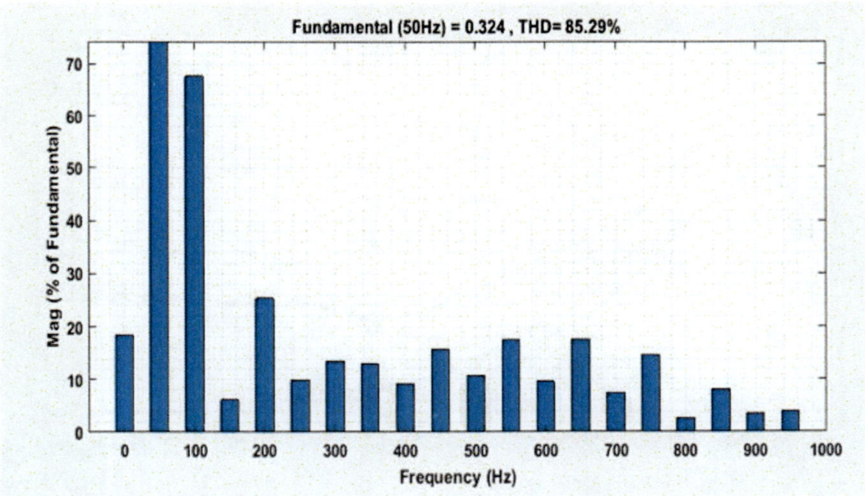

Fig. 9 Total harmonic distortion during PCC disconnect

Table 2 Energy alteration for different scenario

S. No.	Events	ΔE
1	Normal operation	0.0042
2	Islanded operation	0.167
3	Distributed generator trip	0.038
4	Opening of PCC	0.0256
5	DG overloading	0.0063

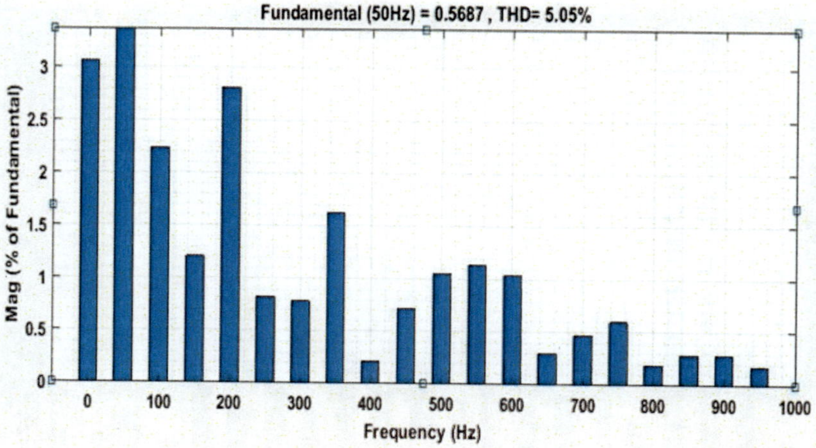

Fig. 10 Total harmonic distortion during DG overloading

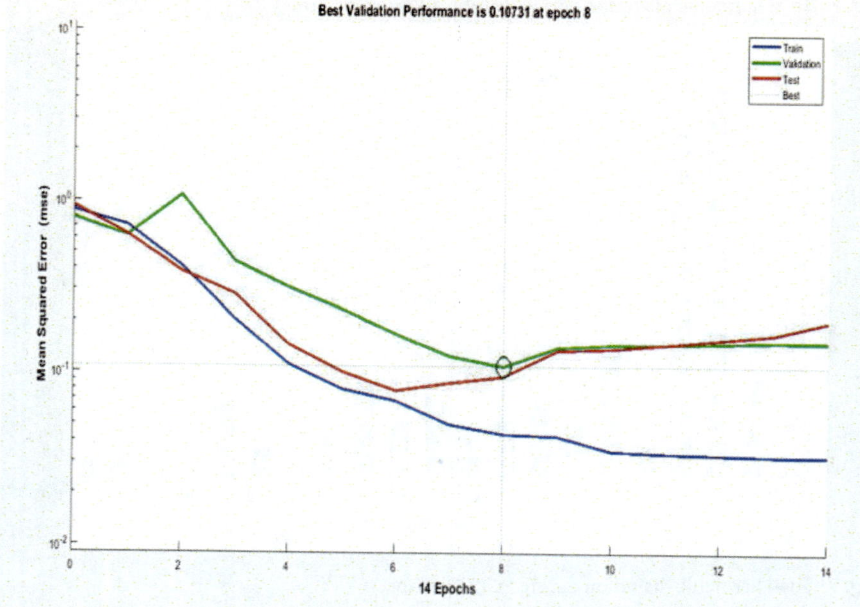

Fig. 11 TPP of the ANN showing mean square error

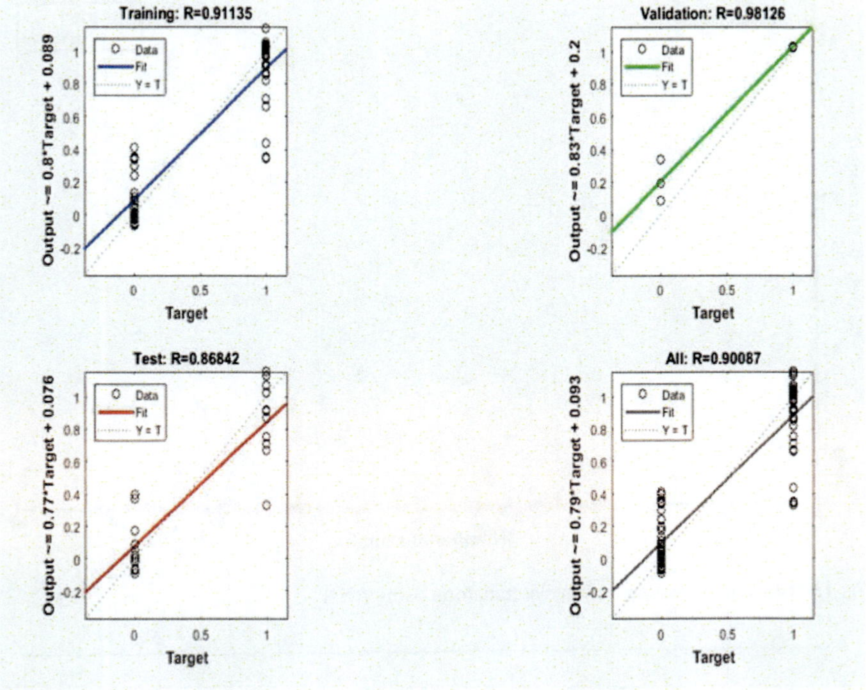

Fig. 12 Regression curves

(90%). Figure 3.9, as shown below, shows the output of ANN when the output of wavelet transform is fed into it.

As can be seen from Fig. 13, when there is no fault in the system and under normal conditions, it shows 0 and when fault occurs under islanding condition, it shows 1.

Result of S-transform

Figure 14 shows the negative sequence voltage having two cycles, one before the fault and one after the fault.

From Fig. 14, it is observed that under normal condition there is no impact of negative sequence voltage, but after islanding has occurred there is negative sequence voltage which can be inferred from the sudden rise in voltage.

Figure 15 shows the plot for retrieved signal samples considered at nominal frequency of 50 Hz with fault given after 4000 samples.

From Fig. 15, it can be observed that islanding condition is detected after 4000 samples accompanied by sudden change in the system frequency.

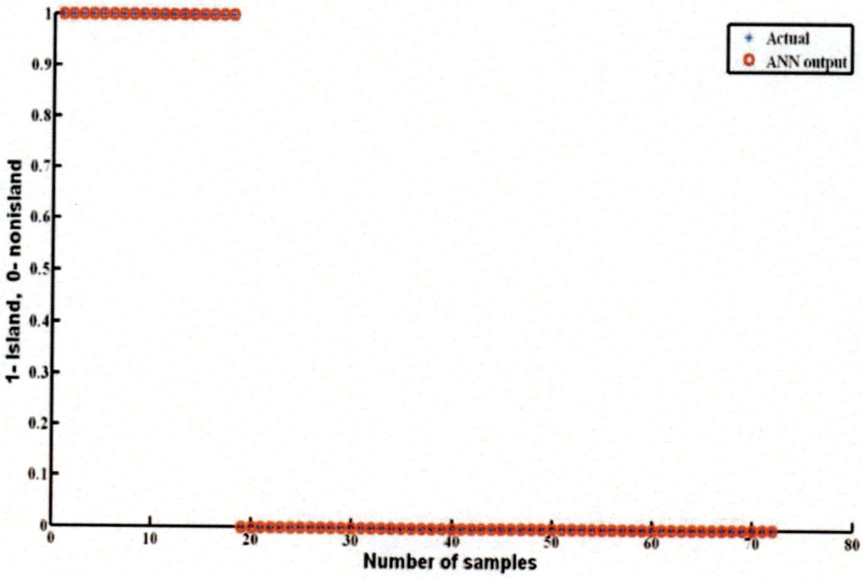

Fig. 13 Islanding detection by wavelet transform using ANN

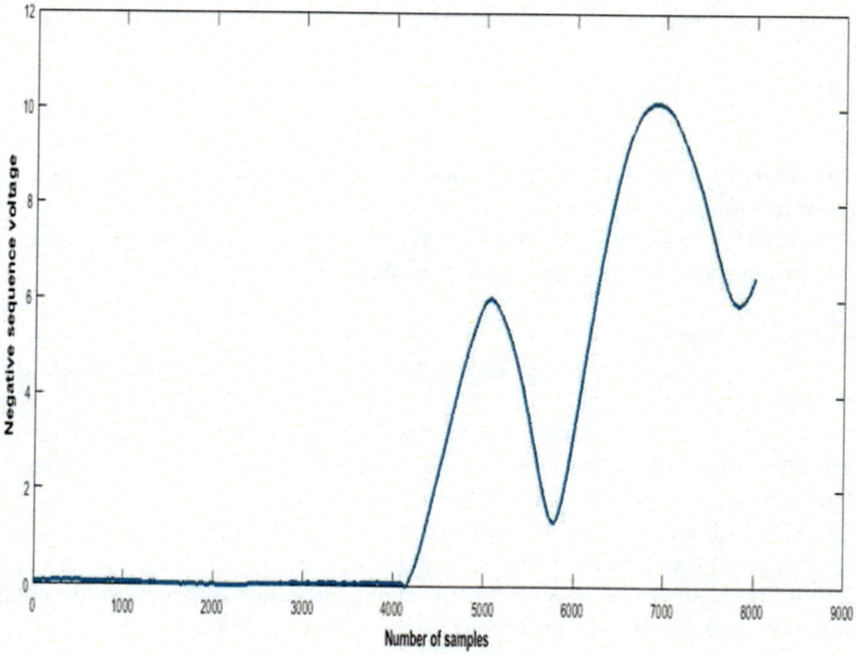

Fig. 14 Figure showing two cycles, one before islanding and one after islanding

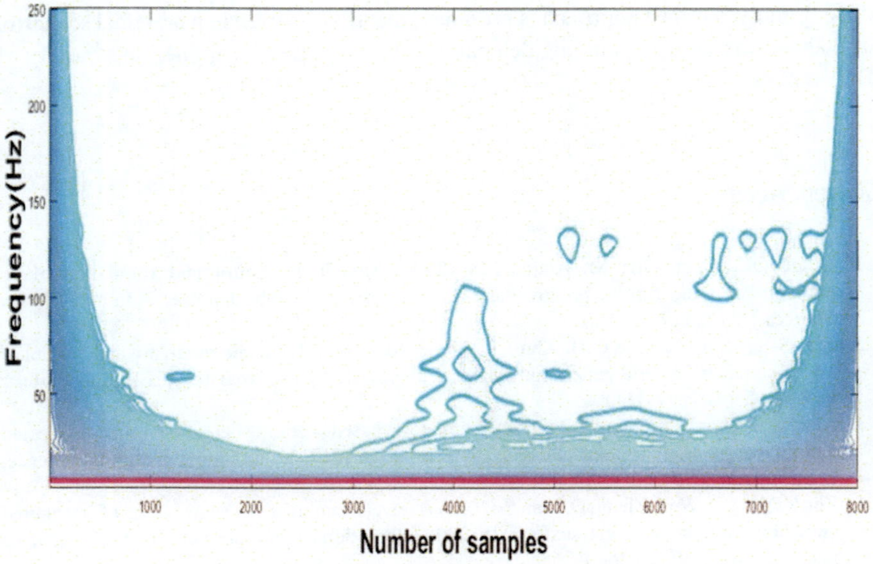

Fig. 15 Detection of islanding using S-transform

5 Conclusions

Two ANN-based islanding detection techniques are discussed in this paper, one using the wavelet transform and the other using the S-transform, which are described based on the literature review. Negative sequence voltage obtained at the target DG site is used in the suggested methods. Considering different cases, the negative sequence voltage signal is decomposed using the wavelet transform, then features are extracted, trained and test data set is created, then the ANN is trained using the selection feature as input, and the trained neural network is tested. Another method that is proposed for islanding detection is to use the CUSUM of the energy content of the voltage signal's negative sequence component. The simulation results show that both techniques detect islanding conditions with 100 percent accuracy and clearly differentiate them from non-islanding conditions. The S-transform technique, on the other hand, provides higher resolution and faster identification of islanding than the wavelet transform approach.

To summarize, this paper focuses on the use of artificial intelligence in conjunction with signal processing to identify islanding. The wavelet transform technique and the S-transform approach have both been studied. These two methods accurately detect the islanding state. We achieve greater resolution with S-transform methods to islanding detection, and simulation time is significantly shorter. The S-transform with ANN takes 1–3 s to simulate, while the wavelet transform takes 30–45 s. Raw data is not used in the approach we have covered; instead, pre-35 data processing is done. Finally, a robust approach is presented that may be used in a practical scenario. The above-mentioned islanding detection techniques may be applied to a variety of DG sources.

References

1. Onara OC, Uzunoglua M, Alama MS (2008) Modeling, control and simulation of an autonomous wind turbine/photovoltaic cell/ultra-capacitor hybrid power system. J Power Sources 185(2):1273–1283
2. Kim S-K, Jeon J-H, Cho C-H, Ahn J-B, Kwon S-H (2008) Dynamic modeling and control of a grid-connected hybrid generation system with versatile power transfer. IEEE Trans Industr Electron 55(4):1677–1688
3. Ray PK, Kishor N, Mohanty SR (2010) S-transform based islanding detection in grid-connected distributed generation-based power system. In: 2010 IEEE International energy conference. IEEE, pp 612–617
4. Zhu Y, Yang Q, Wu J, Zheng D, Tian Y (2008) A novel islanding detection method of distributed generator based on wavelet transform. In: 2008 International conference on electrical machines and systems. IEEE, pp 2686–2688
5. Jang S-I, Kim K-H (2004) An islanding detection method for distributed generations using voltage unbalance and total harmonic distortion of current. IEEE Trans Power Delivery 19(2):745–752
6. Redfern MA, Usta O, Fielding G (1993) Protection against loss of utility grid supply for a dispersed storage and generation unit. IEEE Trans Power Delivery 8(3):948–954
7. Refern MA, Usta O, Fielding G (1993) Protection against loss of utility grid supply for a dispersed storage and generation unit. IEEE Trans Power Delivery 8(3):948–954
8. Pai F, Huang S (2001) A detection algorithm for islanding prevention of dispersed consumer-owned storage and generating units. IEEE Trans Energy Convers 16(4):346–351
9. Kim JE, Hwang JS (2000) Islanding detection method of distributed generation units connected to power distribution system. In: PowerCon 2000. 2000 International conference on power system technology, vol 2. IEEE, pp 643–647
10. Gaouda AM, Salama MMA, Sultan MR et al (2000) Application f multiresolution signal decomposition for monitoring short-duration variations in distribution systems. IEEE Trans Power Delivery 15(2):478–485
11. Prabhakar S, Mohanty AR, Sekhar AS (2002) Application of discrete wavelet transform for detection of ball bearing race faults. Tribol Int 35(12):793–800
12. Serrano EP, Fabio MA (1996) Application of the wavelet transform to acoustic emission signals processing. IEEE Trans Signal Process 44(5):1270–1275
13. Smith GA, Onions PA, Infield DG (2000) Predicting islanding operation of grid connected PV inverters. IEEE Proc Electr Power Appl 147:1–6
14. Menon V, Nehrir MH (2007) A hybrid islanding detection technique using voltage unbalance and frequency set point. IEEE Trans Power Syst 22(1):442–448
15. Ray P, Panigrahi BK, Senroy N (2011) An AI approach for fault distance estimation in series compensated transmission line. In: 2011 International conference on energy, automation and signal. IEEE, pp 1–6
16. Jain AK, Duin RPW, Mao J (2000) Statistical pattern recognition: a review. IEEE Trans Pattern Anal Mach Intell 22(1):4–37
17. Santoso S, Powers EJ, Grady WM (1997) Power quality disturbance data compression using wavelet transform methods. IEEE Trans Power Delivery 12(3):1250–1257
18. Wilamowski BM, Yu H (2010) Improved computation for Levenberg–Marquardt training. IEEE Trans Neural Netw 21(6):930–937

19. Ephraim Y, Malah D (1985) Speech enhancement using a minimum-mean square error short-time spectral amplitude estimator. IEEE Trans Acoust Speech Signal Process 32(6):1109–1121
20. Jang S-I, Kim K-H (2000) An islanding detection method for distributed generations using voltage unbalance and total harmonic distortion of current. IEEE Trans Power Delivery 19(2):745–752

Particle Swarm Optimization Technique for Current Equalization of PV Systems to Achieve Higher GMPP Under PSCs

Rupendra Kumar Pachauri, Hanuman Prasad, Pankaj Kumar Gupta, and Manish Sharma

Abstract The sun irradiations levels are always vary at different times, which affect the performance of solar photovoltaic (PV) systems in terms of power losses (PL). But many natural and un-natural causes exist and behave an obstruction to extract best performance from PV systems. The popular unacceptable causes are partial shading conditions (PSCs), which affects the PV performance highly. To diminish the effect of PSCs, relevant PV module rearrangement method is another good solution. In this paper, particle swarm optimization (PSO) technique demonstrates the current equalization of PV array systems under PSCs. In that method, a particle reaches the optimum value depending on its present velocity, past experience, and the experience of its nearby group members. Also, traditional arrangement techniques such as the series–parallel (SP) and total cross-tied (TCT) configurations have been used. As a result, performance parameters such as power and voltage at global maximum power point (GMPP), PL, and fill factor (FF) have been assessed. This paper displays the methods of implementation of the techniques to avert the effect of PSCs and to improve the energy output of the PV array under non-uniform irradiation circumstances. MATLAB/Simulink modelling is carried out to compare the results of various reconfiguration techniques and to find out which one is the best one under considered PSCs.

Keywords Solar energy · Shading effect · Power loss · Renewable energy · Global maximum power

---

R. K. Pachauri (✉)
Electrical and Electronics Engineering Department, School of Engineering, University of Petroleum and Energy, Dehradun 248007, India
e-mail: rpachauri@ddn.upes.ac.in

H. Prasad
Electrical Engineering Department, Model Institute of Technology, Jammu 181123, India

P. K. Gupta · M. Sharma
School of Computer Science and Application, IIMT University, Meerut 250001, India
e-mail: manishsharma@iimtindia.net

K. Namrata et al. (eds.), *Smart Energy and Advancement in Power Technologies*, Lecture Notes in Electrical Engineering 926,
https://doi.org/10.1007/978-981-19-4971-5_35

1 Introduction

Renewable energy is a type of energy that can be harnessed in a variety of ways, including wind, water, solar, geothermal, and so on. Solar energy is abundantly available on our planet, the Earth, and it will continue to be available for thousands of years to come [1]. So, the rising power demands of the coming generations can also be met through the energy from our sun. Solar power is the type of power source which sustains life in all forms (humans, plants, animals, etc.) in this world. Solar energy is the radiation of the sun that is available everywhere on our planet. The advantage of solar power is that solar power is a clear and eco-friendly power source with no environmental pollutants. The initial installation cost of solar PV systems is very high. But the maintenance cost of the PV arrays is low. So, solar power is an economical source of energy for humans. The limitation of solar energy is that it cannot be stored. It is also not available at night [2]. A mismatch in the electrical curves of the PV panels in a module is one of the main causes that affects the power output of an solar photovoltaic system (SPV). The main causes of the mismatch are production methods utilized, the ageing of PV arrays, and PSCs. Improper maintenance may cause the collection of dirt on the PV modules, which leads to non-uniform acceptance of radiation.

PSCs occur on PV modules at different times of the day due to varying irradiation levels of the sun. The main causes of partial shadowing are falling dry leaves, tree shadows on the PV arrays, nearby phone towers, etc. Multiple peaks in the power-voltage (P–V) curves of solar PV systems are caused by PSCs. Therefore, various rearrangement techniques of the PV arrays, such as the PSO-TCT configuration, SP and TCT configurations, are considered for extensive study. In this way, a greater GMPP is accomplished, and many peaks do not occur in the PV system, and thus, the power yield of the PV arrays is improved compared to before [3, 4].

1.1 Literature Review

In this paper, quality research manuscripts are considered to extract the novelty work. Various techniques such as hybrid reconfiguration and puzzle-based reconfiguration have implemented to do this. In electric reconfiguration, the connection links between arrays are changing to enhance power generation w.r.t. PSCs. It requires switches, sensors, measuring unit, data logger, etc., to do this [5]. In [6], the authors have Magic Square method on 9×9 PV array under long wide shading. It gives 25.03% and 6.38% power enhancement as compare to TCT and Su-do-Ku, respectively. In [7], for instance, a new method for reconfiguring interconnected TCT PV modules based on the Su Do Ku puzzle scheme to distribute the effect of shading over the entire PV array has been proposed. In [8], a project that involved the reconfiguration of 81 photovoltaic modules assembled in 9×9 modules has been realized, with

TCT interconnection, using MATLAB/Simulink. In order to find an optimal reconfiguration for each shadow, the genetic algorithm (GA) has been used to obtain the maximum output power. This study has revealed that the GA provided 34.96% more power than a fixed TCT configuration. Authors in [9] have presented a PV array of 5 × 5 modules interconnected in TCT, in which a dominance square (DS) method for the physical relocation of PV modules has been introduced. The obtained results have indicated that the use of the DS approach may increase the maximum output power compared to other conventional methods. To investigate the execution during 15 distinct random shading profiles, SP, TCT, and BL array configurations of 2 × 4, 2 × 6, 3 × 3, 3 × 4, 4 × 2, 4 × 3, 4 × 4, 6 × 2, and 6 × 4 sizes were considered [10]. For each shade pattern, the electrical performance of various combinations was examined. The series, parallel, SP, and TCT PV array interconnections of three sizes were explored for performance evaluation under various shading conditions [11, 12].

1.2 Novelty of Work

With the motivation available literature, the authors have tried to find the novelty. Following salient points of study are given below and shown novelty of work done in the present paper extensively.

- Conventional SP, TCT configurations are reported in this paper and compared the performance with PSO-based optimal placements of PV modules in array under PSCs.
- PSO-based placement is found best performance under such considered shading environment.
- Performance evaluation is compared in terms of location of GMPP, shading factor analysis, improved FF and minimized PL, etc.

2 PV System Technology and Shading Scenarios

2.1 PV System Modelling

The voltage of the PV modules is expressed in Eq. (1) and electrical equivalent diagram is depicted in Fig. 1 as follows,

$$V_{\text{Cell}} = \frac{AkT_{\text{C}}}{e}\ln\left(\frac{I_{\text{ph}} + I_{\text{DS}} - I_{\text{m}}}{I_{\text{DS}}}\right) - I_{\text{m}}\left(\frac{R_{\text{s}}R_{\text{sh}}}{R_{\text{s}} + R_{\text{sh}}}\right) \quad (1)$$

Where cell voltage stands for V_{cell}, A: ideality factor, k: Boltzmann's constant, I_{ph}, I_{m}, and I_{DS}: photo-current, module, and saturation currents. In addition, series and shunt resistances are represented as R_{S} and R_{sh}, respectively.

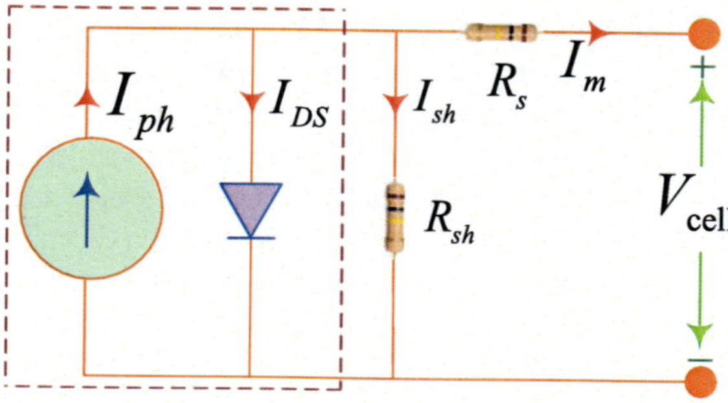

Fig. 1 Electrical circuit of a PV cell [13, 14]

Table 1 PV module specifications at STCs

Index	Specifications
S. C. current (I_{SC})	0.55A
O. C. circuit current (V_{OC})	11.25 V
Max. current (I_m)	0.52A
Max. voltage (V_m)	9.62 V
Max power (P_m)	5 W

For designing PV array systems such as SP, TCT, and PSO-based configurations, a 5 W commercial available PV module (SFTI18P5: Manf. Solar Universe India) is chosen as a single unit in an array. For MATLAB/Simulink modelling of PV system, the required specifications at standard test conditions (STCs) are shown in Table 1 as follows.

2.2 PV Array Systems and Shading Scenarios

(a) ***SP configuration***: For obtaining such a huge power, we should link the required quantity of modules in series and parallel. When some PV arrays are linked in series, the voltage of the solar PV system rises. For improving the current, some PV arrays are linked in parallel. As the series and parallel linking of the PV array rises, the resultant energy of the arrays also gets enhanced. Figure 2 shows the SP structure of PV array as follows,

(b) ***TCT configuration***: In total cross-tied arrangement, the voltage of each individual node and total of currents in distinct junctions are alike. As compared to SP, in TCT, mismatch losses because of partial shading decrease considerably. TCT configuration is shown in Fig. 3 as follows,

(c) ***PSO-TCT configuration***: The PSO technique is dependent on the behaviour of particles in a group or swarm. In this method, a particle reaches the optimum value depending on its present velocity, past experience, and the experience

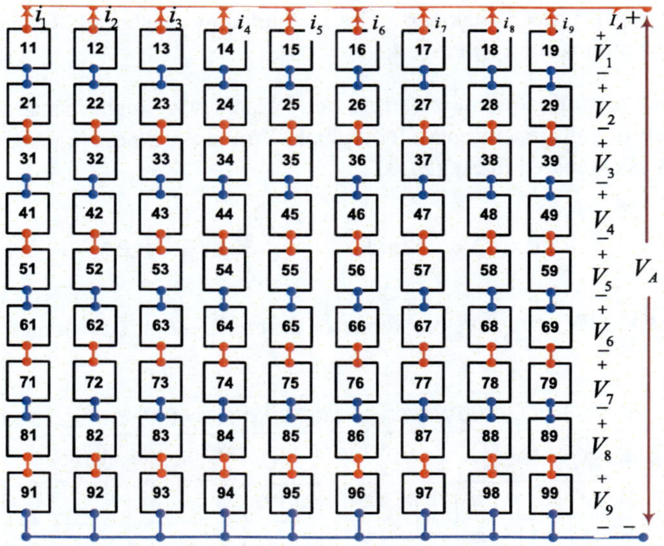

Fig. 2 SP connection of modules in PV array

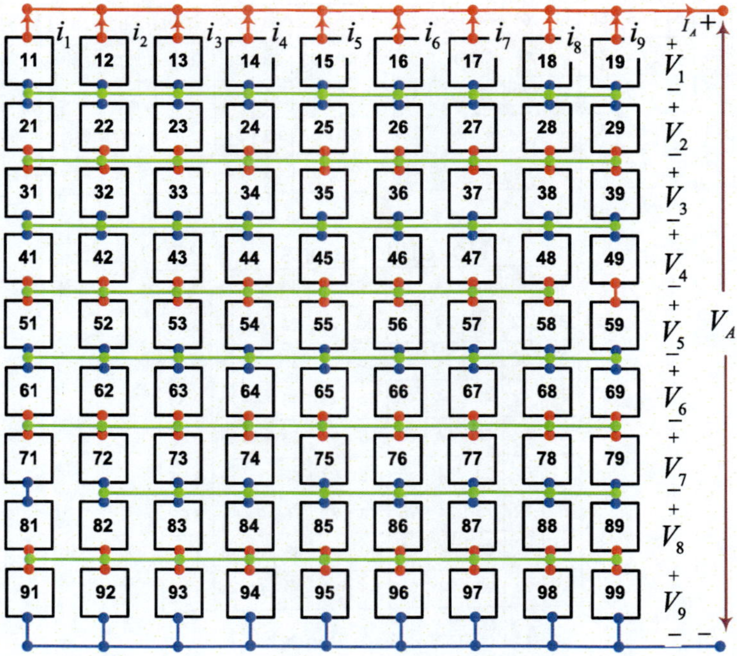

Fig. 3 TCT connection of modules in PV array

of its nearby group members. The methodology to show PSO-based current equalization is shown in Fig. 4.

PV array re-configuration is carried out for achieving higher shade dispersion. On the basis of voltage, current of individual PV module, fitness function is used for evaluation as expressed in Eq. (2).

$$\max(f) = \mathrm{Sum}(P) + \left(\frac{W_e}{E_e}\right) + (W_f + P_a) \tag{2}$$

For 9 × 9 sized PV arrray system, the error difference (E_e) is calculated as,

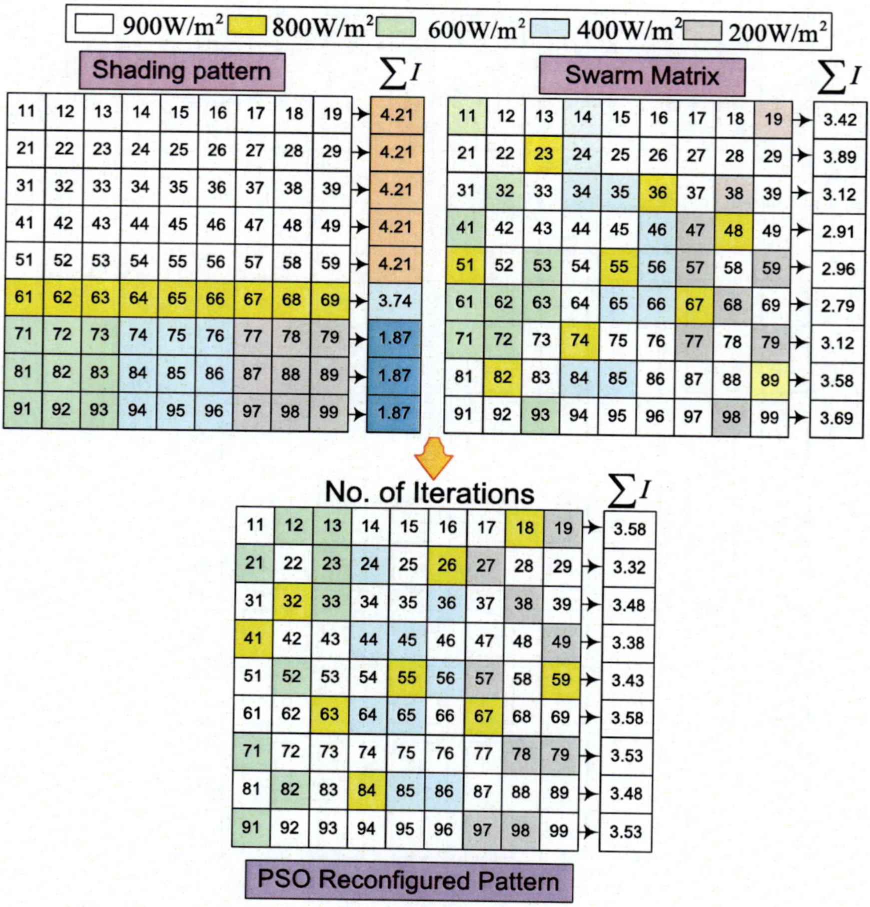

Fig. 4 Shade dispersion techniques using PSO method [15]

$$E_e = \sum_{k=1}^{9} |I_{max} - I_{act}| \tag{3}$$

where I_{max} stands for maximum value, and I_{act} actual values of generated current.

2.3 Shading Scenarios

In this paper, two realistic conditions-based shading scenarios are considered for PV system performance evaluation. In the first shading pattern, distinguish sun irradiance levels are reported such as 900, 800, 600, 400, and 200 W/m^2 for extensive study. In addition, the second shading pattern consists of almost similar irradiation levels such as 900, 800, 700, 400, and 300 W/m^2 for extensive study. Both the shading patterns are shown in Fig. 5a, b as follows.

3 Performance Assessment Parameters

3.1 Power at GMPP

Because of the non-uniform shading conditions, multiple power maxima are observed on P–V curves. The closest/minimum and the highest values of power maxima are referred to as LMPP and GMPP, respectively.

3.2 Power Loss

The difference between MPP at uniform irradiance and MPP at global point is responsive to power loss and is expressed in Eq. (4) as follows,

$$\text{PL} = \text{MPP}_{\text{at uniform irradiation}} - \text{GMPP}_{\text{at PSCs}} \tag{4}$$

3.3 Fill Factor

In first impressions, FF is the most important performance index parameter for defining efficiency and understanding PV performance.

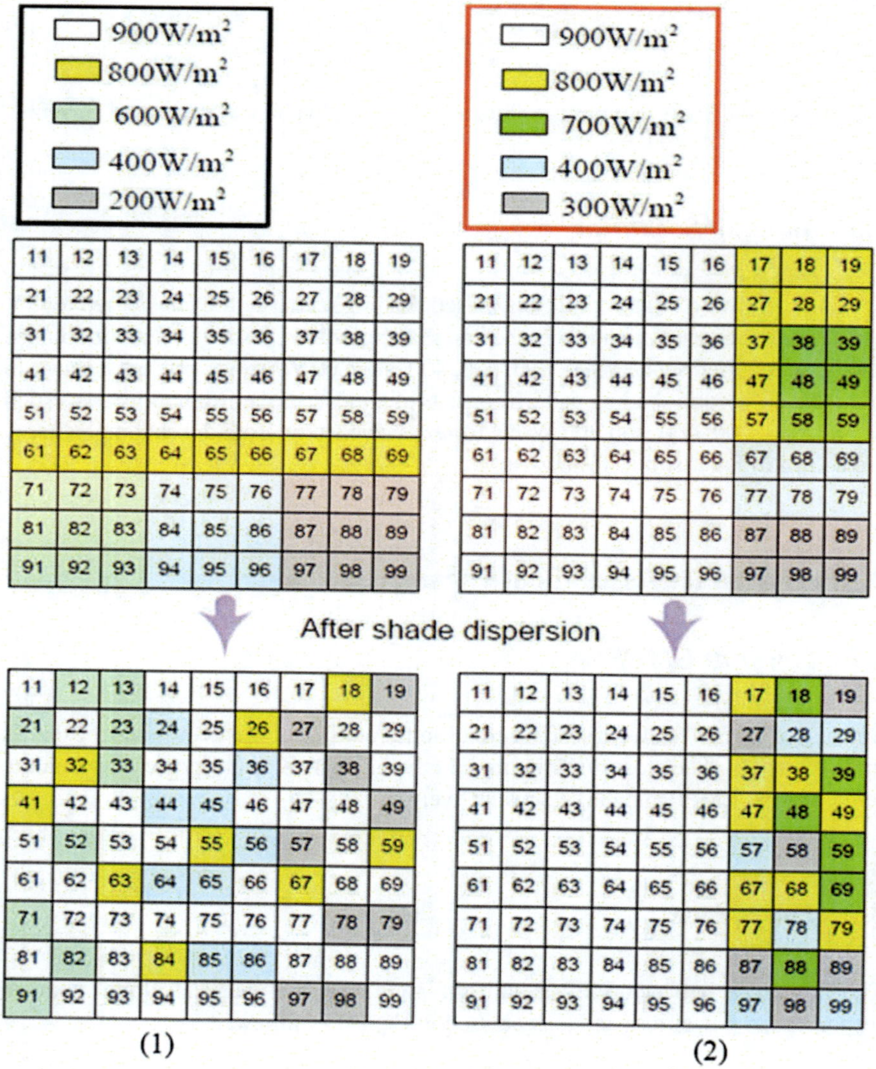

Fig. 5 Shading patterns: (1)–(2)

$$\mathrm{FF} = \frac{V_m \times I_m}{V_{OC} \times I_{SC}} \tag{5}$$

Under shading patterns 1–2, theoretical analysis of current calculation can be performed using SP, TCT, and PSO-based PV array configurations. Equation (6)–(12) are used to calculate the current produced as follows.

Current produced for case 1

$$I_{r1} = I_{r2} = I_{r3} = I_{r4} = I_{r5} = \left(\frac{900}{1000}\right)I_m + \left(\frac{900}{1000}\right)I_m \cdots + \left(\frac{900}{1000}\right)I_m = 8.1I_m \quad (6)$$

$$I_{r6} = \left(\frac{800}{1000}\right)I_m + \left(\frac{800}{1000}\right)I_m \cdots + \left(\frac{800}{1000}\right)I_m = 7.2I_m \quad (7)$$

$$I_{r7} = I_{r8} = I_{r9} = 6 \times \left(\frac{400}{1000}\right)I_m + \left(\frac{200}{1000}\right)I_m + \left(\frac{200}{1000}\right)I_m + \left(\frac{200}{1000}\right)I_m = 3.0I_m \quad (8)$$

Current produced for case 2

$$I_{r1} = I_{r2} = 6 \times \left(\frac{900}{1000}\right)I_m + \left(\frac{800}{1000}\right)I_m + \left(\frac{800}{1000}\right)I_m + \left(\frac{800}{1000}\right)I_m = 7.8I_m \quad (9)$$

$$I_{r3} = I_{r4} = I_{r5} = 6 \times \left(\frac{900}{1000}\right)I_m + \left(\frac{800}{1000}\right)I_m + \left(\frac{700}{1000}\right)I_m + \left(\frac{700}{1000}\right)I_m = 7.6I_m \quad (10)$$

$$I_{r6} = I_{r7} = 6 \times \left(\frac{900}{1000}\right)I_m + \left(\frac{400}{1000}\right)I_m + \left(\frac{400}{1000}\right)I_m + \left(\frac{400}{1000}\right)I_m = 6.6I_m \quad (11)$$

$$I_{r8} = I_{r9} = 6 \times \left(\frac{900}{1000}\right)I_m + \left(\frac{300}{1000}\right)I_m + \left(\frac{300}{1000}\right)I_m + \left(\frac{300}{1000}\right)I_m = 6.3I_m \quad (12)$$

4 Results and Discussion

4.1 P–V and I–V Curves Under Ideal Shading Scenarios

For assessment of shading impact on PV performance, 9×9 size PV system characterization is carried out, and obtained P–V and I-V curves are shown in Fig. 6. Power at GMPP is found as 405 W in ideal conditions, found through P–V curve.

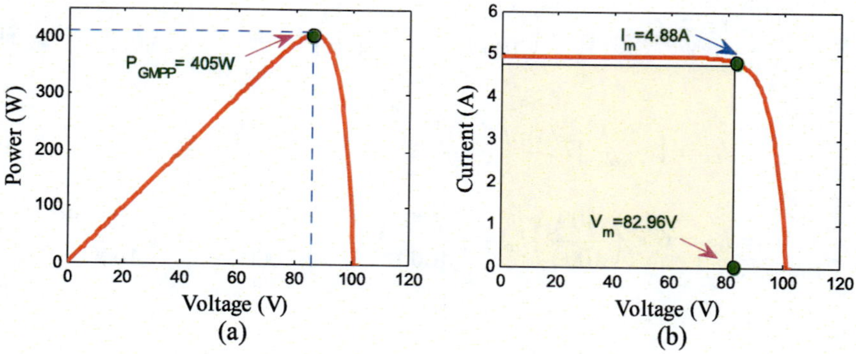

Fig. 6 P–V and I–V curves at ideal irradiations (1000 W/m^2)

4.2 P–V and I–V Curves Under Shading Patterns: 1–2

During PSCs, an extensive study is carried out and found multiple MPP on P–V curve, i.e. GMPP and LMPP. Under both shading patterns 1–2, the GMPP locations for PSO-based PV array configuration are found higher as 317.4 W and 335.9 W compared to SP (220.3 W and 284.4 W) and TCT (227.4 W and 298.8 W) configurations, respectively. The behaviour of P–V and I–V curves is shown in Figs. 7 and 8 as follows.

Quantitative observations are carried pout during investigation under shading scenarios. All the performance indices are reported in Table 2 as follows.

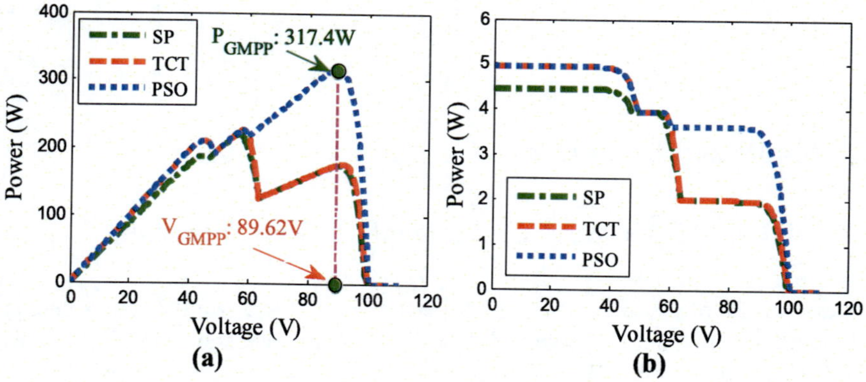

Fig. 7 P–V and I–V curves under shading pattern 1

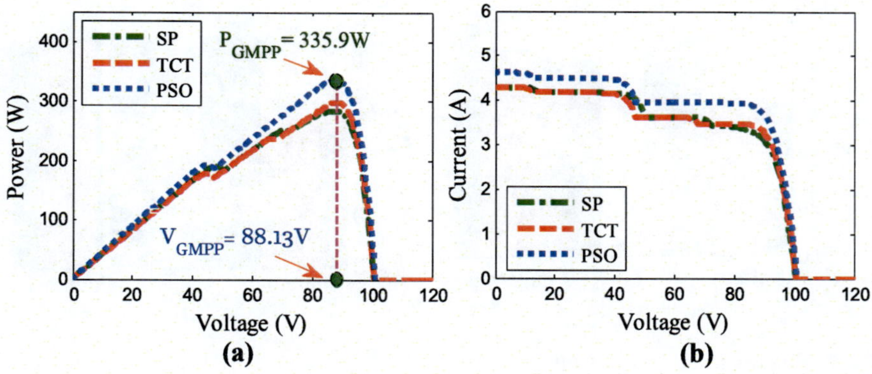

Fig. 8 P–V and I–V curves under shading pattern 2

Table 2 Quantitative performance indices of PV systems under shading scenarios

Performance parameters	Shading pattern 1			Shading pattern 2		
	SP	TCT	PSO	SP	TCT	PSO
$V_{OC}(V)$	100.05	100.35	100.5	100.6	100.6	101.3
$I_{SC}(A)$	4.45	4.95	4.95	4.28	4.28	4.61
$V_m(V)$	56.85	58.6	89.62	87.63	88.52	88.13
$I_m(A)$	3.87	3.88	3.54	3.24	3.37	3.81
$P_{GMPP}(W)$	220.3	227.4	317.4	284.4	298.8	335.9
FF	0.494	0.457	0.638	0.660	0.693	0.719
PL(W)	184.7	177.4	87.6	120.6	106.2	69.1
PG(W)	–	3.12%	30.5%	–	4.81%	11.04%

4.3 *Power and Voltage at GMPP*

Bar chart representation is required to show the effective comparison of PV system under shading scenarios. Performance parameters such as power and voltage at GMPP are observed and compared. Under both shading patterns 1–2, PSO-based configured PV system shows higher power output due to good shade dispersion capability. Power and voltage at GMPP are shown in Fig. 9a, b as follows.

4.4 *FF and PL Analysis*

Assessment of FF is required to estimate the PV system efficiency in a first perception. In this context, FF is calculated under both shading scenarios and found higher for

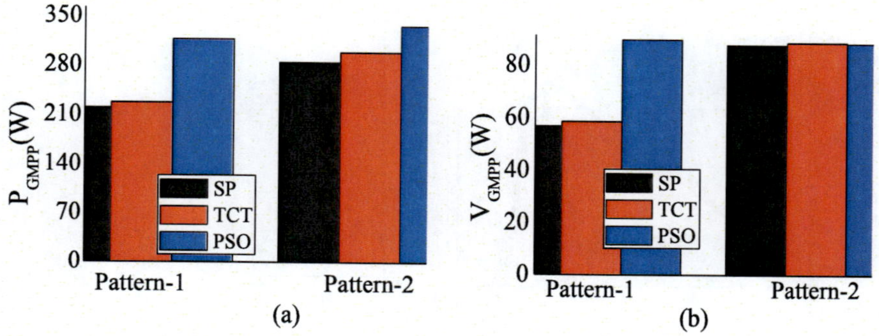

Fig. 9 **a** Power, **b** voltage at GMPP

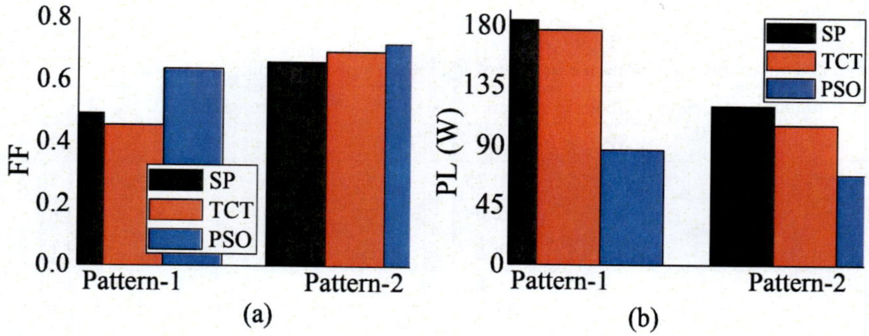

Fig. 10 **a** FF, **b** PL analysis

PSO-based PV array configuration as 0.638 and 0.719 compared to SP (0.494 and 0.660) and TCT (0.459 and 0.693) configurations.

In addition, minimized PL is found also for PSO-based configured PV systems as 87.6 and 69.1 W compared to conventional PV array configurations. Graphically representation is given in Fig. 10a, b as follows.

4.5 Power Gain

PSO technique is used to enhance shade dispersion capability, which is responsible to enhance power out of PV system. In present study, PSO and TCT configured PV systems are having higher PG as 30.5%, 11.04%, and 3.12%, 4.81%, respectively, with respect to SP configuration (Fig. 11).

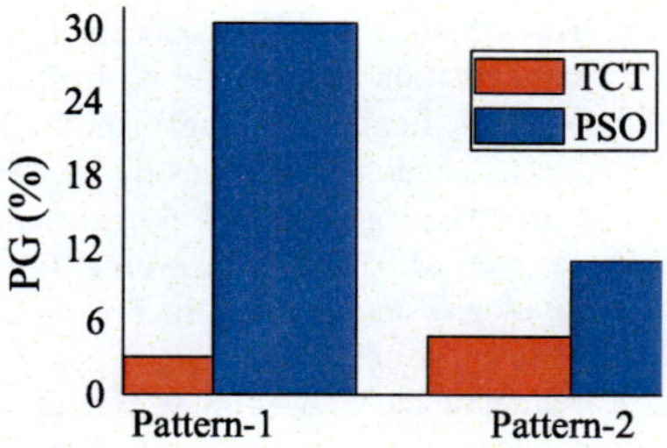

Fig. 11 PG analysis

5 Conclusion

In present study, 9 × 9 size SP, TCT, and PSO techniques-based configurations have been analyse. Moreover, all the obtained results in terms of power, voltage at GMPP, FF, PL, and PG are analysed during shading scenarios 1 and 2. Important salient points of present study are as follows:

- Under shading conditions: 1–2, the highest power at GMPP is found as 317.4 W and 335.9 W, respectively, related to SP and TCT configurations.
- Moreover, in PSO-based configuration, the PL is observed quite less viz. 87.6 and 69.1 W compared to SP (184.7 and 120.6 W), TCT (177.4 and 106.2 W) configurations.
- Furthermore, the improved FF is observed as 0.638 and 0.719 for PSO-based PV configuration compared to conventional configurations.

Overall, present study proved that the higher shade dispersion factor is helped to enhance the performance of PV array configurations. In addition, study can be adopted to develop at the commercial levels.

References

1. Bhadoria VS, Pachauri RK, Tiwari S, Jaiswal SP, Alhelou HH (2020) Investigation of different BPD placement topologies for shaded modules in a series-parallel configured PV array. IEEE Access 8:216911–216921
2. Tao Y, Bai J, Pachauri RK, Sharma A (2020) Parameter extraction of photovoltaic modules using a heuristic iterative algorithm. Energy Convers Manage 224:1–18
3. Bai J, Sun L, Pachauri RK, Wang G (2021) Investigate on photovoltaic array modeling and the MPPT control method under partial shading conditions. Int J Photo Energy 2021:1–16
4. Gupta A, Chauhan YK, Pachauri RK (2016) A comparative investigation of maximum power point tracking methods for solar PV system. Sol Energy 136:236–253
5. Pachauri RK, Alhelou HH, Bai J, Golshan MEH (2021) Adaptive switch matrix for PV module connections to avoid permanent cross-tied link in PV array system under non-uniform irradiations. IEEE Access 9:45978–45992
6. El Iysaouy L, Lahbabi M, Oumnad A (2019) A novel magic square view topology of a pv system under partial shading condition. Energy Procedia 157:1182–1190

7. Rao PS, Ilango GS, Nagamani C (2014) Maximum power from PV arrays using a fixed configuration under different shading conditions. IEEE J Photovoltaics 4(2):679–686
8. Rani BI, Ilango GS, Nagamani C (2013) Enhanced power generation from PV array under partial shading conditions by shade dispersion using Su-Do-Ku configuration. IEEE Trans Sustain Energy 4(3):594–601
9. Jazayeri M, Uysal S, Jazayeri K (2014) A comparative study on different photovoltaic array topologies under partial shading conditions. In: Proceedings of the IEEE PES T&D conference and exposition, 1–5
10. Ramaprabha R (2014) Selection of an optimum configuration of solar PV array under partial shaded condition using particle swarm optimization. Int J Electr Comput Energet Electron Commun Eng 8(1):89–96
11. Storey J, Wilson PR, Bagnall D (2014) The optimized-string dynamic photovoltaic array. IEEE Trans Power Electron 29(4):1768–1776
12. Vijayalekshmy S, Iyer SR, Beevi B (2014) Comparative analysis on the performance of a short string of series-connected and parallel-connected photovoltaic array under partial shading. J Inst Eng India Ser B 95(3):1–10
13. Gupta A, Kumar P, Pachauri RK, Chauhan YK (2014) Effect of environmental conditions on single and double diode PV system: a comparative study. Int J Renew Energy Res 4(4):849–858
14. Singh A, Pachauri RK, Chauhan YK (2016) Comprehensive investigation of PV arrays with puzzle shade dispersion for improved performance. Sol Energy 129:256–285
15. Babu TS, Ram P, Dragicevic T, Miyatake M, Blaabjerg F, Rajasekar N (2018) Particle swarm optimization based solar PV array reconfiguration of the maximum power extraction under partial shading conditions. IEEE Trans Sustain Energy 9(1):74–85

A Review on Photovoltaic Cells

Trushna Prajapati and Abhishek Priyam

Abstract The extraordinary advancement in photovoltaic (PV) technologies over the last 5 years requires a renewed evaluation of their performance and prospective progress in the future. We analyze here comparing PV cell properties across technologies, we evaluate the extent to which any technology can move in the near future. Although accurate or revolutionary advancements cannot be foreseen, cross-fertilization happens between technologies and results in one cell type indicate evolutionary developments in other cells. This transfer of knowledge is crucial since the development of metal halide perovskites helps to connect the hitherto separate technological strands of photovoltaic research.

Keywords Photovoltaic cells · PV technology · Review · Classification of solar cells

1 Introduction

Due to the current scenario-depletion of natural resources for energy generation-solar energy systems have evolved immensely. Particularly, solar photovoltaic systems have seen a rise in usage of 22%. They produce silent and clean energy. The only source they used for intake of energy is sunlight [1].

Photovoltaic cells are compact, thus, can be installed easily in an area where sunlight is in abundance. They can easily be installed on the unoccupied space of roof tops. Apart from cost and irregularity in availability of sunlight one of the major disadvantages include the release of harmful chemicals like cadmium and arsenic. However, their concentrations can be controlled through monitored disposal and their effect on environment is minimal. Considering the bloom in sustainable energy requirements in huge quantities there are many researches that are going on to use solar energy to its full potential.

T. Prajapati (✉) · A. Priyam
Department of Mechanical Engineering, MPSTME, NMIMS University, Mumbai 400056, India
e-mail: Trushnajprajapati@gmail.com

K. Namrata et al. (eds.), *Smart Energy and Advancement in Power Technologies*, Lecture Notes in Electrical Engineering 926,
https://doi.org/10.1007/978-981-19-4971-5_36

1.1 Advantages, Disadvantages and Working of Photovoltaic Cells

Photovoltaic cells have all static parts; therefore electrical energy is formed by Solar Energy. PV systems are reliable, modular and durable and thus the need for regular maintenance is not required. System is compatible with almost all environments. The distribution expenses of energy have been reduced since the equipment can be easily set up at the site of use [2]. Whereas the expenses of installation of PV system are extremely high currently. Efficiency of solar cells is comparatively low. As a result, a greater number of PV cells are to be installed to generate useful power. Storage of solar energy as electrical energy makes the PV system even more expensive [3].

A photovoltaic cell consists semiconductor materials that absorb photons. The semiconductor when struck by photons give out one electron leaving behind a vacant space more commonly referred to as holes. There are two variants of silicon–one is doped with phosphorus and other with boron. As a result of this the part of silicon that is doped as phosphorus atoms becomes electron rich and boron doped silicon becomes electron deficient. The electrons move along the concentration gradient. This works as a battery wherein phosphorus doped silicon is negative terminal and the other part is positive terminal. The moving electrons and holes thus produce electric current. A solar cell consists of a p-type layer of silicone next to an n-type silicon layer (Fig. 1). The n-type layer contains an overabundance of electrons and surplus positive holes exist in the p-type layer (which are vacancies due to the lack of valence electrons). The electrons on either side (n-type layer) migrate to the holes on the other side of the junction near the junction of the two layers (p-type layer). This generates an area around the intersection, which is known as the depletion zone (Fig. 1, closeup). When the entire hole is filled with electrons in the depletion zone, the p-type of the depletion zone (where the hole was initially found) now contains negative ions and the n-type of the depletion zone, where there were electrons, now includes positively charged ions. The existence of these re-laded ions provides an internal electrical field that inhibits electrons from filling poles in the p-type layer in the n-type layer [4].

1.2 Solar Cell Characteristics

1.2.1 I–V Characteristics

This can be mathematically expressed as [5]:

$$I = I_0\left\{\exp\left(\frac{V}{V_T}\right) - 1\right\} \tag{1}$$

When the negative terminal of cell is lit, the characteristic gets changed as:

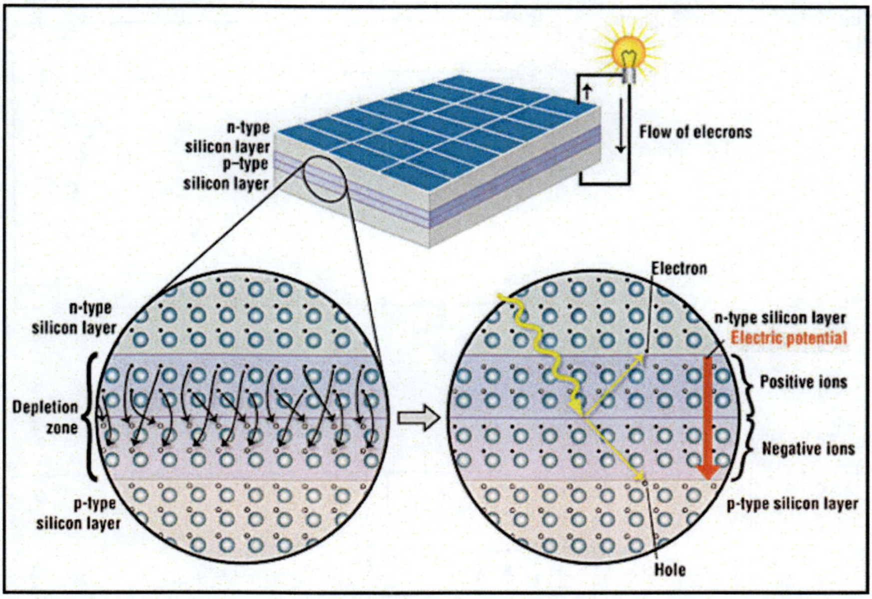

Fig. 1 Working of the photovoltaic cell, illustrating both layers of silicon—silicon doped with boron and silicon doped with phosphorus

$$I = -I_L + I_0\left\{\exp\left(\frac{V}{V_T}\right) - 1\right\} \tag{2}$$

Here I_{L} is known as light generated current

This can also be written as shown below [5]:

$$I = I_{\mathrm{L}} - I_0\left\{\exp\left(\frac{V}{V_{\mathrm{T}}}\right) - 1\right\} \tag{3}$$

Figure 2 shows three important points—short circuit, maximum power and open circuit [6].

1.2.2 Equivalent Circuit

Equivalent circuit of solar cell for practical use is shown in Fig. 3.

The characteristic of for practical cell can be modified as [3]:

$$I = I_{\mathrm{L}} - \left\{I_0\left[\exp\left\{\frac{(V + IR_s)}{V_{\mathrm{T}}}\right\} - 1\right] - \frac{(V + IR_{\mathrm{s}})}{R_{\mathrm{sh}}}\right\}$$

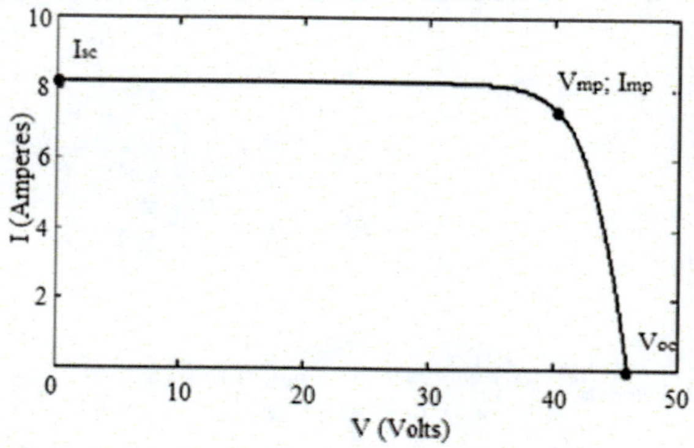

Fig. 2 I–V curve for solar panel

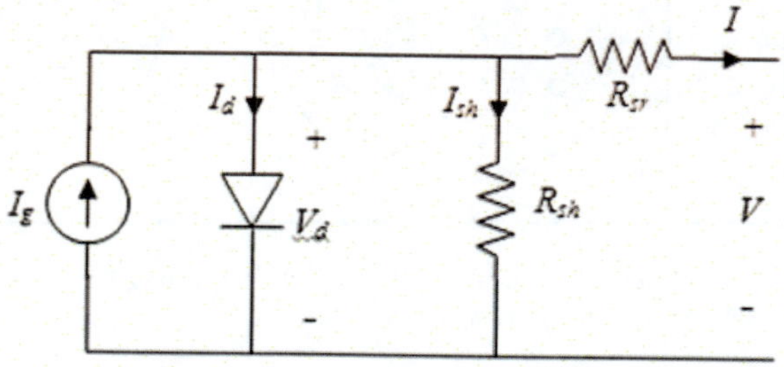

Fig. 3 Equivalent circuit of solar cell [3]

1.2.3 Solar Cell Design Parameters

1. Short Circuit Current: When for every photon each respective electron flows to provide current, the phenomenon is called maximum short circuit current. A material can produce more current when it is in lower band gap, since it can absorb more protons. Therefore, a decrease in energy band gap will increase solar cell current [7].
2. Open Circuit Voltage:

$$V_{\mathrm{oc}} = V_{\mathrm{T}} \ln\left\{\frac{I_{\mathrm{L}}}{I_0} + 1\right\}$$

   This expression shows that increase in band gap, I_0 decreases, thus V_{oc} decreases.
3. Fill Factor: Fill factor can be obtained using the below empirical expression [8]:

$$\mathrm{FF} = \frac{v_{\mathrm{oc}} - \ln(v_{\mathrm{oc}} + 0.72)}{v_{\mathrm{oc}} + 1}$$

4. Efficiency: For an ideal solar cell the maximum efficiency that can achieved is 31% in the optimal band gap of 1.45 eV.

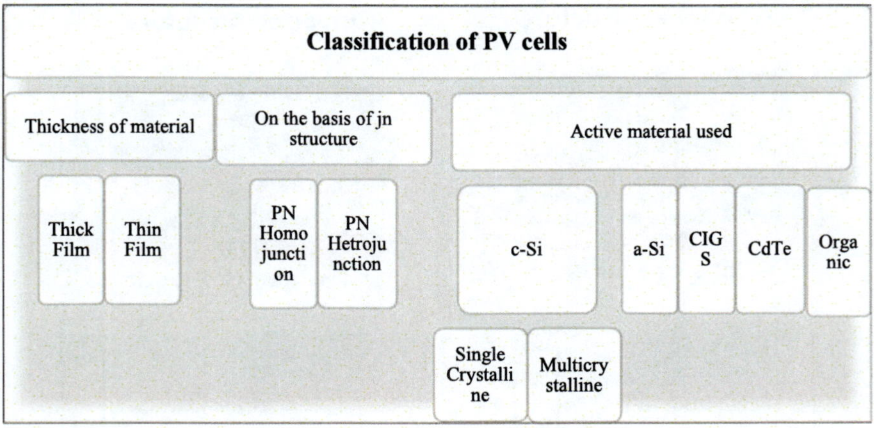

Fig. 4 The flow chart shows classification of PV cells [9]

1.3 Classification of PV Cells

A brief classification of various solar cells on the basis of material thickness, junction structure and the material used has been presented as a flowchart in the Fig. 4.

1.3.1 By Thickness of Material

Thick Film

A thick film solar cell has a layer of paste made from P_2O_5 and B_2O_5. However, due to high reactivity of P_2O_5 with the environment, this method is no longer used commercially. Almost all the cells manufactured today for daily activities are thin film cells. But these cells do provide higher fill factor as compared to thin film cells. A research shows that for a thick film cell of thickness 400 nm the efficiency and fill factor of the same is 11% and 40%, respectively [10].

Thin Film

Thin film cells are used for all commercial applications today. Here, the cells are made by either spraying or printing the photovoltaic material on a metal or a glass surface. This reduces the size of each cell but increases the power to size ratio of the cell. Hence, looking through the manufacturing aspect of the same, the cells are easier and cheaper to manufacture. The photovoltaic material used to manufacture these cells includes the following chemicals—Cadmium Telluride, Amorphous Silicon, Copper Indium diSelendide, Copper Indium Gallium diSelendide. These types of solar cells are also mentioned in the active material category [10].

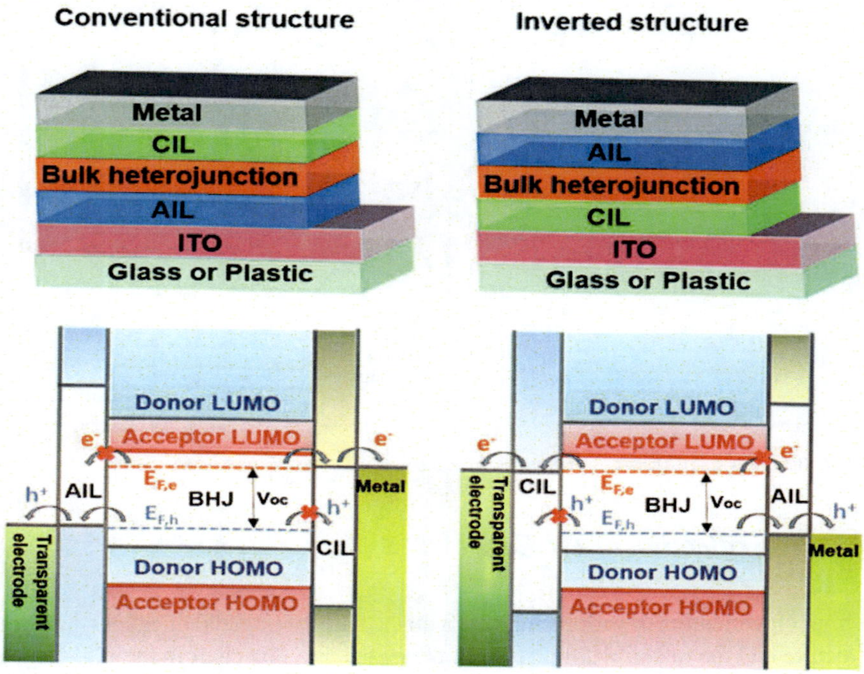

Fig. 5 Heterojunction organic solar cells [12]

1.3.2 By Junction Structure

PN Homojunction

A homojunction is formed when two photovoltaic materials have same band gap but different doping properties. When this phenomenon occurs between n-type and p-type semiconductor material such as silicon a PN Homojunction is formed [11].

PN Heterojunction

On the other hand, a heterojunction is formed between two completely different semiconductors. Here, the semiconductors have different band gaps and different doping properties as well. These materials have complicated manufacturing processes, hence are used for critical specific applications only. Figure 5 illustrates the heterojunction photovoltaic cells [11].

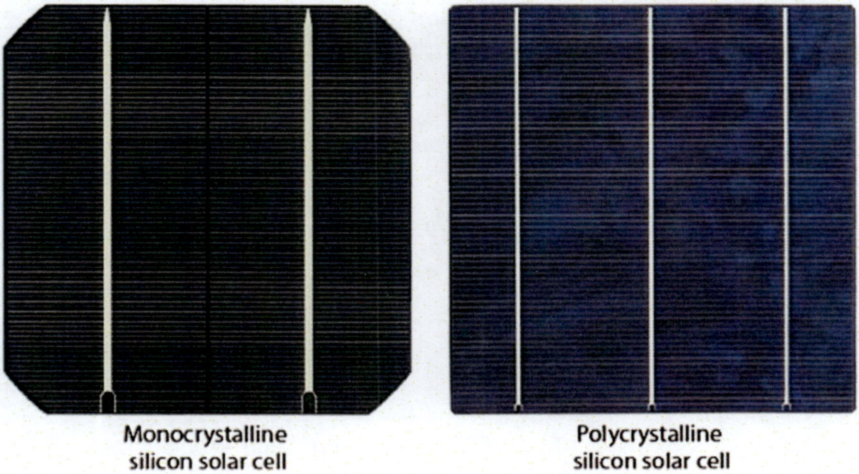

Fig. 6 Monocrystalline and polycrystalline cells [12]

1.3.3 By Active Material

Crystalline Silicon also Known as C-Si

Silicon crystals used for manufacture of photovoltaic cells are of the following types:

1. Single/Mono Crystalline silicon
2. Multi/Poly Crystalline silicon.

Single silicon cells give high efficiency up to 13–19%, however, they are difficult to manufacture. This is because the whole solar cell is to be manufactured from a single silicon crystal. Thus, the process of manufacturing is expensive. On the other hand, multi crystalline silicon cells are not as much efficient as single silicon cells. Multi silicon cells provide a cell efficiency of 9–14%. They are cheaper and easier to manufacture, but due to the decrease in efficiency in multi silicon cells, they are not used commercially. Monocrystalline and polycrystalline cells are represented in Fig. 6.

Amorphous Silicon also Known as A-Si

Amorphous silicon is a non-defined crystal structure unlike crystalline silicon. Thus, the interaction between photons and silicon atoms, this increases the light absorbing capacity of the cell. This allows the cell to be thin film with high absorption capacity. However, the major disadvantage of this type of cells is their efficiency. The highest efficiency achieved using amorphous silicon cells is 13% only. A comparison of the

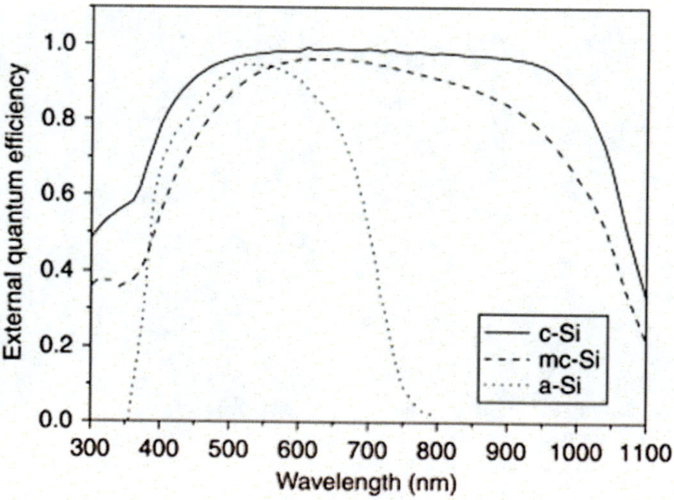

Fig. 7 Comparison of the quantum efficiencies of crystalline, multi crystal and amorphous silicon cells [13]

quantum efficiencies of crystalline, multi crystal and amorphous silicon cells are presented in Fig. 7.

Organic Solar Cells

In organic solar cell the absorbing layer is made of organic semiconductors. The organic semiconductors are placed between the highest occupied molecular orbital and the lowest unoccupied molecular orbital. These terms replace positive and negative side in an inorganic solar photovoltaic cell [13].

Also, a detailed literature survey has been done and tabulated with their major findings.

Author/year	Paper title	Parameters used	Results/ Findings
Cardona et al. [13]	A simple model for sizing stand alone photovoltaic systems	LOLP, dimensionless energy deficit = 0.01, 0.05, 0.1 Array capacity = 0.25–0.9 Battery capacity = 2–9 Array surface tilt, $\beta = 60°, 50°$ Yearly average value = 17.5 MJm^{-2}	• Coefficient of determination for the stand alone photovoltaic installation ranges from 0.94 to 0.98
Wai et al. [14]	High-performance stand alone photovoltaic Generation system	Conversion efficiency = 95–97% Output power = 30–320 W Fcs = 100 Hz PV efficiency = 11.92% Open circuit voltage = 21.61 V Short circuit current = 4.96 A Resistive load = 100–300 Ω	• A high-performance PV Stand Alone system is developed • It has a maximum conversion efficiency of about 96.5% • It is observed that there is an increase in general efficiency by incorporating sun tracking scheme. (this saves the cost of conventional sun tracker)
Jakhar et al., 2017 [15]	Modeling and simulation of concentrating photovoltaic system with earth water heat exchanger cooling	Fluid density = 1000 kg/m^3 Thermal conductivity of absorber plate = 385 W/m K Copper tube diameter = 0.012 m Concentration ratio = 2:6 Reference temperature = 25 °C Collector area = 5 m^2 Edge insulation thickness = 0.025 m	• Power is obtained during the hours of moderate sunlight, thus, during early morning and late evening • It is also observed that as the mass flow rate increases the concentrate photovoltaic power increases • However during hours with maximum sunlight the mass flow rates experience a dip
Green [16]	Radiative efficiency of state-of-the-art photovoltaic cells	AMI 1.5 spectrum range = 600–700 nm External radiative efficiency = 1.26% Efficiency of diode = 20%	• They derived a parameter called ERE form standard solar cell efficiency, which in turn proved to be a useful parameter to compare the performance of the cell

(continued)

(continued)

Author/year	Paper title	Parameters used	Results/ Findings
Abdolzadeh et al. 2009 [17]	Improving the effectiveness of a photovoltaic water pumping system by spraying water over the front of photovoltaic cells	Mean voltage = 14.37 V when not spraying Mean voltage = 16.06 V when spraying Mean current = 6.36 A when not spraying Mean current = 6.52 A when spraying Maximum power = 80 W not spraying Maximum power = 60 W spraying	• Heat removal from the face of the PV cell is proved to improve efficiency of the same • Due to the variations it is observed that effect of spraying water is more observed in lower cells as compared to higher cells • However the efficiency of the overall system is still high
Shrotriya et al. [18]	Transition metal oxides as the buffer layer for polymer photovoltaic cells	Active area of device = 0.105 m^2 Power conversion efficiency = <2% Short circuit density current = 7.8 mA/cm^2 Open circuit voltage = 0.49 V Fill factor = 51% Work function = 4.7–5.2 eV Thickness = 1 nm to 20 nm Thickness of oxide film = 3 nm	• For polymer PV devices efficiency can be increased by incorporating a buffer layer made from transition metal oxides namely, V_2O_5 and MoO_3 • A PV device with a buffer layer of MoO_3 had power efficiency of 3.3% • Open Circuit Voltage of the device is independent of the work function of the metal oxide used
Sambur et al. [19]	Multiple exciton collection in a sensitized photovoltaic system	Semiconductor absorber = 31% Bulk band gap = >3.4 eV Particle diameter = 2.5 to 9.9 nm Absorbance Spectra = 0.85 eV to 1.35 eV	• MEG absorber efficiency of 45% is achieved, thus encouraging further developments of photovoltaic devices that surpass Shockley-Queisser efficiency limit • Power conversion efficiency in thin film can be improved but the limit to which the improvements can be made are unclear

(continued)

(continued)

Author/year	Paper title	Parameters used	Results/ Findings
Águas et al. [20]	Thin film silicon photovoltaic cells on paper for flexible indoor applications	Cells production temperature = 150 °C Fill factor = 41% Open circuit voltage = 0.82 V Short circuit current density = 10.2 mA m^{-2} Cell efficiency = 3.4% Contact angle ≤60°	• Efficiency upto 3% is achieved using optimized n-i-p configuration • Using this device for indoor applications can increase the efficiency upto the range of 6% to 8% • Using technologies like distinct light trapping schemes, the efficiency is predicted to be increased to 10%
Jenekhea et al. [21]	Efficient photovoltaic cells from semiconducting polymer heterojunctions	Collection efficiency = 49% Power conversion efficiency = 1.4% Sunlight illumination = 2% Current density = −1.4 to 1.5 mA/cm^2 Wavelength = 300–400 nm External quantum efficiency = 66%	• For sun intensity of AM 1.5, fill factors in the range 41%–43% where observed • Power conversion efficiency is in the range 1.2% to 0.7% • Hetero junction photovoltaic cells have improvement of factor 2 in fill factor as compared with PPV Schotty barrier devices
Goh et al. [22]	Effects of molecular interface modification in hybrid organic–inorganic photovoltaic cells	Open circuit voltage = 0.25 V Dipole moment = −4 to 5 D Current Density = 0.5 mA/cm^2 Voltage = −0.2 to 0.8 V Wavelength = 400–700 nm Fill Factor = 57–66% Surface roughness = 0.27 nm	• A band angle shift in TiO_2 can be observed by molecular dipoles and acid base interactions • Surface modified TiO_2 has better electron accepting nature as compared to bare TiO_2 • The consequence of dipoles on the band edge shift is unknown. A surface modification is needed to find out

(continued)

(continued)

Author/year	Paper title	Parameters used	Results/ Findings
Gregg et al. [23]	Comparing organic to inorganic photovoltaic cells: theory, experiment and simulation	Excitation radius = 1 nm Binding energy = 0.25 eV Doping density = 1.4×10^5 cm^3 Open circuit = 50 mW/cm^2 Illumination = 2.5 eV	• The paper studies the basic difference between organic and inorganic photovoltaic cells • Open circuit voltage of inorganic cells is higher than electrical potential energy, however, in organic cells the two parameters limit each other
Blakers et al. [24]	High efficiency silicon solar cells	Cell thickness = 110 μm Concentration levels = 100 suns Cell efficiency = 30% Cell area = 4 cm^2 Open circuit voltage = 630 mV	• Back contact design is discussed in the paper • It decreases losses observed due to resistance • Efficiency of 26% to 27% is observed
Chenni et al. [25]	A detailed modeling method for photovoltaic cells	RS = 0Ω, 8Ω, 16Ω Temperature = −25, 0, 25, and 50 °C Insolation = 1000 W/m^2	• Efficiency of this experiment is shown to be based on four parameter model
Hua et al. [26]	Implementation of a DSP-controlled photovoltaic system with peak power tracking	Maximum power = 61.6 W; V_{OC} = 21.1 V I_{SC} = 4.01 A; voltage at load = 17.3 V Current at load = 3.57 A; Thickness = 50 mm Weight = 10 kg; Digital controller = TMS320C25 DSP Solar array = Solarex MSX60	• Maximum Power Point Tracking system is used to make the system work efficiently • The principle of energy conversion is applied to develop system transfer function • By increasing execution speed for improved response of the system

(continued)

(continued)

Author/year	Paper title	Parameters used	Results/ Findings
Park et al. [27]	A novel real-time simulation technique of photovoltaic generation systems using RTDS	RS = 0.031 Ω; RSH = 40 Ω; VOC = 0.57 V Fill factor = 78%; number of cells in panel = 36 cells in series; energy = 1.103 eV; Temperature = 11 °C and 22.72 °C; switching frequency = 10 kHz; electric load = FK series	• It is observed from voltage-current load of 35ohms and 40 ohms– the points for the practical and theoretical situations have a close match • The simulation model is working perfectly
Chai et al. [28]	Black-silicon assisted photovoltaic cells for better conversion efficiencies: a review on recent research and development efforts	Reflectance = 0.4–0.61%; Wavelength = 400–1000 nm; Nanocone height = 600–1000 nm; Nanocone diameter = 200–400 nm; Random Periodicity = 20–23 cones per μm^2; Current density = −10 to 40 mA/cm^2; temperature = 0–500 K; efficiency limit = 29.4%; b-Si solar cell efficiency = 21.4%	• The b-Si solar cell efficiency achieved is 22.1% • b-Si is used to make photovoltaic cells and photodetection camera • Currently, the important step in this research is to focus on light trapping technologies • Photonic crystals are used
Drouiche et al. [29]	Dry magnetic separation on the recovery of metal fragments from Kerf Slurry waste produced during the manufacture of photovoltaic solar cells	Mass % of magnetic materials recovered = 100–15%, drying = 343 K ideal current Intensity = 0.5A	• During the manufacture of Photovoltaic cells, huge amount of waste is either got rid off or incinerated • The authors have suggested a method that is by magnetizing the waste materials. The efficiency of this method is evaluated to be 96%

(continued)

(continued)

Author/year	Paper title	Parameters used	Results/ Findings
Cheng et al. [30]	Enabling high-performance Tandem organic photovoltaic cells by balancing the front and rear subcells	Short circuit current density = 10.3 to 11.7 mA cm^{-2}; conversion efficiency = 15.1%; power conversion efficiency = 7–17%; Wavelength = 400–1000 nm; coherence length = 1.38–1.41 nm; voltage = 0–1.8 V fill factor = 70%	• To increase the conversion efficiency of organic photovoltaic cells, a third component is added to the existing to the system. This is a polymer donor • While, increasing the efficiency of the cell, other parameters like fill factor and open-circuit voltage are kept constant
Karakawaa, et al. [31]	Factors contributing to degradation of organic photovoltaic cells	Active layer thickness = 190 nm; current density = −10 to 15 mA/cm^2; voltage = −1 to 1 V; irradiation time = 0–100 h PCE = 2–5%; absorbance = 0–1.4; wavelength = 300–900 nm; effective cell area = 1 cm^2; relative humidity = 40–50%	• The authors suggest that modifications in EH-IDTBR are vital in improving the efficiency of the photovoltaic cells • By using these changes, the conversion efficiency of the solar cell increases upto 10%

2 Results and Discussion

A review of photovoltaic cells is a demonstrated environmentally benign energy source that continues to photovoltaic research with attractive features. Because existing PV systems continue to be very inefficient and unusual, they are not cost-specific and are only employed on a regular basis if a local power source is not available. Photovoltaic developments will continue to flourish in the areas of thin foil and nanocrystalline material and soon raise photovoltaic efficiency to over 50%. With increased efficiency, PV technology will attract more people, which will cut costs. Because the sun offers 10,000 times more power, photovoltaic advances will one day replace ecologically damaging power plants with a proven and clean source of energy. extended, then efficiencies of around 10% should be within.

3 Conclusion

Efficiency of a PV cell can be improved by incorporating the following:

- Using a technology to continuously track the position of sun called active sun tracking technology.
- Cleaning the surface of the front surface of the PV cell by spraying water on it.
- Incorporating a buffer layer, preferably made from transition metal oxides.
- Using technologies like Distinct light trapping schemes.
- Using thin film photovoltaic cells for indoor applications can increase its efficiency up to 3–6%.

References

1. http://www.energybc.ca/ (2005) [Online]. Available: http://www.energybc.ca/cache/solarpv/www.cetonline.org/Renewables/PV_pro_con.html
2. "Planete Engergies," (2010) [Online]. Available: https://www.planete-energies.com/en/medias/close/how-does-photovoltaic-cell-work. [Accessed 2021]
3. Cubas JPSMC (2014) Explicit expressions for solar panel equivalent circuit parameters based on analytical formulation and Lambert W-function
4. Izawa S, Perrot A, Lee J-H, Hiramoto M (2019) Organic PN homojunction solar cell. Org Electron 45–49
5. MB, Hansen C Single diode equivalent circuit model. PV performance modeling collaborative
6. Khan B Non conventional energy resources. McGraw Hill Education Pvt. Ltd., Chennai
7. AD, Rajapakse MD (2009) Simulation tools for photovoltaic system grid integration studies. In: IEEE Electrical power & energy conference (EPEC)
8. JWXMJGCXXWXZZWF, Zhenghao Hu Z, Semitransparent organic solar cells exhibiting 13.02% efficiency and 20.2% average visible transmittance. J Mater Chem
9. SPALJ-H, Izawa HM (2019) Organic PN homojunction solar cell. Org Electron
10. Rooij DD, sinovoltaics [Online]. Available: https://sinovoltaics.com/learning-center/solar-cells/amorphous-silicon-solar-cells-structure-and-applications/

11. Jestin Y (2012) Down Shifting of incident light for photovoltaic applications. Compressive Renew Energy 563–585
12. Larson B, NREL transforming energy [Online]. Available: https://www.nrel.gov/pv/organic-photovoltaic-solar-cells.html
13. M, Sidrach-de-Cardona MLL (1998) A simple model for sizing stand alone photovoltaic systems. Solar Energy Mater Solar Cells, 199–214
14. Wai R-J, Wang W-H, Lin C-Y (2008) High-Performance Stand-Alone Photovoltaic Generation System. IEEE Trans Ind Electron 240–250
15. Jakhar S, Soni MS, Gakkhar N (2017) Modelling and simulation of concentrating photovoltaic system with Earth water heat exchanger cooling. Energy Procedia 78–85
16. Green MA (2011) Radiative efficiency of state-of-the-art photovoltaic cells. Prog Photovoltaics: Res Appl 472–476
17. M, Abdolzadeh AM (2009) Improving the effectiveness of a photovoltaic water pumping system by spraying water over the front of photovoltaic cells. Renew Energy 91–96
18. VLGYYCC-W, Shrotriya YY (2006) Transition metal oxides as the buffer layer for polymer photovoltaic cells. Appl Phys Lett
19. Sambur JB, Novet T, Parkinson BA (2010) Multiple exciton collection in a sensitized photovoltaic system. Science 63–66
20. HMTVAGDMMJSWA, Águas MR (2015) Thin film silicon photovoltaic cells on paper for flexible indoor applications. Adv Funct Mater 3592–3598
21. SA, Jenekhe YS (2000) Efficient photovoltaic cells from semiconducting polymer heterojunctions. Appl Phys Lett 2635–2637
22. CSSR, Goh MMD (2007) Effects of molecular interface modification in hybrid organic-inorganic photovoltaic cells. J Appl Phys
23. BA, Gregg HMC (2003) Comparing organic to inorganic photovoltaic cells: theory, experiment, and simulation. J Appl Phys 3605–3614
24. Blakers A, Zin N, McIntosh KR, Fong K (2013) High efficiency silicon solar cells. Energy Procedia. 1–10
25. RMMKT, Chenni BA (2007) A detailed modeling method for photovoltaic cells. Energy 1724–1730
26. Hua C, Lin J, Shen C (1998) Implementation of a DSP-controlled photovoltaic system with peak power tracking. IEEE Trans Ind Electron 99–107
27. Park M, Yu I-K (2004) A novel real-time simulation technique of photovoltaic generation systems using RTDS. IEEE Trans Energy Convers, 164–169
28. Chai JY-H, Wong BT, Juodkazis S (2020) Black-silicon-assisted photovoltaic cells for better conversion efficiencies: a review on recent research and development efforts. Mater Today Energy 18
29. Boutouchent-Guerfi N, Boussourdi MA, Lami A, Ould-Hamou M, Drouiche N (2021) Dry magnetic separation on the recovery of metal fragments from Kerf Slurry waste produced during the manufacture of photovoltaic solar cells. Silicon 149–153
30. H-CWRZYZSDZL, Cheng P (2020) Enabling high-performance Tandem organic photovoltaic cells by balancing the front and rear subcells. Adv Mater
31. Karakawa M, Suzuki K, Kuwabara T, Taima T, Nagai K, Nakano M, Yamaguchi T, Takahashi K (2020) Factors contributing to degradation of organic photovoltaic cells. Org Electron 76

Dual-Axis Solar Tracker for an Automated Irrigation System

U. Arjun, L. Gayathri, B. K. Gowri, V. P. Malavika, Ajish Ashok, and C. Sojy Rajan

Abstract The threats of energy depletion, global warming, and water wastage have necessitated an urgent need for harnessing renewable energy resources. Solar photovoltaic (PV) power generation is the most popular renewable energy source as it is a clean source and easily installable. The main shortcoming of this technology is its intermittent nature, which can be overcome to a considerable extent by adopting solar tracking technologies. Numerous researches in this area have proved that it is the best technology for substantially improving the electricity production of photovoltaic systems. Traditional fixed trackers are the most common types, but they are unable to keep up with the shifting orientation of the sun as it moves around the earth, resulting in less energy production. The primary objective of this work is to implement the maximum power point tracking (MPPT) using dual-axis solar tracker and to develop an automated irrigation system which reduces the wastage of excess water. This work also has an AC load side so that the excess energy stored in the battery can be optimally utilised for lighting purpose of the irrigation side as well as to comply with the needs of the farmer. The hardware consists of four LDR sensors used to measure light intensity in all directions. The dual-axis movement is achieved using two DC motors whose rotation is controlled using the Arduino NANO. Output power from the PV panel is fed to the battery, and power to run the motor is taken from the battery. Real-time clock (RTC) is used to set a span of time to run the motor for irrigation purposes. DHT-11 sensors are used for sensing the surrounding temperature-humidity. Integration of DHT and RTC is done in the Arduino board, and LCD displays the panel voltage, output voltage, and temperature-humidity value. This work is expected to be highly beneficial to the agricultural sector as it saves water and utilises renewable energy. The hardware frame work comprises GA sheet and pipes along with spur gear system for attaining the required motion as per the response from the sensors. The PV panel is integrated to this hardware frame work. At the base, the battery and circuitry comprising relays for motor (DC), Arduino, a charge controller, boost circuit, LCD display, and the inverter circuit are installed. The irrigation side comprises submerged type water pump along with a pneumatic

U. Arjun (✉) · L. Gayathri · B. K. Gowri · V. P. Malavika · A. Ashok · C. Sojy Rajan
Mar Baselios College of Engineering and Technology, Trivandrum, Kerala, India
e-mail: arjunu.17ee024@mbcet.ac.in

K. Namrata et al. (eds.), *Smart Energy and Advancement in Power Technologies*, Lecture Notes in Electrical Engineering 926,
https://doi.org/10.1007/978-981-19-4971-5_37

tube and droplet tube system to implement the sprinkler mechanism. The paper also presents MATLAB simulation of a PV and battery system supplying power to a resistive load. The relevant waveforms have also been presented in the study.

Keywords Maximum power point tracking (MPPT) · Perturb and observation (P&O) · Photovoltaic (PV) · State of charge (SoC)

1 Introduction

As the supply of fossil fuels declines and demand for electricity rises, the opportunities to use renewable energy sources expand. The consequences of using non-renewable energy sources are exacerbated by the impact of greenhouse gases and global warming. Solar energy is one of the cleanest types of renewable energy available when compared to the rest. The amount of solar irradiance emitted by the sun is enormous. Solar trackers are now available to use the sun's rays and convert it into useable energy. It is a viable option for enhancing the power generation by increasing the PV panels power efficacy. Traditional fixed trackers will be unable to keep up with the shifting orientation of the sun as it moves around the earth. In comparison with fixed titled PV panel, the suggested dual-axis tracker would be able to provide up to 40% more power.

Remote rural areas lack direct electricity, and so there is an urgent need to electrify these areas. In this case, availability of abundant solar irradiation reveals a feasible solution. An ideal tool for solar positioning which tracks longitude trajectory of the sun and includes dust removal mechanism has been presented in [1]. During day, solar locator is used which effectively tracks the sunlight. Four LDR sensors and two servomotors are used in the system. The amount of solar radiation is continuously monitored by the LDR, and the signal is sent to the microcontroller. According to the instructions from microcontroller, servomotor rotates to position the panel. Adust removal mechanism is also incorporated.

A dual-axis solar tracking system based on LDR and a DC motor with gear configuration is proposed in [2]. The hardware set-up consists of a solar panel, two DC motors with gear configuration, and LDR with an electronic circuit. On a circular plate, four LDR sensors with 900 separations are connected in series. The Arduino UNO is in charge of the tracking system, which sends commands to the DC motors to determine elevation and orientation. The parameters are monitored and supervised on a computer program through Bluetooth module.

A solar-powered off-grid irrigation system that uses solar panels oriented in the direction of sun to provide the power in remote areas is presented in [3]. Here, the solar tracking is bidirectional, and the components used are Arduino UNO, OP-AMP, servomotor, solar panel, and DWT 11 moisture sensor. Arduino receives the signal from the solar panel which compares the panel output voltage and current with reference voltage and current and regulates the motor accordingly. A rechargeable battery is charged from the panel which powers the pump. The moisture sensor implemented has a predefined value, and if the moisture level falls below this value, the pump will start.

A high efficiency solar tracker using Arduino to trace maximum sunlight to power the solar panel has been presented in [4]. It makes use of two servomotors and five LDRs to capture maximum sunlight. The analog input from LDR is converted to digital signal using ADC converter, and the output is sent to servomotor for panel rotation. The output voltage and efficiency of the tracker were evaluated and compared to those of a static panel at different time intervals revealing that the system reacts best at 10 min interval.

A dual-axis solar tracking system that uses incident photons to keep the sun orientation board in correct position has been proposed in [5]. A cost-effective pilot panel is used with the solar panel to capture maximum light intensity and is given to a microcontroller. On the basis of this value, motor rotates for the maximum power point. Different MPPT calculations under similar conditions are done. Various illumination and temperature conditions can be set up. Different tracking voltages were taken with LDR individually and LDR with pilot panel. It was concluded that tracking voltage is higher in LDR with the pilot panel.

A simple automated dual-axis solar tracking system with four LDR sensors for Baghdad city has been introduced in [6]. The hardware side consists of the basic frame work along with the panel, two DC motors, and four LDR sensors. The electronic circuit is designed in such a manner that it works as a comparator to interpret the data from the LDRs and provides the required command (clockwise/anticlockwise) to obtain the required tilt for the panel so that the irradiance is always incident on the panel for obtaining maximum efficiency. An add-on feature to sense the weather conditions has also been incorporated. If the weather is not favourable (like cloudy conditions), then the electronic system will abruptly stop the tracking.

An automatic tracking system has been put-forth in [7]. A machine side and an automatic sun-tracking control circuit comprise automatic sun-tracking. A silicon photo cell assessment circuit, biaxial machinery, tracking system, and battery make up the machine part. Sensor signal processing circuit and microcontroller circuit make up the automatic sun-tracking control circuit. By using the microprocessor AT89C2051 and utilising the data from the individual sensors (horizontal and vertical), automated tracking is obtained. The power required for mechanical working of the tracker is obtained from the panel.

Reference [8] proposes a better PV system based on dual-axis solar tracking and MPPT. The approach used here is a modified variable step size incremental conductance algorithm. It was modelled using MATLAB, and the prototype was also developed for the same. The results indicated an increase in the system's overall efficiency.

The conversion of a fixed single axis system to dual-axis solar tracker utilising MPPT has been proposed in [9]. Here, an Arduino has been used for implementing several MPPT algorithms, namely P&Q method, neural method, and IC method as a case study. By utilising these MPPT techniques, the panel efficiency has been proved to be improved.

In [10], an update on current dual axis tracker technology is proposed. The design of a solar tracker with four LDR sensors and control circuits is discussed in this study. Two LDR sensors align the array inclination angle throughout the day, while the other two sensors align the solar array tilt angle seasonally. The IC MPPT algorithm is utilised to ensure that the PV array delivers maximum power to the load. The DC output voltage from the panel is converted to AC voltage by an inverter, and a PID controller is used in the feedback loop to improve performance.

In [11], an automated solar-powered irrigation system is described. This study describes an embedded system for off-grid irrigation, which is an intelligently regulated mechanism for the most efficient use of irrigation resources. With the use of GSM technology, which provides messages about the current state, the farmer may effortlessly water his fields from any location. The key advantages are the most efficient use of power through water resource management and the reduction of government-supplied electricity. This provides an efficient and environment-friendly approach of the irrigation system which will automate the agriculture sector.

An overview on an efficient automated irrigation system has been presented in [12]. An automated irrigation system has been developed in this work to eliminate the issues of manually turning on and off of pumps and water wastage during irrigation. A wireless sensor section (WSS) and a wireless controller section (WCS) are included in this system. The soil moisture sensor, temperature and humidity sensors, power supply, microprocessor, and GSM module are all part of the WSS. Microcontroller, relay, motor, power supply, and GSM/GPRS module make up the WCS. By utilising the sensor data, smart irrigation is done and communicated via GSM.

Another type of automated irrigation system using solar power has been proposed in [13]. A solar-powered variable rate automatic microcontroller-based irrigation system has been presented. The sensors are installed in the field, and the farmer is informed of the water level in the field. Even from a remote location, the farmer can regulate the system by a cellular phone. During an emergency situation, that is, when the water level is drastically low, motoring starts even without the consent of the farmer, thereby maintaining the water level of the field.

A briefing on engineering quality control of solar-powered intelligent water-saving irrigation has been suggested in [14]. A solar water-saving irrigation system is applied here, which is dependent on parameters such as crop height, water flow, and so on.

MATLAB simulation model of a solar power station with an automated dual-axis solar tracker has been presented in [15].

The primary objective of this work is to implement the maximum power point tracking (MPPT) using dual-axis solar tracker and to develop an automated irrigation system which reduces the wastage of excess water. This work also has an AC load side so that the excess energy stored in the battery can be optimally utilised for lighting purpose of the irrigation side. The study also includes a MATLAB simulation of a PV and battery system that supplies electricity to a resistive load, as well as an analysis of the waveforms.

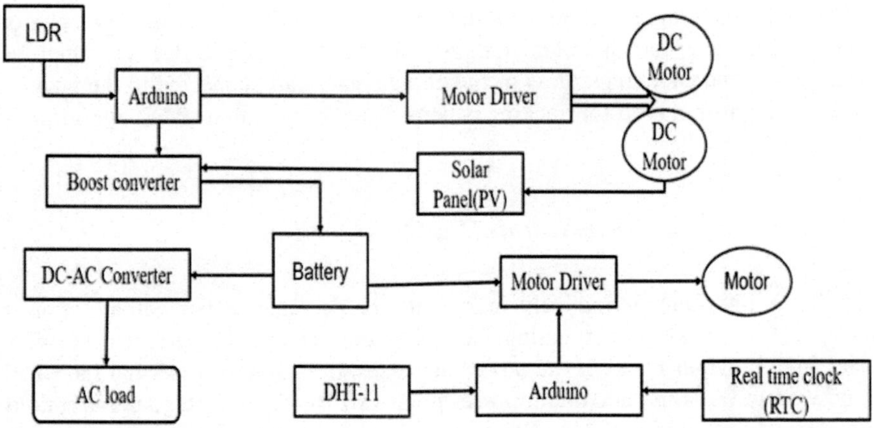

Fig. 1 Block diagram of the proposed system

2 Block Diagram of the Proposed System

The proposed dual-axis solar tracker utilises MPPT algorithm to extract maximum power from solar irradiance. The block diagram of the proposed system is as shown in Fig. 1. The PV panel comprises two pairs of LDRs that track the sun's exact location along the inclination axis as well as the azimuth axis. For effective tracking, a funnel shaped arrangement is adopted for the LDRs in order to avoid shadow effect. The output from LDR channels is given to Arduino Nano, and they provide digital signals to the motor driver to drive the DC motors for dual-axis tracking. Three DC motors are used for tracking. Two DC motors are used for the up-down movement and one for 360° rotation of the solar panel. Output voltage from the panel is boosted using a boost converter. The MOSFET switches of the boost converter operate according to the digital PWM pulses from the Arduino. The boosted voltage is used to charge the battery.

The power stored in the battery is used to supply both the AC load and DC load. At the DC side, there is a motor for irrigation purpose. The control of the motor is done using sensors. To detect temperature and humidity, DHT-11 is adopted, and a clock signal RTC (real-time clock) is also set for a short duration during the morning and evening time. The motor operates according to the command from the Arduino. The excess stored power will supply the AC load also. The DC-AC converter converts the DC voltage out from the solar panel to AC voltage.

3 Maximum Power Point Tracking (MPPT)

Maximum power point tracking (MPPT) is a photovoltaic (PV) inverter algorithm that effectively controls the impedance seen by the solar array in order to keep the

PV system operating at, or close to, the peak power point of the panel in spite of varying conditions such as solar irradiance, temperature, and load. It will improve efficiency of the panel regardless of the climate and load connected to the system. Here, the algorithm used for tracking is perturb and observation (P&O) algorithm.

3.1 Perturb and Observation (P&O) Algorithm

The P&O algorithm periodically increments or decrements the output terminal voltage of the array by comparing the power obtained in the current cycle with that of the previous cycle. If the power increases, it is an indication that the panel is moving towards the maximum power point (MPP). Thus, voltage perturbations move the panel more towards MPP and vice versa, as shown by the flow chart in Fig. 2.

The flow chart explains the working of P&O algorithm, i.e. the voltage and current are measured initially in order to calculate the power. The prime step is the comparison of this calculated power with the power that was calculated earlier [$P(n-1)$]. If the difference is zero, then it is the actual null point, i.e. the panel is in the maximum

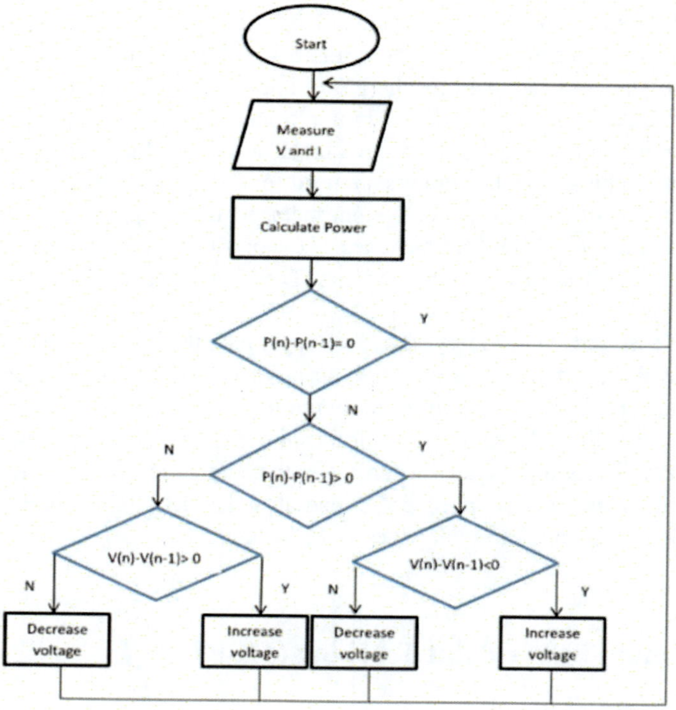

Fig. 2 Flow chart of P&O algorithm

power point, and there is no need for further movement. If the initial condition is not true, then the power difference is again taken to check whether it is greater than or less than zero. If the value is greater than zero, the corresponding voltages of the respective powers are compared. If their difference is less than zero, then the command is to "increase voltage" that is to move the panel to the next position to attain the null point. If the difference is greater than zero, then the condition is "decrease voltage" that is to shift to the earlier position in order to attain the MPPT. Similar process takes place for the other expression $[V(n) - V(n-1)]$. The orientation of the motor decides the movement of the panel to attain the MPPT (right or left to up or down).

3.2 *RTC-DHT Flow Chart*

The flow chart in Fig. 3 explains the working of the motor depending upon the timing that will be set on the RTC module, connected to the LCD display. When the mode button is long-pressed, it will loop between four values, i.e. if mode is 1 then it will display the current time, if mode is 2 the input and output voltages will be displayed, if mode is 3 then it shows the temperature and humidity values, and if mode is 4 the ON/OFF alarms for the motor can be set. On the other hand, if cancel button is long-pressed, the timer for switching the motor will be turned OFF. Then, it again checks if the timer is triggered, and according to that, it decides whether to turn the pump ON/OFF according to the triggered timer.

4 Hardware Implementation

Figure 4 shows the hardware prototype of the proposed system. Four LDR sensors are placed on the panel with LDRs covered with a PVC ring in order to avoid shading effect. The LDRs are placed as two channels, one channel for up and down movement and other for 360° rotation. A DC motor with a gear system is used to control the speed of the motor and to obtain 360° axial movement, and two DC motors are placed on top of two arms, for X–Y rotation of the solar panel, i.e. the up and down movement. The motor will work automatically when timer in set on the LCD panel. Depending on the thrust, the rotatory motion develops, and water is sprinkled.

5 MATLAB Simulation of PV and Battery System

In this section, MATLAB simulation of a PV and battery system (Fig. 5) supplying power to a resistive load has been presented to plot the possible PV and battery characteristics. In this model, a PV panel is integrated with a lithium-ion battery through a boost converter and bidirectional inverter. The panel parameters were set

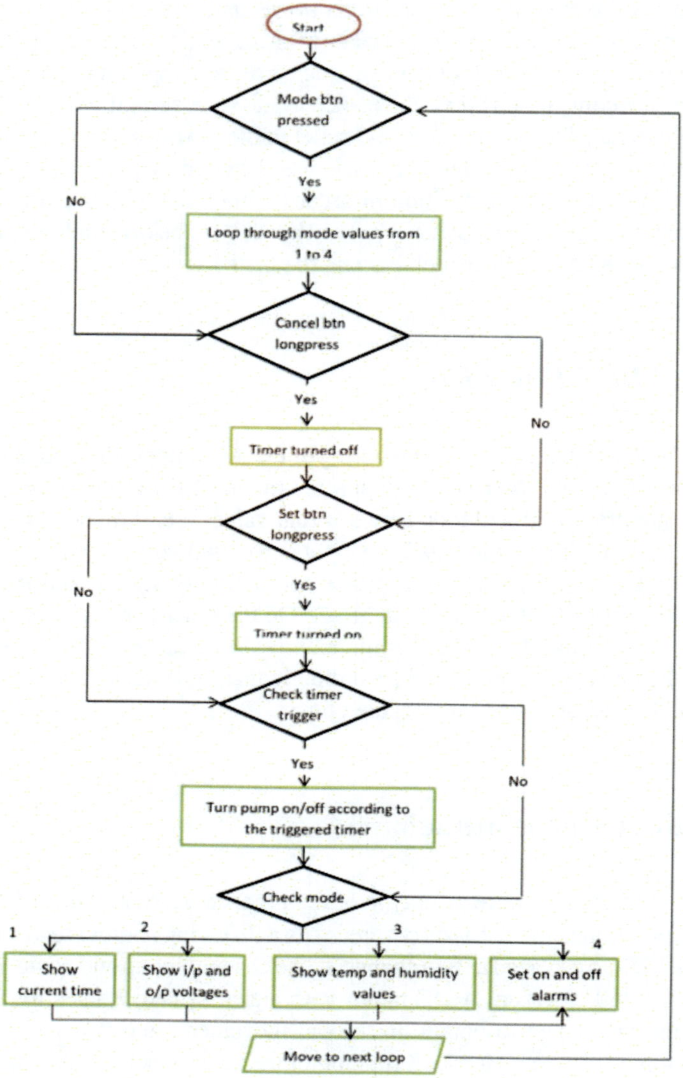

Fig. 3 Flow chart of RTC-DHT

in such a way that maximum power output from panel is 1 kW. The output of boost converter is the coupling point for PV and battery. The bidirectional converter allows both charging and discharging of the battery. The energy thus stored in the battery can be used for supplying the load. The parameters of all the components have been shown in Table 1.

Fig. 4 Hardware prototype

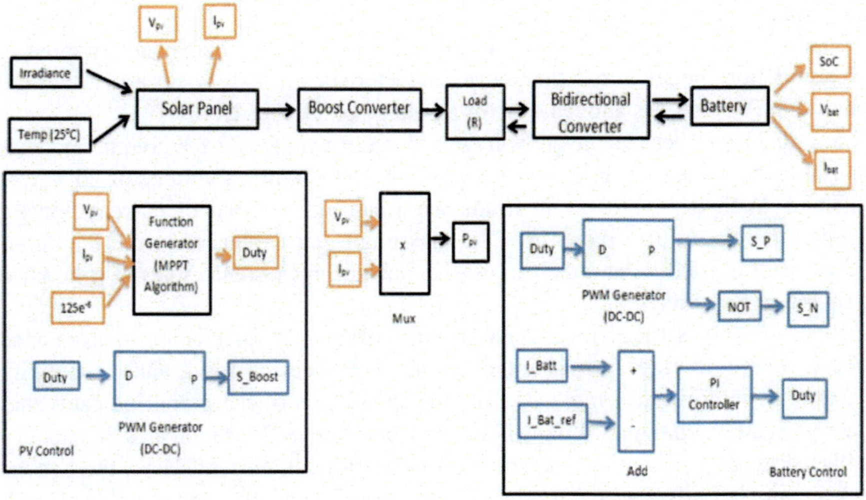

Fig. 5 Simulation block diagram

Table 1 Specifications

Components	Specification
Solar panel	Soltech 15TH-215-P PV module V_{OC}—36.3 V I_{SC}—7.84 A
Boost converter	C—100 μF, 3300 μF R—0.01 mΩ, 0.1 Ω L—5 mH Switch—IGBT
Load	R—6 Ω
Bidirectional converter	L—0.5 mH C—1000 μF R—0.1 Ω, 0.1 mΩ Switch—IGBT (for both positive and negative switching)
Battery	Li-ion battery SoC—45% (initial) Nominal voltage—24 V Rated capacity—50 Ah

6 Simulation Results and Discussions

MATLAB simulation results obtained for the PV array and battery integrated model have been presented in this section.

Figure 6 shows the P–V and I–V characteristics of the solar panel adopted in the simulation model. From the above characteristics, it can be observed that the maximum PV power is about 1 kW for a voltage of 29 V at 25 °C.

Figure 7 describes changes in voltage and current values for the gradual increase of irradiance value. For an irradiance value of 1000 W/m^2, a voltage of 20 V was obtained. With the increases in irradiance value, the amount of current flowing through the circuit also increases. The maximum power obtained from the simulation result is 1000 W, and the variation in power for different values of irradiance is shown in P_{PV}-time graph.

Figure 8 shows the variations of battery parameters for change in irradiance values. SoC-time graph provides the information about charging and discharging of battery according to irradiance. The negative slope of the graph indicates that battery is getting charged from the PV array, and positive slope indicates that battery is discharging. I_{batt}–Time graph shows the variations in current. Here, the battery current was initially about 20A and starts to decrease as the battery gets charged. $V_{(batt)}$–time graph depicts the increase in battery voltage with irradiance. The increase in battery voltage indicates that the battery is charging.

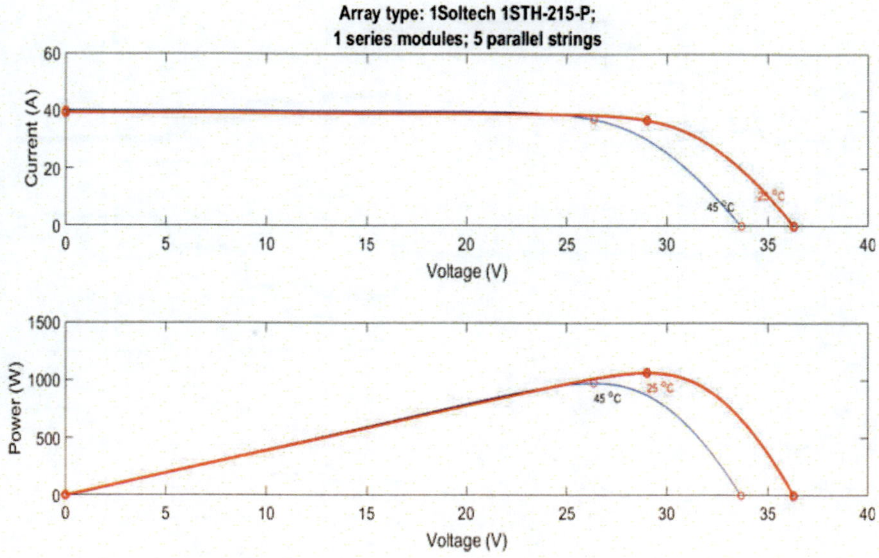

Fig. 6 P–V and I-V characters of PV array

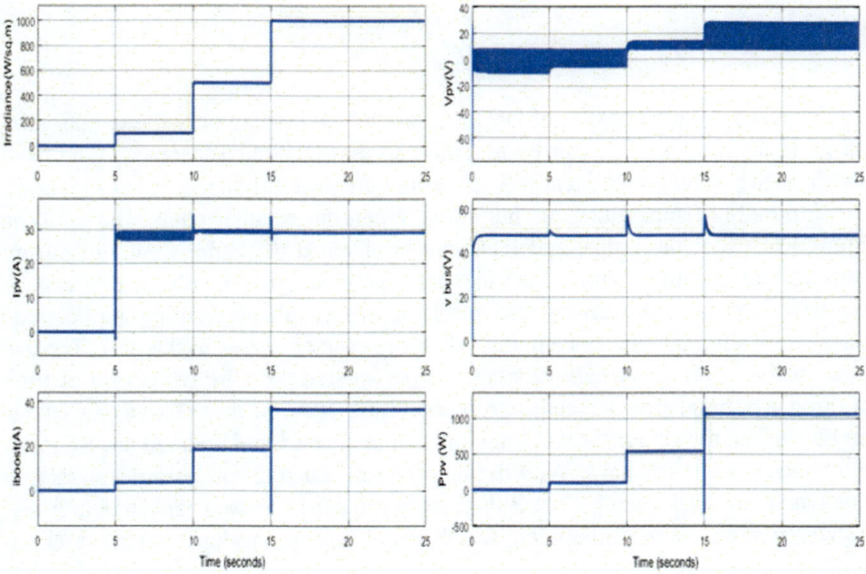

Fig. 7 Simulation results of PV array

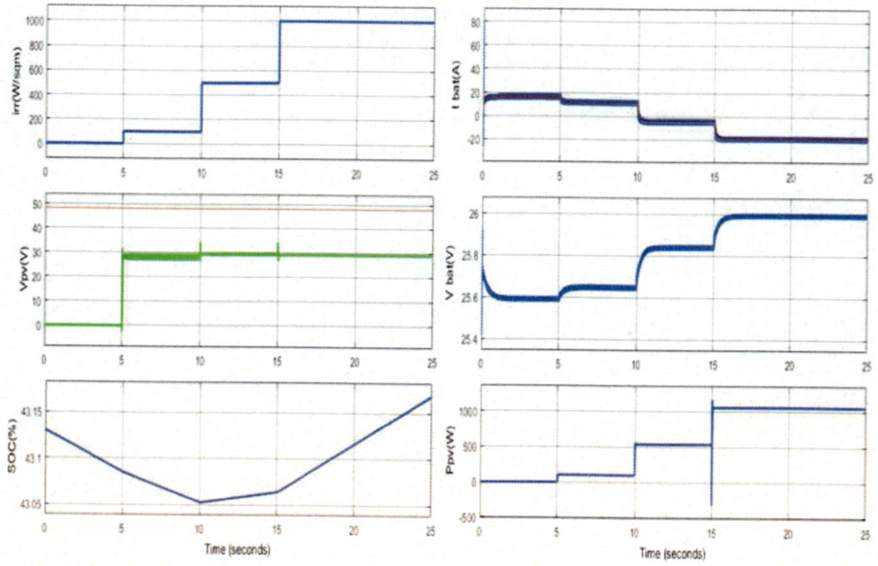

Fig. 8 Simulation results of battery

7 Future Scope

The proposed system consists of DHT sensors which display only the temperature and humidity values. This can be modified so that DHT-11 sensors may be used to reduce the watering frequency. The sensors are currently used only to measure the surrounding temperature and humidity. When the weather cools, less water is required to water the crops, and RTC can be given a particular time slot during morning and evening hours. Since irrigation application is involved, this can be converted into a smart farming system which will handle all farming applications and entirely automate the system, thus allowing ease of access to the user. The use of smart irrigation system allows plants to be watered for a longer period of time resulting in optimal growth conditions. It saves time, and time delay based on weather can be applied for automatic watering and can be altered and updated according to the changes in environment. With the help of soil sensors, soil parameters can be displayed in the LCD panel. The fertility of the soil can be sensed, and an output will be given in terms of recommending which crop is best to produce making it a crop prediction application.

8 Conclusion

A dual-axis solar tracking system was developed in the research, which will aid in enhancing the performance of a solar PV system. An automatic tracking system has been proposed to maintain vertical contact between the solar panel and the sun, as well as an automatic irrigation system to irrigate a small area and encourage productive farming. The battery's excess energy can be utilised to meet the AC load needs. It is projected to increase solar energy use and photovoltaic power generation system efficiency. With the optimal utilisation of solar energy and dual-axis tracking, this technology will assist to alleviate the energy crisis to a large extent. Because the hardware prototype is RTC-integrated, the requisite irrigation interval can be met. This system is projected to benefit the agricultural sector in particular because it uses an environmentally friendly dual-axis tracked PV source to power an automated watering system. The system, which consisted of a PV array and a battery, was simulated in MATLAB.

References

1. Chang KC, Zhou Y, Sajid T (2020) Study of low cost and high efficiency intelligent dual axis solar panel system. In: IEEE International conference on artificial intelligence and computer applications
2. Mustafa FI, Al Ammri AS (2017) Direct and indirect sensing two-axis solar tracking system. In: International renewable energy congress (IREC 2017)
3. Dutta PK, Mallikarjuna K (2017) Sensor based solar tracker system for moisture detection and auto-irrigation. In: IEEE International conference on power control signals and instrumentation engineering
4. Zolkapli M, AI-Junid S A, Othman Z, High efficiency dual axis solar tracking development using Arduino. In: IEEE International conference on technology, management, engineering and environment
5. PushpaValli M, Sivagami P, Abhirami P (2018) Solar panel tracking and power generation automatic system. In: IEEE 4th International symposium in robotics and manufacturing automation (ROMA)
6. Mustafa FI, Mustafa FF, Shakir S (2018) Simple design and implementation of solar tracking system two axis with four sensors for Bagdad city. In: IEEE 9th International renewable energy congress
7. Zhang B, Gao G, Zhu Y (2010) Designment of automatic tracking system of solar energy system. In: IEEE, 2nd International conference on industrial mechatronics and automation (ICIMA), pp 19–29
8. Li H, Zhao C, Wang H, Luo J, Xie S (2014) An improved PV system on dual axis solar tracking and MPPT. In: IEEE International conference on mechatronics and automation
9. El-monem Salama HA, Taha ATM (2018) Practical implementation of dual axis solar power tracking system. In: IEEE International Middle-East power system conference
10. Allamehzadeh H (2019) An update on solar energy and Sun tracker technology with a dual axis Sun tracker application. In: 46th IEEE PVSC conference
11. Yalla SP, Rajesh Kumar KV, Ramesh B (2013) Energy management in an automated solar powered irrigation system. In: IEEE International conference on information communication and embedded system

12. Patil B, Gabhane SK, Repal SS (2018) Efficient automated irrigation system. In: IEEE International conference on IoT in social, mobile, analytics and cloud
13. Uddin J, Taslim Reza SM, Newaz Q, Uddin J, Islam T, Kim JM (2012) Automated irrigation system using solar power. In: IEEE 7th International conference on electrical and computer engineering
14. Liu X, Wu H, Ling J, Tao J, Yao L (2010) Engineering quality control of solar-powered intelligent water-saving irrigation. In: IEEE 2nd International Asia conference on informatics in control, automation and robotics
15. Mitrofanov SV, Baykasenov DK, Suleev MA (2018) Simulation model of autonomous solar power plant with dual-axis solar tracker In: IEEE International ural conference on green energy

Electricity Theft Detection Methods and Analysis Using Machine Learning: Overview

Ranbirjeet Kaur and Garima Saini

Abstract Losses are involved from generation to transmission and distribution (T&D) of electricity. The score of these losses is rising in many countries severely. The main issue in electricity usage is the theft of electricity, which is dangerous for power suppliers and has caused budget losses. Detection and control of electricity theft is a challenge that involves a variety of factors such as economic, social, regional, administrative, political, infrastructure, level of education, etc. Electricity theft detection is very much important to make the power system reliability. Fictitious use of electricity lowers the quality of supply, increases a load of production, causes certain consumers to pay an extra amount of bills, and affects the overall rollback. Non-technical losses occur due to improper and illegal measurements of energy consumed by energy meters. Non-technical losses are a big problem for the occurrence of security risks and immeasurable financial losses. Sometimes it is difficult to locate the tampered meters, damaged/broken meter terminals, and/or unlawful applications that cannot be traced out at the stage of checking's done by the flying squad. IoT base smarts meters are used nowadays to overcome this problem, as they involve two-way communications. This meter is also intelligent enough to detect excessive energy use and warn the consumer to reduce consumption or stop the supply process automatically. This review contains comparisons and research among the various ways to detect theft. Accuracy is compared and analyzed with complete literature reviews. In this paper, we have studied various machine learning algorithms used to detect the theft of electricity.

Keywords Losses · Electricity theft · Tempered meters · Machine learning · IoT

R. Kaur (✉) · G. Saini
Department of Electronics and Communication, National Institute of Technical Teacher's Training & Research Chandigarh, Chandigarh, UT, India
e-mail: ranbirjeetkaur88@gmail.com

G. Saini
e-mail: garima@nitttrchd.ac.in

K. Namrata et al. (eds.), *Smart Energy and Advancement in Power Technologies*, Lecture Notes in Electrical Engineering 926,
https://doi.org/10.1007/978-981-19-4971-5_38

1 Introduction

Power losses are categorized into two types that is technical losses (TLs) and non-technical losses (NTLs) [1] Technical losses occur due to joules effect on power lines and loss of transformer during electrical transport. The technical losses calculation is much difficult, because of the difficulty of locating the point of losses and estimation of energy lost. Technical losses can't be eliminated, but can be decreased by using certain conversion methods within the power system. Non-technical losses can only be defined as the total loss minus technical losses [2] in electricity consumption, the fewer technical losses occur as a result of the infrastructure or system and these losses are taken into consideration to be 5% of total energy consumption. The main reason of NTL is payment delays and irregularities, power theft, energy meters having fault, fraud and defaulting amount [3]. Many investigators have reported a separate loss, called gratis, that occurs when electricity is given free of cost to different categories like SC/ST, BPL, etc. [4]. Nowadays, cyberbullying is a vicious cycle to changes data usage patterns, which lowers consumer bills [5]. Even so, power losses are caused by a small number of consumers, reducing profits and power efficiency of operating companies, which has led to an increase in the cost of all users and created similar problems such as deregulation, disruption of normal industrial processes, and inflation. NTL. Mischievous consumers who consume electricity in various ways are called non-technical loss (NTL). NTLs are the loss of power during the process of generating, transmitting, and distributing electricity, and these losses are majorly caused by the theft of electricity on the consumer side [4]. NTL behavior often involves skipping an electric meter, interrupting meter readings, or hacking a meter. With increasing global demand for energy, the value of consumption has increased equally. In many countries, electricity theft loss (ETLs) causes a large proportion of the total electrical losses [6] and occurs mainly in low electricity grids. Electricity theft loss can cause major issues, such as loss of energy suppliers' income, less stability protection and authenticity of the grid system and also improve insignificant use of sources. In India, ETLs had been predicted at $4.5 billion [7] which keep on increasing year after year. ETLs are said to account for 39.9% of total energy losses in various countries like Malaysia, Brazil, and Lebanon [8]. Electricity theft loss in a few states in China has reached approximately 2 hundred million kWh, with a complete cost of 1 hundred million in Wuhan city. As stated in [9], losses due to energy theft are about 1 hundred million Canadian bucks yearly with power losses that could be provided to 77,000 families in one year. Annual revenue loss as a result of electricity theft in America has been 0.5–3.5% of total revenue [10, 11]. Consequently, research to improve electricity theft recovery strategies has come to be increasingly more needed due to its significance in energy conservation and reduced intake [12].

The overview of total T&D losses for one year in different countries is shown in Fig. 1. It is clear that in many nations, trillions of kWh of electricity are pilfered every year.

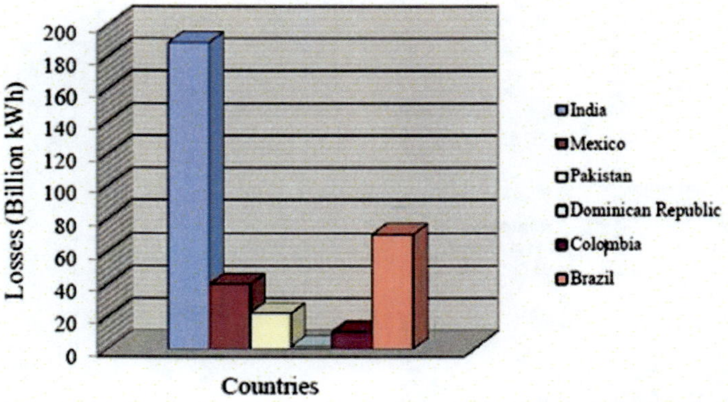

Fig. 1 Overview of T&D losses (for an entire year) in different nations [10]

The smart grid has to turn out to be the best way to monitor power systems in the future. This system may be defined as a whole electrical network that collaborates the framework of the electrical and pc system to control and screen power consumption, and a sensible screening system that will screen the utilization pattern and overall performance mode of all system-related consumers [13]. The smart grid offers "facility of clients" area screening, controlling, and predicting electricity consumption through integrating the contemporary digital devices with the existing power system. In this system, the device at the consumer's site will provide utilization to energy at the workspace via using the internet, and the power transfer corporation will make payment methods according to these consumed units calculated from previous and present readings. The application center takes consumer readings from the scheduled client updates thru the internet (Wi-Fi). The primary aim is to decrease the %age of losses because of wastage of power and supply of efficient, fee-effective, and protective energy [14]. The device which is used to report energy usage is known as a smart meter and is an old-fashioned computer version of the meter. Processor, limited storage, and communication facilities, and maintenance capacity increased customer power expansion, making smart meters a fundamental part of smart grid systems.

1.1 Methods of Stealing Electricity

Taping energy directly from that of an overhead distribution feeder is the most common and easiest method of pilfering electricity. The misuse of energy meters used to record and bill manufacturing, commercial and residential energy use is the next most popular form of electricity theft. Although there are several tampering techniques with such meters, some of these might include [15]:

- Exposing meters of strong magnetic fields to clear memory.

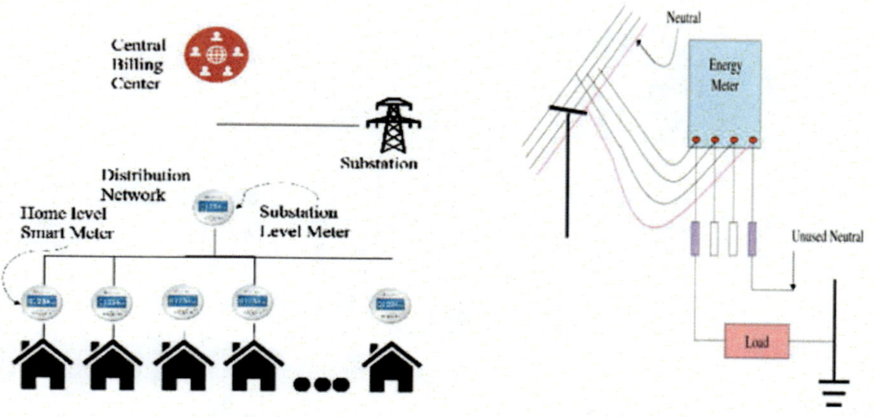

Fig. 2 Home distribution network and energy theft situation

- Inserting of film or exposure to the high viscous fluid to disrupt the disk rotation.
- Implementing emerging technologies including remote sensor devices.
- Crystal frequency deception of integrated circuits.
- Building connections between the power cord controls lines will be permanent times redirect the current reading to the meter representing zero reading.
- RF devices are set to electric meters to impact meter accuracy.
- Installation of a shunt between incoming and outgoing meter tables.
- Changing incoming and outgoing meter terminals.
- Causes damage to the meter pressure coil.
- Reset meter reading.
- Making undesirable harmonics.
- Exposure of radiation to shock mechanics.
- The voltage is controlled from the meter connection so that the reading value is less than the original use.

Figure 2 shows an example of a home distribution network and energy theft situation.

1.2 Smart Meters for the Power Grid

A smart meter is an automatic power tool that monitors consumer power consumption as well as reports you and the details related to that service in private. The smart meter enables data collection resources (due to its bidirectional capability) to collect data from consumer premises to collect energy consumption and also energy returned to the power grid in case of the solar system installed at the consumer's site. Smart meters can run command signals directly as well as virtually, about protected communication. So smart meters can be used in the customer's environment just to monitor and track all household appliances and devices. They can collect diagnostic data about distribution systems, consumer electronics; therefore, they can exchange this data within their control with other meters. They calculate electricity consumption, mobilize resources to authorize and install devices for energy storage. Distributed sources of electricity generation are a vital part of future lifestyles. All additional services and demand management methods enable services to combine a large amount of data in real time. Smart meters can be installed in such a way that in some way electricity is consumed from electricity grid charged, except electricity used from individual manufactured customer resources including storage devices. Moreover, because electricity is supplied to the grid, the credit will be given to consumers. Usually, data is exchanged by a smart meter is a combination of symbols such as a specific meter code, time stamp, and rapid rate of energy consumption. Smart energy meters can regulate the maximum usage of energy, control household types of equipment and therefore can remotely disconnect or restore the electricity supply toward any consumer in the situation of neighborhood fault or event. Smart meters play a major role in tracking grid efficiency as well as features of consumer power consumption in a smart grid system. Regular screening of energy consumption of all the consumers makes it very effective to find and control increased electricity demand. Home power management strategies are often informative for consumers about the cost-effective usage of household appliances without installing any special meters. Smart meters can be used to monitor light, heat, cool air, and more gadgets in the context of this. In a given schedule, smart meters can be programmed to operate home appliances. In addition, smart meter integration enables utilities to detect illegal use and energy theft to increase the efficiency of distribution and quantity of power supplied [16].

Figure 3 shows a smart energy metering system with its unique modules. A meter is smart if it has had the following capabilities.

- It must capture and record the consumer's real-time electricity consumption.
- Using cloud computing, there should be remote access through the mobile device application.
- Must be able to record events such as device status measurement.
- It should be able to detect and eliminate wave harmonics.

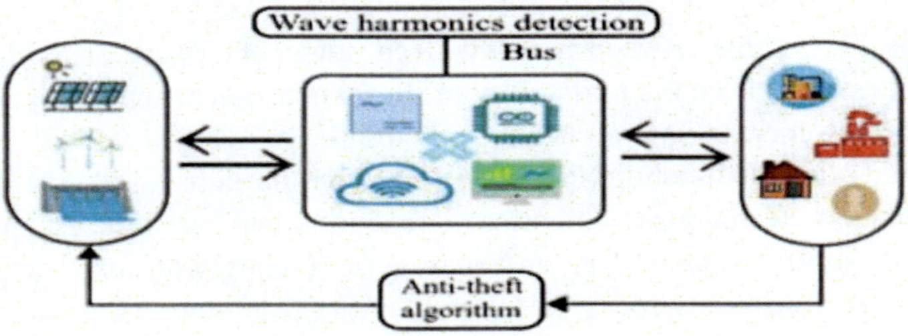

Fig. 3 Smart energy metering system

- There must be an anti-theft algorithm.
- Should be able to perform forecasting from time to time.

2 Machine Learning Methods for Detecting Theft of Electricity

In the past, the installation of electric meters, professional electric meters, and traditional bidirectional metering techniques has been followed to deal with electricity theft [17, 18]. These days, energy theft detection techniques depend on the classification of the statistics that are collected through the measurement system of the smart meter. Electricity theft classification and ordinary behaviors are carried out via statistics evaluation [19]. The current techniques for the detection of electricity theft specifically consist of three methods mainly. First is known as state-based analysis, second is a game theory, and third is known as the classification method [20].

State-based analysis techniques use unique methods to offer better performance in terms of accuracy. State-based detection schemes will offer the extra funding required for tracking applications, along with equipment expenses, program start-up prices, software program charges, and operating/training expenses. In [21], it was found that power theft detection in microgrids, considering the physical model of the microgrid power system and consumer privacy protection. It has advised two algorithms for detecting ability theft to successfully perceive energy thieves. One set of rules is known as the centralized state-based estimation algorithm which works on the principle of the Kalman filter (SEK), and central Kalman filter is employed. But, this filter couldn't keep consumer privacy and also lacked numerical stability in massive structures with errors of high percentage. Another algorithm, which is called the privacy-preserving bias estimation (PPBE) algorithm, works on the phenomenon of freely integrated filters and might maintain customer's privacy through keeping their electricity measurement system operators secret, other customers. But, state-based detection methods used tools that give a high percentage of accuracy that comes however at an extra investment needed for a tracking system which includes device expenses, implementation charges, software fees, and training/operating expenses.

Another manner to detect theft is primarily based on the concept of the game theory. The authors in Reference [22] formulate a detection of electricity theft problem just like a game between an unlawful consumer and a power supplier. Power supplier seeks to enhance the availability of illegal consumers and at the same time as unlawful customers try it to reduce the possibilities of being busted by converting the probability density function (PDFs) of unit's consumption.

Classification techniques comprise strategies designed for segmentation consisting of skillful systems, i.e., an expert system along with machine learning. Expert programs primarily focus on trained computer systems via human experts to cope up with complicated issues and illustrate identical results as an expert [23]. In [24], the author determined the probabilities that exist in the application of machine learning methods of non-technical losses to consumers. The evaluation was drawn based on collaborative work. We can analyze the daily consumption pattern of consumers with the help of machine learning algorithms that are used to establish classification models. The most commonly used are random forest, decision tree, support vector machine (SVM). Here the author used the combination of SVM along with a decision tree [25]. In [26–28] clustering and classification using SVM were used to find the irregularities in the dataset. To make this technique more effective, author applied clustering for both the primary and secondary steps to find the electricity usage. The collaboration of 1-class SVM, decision tree, and optimum path forest is used in [29]. Fuzzy logic and SVM [30], also generic algorithm and support vector machine, i.e., GA-SVM, are used, which is effective to enhance accuracy for detection of theft of electricity.

3 Literature Review

In this section, we have discussed a literature review related to energy theft and methods to detect the theft.

Nazmul Hasan et al. [31] Author analyzed that power theft is also extremely high dangerous consequences amid the non-technical loss of the power provider. Fraud energy consumption reduces the level of supply, raises the load of generation, causes unsustainable electricity bills to be charged by legal customers, and affects the general economy. Adapted to smart grids will greatly reduce these losses. A large amount of data is made into an intelligent grid system, including power consumption by individual users. Machine learning and deep learning methods can reliably distinguish the consumers doing electricity theft using this information. In this paper based on a convolution neural network (CNN) network and a very short period memory (LSTM), a system to detect electrical theft was proposed. Convolution neural network is an automated tool for feature extraction as well as segmentation process. However, the details of the time series were its signature of power use, based on CNN. An LSTM (CNN-LSTM) smart grid data allocation framework has been developed. The proposed model achieved 89% accuracy, precision 90%, and recall 87% as it was checked for 17,120 no. of consumers.

Punmiya et al. [32] In this paper, author finds that most current machine learning (ML) techniques focus on classifier hyperparameters, ML algorithm, GBTD that is gradient boosting theft detector, based on previous performance according to optimization of feature, improves the efficiency of detection and the complexity of time. By production stochastic features such as SD, known as standard deviation, mean, minimum, and maximum values of electricity used on daily basis, GBTD improves the rate of detection known as DR and FPR that is a false positive rate of those gradient boosting classifiers (GBCs). By weighted feature-based extraction methods, GBTD is also declining the complexity of the classifier. The actual implementation of the proposed ML for theft discovery was emphasized by the reduction of FPR and the removal of the storage area once improves the time complexity of GBTD class models. The proposed GTBT model has a detection rate i.e. DR 95–97% in terms of average while the existing consumption pattern-based electricity theft detector (CPBETD) has DR 88%. For FPR also the GTBT model performed well having FPR 5–7% in comparison to the exiting model having FPR 15%.

Zheng et al. [33] Author analyzed that Smart grid data analysis helps detect power theft due to unusual use of electrical patterns by electricity thieves. However, while most of them were made with one-dimensional (1-D) data but failed to take the electricity consumption periodicity, current methods have low accuracy of detection of electrical theft. Throughout the text, the author originally introduced something new regarding electrical theft detection which is based on a large and deep convolution neural network (CNN) to resolve the issues mentioned above. Large and deep CNN consists of, in particular, two parts: a wide component and a deep CNN component. Depending on the 2-D specifications in power consumption, the deep part of CNN can accurately classify non-periodicity of theft of electricity and the periodicity of general electricity consumption. In the meantime, a large part of it can capture the global aspects of data in 1-D power consumption. As a result, a larger and deep CNN model can achieve excellent efficiency in detecting electrical theft. Based on a physical database basis, research shows that CNN's larger and deep model surpasses other existing methods.

Ballal et al. [34] It has been reported that with the infrastructure available in the region, real-time electricity theft detection and prevention scheme (ETDPS) was developed. Recommended ETDPS focused on programmable control of logic determines the pilfering data positions and also determines the stolen power of unauthorized users. In the lab, the model has been tested and the findings show that ETDPS works very satisfactorily within diverseness of operating environments. After the implementation of the above model, it was found that the success rate for theft detection was >70%, as out of 219 theft cases it performed well for 158 cases and calculated revenue loss was 8.98% approx.

Li et al. [35] In this paper first, the convolutionary neural network (CNN) is built on this model to read the characteristics of large and flexible smart meter data among unique hours of the day on various days on the effectiveness of convolution and down sampling operations. In addition, to delay the possibility overfitting, a drop-out layer was introduced a back-propagation (BP) algorithm was performed in the training process to switch the network parameters. Also, random forest (RF) was equipped to determine if the user was stealing electricity according to the characteristics obtained. Finally, detection is made based on real power consumption data, and results show that alternatives to accuracy are better than the suggested discovery model. The classification score in terms of average shows that CNN-RF has precision, recall, and F-1 score is 97%. In terms of area under the curve, the results for the proposed hybrid CNN-RF method is 0.99, but the AUC value for CNN-GBDT is 0.97, for CNN-LSTM is 0.98, for CNN-SVM is 0.93, for CNN is 0.77, for SVM is 0.91, for RF is 0.63, for LR is 0.77.

Chandel et al. [36] This study aims to address the problems in the detection of theft of electricity, which is linked to the Bidirectional Long Short-Term Recurrent Neural Network based on memory (CNN-RNN-BiLSTM). 1-D global data variables are recorded by the CNN and define the periodicity and non-periodicity of 2-D data records of consumption of electricity. RNN-BiLSTM extends the ability to store neural memory with the flow of information in both directions, which allows the process of correlation in series of learning sequences. The results of the proposed system for the theft of electricity show a better accuracy rating with reduced training time, and no of epochs were also reduced. After comparison, the proposed method with the current support vector machine (SVM) and SVM multi-class models, validation of the proposed study was performed. Comparative results show that the enhanced electricity theft detection model is CNNRNN- BiLSTM works brilliantly. The author tested the model of 6000 consumers, and it achieves 99% accuracy in the training phase and 97.12% accuracy in the testing phase.

Zhengwei Qu et al. [37], A new approach to detecting theft of electricity which focused on improved synthetic minority oversampling technique (SMOTE) process & improved randomly forestry (RF) process was implemented by the author to increase the effectiveness of the study of grid companies. Positive data (PD) and negative-ND (data), respectively, are labeled as general consumer data and electricity theft consumers. The total of ND was very low than that of PD, which has led to a database containing these two types of data being incompatible. To stabilize the database, an improved SOMTE based on K means algorithm (K-SMOTE) was introduced. Through the K-means process, the ND cluster center was calculated. After that, on the basis of the collection, SMOTE normalized ND to stabilize all dataset. At the final stage, RF classification was done with the measured data and the maximum number of decision trees on the RF is determined based on the out-of-bag data error (OOB error). Professional RF Editor described ways to steal electricity from the user side. After implementation of the above algorithm, it was observed that unbalanced data were processed with and without K-SMOTE and the mean value for accuracy in the RF model was 85.3% and 94.53% respectively. Also, the ROC curve was larger with K-SMOTE and was close to 0.1, whereas in the case of SVM and BPN, it was 71.26% and 84.82% respectively.

Rafia Nishat Toma et al. [38] Author analyzed that the SVM is a widely used machine learning technique used in the analysis of the training data which is collected from (SM) smart meters and measure the accuracy of predictions data. The key component is to reduce the dimensions of data so that the tutoring process could have less complexity. Then grid search method is applied to find the most accurate meta-parameters in SVM are selected where approximately 90% accuracy, 89% precision,90% recall, 89% f1 score are reached. The results of this classification regarding different SVM meta-parameters (standard parameters C and γ) were introduced. The value of C was fixed 100, and value γ was taken 0.001 to achieve high accuracy and low FPR, i.e., 9%. In addition, the results obtained are clear that the techniques used have high accuracy and a low level of falsehoods for real-time consequences.

Li et al. [39] In this paper, it appears that the author is developing a theft detection device called the SETS (Smart Energy Theft System) which uses machine learning and mathematical models. It has three phases of decision-making modules; the first phase has a prediction model that uses a multidisciplinary forecast system. This program incorporates a different kinds of machine learning models into a single power consumption forecast system. The second step is the primary decision-making process that performs decisions making using simple moving average (SMA) for different filters. The final phase takes the second phase which is the last phase of the power theft decision. At stage 1 the model has a theft detecting accuracy of only 56.39%, but with the addition of stage to there is a big change inaccuracy which rises to 99.89%. And with the last and final stage, the algorithm achieves an accuracy of 99.89% which improves a smart home based on IoT security.

Hossain et al. [40] The author conducts extensive research into big data usage and machine learning, as introduced by the evolution of future generation energy system which is called smart grid system (SG). The main context of the new grid infrastructure is connectivity that is furnished by (IoT) Internet of Things. Connectivity, as well as uninterrupted bidirectional communication in the system, also produced a large amount of data that required much higher strategies than conventional methods of proper analysis and decision-making. An integrated SG system can provide data prediction strategies and data acquisition strategies, as well as cost-effectiveness. Big data analysis and machine learning crucial critical issues; IoT equipment and their data become a major objective for the attack.

Ullah et al. [41] Here the author proposed HDNN that is hybrid deep neural network for this project, in which convolution neural network (CNN) is combined with gated recurrent unit (GRU) and particle swarm optimization (PSO), the combination is known as CNN-GRU-PSO-HDNN. The data is taken from the real-world electricity users that can be easily accessed from an online source, known as the State Grid Corporation of China (SGCC). The actual database contains the actual values and the missing values. Initially to filter the data, pre-processing steps were performed. Then extraction and feature selection is done with the help of CNN. In addition, the classification of the legal and illegal consumers is performed with the help of the GRU-PSO process. The performance of the given HDNN model is compared to other measurement techniques such as support vector machine (SVM), logistic regression (LR), GRU, and long short-term memory (LSTM). The performance of the proposed model is verified for different factors such as area under the curve (AUC), precision, recall accuracy, and F1-score. The results of the proposed model are much better than existing ETD strategies. Here the total 5000 consumers were considered as data set where 75% of data was trained and the remaining 25%, i.e., 1250 consumers were left for testing purposes, and it was observed that the proposed model has a high AUC value, i.e., 0.89.

Based on the literature survey, we have compared the different techniques based on the dataset used to analyze the different performance parameters (Table 1).

4 Inferences Drawn from Literature Review

The literature study in the previous section can make the researcher understand the various methods of electricity theft detection. This section discusses the major inferences from the literature and is expressed below.

Table 1 Summary and comparison of different techniques for various performance parameters

References	Technique used	Number of consumers in data set	Accuracy %	Precision %	Recall %	F1 score %	DR %	FPR %
[42]	Wide and deep CNN	42,372	–	94.04	–	–	–	–
[25]	DT coupled SVM	NA	92.5	–	–	–	–	–
[28]	SVM	36,170	60	–	–	–	–	–
[29]	SVM, OPF, decision tree	NA	86.2	54.4	64	–	–	–
[30]	SVM-FIS	36,170	72	–	–	–	–	–
[31]	CNN-LSTM	17,120	89	90	87	–	–	–
[32]	ML-GTBT	5000	–	–	–	–	97	7
[35]	CNN-RF	7584	–	97	97	97	–	–
[36]	CNN-RNN-BiLSTM	6000	–	97.12	–	–	–	–
[37]	SMOTE-RF	NA	–	94.53	–	–	–	–
[38]	SVM	10,000	90	89	90	89	–	9
[39]	SMA with MLP-RNN-LSTM-GRU	NA	99.89	–	–	–	–	–
[41]	CNN-GRY-PSO	5000	89	–	–	–	–	–
[43]	Genetic-SVM	186,968	62	–	–	–	–	–
[44]	Fuzzy classification	NA	74.5	–	–	–	–	–
[45]	Fuzzy logic	NA	55	–	–	–	–	–
[46]	Neural networks (NN)	NA	83.5	–	–	–	–	–
[47]	Neuron-fuzzy	4159	68.2	51.2	–	–	–	–
[48]	CNN-LSTM-Autoencoder	12,180	96.9	–	–	–	–	–

- Electricity theft detection is an important function of maintaining high-performance systems. There are many methods and techniques used for the detection of the theft of electricity.
- The normal process is designed using a convolution neural network. This network provides a platform for easy access to theft. A common source of theft could be meter interference or another issue. CNN allows deal with the issue because of its most beneficial features.
- Periodicity of electric usage, non-periodicity of loss of electricity, false positive rate, and deviations are several factors considered in different studies.
- Logistic regression (LR) and support vector machine (SVM) are most commonly used for electricity theft detection. But these techniques work only when the dataset is small, but in the case of imbalanced data and large dataset this technique is not much effective.
- Random forest is an implementation of numerous no. of decision trees and can maintain significant control on overfitting in comparison to a single decision tree. It can deal with high-dimensional data and maintain computational efficiency very high [35].

- Other methods involve gradient boosting, which is useful but less effective. GB is an iterative decision tree algorithm having multiple no. of trees for decision making. In the end, to get the final output, all outcomes or weights are collected in the GBDT.
- Finally, it is clear that CNN is used for detection purposes; however, performance can change due to the use of various classifiers and pre-processing categories. Convolution neural network (CNN) along with long short-term memory (LSTM) is implemented untidily to resolve the job of classification of data. Feature extraction from the available dataset is done with the help of CNN and LSTM work on sequential data. The combination of LSTM along with CNN is used in various fields that include sentiment analysis, text from image processing or text form video processing, and is also used for language processing [31].
- CNN is not capable to learn temporal and context features, especially long-term dependency between two entities.
- CNN does not take responsibility for unsafe data on systems; as a result of the operation of the system is deteriorating.

The accuracy and performance of the current methods of power theft detection are reasonable. But, these techniques have certain limitations, mentioned here:

- Traditional electricity theft detection contains manual techniques that are human-based meter reading and testing of direct connections to transmission lines. These strategies involve a supplementary amount of employing of inspectorial teams.
- LDR and FPR that is a low detection rate and the false positive rate is achieved through game theory-based techniques [49].
- High cost is offered in the case of state-based solutions as they need an extra expense for the implementation of hardware [50].
- The major issue in electricity theft detection techniques using machine learning algorithms to handle the unbalanced dataset. But with the help of SMOTE and RUS methods imbalanced data is handled, but the problem of overfitting occurs.
- Due to erroneous values in available data classification accuracy is reduced [51].
- The conventional machine learning techniques like SVM and logistic regression have low accuracy to classify the massive dataset [51].

4.1 Advantages and Disadvantages of LSTM

Advantages

- LSTM works very nice over a wide range of parameters like learning rate, input gate bias, output gate bias. So there is no need to do fine adjustments. The complexity of updating each weight in backpropagation through time, i.e., BPTT is reduced with the usage of LSTM.
- Long short-term memory is a form of RNN that may learn from the order dependence among various objects in a sequence. LSTMs are capable of learning the

context needed to make predictions in the time series forecasting issues, rather than having this context pre-specified and fixed.
- LSTM performs very well to classify, process, and predict the time series which gives time lags of unknown duration. Also, LSTM performs better over RNN or other sequence learning methods and has insensitivity to gap length.
- While detecting theft of electricity LSTM can be used to classify the dataset into honest and dishonest consumers.
- If LSTM is used along with CNN and RNN both that is CNN-RNN-BiLSTM, the combination performs very well in an extension of the memory storage capacity of the neural network and also allows the bidirectional flow of information.

Disadvantages

- LSTMs have become famous due to the fact they resolve the trouble of vanishing gradients. The problem of gradient vanishing is reduced by using LSTM along with GMM, i.e., Gaussian Mixture Model [52]. However it seems, it fails to eliminate the problem. The basic problem is that the statistics have to move from one cell to another cell for its evaluation. Furthermore, the cell has to turn out to be quite complicated with some additional functions (including forget gates) being introduced in the picture. So the model becomes quite complex.
- LSTM needs numerous sources and much time for training and emerges as equipped for applications in the real world. In technical terms, they require extra memory bandwidth due to linear layers found in every cell and the system fails to offer so. Accordingly, hardware-sensible, LSTMs grow to be pretty inefficient.
- They are affected by weight initializations done randomly and subsequently behave similarly to that of feed-forward NN.
- LSTM can't solve the problem of imbalanced dataset, i.e., units consumption of consumers. But the problem can be reduced using SMOTE [53].
- LSTMs are liable to overfitting, and it is difficult to implement the dropout algorithm to lower this overfitting difficulty. Dropout algorithm is a regularization method where input unit and recurrent unit to LSTM network are probabilistically excluded from activation function and weight updates at the same time as training a network.

5 Security Aspects of Machine Learning in Terms of Detection of Theft of Electricity

As it is well known that machine learning is widely used in many applications due to its technical breakthrough in current years. But machine learning models are insecure because of various attacks due to which the security of the system is compromised and further its applications. In literature [54], security issues are analyzed and different security evaluations have been done from the training to the testing phase. Figure 4 shows the overview of the machine learning system.

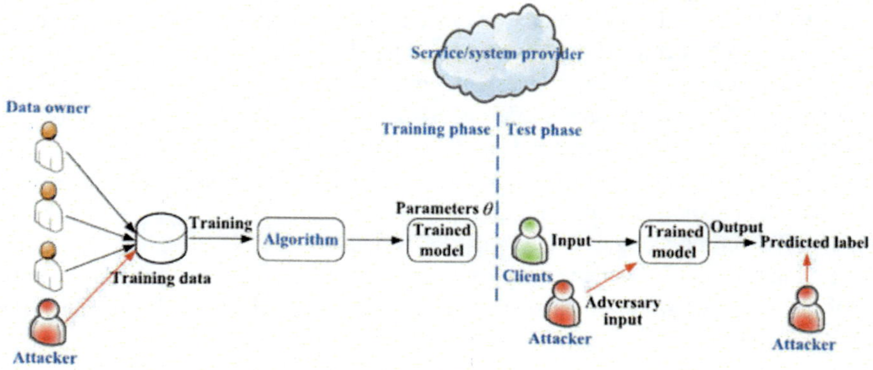

Fig. 4 Overview of machine learning system

Security threats to machine learning systems are mainly divided into two parts according to the learning model's training status. These two types are,

1. Threats before or during the training of the model.
2. Threats after the machine learning model has been trained.

Figure 5 shows the flow diagram for the security threats to machine learning systems and its types.

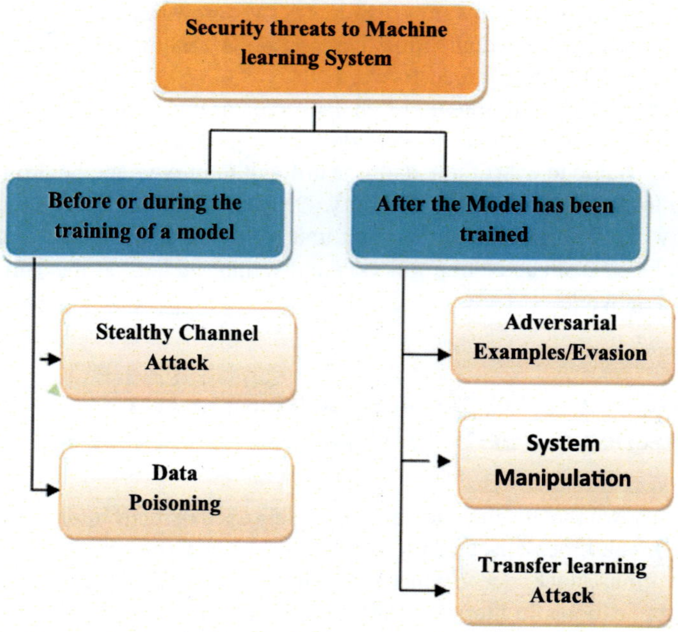

Fig. 5 Security threats to machine learning system

- **Stealthy Channel Attack**

Good quality of data plays important role in creating the machine learning model. So collecting the relevant data is a challenging task. As in detection of theft of electricity collection of units consumed by a consumer is major work to do, and it is collected from various sources, here the fraudulent data can be inserted which will badly affect the accuracy of the system. So, it's a hard duty of data collectors to be diligent while collecting the data. Data imbalance problem also occurs during the collection data as the numbers of normal and abnormal consumption of units are not in proportion. Sometimes, abnormal samples or theft samples are rare or do not exist in the dataset, due to this lack of exhaustive dataset affects the detection rate.

- **Data Poisoning**

Machine learning models are highly dependent on the data for the learning process, i.e., the consumption of the consumers. So it is very important to make sure regarding the integrity, reliability, and security of the data to be used for the learning process. Otherwise, the model will be unable to predict honest and dishonest consumers. The data is corrupted in such a way that the whole system down to its knees.

Another way of data poisoning is by gradient descent attack. If the data used by the model is poisoned, it can push the model toward an infinite loop and the model will be stuck into the iteration process only and will not reach the required results. Sometimes the model is trained on inaccurate parameters, like the consumer may change his/her appliances which will affect the detection rate and the model will falsely believe that predicted dishonest consumers are the right answer. The overview of poisoning attacks is shown in Fig 6.

- **Adversarial Examples/Evasion Attack**

In these kinds of attacks input or testing, data is manipulated in such a way that machine learning system makes false predictions due to malicious inputs. Simply, for example, if some consumers have gone on vacations for one month and their bill will be generated on basis of average due to "Lock". But if the model is trained that "Lock" means dishonest, and this picture does not exist in the real world and the model will be forced to make the decision based on that.

- **System Manipulation**

A good machine learning system keeps on learning from the environmental feedback and keeps on improving it like reinforcement models. For example, the consumption of industrial consumers may vary due to lockdown in the COVID-19 period, or for domestic consumers, the consumption will vary for summer and winter, but if the feedback system of the model is attacked then the results will go in the wrong direction.

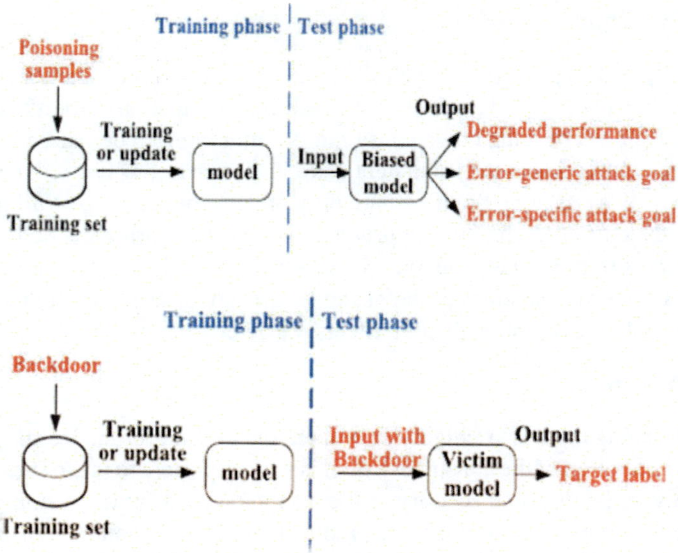

Fig. 6 Data poisoning attack on the machine learning model

- **Transfer Learning Attack**

Sometimes for fast production machine learning models which are pre-trained are used. Which saves the time to train a large amount of training data? The conventional machine learning models are tweaked to meet particular functions by offering specialized training. At this stage of transfer learning, an attack may lethal. If the selected model is popular attackers can launch attacks that can even fool your project-specific machine learning model.

- **Output Integrity Attack**

If the attacker gets in between the model and interface accountable for showing the outcomes, then manipulated outcomes may be displayed. This form of attack is known as the output integrity attack. Due to a lack of awareness of the real internal working of the machine learning system theoretically, it will become hard to find the actual results. Consequently, while the output has been displayed by using the system it is taken at face value. This naivety may be exploited by the attackers by compromising the integrity of the output.

6 Conclusion

Literature research has been done and analyzed that machine learning algorithms use smart meters to detect the abnormal consumption behavior of the consumers to find the dishonest consumers, but it does not perform well when the data is in an imbalanced form. A model is proposed to detect electricity theft based on a convolution neural network (CNN) along with (LSTM) which is a long short-term memory algorithm. The convolution neural network is implemented to extract the features *while LSTM used the suita*ble features for the classification of the records into legal users and theft users. To resolve the trouble of the dataset which is not balanced data set, the Synthetic Minority Over Sampling method (SMOTE) is applied. It has accomplished accurate results in terms of accuracy, recall, and precision, but the overfitting problem is not taken into consideration, that's resulting from the addition of duplicate records via SMOTE. The model provided practical accuracy, but after that, a study of other texts reveals that there is still room for improvement.

References

1. Lo YL, Huang SC, Lu CN (2012) Non-technical loss detection using smart distribution network measurement data. In: Proceedings of the IEEE PES innovative smart grid technologies, Tianjin, China, 21–24 May 2012, pp 1–5
2. Agüero JR (2012) Improving the efficiency of power distribution systems through technical and non-technical losses reduction. In: Proceedings of the PES T&D 2012, Orlando, FL, USA, 7–10 May 2012, pp 1–8
3. McLaughlin S, Holbert B, Fawaz A, Berthier R, Zonouz S (2013) A multi-sensor energy theft detection framework for advanced metering infrastructures. IEEE J Sel Areas Commun 31:1319–1330
4. Smith TB (2004) Electricity theft: a comparative analysis. Energy Policy 32:2067–2076
5. Cabral JE, Gontijo EM (2004) Fraud detection in electrical energy consumers using rough sets. In: Proceedings of the 2004 IEEE international conference on systems, man and cybernetics (IEEE Cat. No. 04CH37583), The Hague, The Netherlands, 10–13 Oct 2004; IEEE, Piscataway, NJ, USA, pp 3625–3629
6. Huang SC, Lo YLCN (2013) Non-technical loss detection using state estimation and analysis of variance. IEEE Trans Power Syst 28:2959–2966
7. Bhavna B, Mohinder G (2004) Reforming the power sector, controlling electricity theft and improving revenue. Public policy for the private sector. Available online: http://rru.worldbank.org/PublicPolicyJourna. Accessed 16 Dec 2019
8. Soma SSRD, Wang L, Vijay D, Robert CG (2013) High-performance computing for detection of electricity theft. Int J Electr Power Energy Syst 47:21–30
9. Smart Meters Help Reduce Electricity Theft, Increase safety. BCHydro. Available online: https://www.bchydro.com/news/conservation/2011/smart_meters_energy_theft.html. Accessed 16 Dec 2019
10. Dzung D, Naedele M, Von Ho TP, Crevatin M (2005) Security for industrial communication systems. Proc IEEE Secur Ind Commun Syst 93:1152–1177
11. Krebs B (2019) FBI: smart meter hacks likely to spread. Available online: ttp://krebsonsecurity.com/2012/04/fbismart- meter-hacks-likely-to-spread. Accessed 16 Dec 2019
12. Carlos L, Félix B, Iñigo M, Juan I, Guerrero JB, Rocío M (2011) Integrated expert system applied to the analysis of non-technical losses in power utilities. Expert Syst Appl 38:10274–10285
13. Yu X, Cecati C, Dillon T, Simoes MG (2011) The new frontier of smart grids. IEEE Ind Electron Mag 5:49–63

14. Mavridou A, Papa M (2011) A situational awareness architecture for the smart grid. In: Global security, safety and sustainability & e-Democracy. Springer, Berlin/Heidelberg, Germany, pp 229–236
15. "Pilferage of electricity—issues and challenges," power sector news, KSEB Officers' Association, [Online]. Available: http://www.kseboa.org/news/pilferage-of-electricity-issues-and-challenges.html
16. Depuru SSSR, Wang L, Devabhaktuni V (2011) Electricity theft: overview, issues, prevention and a smart meter based approach to control theft. Energy Policy 39:1007–1015
17. Han W, Xiao Y (2014) NFD: a practical scheme to detect non-technical loss fraud in smart grid. In: Proceedings of 2014 IEEE international conference on communications (ICC), Sydney, NSW, Australia, 10–14 June 2014, pp 605–609
18. Grochocki D, Huh JH, Berthier R (2012) AMI threats, intrusion detection requirements, and deployment recommendations. In: Proceedings of the IEEE third international conference on smart grid communications, Tainan, Taiwan, China, 5–8 Nov 2012, pp 395–400
19. Hao R, Ai Q, Xiao F (2017) Architecture based on multivariate big data platform for analyzing electricity consumption behavior. Electr Power Autom Equip 37:20–27
20. Jiang R, Lu R, Wang Y, Luo J, Shen C, Shen X (2014) Energy-theft detection issues for advanced metering infrastructure in smart grid. Tsinghua Sci Technol 19:105–120
21. Salinas S, Li P (2016) Privacy-preserving energy theft detection in microgrids: a state estimation approach. IEEE Trans Power Syst 31:883–894
22. Cárdenas AA, Amin S, Schwartz G, Dong R, Sastry S (2012) A game theory model for electricity theft detection and privacy-aware control in AMI systems. In: Proceedings of the 50th annual Allerton conference on communication, control, and computing (Allerton), Monticello, IL, USA, 1–5 Oct 2012, pp 1830–1837
23. O'Leary DE (2016) Summary of previous papers in expert systems review. Intell Syst Account Financ Manag 1:3–7
24. Coma-Puig, B.; Carmona, J. Bridging the Gap between Energy Consumption and Distribution through Non-Technical Loss Detection. Energies **2019**, 12, 1748
25. Jindal A, Dua A, Kaur K, Singh M, Kumar N, Mishra S (2016) Decision tree and SVM-based data analytics for theft detection in smart grid. IEEE Trans. Ind. Inform. 12:1005–1016
26. Hodge V, Austin J (2004) A survey of outlier detection methodologies. Artif Intell Rev 22:85–126
27. Jokar P, Arianpoo N, Leung VC (2015) Electricity theft detection in AMI using customers' consumption patterns. IEEE Trans Smart Grid 7:216–226
28. Nagi J, Mohammad A, Yap KS, Tiong SK, Ahmed SK (2008) Non-technical loss analysis for detection of electricity theft using support vector machines. In: Proceedings of the 2008 IEEE 2nd international power and energy conference, Johor Bahru, Malaysia, 1–3 Dec 2008; IEEE, Piscataway, NJ, USA, pp 907–912
29. Di Martino M, Decia F, Molinelli J, Fernández A (2012) Improving electric fraud detection using class imbalance strategies. In: Proceedings of the international conference on pattern recognition applications and methods (ICPRAM), Vilamoura, Portugal, 6–8 Feb 2012, pp 135–141
30. Nagi J, Yap KS, Tiong SK, Ahmed SK, Nagi F (2011) Improving SVM-based nontechnical loss detection in power utility using the fuzzy inference system. IEEE Trans Power Delivery 26:1284–1285
31. Nazmul Hasan M, Toma RN, Nahid A-A, Manjurul Islam MM, Kim J-M (2019) Electricity theft detection in smart grid systems: a CNN-LSTM based approach. Energies 12:3310. https://doi.org/10.3390/en12173310
32. Punmiya R, Choe S (2019) Energy theft detection using gradient boosting theft detector with feature engineering-based preprocessing. IEEE Trans Smart Grid 10(2):2326–2329. https://doi.org/10.1109/TSG.2019.2892595
33. Zheng Z, Yang Y, Niu X, Dai H, Zhou Y (2018) Wide and deep convolutional neural networks for electricity-theft detection to secure smart grids. IEEE Trans Industr Inf 14(4):1606–1615. https://doi.org/10.1109/TII.2017.2785963

34. Ballal MS, Suryawanshi H, Mishra MK, Jaiswal G (2020) Online electricity theft detection and prevention scheme for smart cities. IET Smart Cities 2(3):155–164. https://doi.org/10.1049/iet-smc.2020.0045, https://doi.org/10.1049/iet-smc.2020.0045
35. Li S, Han Y, Yao X, Yingchen S, Wang J, Zhao Q (2019) Electricity theft detection in power grids with deep learning and random forests. J Electr Comput Eng, 12p, Article ID 4136874. https://doi.org/10.1155/2019/4136874
36. Chandel P, Thakur T (2019) Smart meter data analysis for electricity theft detection using neural networks. Adv Sci, Technol Eng Syst J 4(4):161–168. https://doi.org/10.25046/aj040420
37. Qu Z, Li H, Wang Y, Zhang J, Abu-Siada A, Yao Y (2020) Detection of electricity theft behavior based on improved synthetic minority oversampling technique and random forest classifier. Energies. 20:2039. https://doi.org/10.3390/en13082039
38. Toma RN, Hasan MN, Nahid A, Li B (2019) Electricity theft detection to reduce non-technical loss using support vector machine in smart grid. In: 2019 1st International conference on advances in science, engineering and robotics technology (ICASERT), Dhaka, Bangladesh, 2019, pp 1–6. https://doi.org/10.1109/ICASERT.2019.8934601
39. Li W, Logenthiran T, Phan V-T, Woo WL (2019) A novel smart energy theft system (SETS) for IoT based smart home. IEEE Internet of Things J 6:5531–5539. https://doi.org/10.1109/JIOT.2019.2903281
40. Hossain E, Khan I, Un-Noor F, Sikander SS, Samiul Haque Sunny M (2019) Application of Big Data and machine learning in smart grid, and associated security concerns: a review. https://doi.org/10.1109/ACCESS.2019.2894819, IEEE Access
41. Ullah A, Javaid N, Samuel O, Imran M, Shoaib M (2020) CNN and GRU based deep neural network for electricity theft detection to secure smart grid. In: 2020 International wireless communications and mobile computing (IWCMC). https://doi.org/10.1109/IWCMC48107.2020.9148314
42. Zheng Z, Yang Y, Niu X, Dai H-N, Zhou Y (2018) Wide and deep convolutional neural networks for electricity-theft detection to secure smart grids. IEEE Trans Ind Inform 14:1606–1615
43. Nagi J, Yap KS, Tiong SK, Ahmed SK, Mohammad A (2008) Detection of abnormalities and electricity theft using genetic support vector machines. In Proceedings of the TENCON 2008–2008 IEEE Region 10 conference, Hyderabad, India, 19–21 Nov 2008. IEEE, Piscataway, NJ, USA, pp 1–6
44. Angelos EWS, Saavedra OR, Cortés OAC, de Souza AN (2011) Detection and identification of abnormalities in customer consumptions in power distribution systems. IEEE Trans Power Deliv 26:2436–2442
45. Nagi J, Yap KS, Nagi F, Tiong SK, Koh S, Ahmed SK (2010) NTL detection of electricity theft and abnormalities for large power consumers in TNB Malaysia. In: Proceedings of the 2010 IEEE student conference on research and development (SCOReD), Putrajaya, Malaysia, 13–14 Dec 2010, IEEE, Piscataway, NJ, USA pp 202–206
46. Muniz C, Figueiredo K, Vellasco M, Chavez G, Pacheco M (2009) Irregularity detection on low tension electric installations by neural network ensembles. In: Proceedings of the 2009 international joint conference on neural networks, Atlanta, GA, USA, 14–19 June 2009. IEEE, Piscataway, NJ, USA, pp 2176–2182
47. Muniz C, Vellasco MMBR, Tanscheit R, Figueiredo K (2009) A neuro-fuzzy system for fraud detection in electricity distribution. In Proceedings of the IFSA/EUSFLAT conference, Lisbon, Portugal, 20–24 July 2009, pp 1096–1101
48. Bhat RR, Trevizan RD, Sengupta R, Li X, Bretas A (2016) Identifying non-technical power loss via spatial and temporal deep learning. In: Proceedings of the 2016 15th IEEE international conference on machine learning and applications (ICMLA), Anaheim, CA, USA, 18–20 Dec 2016. IEEE, Piscataway, NJ, USA, pp 272–279
49. Amin S, Schwartz GA, Cardenas AA, Sastry SS (2015) Game-theoretic models of electricity theft detection in smart utility networks: providing new capabilities with advanced metering infrastructure. IEEE Control Syst Mag 35:66–81
50. Leite JB, Mantovani JRS (2016) Detecting and locating non-technical losses in modern distribution networks. IEEE Trans Smart Grid 9:1023–1032

51. Wang S, Chen H (2019) A novel deep learning method for the classification of power quality disturbances using a deep convolutional neural network. Appl Energy 235:1126–1140
52. Ding N, Ma H, Gao H, Ma Y, Tan G (2019) Real-time anomaly detection based on long short-term memory and Gaussian mixture model. Comput Electr Eng 70:106458
53. Aslam Z, Javaid N, Ahmad A, Ahmed A, Gulfam SM (2020) A combined deep learning and ensemble learning methodology to avoid electricity theft in smart grids. Energies 13:5599. https://doi.org/10.3390/en13215599
54. Xue M, Zhang Y (2020) Machine learning security: threats, countermeasures, and evaluations 32. IEEE Access 8:74720–74742
55. "World electricity distribution losses," International Energy Annual 2006, U.S. Energy Information Administration, [Online]. Available: http://www.eia.doe.gov/pub/international/iealf/tables5.xls

Monitoring of Grid Connected 1.43 MWp Rooftop Solar PV Plant by Internet of Things (IoT)

Santu Hore, Raja Kumar Sakile, and Umesh Kumar Sinha

Abstract An installation of rooftop solar PV system is the most promising option to reduce the cost of power followed by manufacturing cost as well as mitigation of renewable power obligation for industries, but there are challenges to monitor the real time system parameters, actual power generation, plant efficiency, and working status of equipment of the plant due to its location on rooftop. In this paper, detailed engineering has been done for implementation of IoT to monitor the 1.43 MWp rooftop solar PV plant installed on industrial shed. The flow diagram, hardware requirement and software platform for implementation of IoT are discussed in details. This will facilitate historical analysis of plant performance, performance evaluation, fault detection, and real-time analysis of the plant. This paper also contains technical specifications of each component of rooftop PV plant and IoT platform to monitor the plant.

Keywords Internet of Things (IoT) · IoT architecture · Rooftop solar PV

1 Introduction

As a part of 'World's largest Renewable energy programme', India has taken of 175 GW renewable energy installation target till 2022. To meet the target, rooftop solar PV installations will be the key focus area considering rooftop solar installations can grow 13 GW or more by 2022 [1]. India is the solar richest country, and Indian industries are having large scale of industrial sheds where rooftop solar PV power plants can be installed easily. So, Indian industries are thinking differently to generate

S. Hore (✉) · R. K. Sakile · U. K. Sinha
Department of Electrical Engineering, NIT Jamshedpur, Jamshedpur, India
e-mail: horeaec@gmail.com

R. K. Sakile
e-mail: 2018rsee006@nitjsr.ac.in

U. K. Sinha
e-mail: uksinha.ee@nitjsr.ac.in

K. Namrata et al. (eds.), *Smart Energy and Advancement in Power Technologies*, Lecture Notes in Electrical Engineering 926,
https://doi.org/10.1007/978-981-19-4971-5_39

the solar power in large scale to meet their energy demand, reduce manufacturing cost, and mitigation of renewable power obligation (RPO).

Rooftop solar PV plant mean the complete installation of solar PV panels located on the roof top of buildings or industrial sheds. Total power generation from the PV plant can be effected by several factors such as average duration of daylight hours, panel temperature, tilt angle, intensity, and dust accumulation on panels, inverter efficiency, etc. [2]. To ensure maximum efficiency and optimum power output of the solar PV plant, sufficient measure needs to be taken and monitored, but due to location of the plant it always may not be possible to monitor the system by conventional method including manual inspection and hence the application of IoT comes.

The internet is a simple communication network that connects individuals to information banks where we can describe Internet of Things (IoT) as an interconnected system between differently addressed physical components with processing, sensing and evaluating capabilities and communicate through internet to a central platform. In other words, we can say IoT is an information sharing environment where objects are connected through wireless or wired network. The IoT applications are using in various sectors like health care, smart cities, energy system, security management, education, transport consumer electronics, etc. [3].

In this paper, detailed engineering has been discussed for implementation of IoT to monitor the MW level rooftop solar PV plant installed on industrial shed. The proposed flow diagram, hardware requirement and software platform for implementation of IoT has also been discussed in details. This will facilitate historical analysis of plant performance, performance evaluation, fault detection, and real-time analysis of the plant. In this paper detail technical specification of each component of rooftop PV plant and IoT platform to monitor the plant has also been discussed.

1.1 Grid Connected Rooftop Solar PV Plant

In current pandemic (COVID-19) situation, industries are taking initiatives for manufacturing cost reduction and increase of fossil fuels cost is the biggest challenge to achieve the same. To reduce the power cost followed by manufacturing cost, industries are looking differently to resource the power from cheapest sources or to generate power from renewable resources.

Figure 1 shows that industrial sector is the bulk power consumer and it is consuming approx. 43% of the total energy consumption [4]. Considering the sharp depletion of conventional resources it is therefore required to move forwards for non-conventional energy like wind, solar, biogas, etc. Among all renewable sources, solar energy is the easy and profitable resource for green energy generation.

Industries are having large no. of industrial sheds, roof tops of which can easily use for installation of solar modules to reduce the cost of land required for solar power plants. Solar energy is profitable as it does not require any fuel. Due to fluctuating in nature and non-availability in night hours, solar power plant can be designed as

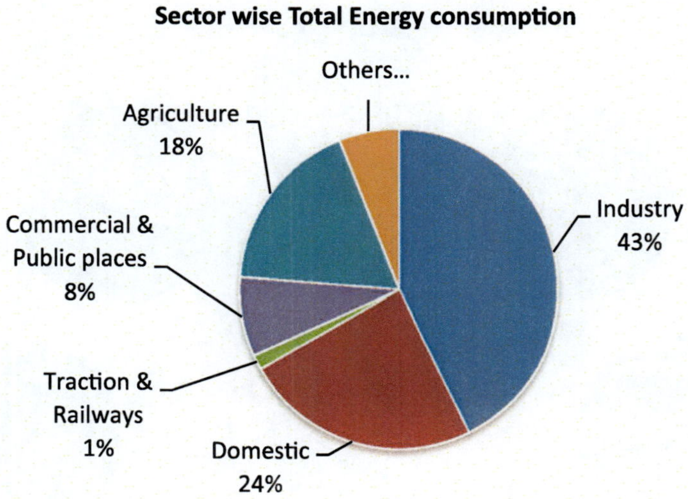

Fig. 1 Sector wise total energy consumption (*Source* Energy Statistics India 2021)

storage type and or grid connected type among which grid connected solar power plant is preferable for industries to ensure reliable power supply to the loads [5].

1.2 *The Internet of Things*

In the year 1999, Kevin Ashton was first instituted the phrase "Internet of Things". In these days, everyone is aware about the solar PV technology and about solar power generation system, but now the main challenge is how to measure and monitor the performance of solar PV plants located at remote locations or on roof tops [6]. With technology advancement and low cost solution, solar PV plants are installing in large scales across the world and these PV plants need to be monitored remotely to ensure maximum power output and efficiency. With the help of Internet of Things (IoT), performance of solar plants can be monitored easily from a central location [7].

Internet is the network between computer to computer or computer to data base, where we can describe IoT as a network between various devices, devices to computers to networks. Through the IoT, we can manage education to health-care system, agriculture to power g0rid management, every area can be monitor and control by IoT from our smartphone to available smart devices [8] (Fig. 2).

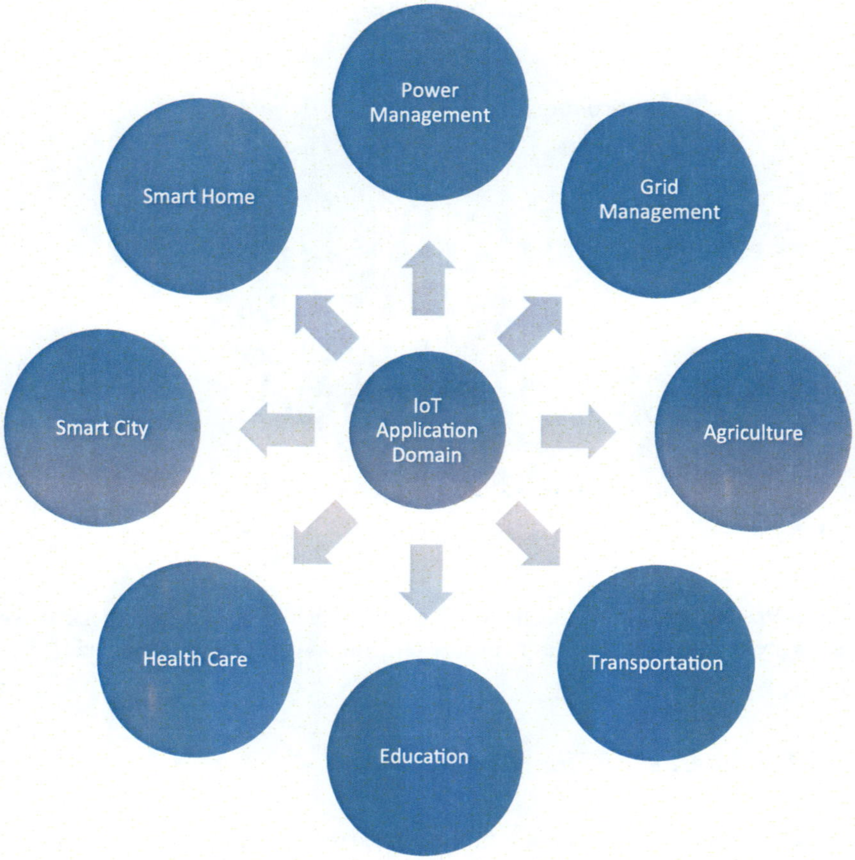

Fig. 2 IoT application domain

2 Plant Description

Design of a rooftop solar PV plant in MW level is always challenging for designer. In the current trend of frequent technology upgradation, designer shall need to know about the technical details of the plant including the communication system through IoT [9]. Communication system through IoT need to be designed for real time monitoring of solar power generation, irradiance, temperature of solar cells, working condition of ON grid inverters, winding temperature of transformers and other operational parameters [10].

Plant Location

The proposed implementation of IoT is carried out on 1.43 MWp rooftop solar PV plant located at 22.8°N, 86.3°E. The complete plant was installed on roof top of an industrial shed having tilt angle of 19°. Detail about the plant is depicted in Table 1.

Table 1 Details of plant location

Location details		
S. No.	Parameters	Specification
1	Latitude and longitude	22.8°N, 86.3°E
2	Facing	South
3	Glob Hor	1810.80 (kWh/m^2/year)
4	Tilt angle	19°
5	Location of plant	Rooftop
6	Distribution grid voltage	6600 V

Table 2 Technical parameters of PV module

PV panel specification		
S. No.	Parameters	Specification
1	PV panel type	72 cell polycrystalline
2	No of cells in each panel	72 Nos
3	Peak power (P_{max})	320 Wp
4	Module efficiency	16.40%

Solar PV Panels

Efficiency, cost, and availability are top factors that decide the selection of the solar PV panels. The 1.43 MWp rooftop solar PV plant was designed by 320 Wp Polycrystalline module as depicted in Table 2.

ON Grid Inverters

To convert the DC output of the solar module to AC, two nos. 630 KW each grid tied inverter has been installed with overloading consideration of 40% as depicted in Table 3.

Table 3 Technical parameters of ON grid inverters

Inverter specification		
S. No.	Parameters	Specification
1	Type	Grid-connected PV inverter
	DC- input	
2	V_{max} PV	1000 V
3	Max. input current	1356 A
	AC-output	
4	Rated output power	630 kW
5	Rated output voltage	3–360 V
6	Rated output frequency	50 Hz

Step-up Transformer

To synchronize the solar PV generation with the 6.6 kV power distribution grid one no. 1250 KVA, 0.36/6.6 kV, 3 winding cast resin transformer has been installed. The vector group of solar transformer is selected YNd11d11) as depicted in Table 4.

Energy Meter

Total generation of the solar power plant is monitoring through three phase, four wire energy meter. The energy meter is having the communication port RS 485 for network connectivity through which real-time generation of solar plant can be monitored with the implementation of IoT. The detail technical parameters of energy meter used for this plant as depicted in Table 5 (Fig. 3).

Table 4 Technical parameters of solar transformer

Transformer specification		
S. No.	Parameters	Specification
1	Power rating	1250/625–625 KVA
2	Vector group	YNd11d11
3	Phase	3
4	No load voltage	HV: 6.6 kV
		LV-1: 0.360 kV
		LV-2: 0.360 kV
5	Current	HV: 109.35 A
		LV-1:1002.34 A
		LV-2: 1002.34 A
6	Communication port	RS 485

Table 5 Technical parameters of smart energy meter

Energy meter specification		
S. No.	Parameters	Specification
1	Type	3 phase, 4 wire
2	Energy parameters	MWh, MVArh, MVAh, MW, MVA
3	Voltage	6.6 kV/$\sqrt{3}$/110 V/$\sqrt{3}$
4	Ib	5 A
5	Imax	10 A
6	Frequency	50 Hz
7	Class	0.2 s
8	Communication port	RS 485

Fig. 3 Actual site photographs of 1.43 MWp rooftop solar PV plant

3 Implementation of IoT

The proposed IoT application for this 1.43 MWp solar PV plant is designed using four layer architecture named as field and control, data communication, and IoT platform and analytics [11]. The field is designed with smart sensors, smart meters, measuring, and monitoring devices to capture the desired data. The field devices are then connects to IoT platform with communication cables and through communication port of devices. After processing and cleaning of data, the IoT application transform the data from storage device to cloud for access from smart application devices, mobile apps, etc. [12] (Fig. 4).

Hardware Setup

The hardware part of the IoT system for 1.43 MWp rooftop solar PV plant designed by solar log module, network switch, smart meters, sensors, hard disk for data storage, and GPRS SIM. The smart energy meters and sensors are connected through communication cable to the network switch. The network switch further connected to solar log module by single pair communication cable [13]. The solar log module then connected to the cloud via GPRS SIM and the processed values can be monitored through desktop/mobile via app for real-time monitoring and analysis purposes [14]. Detail hardwires used in this project is summarized in Table 6 (Figs. 5, 6, 7 and 8).

Software Setup to IoT Platform

Implementation of IoT is carried out on 1.43 MWp rooftop solar PV plant through the designed software platform by solar log. The software is having inherent system to process and display of PV system performance at a glance. The system can also be easily accessed in mobile through the App.

Work Flow

Figure 9 represents the implementation of IoT process to monitor the performance and system parameters of 1.43 MWp rooftop plant. Real-time parameters of DC system

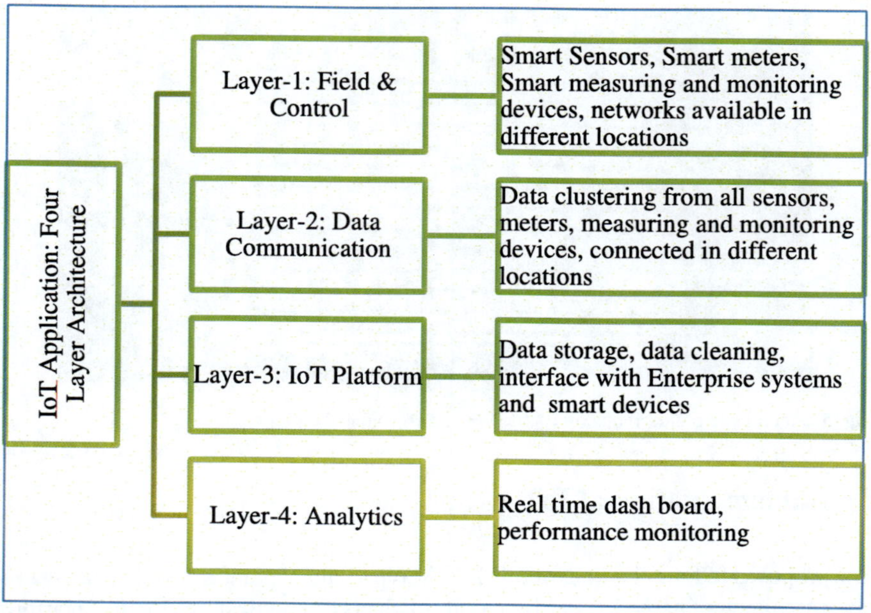

Fig. 4 Four layer architecture of IoT implementation

Table 6 Hardware details used in this project

S. No.	Components	Specifications
1	Solar log module	Operating parameters: 1 A, 12 V DC Connectivity: GPRS Communication port: RS 485
2	Adapter for solar log	Input: 220 V AC; output: 12 V DC
3	Network switch	8-channel network switch
4	Smart energy meters	Suitable to communicate with RS 485 port
5	Hard disk	500 GB
6	Communication cable	Single pair, 7 × 0.2 mm conductor stranding, tinned copper materials
7	Communication media	GPRS SIM

first sense the parameters and send to solar log for further process. The solar log sense the values, process for display and upload to the cloud through GPRS system [15]. Then the user can login into the web portal and mobile app for monitoring, record, and analysis the parameters.

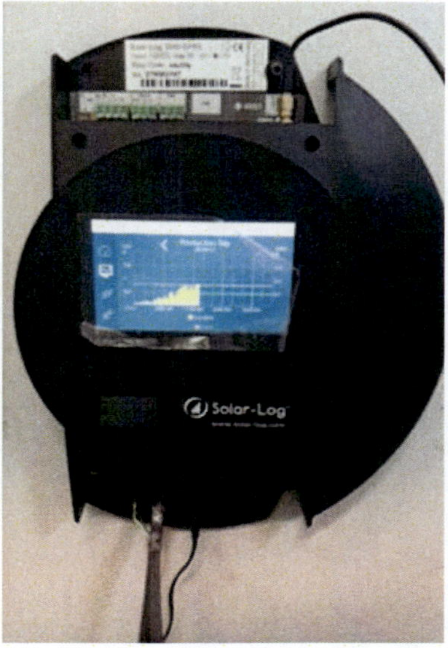

Fig. 5 Solar log module

Fig. 6 Channel network switch

4 Results and Discussions

The 1.43 MWp rooftop solar PV plant is installed through 4480 nos. solar PV panels in combination of 224 strings on rooftop of industrial shed. After installation of the PV plant, toward monitoring of the real-time parameters and performance of the solar PV plant, IoT system has also been installed.

Fig. 7 Smart meters with RS 485 port communication

Fig. 8 Portable antenna for GPRS

From the IoT platform, it can easily be monitored the day wise solar power generation for 12 h. as well as for 24 h. for any typical day. Sample power generation graph for different days are showing in Figs. 10 and 11.

This PV plant is consisting two nos. 630 KW Inverter and through the IoT implementation, inverter (Inverter-I & Inverter-II) wise power generation can also be monitored. Inverter wise solar power generation for any typical days are showing in Fig. 12.

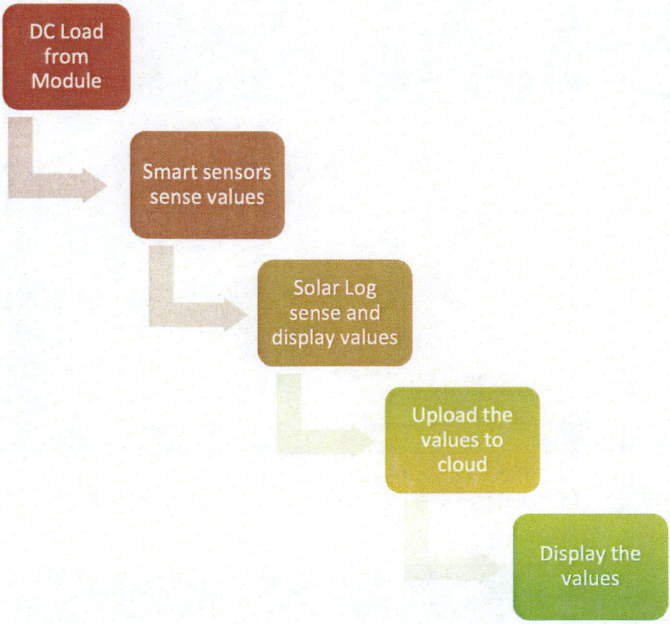

Fig. 9 Work flow of Iot implementation in rooftop solar PV plant

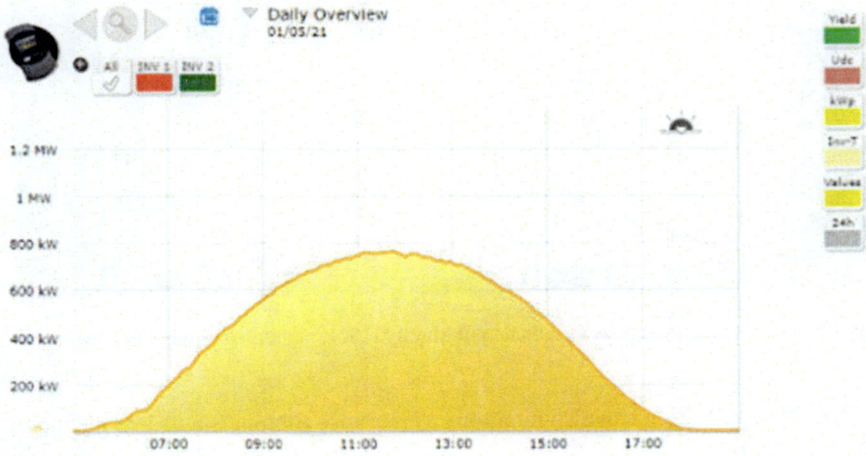

Fig. 10 Day (12 h) wise solar power generation of May, 1, 2021 monitored through IoT application

Through the IoT application it can also be monitor the yield history on daily, monthly, and yearly basis. The yield history of the plant is showing in Figs. 13, 14, 15 and 16.

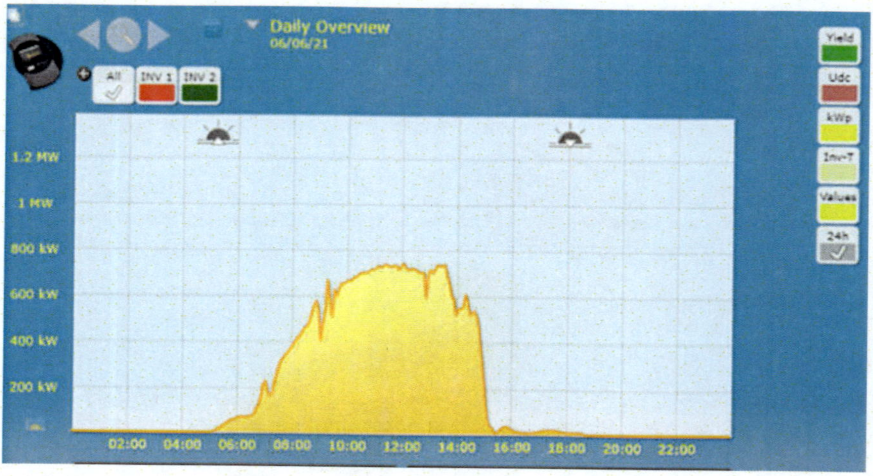

Fig. 11 Day (24 h) wise solar power generation of June 6, 2021, monitored through IoT application

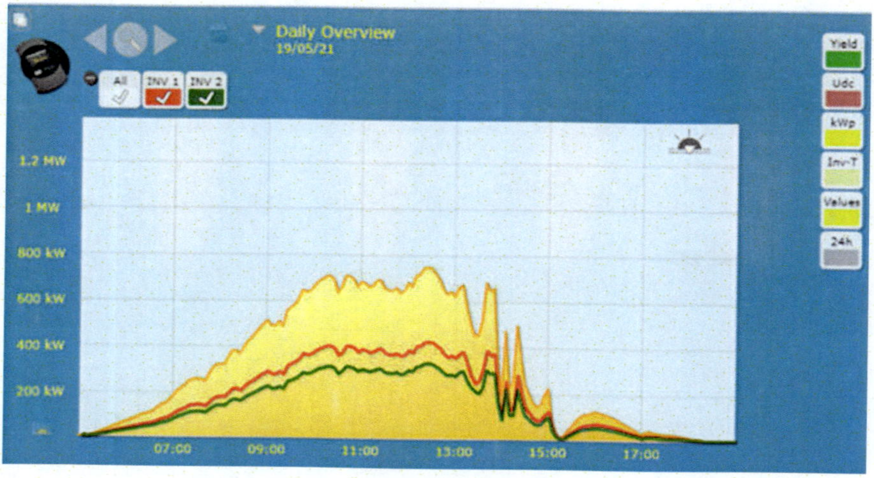

Fig. 12 Inverter wise solar power generation of May 19, 2021, monitored through IoT application

The IoT application can also monitor inverter wise running status as showing in Figs. 17 and 18.

5 Conclusion

To meet the manufacturing cost reduction industries are looking for renewable energy resources, and solar PV is the most promising options to meet the target. Industries

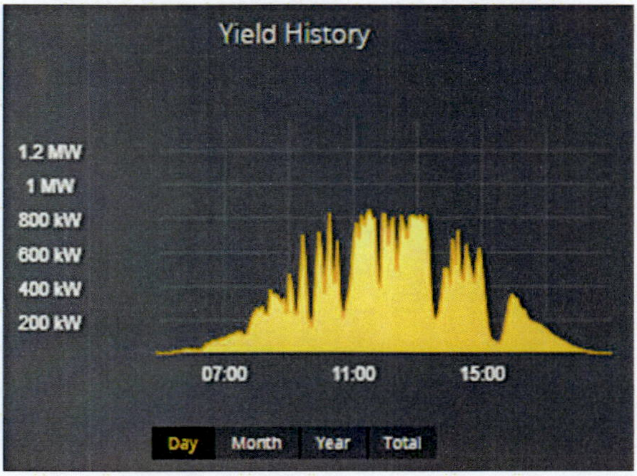

Fig. 13 Day wise yield history

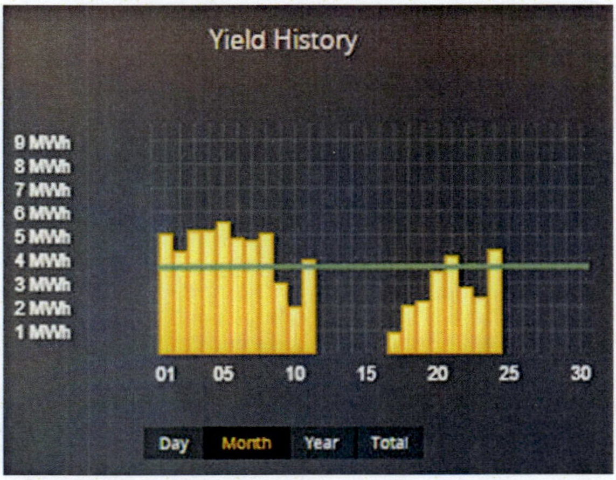

Fig. 14 Month wise yield history

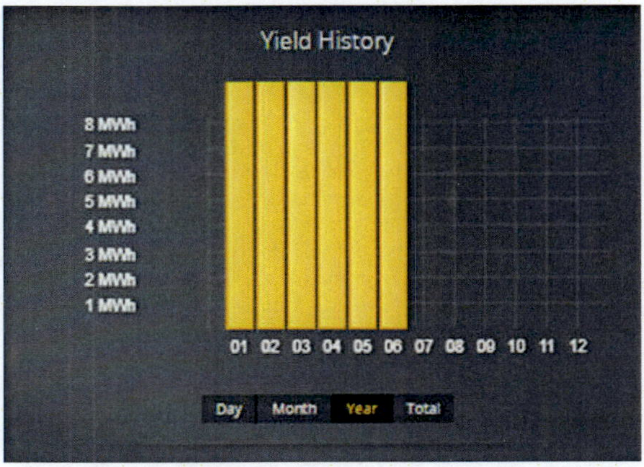

Fig. 15 Yearly (Jan-21 to June-21) yield history monitored through IoT

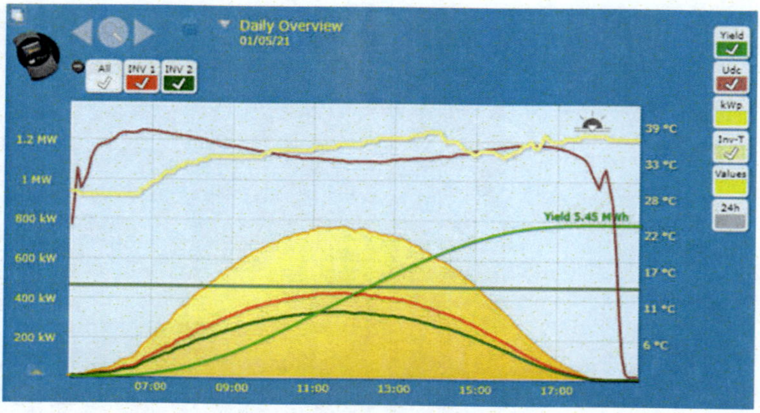

Fig. 16 Web based monitoring of inverter wise generation, yield, inverter temperature, input voltage

Inverter	Date	Status	Error
1	24.06.2021 18:34 -	Stand-by	
1	24.06.2021 05:21 - 24.06.2021 18:34	Alarm	
1	24.06.2021 05:20 - 24.06.2021 05:20	Run	
1	24.06.2021 05:15 - 24.06.2021 05:19	Stand-by	
1	24.06.2021 05:06 - 24.06.2021 05:15	Alarm	
1	24.06.2021 05:05 - 24.06.2021 05:06	Run	
1	24.06.2021 04:48 - 24.06.2021 05:05	Initial Stand-by	
1	23.06.2021 18:35 - 24.06.2021 04:48	Stand-by	
1	23.06.2021 05:26 - 23.06.2021 18:34	Alarm	
1	23.06.2021 05:26 - 23.06.2021 05:26	Run	
1	23.06.2021 05:25 - 23.06.2021 05:25	Start-up	
1	23.06.2021 05:21 - 23.06.2021 05:25	Stand-by	
1	23.06.2021 05:12 - 23.06.2021 05:21	Alarm	
1	23.06.2021 05:11 - 23.06.2021 05:12	Run	
1	23.06.2021 04:51 - 23.06.2021 05:11	Initial Stand-by	
1	22.06.2021 18:33 - 23.06.2021 04:51	Stand-by	
1	22.06.2021 05:15 - 22.06.2021 18:33	Alarm	
1	22.06.2021 05:14 - 22.06.2021 05:15	Run	
1	22.06.2021 05:09 - 22.06.2021 05:14	Stand-by	
1	22.06.2021 05:00 - 22.06.2021 05:09	Alarm	
1	22.06.2021 04:59 - 22.06.2021 05:00	Run	
1	22.06.2021 04:54 - 22.06.2021 04:59	Initial Stand-by	
1	22.06.2021 04:54 - 22.06.2021 04:54	Start-up	
1	22.06.2021 04:47 - 22.06.2021 04:54	Initial Stand-by	
1	21.06.2021 18:21 - 22.06.2021 04:47	Stand-by	
1	21.06.2021 18:12 - 21.06.2021 18:21	Alarm	
1	21.06.2021 18:11 - 21.06.2021 18:12	Run	
1	21.06.2021 18:05 - 21.06.2021 18:11	Stand-by	
1	21.06.2021 05:12 - 21.06.2021 18:05	Alarm	
1	21.06.2021 05:11 - 21.06.2021 05:12	Run	
1	21.06.2021 05:06 - 21.06.2021 05:11	Stand-by	
1	21.06.2021 04:57 - 21.06.2021 05:06	Alarm	
1	21.06.2021 04:56 - 21.06.2021 04:57	Run	
1	21.06.2021 04:45 - 21.06.2021 04:56	Initial Stand-by	
1	20.06.2021 18:22 - 21.06.2021 04:45	Stand-by	
1	20.06.2021 05:22 - 20.06.2021 18:22	Alarm	

Fig. 17 Running status of inverter-I

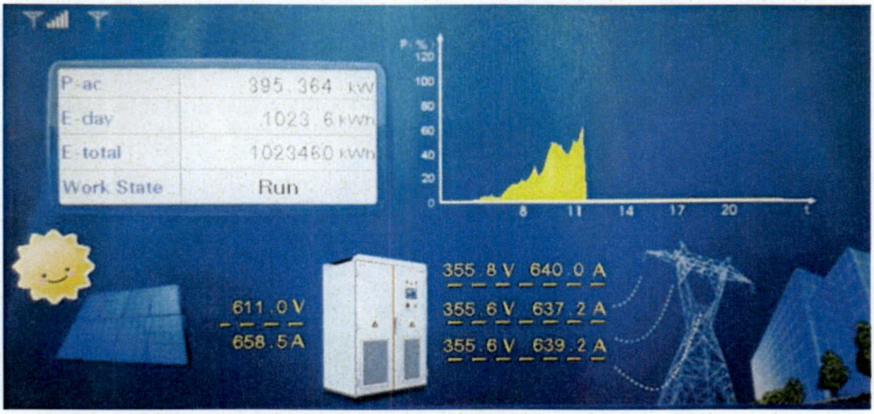

Fig. 18 Display of real-time power generation in web-based application

are having large rooftops and installation of solar PV plants on rooftop is more cost effective solution. But, monitoring of rooftop solar PV plant is not a very easy task due to its location. In view of this, it was decided to implement the IoT system to monitor the generation of the plant, system parameters, running status of equipment of the plant and accordingly IoT implementation has been done with the help of solar log system. All the communication has been established by communication cable through RS 485 port of smart devices to solar log module. Solar log module then process and transfer the data to cloud via GPRS SIM, and the parameters are visible to desktop and smart mobile application.

References

1. IEEFA report: https://ieefa.org/wp-content/uploads/2019/05/IEEFA-India_Vast-Potential-of-Rooftop-Solar-In-India.pdf
2. Priharti W, Rosmawati AFK, Wibawa IPD (2019) IoT based photovoltaic monitoring system application. In: Innovative researches international conference on engineering, technology and innovative researches J Phys Conf Ser 1367
3. Hussein ARH (2019) Internet of things (IOT): research challenges and future applications. (IJACSA) Int J Adv Comput Sci Appl 10(6)
4. Energy Statistics India 2021, Ministry of Statistics and Programme Implementation (MOSPI): http://mospi.nic.in
5. Vasita J, Shakhiya Q, Modi J (2017) Feasibility study and performance evaluation of a grid-connected rooftop solar PV system. In: IEEE, international conference on information, communication, instrumentation and control (ICICIC-2017)
6. Tayir Gulzar naqishbandi, Shavet Sharma (2018) IoT based solar power monitoring system. JASC J Appl Sci Computations 5(10). ISSN No: 1076-5131
7. Tellawar MP, Chamat N (2019) An IOT based smart solar photovoltaic remote monitoring system. Int J Eng Res Technol (IJERT) 8(09). ISSN: 2278-0181

8. Adhya S, Saha D, Das A, Jana J, Saha H (2016) An IoT based smart solar photovoltaic remote monitoring and control unit. In: 2016 2nd international conference on control, instrumentation, energy & communication (CIEC)
9. Mishra SS Design and installation of grid connected roof top solar Pv system. In: International conference on recent innovations in electrical, electronics & communication engineering (ICRIEECE)
10. Bhoye H, Sharma G (2014) An analysis of one MW photovoltaic solar power plant design. Int J Adv Res Electr Electron Instrum Eng 3(1)
11. Srinivasan KG, Vimaladevi K, Chakravarthi S, Solar energy monitoring system by IOT. Special Issue Published in Int J Adv Network Appl (IJANA)
12. Deshmukh NS, Bhuyar DL, Jadhav AT (2018) Review on IoT based smart solar photovoltaic plant remote monitoring and control unit. Int J Adv Sci Res Eng Trends 3(3)
13. Sathiyanathan N, Selvakumar S, Selvaprasanth P (2020) A brief study on IoT applications. Int J Trend Sci Res Dev (IJTSRD) 4(2). e-ISSN:2456–6470
14. Lokesh Babu RLR, Rambabu D, Rajesh Naidu A, Prasad RD, Gopi Krishna P, IoT enabled solar power monitoring system. Int J Eng Technol. Website: www.sciencepubco.com/index.php/IJET
15. Kumar NM, Atluri K, Palaparthi S, Internet of Things (IoT) in photovoltaic systems. In: 2018 national power engineering conference (NPEC)

Frequency Control of Wind Integrated Isolated Power System with I-PD Controller

K. Appala Naidu and Binod Shaw

Abstract To improve the frequency profile in an interconnected power system, the parameter gains of the I-PD controller are achieved with the crow search optimization (CSO) algorithm. Comparative performances of the system in presence of load disturbances, inertia, and droop control of wind are carried out with a PID controller. The improvements in terms of the fitness function, overshoot, and undershoot are observed in isolated power systems integrated with wind power, using the I-PD controller proposed in this paper. Three conventional power generating units-thermal, hydro, and gas are controlled by this I-PD controller when the system is integrated with HVDC tied wind plant. Two control mechanisms are adopted for wind known as inertia and droop control. The optimal settling time values show the superiority of the I-PD controller over the PID controller. Simulation investigations are carried out in MATLAB-SIMULINK.

Keywords Isolated power system · Wind · PID controller · I-PD controller

1 Introduction

Integration of wind and solar electrical power generating units influences the system frequency profile whenever wind speed and irradiance vary with time, respectively, along with regular load disturbances [1]. Frequency control studies for such systems are crucial compared to conventional or non-renewable interconnected power systems [2]. Several works are carried out in this area for isolated [3–7], interconnected [8–10], and hybrid power systems [11–14] which are integrated with renewables.

K. Appala Naidu (✉) · B. Shaw
Department of Electrical Engineering, National Institute of Technology, Raipur, India
e-mail: karanam2010@gmail.com

K. Appala Naidu
Department of Electrical Engineering, Vignan's Institute of Information Technology, Visakhapatnam, India

K. Namrata et al. (eds.), *Smart Energy and Advancement in Power Technologies*,
Lecture Notes in Electrical Engineering 926,
https://doi.org/10.1007/978-981-19-4971-5_40

Intelligent controllers provided better control for microgrid scenarios [3]. Fuzzy assisted with particle swarm optimization (PSO) approach is presented in [3] to minimize the frequency deviations of the AC microgrid during load, wind, and PV power changes that cause frequency disturbances. In [4], a PI controller with a genetic algorithm (GA) is proposed for an isolated power system with smart loads. The systems available in [3, 4] consist of diesel generating units that cause pollution. Therefore, alternative systems are available in the literature. In this line, a complete renewable-based microgrid system is presented in [5] with biodiesel and biogas units for remote end-users. The authors used the grasshopper optimization algorithm (GOA) for finding nominal gain values of the controller to mitigate the deviations in frequency error and thereby improving the frequency profile of the remote microgrid system. In [6], the crow search optimization (CSO) algorithm is used for finding optimal parameters of the PI and PID controllers with and without primary regulation concept. Apart from these intelligent and optimized approaches, H-infinity and μ synthesis approaches are also tested for isolated microgrids to improve the stability and control during system uncertainties [7]. All the methodologies are applied for isolated microgrid environments with PV and wind presence. However, research works are carried out for interconnected power system cases in presence of renewables [8–10]. Further, few works are available in hybrid power system frameworks. In [11–14], different algorithms and controllers are applied for isolated power systems to mitigate the frequency deviations. Apart from the PID controller, other controllers increase program burden and required additional data from the system to get superior results.

In this paper, the I-PD controller is adopted to improve the frequency profile of the isolated power system, and the CSO algorithm is utilized for identifying the optimal parameter gains of the controller. For comparison, the PID controller has opted in this paper. Improvements are presented in terms of AGC and optimization perspectives. The rest of the paper is organized as follows: Isolated power system simulation model details are reported in Sect. 2 followed by the controller in Sect. 3. Section 4 consists of simulation results, and finally, conclusions are available in Sect. 5.

2 Isolated Power System Simulation Model

To study the frequency control with wind integration, an isolated power system is considered in this paper whose simulation model block diagram is shown in Fig. 1. This isolated power system consists of four plants known as thermal, hydro, gas, and wind. Among the units, thermal, hydro, and gas plants are controlled by both primary and secondary control mechanisms of AGC. The wind unit is controlled either by droop or by inertia or by both strategies. This wind plant is integrated with the HVDC link. The participation factors of thermal, hydro, gas, and wind plants are given by K_{TH}, K_{HY}, K_{G}, and K_{WF}, respectively.

These gain parameters are also adjustable. For the secondary AGC, two controllers are used in this paper. The controllers are proportional-integral-derivative (PID)

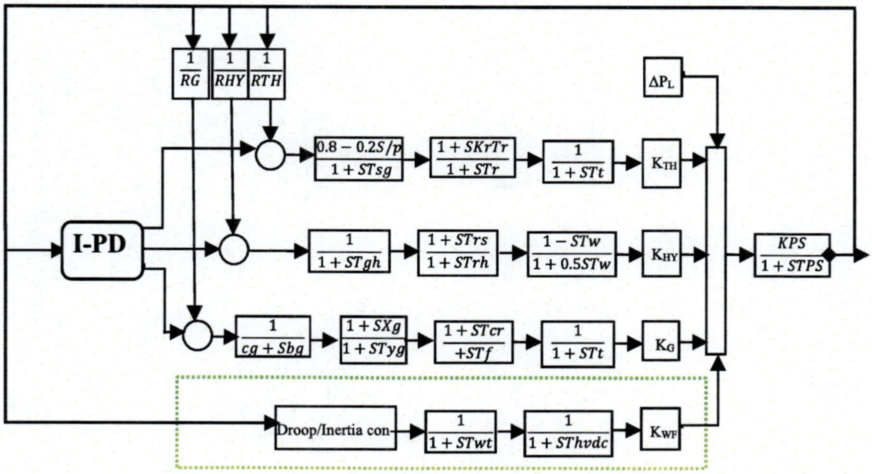

Fig. 1 Block diagram representation of simulated power system

controller and integral-proportional-derivative (I-PD) cascade controller. These controllers are placed in a secondary AGC loop with a centralized mechanism so that a simultaneous control output is provided by PID and I-PD controller to minimize the frequency deviations of the power system during load and generation power changes. The system parameter values are available in [8].

3 PID and I-PD Controllers

PID controller is extensively used for secondary AGC to minimize the frequency deviations during load perturbations. The schematic block diagram of the PID controller used in this paper is shown in Fig. 2.

In the case of an isolated hybrid power system, the reference input signal $r(s)$ is Δf, and the output of the controller is given by

Fig. 2 Schematic block diagram of PID controller for AGC studies

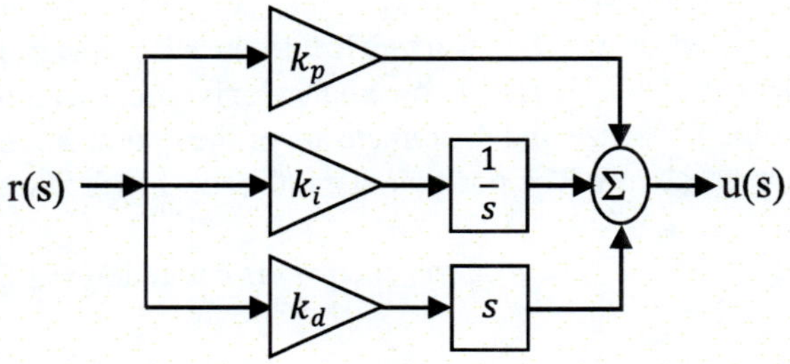

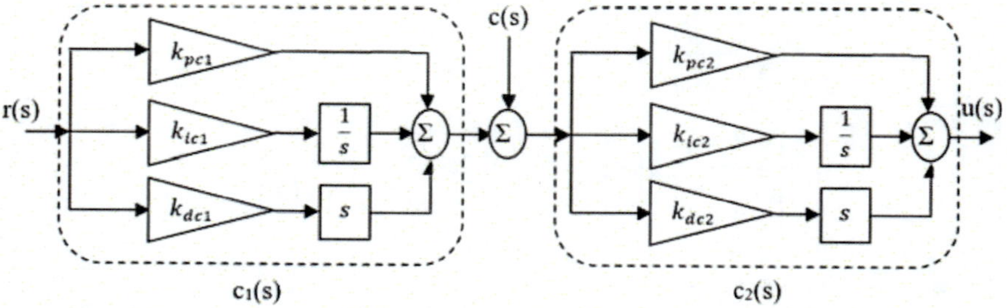

Fig. 3 Schematic block diagram of cascade PID controller

$$u(s) = \left\{k_p + \frac{k_i}{s} + k_d s\right\}\Delta f \tag{1}$$

To enhance the performance of AGC mechanism, PID controller is replaced by the I-PD controller. It is designed based on a cascade PID controller scheme without increasing the number of decision variables in the control problem. Figure 3 shows the cascade PID controller structure in which three parameters are set to be zero to make it an I-PD controller.

The output of this I-PD controller is given by

$$u_i(s) = \left\{\left(\frac{k_i}{s}\right) - 1\right\}(k_\mathrm{p} + k_\mathrm{d}s)\Delta f \tag{2}$$

In Eqs. (1) and (2), k_p denotes the proportional gain parameter, k_i denotes the integral gain parameter, and k_d denotes the derivative gain parameter. The optimal parameter gains of the controllers are obtained by using crow search algorithm with an objective function framed with frequency deviations of the system known as the integral square error (ISE) given by

$$J = \int_0^t (\Delta f)^2 \mathrm{d}t \tag{3}$$

The decision variables of both PID and I-PD controllers with upper and lower limits are given by $k_\mathrm{p}, k_i, k_\mathrm{d} \in [k_{\min} k_{\max}]$. In these limits, the optimal values are achieved by CSO. Askarzadeh introduced a powerful algorithm based on meta-heuristics in 2016 [15]. Crow's behavior is mimic in the algorithm. Initial assumptions of the flock size and the dimension of the optimization problem are taken as N and n, respectively. The position of an ith crow in kth iteration is given in Eq. 4.

$$X_{i,k} = \text{location of crow } i \text{ in } k \text{ iteration} = [x_{i,k}^1, x_{i,k}^2, \ldots, x_{i,k}^n] \tag{4}$$

where i indicates the number of crow in N population, $i \in (1, N)$ and iteration $k \in (1, k_{\max})$. The food hiding location of a crow is recollected by a crow and is retained for revival in $M_{i,k}$

$$M_{i,k} = \left[m_{i,k}^1, m_{i,k}^2, \ldots, m_{i,k}^n\right] \quad (5)$$

New updated positions are achieved in two ways based on crow's interactions

$$X_{i,k} = X_{i,k} + r_i \times fl_{i,k} \times \left(M_{i,k} - X_{i,k}\right) \quad (6)$$

$$X_{i,k+1} = \begin{cases} X_{i,k} + r_i \times fl_{i,k} \times \left(M_{i,k} - X_{i,k}\right) & r_i \geq AP \\ \text{random} & \text{otherwise} \end{cases} \quad (7)$$

where r_i denotes a random value 1 or 2, $fl_{i,k}$ denotes the length of flight of crow i during iteration k. AP denotes the crow's awareness probability. This procedure repeats for all particles until convergence criteria are satisfied.

4 Simulation Results

The enhancement in the frequency profile of the isolated power system is observed in three control mechanisms of the wind unit known as droop control, inertia control, and droop-inertia control. In all cases, 1% load decrease condition is simulated as load disturbance to study the impact of both PID and I-PD controllers. The optimal gain parameters of both controllers are achieved using CSO and reported in Table 1.

In the case of both droop and inertia control mechanisms of wind, the frequency deviations of the system are controlled by PID and I-PD controllers for a load change of 0.01 pu. The comparative performances of both controllers are reported from Figs. 4, 5 and 6. In Fig. 4, the frequency changes from its nominal value are plotted. Figure 5 shows the change in wind power output, and controller output signals are presented in Fig. 6. The fitness function values of the system with CSO are 0.0252 for PID and 0.0244 for I-PD controllers. These values provide evidence for the improvements of I-PD over PID.

Further investigations are carried out in presence of individual droop and inertia control of the wind system. In the case of droop control, the frequency profile information is presented in Figs. 7 and 8 for inertia control. In both cases, superior results are obtained with the I-PD controller than the PID controller. Irrespective of the control mechanism of wind, the over peak shoot, settling time, and steady-state error

Table 1 Controller parameter gains of the test system

Case	Proportional gain	Integral gain	Derivative gain
PID	0.9975	0.9995	0.9857
I-PD	0.8202	0.8002	0.8447

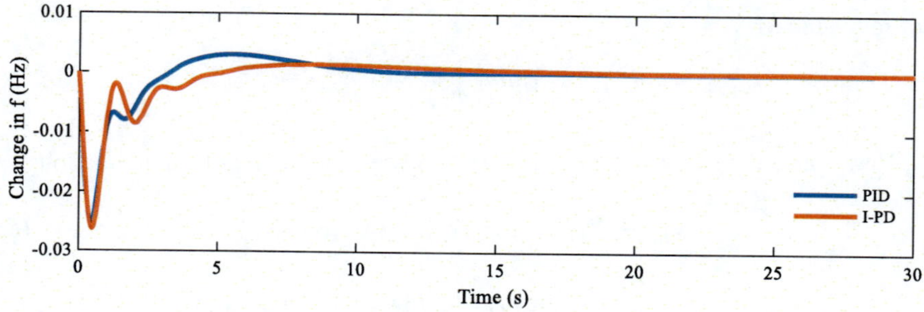

Fig. 4 System frequency deviations for the droop-inertia control case

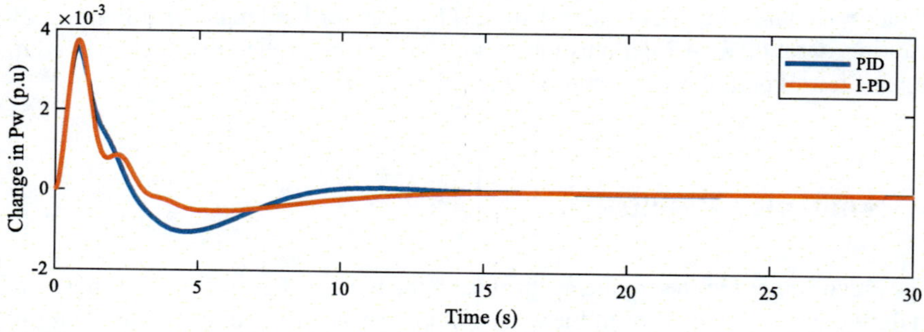

Fig. 5 Wind output power deviations for the droop-inertia control case

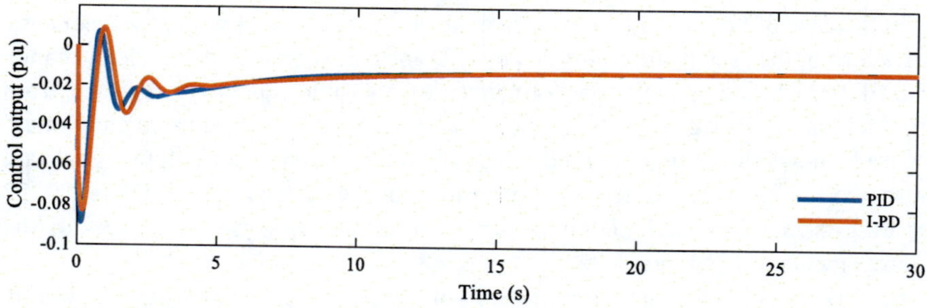

Fig. 6 Controller outputs for the droop-inertia control case

values are reduced with the proposed I-PD controller. However, the peak error change is comparatively closer with PID results.

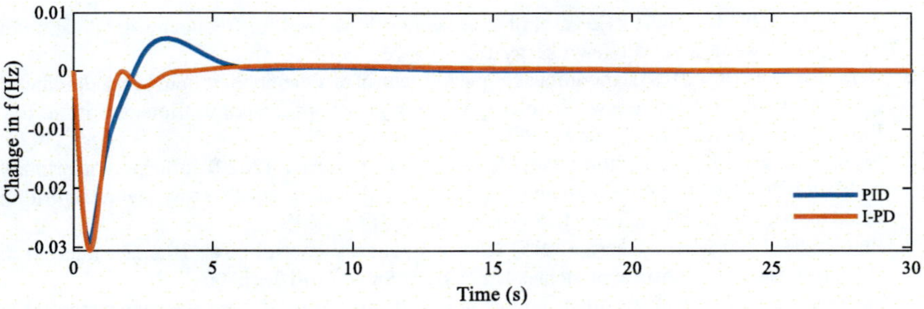

Fig. 7 System frequency deviations for droop control

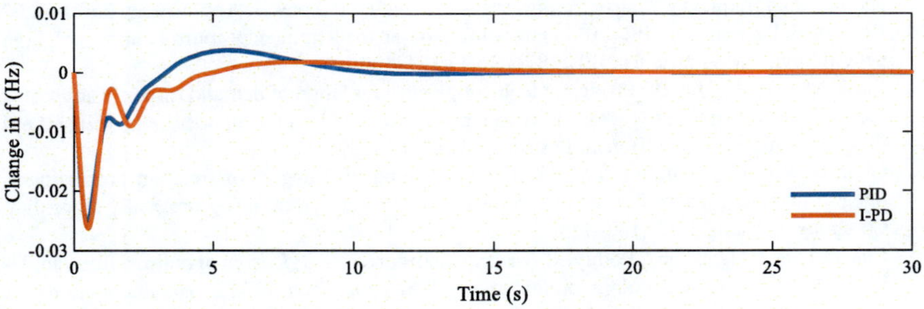

Fig. 8 System frequency deviations for the inertia control case

5 Conclusions

CSO tuned I-PD controller is suggested in this paper to mitigate the deviations in frequency error of the isolated power system with diverse power generating units integrated with the wind system. The wind plant is integrated with the grid through an HVDC link to enhance the power transfer capacity. The comparative results with the PID controller provided superiorities of the I-PD controller in terms of performance specifications. Along with AGC objectives, the I-PD controller required the same number of decision variables like PID, optimization problem needs not be modified, and no additional information is required from the system.

References

1. Bevrani H (2009) Robust power system frequency control. Springer
2. Kayicki M, Milanovic JV (2009) Dynamic contribution of DFIG-based wind plants to system frequency disturbances. IEEE Trans Power Syst 24(2):859–867
3. Bevrani H, Habibi F, Babahajyani P, Watanabe M, Mitani Y (2012) Intelligent frequency control in an AC microgrid: online PSO-based fuzzy tuning approach. IEEE Trans Smart Grid 3(4):1935–1944

4. Nandar CS (2013) Robust PI control of smart controllable load for frequency stabilization of microgrid power system. Renew Energy 1(56):16–23
5. Barik AK, Das DC (2018) Expeditious frequency control of solar photovoltaic/biogas/biodiesel generator based isolated renewable microgrid using grasshopper optimisation algorithm. IET Renew Power Gener 12(14):1659–1667
6. Patan MK, Raja K, Azaharahmed M, Prasad CD, Ganeshan P (2020) Influence of primary regulation on frequency control of an isolated microgrid equipped with crow search algorithm tuned classical controllers. J Electr Eng Technol 16(2):681–695
7. Bevrani H, Feizi MR, Ataee S (2015) Robust frequency control in an islanded microgrid: H-infinity and μ synthesis approaches. IEEE Trans Smart Grid 7(2):706–717
8. Tavakoli M, Pouresmaeil E, Adabi J, Godina R, Catalão JP (2018) Load-frequency control in a multi-source power system connected to wind farms through multi terminal HVDC systems. Comput Oper Res 96:305–315
9. Shang-Guan X, He Y, Zhang C, Jiang L, Spencer JW, Wu M (2020) Sampled-data based discrete and fast load frequency control for power systems with wind power. Appl Energy 259:114202
10. Bevrani H, Daneshmand PR (2011) Fuzzy logic-based load-frequency control concerning high penetration of wind turbines. IEEE Syst J 6(1):173–180
11. Barik AK, Das DC (2019) Proficient load-frequency regulation of demand response supported bio-renewable cogeneration-based hybrid microgrids with quasi-oppositional selfish-herd optimisation. IET Gener Transm Distrib 13(13):2889–2898
12. Guha D, Roy PK, Banerjee S (2018) Optimal tuning of 3 degree-of-freedom proportional-integral-derivative controller for hybrid distributed power system using dragonfly algorithm. Comput Electr Eng 1(72):137–153
13. Chaiyatham T, Ngamroo I, Pothiya S, Vachirasricirikul S (2009) Design of optimal fuzzy logic-PID controller using bee colony optimization for frequency control in an isolated wind-diesel system. In: 2009 transmission & distribution conference & exposition: Asia and Pacific 2009 Oct 26. IEEE, pp 1–4
14. El-Fergany AA, El-Hameed MA (2017) Efficient frequency controllers for autonomous two-area hybrid microgrid system using social-spider optimiser. IET Gener Transm Distrib 11(3):637–648
15. Askarzadeh A (2016) A novel metaheuristic method for solving constrained engineering optimization problems: crow search algorithm. Comput Struct 169:1–12

LFC Technique of an Interconnected Hybrid Grid System with the Forecasting of Wind Power

P. Suresh Kumar, B. Arundhati, Y. Srinivasa Kishore Babu, and Madisa V. G. Varaprasad

Abstract As the wind power is random and strong, higher efforts have been incorporated into the load frequency (LF) controlling techniques. The research objective is we consider an interconnected power grid unit with a wind power system; the LF control approach has been developed based on the prediction of the wind by Kalman's filter algorithm techniques which influence the change in system frequency due to wind, power. The Kalman's filter algorithms are used for the prediction of wind power that has been generated in the starting and also the wind power forecasted for the purpose LF controller and designing. Further a LF control model interconnected in a power system network is been established. This adaptive control technique is been applied to a four-area interconnected distributed grid system with three thermal plant units integrated with a wind power unit. The simulation analysis is carried in MATLAB/Simulink which shows that the suggested LF control approach based on the Kalman's filter algorithm with wind system prediction can beneficially reduce the occurred fluctuation in frequency and keep minimize the fluctuations in frequency. Simulation results are further compared with traditional proportional–integral–derivative controllers based on load frequency controller with a conventional wind system.

Keywords Kalman's filter algorithm · Wind power forecast · Wind power · Load · Frequency control (LF control)

1 Introduction

Due to the high uncertainty and continuous fluctuations of the wind speed, it will be affecting the safety and stability of the system [1–3]. Depending on the balance of

P. Suresh Kumar (✉) · B. Arundhati · M. V. G. Varaprasad
Vignan's Institute of Information Technology, Visakhapatnam, India
e-mail: suresh0260@gmail.com

Y. Srinivasa Kishore Babu
JNTUK University College of Engineering, Dwarapudi, India

K. Namrata et al. (eds.), *Smart Energy and Advancement in Power Technologies*, Lecture Notes in Electrical Engineering 926,
https://doi.org/10.1007/978-981-19-4971-5_41

active power component and integration of intermittent, wind power shows a significant impact on the system frequency dynamic characteristics and also maintains frequency stability. Therefore, large power generation by wind into an interconnected system will create a greater challenge to LF controller in a distributed interconnected power system network [4].

In the latest research area, researchers have been studying on the frequency control approach of wind power unit systems. Few adapt to artificial intelligence techniques in order to optimize the parameters of a traditional proportional–integral–derivative controller and study the effect of frequency control. For example in [5], the parameters of a traditional proportional–integral–derivative controller are calculated and optimized by neural networks algorithms. This not only makes the LF controller have a universal application of proportional–integral–derivative universal controller but also improve the adaptive capacity of LF controller. In the optimization of neural networks, it is easy to develop an optimal solution. An energy-storage device is added in wind farm unit to suppress and steady the fluctuations produced due to wind power, but it increases the difficult in construction and the cost of investment. A frequency control link is introduced in the wind turbine for adjusting, the active power of the wind power turbine and involves in system's frequency adjustment [6, 7]. In this, the output power of the wind turbine must meet load reduction by reserved demand capacity. Hence, this will not covenant the MPPT of the wind turbine and reduce the rate of utilization of actual wind power. In References [8–10], we study the improvement in frequency and effect of modulation in frequency by the unit speed governing system and automatic generation control device if a wind power unit is interconnected with the grid, but, frequency, regulation capacity of systems unit is limited. And further to maintain the frequency stability and maximize the utility of wind power generated, it is very important to maintain the frequency fluctuations in the existing interconnected system. Hence, an intelligent LF controller has been designed as an effective solution to minimize the frequency fluctuations.

In this paper, multiple-area interconnected power system network with the integration of wind power unit is taken as research objective. Firstly, we use Kalman's filter to forecast the wind power, and a traditional LF controller has been designed depending on the predicted wind power, meanwhile a phase-compensation technique has been incorporated into the traditional controller so as to improve the dynamic response of the frequency. Hence, an adoptable LF control technique has been proposed for a wind power unit and is applied for a four-area distributed interconnected power grid system. Results based on, MATLAB/Simulink will verify the benefit of the control approach.

2 Forecasting of Wind Power Depending on Kalman's Filter Algorithm

2.1 *Actual Power of Wind Forecasted by Kalman's Filter Algorithm*

The basic model of forecasting has been constructed based on the Kalman's filter algorithm technique. By using the state equations, the next state of wind power has been forecasted by using the present situation of wind power. In this, we assume that wind power at present time "t" as a studied value and the forecasted wind power which is expressed as

$$P_y\left(t + \frac{\Delta t}{t}\right) = A_t P_z(t/t) \tag{1}$$

where $P_y\left(t + \frac{\Delta t}{t}\right)$ is wind power at a time instant of $(t + \Delta t)$ which is forecasted from the actual wind power unit at time "t", it is considered, as a one-step forecast power. $P_z(t/t)$ is the optimal estimation, of the power at a time of "t" in a state transition matrix. The covariance of the matrix is updated after obtaining the actual wind power at $(t + \Delta t)$.

$$C_y\left(t + \frac{\Delta t}{t}\right) = A_t C_z(t/t) A_t^T + Q \tag{2}$$

The power at a time instant of $(t + \Delta t)$ is obtained. According to Kalman's filter algorithm [10], optimal estimation power is calculated at a time of $(t + \Delta t)$ and is given as

$$P_z\left(t + \frac{\Delta t}{t} + \Delta t\right) = P_y\left(t + \frac{\Delta t}{t}\right) + K_{gt+\Delta t}\left[P_c(t + \Delta t) - H_t + x\left(t + \frac{\Delta t}{t}\right)\Delta t\right] \tag{3}$$

where $P_z\left(t + \frac{\Delta t}{t} + \Delta t\right)$ is optimal power estimation at $(t + \Delta t)$ and $K_{gt+\Delta t}$ is kalman's filter algorithm gain. $H_{t+\Delta t}$ is measurement matrix, and $x\left(t + \frac{\Delta t}{t}\right)$ is state matrix. Similarly, the covariance is updated correspondingly.

$$C_y\left(t + \frac{\Delta t}{t} + \Delta t\right) = \left[I - K_{gt+\Delta t} H_t + \Delta t\right] C_y\left(t + \frac{\Delta t}{t}\right) \tag{4}$$

The Kalman's filter gain $K_{gt+\Delta t}$ is given as

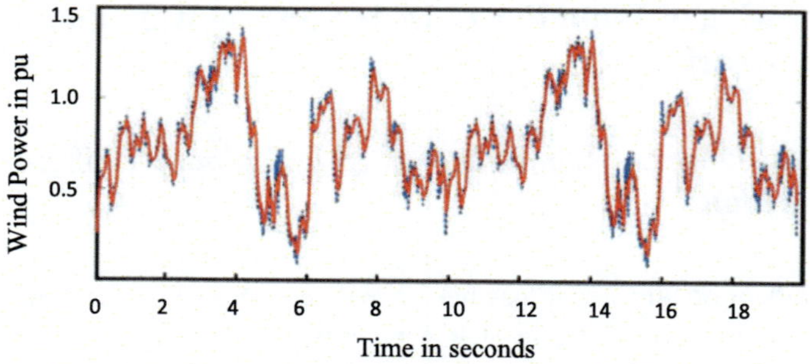

Fig. 1 Actual wind power and forecasted wind power curve

$$K_{gt+\Delta t} = C_y\left(t + \frac{\Delta t}{t}\right)H_t\left[H_{t+\Delta t}C_y\left(t + \frac{\Delta t}{t}\right)H^T_{t+\Delta t} + R\right] \tag{5}$$

R is measurement-noise variance.

2.2 *Verifying the Optimal Forecast Results of Wind Power*

By considering the speed of the wind and combining with the engineering practices, the actual starting values have been considered as: $x(0/0) = x_0$,$C_y(0/0) = 10I$. And let $Q = 1$ and $R = 1$. A basic set of wind power unit data mentioned in Reference [9] has been forecasted by the Kalman's filter algorithm [11]. Figure 1 represents the actual wind power curve and the actual forecasted wind power curve. These results verify the feasibility and validity of Kalman's filter technique in the wind- power unit forecasting.

The blue color dotted line represents the actual power of wind unit, and the red color dotted line represents the forecasted wind power.

3 LF Control Approach Considered with Wind Power Forecast

3.1 *Multiple-Area LF Control (LFC) Model*

LF control models are expressed as a linear model when there is no remarkable change in the load. Figure 2 shows the ith control area model from a multiple-area system [12, 13].

A LF control model has turbine, speed governor unit, generator, controller, and tie lines. The parameters are.

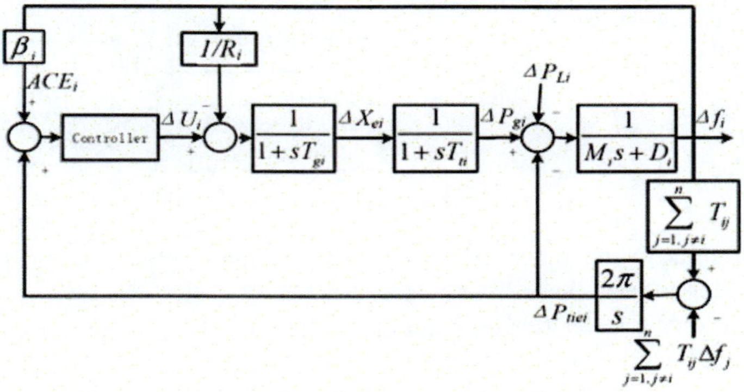

Fig. 2 LF control of multiple-area system

T_{gi} = time constant of speed governor
D_i = damping coefficient
β_i = frequency deviation coefficient
M_i = unit inertia constant
T_{Ti} = turbine time constant of the ith area
ACE = area control area error
Δf_i = frequency error
ΔP_L = disturbance in the load
ΔX_{ei} = change in valve adjustment position
ΔP_{tie} = tie line power between two areas in a system.

3.2 *Design of a LFC Approach Including Wind Power Forecast*

This paper discusses LF control technique considering an actual power of wind forecasted for the fluctuations in frequency produce by the integration of wind system. This technique makes use of the wind power which is forecasted in advance so as to modify the output power of a synchronous alternator so as to reduce the frequency fluctuations of the system which are caused due to the addition of wind power [14].

A traditional controller has been suggested in this paper which comprises of a decided value of proportional–integral–derivative controller and a specified signal. These values contain phase, inversion and compensation of the wind power which are forecasted by the wind generation. The block diagram of the controller unit is shown in Fig. 3.

Active power component generated from wind power is considered as the load of the power system. The designed controller is utilized to apply the inverted output power that is forecasted from actual wind power which acts as a general controlled

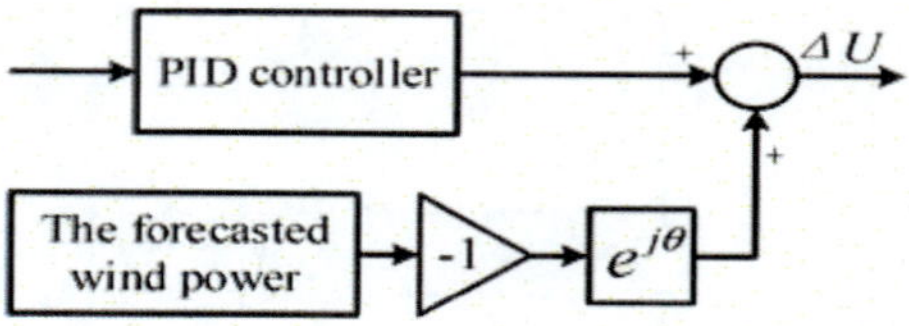

Fig. 3 LFC considering the forecast of wind power

command input to the generator, so that the generator output can be inverted in corresponding to wind power unit output.

The governor transfer function and the turbine transfer function of a first-order inertia link are represented as follows

$$G_t(s) = \frac{1}{(ST_{ti} + 1)} \tag{6}$$

$$G_g(s) = \frac{1}{(ST_{gi} + 1)} \tag{7}$$

The generator in the system will be responding according to the controlled power from the LFC controller. However, the generator output will be having certain lag in phase in relation to control power. Hence, it is required to determine the lag in phase angle with respect to frequency response characteristics and the fluctuations involved in wind power system. In Fig. 3, the phase link compensation $e^{j\theta}$ was added in order to invert the signal of forecasted wind power energy in LFC. It is also necessary for improving the dynamic response of frequency. Figure 1 shows the design of a phase plot of the compensated link intricate in the example which was integrated to single-area system; now, let us assume the tie constant of the governor and the turbine to be 0.2 and 0.3 s for a single-area system, respectively. The spectrum inspection done by fast Fourier transforms (FFT) has been used in order to determine the frequency, and the results are briefly shown in Figs. 4 and 5.

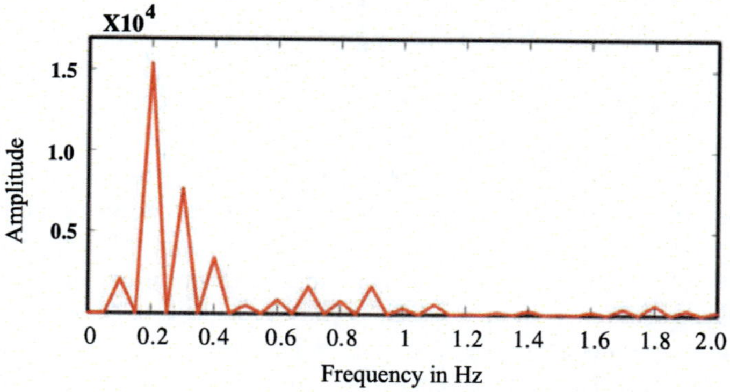

Fig. 4 Wind power spectrum analysis

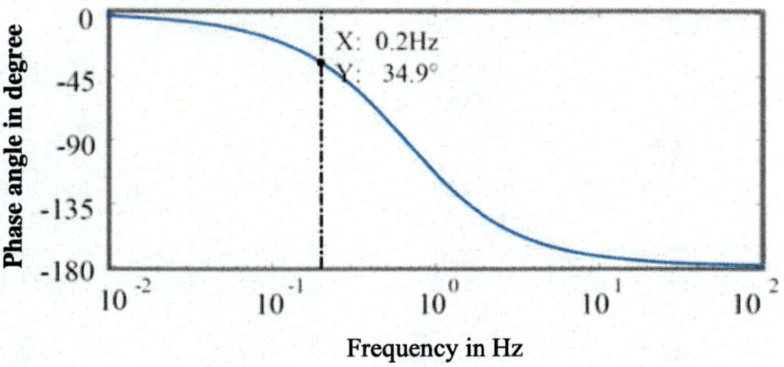

Fig. 5 Frequency response of the generator phase curve in a single-area system

From the above result, it can be resolved, that the lag in the phase is derived from generator t/f function with wind power frequency to be around 0.2 Hz at a phase of 34.9°. Hence, the compensated phase angle of a compensated phase link that has been added to the LFC controller is 34.9°. This is achieved by introducing compensation link $e^{j\theta}$.

3.3 Simulation, Results and Analysis

For verifying the effectiveness of the LFC approach while considering a wind power area forecast in an interconnected system, MATLAB/Simulink simulation is carried out for a four-area interconnected system, i.e., $i = 4$. The diagrammatic representation of a four-area interconnected system is shown in Fig. 6, and parameters of the system are shown in Table 1 (Fig. 7).

The blue color dotted line represents the actual power of wind unit, and the red color dotted line represents the forecasted wind power (Fig. 8).

The blue color dotted line represents the actual power of wind unit, and the red color dotted line represents the forecasted wind power.

Fig. 6 Four-area interconnected system

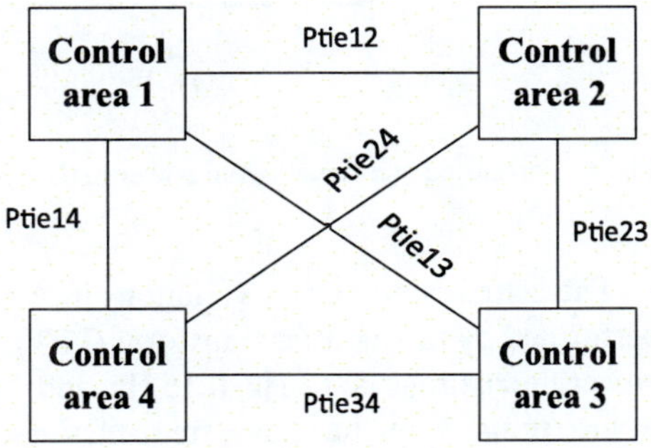

Table 1 Parameters of the system

*I*th area	1	2	3	4
R_i	12/5	27/10	5/2	2
β_i	17/40	17/40	17/40	17/40
T_{gi}	8/100	9/125	7/100	17/200
T_{ti}	3/10	33/100	35/100	3/8
M_i	0.167	0.223	0.16	0.130
D_i	1/120	1/112.5	1/125	1/115
T_{ij}	109/200	109/200	109/200	109/200

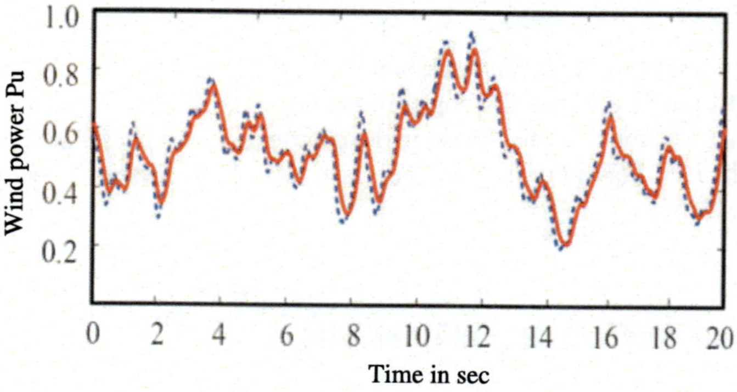

Fig. 7 Forecasted and actual, wind power in the second area

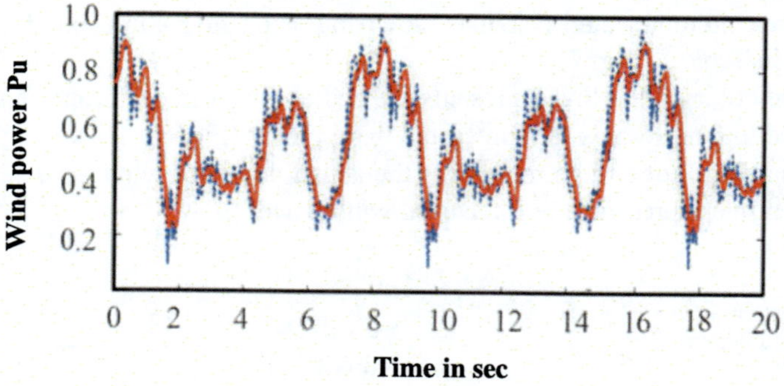

Fig. 8 Forecasted and actual, wind power in the fourth area

The actual wind power, variations in Areas 1, 2, and 4 by spectrum analysis are performed by fast Fourier transform (FFT). It is observed that frequency of a wind power fluctuations is 0.2 Hz, 0.15 Hz, and 0.25 Hz, respectively, in Areas 1, 2 and 4. In the similar way, the phase, frequency response characteristics of the wind power fluctuate frequency and the function of generator in each respective area. And the generator *t/f* can also be determined. Based on the compensated phase angle θ_i, phase computation link $e^{j\theta i}$ can be controlled. The compensation phase angle values of the respective Area 1, Area 2, and Area 4 are 31.04°, 20.1°, and 33.8°, respectively.

The blue color dotted line represents the output with proportional–integral–derivative controller, and the red color dotted line represents the proposed optimal LFC approach.

The blue color dotted line represents the output with proportional–integral–derivative controller, and the red color dotted line represents the proposed optimal LFC approach.

The blue color dotted line represents the output with proportional–integral–derivative controller, and the red color dotted line represents the proposed optimal LFC approach.

The simulation, results shown in Figs. 9, 10, 11 and 12 represent the control effect of a traditional proportional–integral–derivative controller and can be easily differed due to wind power unit fluctuations which are connected in a distributed grid system. There is a significant frequency deviation which is exceeding standard frequency fluctuations of ± 0.2 Hz. Hence, it cannot guarantee the frequency of the assumed system which is in an operation range. By adopting LFC controller approach, the deviation in frequency is minimized in each area significantly. And also the frequency dynamic response of each area is better than a traditional proportional–integral–derivative controller. From the above, simulation results, the proposed LFC approach has better control effect than traditional proportional–integral–derivative controller in a multiple-area LFC system.

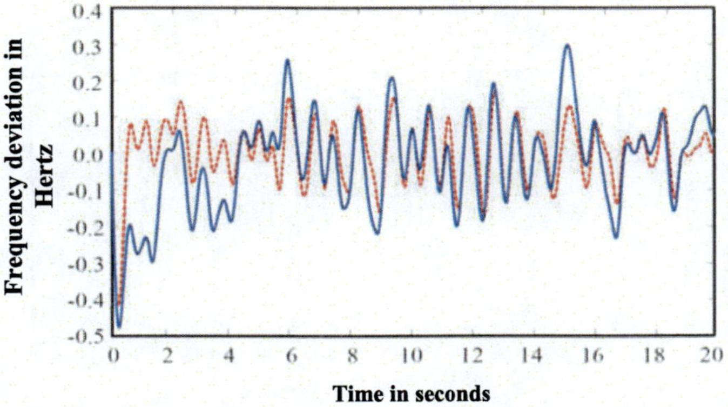

Fig. 9 Deviation of the frequency in the first area

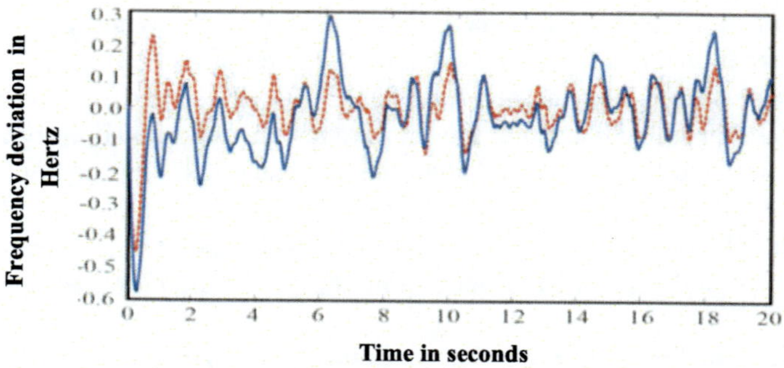

Fig. 10 Deviation of the frequency in the second area

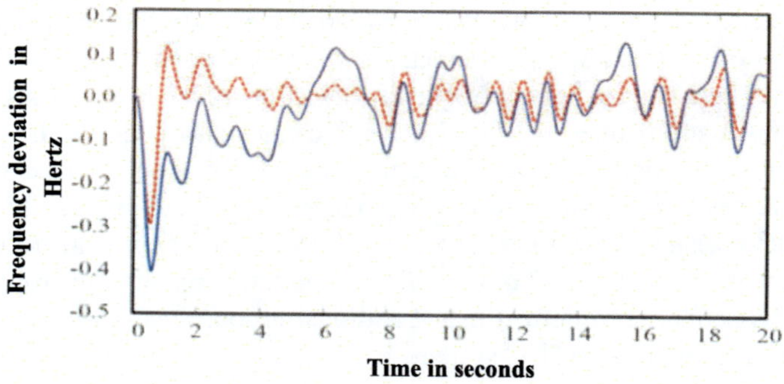

Fig. 11 Deviation of frequency in the third area

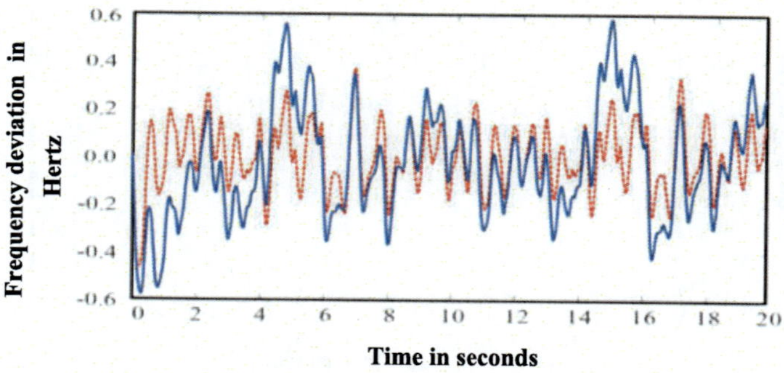

Fig. 12 Frequency deviation in the fourth area

4 Conclusion

In this paper, a LF control (LFC) technique has been considered along with forecasted wind power. Hence, to minimize the fluctuations of wind power in an interconnected system, the Kalman's filter algorithm has been inverted in, order for compensating, the phase lag which has been caused by introducing the dynamic link, of the generator in the respective area. The compensated and inverted output has been used as a signal command for the LF controller, which can beneficially reduce fluctuations caused by wind power unit in an integrated system. Meanwhile, this approach can beneficially compensate the shortage of a traditional and traditional proportional–integral–derivative controller in an interconnected power grid system with wind unit, and it also improves the dynamic response characteristics of frequency. Based on the obtained results, the effectiveness of the proposed control approach is verified.

References

1. Zhang W, Fang K (2017) Controlling active power of wind farms to participate in load frequency control of power systems. IET Gener Transm Distrib 11:2194–2203
2. Junrui H, Yuchun H, Wen T (2012) Load frequency control of wind diesel hybrid power systems via predictive control. In: Proceedings of the 31st Chinese control conference, pp 6698–6703
3. Chen W, Tan W (2014) Load frequency control for power systems with wind turbines. In: Proceedings of the 33rd Chinese control conference
4. Senjyu T et al (2008) A new control methodology of wind farm using short-term ahead wind speed prediction for load frequency control of power system. In: 2008 IEEE 2nd international power and energy conference
5. Miller N, Loutan C, Shao M, Clark K (2013) Emergency response: U.S. system frequency with high wind penetration. In: IEEE power and energy magazine, vol 11(6), pp 63–71
6. "IEE Colloquium on 'Kalman filters: introduction, applications and future developments' (Digest No.27)," IEE Colloquium on Kalman filters: introduction, applications and future developments (1989)
7. Khazraj H, Faria da Silva F, Bak CL (2016) A performance comparison between extended Kalman Filter and unscented Kalman Filter in power system dynamic state estimation. In: 51st international universities power engineering conference (UPEC)
8. Kong L, Xiao L (2007) A new model predictive control scheme-based load-frequency control. In: 2007 IEEE international conference on control and automation
9. Luo C, Ooi B (2006) Frequency deviation of thermal power plants due to wind farms. IEEE Trans Energy Convers 21:708–716
10. Khazraj H, Faria da Silva F, Bak CL (2016) A performance comparison between extended Kalman Filter and unscented Kalman Filter in power system dynamic state estimation. In: 2016 51st international universities power engineering conference (UPEC), 2016, pp 1–6. https://doi.org/10.1109/UPEC.2016.8114125
11. Li Q, Li R, Ji K, Dai W (2015) Kalman filter and its application. In: 2015 8th international conference on intelligent networks and intelligent systems (ICINIS), pp 74–77. https://doi.org/10.1109/ICINIS.2015.35
12. Mi Y, Yang Y, Zhang H, Yu A, Wang L, Ren L (2014) Sliding mode based load frequency control for multi-area interconnected power system containing renewable energy. In: 2014 IEEE conference and expo transportation electrification Asia-Pacific (ITEC Asia-Pacific), pp 1–6. https://doi.org/10.1109/ITEC-AP.2014.6941131
13. Kumtepeli V, Wang Y, Tripathi A (2016) Multi-area model predictive load frequency control: a decentralized approach. In: 2016 Asian conference on energy, power and transportation electrification (ACEPT), pp 1–5. https://doi.org/10.1109/ACEPT.2016.7811530
14. Han Z et al (2011) Study on calculation methods of wind farm's abandoned energy. In: International conference on advanced power system automation and protection, pp 1996–1999. https://doi.org/10.1109/APAP.2011.6180767

Determination of Critical Contingency Based on L-Index and Impact Assessment on Power System

Mehebub Alam, Shubhrajyoti Kundu, Siddhartha Sankar Thakur, and Sumit Banerjee

Abstract The outage of transmission line creates contingency in power system. Therefore, determination of critical contingency is essential for stability analysis purpose. In this article, the critical bus is identified using well-known L-index method, and thereafter, the critical contingency is found through maximum L-index value considering all contingency scenarios. Moreover, the deviation of the bus voltage profile is presented through a new index, i.e., voltage deviation index (VDI). Furthermore, the deviation of the post-contingency power loss is evaluated through proposed loss deviation index (LDI). The case studies are performed on IEEE 6 and 30 bus systems in MATLAB environment. The analysis of the test results will be helpful for contingency-constrained voltage stability study.

Keywords Power system · Contingency · L-index

1 Introduction

In the recent years, the transmission network experiences a lot of stress due to deregulation of entire power grid network. In the deregulated network, each entity, i.e., transmission, generation and distribution operates independently to promote the overall efficiency and flexible operation [1] of the system. However, if contingency occurs in power system due to outage of generator, transmission line or transformer, the system becomes more and more stressed. This stressed network is more vulnerable to collapse. Therefore, power system stability analysis is required to find the critical bus and critical line so that the power engineers and planners can take appropriate action to avoid the system breakdown.

M. Alam (✉) · S. Kundu · S. S. Thakur
Department of EE, National Institute of Technology (NIT) Durgapur, Durgapur, West Bengal 713209, India
e-mail: mehebubjgec1990@gmail.com

S. Banerjee
Department of EE, Dr. B.C. Roy Engineering College, Durgapur, West Bengal 713206, India

K. Namrata et al. (eds.), *Smart Energy and Advancement in Power Technologies*, Lecture Notes in Electrical Engineering 926,
https://doi.org/10.1007/978-981-19-4971-5_42

The researchers have reported several methods in literature [2–6] for stability analysis purpose. In [2], authors proposed Lmn index for line stability analysis. The value of Lmn index must be less than 1 for stable system. The fast voltage stability index (FVSI) is presented in [3] for determination of critical line. The FVSI value nearest to 1 represents the critical line. Sekhawat et al. [4] analyzed the voltage stability using FVSI and Lmn indices. The weak bus identification and maximum loadability determination are done using line stability indices. Authors proposed novel voltage stability index (NVSI) [5] which takes into account both active and reactive power. Unlike FVSI and Lmn, proximity of the value of the NVSI to unity indicates that the system is approaching toward instability. Three line stability indices, i.e., FVSI, Lmn, NVSI and one bus stability index, i.e., ratio index are used in [6] for stability analysis. Additionally, the shunt compensators are used for reactive power support. Furthermore, the L-index is proposed in [7] as a stability measuring tool and used in [8] for determination of critical contingency. The authors in [9] used the performance index as a tool for contingency ranking. The real power, reactive power and voltage are considered combinedly for ranking evaluation. The popular fast decoupled technique is used in [10] for contingency selection. The impact of contingency due to line outage is presented in [11] through PV curve. It is also observed that bus voltages decrease with the increased loading scenario. Further, the optimization of weighting factor is employed in [12] for improving the contingency ranking. The authors proposed voltage and reactive power-related index in [13] for screening and ranking of contingency. The authors in [14] identified the contingency due to line outage using the synchronized measurements at generator buses. A probabilistic performance index is used for ranking of the contingency in [15], and the technique is applied up to second-level contingency. In this study, the critical bus and critical line of the system are determined. Furthermore, the impact of contingency on power grid network is also evaluated in terms of voltage deviation and power loss deviation.

2 Methodology

The whole study is mainly divided into two parts.

- Voltage stability analysis
- Contingency analysis

In voltage stability part, the weak bus and line of the system are identified. The identification of the weakest bus and weakest line is utmost required in order to avoid unpredictable voltage collapse. Further, the weakest bus is to be determined for placing suitable compensating devices for improving the performance of the whole system.

In contingency analysis part, the comparison of the load bus voltage and system loss between base case and critical contingency case is outlined. The detailed description of the whole methodology is highlighted in the following section.

2.1 Analysis of Voltage Stability

The analysis of voltage stability is conducted to find the critical bus and critical line. It is of paramount importance to find the weak bus of the power network. The weak bus is to be monitored properly to avoid the voltage collapse. Furthermore, the compensating action for reactive power support is to be done at the weak bus in order to get optimum economic benefit. Although several methods are available to identify the weak bus, the popular L-index [7]-based method is adopted here. The L-index for k-th load bus is represented by Eq. (1)

$$L_k = \left| 1 - \sum_{i=1}^{n_g} F_{ki} \frac{V_i}{V_k} \right|, \quad k = 1, 2, \ldots, n_l \tag{1}$$

Here, n_g is the number of generator bus, and n_l is the number of load bus. The voltage at i-th and k-th bus is represented by V_i and V_k, respectively. Further, the matrix F_{ki} can be expressed by following Eq. (2)

$$F_{ki} = -[Y_{LL}]^{-1}[Y_{LG}] \tag{2}$$

Here, Y_{LL} is the sub-matrix of bus admittance matrix, and the elements are associated with only load buses. Similarly, Y_{LG} is the sub-matrix of bus admittance matrix, and the elements are associated with load buses and generator buses.

The L-index value lies between 0 and 1. The maximum L-index value corresponds to the critical value. The L-index value close to 0 represents that the system is secured and stable. For zero loading condition, system bus voltages are 1 p.u., and thus, the L-index for load buses will be close to zero which indicates the stable system.

2.2 Contingency Analysis

The second part of our study is finding out the critical contingency and its impact assessment. This task is performed by identifying the critical line. Now, the critical line is identified through repeated load flow study for all possible contingency scenarios. For each contingency scenario, the L-indices of the load buses are calculated and maximum L-index value represents the critical contingency. Therefore, the corresponding line due to which that particular contingency occurs is designated as critical line.

Now, it is interesting to note that each contingency results into the deviation of the system voltage profile and system losses. Deviation of voltage is to be measured to avoid voltage collapse and maintain proper voltage stability. On the other hand, the deviation of system loss is to be measured for loss minimization purpose or to ensure maximum usable power. Therefore, the quantification of the deviation of the system voltage and deviation of the system loss is very vital. In this context, a new index, i.e., voltage deviation index (VDI) is introduced here to quantify the post-contingency voltage deviation. The VDI is expressed by Eq. (3)

$$\mathrm{VDI} = \frac{\sum_{i=1}^{n_l} V_i - \sum_{i=1}^{n_l} V_i^k}{\sum_{i=1}^{n_l} V_i} \quad (3)$$

Here, V_i and V_i^k represent the base case voltage of i-th bus and voltage of i-th bus after k-th contingency.

Furthermore, another index, i.e., loss deviation index (LDI) is introduced to measure the loss deviation. The LDI is written by Eq. (4)

$$\mathrm{LDI} = \frac{\mathrm{Loss} - \mathrm{Loss}^k}{\mathrm{Loss}} \quad (4)$$

Here, in Eq. (4), Loss and Loss^k represent the base case system loss and system loss after k-th contingency. Now, the total system loss can be calculated as

$$\mathrm{Loss} = \sum_{i=1, i \neq j}^{L} L_{ij} \forall \text{ lines} \quad (5)$$

Here, L denotes total number of lines, and L_{ij} denotes power loss at line connecting between i-th bus and j-th bus. Further, the loss at any line can be computed by using following Eq. (6)

$$L_{ij} = P_{ij} + P_{ji} \quad (6)$$

Here, P_{ij} and P_{ji} denote the real power flows from i-th bus to j-th bus and j-th bus to i-th bus, respectively. The real power flows from i-th bus to j-th bus can be written by Eq. (7)

$$P_{ij} = \mathrm{Re}\left[V_i\left(I_{ij}^*\right)\right] \quad (7)$$

Here, I_{ij} denote the current flows from i-th bus to j-th bus.

By stepwise, the whole computational process is demonstrated below:

1. Read the system data.
2. Run the load flow program.
3. Calculate the L-index value for load buses using (1) and (2).
4. Determine the maximum L-index value, and the corresponding bus is designated as critical bus.
5. Simulate the contingency one by one, and for each contingency, the maximum L-index value is stored.
6. Sort the L-index value for different contingency scenarios.
7. Find out the maximum value of L-index among all contingency scenarios.

8. The maximum L-index value corresponds to the critical contingency, and designate the critical line.
9. Calculate the VDI and LDI for critical contingency.
10. Display results.

3 Numerical Results and Discussion

The presented scheme is designed in MATLAB, and the logical program is executed. The simulation is done with MATLAB 7.10.0 (R2013a) platform loaded in a computer having Intel core-i3 processor.

3.1 *IEEE 6 Bus Study*

The whole case study is conducted on IEEE 6 bus network (Fig. 1) which has eleven lines, three load buses and two generator buses (buses 2 and 3).

The L-index value for load buses 1, 2 and 3 (buses 4, 5, 6) is found to be 0.2162, 0.2958 and 0.2506, respectively. The computed L-index value for base case is presented in Fig. 2 which shows that the maximum L-index is found for load bus 2, i.e., bus 5. Hence, bus 5 is designated as critical bus. Now, the L-index value for different contingency scenarios is also calculated, and the maximum value for each contingency scenario is presented by Fig. 3. The maximum L-index value among all

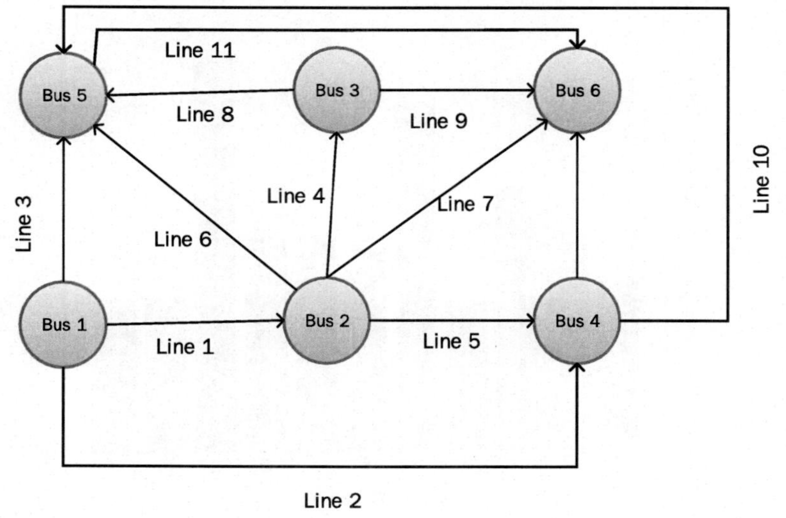

Fig. 1 Schematic of IEEE 6 bus system

contingencies is found 0.8114 which refers to the contingency 9, i.e., outage of line 9. Therefore, line 9 is designated as critical line.

Now, the VDI is calculated for all possible contingencies, and the VDI is represented in Fig. 4. It is seen from Fig. 4 that almost all the VDIs are positive. The positive VDI indicates the degradation of the voltage profile after contingency. Further, the

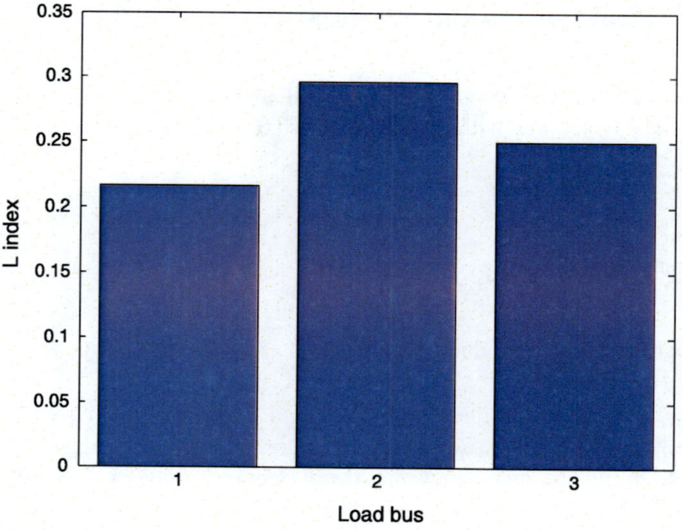

Fig. 2 L-index value for load buses of IEEE 6 bus network

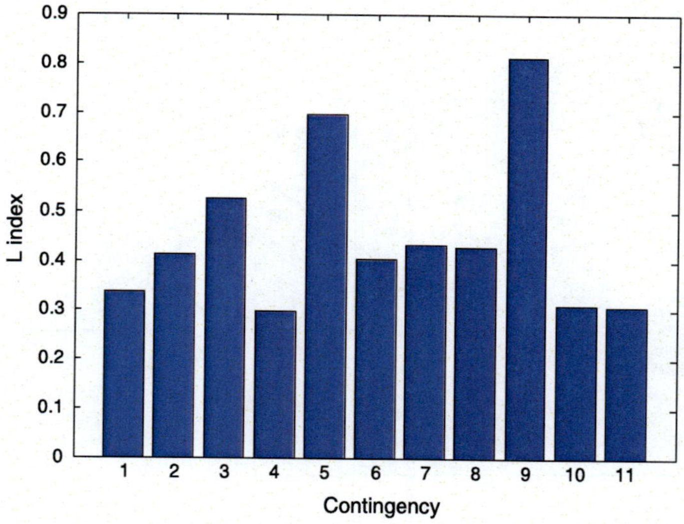

Fig. 3 Maximum L-index value for different contingencies of IEEE 6 bus network

LDIs for all contingencies are displayed in Fig. 5. It is observed that almost all the LDIs are negative. The negative LDI indicates the increase of loss after contingency. It is seen that for critical contingency, i.e., contingency 9, the VDI is positive and LDI is negative.

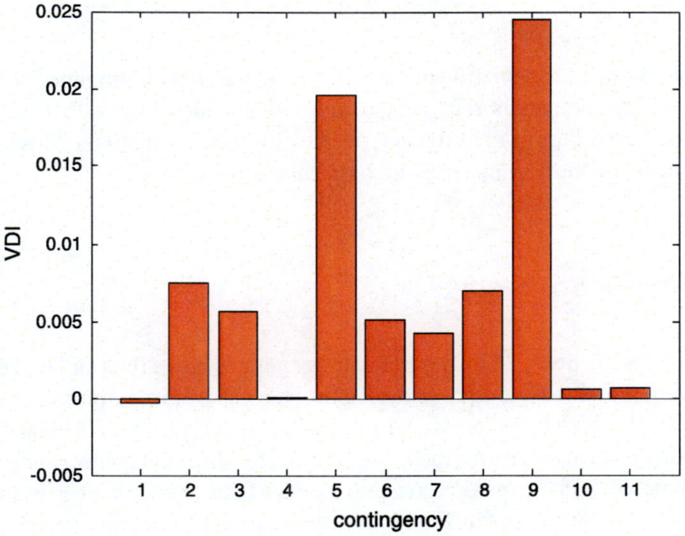

Fig. 4 VDI for different contingencies of IEEE 6 bus network

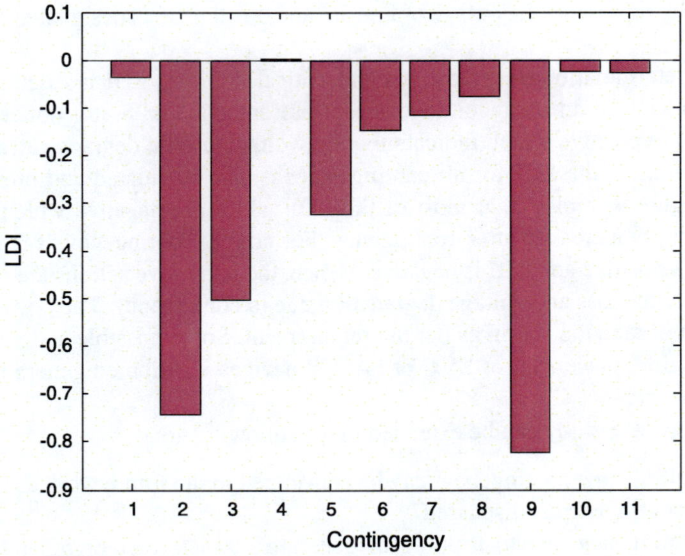

Fig. 5 LDI for different contingencies of IEEE 6 bus network

Table 1 Bus voltage comparison between base case and critical contingency case

Bus no.	Base case voltage (p.u)	Critical contingency voltage (p.u)
4	0.9864	0.9824
5	0.9797	0.951
6	1.0014	0.8834

In order to show the deviation of the voltage profile after contingency, the comparison of the bus voltages is represented in Table 1 which shows that the load bus voltages decrease during contingency. Here, only the comparison of bus voltages between critical contingency case and base case is shown.

3.2 IEEE 30 Bus Study

This system has 41 lines, 24 load buses and 5 generator buses (bus and line data given in Appendix). For this system, considering the base case, the maximum L-index value is found 0.0846 which corresponds to load bus 24 (Fig. 6), i.e., bus 30. Hence, bus 30 is designated as critical bus. Now, the L-index value for different contingency scenarios is also computed, and the maximum value for each contingency scenario is displayed in Fig. 7. It is observed from Fig. 7 that the maximum L-index value among all contingencies is 0.3068 which corresponds to the contingency 33. It is worth noting that load flow solution diverges during outage of line 13, 16 and 34. Therefore, these three contingencies are not considered in the simulation study. Hence, contingency 33 basically represents the outage of line 36, and this line 36 is designated as critical line.

After determination of critical bus and critical line, the VDI is calculated. The computed VDI for different contingencies is presented in Fig. 8. It is noticed that all the VDIs are positive which indicate that the voltage profile degrades after contingency. Similarly, the LDI for all contingencies is also calculated and presented in Fig. 9. Figure 9 implies that most of the LDI values are negative which indicate that the loss is increased after contingency. For critical contingency, it is observed that VDI is positive and LDI is negative. Hence, the corrective actions are needed to compensate the loss and voltage degradation due to contingency. The power system planners will take the action as per the requirement. Some possible actions may be load reduction, placement of SVC or FACTS devices, distributed generation (DG) penetration, etc.

The presented study can be extended in the following area:

- The multiple line contingency may be considered in the future work as here only $N-1$ contingency is discussed
- The contingency due to tripping of generator, transformer or other than line element may be taken into account as future scope of this work

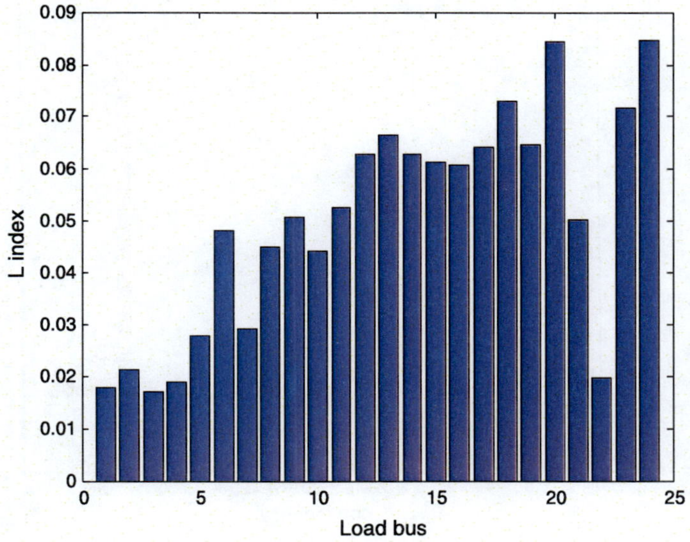

Fig. 6 L-index value for load buses of IEEE 30 bus network

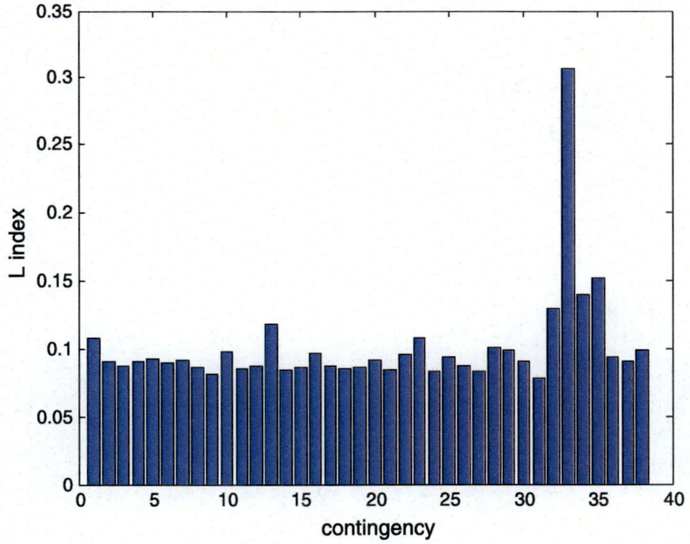

Fig. 7 Maximum L-index value for different contingencies of IEEE 30 bus network

- The appropriate action like placing of FACTS devices may be considered for improving the voltage stability and performance of the network. In this context, the determination proper location and the sizing of the devices will be another future scope of this study

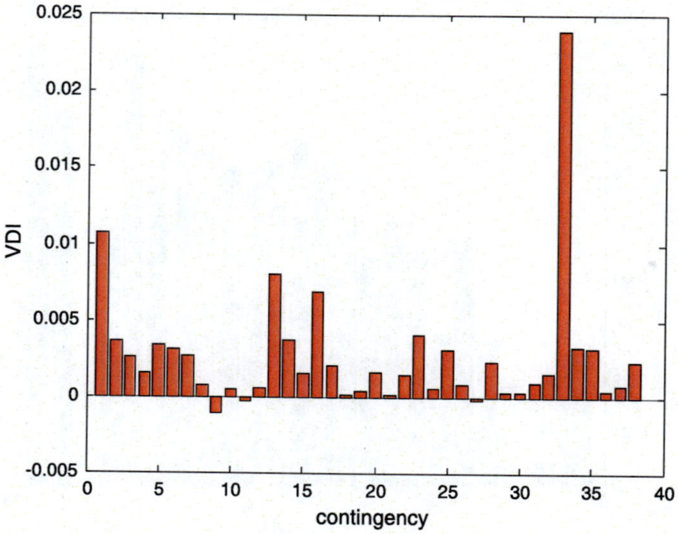

Fig. 8 VDI for different contingencies of IEEE 30 bus network

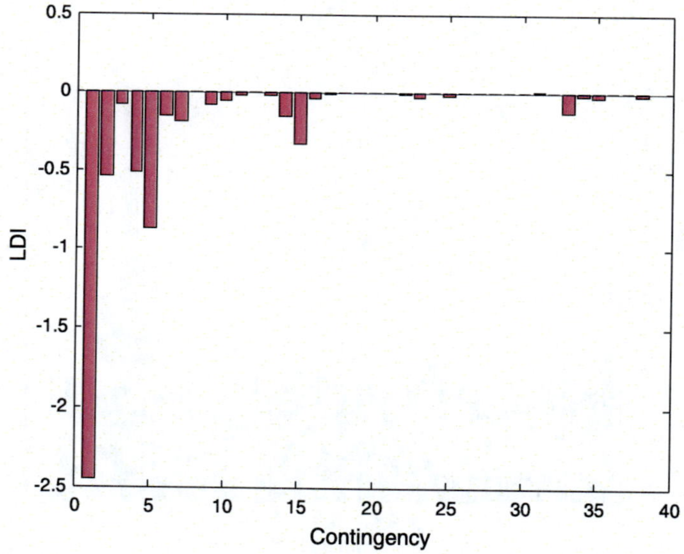

Fig. 9 LDI for different contingencies of IEEE 30 bus network

- The large network such as IEEE 57 bus, 118 bus may be considered as the test bed.

4 Conclusion

In this work, the voltage stability analysis as well as contingency analysis are conducted. The critical bus is identified through L-index value, and thereafter, the critical contingency is also determined. The impact of the contingency is evaluated through two novel indices, i.e., VDI and LDI. In majority of cases, the VDI found positive and LDI found negative. Therefore, it can be concluded that contingency generally degrades the voltage profile of the system and increases the power loss in the system. This study will be helpful to the power system planners for contingency analysis. The information regarding critical bus and critical line may be used for monitoring purpose or installation of suitable devices in order to improve the system stability or transfer capability. In the near future, the work may be extended considering the load variation scenario which is evident in real power network.

Appendix

See Tables 2 and 3.

Table 2 Bus data of IEEE 30 bus network (Type: 3—slack bus, 2—PV bus, 1—load bus)

Bus no.	Type	V	PL	QL	Pg	Qg	Qmax	Qmin	G	B
1	3	1.06	0	0	0	0	0	0	0	0
2	2	1.043	21.7	12.7	40	48.9	50	−40	0	0
3	1	1	2.4	1.2	0	0	0	0	0	0
4	1	1	7.6	1.6	0	0	0	0	0	0
5	2	1.01	94.2	19	0	36.6	40	−40	0	0
6	1	1	0	0	0	0	0	0	0	0
7	1	1	22.8	10.9	0	0	0	0	0	0
8	2	1.01	30	30	0	41.2	40	−10	0	0
9	1	1	0	0	0	0	0	0	0	0
10	1	1	5.8	2	0	0	0	0	0	0.19
11	2	1.082	0	0	0	16.3	24	−6	0	0
12	1	1	11.2	7.5	0	0	0	0	0	0
13	2	1.071	0	0	0	10.3	24	−6	0	0
14	1	1	6.2	1.6	0	0	0	0	0	0
15	1	1	8.2	2.5	0	0	0	0	0	0
16	1	1	3.5	1.8	0	0	0	0	0	0
17	1	1	9	5.8	0	0	0	0	0	0
18	1	1	3.2	0.9	0	0	0	0	0	0

(continued)

Table 2 (continued)

Bus no.	Type	V	PL	QL	Pg	Qg	Qmax	Qmin	G	B
19	1	1	9.5	3.4	0	0	0	0	0	0
20	1	1	2.2	0.7	0	0	0	0	0	0
21	1	1	17.5	11.2	0	0	0	0	0	0
22	1	1	0	0	0	0	0	0	0	0
23	1	1	3.2	1.6	0	0	0	0	0	0
24	1	1	8.7	6.7	0	0	0	0	0	0.043
25	1	1	0	0	0	0	0	0	0	0
26	1	1	3.5	2.3	0	0	0	0	0	0
27	1	1	0	0	0	0	0	0	0	0
28	1	1	0	0	0	0	0	0	0	0
29	1	1	2.4	0.9	0	0	0	0	0	0
30	1	1	10.6	1.9	0	0	0	0	0	0

Table 3 Line data of IEEE 30 bus network

Line no.	From bus	To bus	R	X	Ysh	Tap
1	1	2	0.0192	0.0575	0.0528	1
2	1	3	0.0452	0.1652	0.0408	1
3	2	4	0.057	0.1737	0.0368	1
4	3	4	0.0132	0.0379	0.0084	1
5	2	5	0.0472	0.1983	0.0418	1
6	2	6	0.0581	0.1763	0.0374	1
7	4	6	0.0119	0.0414	0.009	1
8	5	7	0.046	0.116	0.0204	1
9	6	7	0.0267	0.082	0.017	1
10	6	8	0.012	0.042	0.009	1
11	6	9	0	0.208	0	0.978
12	6	10	0	0.556	0	0.969
13	9	11	0	0.208	0	1
14	9	10	0	0.11	0	1
15	4	12	0	0.256	0	0.932
16	12	13	0	0.14	0	1
17	12	14	0.1231	0.2559	0	1
18	12	15	0.0662	0.1304	0	1

(continued)

Table 3 (continued)

Line no.	From bus	To bus	R	X	Ysh	Tap
19	12	16	0.0945	0.1987	0	1
20	14	15	0.221	0.1997	0	1
21	16	17	0.0524	0.1923	0	1
22	15	18	0.1073	0.2185	0	1
23	18	19	0.0639	0.1292	0	1
24	19	20	0.034	0.068	0	1
25	10	20	0.0936	0.209	0	1
26	10	17	0.0324	0.0845	0	1
27	10	21	0.0348	0.0749	0	1
28	10	22	0.0727	0.1499	0	1
29	21	22	0.0116	0.0236	0	1
30	15	23	0.1	0.202	0	1
31	22	24	0.115	0.179	0	1
32	23	24	0.132	0.27	0	1
33	24	25	0.1885	0.3292	0	1
34	25	26	0.2544	0.38	0	1
35	25	27	0.1093	0.2087	0	1
36	28	27	0	0.396	0	0.968
37	27	29	0.2198	0.4153	0	1
38	27	30	0.3202	0.6027	0	1
39	29	30	0.2399	0.4533	0	1
40	8	28	0.0636	0.2	0.0428	1
41	6	28	0.0169	0.0599	0.013	1

References

1. Streimikiene D, Siksnelyte I (2016) Sustainable assessment of electricity market models in selected developed world countries. Elsevier Renew Sustain Energy Rev 57:72–82
2. Berizzi A, Finazzi P, Dosi D, Marannino P, Corsi S (1998) First and second order methods for voltage collapse assessment and security enhancement. IEEE Trans Power Syst 13(2):543–551
3. Musirin I, Rahman TKA (2002) Novel Fast Voltage Stability Index (FVSI) for voltage stability analysis in power transmission system. In: Student conference on research and development proceedings 2002, 17 July
4. Shekhawat N, Gupta AK, Kumar Sharma A (2018) Voltage stability assessment using line stability indices. In: 2018 3rd international conference and workshops on recent advances and innovations in engineering (ICRAIE) Jaipur, India, pp 1–4
5. Kanimozhi R, Selvi K (2013) A novel line stability index for voltage stability analysis and contingency ranking in power system using fuzzy based load flow. J Electr Eng Technol 8(4):694–703
6. Jirjees MA, Al-Nimma DA, Al-Hafidh MSM (2019) Selection of proper voltage stability index for real system loading. In: 2019 2nd international conference on electrical, communication,

computer, power and control engineering (ICECCPCE), Mosul, Iraq
7. Kessel P, Glavitsch H (1986) Estimating the voltage stability of a power system. IEEE Trans Power Delivery 1(3):346–354
8. Kundu S, Alam M Thakur SS (2018) State estimation with optimal PMU placement considering various contingencies. In: 2018 IEEE 8th power India international conference (PIICON) Kurukshetra, India, pp 1–6
9. Adewolu BO, Saha AK (2020) Evaluation of performance index methodology for power network contingency ranking. In: 2020 international SAUPEC/RobMech/PRASA conference Cape Town, South Africa, pp 1–6
10. Rani Gongada S, Rao TS, Rao PM, Salima S (2016) Power system contingency ranking using fast decoupled load flow method. In: 2016 international conference on electrical, electronics, and optimization techniques (ICEEOT), Chennai, India, pp 4373–4376
11. Alam M, Mishra B, Thakur SS (2018) Assessment of the impact of line outage in modern power system. In: 2018 international conference on current trends towards converging technologies (ICCTCT), Coimbatore, pp 1–6
12. Dwivedi M, Dhandhia A, Pandya V (2017) Optimization of weighting factors of performance index to improve contingency ranking. In: IEEE international conference on power systems (ICPS) pp 319–322
13. Cruz EFD, Mabalot AN, Marzo RC, Pacis MC, Tolentino JHS (2016) Algorithm development for power system contingency screening and ranking using voltage reactive power performance index. In: IEEE region 10 conference (TENCON) pp 2232–2235
14. Alam M, Kundu S, Thakur SS, Banerjee S (2020) A new cost effective algorithm for online identification of line outage contingency using current phasor of PMU. Sustain Energy Grids Netw 23:1–12
15. Al. Shaalan AM (2020) Contingency selection and ranking for composite power system reliability evaluation. J King Saud Univ Eng Sci 32:141–147

Energy-Efficient Railway Lighting Design—A Case Study

Shubhankar Sardar, Arnab Ganguly, Amartya Roy, Srijan Banerjee, and Sajjan Kumar

Abstract In this paper, an energy-efficient lighting design scheme has been developed for coach lighting of an EMU train. The design is proposed by aiming to achieve energy efficiency by achieving the desired outcomes of lighting parameters and maintaining the aesthetic beauty of lighting. The change in the lighting design scheme has been proposed on the basis of a survey work done at Eastern Railway, Howrah, EMU division. This survey work was done to get the technical data of the existing lighting design. Here, instead of fluorescent lamps, LED lamps are used as an artificial light source in the proposed design to make the betterment from existing design scheme in terms of lighting outcomes and energy efficiency. The proposed scheme has been designed in DIALux 4.13 software. From this proposed design scheme, the uniformity and average illumination level can be improved; at the same time, the energy consumption per day per coach can be reduced drastically as compared to existing design.

Keywords Railway lighting · Artificial lighting · Energy efficient · LED

1 Introduction

Illumination engineering is a dynamic field of application where designing of a lighting scheme involves a number of lighting parameters [1]. The design can be done in various methods by varying the design variables in a continuous or discrete way [2, 3]. Lighting design is such a field where to illuminate the object is not only the concern. With achieving the desired outcome for lighting parameters, it is also obvious to maintain the aesthetic beauty that actually created an appropriate luminous environment, which in turn helps to make the mood [4, 5]. Energy efficiency is another real concern that takes an important role to play in creating an appropriate design

S. Sardar · A. Ganguly (✉) · A. Roy · S. Banerjee · S. Kumar
Electrical Engineering Department, Gargi Memorial Institute of Technology, Baruipur, Kolkata, West Bengal, India
e-mail: arnab.ee_gmit@jisgroup.org

K. Namrata et al. (eds.), *Smart Energy and Advancement in Power Technologies*, Lecture Notes in Electrical Engineering 926,
https://doi.org/10.1007/978-981-19-4971-5_43

scheme [6]. So the main goal to propose a new lighting design is always to betterment of the existing design scheme in terms of achieving the lighting parameters, creating aesthetic beauty, and also decreasing the electrical energy consumption [7, 8]. It can make a lighting design efficient. Railway lighting is an application-specific lighting design where the designer has to ensure the fulfilment of some other technical requirements with maintaining the desired lighting outcomes [9, 10]. Here, the coach lighting is considered for the case study. Achieving the lighting parameters like average maintained illuminance and overall uniformity, energy consumption also has to be considered for the installation [11, 12]. Passenger safety is a big issue where lighting can play a role. Railway lighting consists of emergency lighting in every EMU coach that mainly required when the train passes through a neutral section, as on that point all the other artificial light sources are in off condition [13]. So for lighting installation, the designer has to keep this in mind. In this paper, we have proposed a new lighting design scheme for EMU coach where LED lamps are considered as the artificial light source by replacing the existing T8/T5 fluorescent lamp. The proposed design is prepared using DIALux 4.13 software where aims are to create a more energy-efficient system is with achieving the desired lighting parameters [14, 15].

2 Existing Design Scheme

In the existing design scheme of EMU coaches under Eastern Railway, Howrah, EMU division, this is noticed that it consists of two types of EMU coaches, i.e. old coach and a new coach. Here, it is already mentioned that we are mainly considering the lighting design aspect, and from that perspective, there are some differences between the old EMU coach and the new EMU coach. An EMU is abbreviated as "electric multiple unit", and this type of multiple-unit train mainly consists of motor-coach and tailor coach. In the old EMU coach, mainly T12 fluorescent lamps are used inside the train compartment as the main light source and GLS lamps are used for emergency lighting in front of the gate of each compartment. But in the new EMU coach, T12 fluorescent lamps are replaced with T8/T5 fluorescent lamps, and for emergency lighting purposes, GLS lamps are also replaced with fluorescent lamps. In the old EMU coach, each luminaire consists of a single FTL but in the new EMU coach, each luminaire consists of two numbers of FTLs. The luminaire arrangements of old coach and new coach of EMU are shown in Fig. 1. The total number of light fixtures also differs from old coach to new coach. Here, a comparison between the old coach and the new coach is given in Table 1 as per Indian Railway Datasheet.

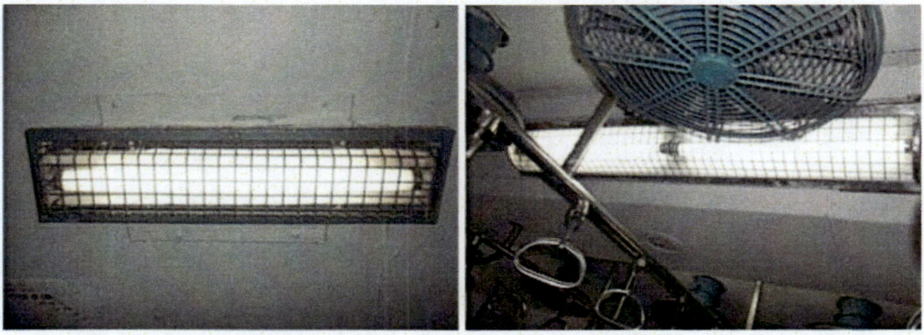

Fig. 1 Lighting fixture of old and new EMU coach

Table 1 Comparison chart between old coach and new coach

Name of the coach	Numbers of lamp	
	Old coach	New coach
Motor	24/26	30
Tailor	26	36

3 Calculation of Per Day Energy Consumption of Existing Scheme

Energy consumption is a very important aspect of any electrical installation. In this project, we have done a calculation regarding per day energy consumption of the existing design scheme. Here, the new EMU coach is considered for this calculation as gradually, the number of old EMU coaches are decreasing with the prominence of new coaches used in Eastern Railway, Howrah, EMU division. The data regarding the number of lamps in a particular tailor coach of EMU is collected from the survey work. With this data, per day energy consumption of a specific new EMU coach is prepared, where it is considered as the artificial lamps are used for 6 h every day (5.00 PM–11.00 PM). The mentioned energy consumption data of existing new EMU coach is given in Table 2.

Table 2 Per day energy consumption data of existing new EMU coach

Coach type	No of lamp/coach (excluding emergency lamp)	Wattage/lamp (W)	Total lamp load/coach (W)	Burning hour/day (h)	Total energy consumption (KWh)
New coach (existing)	28	18	504	6	3.02

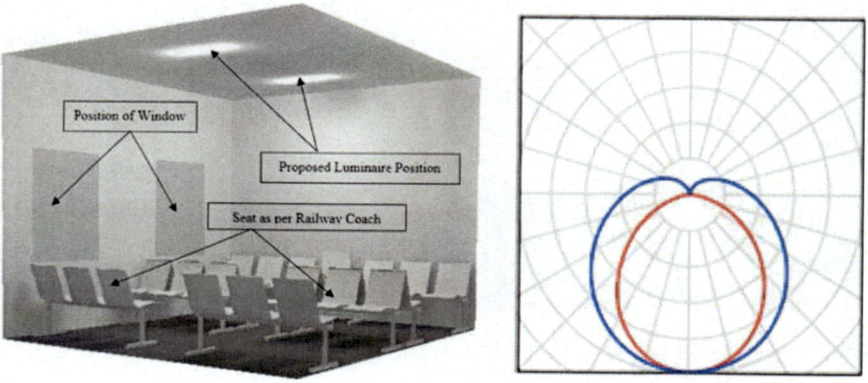

Fig. 2 3D view of the proposed scheme and photometry of proposed LED luminaire

4 Proposed Design Scheme

In this paper, we have prepared a proposed lighting design scheme for EMU coach, where it is proposed to replace the existing fluorescent lamps with LED lamps (Ecopoint T-series compact linear fitting) of the same dimension (600 mm width). DIALux 4.13 software is used to prepare and check the feasibility of the proposed design. 3D view of the proposed design scheme and the photometry of the proposed LED luminaire are shown in Fig. 2. This change is proposed for the overall betterment of the design in terms of illumination and electrical energy consumption point of view. Here a small section of an EMU coach, dimension of (3.65 m × 4 m × 3 m) is considered for the proposed design scheme. Seating arrangements have done as per the arrangements of the existing new EMU coach, which we surveyed in Eastern Railway, Howrah, EMU division. Algorithms of our proposed scheme are given bellow,

Step 1: Measurement of existing lighting parameter (E_{avg}, E_{max}, E_{min}, U_0) of new coach;
Step 2: Simulate the existing parameter (E_{avg}, E_{max}, E_{min}, U_0) in DIALux 4.13;
Step 3: Proposed energy-efficient lighting design in DIALux 4.13;
Step 4: Comprising between existing and proposed scheme of new coach.

5 Calculation of Per Day Energy Consumption of Proposed Scheme

Energy consumption of the proposed design scheme is done on a per daily basis. The total number of lamps for an EMU coach can be calculated from the proposed design done in DIALux 4.13 software as it obviously can be considered as a symmetrical design throughout the entire coach. The design symmetry is also noticed in the

Table 3 Per day energy consumption data of proposed EMU coach

Dimension	No. of lamp (excluding emergency lamp)	Wattage/lamp (W)	Total lamp load (W)	Burning hour/day (h)	Total energy consumption (KWh)
3.65 m × 4 m × 3 m	2	11	22	6	0.132
Entire coach	14	11	154	6	0.924

time of survey to an EMU coach under Eastern Railway, Howrah, EMU division. The burning hour of the proposed LED lamps is considered as same (6 h) as the existing scheme, designed by fluorescent lamps to draw a comparison between the existing and the proposed design scheme. The mentioned energy consumption data of proposed EMU coach is given in Table 3.

6 Result Analysis

This project has two parts. Firstly, the survey work done in the EMU coach under Eastern Railway, Howrah, EMU division, and next, the proposed change in design made in DIALux 4.13 software. So here, we have two results to show the outcome of survey work and the software design, respectively. It will be followed by a comparison between the existing and proposed design scheme. The software design is done for the particular area of an EMU coach whose dimension is (3.65 m × 4 m × 3 m). This mentioned area dimension is the same as the area dimension for which the survey work was done in an EMU coach. From a lighting point of view, the measured parameters are the averaged maintenance horizontal illuminance and the overall uniformity. The number of lamps required to achieve the standard illuminance level mentioned in the Indian Railway Standard is also an important parameter. The average maintained illuminance level for passenger coach is (30–60 lx), as mentioned in Indian Railway Standard. So, the first and foremost criteria is to achieve the average illuminance value with maintaining proper uniformity level. The illuminance values are the horizontal illuminance outcome and real-time values, taken by Luxmeter. The Luxmeter was fixed in the seats of the EMU coach in horizontal mode to take the values. Here, the illumination-oriented result outcome from the survey work is given in Table 4. The ISOLUX contour of the existing lighting design is given in Fig. 3.

The proposed design for the betterment of the existing scheme is prepared in DIALux 4.13 software. The workspace is taken at the height of 0.750 m, the same as the height of the seat of the existing coach, considered at the time of the survey. The final outcome of the lighting parameter of proposed coach is given in Table 5. The ISOLUX contour of this proposed lighting design is given in Fig. 4.

Now, a comparison can be done between existing coach and the proposed design scheme from the lighting point of view. Details of comparison are given in Table 6.

Table 4 Illuminance values under artificial light in new coach (existing)

Horizontal illuminance values (lux)						
44	65	75.6	*Gap between row*	75.4	65.5	44.3
Gap between seat				*Gap between seat*		
43.8	65.4	73.8		73.2	65.3	43.6
44.2	64.7	73		72.7	64.3	44.1
Gap between seat				*Gap between seat*		
43.2	64.2	74.6	75.3	74.8	64.4	43

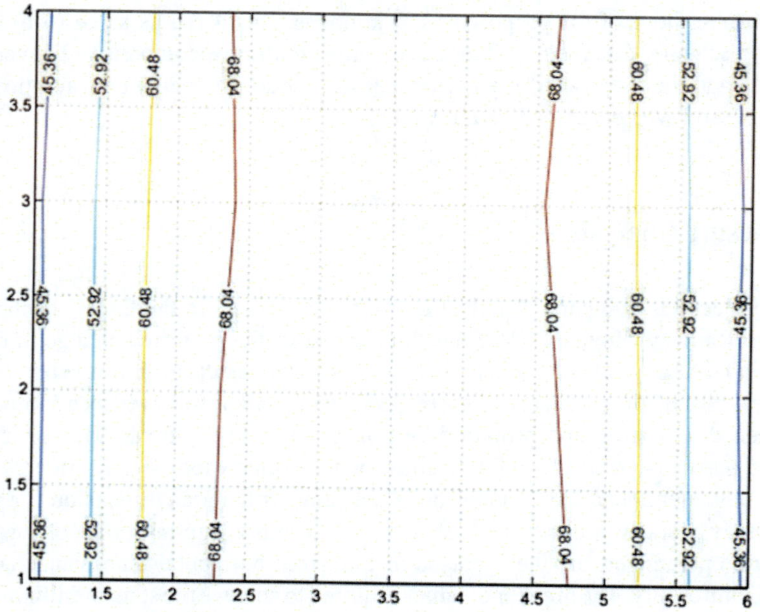

Fig. 3 ISOLUX contour of the existing lighting design

Table 5 Illuminance values under artificial light for proposed coach

Maximum illuminance E_{MAX} (lux)	Minimum illuminance E_{MIN} (lux)	Average illuminance E_{AVG} (lux)	Overall uniformity (U_0)
80	51	69	0.732

It is obvious that the proposed lighting design gives a better outcome than the existing one from an illumination point of view. It gives better average maintained illuminance and better overall uniformity, which in turn gives a better luminous environment.

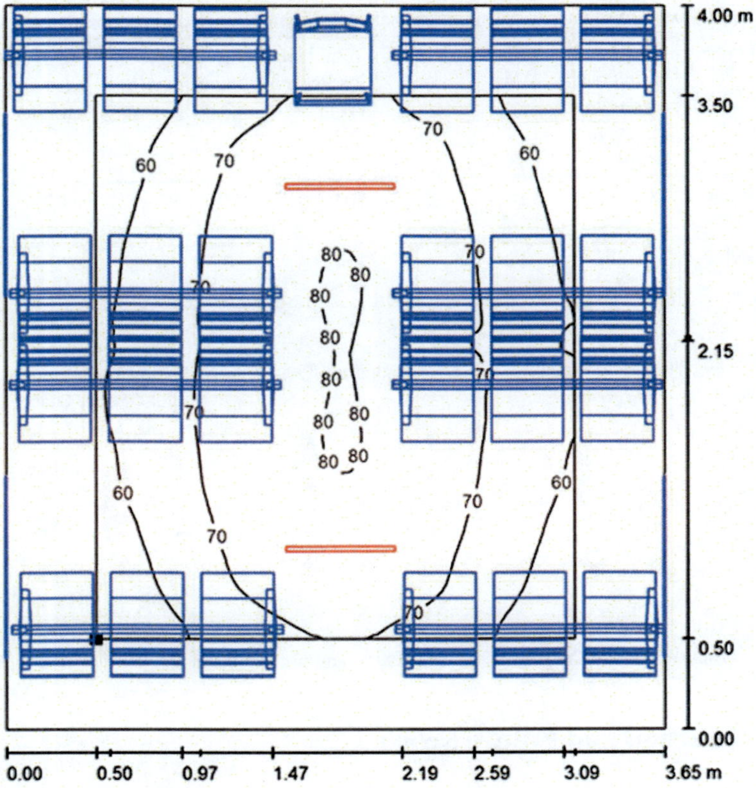

Fig. 4 ISOLUX contour of this proposed lighting design

Table 6 Comparison of lighting parameters between existing coach and proposed coach

Coach type	Maximum illuminance E_{MAX} (lux)	Minimum illuminance E_{MIN} (lux)	Average illuminance E_{AVG} (lux)	Overall uniformity (U_0)
Proposed coach	80	51	69	0.732
Existing coach	75.6	43	61.49	0.699

Another comparison between the existing design and the proposed design can be drawn from an energy consumption point of view. The mentioned comparison is given in Table 7.

It can be said that the proposed design scheme using LED lamps consume less energy than the exiting design consists of fluorescent lamp. The conclusion can be drawn from the result analysis that the proposed lighting design of EMU coach using LED lamps will give better lighting ambience and less energy consumption than the existing one. Comparison of all the parameters of existing and proposed scheme is also shown in Fig. 5.

Table 7 Comparison of energy consumption between existing coach and proposed coach

Coach type	No. of lamp/coach (excluding emergency lamp)	Wattage/lamp (W)	Total lamp load/coach (W)	Burning hour/day (h)	Total energy consumption (KWh)
New coach (existing)	28	18	504	6	3.02
Proposed coach	14	11	154	6	0.924

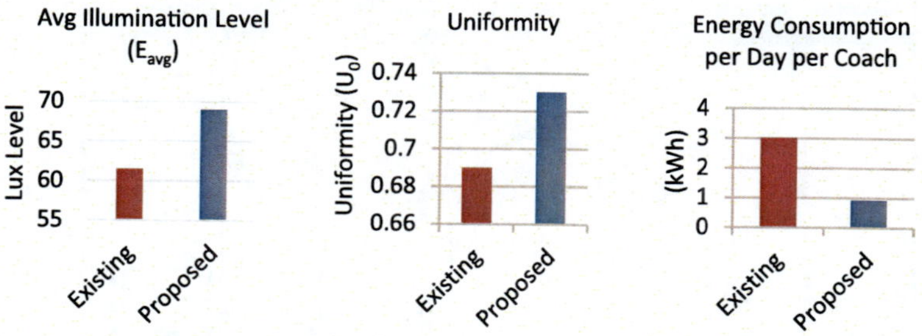

Fig. 5 Comparison of all the parameters of existing and proposed scheme

7 Conclusion and Future Scope

From the paper, it can be concluded that there will be a 50% reduction in the number of lights if we use LED in place of conventional lights in a coach. This is done by the efficient design of the layout for installing LED tubes. The proposed scheme improves the overall maintaining average illuminance and uniformity. From a saving point of view, an amount of 2.1 kW can be saved every day per coach. In this proposed scheme, we can also think about the daylight integration, so that based on the availability of daylight, we can make more energy-efficient system by using dimmable artificial lamps where the dimming will depend on the availability of natural light in day-time.

References

1. Rufer A (2010) Energy storage for railway systems, energy recovery and vehicle autonomy in Europe. In: Proceedings of 9th international power electronics conference (IPEC 2010), Singapore, 21–24 Jun, pp 3124–3127
2. Mandal P, Dey D, Roy B (2021) Optimization of luminaire layout to achieve a visually comfortable and energy-efficient indoor general lighting scheme by particle swarm optimization. LEUKOS J Illumination Eng Soc 17(1):91–106
3. Villa C, Labayrade R (2013) Multi-objective optimisation of lighting installations taking into account user preferences—a pilot study. Lighting Res Technol 45:176–196

4. Mendes LA, Freire RZ, Coelho LS, Moraes AS (2017) Minimizing computational cost and energy demand of building lighting systems: a real time experiment using a modified competition over resources algorithm. Energy Build 139:108–123
5. Ramirez JAR, Llano CA (2012) Guide for interior lighting installations design using Dialux Technological University of Pereira
6. Chew I, Kalavally V, Oo NW, Parkkinen J (2016) Design of an energy-saving controller for an intelligent LED lighting system Energy. Build 120:1–9
7. Steiner M, Klohr M, Pagiela S (2007) Energy storage system with ultracaps on board of railway vehicles. In: Proceedings of 12th European conference power electronics application (EPE 2007), Aalborg, Denmark, 11–14 Sep, pp 1–10
8. Coppola M, Del Pizzo A, Iannuzzi D (2012) A power traction converter based on modular multilevel architecture integrated with energy storage devices. Presented at 2nd international conference electronics system aircraft railway ship propulsion (ESARS 2012), Bologna, Italy, 16–18 Oct
9. Battistelli L, Fantauzzi M, Iannuzzi D, Lauria D (2012) Energy management of electrified mass transit systems with energy storage devices. In: Proceedings of 21st international symposium power electronics electronic drives automation motion (SPEEDAM 2012), Sorrento, Italy, pp 1172–1177
10. Gil R, Gil S and Iannelli L (2015) 1, 5 GW saving in the peaks of electricity consumption-LED Lighting Petrotecnia
11. Iannuzzi D, Tricoli P (2010) Optimal control strategy of onboard super capacitor storage system for light railway vehicles. In: Proceedings of IEEE international symposium industrial electronics (ISIE 2010), Bari, Italy, 4–7 Jul, pp 280–285
12. Iannuzzi D, Lauria D, Tricoli P (2012) Optimal design of stationary supercapacitors storage devices for light electrical transportation systems. J. Optim Eng 13(4):689–704
13. Mukherjee P (2016) An overview of energy efficient lighting system design for indoor applications of an office building. Key Eng Mater 692:45–53
14. Kamaruddin MA, Arief YZ, Ahmad MH (2016) Energy analysis of efficient lighting system design for lecturing room using DIAlux Evo 3. Appl Mech Mater 818:174–178
15. Ciccarelli F, Iannuzzi D, Lauria D (2012) Stationary ultracapacitors storage device for improving energy saving and voltage profile of light transportation networks. Transp Res C Emerg Technol 21(1):321–337

Design of 20 kwp Solar PV System with Different Tracking Systems Using PVsyst and Sketch-Up

K. Mahesh, V. Joshi Manohar, Devineni Gireesh Kumar, M. Prameela, K. Ramakrishna, S. Saravanan, DSNMRAO, and P. Chandra Babu

Abstract The development of renewable energy technology in particular the use of solar energy in the current energy market resulted from the technical, economic, and environment policies at world level. Since it became an integral component of the supplier to the reach for planning and analyzing the solar PV system before realistic plants. So, installer must design the PV system before installing it practically to minimize the cost and to reduce losses in the system after installation. In this study, performance comparison of single axis and dual axis tracker controlled solar photovoltaic power system is done with fixed tilt solar photovoltaic system with and without shading using PVsyst V6.7 and SketchUp with a plug-in specific to develop photovoltaic generation is called Skelion. SketchUp is used to build a three-dimensional model design of the photovoltaic project. This study also includes simulated performance evaluation of 20 kWp grid connected Si-poly photovoltaic system at BVRITN. For the efficiency of the photovoltaic device, the simulation effects have been studied. This involves assessing the PV array's effective energy efficiency, grid injecting energy, standardized output by kWp, and performance ratio. Shading of solar panels is an important observation in our study as because it reduces effective output of PV array.

Keywords PVsyst · Dual axis tracking · Single axis tracking · SketchUp · Fixed tilt · Shading analysis

K. Mahesh · D. G. Kumar (✉) · M. Prameela · K. Ramakrishna · S. Saravanan · P. Chandra Babu
Electrical & Electronics Engineering, BV Raju Institute of Technology, Narsapur, Telangana, India
e-mail: gireesh218@gmail.com

V. Joshi Manohar
Electrical & Electronics Engineering, Presidency University, Bangalore, Karnataka, India

DSNMRAO
Electrical & Electronics Engineering, Gokaraju Rangaraju Institute of Engineering & Technology, Hyderabad, Telangana, India

K. Namrata et al. (eds.), *Smart Energy and Advancement in Power Technologies*, Lecture Notes in Electrical Engineering 926,
https://doi.org/10.1007/978-981-19-4971-5_44

1 Introduction

Energy is an essential part of our fast-growing economies. There are numerous energy resources. Human beings started to extract energy from fossil fuels but growing population and increasing demands in energy has led to increase in fuel prices accordingly and depletion of fossil fuels [1]. With growing increase in energy demand and rapidly depleting fossil fuels the energy resources must be renewable and sustainable. Solar energy is one of the cleanest renewable sources of energy [2]. Numerous technologies have been developed to generate electricity from solar energy. There are different types of PV systems depending on the type of load and grid configuration. Grid connected system is most common in the case of large-scale PV systems [3, 4]. The PV modules are affixed at such tilt angles that maximum of sunlight is incident on the module surface. But the path of sun over a specific area where modules are installed is not same throughout a year [5]. To track the sun for maximum sunlight PV modules are adjusted in orientation and tilt angles. Tracker systems are being used for this purpose [6]. Based on the direction and tilt angle following are the main types of PV systems.

1.1 Fixed Tilt System

PV modules are oriented to collect maximum possible sunlight. The PV modules are tilted perpendicular to the sun rays for maximum efficiency. Since the angle of sun varies all year so the tilt angel is different for summer and winter [7, 8]. Increasing the tilt angle increases energy generation in winter and decreasing the tilt angle is increases energy generation for summer season. PV modules in fixed tilt system are tilted at a fixed angle which is calculated as an optimum angle for the whole year. Figure 1 shows a PV module oriented toward south and tilted at fixed angle.

1.2 Single Axis Tracking System

Tracker controlled PV systems are installed to track the sun by changing the orientation all day making sure that maximum energy is absorbed by the PV modules. The power consumption of tracker systems is 2–3% of the increased output. The solar trackers, one axis and the two axis trackers, are two major groups [9, 10]. Solar trackers with a single axis will rotate on a single fixed axis in one direction. The various kinds of axis trackers such as inclined, horizontal, vertical, and polar aligned are available, depending on the rotation. The single axis tracker is seen on the tilted axis in Fig. 2.

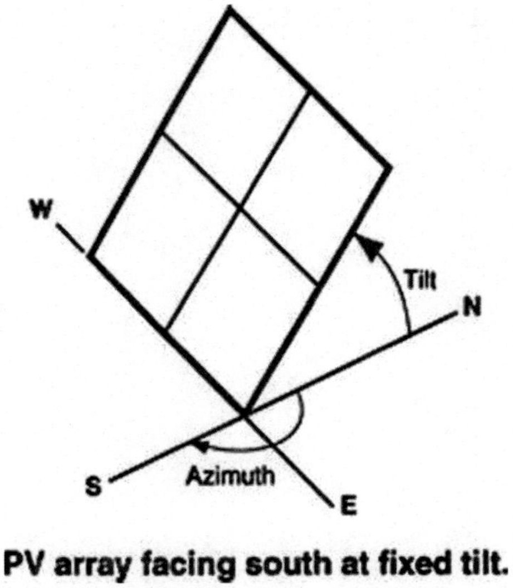

Fig. 1 Fixed tilt system

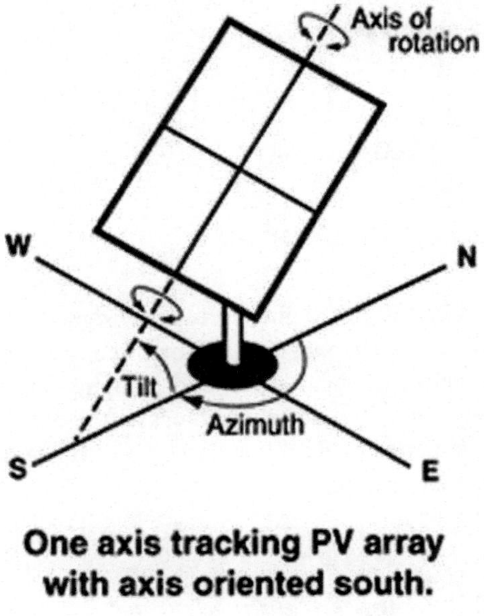

Fig. 2 Single axis tracking

1.3 Dual Axis Tracking System

Dual axis solar trackers can move in both directions vertically and horizontally and hence generating more output power as compared to single axis tracker systems. The dual axis tracker system is designed in such a manner that it always faces the sun and

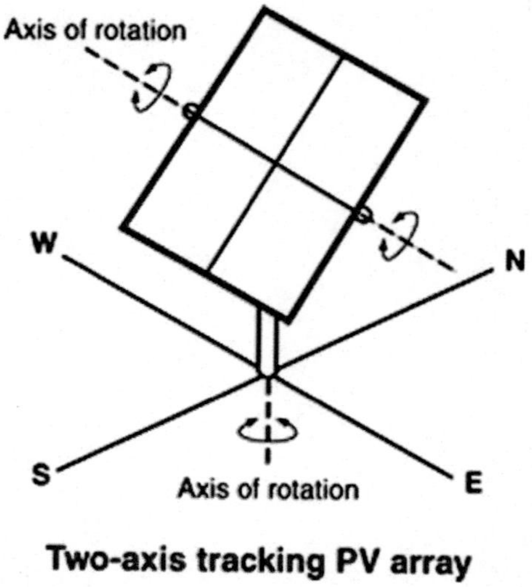

Fig. 3 Dual axis tracking

sunlight is incident to the collector's surface [11]. A dual axis solar tracker system is shown in Fig. 3.

2 Methodology

2.1 PVsyst Software

PVsyst is a modeling software designed to forecast solar energy efficiency. Meteor data and personal data are imported from several sources. The performance of grid-connected, stand-alone, and pumping systems will be assessed using the module selection. The software forecasts the device output calculated with precise hourly simulation data accurately [12, 13]. It needs to have some inputs on the software to achieve results.

2.2 Inputs Required

The configuration of PV system would rely on the environment since the amount of solar radiation at every location is different. It's because of the location in relation to the sun. This disparity in direction is seen by the special set of parameters such as latitude, longitude, and location altitude.

Table 1 Location details

Name of the site		Narsapur (BVRITN)
Latitude		17.72°N
Longitude		78.25°E
Altitude		540 m
Optimum tilt angle	Summer	2°
	Winter	32°
Azimuth		0°
Tilt angle		17°

2.3 Location

Table 1 shows the location details of Narsapur (BVRITN) for the study.

2.4 Tilt Angle

The angle of inclination is known as the slope of solar panels installed to face the sun. The location of sun varies about earth every day, and the mounting angles of panels continue to change as well. Tilt angle is generally assumed to be equivalent to the latitude of the position considered [14]. To achieve full solar power on the panels, optimum tilt angle values are required [15, 16].

2.5 Solar PV Module

Table 2 shows the specifications of solar PV module used at Narsapur (BVRITN).

Specifications are at STC: 1000 W/m^2 Insulation, AM 1.5, Cell Temperature 25 °C.

Table 2 Polycrystalline based Solar PV module specifications (SS-250)

Name of manufacturer	Sirius solar energy systems
Model	SS-250
Type	Polycrystalline
Rated power (P_{max})	250 Wp
Open circuit voltage (V_{oc})	37.56 V
Short circuit current (I_{sc})	8.53 A
Rated voltage (V_{max})	31.18 V
Rated current (I_{mp})	8.02 A
Power tolerance	± 3%

Table 3 Polycrystalline based Solar PV module specifications (6Tp-250)

Name of manufacturer	Helios USA
Model	6Tp-250
Type	Polycrystalline
Rated power (P_{max})	250 Wp
Open circuit voltage (V_{oc})	37.80 V
Short circuit current (I_{sc})	8.650 A
Rated voltage (V_{max})	30.70 V
Rated current (I_{mp})	8.140 A
Power tolerance	± 3%

As the data base of Sirius Solar Energy Systems is not included in PVsyst an approximate characteristic panel with similar ratings has been chosen so that it can operate in the same way as Sirius Model No. SS-250. Table 3 shows the specifications of solar PV model used for study purpose.

Specifications are at STC: 1000 W/m^2 Insolation, AM 1.5, Cell Temperature 25 °C.

2.6 Inverter

Table 4 shows inverter specifications used at Narsapur (BVRITN) and for study purpose. Inverter used is REFUsol 020 K. It is a 20 Kw inverter.

Table 4 Inverter specifications

DC data	
DC max. input voltage	1000 V
DC MPP range	480–850 V
DC max. input current total	41.0 A
Operating temperature range	−25 … +55 °C
AC data	
AC nominal operating voltage	3AC 400 V + N
AC nominal operating frequency	50/60 Hz
AC rated power	19.2 kVA
AC max. output current	3 × 29.0 A

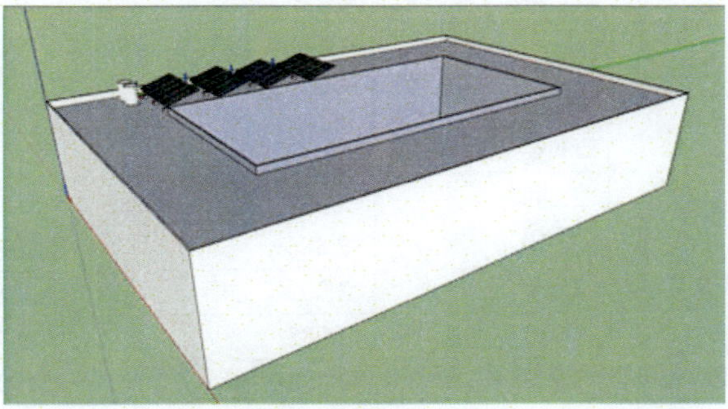

Fig. 4 Three-dimensional model built in SketchUp

2.7 SketchUp Software

In SketchUp, we can create three-dimensional models of any structure. So, it is widely applied in the field of architecture to model three dimensional products. In addition, this program is also used in civil engineering and design. The strong, real-time Shadow Engine of SketchUp allows you to carry out accurate studies on PV modules. To measure the cast shadow location, you just need to know the year, period, and place of the module on the globe. SketchUp makes geographical location of the module at quick. Most of the shadow configurations panel does. Only slide the sliders Time and Date to see cast shadows in real time.

3 Data Collection

3.1 SketchUp

With three-dimensional modeling offered by software SketchUp it can be seen the structure where the photovoltaic panels are to be installed. The three-dimensional digital reproduction of the installation site provides more precisely the project measures, and better visualization of architectural design. Figure 4 shows the illustrated structure digitally reproduced with SketchUp.

After the three-dimensional modeling done in SktechUp, Skelion was used for installing solar panels on the mounted structure.

3.2 Energy Yield Analysis of Fixed Tilt, Single Axis Tracking, and Dual Axis Tracking

In this study, a single axis solar tracker system and dual axis solar tracker system is simulated, and energy generation of complete year is calculated, while fixed tit

Table 5 Simulation results

Month	Fixed tilt without shading	Fixed tilt with shading	Single axis tracker	Dual axis tracker
	Energy injected into grid (MWh)			
Jan	2.889	2.861	3.200	3.836
Feb	2.910	2.886	3.078	3.895
Mar	3.090	3.062	3.106	3.925
Apr	3.013	2.985	3.067	3.899
May	2.929	2.898	3.199	4.011
Jun	2.350	2.313	2.579	2.989
Jul	2.073	2.039	2.242	2.579
Aug	2.145	2.109	2.210	2.523
Sep	2.519	2.488	2.522	3.101
Oct	2.837	2.806	2.896	3.477
Nov	2.623	2.595	2.804	3.288
Dec	2.855	2.823	3.223	3.776
Year	32.233	31.863	34.124	41.298

system is physically installed. Simulation results of all the systems are shown in Table 5.

3.3 Meteorological Data from Meteonorm and Performance Ratio of Fixed Tilt, Single Axis Tracking and Dual Axis Tracking

Table 6 shows the summarized results of Meteorological data and Performance ratio of various systems using simulation.

Table 6 Meteorological data and performance ratio

	Fixed tilt without shading	Fixed tilt with shading	Single axis tracker	Dual axis tracker
GlobHor (Kwh/m^2)	1949.6	1949.6	1949.6	1949.6
DiffHor (Kwh/m^2)	824.23	824.23	824.23	824.23
T Amb (°C)	26.91	26.91	26.91	26.91
GlobInc (Kwh/m^2)	2056.3	2056.3	2185.7	2668.4
GlobEff (Kwh/m^2)	1994.5	1969.4	2124.1	2625.8
Performance ratio (PR)	0.784	0.775	0.781	0.774

Table 7 Simulation results of normalized productions per day

	Fixed tilt without shading	Fixed tilt with shading	Single axis tracker	Dual axis tracker
Produced useful energy Y_f (kWh/kWp/day)	4.42	4.36	4.67	5.66
Collector loss L_c (kWh/kWp/day)	1.12	1.17	1.21	1.53
System loss L_s (kWh/kWp/day)	0.1	0.1	0.1	0.12

3.4 Normalized Productions Per kWp

Table 7 shows normalized productions per kWp for fixed tilt, Single axis tracking and Dual axis tracking.

4 Results and Discussions

4.1 Comparison of Excepted Energy Yield of Fixed Tilt, Single Axis Tracking and Dual Axis Tracking

See Fig. 5; Table 8.

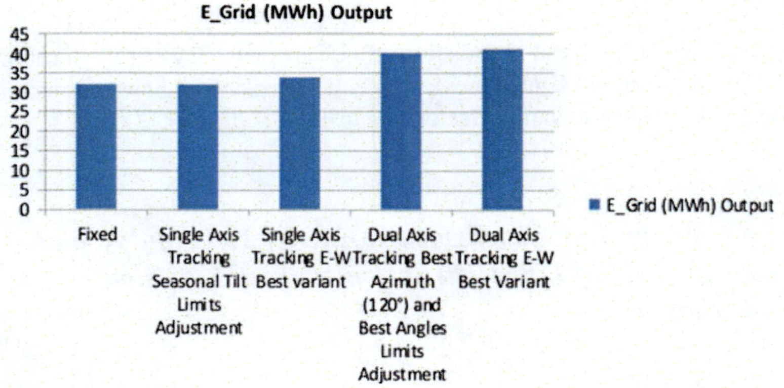

Fig. 5 Simulated energy yield comparison

Table 8 Excepted energy yield for various systems

Type of the system	Tilt	E_grid (MWh) output
Fixed	17°	32.23
Single axis tracking seasonal tilt limits adjustment	2°, 32°	32.34
Single axis tracking E-W best variant	−60°, 60°	34.12
Dual axis tracking best Azimuth (120°) and best angles limits adjustment	−7°, 41°	40.24
Dual axis tracking E-W best variant	−60°, 45°	41.30

Table 9 Variation of incident radiation and losses

Type of the system	Global incident in collector plane (%)	Loss due to IAM factor (%)	Loss due to temperature (%)
Fixed	+5.5	−2.9	−13.9
Single axis tracking seasonal tilt limits adjustment	+5.9	−3.0	−13.8
Single axis tracking E-W best variant	+12.1	−2.8	−14.4
Dual axis tracking best Azimuth (120°) and best angles limits adjustment	+33.3	−1.8	−15.9
Dual axis tracking E-W best variant	+36.9	−1.6	−16.2

4.2 Variation of Incident Radiation and Losses

Average Global Horizontal Irradiation for Narsapur (BVRITN) location is 1949.6 kWh/m^2/year and Global incident in collector plane is shown as percentage increase and losses due to various factors are in Table 9 (Fig. 6).

4.3 Comparison Between Excepted Energy Yield and Actual Energy Yield of Fixed Tilt System with and Without Shading

See Fig. 7; Table 10.

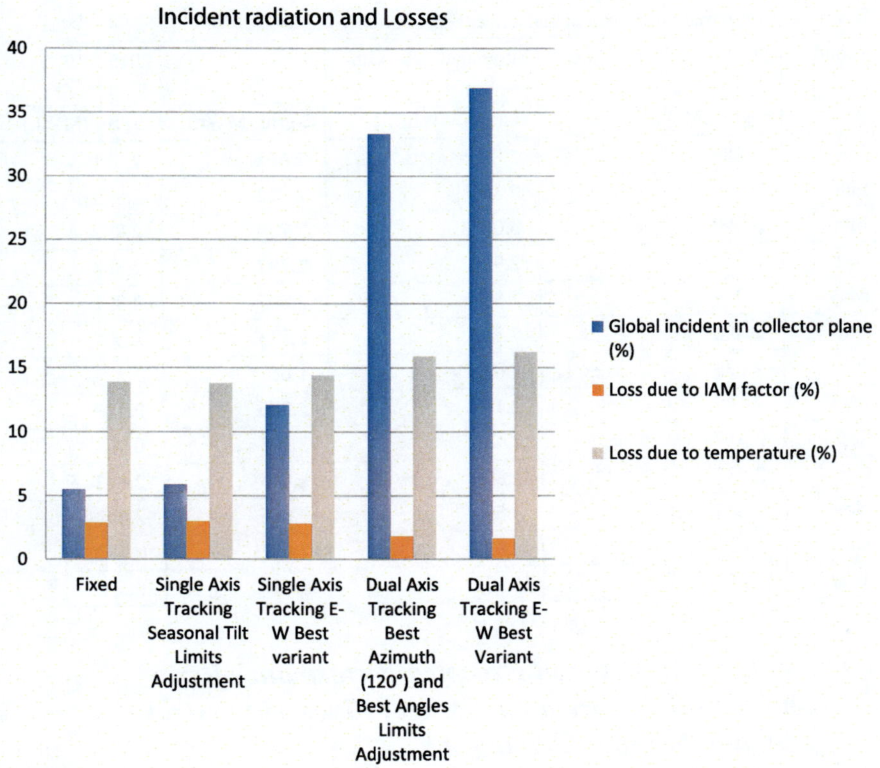

Fig. 6 Incident radiation and losses

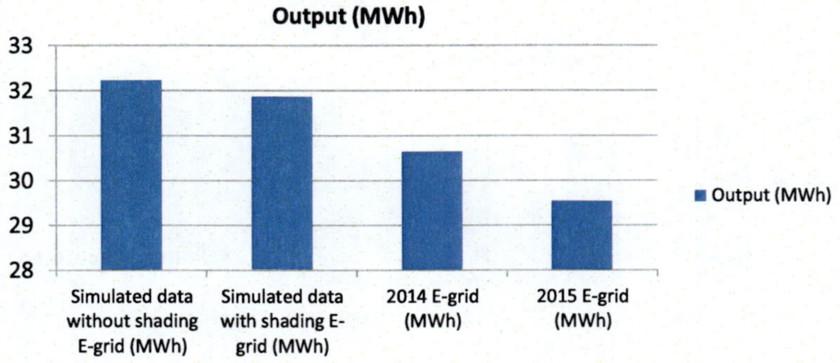

Fig. 7 Simulated and actual energy yield (Yearly)

Table 10 Comparison between actual and excepted energy yield

Month	Simulated data without shading	Simulated data with shading	2014	2015
	E-grid (MWh)	E-grid (MWh)	E-grid (MWh)	E-grid (MWh)
Jan	2.889	2.860	2.662	2.851
Feb	2.910	2.886	2.712	2.899
Mar	3.090	3.061	2.739	2.892
Apr	3.012	2.984	2.797	2.898
May	2.929	2.898	2.803	2.891
Jun	2.350	2.313	2.573	2.064
Jul	2.073	2.038	1.669	2.224
Aug	2.145	2.109	2.276	2.123
Sep	2.519	2.488	2.395	0.763
Oct	2.837	2.806	2.675	2.672
Nov	2.623	2.595	2.666	2.493
Dec	2.855	2.822	2.687	2.782
Year	32.231	31.861	30.654	29.552

4.4 Comparison Between Excepted Performance Ratio and Actual Performance Ratio of Fixed Tilt System with and Without Shading

See Fig. 8 and Table 11.

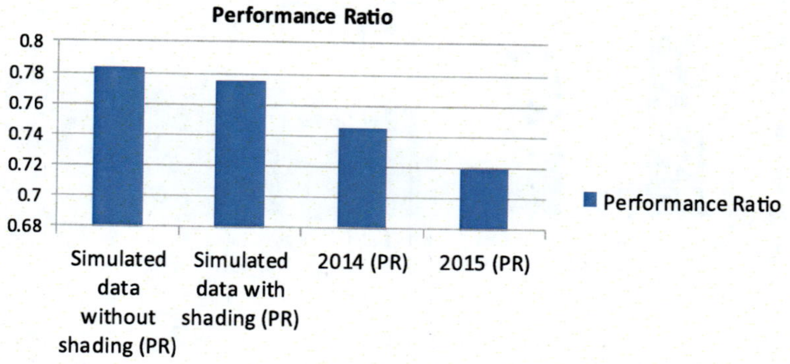

Fig. 8 Simulated and actual performance ratio (yearly)

Table 11 Comparison between actual and excepted performance ratio

Month	Simulated data without shading	Simulated data with shading	2014	2015
	PR	PR	PR	PR
Jan	0.796	0.778	0.765	0.754
Feb	0.756	0.793	0.801	0.808
Mar	0.793	0.790	0.795	0.801
Apr	0.784	0.788	0.771	0.758
May	0.747	0.748	0.780	0.787
Jun	0.794	0.783	0.781	0.786
Jul	0.791	0.775	0.733	0.725
Aug	0.678	0.700	0.724	0.868
Sep	0.645	0.857	0.754	0.744
Oct	0.808	0.754	0.745	0.785
Nov	0.775	0.716	0.725	0.746
Dec	0.696	0.859	0.799	0.240
Year	0.743	0.755	0.780	0.719

5 Conclusion

In this paper, compared different type of system that will maximize power generation through tracking the sun. In this study performance of a 20 kWp photovoltaic system single and dual axis tracker systems are compared with fixed tilt system with identical parameters. The annual electricity yield is 32.23 MWh for fixed tilt system, 34.12 MWh for single axis tracker system and 41.30 MWh for dual axis tracker system. The single axis horizontal EW tracker generates 5.8% more energy than a fixed tilt system with same specifications, while a dual axis horizontal EW tracker system generates 21% more energy than single axis tracker system and 28.1% more energy than fixed tilt system. We have also compared fixed tilt system with and without shading, and it was observed that fixed tilt system without shading (Exact location with similar conditions) generates 1.6% more energy than a fixed tilt system with shading [8]. Energy production of practical data and simulated data with shading was compared and was observed that there was 3.78% decrease in energy production. Table 10 shows that energy generation decreased from 2014 to 2015. This is due to solar panel degradation [9, 10], as well as other factors such as soiling, ohmic losses, module output loss, array mismatch loss, and bird fall. The system's annual performance ratio is calculated to be 0.775, which is appropriate.

Acknowledgements The authors would like to thank management of B.V. Raju Institute of Technology, Narsapur, Telangana, India for providing all facilities to write this paper on Performance Analysis of 20 kWp PV System.

References

1. Gielen D, Boshell F, Saygin D, Bazilian M, Wagner N, Gorini R (2019) The role of renewable energy in the global energy transformation. Energy Strategy Rev 24:38–50
2. Elhodeiby AS, Metwally HMB, Farahat MA (2011) Performance analysis of 3.6 kw rooftop grid connected photovoltaic system in Egypt. In: International conference on energy systems and technologies (ICEST 2011), 151–157. Retrieved 2011–07–21
3. Shah R, Mithulananthan N, Bansal R, Ramachandara Murthy V (2015) A review of key power system stability challenges for large-scale Pv integration. Renew Sustain Energy Rev 41:1423–1436
4. Omran WA, Kazerani M, Salama M (2011) Investigation of methods for reduction of power fluctuations generated from large grid connected photovoltaic systems. IEEE Trans Energy Convers 26(1):318–327
5. Tan W-S, Hassan MY, Majid MS, Rahman HA (2013) Optimal distributed renewable generation planning: a review of different approaches. Renew Sustain Energy Rev 18:626–645
6. Ellis A, Nelson R, Von Engeln E, Walling R, MacDowell J, Casey L, Seymour E, Peter W, Barker C, Kirby B et al (2012) Reactive power performance requirements for wind and solar plants. Power and energy society general meeting. IEEE, pp 1–8
7. Ahmed T, Khalid W, Shah IA (2018) Comparison of single and dual axis tracker controlled with fixed tilt solar PV system in Pakistan. Int J Power Eng Energy (IJPEE) 9(2):832–837
8. de Oliveira HM, de Aguiar Sodré E (2018) Study of a grid-tied photovoltaic system in Caruaru using PVsyst and Skelion. IEEE. 2018 Simposio Brasileiro de Sistemas Eletricos (SBSE). 12–16 May 2018
9. Kumar NM, Kumar MR, Ruth Rejoice P, Mathew M (2017) Performance analysis of 100 kWp grid connected Si-poly photovoltaic system using PVsyst simulation tool. In: 1st international conference on power engineering, computing and control, PECCON-2017, 24 Mar 2017
10. Anand VP, Priyan OB (2014) Effect of shading losses on the performance of solar module system using MATLAB simulation. In: 2014 IEEE 2nd international conference on electrical energy systems (ICEES), 61–64
11. Ibrahim D (1995) Optimum tilt angle for solar collectors used in Cyprus. Renew Energy 6(7):813–819
12. Yadav AK, Chandel SS (2013) Tilt angle optimization to maximize incident solar radiation: a review. Renew Sustain Energy Rev 23:503–513
13. https://www.solarpaneltilt.com
14. Teo JC, Tan RHG, Mok VH, Ramachandaramurthy VK, Tan CK (2018) Impact of partial shading on the P-V characteristics and the maximum power of a photovoltaic string. Energies
15. http://businessfeed.sunpower.com/articles/what-to-know-about-commercial-solar-panel-degradation
16. Jordan DC, Kurtz SR (2012) Photovoltaic degradation rates-an analytical review. NREL

Performance Assessment and Improvement of Classifiers Using Error Correcting Output Code for Islanding Detection in Microgrid

Dhruba Kumar

Abstract Islanding detection has become an important issue for interconnected power distribution systems. The main objective of the paper is to develop a new error correcting method which is reliant on the results of the pre-established classifiers. The measurements from different locations of the laboratory-based distribution system are collected, and the dimensionality is reduced using any one of the dimension reduction algorithms, i.e. principal component analysis, probabilistic principal component analysis, kernel principal component analysis and Andrews method. After dimension reduction, comparative analysis of classifiers for islanding detection in a laboratory-based microgrid has been established. The algorithms show unique characteristics when tested on the basis of decision surface, probability of classification and confusion matrix. The dimension reduction times and the classification training times are compared. Eventually, a novel error correcting output code has been implemented. The dimension reduction process used in this error correcting codes is a combination of kernel principal component and Andrews method. This error correcting method has been proved to be much beneficial for improving the accuracy of any type of classifiers for islanding detection.

Keywords Classification · Dimension reduction · Error correcting output code · Power system analysis · Microgrid · Islanding detection

1 Introduction

A microgrid is consisting of domestic, industrial loads and small-scale generation with distributed resources, such as solar and wind. The distributed generator can be operated either in standalone mode or in grid connected mode. Depending upon these two modes, the microgrid cannot be or can be connected to the grid. The changeover from grid connected mode to standalone mode needs islanding detection. Islanding is a condition in which the distributed generator feeds the load while the

D. Kumar (✉)
Department of Electrical Engineering, National Institute of Technology Patna, Patna, India
e-mail: dhrubakumar22@gmail.com

K. Namrata et al. (eds.), *Smart Energy and Advancement in Power Technologies*, Lecture Notes in Electrical Engineering 926,
https://doi.org/10.1007/978-981-19-4971-5_45

grid supply is discontinued. If the islanding is not detected, the distributed generator will continue to supply the local loads in an undesirable manner causing damage to the equipment. The islanding can be detected either by observing any change in the measured parameters or by inserting external signals. Classifiers have been used in this paper for identifying any change in the measured parameters which are supposed to be similar to the measurements occurred during islanding.

Machine learning and their overbearing aspects like classification algorithms are becoming major importance in smart grid. Naïve Bayes classifier has been implemented in [1] for islanding detection in IEEE 34 bus system. The same 34 bus system has been implemented in a real-time digital simulator for classification of islanding and non-islanding events using random forest in [2]. Several classification methods have been compared to each other by Hussein Al-Bataineh et al. in [3] for islanding detection in a small 13 bus system. Based on the Hurricane Gustav incident in the year 2008, the islanding contingency detection has been focused on using decision tree in [4]. An IEEE 9-bus transmission system is used in [5] for the classification of symmetrical faults. Twelve features can be obtained from measurements, i.e. voltage, current, resistance, reactance, active and reactive power etc. from 9-bus system. Wavelet packet transform (WPT) can be used for feature extraction as mentioned in [6]. Two different classifiers have been compared to each other for microgrid fault detection after the application of WPT.

Support vector machine-based islanding detection methodology as experimented in [7] is a superior approach to classify the fault when a voltage surge relay fails. The same methodology has been discussed in [8] for multi-distributed generation system. Another application of support vector machine-based islanding detector has been implemented in [9]. The robustness of the classifier is tested for different loading circumstances, sorts of a grid disturbance, etc.

The signals taken from the grid are processed using discrete wavelet transform, and the probability of islanding is evaluated using artificial neural network classifier in [10]. A modified ensemble classifier as mentioned in [11] can detect islanding in a short time. Artificial neural network-based classifiers of different hidden layer sizes are publicized in [12] for islanding detection in IEEE 9-bus system. Radial basis function-based neural network has been compared to multilayer perceptron in [13] for different numbers of hidden nodes for microgrid islanding. The features of artificial neural network classifier can be minimized using particle swarm optimization as authenticated in [14] for islanding detection. An added predictive algorithm can be developed for islanding detection depending upon the output of neural network classifier as prescribed by Kumar et al. in [15]. This method has been used alternatively in [16] for islanding detection in laboratory-based system during interrupted communication.

Furthermore, the background of error correcting code is given by Sect. 2. In Sect. 3, some basic results and descriptions are provided. The improvement of classification accuracy using error correcting code is discussed and compared in Sect. 4. Finally, the paper has been concluded in Sect. 5.

Table 1 Coding design matrix

	Learner 1	Learner 2
Islanding	1	0
Non-islanding	0	1

2 Background of Error Correcting Output Codes Model (ECOC)

For classification of islanding and non-islanding events, the error correction output code model uses two separate binary learners. Learner 1 considers islanding as a positive class and non-islanding as a negative class, whereas learner 2 treats non-islanding as positive class and islanding as negative class. The learners are SVM-based learners (ECOC-aided SVM). The matrix shown in Table 1 is called coding design matrix. There are several methods of generating coding design matrix, e.g. one versus all, one vs one, ordinal and ternary complete. One vs one coding is the easiest to implement.

After obtaining results from all the classifiers, the next objective is to find out which one of the results is approximately the same as given training data. This procedure is called decoding. There are a number of techniques for decoding, e.g. hamming decoding, inverse hamming decoding, Euclidean decoding, etc. Hamming decoding is most useful for decoding.

For the improvement of classifiers, error correcting output code needs certain adjustment which is dependent on the outcome of the other algorithms. So, the procedure of improvement of classifiers using ECOC has been established in Sect. 5 after getting results from other algorithms.

3 Theoretical Discussion and Results

In this section, all the classification methods are evaluated prior to the application of dimension reduction algorithms.

3.1 Description of Feature Extraction from the Test System

The layout of the test system is depicted in Fig. 1a. Islanding is created by disconnecting the grid supply, and any one or two of the DGs remains running. Chroma programmable electronic load is used to create different loading condition, as mentioned in UL1741 standards. NI9225 and NI9227 data cards are used for measuring voltage and current waveforms respectively from different locations. From the signals, voltage phasors, current phasors, frequencies and total harmonic distortions are calculated and send as features to the dimension reduction algorithms for further analysis. The hardware arrangement is shown in Fig. 1b. The data points after dimension reduction are shown in Fig. 2. The islanding and non-islanding data points are separated by colours as well as shapes. The lower-left coordinate axis is for real-time data and the upper-right axis is for simulated data.

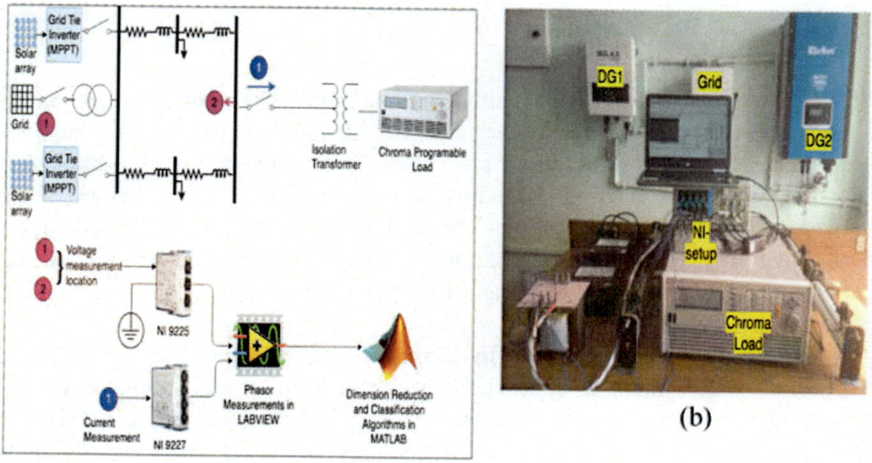

Fig. 1 Test system **a** layout of test system, **b** hardware arrangementThe real-time and simulated data points are almost matching

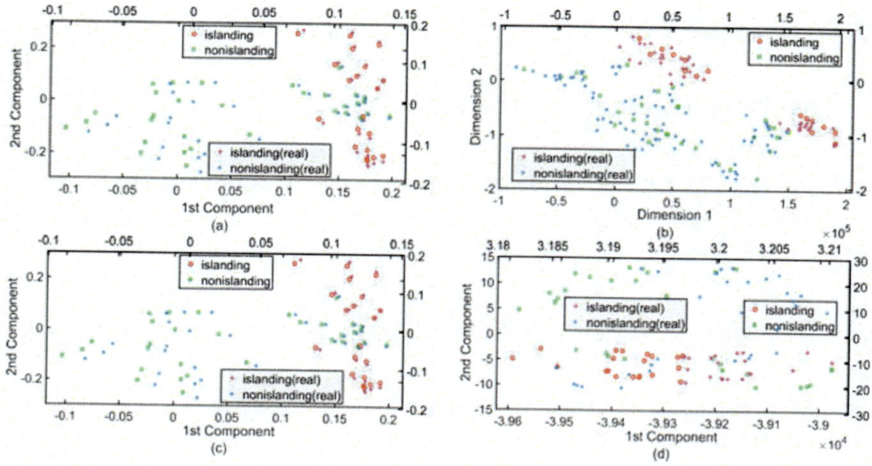

Fig. 2 Simulated and real-time (real) data points after dimension reduction using **a** PCA, **b** AM, **c** PPCA, **d** KPCA

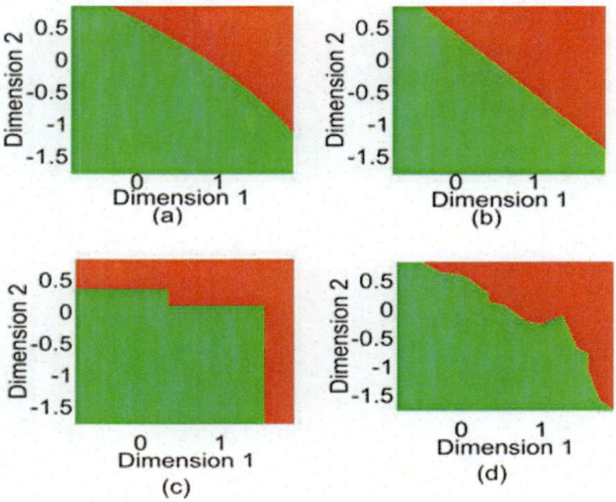

Fig. 3 Decision surface (best case) for islanding (red) and non-islanding (green) using AM for **a** NB, **b** DA, **c** RF, **d** NN

3.2 *Analysis of Classification Algorithms Using Decision Surface*

A decision surface covers the complete region for all possible data points in the reduced dimension. The non-islanding and islanding data points are coloured green and red respectively in the decision surface. Decision surfaces are shown in Fig. 3.

3.3 *Analysis of Classification Probability*

If a data point is completely sure to be classified (probability is 1) under islanding or non-islanding category, it is marked yellow, and if the data point is completely uncertain to be classified (probability is 0) under any category, it is coloured blue in the classification probability plot. The colour in between yellow and blue is determined by the corresponding probability of the data point to be in class islanding or non-islanding. Classification probabilities is shown in Fig. 4.

3.4 *Estimating Confusion Matrix*

Confusion matrix is used to know the actual number of misclassified and correctly classified data points for a classifier. The diagonal numbers in the matrix represent correctly classified events. The off-diagonal numbers are wrongly classified events. The higher are the diagonal numbers, the better is the accuracy of the classifier. Confusion matrices are shown in Figs. 5 and 6.

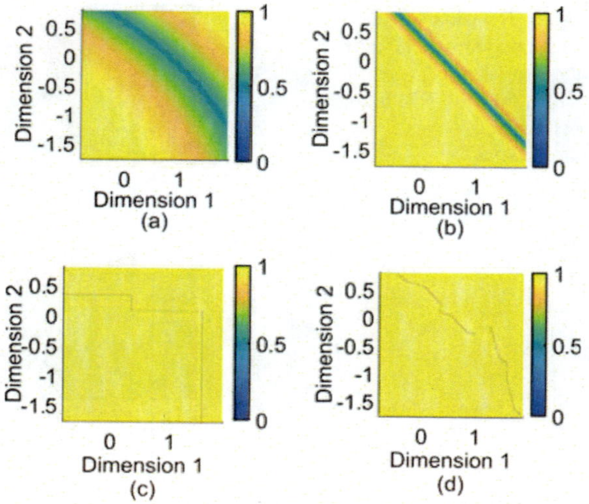

Fig. 4 Probability of classification using AM for **a** NB, **b** DA, **c** RF, **d** NN

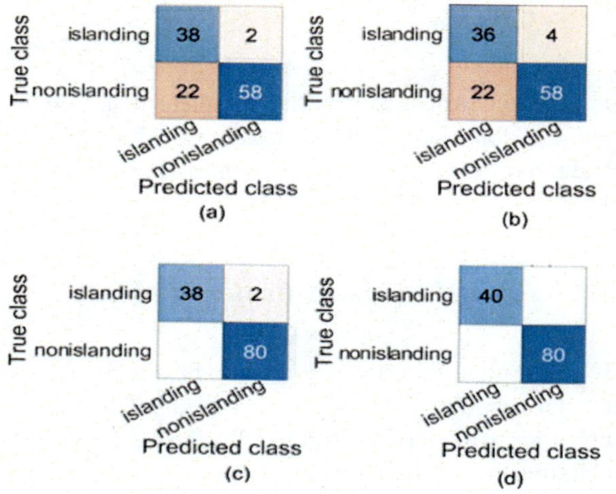

Fig.5 Confusion matrix (best case) using PCA/PPCA for **a** NB, **b** DA, **c** RF, **d** NN

3.5 *Validating Simplicity of Andrews Method by Tree Classifier*

The classification trees is shown in Fig. 7 for AM-based method. The tree is having 1 root, 2 nodes and 4 leaves. Other classification tress are more complicated. The KPCA tree is having 1 root, 4 nodes and 6 leaves. The PCA or PPCA-based tree is having 1 root, 5 nodes and 7 leaves. So, the algorithms can be arranged as AM, KPCA and PCA/PPCA in decreasing order of simplicity.

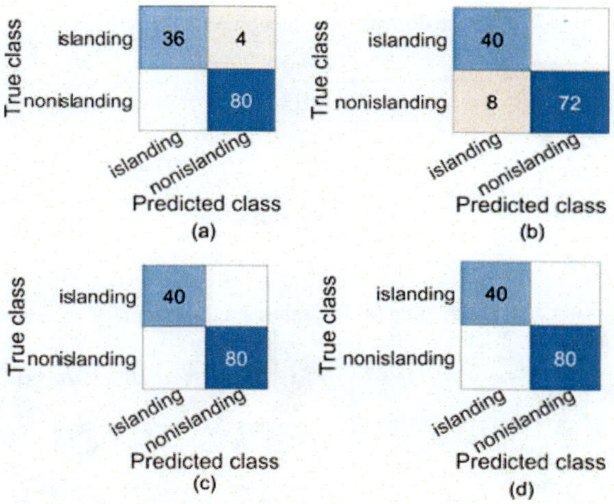

Fig.6 Confusion matrix (best case) using AM for **a** NB, **b** DA, **c** RF, **d** NN

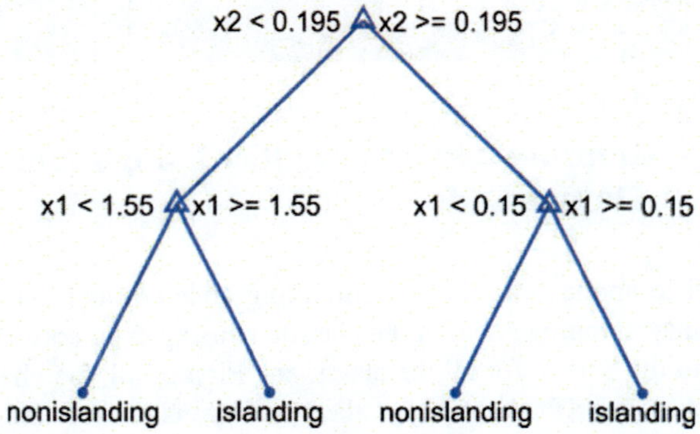

Fig. 7 Best classification tree using AM

3.6 *Comparison of Time Consumptions and Accuracy of the Algorithms*

The time consumption is tested in a setup (specifications: Core i7-7500U 7th generation × 64-bit processor, 2.7–2.9 GHz, 8 GB RAM). The average times of dimension reduction by PCA, PPCA, KPCA and AM are shown in Table 2 separately for 60 × 7 and 120 × 7 data. Time taken by AM method is comparatively smaller than the other methods. The time of dimension reduction increases with the increase in data size. The reverse is true for AM where dimension reduction time decreases with the increase in data size up to a certain limit.

Four types of information are attained by each cell in Table 3. The classification algorithms and dimension reduction algorithms are represented by columns and rows respectively. The value inside the cell represents time in second for training the classifier and the colour represents accuracy in percentage. The colour code from [*r g b*] = [51 204 204] to [*r g b*] = [0 0 153] outlines 66.7% accuracy to 100% accuracy.

Table 2 Average dimension reduction time (s)

	For 60 × 7 data	For 120 × 7 data
PCA	0.0013	0.0014
PPCA	0.06	0.74
KPCA	0.008	0.013
AM	6.89×10^{-5}	4.55×10^{-5}

Table 3 Average time (s) for training and percentage accuracy of the classifiers for 120 × 7 data and 60 × 7 data

Colour code for accuracy in %:	66.7	73	80	87	91	92	94	97	98	99	100

	NB	DA	RF	NN	SVM	ECOC aided SVM	ANN of hidden layers 1 & 2	
PCA	0.009	0.011	0.007	0.005	0.006	0.013	0.002	0.002
KPCA	0.013	0.012	0.009	0.007	0.2962	0.3032	0.009	0.002
PPCA	0.013	0.011	0.009	0.007	0.0108	0.0167	0.003	0.003
AM	0.012	0.013	0.009	0.007	0.011	0.019	0.003	0.003

4 Improvement of Classifier Using Error Correcting Output Code

The application of error correcting code so far has been limited to SVM only. A major improvement of the classifiers using error correcting code has been discussed in the Sect. 4 for all the classifiers. First of all, the phasor data are processed using AM and KPCA, and four class problem is put forward to the ECOC where number of learners, classes and classification accuracy using KPCA and AM are predefined. The learners of classifiers are trained after defining the coding design matrix and priority-based coding design matrix. These trained learners can be used for testing purpose. The output classes are obtained after decoding. The detailed description of four class problem, coding design, priority-based coding design, decoding, etc. is discussed hereafter in this section.

4.1 Defining Four-Class Problem

Almost all the data points, whether these are islanding or not, are well supported by AM method. But a few data points are missed by AM method. The classification accuracy of KPCA is next to AM. The data points obtained from KPCA and AM can be combined together to define a multiclass problem where islanding points obtained from AM and KPCA are considered as two subclasses (1 and 3, respectively), and non-islanding points from AM and KPCA are considered as another two subclasses

Table 4 Binary coding design matrix for four-class problem

Class number	Binary codes for learners 1 to 7						
	1	2	3	4	5	6	7
1	1	1	1	1	1	1	1
2	0	0	0	0	1	1	1
3	0	0	1	1	0	0	1
4	0	1	0	1	0	1	0

(2 and 4, respectively). An error correcting output code can now be applied to this four-class problem. Note that error correcting output code does not produce better results in two-class problem as observed in ECOC-aided-SVM.

4.2 Binary Coding Design for Four-Class Problem (Simplified ECOC)

The binary coding design matrix is shown in Table 4. A total of 7 learners can be formed for four classes. For learner 1, only class 1 is positive and all other classes are negative. For learner 2, classes {1, 4} are positive and {2, 3} are negative. For learner 3, classes {1, 3} are positive and {2, 4} are negative. For learner 4, all the classes are positive except class {2}. For learner 5, the positive classes are {1, 2} and negative classes are {3, 4}. The positive classes are {1, 2, 4} for learner 6, only class {3} is negative. For learner 7, the positive and negative classes are {1, 2, 3} and {4}, respectively. All the classes have different codewords, and all the outputs are different for different learners. For example, class 3 has a codeword {0, 0, 1, 1, 0, 0, 1}.

4.3 Modification of Correcting the Codes Depending upon the Accuracies of AM and KPCA (Priority-Based ECOC)

It is not recommended to consider codewords of the four classes equally. Codewords for classes 1 and 2 are more effective than codewords for classes 3 and 4. The effectiveness can be determined precisely by the accuracies of AM and KPCA method. Let the ratio of accuracy of AM and KPCA is k for a particular classifier. Also let {0, 0, 0.1, 0.5, 0.3, 0.3, 0.9} is an output code. The Euclidian distances of this code is $\sqrt{5/2}$ and $\sqrt{5/2}$ from typical classes 2 and 3, respectively. But codeword of class 2 is k times more significant. So, the Euclidian distance can be modified as $(\sqrt{5/2})k$ and

$\sqrt{5}/2$ from standard classes 2 and 3, respectively. As k is greater than 1 the output code {0, 0, 0.1, 0.5, 0.3, 0.3, 0.9} is nearer to class 2.

Note that, without using priority-based ECOC, the Euclidian distance was same ($\sqrt{5}/2$). This is unacceptable for good classification.

4.4 Validation of the Improvement of Classifier by ECOC (Mathematically)

Hamming distance can be calculated between any pair of codewords. For example, codewords {0, 0, 1, 1, 0, 0, 1} and {0, 1, 0, 1, 0, 1, 0} have a hamming distance equal to 4. The minimum of all hamming distances between any pair of codewords in Table 4 is 3. So, the design matrix can correct at least $\lfloor(3 - 1)/2\rfloor = 1$-bit error. Even if the error is slightly more than $\lfloor(3 - 1)/2\rfloor$, the nearest codeword remains the same. Correcting 1-bit error is equivalent to improving $100/(4 \times 7) = 3.57\%$ accuracy. The priority-based ECOC can correct more error which is equivalent to improving $3.57\% \times k$ accuracy.

4.5 Validation of the Improvement of Classifier by ECOC (in Simulation and Real Time)

The increase in accuracy from the simplified four-class problem has been shown in Fig. 8. A colour-bar is associated with each subplot. The increase in accuracy (%) has been described by the colour bar. For all subfigures, it is observed that NN accuracy is not improving because it is already extreme. As the SVM accuracy has much improved for all cases, the usefulness of ECOC is obvious.

The estimated accuracies in real time as well as in simulation using both simplified ECOC and priority-based ECOC are not equal to the mathematical accuracies because the ECOC is correcting more than 1-bit error as established in Table 5. While using simplified ECOC, the accuracies are higher than ordinary four-class problem. The accuracies can further be improved using priority-based ECOC.

4.6 Comparing Difference in Improvement Between Simplified and Priority-Based ECOC

In few cases, the improvements can be higher using simplified ECOC than priority-based ECOC. Two separate comparisons are shown in Figs. 9 and 10 for simulation and real time, respectively. In Fig. 9, the difference in accuracies between simplified and priority-based ECOC is shown for simulated data. The similar difference in

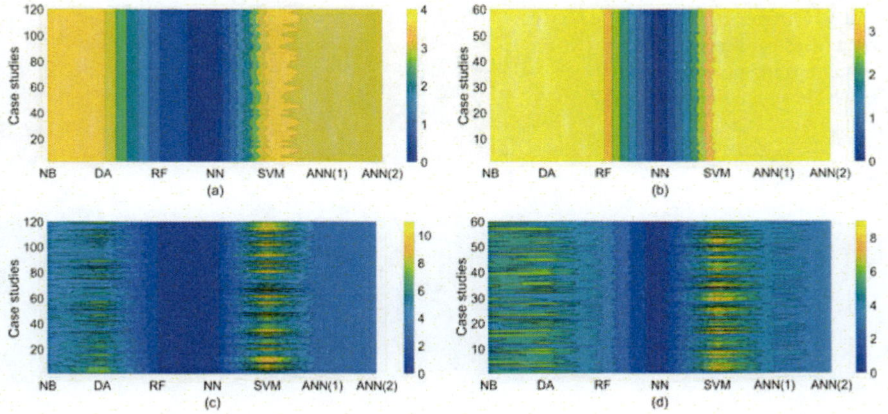

Fig. 8 Improvement of accuracy (for all cases) with respect to four-class problem with **a** simplified ECOC in real time, **b** simplified ECOC in simulation, **c** priority-based ECOC in real time, **d** priority-based ECOC in simulation

Table 5 Improvement of accuracy (average) of the classifiers by the application of proposed method (ECOC)

	Four-class without ECOC	Four-class with ECOC			
		Simplified ECOC		Priority-based ECOC	
		Mathematical	Actual	Mathematical	Actual
NB	94.5	98.07	98.07	98.26	98.9
DA	90.5	94.0	n94.08	94.35	96.5
RF	99	100	100	100	100
NN	100	100	100	100	100
SVM	86.5	90.07	90.34	91.39	94.4
ANN(1)	97	100	100	100	100
ANN(2)	97	100	100	100	100

real time is shown Fig. 10. The differences obtained from a particular classifier are represented by a box. Some useful statistical information, such as median and quartiles is observed from a box. It is mathematically disagreeable for the data points in a box to go above zero. This disagreeable situation occurs in few cases.

From Fig. 9, it is observed that the change in accuracies is negligible for NB and NN. The change in accuracies is 2.5% and 5% for RF and ANN, respectively. For the classifiers like DA and SVM, the changes are variable.

The change in accuracy is negligible for NN as shown in Fig. 10. The change in accuracies is concentrated in the neighbourhood of 1% and 3% in case of ANN and RF classifiers, respectively. A wide variety of changes are observed for SVM classifier.

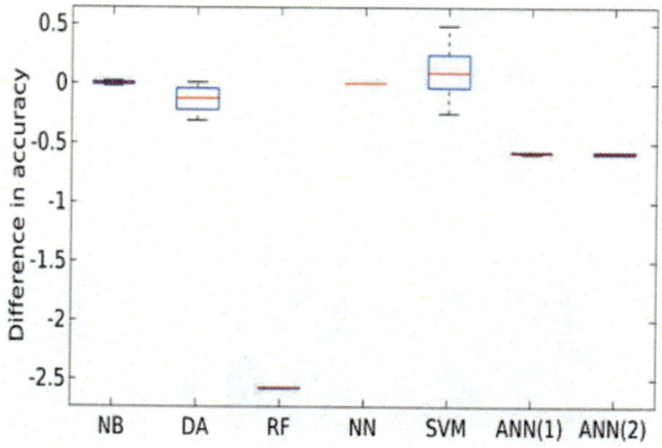

Fig. 9 Difference of accuracies between simplified and priority-based ECOC in simulation

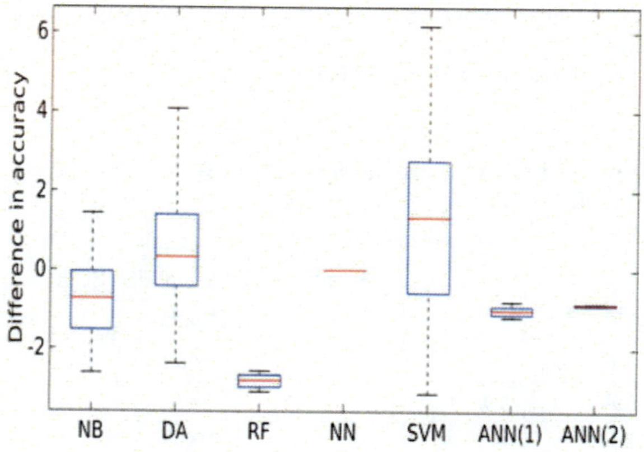

Fig. 10 Difference of accuracies between simplified and priority-based ECOC in real time

4.7 Comparison of Training Time and Detection Time

The training time, as well as detection time, can be visualized in Table 6. SVM method is having much higher training and detection time which are lower for all possible ANN classifiers. While using ECOC, the detection time is smaller than four-class problem but training time is greater. The ECOC is engaged in almost same training and detection time for both simplified and priority-based ECOC.

4.8 Comparison with Other Methodologies

The proposed priority-based ECOC has been compared to the existing islanding methodology in Table 7. The existing methodologies in [1, 3, 4, 26–28], etc. are good but not superior to the proposed methodology. Any kind of classifier can achieve better accuracy while implementing the proposed priority-based error correcting

Table 6 Average training time (s) and average detection time (ms) of the classifiers

Classification Algorithm	Colour code for detection time (ms)		Training time (s)		
		70 ms	4-class without ECOC	4-class with ECOC	
				Simplified ECOC	Priority based ECOC
NB		60 ms	0.013	0.019	0.019
DA		50 ms	0.013	0.019	0.019
RF			0.008	0.011	0.011
NN			0.007	0.011	0.011
SVM		40 ms	1.51	1.52	1.52
ANN(1)			0.007	0.003	0.003
ANN(2)			0.005	0.002	0.002

output code method. The accuracies of NB, DA, RF, NN, SVM and ANN are 98.98%, 96.56%, 100%, 100%, 94.46% and 100%, respectively as shown in Table 7.

Table 7 Comparison with existing methodology

Methodology	References	Whether error correcting method implemented	Classification accuracy
Nearest neighbour	3	No	86.6
Random forest	3	No	86
Decision tree	4	No	98.13 (maximum)
ESPRINT	1	No	95.56
Naïve Bayes	Proposed	Yes	98.98
Discriminant analysis	Proposed	Yes	96.56
Random forest	Proposed	Yes	100
Nearest neighbour	Proposed	Yes	100
Support vector machine	Proposed	Yes	94.46
ANN	Proposed	Yes	100

5 Conclusion

Four types of dimension reduction algorithms together with their effects on six different classification algorithms have been studied in this paper which not only highlights the assessments of classification algorithms but also improves the classification accuracies by correcting the errors of output codes.

From the results, it is observed that some classification algorithms have higher accuracy; some classification algorithms have less training time, etc. These performances of the classification algorithms also depend on dimension reduction algorithms. So, the algorithms can be sorted with respect to accuracy, training time and time consumption. All the classification algorithms are trained faster while the classifiers are running with AM. The accuracy using AM method is being followed by KPCA and PCA. Accuracies of PCA and PPCA method are same but PCA is faster than PPCA. The SVM runs very slow while KPCA is used with it. The accuracy is good for NN classifier followed by the accuracies of RF and ANN. The time consumption for training the ANN classifier is the least; the training time is highest for SVM.

Finally, after enhancing the algorithms with proposed ECOC, the accuracies of all classifiers reach more than 95%. Even the most inefficient SVM method has been performing satisfactorily using this proposed ECOC. The implementation of ECOC with other algorithms is helpful for the researchers in any other topics related to classification.

Acknowledgements The work has been supported by Science and Engineering Research Board (SERB), Department of Science and Technology (DST), Government of India, under project file no. EEQ/2016/000347. The authors are grateful to DST for funding this project.

References

1. Najy WKA, Zeineldin HH, Alaboudy AHK, Woon WL (2011) A Bayesian passive islanding detection method for inverter-based distributed generation using ESPRIT. IEEE Trans Power Deliv 26(4):687–2696
2. Adari S, Bhalja BR (2016) Islanding detection of distributed generation using random forest technique. In: Proceedings of 6th international conference on power systems (ICPS), New Delhi, IEEE, pp 1–6
3. Al-Bataineh H, Kavasseri RG (2017) Islanding detection with data mining methods—a comparative study. In: Proceedings of ninth annual IEEE green technologies conference (GreenTech). IEEE, pp 104–109
4. Sun R, Wu Z, Centeno VA (2011) Power system islanding detection & identification using topology approach and decision tree. In: Proceedings of IEEE power and energy society general meeting. IEEE, pp 1–6
5. Lwin M, Min KW, Padullaparti HV, Santoso S (2017) Symmetrical fault detection during power swings: an interpretable supervised learning approach. In: Proceedings of IEEE power & energy society general meeting. IEEE, pp 1–5

6. Ranjbar S, Jamali S (2019) Fault detection in microgrids using combined classification algorithms and feature selection methods. In: Proceedings of international conference on protection and automation of power system (IPAPS). IEEE, pp 17–21
7. Alam MR, Muttaqi KM, Bouzerdoum A (2014) A multifeature-based approach for islanding detection of DG in the subcritical region of vector surge relays. IEEE Trans Power Deliv 29(5):2349–2358
8. Alam MR, Muttaqi KM, Bouzerdoum A (2014) An approach for assessing the effectiveness of multiple-feature-based SVM method for islanding detection of distributed generation. IEEE Trans Ind Appl 50(4):2844–2852
9. Matic-Cuka B, Kezunovic M (2014) Islanding detection for inverter-based distributed generation using support vector machine method. IEEE Trans Smart Grid 5(6):2676–2686
10. Kermany SD, Joorabian M, Deilami S, Masoum MAS (2017) Hybrid islanding detection in microgrid with multiple connection points to smart grids using fuzzy-neural network. IEEE Trans Power Syst 32(4):2640–2651
11. Khamis Y, Xu ZY, Dong, Zhang R (2018) Faster detection of microgrid islanding events using an adaptive ensemble classifier. IEEE Trans Smart Grid 9(3):1889–1899
12. Kumar D, Bhowmik PS (2018) Artificial neural network and phasor data-based islanding detection in smart grid. IET Gener Transm Distrib 12(21):5843–5850
13. Hashemi F, Mohammadi M, Kargarian A (2017) Islanding detection method for microgrid based on extracted features from differential transient rate of change of frequency. IET Gener Transm Distrib 11(4):891–904
14. Raza S, Mokhlis H, Arof H, Naidu K, Laghari JA, Khairuddin ASM (2016) Minimum-features-based ANN-PSO approach for islanding detection in distribution system. IET Renew Power Gener 10(9):1255–1263
15. Kumar D, Bhowmik PS (2019) Hidden Markov model based islanding prediction in smart grids. IEEE Syst J 13(4):4181–4189
16. Kumar D (2021) Islanding detection in microgrid compromising missing values using NI sensors. IEEE Syst J (Early access)

Present Wind Energy Market Scenario in India

Sandeep Gupta, Preeti Singh, and Raju Kumar Swami

Abstract Presently, India has the 5th worldwide position for the total installed renewable power capacity. India has planned the retinue of new renewable sources of energy (NRSE) schemes under the ministry of non-conventional energy by the ninth plan. Wind power has seen consistent development in India for around three decades. Wind power is a pollution-free renewable energy source. Wind energy is available at no cost, and with new technology, it can be captured efficiently. Therefore, this paper shows the overall present scenario and progress of wind energy in India. Present government policies and goal-oriented objectives are also explained in this paper.

Keywords Environmental challenges · Non-conventional sources · Wind energy · Government policies

1 Introduction

The source that does not exhaust or can be refilled within a human lifetime is termed renewable energy. Productivity is rapidly increasing in industrial and farming areas [1–3]. It is supreme to give a cost-effective, well-managed alternative to electric power production to society. Renewable energy could be the only solution to the energy crisis in this new prosperity. It can enhance the economy of the nation. In our daily life, the role of electrical energy is increasing epidemically and can be safely managed at an individual level by society. There is no logic as to why 100% of

S. Gupta (✉)
Department of Electrical & Electronics Engineering, EKLAVYA University, Damoh (Madhya Pradesh), India
e-mail: jecsandeep@gmail.com

P. Singh
ECE Department, University Institute of Engineering and Technology, CSJM Kanpur (U.P.), Kanpur, Uttar Pradesh, India

R. K. Swami
EE Department, Pacific University, Udaipur, India

K. Namrata et al. (eds.), *Smart Energy and Advancement in Power Technologies*, Lecture Notes in Electrical Engineering 926,
https://doi.org/10.1007/978-981-19-4971-5_46

our power needs could not be met by renewable energy sources. India, being a rising nation, has observed an acute growing power demand due to speedy industrialization growth and an increasing demographic framework.

The wind power age limit in India has fundamentally expanded lately. Starting on February 29, 2020, the all-out introduced wind power limit was 37.669 GW, the fourth biggest introduced wind power limit in the world [4–8]. The wind power limit is for the most part, spread over the Southern, Western, and Northern locales. Wind power alone can deal with the entire power prerequisite of India is not so distant hundreds of years. With the help of this technology advancement and offshore potential, there could be increased power generation in all areas [9–14].

Wind power has seen consistent development in India for around twenty years. As of now, wind power is one of the key sustainable power hotspots for the power age in India. Presently, wind energy has spread over the South, West, and North areas of India. The Indian government reported an aspiring objective of introducing 175 GW of environmentally friendly power (RE) by December 2022 [15]. The measure of total energy consumption is increasing, as shown in Figs. 1 and 2 [16, 19–22]. Figure 1 shows that currently, renewable energy has the second position for power generation [17, 18, 23]. So, day by day, the utilization of this technology is increasing. Figure 2 shows the quarterly development in power generation capacity with the help of different energy sources.

Therefore, this paper completely presents a short survey of the arrangements, improvement projects, and the current status of the wind market in the nation. This paper is organized as follows: Sect. 2 provides the details of various sources of renewable energy; Sect. 3 describes the Indian wind energy potential; Sect. 4 explains the challenges of wind energy; Sect. 4.3 presents the existing and future scenario. The paper concludes in Sect. 5.

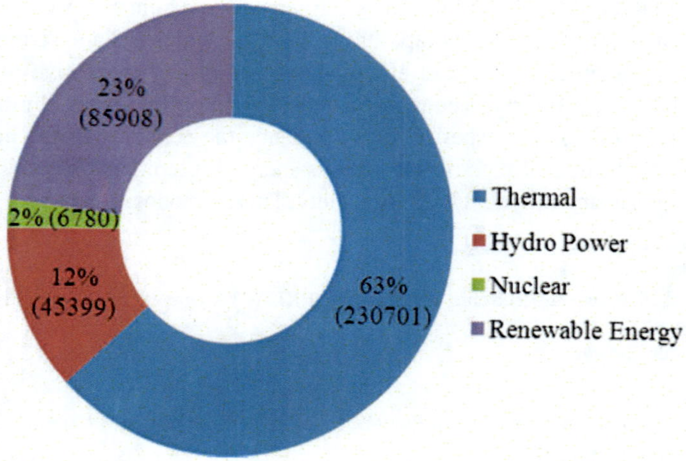

Fig. 1 Various sources-based installed power generation capacity (MW) as on 31.12.2019

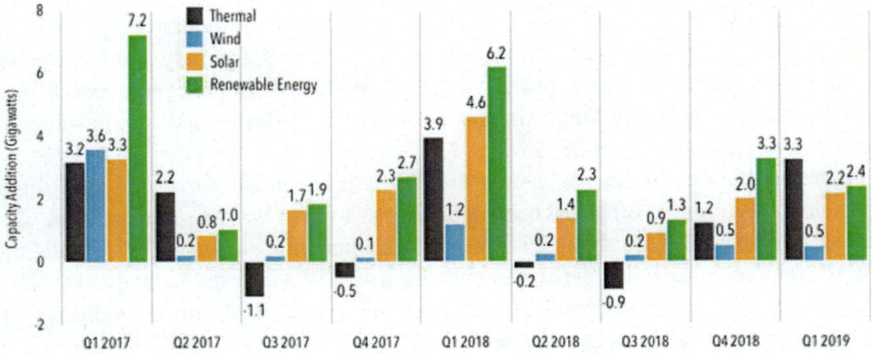

Fig. 2 Indian power installation capacity development through different sources

2 Indian Wind Energy Potential

India has expected offshore wind vitality potential to produce around 70,000 megawatts (MW) of power, the majority of that recognized in seaside Gujarat and Tamil Nadu, as indicated by the administration of new and practical force source (MNRE). "Initial examinations exhibit toward the ocean wind imperativeness capacity of around 70 gigawatts (GW) inside the recognized zones in the banks of Gujarat and Tamil Nadu, in a manner of speaking," India recognizes offshore wind vitality capability of 70,000 MW along with Gujarat, TN coasts [24, 25].

Wind power has seen consistent development in India for around three decades (1985–2015). The nation, as of now, positions fourth on the planet in wind power, with 37.5 gigawatts (GW) of limit introduced [26, 27]. The greater part of which was driven by motivators, for example, quickened deterioration and age-based instalments and alluring feed-in duties (FiT) as shown in Fig. 3.

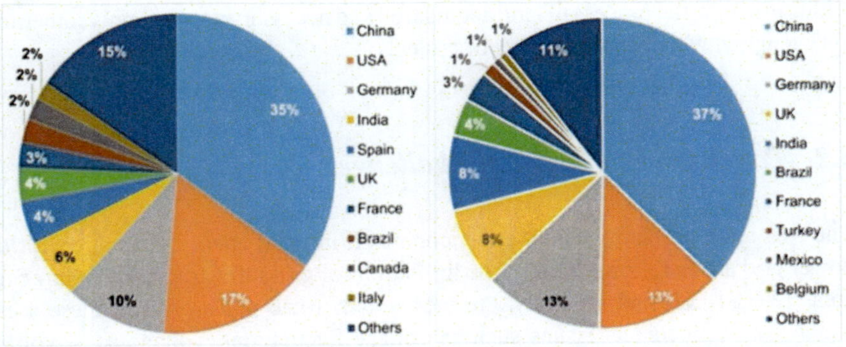

(a) Wind energy capacity at till December 2017 **(b)** New installation in 2017

Fig. 3 Power production situation through the wind in world (As on December 2017) [28]

In 2015, India declared a goal-oriented objective of introducing 175 GW of sustainable power source (RE) by December 2022. Be that as it may, it concurred to some degree with the humble objective of 60 GW to twist as the center moved into sunlight-based force. By then, the residential breeze industry had just developed, with an introduced limit of 25 GW.

The breeze area has overwhelmed the RE limit expansion for very nearly three decades. However, its offer has been declining as of late. Wind limit expansion was created in 2016–17, with about 5.5 GW of establishments as shown in Fig. 3. Utilizing this development, an objective of accomplishing 60 GW wind establishments by 2022 required 5 GW expansions for the following seven years, which was unambitious for the development the business was seeing [29, 30].

3 Challenges in Wind Energy

Nowadays, wind energy is one of the popular growing renewable energy sources which reduces the dependency on fossil fuels and helps to combat global warming. But, several challenges are linked with harnessing this electrical power for grid application, mostly due to its unpredictable nature. The major challenges are as follows:

3.1 *Location* Challenge

As construction of wind power plants requires large areas, these areas should be free from building obstructions and topology obstacles (because the wind speed is influenced by these obstacles). Therefore, these plants are usually located in rural or remote areas at a sufficient height where there is adequate availability of land. In addition to this, the process of land acquisition and necessary paperwork is sometimes very lengthy, which causes delays in projects.

3.2 *Power Grid Connection Challenge*

The two main problems which are encountered in wind energy generation for the power grid are as follows: Firstly, the limitations of grid infrastructure in many rural areas [29, 31, 32] and secondly, the presence of weak power grid in these areas, which lead to technical issues such as voltage fluctuations, voltage sags, voltage unbalances, faults, etc., due to the variable nature of wind [33, 34]. Due to the limited grid infrastructure, the amount of electrical energy produced by windmills cannot be efficiently transmitted to the consumers, which leads to the wastage of energy.

3.3 Environmental Challenges

Although wind energy is a clean source of energy and has less effect on the environment as compared to conventional power plants, sound and visual impacts are the two main health challenges associated with wind energy. Sound or noise pollution is the most critical environmental challenge to harnessing wind energy. Because when the wind turbines operate, they produce noise that can be quite distracting to nearby people [35, 36]. Due to the noise, property values in that area also go low.

Wind energy has visual or aesthetic impacts on the landscape. It depends upon the size, shape, and layout of the wind turbine. These impacts are specific to a particular site.

3.4 Cost Challenge

The initial investment cost (approx. 80% of the total project) of wind power is significantly higher as compared to conventional energy sources because of the wind turbine construction costs, transportation costs of wind turbines to remote areas, packing and storage of all associated components are very high. Wind power plants have high production costs. Wind energy is intermittent, which leads to a lower plant capacity factor. A lower plant capacity factor means less output power, which increases the overall production cost.

3.5 Turbine Design Challenge

Proper design of the wind turbine is also an important challenge in harnessing wind energy [36, 37]. The wind turbines must be properly designed for blade loading and aerodynamic stability.

3.6 Variable Output Power

The wind is variable (intermittently) due to weather patterns and cycles of day and night. Therefore, wind power generated through wind also varies accordingly. Therefore, wind power plants cannot be used as base-load power plants.

3.7 Major Technical Issues

The integration of energy generated by windmills produces several technical issues due to the variable nature of the wind. This affects the power quality and power reliability [33]. The major parameters that affect power quality are as follows:

(i) Voltage fluctuations
(ii) Low power factor
(iii) Electromagnetic interference
(iv) Synchronizations
(v) Power system transients and harmonics
(vi) Reactive power etc.

4 Wind Energy Present and Future

4.1 Wind Policies

The largest state in the country, Rajasthan, has come up with a draft of a solar-wind hybrid policy that aims to achieve 2000 MW of wind power capacity. The target is set for the fulfillment of the Renewable Purchase Obligation (RPO) of State DISCOMs in respect of wind energy as determined by the RERC up to 2023–24. The National Institute of Wind Energy (NIWE) has assessed a wind power potential of 18,770 MW in Rajasthan. Figure 4 shows the wind energy-based capacity in different states [16, 17]. In this figure, Tamil Nadu has the maximum installed capability in the nation with around 9.3 GW, and its state regulations are very favorable for wind power generation. Wind power yearly generation is shown in Fig. 5. From this figure, we can say that the growth of wind power generation is continuously increasing yearly.

The state will allow the setting up of projects that were already approved before 31.03.2016 (31st March 2016) by the State Level Empowered Committee under the Wind Policy, 2012. For the sale of power to DISCOMs of Rajasthan at the weighted average tariff determined through competitive bidding from the first lot of wind power projects for the fulfillment of the Renewable Purchase Obligation (RPO) target. Rajasthan state wants to promote wind power projects with storage devices to decrease the variability of wind power outcomes on the grid and to ensure the accessibility of firm power. Therefore, 5% of the RPO targets in MW include solar/wind-solar hybrid with storage and will be in addition to the RPO target.

Quarterly wind power installations are shown in Fig. 6 [38]. For the already functioning wind energy projects, the state will promote the repair of existing wind turbines of capacity below 1 MW which have remained in operation for at least 10 years. For the wind power projects with storage systems, additional land will be allotted as per the rules of the Revenue Department, Govt. of Rajasthan. In the case of land allotment, submission of a cash security deposit of Rs. 3 lakh/MW will be required. The state will also look into private lands for setting up wind projects.

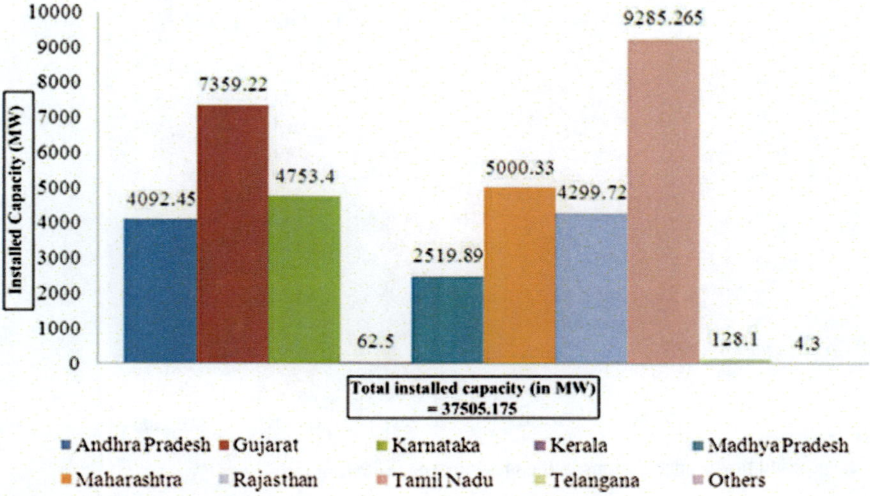

Fig. 4 Wind power-based installation capacity as of December 2019 in the different states

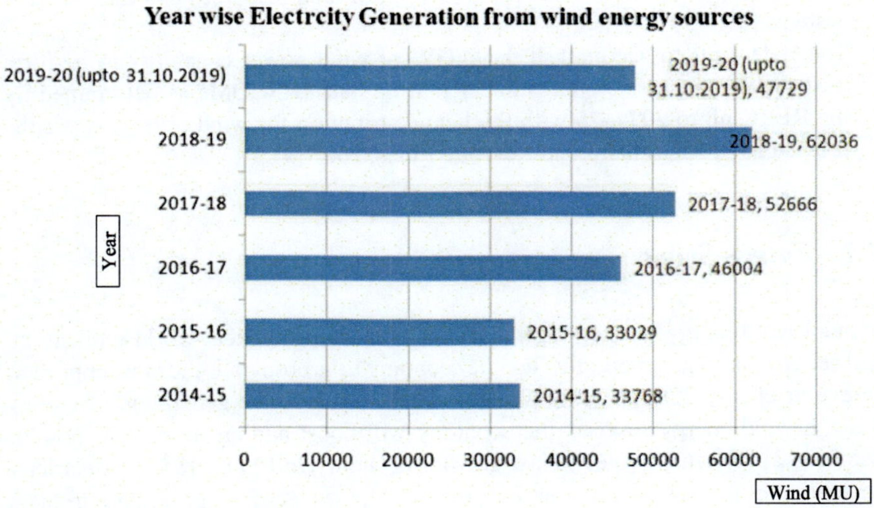

Fig. 5 Electricity generation yearly from wind power

4.2 Wind Power Vision of Government

- To promote wind power projects and required storage systems.
- Promotion of "Repowering" of wind power projects.
- Promotion of Wind Resource Assessment Program.
- To promote industries in the manufacturing of wind energy equipment.

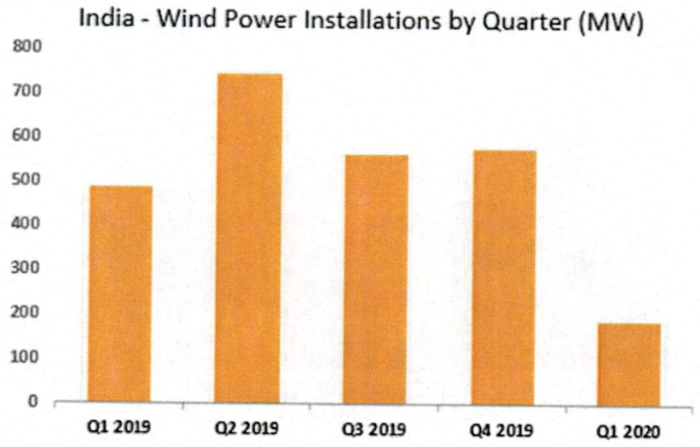

Fig. 6 Wind power installations in the previous year with quarterly

- Hybridization of wind and solar technologies to meet the challenge of grid security and stability along with optimum utilization of land resources and transmission systems [39, 40].
- The state aims to accomplish the 2 GW of wind power capability to perform a Renewable Purchase Obligation (RPO) of state DISCOMs as determined by the RERC up to 2024–25 with the land distribution for wind energy as per the provisions of Rajasthan Land Revenue Rules, 2007 [41].

4.3 Future Scope

In India, wind energy has scope for innovation, translating into real-world application, and better economic opportunity. For the economy's purposes, it is the most important source of energy. If India takes a major step to improve the generation of energy through wind energy sources, the economy will boost and create decent jobs for people and play a major role in a fast-growing economy. In 1950–1985, the electricity generation in India was very slow, but since 1990, the electricity generation in India has been very fast. The biggest advantage in the field of wind energy generation in India is that it has a large coastline which is approximately 7600 km. That is very useful for the wind generation system to get a major amount of fresh air. India is the third-largest energy consumer in the world and is the backbone of the economy. The IEA predicts that India's consumption rate of energy in 2040 will be more than double the total present consumption of energy. It is more important for India as our economy continues to evolve, and we must ensure every Indian has access to opportunity, a decent job, and a livelihood. India plans to add 60 gigawatts (GW) of wind power installed capacity by the year 2022.

The recent survey indicates that a gross wind power potential based on 302 GW at 100 m above ground level is mostly used in only seven states of the country. These are Gujarat, Rajasthan, Maharashtra, Tamil Nadu, Madhya Pradesh, Karnataka, and Andhra Pradesh. The government is promoting the wind power projects in the entire country through private sector investment by providing financial support like subsidies and many types of benefits like concessional custom duties exemption on certain components of wind electric generators.

The government through NIWE has installed over 800 wind monitoring stations all over the country. The main future scope for India is to provide a clean energy generation and energy security field of electricity or energy generation. Promote the benefits of wind energy and the big role it can play to provide a sustainable path for economic and social development in the country.

Make sure that the industry develops sustainably by protecting the interest of all stakeholders. To make, the wind energy is the primary source of energy for the people of India. In the future, the main focus of energy generation will be on renewable energy sources in India because India has some favorable geological conditions, and they play a major role in the economy of India to grow up faster with better results.

5 Conclusion

A wind turbine converts the wind's kinetic energy into mechanical energy, which is then turned into electrical energy by an induction generator in a wind power plant. Wind energy reduces carbon emissions and offers a large potential for meeting energy demands. It may also be a lucrative source of employment. However, wind energy initiatives come with a number of difficulties. As a result, in order to utilize wind energy effectively, these problems must be addressed by research and development, innovations & technological breakthroughs in wind turbines, grid infrastructure strengthening, storage method improvements, and so on. If India makes significant steps to enhance energy generation using wind energy sources, the economy would be stimulated, creating good employment for people and playing a significant part in a fast-growing economy. Renewable energy sources will be the major focus of energy generation in India in the future. Because India has some advantageous, such as geological characteristics, which have a significant influence on the Indian economy's rapid growth and improved results.

References

1. Nazir MS et al (2020) Potential environmental impacts of wind energy development: a global perspective. Curr Opin Environ Sci Health 13:85–90
2. Wang HKH (2020) Renewable energy management in emerging economies: strategies for growth. Taylor and Francis, Routledge. ISBN: 9781351061582

3. Shukla RD, Tripathi RK, Gupta S (2010) Power electronics applications in wind energy conversion system: a review. In: 2010 IEEE international conference on power, control and embedded systems, pp 1–6
4. Herbert GMJ, Iniyan S, Sreevalsan E, Rajapandian S (2007) A review of wind energy technologies. Renew Sustain Energy Rev 11(6):1117–1145
5. Singh AK, Idrisi AH (2020) Evolution of renewable energy in India: wind and solar. J Inst Eng (India) Ser C 101(2):415–427
6. Tyagi R Atul A, Shaikh SA (2015) Indian renewable energy act 2015: a step towards reducing carbon footprint. Indian J Power River Valley Dev 68(9/10):145–151
7. Gupta S, Sharma A (2018) Global scenario of solar photovoltaic (SPV) materials. In: Advanced computational and communication paradigms. Springer, Singapore, pp 126–133
8. Satpute AV, Vijay Kumar E (2020) Current scenario of wind power in India, government policies, initiatives, status and challenges. Int J Energy Sector Manag 15(1):209–226
9. Ibrahim D (2000) Renewable energy and sustainable development: a crucial review. Renew Sustain Energy Rev 4(2):157–175
10. Gupta A, Bonds G (2020) Financing India's renewable energy vision. ISBN: 978-93-89622-35-5. Available: http://envecologic.com/wp-content/uploads/2020/05/ORF_IssueBrief_336_RenewableEnergy.pdf
11. Madan D, Mallesham P, Sagadevan S, Veeramani C (2020) Renewable energy scenario in Telangana. Int J Ambient Energy 41(10):1110–1117
12. Panwar NL, Kaushik SC, Kothari S (2011) Role of renewable energy sources in environmental protection. Renew Sustain Energy Rev 15(3):1513–1524
13. Burton T, Jenkins N, Sharpe D Bossanyi E (2011) Wind energy handbook. Wiley, New York. ISBN: 978-0-470-69975-1
14. Tripathi L, Mishra AK, Dubey AK, Tripathi CB, Baredar P (2016) Renewable energy: an overview on its contribution in current energy scenario of India. Renew Sustain Energy Rev 60:226–233
15. Chaurasiya PK, Warudkar V, Ahmed S (2019) Wind energy development and policy in India: a review. Energy Strat Rev 24:342–357
16. Perumal R, Raja Singarayar R (2019) Renewable energy market in India-an overview, opportunities & the challenges ahead. ZENITH Int J Multidiscip Res 9(3):166–169
17. Raghuwanshi SS, Arya R (2019) Renewable energy potential in India and future agenda of research. Int J Sustain Eng 12(5):291–302
18. Vishnupriyan J, Manoharan PS (2018) Multi-criteria decision analysis for renewable energy integration: a southern India focus. Renew Energy 121:474–488
19. Swami RK, Samuel P, Gupta R (2016) Power control in grid-connected wind energy system using diode-clamped multilevel inverter. IETE J Res 62(4):515–524
20. Jain P (2011) Wind energy engineering. McGraw-Hill, New York
21. Mohtasham J (2015) Renewable energies. Energy Procedia 74:1289–1297
22. Singh S, Bhatti TS, Kothari DP (2004) Indian scenario of wind energy: problems and solutions. Energy Sourc 26(9):811–819
23. Gupta S, Kumar TK, Shivaji RM, Nachimuthu K (2020) Wind energy potential, challenges with major technical issues. J Green Eng 10(12):12973–12987
24. Elavarasan RM, Shafiullah GM, Padmanaban S, Kumar NM, Annam A, Vetrichelvan AM, Mihet-Popa L, Bo Holm-Nielsen J (2020) A comprehensive review on renewable energy development, challenges, and policies of leading Indian states with an international perspective. IEEE Access 8:74432–74457
25. Madhu S, Payal S (2014) A review of wind energy scenario in India. Int Res J Environ Sci 3(4):87–92
26. Upadhyay DK, Mishra M (2020) Blue economy: emerging global trends and India's multilateral cooperation. Maritime Affairs. J Nat Maritime Found India 16(1):30–45
27. Sholapurkar RB, Mahajan YS (2015) Review of wind energy development and policy in India. Energy Technol Policy 2(1):122–132

28. Dawn S, Tiwari PK, Goswami AK, Singh AK, Panda R (2019) Wind power: existing status, achievements and government's initiative towards renewable power dominating India. Energ Strat Rev 23:178–199
29. Rehman S, Natarajan N, Vasudevan M, Alhems LM (2020) Assessment of wind energy potential across varying topographical features of Tamil Nadu, India. Energy Explor Exploit 38(1):175–200
30. Veers P, Dykes K, Lantz E, Barth S, Bottasso CL, Carlson O, Clifton A et al (2019) Grand challenges in the science of wind energy. Science 366(6464)
31. Pachauri RK, Chauhan YK (2012) Assessment of wind energy technology potential in Indian context. Int J Renew Energy Res (IJRER) 2(4):773–780
32. Singh AK, Parida SK (2013) Evaluation of current status and future directions of wind energy in India. Clean Technol Environ Policy 15(4):643–655
33. Agarwal T, Verma S, Gaurh A (2016) Issues and challenges of wind energy. In: 2016 IEEE international conference on electrical, electronics, and optimization techniques (ICEEOT), pp 67–72
34. Wood DH (2020) Grand challenges in wind energy research. Front Energy Res 8:337
35. Gupta S, Kailash K et al (2019) An IoT based approach to minimize air pollution. Int J Recent Technol Eng (IJRTE) 7(6S5):596–599
36. Gandhar S, Ohri J, Singh M (2020) A critical review of wind energy based power generation systems. Asian J Water Environ Pollut 17(2):29–36
37. Muyeen SM ed (2012) Wind energy conversion systems: technology and trends. Springer Science & Business Media
38. Ranjan R (2020) 2.07 GW of wind power capacity installed in India during FY 2019–20. Mercom India. Retrieved 18 May (2020). Available: https://mercomindia.com/2-07-wind-power-capacity-installed-india/.
39. Sharma S, Sinha S (2019) Indian wind energy & its development-policies-barriers: an overview. Environ Sustain Ind, pp 1–34
40. Khare V, Nema S, Baredar P (2013) Status of solar wind renewable energy in India. Renew Sustain Energy Rev 27:1–10
41. Bansal N, Srivastava VK, Kheraluwala J (2019) Renewable energy in India: policies to reduce greenhouse gas emissions. In Greenhouse gas emissions. Springer Singapore, pp 161–178. ISBN 978-981-13-3271-5

Collaborative Control of Voltage and Frequency of an Interconnected Power System Using MFO-Optimized Cascaded PI-PDF Controller

B. V. S. Acharyulu, S. K. Swamy, B. Seshasai, and K. Laxmana Rao

Abstract An interconnected non-reheat two area thermal system with the amalgamated exploration of automatic voltage regulation (AVR) and automatic load frequency control (ALFC) is presented in this paper. A modish trail has been made to apply a cascade controller, namely proportional integral proportional derivative with filter (PI-PDF) in LFC and AVR. A recently developed naturally inspired optimization technique known as the moth flame optimization (MFO) technique is used to concurrently improve controller gains. The analysis of conventional controllers which include proportional integral derivative filter (PIDF), proportional integral derivative (PID) and the proportional integral (PI) is discussed and compared to the PI-PDF cascade controller to know their effectiveness. Several observations are noted where the PI-PDF cascade controller is compared with the other controllers. As per the sensitivity analysis, the MFO-optimized PI-PDF cascade controller parameters acquired are resistant to large changes in system settings and do not require resetting. In area-1, the system's dynamic performance is inspected using a 20% step load.

Keywords Moth flame optimization (MFO) · Automatic voltage regulator (AVR) · Proportional integral derivative (PIDF) controller · Automatic load frequency control (ALFC)

1 Introduction

It is very much essential to provide electricity at rated amdrequency and voltage. The voltage and amdrequency are two main important factors to decide the stability of the power system network. But on account of load aberration and reactive power variations will lead to change in voltage, amdrequency and interline power exchange

B. V. S. Acharyulu (✉) · B. Seshasai · K. L. Rao
Department of EEE, LIET, Vizianagaram, Andhra Pradesh, India
e-mail: acharyulu201@yahoo.com

S. K. Swamy
Department of EEE, AITAM, Tekkali, AP, India

K. Namrata et al. (eds.), *Smart Energy and Advancement in Power Technologies*, Lecture Notes in Electrical Engineering 926,
https://doi.org/10.1007/978-981-19-4971-5_47

in the interconnected power system. In order to change the amdrequency and voltage, the generator has ALFC loop and AVR loop. The amdrequency of output power can be changed by ALFC loop, whereas the voltage can be controlled by AVR loop [1, 2]. The amdrequency and the voltage in power system are changed due to load change. Hence, an attempt has been made in the paper to study the combined model of ALFC and AVR for the interconnected system to control voltage and amdrequency simultaneously.

Many authors look in to the consequences of ALFC and AVR for the interconnected power system [3–9]. Gupta et al. [10] presented voltage and amdrequency control of single area power system. Optimum control of voltage and amdrequency achieved by PSO-optimized PID controller [11]. Chandrakala Vijaya et al. [12] presented minimization of amdrequency and voltage variations simultaneously for the interconnected power system. Therefore, it is noticed that there are hardly any studies that have been done on the interconnected system of an ALFC in collaboration with AVR. Hence, further research is needed in this area. In the past literature, the combined study was done using the controllers such as I, PI and PID. Astonishingly, PI-PDF is never put forward in a collaborative model of AVR and ALFC. This requires more inquisition.

The following are the important objectives of present work.

- To develop and establish a PI-PDF controller. The structure of PI-PDF is shown in Fig. 3.
- To develop joint model of ALFC and AVR for non-reheat two area thermal system.
- To optimize parameters and control via MFO.
- To study the performances of PI-PDF controller and compare with other controllers.
- To study controller robustness for parametric changes + and – 25%.

2 System Modeling

The simulation reports have been carried on non-reheat two area thermal power system. The impulse response of the system or its transfer function model [13] is shown in Fig. 1. Each area has a rating of 2000 MW and has a nominal load of 1000 MW. In Fig. 1, the governor time constants are denoted by T_{g1} and T_{g2} in sec; the turbine time constants are T_{t1} and T_{t2} are in sec; the power system gains are K_{ps1} and K_{ps2}; power system time constants are T_{ps1} and T_{ps2} in sec; the amdrequency bias constants are B_1, B_2; the speed regulation constants of area-1and area-2 are R_1 and R_2; T_{12} is the synchronizing time constant; ΔP_L is the load disturbance; the amdrequency aberrations are ΔF_1 and ΔF_2 at area-1 and area-2, respectively, and the tie line power aberration is given as ΔP_{tie}. The nominal values of the system are presented in [13]. The MFO is used to optimize the controller parameters of the system. To validate the performance of controller and algorithm, an error criterion integral square error (ISE) is observed in this paper for comparison [13] and given in following Eq. (1):

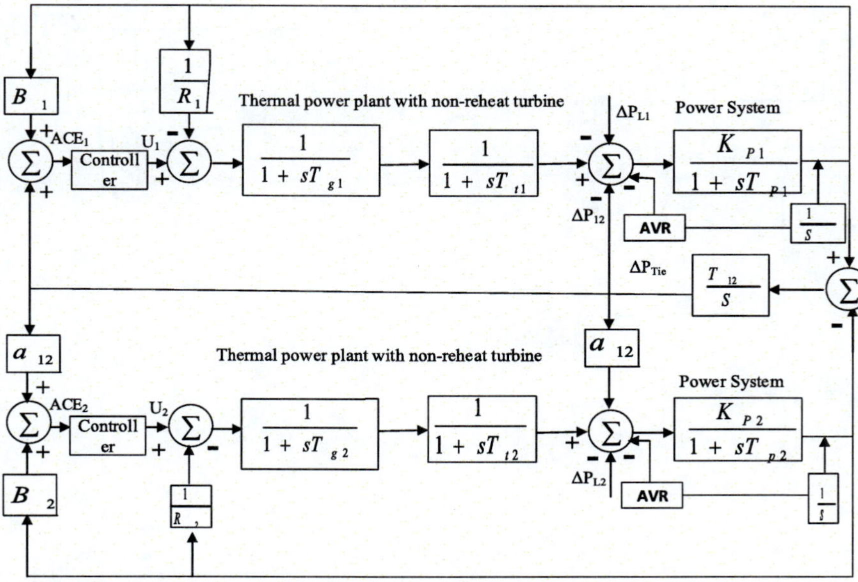

Fig. 1 Non-reheat two area thermal system [13]

$$\mathrm{ISE} = \int_{0}^{\mathrm{tsim}} (\Delta f_1)^2 + (\Delta f_2)^2 + (\Delta P_{\mathrm{tie}})^2 \cdot \mathrm{d}t \tag{1}$$

AVR Loop Modeling

The constant voltage from the alternator is superintended by AVR. Figure 2 shows the cross-coupling coefficients model of AVR system. The error voltage (ΔV) is calculated by constantly contrasting output voltage (V) with the reference signal (Vref). Now, to change the generator filed excitation, the error signal is amplified and is given to the exciter, therefore, the aberrations in terminal voltage be compensated to make system stable (Fig. 3).

3 Moth Flame Optimization (MFO)

SeyedaliMirjalili developed this natural-inspired algorithm called moth flame optimization. The flexure alignment behavior of moths in nature influenced this algorithm. A moth's mechanism for traveling over a long distance in a straight path is very effective. It can be seen moving in a straight line as well as spiraling around the lights. Inevitably, the moths fall on the flames. In this paper, the data point is taken as the moth, and the variable quantities are the moth's location in space. The moth is defined by an m * d matrix with m number of moths and a dimension of d

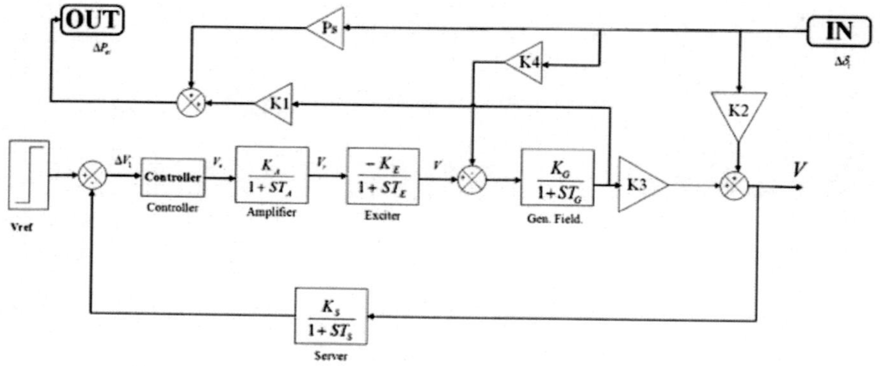

Fig. 2 AVR model

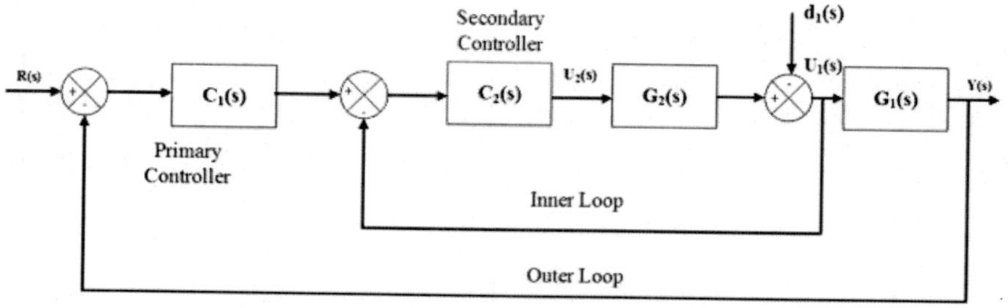

Fig. 3 Cascaded PI-PDF controller structure

represented by N.

$$N = \begin{bmatrix} n_{11}\ n_{12}\ \ldots\ n_{1d} \\ n_{21}\ n_{22}\ \ldots\ n_{2d} \\ \vdots\ \vdots\ \vdots\ \vdots \\ n_{m1}\ n_{m2}\ \ldots\ n_{md} \end{bmatrix} \tag{2}$$

The fitness function is determined by

$$\text{FIT} = \begin{bmatrix} \text{FIT}_1 \\ \text{FIT}_2 \\ \vdots \\ \text{FIT}_m \end{bmatrix} \tag{3}$$

Moths always have their best position due to the release of flag during searching the flame. Moths update their location in relation toward flame according to the following equation,

$$\mathrm{MOFL}_{i_j} = \mathrm{SP}\big(\mathrm{MO}_i, \mathrm{FL}_j\big) \tag{4}$$

Here, MOFL_{i_j} is the updated position of moth, SP is the function called spiral function, MO_i is ith moth and FL_j is jth flame. Further, spiral function is defined as

$$\mathrm{SP}\big(\mathrm{MO}_i, \mathrm{FL}_j\big) = D_{i_j} * e^{\mathrm{at}}\cos(2\pi t) + \mathrm{FL}_j \tag{5}$$

where a is called shape constant of spiral movement and the random variable t is within [−1,1], D_{i_j} is the distance from ith moth to jth flame.

$$D_{i_j} = \big|\mathrm{MO}_i - \mathrm{FL}_j\big| \tag{6}$$

The updation of flame is done by using the following equation

$$\mathrm{FL} = \mathrm{roundup}\left(\mathrm{FL}_{\mathrm{max}} - I * \frac{\mathrm{FL}_{\mathrm{max}} - 1}{I_{\mathrm{max}}}\right) \tag{7}$$

where $\mathrm{FL}_{\mathrm{max}}$ is the max value of flame, I. is present iteration value and I_{max} is max iteration value.

4 Result and Discussions

MATLAB/SIMULINK software is used to design a non-reheat two area thermal system with AVR, and an algorithm for moth flame optimization is written in mfile or script file. A maximum of 100 iterations and 30 number of moths are taken into consideration for this evaluation. The examined system is fed with one controller at a time of all PI, PID, PIDF and PI-PDF controllers. These controllers are employed to diminish amdrequency oscillations in the system. The best parameter gains are obtained by creating 1% disturbance at area-1 for PI-PDF controller $kp_1 = 1.0$; $ki_1 = 0.550$; $kp_2 = 0.980$; $kd_1 = 0.90$; $N = 99.0$; PIDF controller $kp_1 = 0.990$; $ki_1 = 0.5401$; $kd_1 = 0.10$; $N = 100.0$; PID controller $kp_1 = 0.989$; $ki_1 = 0.541$; $kd_1 = 0.101$; PI controller $kp_1 = 0.97$; $ki_1 = 0.53$; using these optimum gains, each controller's dynamic behaviors is juxtaposed as well as the response attribute values are mentioned in Table 1. It is observed from the responses and Table 1 that the considered PI-PDF regulator offers better improvement in terms of OS, ST, US and OBJ. Each controller performances for LFC and AVR is portrayed in Figs. 4a–c, and 5a, b.

Table 1 Comparison of dynamic performances MFO-Optimized PI-PDF

Performance index	ISE	PI	PID	PIDF	PI-PDF
		0.0045	0.0042	0.0040	0.0038
ΔF_1	ST	28.31	25.12	13.21	8.8
	US	− 0.0265	−0.026	−0.025	−0.016
	OS	0.011	0.008	0.005	0
ΔF_2	ST	28.91	26.01	14.34	12.02
	US	−0.0225	−0.0201	0.0181	− 0.0103
	OS	0.013	0.008	0.005	0
ΔP_{tie12}	ST	66.12	66.03	65.56	65.14
	US	− 0.0072	− 0.0071	− 0.0067	− 0.0066
$V1$	ST	6.29	6.21	6.01	5.81
	OS	1.311	1.301	0.089	0
$V2$	ST	6.29	6.21	6.01	5.81
	OS	1.311	1.301	0.089	0

5 Sensitive Analysis

Figure 6a–c depicts the proposed controller's stability review. For the proposed scheme, the variables Tg, Tt, and T_E are varied from + 25 to − 25% [6], and the resulting STs and overshoots are furnished in Table 2. The proposed controller's capacity is shown by the amdrequency variance simulation findings at first area.

6 Conclusion

A novel potent evolutionary natural-inspired computing technique known as the moth flame optimization (MFO) is employed to concurrently improve several parameters like gains in that regulator. A new attempt is made first time to apply a cascade controller, namely PI-PDF in LFC and AVR. According to the observations, the PI-PDF cascade controller significantly surpasses all of the other controllers in this comparison. In area-1, the system's vigorous performance is inspected using a 20% step load. According to the sensitivity analysis, the MFO-optimized PI-PDF cascade controller parameters obtained are resilient to immense changes in system settings.

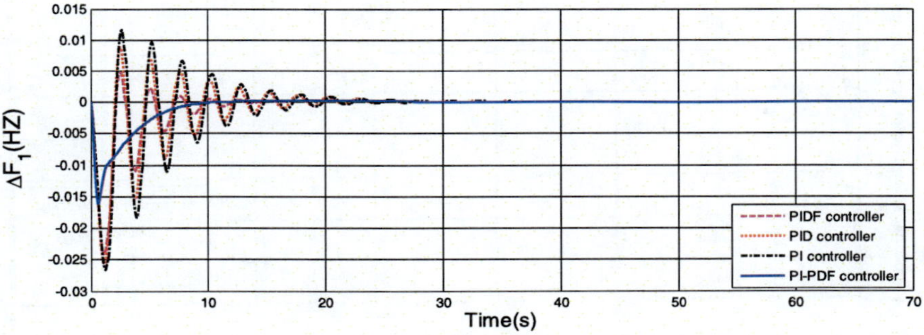

(a)

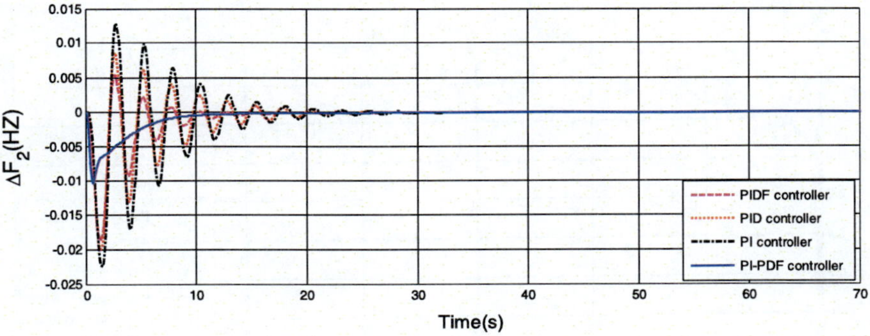

(b)

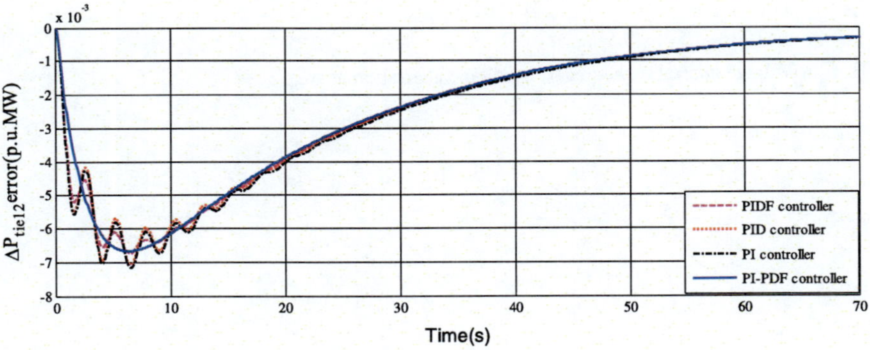

(c)

Fig. 4 **a** Amdrequency aberration of area-1 **b** Amdrequency aberration at area-2 **c** tie line power interchange between area-1 and 2

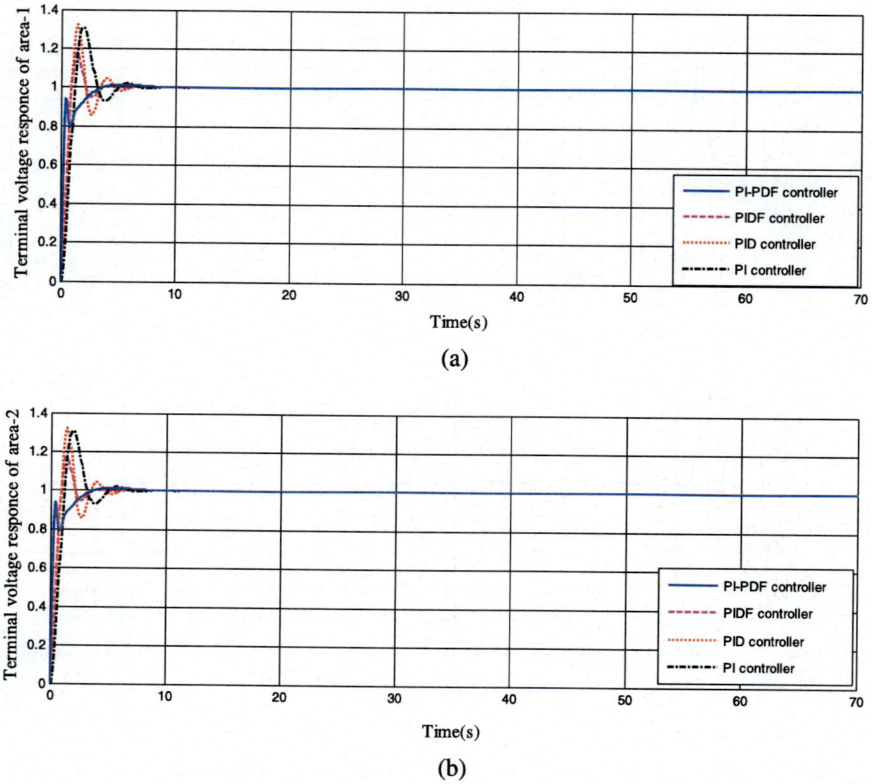

Fig. 5 **a** Voltage aberration at area-1. **b** Voltage aberration at area-2

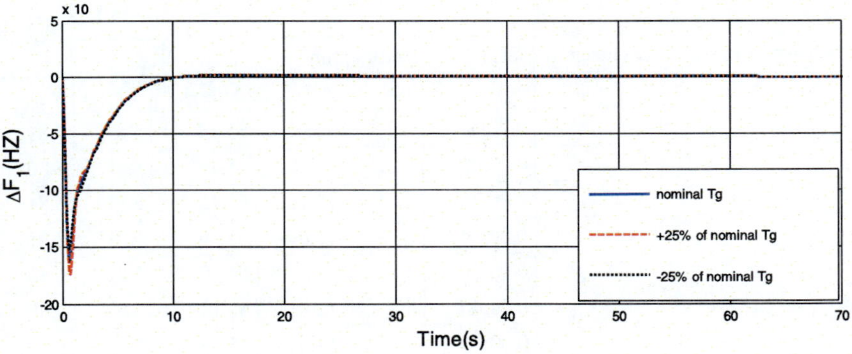

(a)

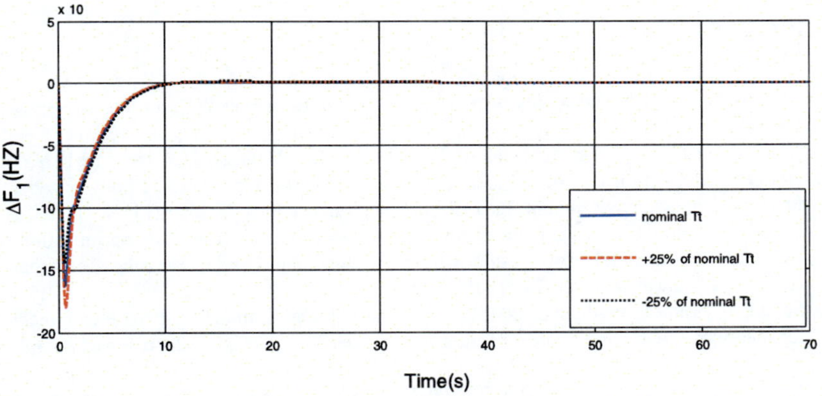

(b)

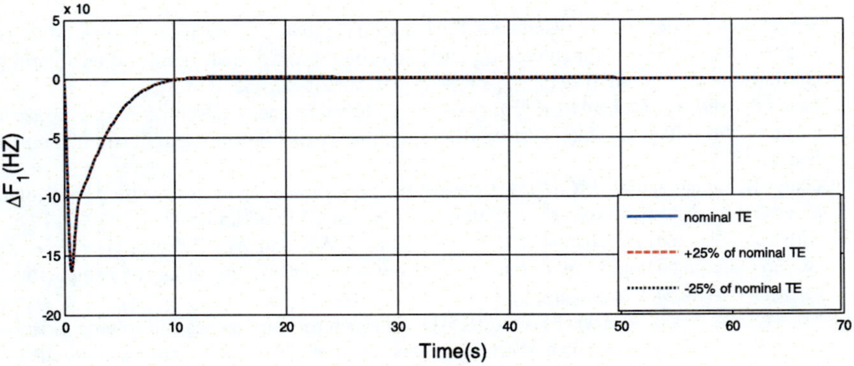

(c)

Fig. 6 **a** Aberration of amdrequency of area-1 for SLP of 20% with variation in Tg. **B** Aberration of amdrequency of area-1 for SLP of 20% with variation in Tt. **c** Aberration of amdrequency of area-q for SLP of 20% in with variation in T_E.

Table 2 Sensitivity analysis

Parameter variation	% change	ISE	Settling time				
			ΔF_1	ΔF_2	ΔP_{tie12}	V_1	V_2
Nominal	0	0.0038	8.8	12.02	65.14	5.81	5.81
Tg_1	− 25	0.00381	8.82	12.12	65.16	5.82	5.80
	+ 25	0.0038	8.81	12.10	65.15	5.81	5.81
Tt_1	− 25	0.00379	8.79	12.09	65.04	5.81	5.81
	+ 25	0.00382	8.82	12.05	65.09	5.81	5.81
T_{E1}	− 25	0.00381	8.8	12.03	65.15	5.82	5.83
	+ 25	0.0038	8.81	12.02	65.14	5.80	5.83

References

1. Elgerd OI (2000) Electric energy systems theory—an introduction, 2nd edn. Tata McGraw-Hill, New Delhi, India
2. Kundur P (1994) Power system stability and control. Tata McGraw-Hill, New York, NY, USA
3. Hota PK, Mohanty B (2016) Automatic generation control of multi-source power generation under deregulated environment. Int J Electr PowerEnergy Syst 75:205–214
4. Mohanty B, Hota PK (2015) Comparative performance analysis of fruit fly optimization algorithm for multi-areamulti-source automatic generation control under deregulated environment. IET Gener Transm Distrib 9(14):1845–1855
5. Sahu RK, Panda S, Pradhan S (2015) A hybrid firefly algorithm and pattern search technique for automatic generation control of multi area power systems. Int J Electr PowerEnergy Syst 64:9–23
6. Mohanty B, Acharyulu BVS, Hota PK (2017) Moth-flame optimization algorithm optimized dual-mode controller for multiarea hybrid sources AGC system. Optim Control Appl Meth 1–15
7. Simhadri KS, Mohanty B, Panda SK (2018) Comparative performance analysis of 2DOF statefeedback controller for automatic generation control using whale optimization algorithm. Optim Control Appl Meth 1–19. https://doi.org/10.1002/oca.2462
8. Panda S, Sahu BK, Mohanty PK (2012) Design and performance analysis of PID controller for an automatic voltage regulator system using simplified particle swarm optimization. J Franklin Inst 349(8):2609–2625
9. Gozde H, Taplamacioglu MC (2011) Comparative performance analysis of artificial bee colony algorithm for automatic voltage regulator (AVR) system. J Franklin Inst 348(8):1927–1946
10. Gupta A, Chauhan A, Khanna R (2014) Design of AVR and ALFC for single area power system including damping control. In: 2014 recent advances in engineering and computational sciences (RAECS), Chandigarh, pp 1–5
11. Kouba NEY, Menaa M, Hasni M et al (2015) Optimal control of frequency and voltage variations using PID controller based on particle swarm optimization. Fourth international conference on systems and control (ICSC), Sousse, pp 424–429
12. Chandrakala Vijaya KRM, Balamurugan S (2016) Simulated annealing based optimal frequency and terminal voltage control of multisource multiarea system. Int J Electr Power Energy Syst 78:823–829
13. Sahu RK, Panda S, Biswal A, Chandra Sekhar GT (2016) Design and analysis of tilt integral derivative controller with filter for load frequency control of multi-area inter connected power systems. ISA Trans 61:251–264

Performance Study of Solar PV System with Bifacial PV Modules

Kiran Jose, S. Sheik Mohammed, and O. Mohammed Mansoor

Abstract Bifacial photovoltaic (PV) is regarded as a promising technology due to its potential to attain higher annual energy yield compared to conventional PV systems. Higher annual energy yield is a significant factor resulting in reduced balance of system (BOS) costs, thereby achieving lower levelized cost of energy (LCOE). In this paper, a solar PV system with maximum power point tracking (MPPT) controller is presented and discussed. The single diode model of a bifacial PV module based on its mathematical equations is developed in MATLAB. This model can be utilized to analyze and validate the performance of any commercially available bifacial PV module. A standalone PV system consisting of bifacial PV module, DC–DC buck converter, and perturb and observe (P&O) MPPT controller is built and simulated in MATLAB/Simulink environment. A detailed analysis of the performance of PV module and the system for different conditions are validated by obtained simulation results.

Keywords Bifacial · Monofacial · Buck converter · Modeling · Perturb and Observe (P&O) algorithm · Maximum power point tracking (MPPT)

Nomenclature

A	Diode ideality factor (1.2)
D	Duty cycle
F	Switching Frequency (kHz)
G	Irradiance (W/m^2)
G_f	Front irradiance (W/m^2)
G_{nom}	Nominal irradiance (1 kW/m^2)

K. Jose · S. Sheik Mohammed (✉) · O. Mohammed Mansoor
Faculty of Engineering, Universiti Teknologi Brunei, Gadong A BE1410, Brunei Darussalam
e-mail: sheik.sulthan@utb.edu.bn

O. Mohammed Mansoor
e-mail: mansooreee@tkmce.ac.in

K. Namrata et al. (eds.), *Smart Energy and Advancement in Power Technologies*, Lecture Notes in Electrical Engineering 926,
https://doi.org/10.1007/978-981-19-4971-5_48

G_r	Rear irradiance (W/m^2)
I	Net current of a PV cell (A)
I_{mpp}	Maximum output current (A)
I_o	Output current (A)
I_{ph}	Photo-generated current (A)
I_s	Saturation current of diode (A)
I_{sc}	Short circuit current (A)
k	Boltzmann const. (1.38e^{-23} J/K)
K_i	Temperature coefficient of I_{sc}
N_s	Number of modules in series
P_{front}	Front-side power (W)
P_{max}	Peak power of module (W$_p$)
P_{mpp}	Maximum output power (W)
P_{rear}	Rear side power
q	Charge of an electron (1.6e^{-19} C)
R_s	Series resistance (Ω)
T_c	Cell temperature (°C)
T_r	Reference temperature (25 °C)
V_i	Input voltage (V)
V_{mpp}	Maximum output voltage (V)
V_o	Output voltage (V)
V_{oc}	Open-circuit voltage (V)
ΔI_L	Inductor ripple current (%)
ΔV_o	Output voltage ripple (%)

1 Introduction

Over the past few years, there has been a rise in demand toward the bifacial PV technology, due to its better performance rates and lower LCOE in contrast with the conventional PV systems [1]. The bifacial technology has been under research since early 1970s [2], and the first work can be found to be a patent published in 1966 [3]. This was followed by few studies regarding its efficiency and practical applications [4]. Over the past few decades, researchers began studying this technology worldwide. In some of these studies, mathematical models were investigated, and different configurations such as heights, tilts, and albedo were also experimented [5, 6].

Bifacial systems produce more power when compared to its monofacial counterparts having the same front-side capacity, as it utilizes solar irradiance falling on both front and rear sides of the cell. This results in the installation of fewer number of panels to achieve the required capacity, which further results in lesser BOS components. This reduces the possibilities for system failures and the entire BOS and plant cost [7]. Also, the distinct properties of bifacial technology provide 2–6% of reduced LCOE when the latitude is greater than 30° [1]. Figure 1 shows the cell structure of a bifacial cell, and Fig. 2 depicts various illumination elements collected by a bifacial PV module.

The bifacial efficiency is determined by bifacial factor (BF), or bifaciality, which is described as relative efficiency of rear side of the module in reference to the relative efficiency of front side of the module. The relative efficiency is also characterized as relative power under standard test conditions (STC) of 25 °C and irradiation level of 1000 W/m^2 [8], which is expressed as in Eq. (1):

$$\mathrm{BF}\,(\%\,) = \frac{P_{\mathrm{rear}}\ @\ \mathrm{STC}}{P_{\mathrm{front}}\ @\ \mathrm{STC}} \times 100 \tag{1}$$

The additional yield in energy produced by the rear side in a bifacial module is calculated as bifacial gain in energy (BG_E) [8]. BG_E is amount of energy produced by the rear side to the amount of energy produced by the front side of the bifacial module, quantified as in Eq. (2):

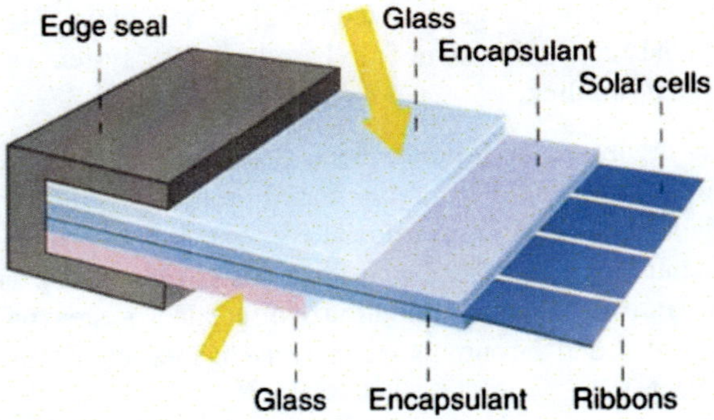

Fig. 1 Cell structures of bifacial PV system [8]

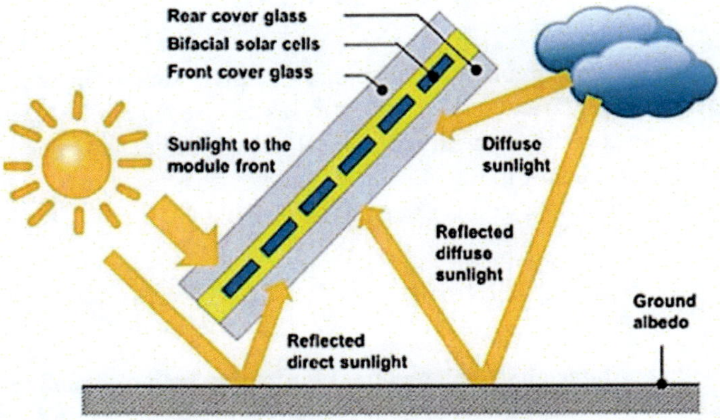

Fig. 2 Illumination elements of bifacial PV module. (© T.Ü.V. Rheinland Energy)

$$\mathrm{BG}_E = \frac{\text{Rear side (kWh)}}{\text{Front side (kWh)}} \tag{2}$$

Majority of the existing literatures associated with mathematical modeling and performance analysis of PV systems focused mainly on the monofacial PV modules. The mathematical modeling of conventional PV module interfaced with a boost converter is discussed in [9]. Implementation of P&O and InC MPPT algorithm on solar controller circuit in MATLAB/Simulink were discussed by the authors in [10]. The performances of PV modules under partial shading conditions are studied using P&O and particle swarm optimization (PSO) MPPT algorithms in [11].

As mentioned earlier, the previous researches do not involve mathematical modeling or performance analysis of a bifacial PV system. To the best of our knowledge, a generalized model for a bifacial module is presently not published in any literatures. Such a model helps in analyzing and validating performance characteristics of commercially available bifacial modules. The contribution of this work is outlined below:

- Mathematical modeling of a generalized bifacial PV module.
- Performance analysis of bifacial module-based solar PV system with P&O MPPT using MATLAB/Simulink.

This paper is organized as follows: Sect. 2 discusses on the development of single diode model of bifacial PV module based on mathematical equations using MATLAB. In Sect. 3, the developed bifacial PV system is explained in steps. This includes fundamentals of P&O MPPT algorithm, design of buck converter, and incorporation of MPPT into the system. Simulation of bifacial PV system with P&O MPPT algorithm is performed, and the outputs are analyzed in Section 4. The presented study is concluded in Section 5.

2 Single Diode Model of Bifacial PV Cell

The single diode PV cell circuit model comprises a current source (I_{pv}) connected in anti-parallel to a diode (D) and series resistor (R_s) as depicted in Fig. 3. A generalized I–V and P-V characteristic for a particular level of insolation and temperature of a PV cell is also shown below in Fig. 4.

The net current of the PV cell is given as in Eq. (3):

$$I = I_{\text{pv}} - I_s\left\{\exp\left(\frac{q}{\text{Ak}T_c N_s}V + \text{IR}_s\right) - 1\right\} \tag{3}$$

The modeling of bifacial PV cell requires considering both front and rear irradiances, unlike their monofacial counterpart. Therefore, irradiance (G) in I_{pv} equation is further modified. The photo-generated current, which mainly depends upon the solar insolation and cell temperature [9], is described as in Eq. (4):

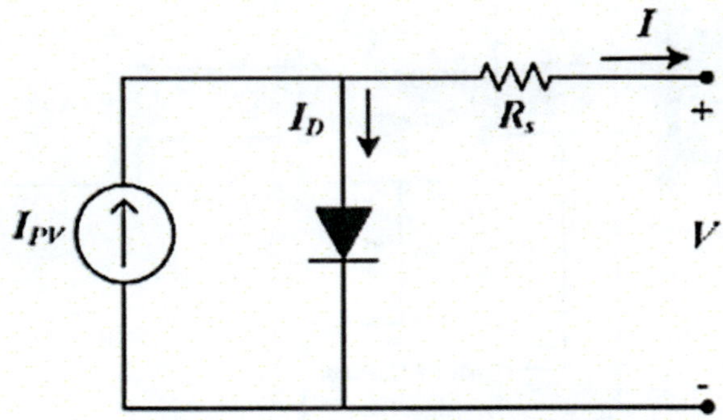

Fig. 3 Single diode model of PV cell

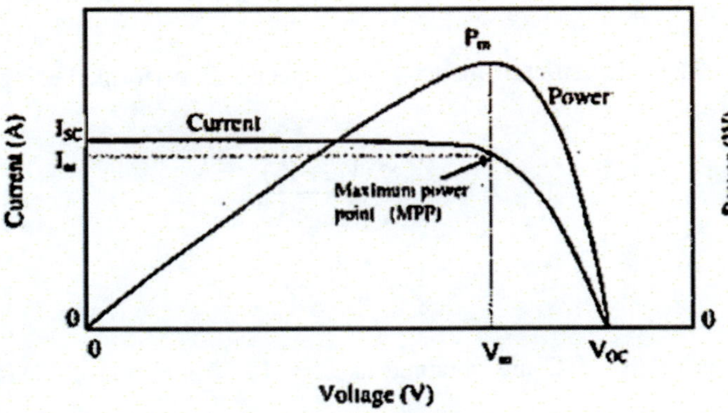

Fig. 4 I–V and P-V characteristics of PV cell [9]

$$I_{pv} = [I_{sc} + K_i(T_c - T_r)]\frac{G_f + G_r}{G_{nom}} \quad (4)$$

3 Architecture of Solar PV System

The MPPT mechanism utilizes an algorithm and a power electronics circuitry where load and source impedance are to be matched to extract MPP. This matching is attained utilizing a DC–DC converter. The converter adjusts the switching duty cycle, thereby regulating the voltage generated. Block diagram of the system is shown in Fig. 5.

This work uses a buck converter in order to step down module voltage in grid connected applications. The basic configuration of buck converter is depicted in Fig. 6, and the parameters of the converter used in this work are listed as $V_i = 81$ V, $P_{mpp} = 770$ W, $V_o = 48$ V (< 2% ripple), $I_o = 16$A (< 10% ripple), $D = 0.6$, and

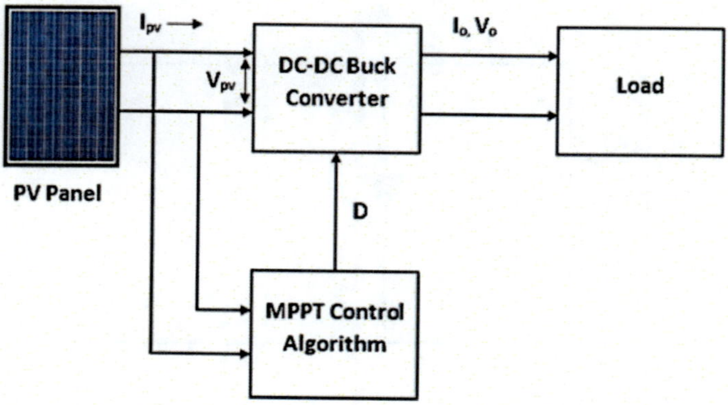

Fig. 5 Block diagram of P&O MPPT-based PV system

$f = 10$ kHz. The calculation of inductor and capacitor is performed using Eqs. (5) and (6) [12].

$$L = \frac{V_o \times (V_i - V_o)}{\Delta I_L \times f \times V_i} \tag{5}$$

$$C = \frac{\Delta I_L}{8 \times f \times \Delta V_o} \tag{6}$$

The values of L and C are obtained as $L = 19.55$ mH and $C = 15.57$ µF, respectively.

In order to extract P_{MPP} from the PV, MPPT capability is used. P&O MPPT algorithm is a commonly used MPPT control technique, where a slight perturbation is applied to the system voltage to the cause variation in the power output of the PV module. This output is measured and compared periodically with the previous power output. If the power is increased, the process is repeated; otherwise perturbation is inversed [13].

Fig. 6 Buck converter

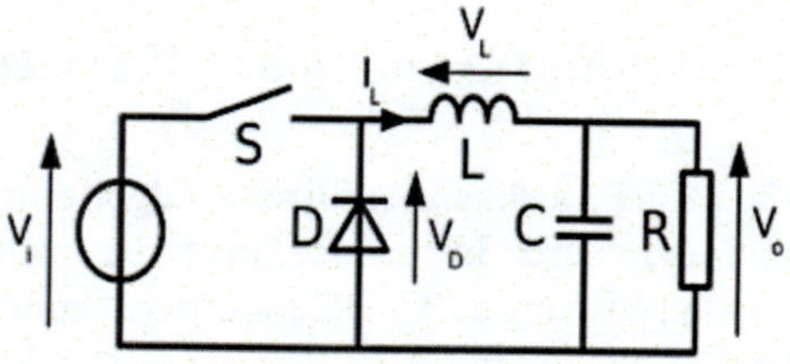

4 Simulation and Analysis

4.1 Simulation of Bifacial PV Module-Based System

The simulation model of solar PV system with bifacial PV string controlled by P&O MPPT controller is shown in Fig. 7. The bifacial PV module is developed based on the mathematical model which is discussed in Section II and is shown in Fig. 8.

The PV module chosen for validating the developed model is Adani Solar ASB-7-AAA bifacial PV module [14]. The maximum power output in monofacial condition is 385 W and can go up to 500 W when used as bifacial model. The electrical data needed to simulate the system is obtained from the datasheet [14] as depicted in Table 1.

The subsystem of P&O MPPT controller is depicted in Fig. 9. Developed system is simulated for different conditions, and observed results are discussed in the next section.

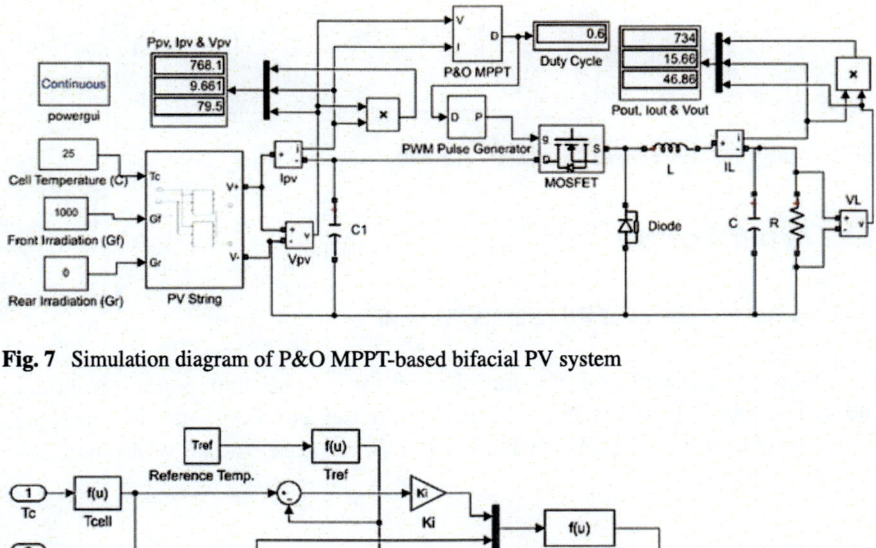

Fig. 7 Simulation diagram of P&O MPPT-based bifacial PV system

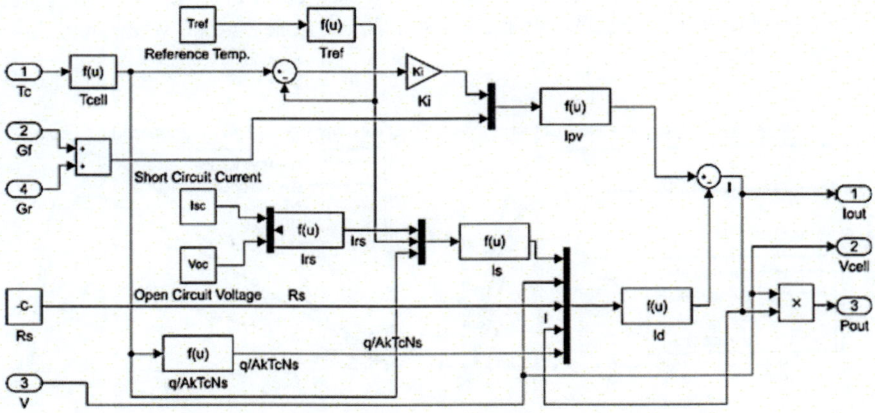

Fig. 8 Subsystem of developed generalized bifacial PV module

Table 1 Electrical data for Adani solar ASB-7-AAA bifacial PV module (at STC)

Electrical specifications	Monofacial	Bifacial (reference 385 W_p front)		
G_r	0%	10%	20%	30%
P_{max} (W_p)	385	420	460	500
V_{mpp} (V)	40.5	40.51	40.52	40.53
I_{mpp} (A)	9.53	10.40	11.39	12.20
V_{oc} (V)	49.2	49.21	49.22	49.23
I_{sc} (A)	10.07	10.92	11.96	12.81
K_i	0.065%/°C			
N_s	72			

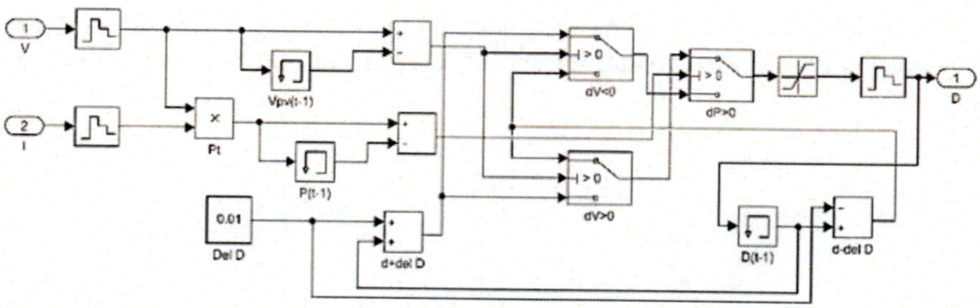

Fig. 9 Subsystem of developed P&O MPPT controller

4.2 *Result Analysis*

4.2.1 Characteristics of Bifacial PV Module

The model is simulated as a monofacial setup for various illumination conditions, and its I–V and P-V curves at $T_c = 25$ °C are presented in Fig. 10. Test cases are: Case 1—1000 W/m^2 (i.e., STC), Case 2—800 W/m^2, Case 3—600 W/m^2, and Case 4—400 W/m^2.

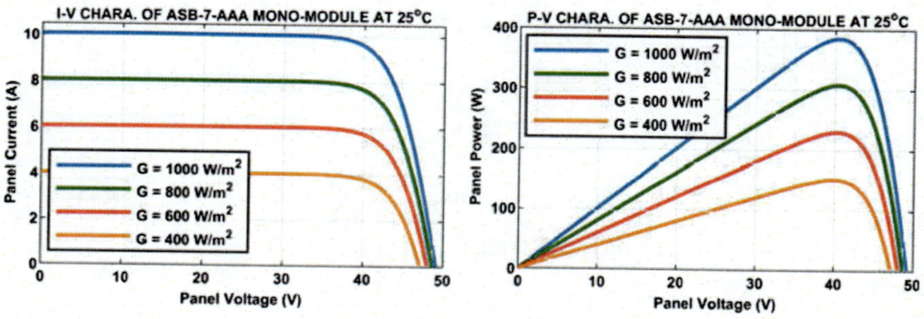

Fig. 10 I–V (left) and P-V characteristics (right) of ASB-7-AAA monofacial module

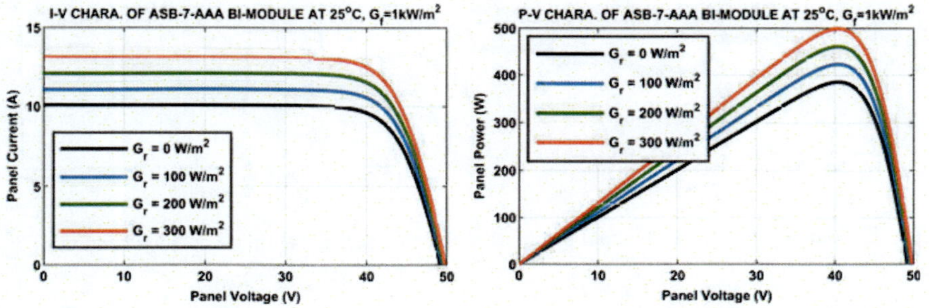

Fig. 11 I–V (left) and P-V characteristics (right) of ASB-7-AAA bifacial module

Further, the model is simulated in bifacial conditions for different values of G_r, where $G_f = 1000$ W/m^2 and $T_c = 25$ °C. Test cases are: Case 1—0 W/m^2 (i.e., monofacial condition)., Case 2—100 W/m^2, Case 3—200 W/m^2, and Case 4—300 W/m^2. Figure 11 depicts I–V and P-V behavior for the above test conditions.

4.2.2 P&O MPPT Analysis

The MPPT analysis is performed for both monofacial and bifacial models. Figure 12 shows G_f provided in both conditions, while $G_r = 300$ W/m^2 for bifacial condition. T_c was maintained at 25 °C.

In Case 1, the simulation is performed for the monofacial setup, where the $P_{\max} =$ 770 W. The D value of the converter generated by P&O MPPT algorithm and output power for different irradiation levels is depicted below in Fig. 13.

In case 2, the bifacial model with $P_{\max} = 1000$ W is simulated with different irradiation levels. Duty cycle and output power for the simulated converter are shown in Fig. 14.

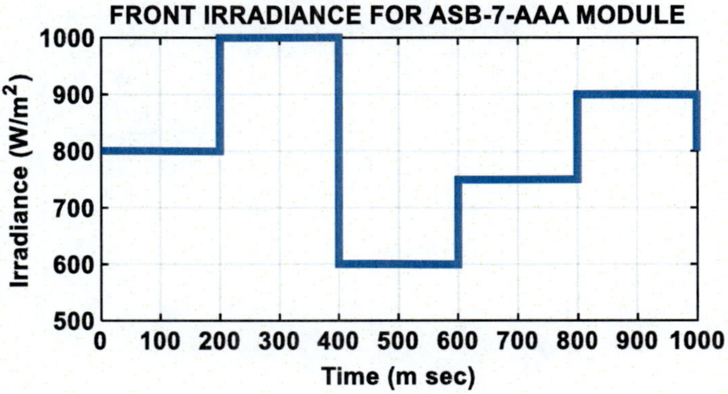

Fig. 12 Front irradiation levels for test conditions

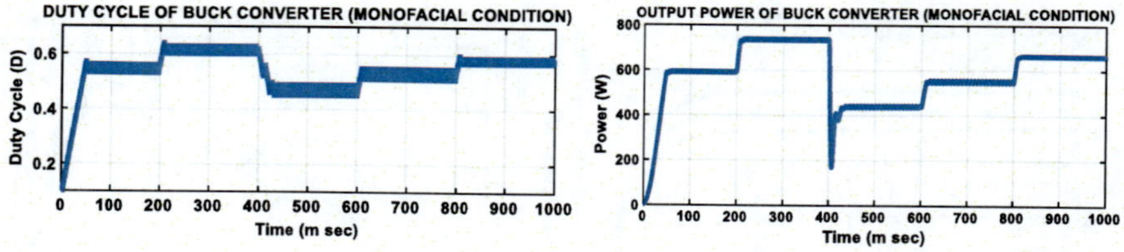

Fig. 13 Duty cycle (left) and output power of buck converter (right) (in monofacial condition)

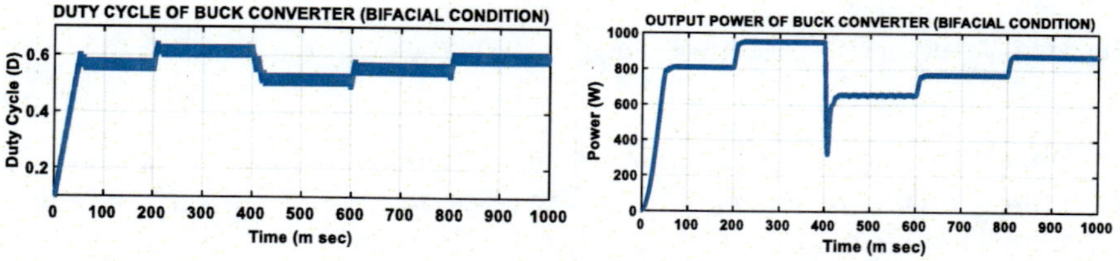

Fig. 14 Duty cycle (left) and output power of buck converter (right) (in bifacial condition)

The average maximum power output extracted by the MPPT controller for Case 1 and 2 is compared in Table 2. The results show that bifacial installations can produce higher power output from its similar rated monofacial counterparts for different levels of G_f values.

The efficiency of the panel for different rear irradiation conditions is simulated, and the output obtained is presented in Table 3. The power values obtained are matching up to the ratings provided by the manufacturer which is depicted in Table 1. The duty cycle of the converter is nearly 60%. The efficiency of the converter is calculated to be around 96% for all test conditions.

Table 2 Comparison of maximum power output in Cases 1 and 2

G_f (W/m^2)	Monofacial power output (W)	Bifacial power output (W)
800	590.2	807
1000	734.9	948.1
600	439	657.5
750	551.1	768.5
900	665.5	876.4

Table 3 Tracking efficiencies of ASB-7-AAA module under different G_r conditions

Test conditions	PV output power (W)	Converter output power (W)	Efficiency (%)
$T_c = 25$ °C, $G_f = 1000$ W/m^2, $G_r = 0$ W/m^2	768.4	734.9	95.64
$T_c = 25$ °C, $G_f = 1000$ W/m^2, $G_r = 100$ W/m^2	838.6	802.9	95.74
$T_c = 25$ °C, $G_f = 1000$ W/m^2, $G_r = 200$ W/m^2	916.6	877.3	95.71
$T_c = 25$ °C, $G_f = 1000$ W/m^2, $G_r = 300$ W/m^2	989.8	948.1	95.85

5 Conclusion

The modeling of bifacial PV module using MATLAB and performance analysis of bifacial module-based solar PV system with P&O MPPT is presented. The bifacial PV model presented in this paper is suitable for analyzing the performance characteristics of commercially available bifacial PV modules and also the solar PV systems. The performance characteristics of the developed bifacial PV module model are tested for ASB-7-AAA panel under different irradiation and gain values. The generated results are on par with the behavior given by the manufacturer in the data sheet.

Simulation of P&O MPPT-based PV system with buck converter is also performed. The developed system is simulated for different irradiation levels, and the outputs are analyzed. The PV system has an efficiency close to 96% for all test conditions.

References

1. Gu W et al (2020) (2005) A comprehensive review and outlook of bifacial photovoltaic (bPV) technology. Energy Convers Manage 223:113283
2. Cuevas A (2005) The early history of bifacial solar cells. In: Proceedings of the 20th European photovoltaic solar energy conference, pp 801–805
3. Hiroshi M (1966) Radiation energy transducing device, US Patent
4. Riaz MH et al (2021) Module technology for agrivoltaics: vertical bifacial versus tilted monofacial farms. IEEE J Photovoltaics 11(2):469–477
5. Rouholamini M et al (2021) modeling, configuration, and grid integration analysis of bifacial PV arrays. IEEE Trans Sustain Energy 12(2):1242–1255
6. Lu W (2021) Power generation characteristics of bifacial photovoltaic modules under different backgrounds. In: 2021 IEEE international conference on artificial intelligence and computer applications (ICAICA), pp 470–473

7. Ayadi O et al (2021) An experimental comparison of bifacial and monofacial pV modules. In: 2021 12th international renewable engineering conference (IREC), pp 1–8
8. Asgharzadeh A (2019) Bifacial photo-voltaic (PV) system performance modeling utilizing ray tracing. PhD Dissertation, Department of Electrical Engineering, University of Iowa
9. Mohammed SS, Devaraj D (2014) Simulation and analysis of stand-alone photovoltaic system with boost converter using MATLAB/Simulink. In: 2014 International Conference on Circuits, Power and Computing Technologies (ICCPCT), pp 814–821
10. Verma N et al (2021) Maximum power point tracking methods for photovoltaic modules. In: 2021 international conference on advance computing and ingenious technology in engineering science (ICACITE), pp 223–227
11. Lakhdara A, Bahi T, Moussaoui A (2021) MPPT techniques of the solar PV under partial shading. In: 2021 18th international multi-conference on systems, signals & devices, pp 1241–1246
12. Morcillo Bastidas JD et al (2021) Simulation and analysis of complex behaviors in a DC–DC buck converter. IEEE Latin Am Trans 19(1): 68–74
13. Sheik Mohammed S, Devaraj D, Imthias Ahamed TP (2016) A novel hybrid maximum power point tracking technique using perturb & observe algorithm and learning automata for solar PV system. Energy 112:1096–1106
14. Adani ELAN-72 P-Type-PERC ASB-7-AAA (AAA=370–395) bifacial PV module datasheet. https://www.adanisolar.com/Downloads#2

Peak Time Energy Management System for Household Load Devices Under Real-Time Pricing

Jigyasa Sharma and Sandeep Bhongade

Abstract This paper describes how to schedule the home load types like air conditioner (AC), electric water heater (EWH), clothes dryer (CD) and electric vehicle (EV), taking into account end user's ease level that is flexibility to choose the operation period of the devices. The real-time pricing (RTP) is accounted so that peak demand will shift to the other time which is having lower pricing hours and due to this user save in the electricity bill. The utility has set a demand threshold level (DL) if user breaks the limit at any time of the billing period, then the supplementary charges will be levied on consumer for drawing more power. This paper examines the comparison study of three different scheduling methods. The scheduling of load devices has been formulated one with home energy management algorithm accompanying load predilections and the other two optimization methods: binary particle swarm optimization (BPSO) and proposed binary salp swarm algorithm (BSSA). The proposed optimization arrived at the conclusion that the solution it has provided surpassed from the rest of the added two methods solution. In comparison with the rest of the methods, the reduction in load factor and the saving from BSSA is very much in consumer's bill. The convergence is also faster than the BPSO.

Keywords Demand side management (DSM) · Binary salp swarm algorithm (BSSA) · Demand response (DR) · Real time pricing (RTP)

1 Introduction

The demand side management (DSM) described as development and implementation of those electric utility actions aimed to regulate the end users electricity consumption pattern which creates desirable modification in electric utility's load

J. Sharma
Department of Electrical Engineering, Indore, Madhya Pradesh, India
e-mail: jigyasa994@gmail.com

S. Bhongade (✉)
Shri G.S Institute of Technology and Science, Indore, Madhya Pradesh, India
e-mail: bhongadesandeep@gmail.com

K. Namrata et al. (eds.), *Smart Energy and Advancement in Power Technologies*,
Lecture Notes in Electrical Engineering 926,
https://doi.org/10.1007/978-981-19-4971-5_49

profile. When the demand for energy is very high that is during peak hours then at that time utility entices the end user to reduce the energy demand or to shift it to another time. With the purpose of meeting the electricity demand in peak period, the electric utility company interrupts the electricity or a blackout occurs. By reducing peak power demand, the potential for blackouts is reduced. Demand response (DR) manage peak demand occurrence and avoid network congestion, as it supports the adaptability needed to change the timing of loads [1]. DSM exists as ultimate influential functions in a smart grid that supports power providers to lower the peak load requirements and to redesign the load outline. The proposed next day load change approach is formulated computationally as minimization problems. An evolutionary algorithm with a heuristic base that is easy to adapt to the set of rules that intended to increase the probability of solving the mentioned problem [2]. DSM involves many constituents like smart devices, inclusion of renewable sources allowing load structuring. Home area network (HAN) plays an important bidirectional communication channel between utility and consumer. It turned on/off the consumer devices as per demand. Due to the following factors, the capital investment required to satisfy the peak demand of the power plants has been reduced [3].

In this paper, binary particle swarm optimization (BPSO) proposed. The optimization adequately plans devices on usage pattern and users performance. BPSO optimizes the scheduling period, which leads to a delay in peak hours and a dynamic curve using dynamic rate [4]. This BPSO provides a planning strategy for the interruptible loads taking into account boundary conditions, manages financial penalization and several consumer ease levels. Objective function is under the control of weight factor [5]. Salp swarm algorithm (SSA) is a new meta-heuristic algorithms supported the manner of deep sea salpidae. The paper includes a binary variant of the SSA called binary salp swarm algorithm (BSSA). The transfer function supports in the change of continuous value into binary value and helps in determining the global optimum solution. The comparative investigation of various transfer function is done with the help of several standard functions [6].

Mostly, above-mentioned methods are primarily designed for continuous problems thereafter convert them to binary optimization. For optimal scheduling of considered load types requires binary search space because load devices works in only two operating state, it is either on or off. The proposed binary salp swarm algorithm (BSSA) presents in this paper directly works in binary search space since the very first step. With the binary values, it keeps updating the leader's position and in the end provides a global optimum solution which gives user the most saving in bill compare to other optimization methods. Due to the binary initialization of swarm, its exploration characteristics enhance and its computation time is also reduced.

In this research paper, building of scheduling method with and without optimization manages to mitigate the peak load demand in accordance with the RTP. The foundation allegiance of this paper is reported by:

- The load types are detached into four classification: Thermostatically load like AC and electric water heater (EWH) serves on thermostat standard. Uninterruptible load that is cloth dryer (CD) once turned on will not turn off until it has finished its job. Shiftable load that is electric vehicle (EV) whose operation can shift at some specified temporal length of event without compromising consumer favorable setting and even stopped in the middle of the working of EV. The indispensable load as refrigerators, cooking food appliances and entertainment gadgets are neither relocated nor halted it conducted in accordance to the user necessity.

- The load types considered in this paper perform substantial role in growing peak demand. The development of the algorithm commands the domestic load types. First the modeling of the loads indicated the devices consumption of the power and also considered the load predilections and user satisfaction.
- In this paper, the physical modeling of all home appliances is done according to [7].

This paper propound binary version of salp swarm algorithm (BSSA) that minimizes electricity costs in households and compares it to Binary Particle Swarm Optimization (BPSO) and Home Energy Management Algorithm. MATLAB 2015(a) is used for the development of the algorithm.

This paper is structured as: Sect. 2 Objective Function and boundary conditions, Sect. 3 Proposed optimization, Sect. 4 Result and Discussion, Sect. 5 Conclusion and future work.

2 Problem Conceptualization

2.1 *Requisites of Home Energy Management (HEM) Structure*

For execution of DR scheme, an advance metering foundation is desired, that is a bidirectional interaction channel between utility and user load devices. The specified limit for the power demand has set by the utility. Home domain interface unit is the prime governing unit which constantly checked the energy absorption of all the domestic loads under real time environment. The DR scheme rooted in Home domain interface unit [8]. The smart devices are IP accessible, intelligent, energy efficacious, self-regulating and communicating according to consumer comforts and predilection. They acquire command indication and notify its status to the interface unit.

2.2 *The Demand Response (DR) Scheme*

The scheme initiated once the home total consumed power ($P_{h,i}$) surpasses the demand threshold level (DL_i). In this paper, load types are classified into 2 units: energy-consuming non-indispensable loads such as AC, WH, CD and EV all are controlled loads and indispensable loads such as light, freeze and distinct charging sockets. DR compatible load models are classified accordingly [9]. Load precedence, predilection and the comforts of the consumer have been taken care. According to end user given precedence to the load, the DR scheme chooses the rank of the load devices among the two classified units. Assuming the given precedence of the load and comfort predilection for our targeted house is shown in Table 1.

Working Home Energy Managing Algorithm: The given precedence to the load decides the rank of the load type under DR program. If energy management system gets a request to turn on any load type and when no DL set by the utility, then the system issued a command indication and activate the load. But when the DL is set, the system will first check the total home demand. If it is more than the DL, the system will relocate the loads according to the pre-established precedence. The load which is of low precedence will be relocate first to other time slots where the DL magnitude is high. As shown that the EV load which is of lowest precedence is the first one to shift then CD, AC and the EWH which is of highest precedence will shift to the last. The algorithm optimally relocates the loads preserving the total demand lower than the DL. The home command function is given by the formula [8]:

$$P_{L,i} + P_{c,i} \leq \mathrm{DL}_i \tag{1}$$

wherein,

$P_{1,i}$ is the consumption of power of all non-indispensable load at the moment i in KW.

$$P_{l,i} = \sum_{j=1}^{N} P_{lj,i}\left(L_j = \text{Controllable loads}\right) \tag{2}$$

$Pc_{,i}$ is the consumption of power of all indispensable load at the moment i, in KW.

$$P_{c,i} = \sum_{j=1}^{K} P_{Ck,i}(C_k = \text{Indispensable load}) \tag{3}$$

DL_i is the demand threshold level in KW set by power utility at a time interval i.

Once the demand threshold level is higher from the total home consumed power, then the DR program will end and all the non-indispensable load will start operating

Table 1 Pre-established load precedence

Load type	Load precedence
EV	Lowest precedence
CD	3rd precedence
AC	2nd precedence
WH	Highest precedence

at their normal level. This is what it called as scheduling of the load and shifting the peak.

2.3 Proposed Objective Function and Its Constraints

The objective function is minimizing the cost of energy consumption, taking into account the limitations and convenience of the end user. Objective function minimization is in accordance with the Eq. (4).

$$\text{Objective Function} = \min(f) \tag{4}$$

$$f = \text{Total price (T)} + \text{Penalization function } (P) \tag{5}$$

$$\text{Total power consumption} = E_{\text{AC}}(t) + E_{\text{EV}}(\text{t}) + E_{\text{WH}}(\text{t}) + E_{\text{CD}}(\text{t}) + E_{\text{Cr}}(\text{t}) \tag{6}$$

$$\begin{aligned}\text{Difference in Power } (P_{\text{diff}}) = {} & \text{Total power consumption exceeding threshold level} \\ & - \text{Threshold level specified by utility (DL(t))}\end{aligned} \tag{7}$$

$$\text{Total price (T)} = \sum_{t=1}^{T} \text{Pr}(t)[E_{\text{AC}}(\text{t}) + E_{\text{EV}}(\text{t}) + E_{\text{WH}}(\text{t}) + E_{\text{CD}}(\text{t}) + E_{\text{Cr}}(\text{t})] \tag{8}$$

$$\text{Penalization function } (P) = \sum_{(t=1)}^{T} \text{Pr}(t) P_{\text{diff}} \tag{9}$$

where

f	total per day electricity cost function
$P_{\text{C}}(\text{t})$	total power consumption of all the devices at time interval t
$E_{\text{AC}}(\text{t})$	AC energy consumption at time interval t
$E_{\text{EV}}(\text{t})$	EV energy consumption at time interval t
$E_{\text{WH}}(\text{t})$	WH energy consumption at time interval t
$E_{\text{CD}}(\text{t})$	CD energy consumption at time interval t
$E_{\text{Cr}}(\text{t})$	indispensable load (Cr) energy consumption at time interval t
Pr(t)	real-time price at time interval t
DL(t)	demand threshold level at time interval t

The sum of total price (T) of 24 h and the penalization function is the objective function. Penalization function is described for the total consumed power for every min. Examine the consumption of power with the demand threshold level (DL). DL varies over time and that utility sets. At the time of low DL, the consumer is still drawing power even after the total power consumption exceeds the DL; then, penalization will be included by the way of price in the objective function.

2.4 *Electricity Bill*

The MATLAB code is constructed for a min time slot. In RTP, the algorithm first measured the amount of power being consumed per hour and the unit price of that hour, both their 24 h multiplication is the electricity cost. Demand charges are the additional charges that the utility levied on users bill for peak demand (KW). Supposed peak demand appeared on chosen day and that multiply by price of peak demand.

$$\begin{aligned}\text{Bill for a month} = {} & \text{total per day electricity cost function } (f) \text{ * total day in a month} \\ & + \text{peak demand of a particular day (KW)} \\ & \text{* price of the peak demand}\end{aligned} \quad (10)$$

2.4.1 Boundary Conditions

The ambient temperature of AC is $T_{AC}(t)$, and minimum value of $T_{AC}(t)$ is 64.5°F and maximum value of $T_{AC}(t)$ is 71.5°F. The discharging water temperature of EWH is $T_{WH}(t)$ limited in the middle of the bottom value which is 41.5 °C and utmost value which is 47.5 °C. The SOC of an electric vehicle battery changes from the primary charge which is 40% to the peak accumulated charge which is 100%. The drying time (min) is shorter than the amount of time it takes /total time span of CD drying process (min) which is 90 min. Several constraints are given for the minimization of the objective function in accordance with the comfort of the consumer and are shown in Table 2.

$$T_{AC}^{min}(t) < T_{AC}(t) < T_{AC}^{max}(t) \quad (11)$$

$$T_{WH}^{min}(t) < T_{WH}(t) < T_{WH}^{max}(t) \quad (12)$$

$$SOC_{EV}^{min}(t) < SOC_{EV}(t) < SOC_{EV}^{max}(t) \quad (13)$$

$$T_{Accumulated} < T_{required} \quad (14)$$

The user sets up the ease setting of the devices such as indispensable load turn on at any moment of the time unaffected by the algorithm. CD completed its task in 70 min in the middle of 7–10 a.m. and 7–10 p.m. If once it is turned on in any

Table 2 User predilection setting

Types	Residential appliances	$I_d \sim E_d$	O_d (1 slot = 1 min)	Characteristics and requirement
1st type	Indispensable loads	1–24	24 h (1440 min)	All day power consumption pattern of non-shiftable loads
2nd type	Cloth dryer (CD)	7–10 19–22	70 min	Original hours of working:70 min operate between [7 and 10 pm] without intervention
3rd type	Electric water heater (EWH)	7–9 20–22	50 min	In a day EWH switched on for four instance of time Original hours of working [7–7:10, 8:00–8:10, 20–20:15, 21–21:15]
	Air conditioner (AC)	1–24	24 h	Room temperature is retained in the middle of specified limit for a time period of 24 h
4th type	Electric vehicle (EV)	18–6	4 h (240 min)	Original on/off charging cycle in the middle of 6 p.m.–3a.m. and charging of EV will complete in four hours

specified time, then CD will not turn off until it finishes its work. User will use the EWH for four instances of time. AC retained the room temperature in the middle of particular limits for a 24-h period. EV fully charges its battery in four hours and can schedule at any moment in the middle of 6 pm to 6 am. Modeling of every residential appliances has been done with the use of three parameters I_d, E_d, O_d, where [I_d, E_d] shows the permissible working period during which appliance d can be turned on and also defined by Eq. (16). O_d defines as total working period requested by user which is showed by Eq. (15) [8].

$$\sum_{t=1}^{T} s_d(t) = O_d \quad \forall d \in (1 \text{ to } D) \tag{15}$$

$$s_d(t) = 0 \quad \forall t \langle I_d \text{ or } \forall t \rangle E_d \tag{16}$$

where

s_d on/off status of the appliance d
D total appliances

T total time slot.

3 Binary Salp Swarm Optimization (BSSA)

In BSSA, sigmoidal transfer function (TF) is proposed to convert the continued value into binary. BSSA optimally shifts the peak demand to the low pricing interval [10].

- Sigmoidal transfer function (TF): This is used for mapping the continuous value into binary value. It can be expressed as [10]:

$$T\left(x_i^j(t+1)\right) = \frac{1}{1 + \mathrm{e}^{-x_i^j(t+1)}} \tag{17}$$

The location of the leader is revised according to below expression:

$$\Delta_i^j(t+1) = \begin{cases} 1 \text{ if } T\left(x_i^j(t+1)\right) > R \\ 0 \text{ otherwise} \end{cases} \tag{18}$$

where R is the random number evenly allotted over 0 and 1.

3.1 Salp Swarm Algorithm (SSA)

- The location of the salps is specified in d spatial search space, where d is the no of variables.
- Food source F which is the swarm's target.
- The leader position updated according to Eq. (19)

$$x_j^1 = \begin{Bmatrix} F_j + c_1\left((ub_j - lb_j)c_2 + lb_j\right)c_2 \geq 0 \\ F_j - c_1\left((ub_j - lb_j)c_2 + lb_j\right)c_2 < 0 \end{Bmatrix} \tag{19}$$

where x_j^1 is the location of the leader sets the salp in jth dimension, F_j is the food source, ub_j and lb_j—upper and lower limits.
- c_1 plays an influential role in SSA. It maintain exploration and exploitation.
- Exploration focuses on discovering best solution by survey a search space correctly. Exploitation focuses on handling the data in local area. The estimation of c_1 is appeared as:

$$c_1 = 2\mathrm{e}^{-\left(\frac{4k}{K}\right)^2} \tag{20}$$

where k is the existing iteration and K is the maximal number of iterations. c_2 and c_3 are haphazard numbers created into the interval [0,1].

- The follower updated its location in accordance with Newton's law of motion and is expressed in Eq. (21)

$$x_j^i = \frac{1}{2}at^2 + v_o t \tag{21}$$

whereas $i \geq 2, x_j^i$ is the location of ith follower in jth dimension, t is the time. v_o is the starting speed and $a = \frac{v_{\text{final}}}{v_o}$ as long as iteration is fixed as 1 and $v_o = 0$. Equation (21) could be rewrite as

$$x_j^i = \frac{1}{2}\left(x_j^i + x_j^{i-1}\right) \tag{22}$$

Variables are the status of the devices over 24 h duration. Each salp is a matrix of length [A * T] where A is the no of appliances and T is the time in minute. In BSSA algorithm, each salp possesses position which of length [N * A * T] where N is the search agent number or salp number. Every salp position contains the status of each device for 1440 min. Fitness function is calculated for every salp position. It runs for 100 iterations and status of the devices updated according to Eqs. (19) and (22) but in binary version, sigmoidal function compares with random number and gets salp position value which is shown in Eq. (17). The parameter values shown in Table 3 for the BSSA algorithm are tested widely by empirical observation.

The BSSA has three important parts of operation:

- The first part is initializing randomly the position of the salps which is on/off status of the devices.
- The second part is assessment in which solution are evaluated to find the best result among all the leaders. The salps find the time of the devices which is the food positions and calculate the power consumption. The consumption of the devices directs the energy cost which is the food fitness function. The fitness obtained by all the salps set as F. The c_1 in Eq. (20) explores the search space and updates the leader position based on upper and lower bound of variables. The best food fitness, i.e., energy cost determines among all the leader position.
- The third part is updating the proposed BSSA parameters as well as solution. The upgrading of the solution is done with some threshold value. The further

Table 3 Parameter values for optimum solution of energy cost by BSSA

Initialization parameter	Values
Search Agents no N	10
Device no	4
Maximum_iteration k	100
Dimension (dim)	24 h

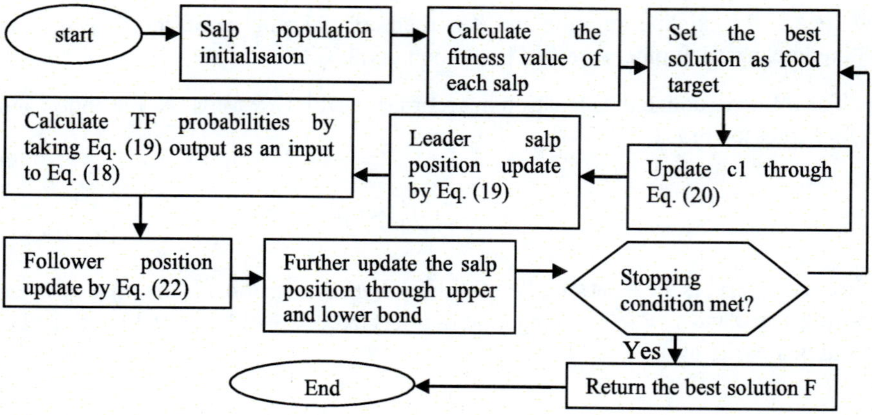

Fig. 1 Flowchart of BSSA [11]

updating of the solution is based on the TF and checks the boundary condition which is Eq. (19) and Eq. (22) until it reaches the maximum iterations, and it returned to the best food fitness function. The proposed algorithm converges the best global optimum solution and gives the minimum energy cost which reduces the consumers electricity bill (Fig. 1).

4 Simulation Results and Discussions

The outside temperature T_o °C of 24 h considered after foregoing research [12]. US real-time prices [7] were picked for research strategy.

A MATLAB program for proposed load types has been developed with the consideration of DL (which is issued by the utility) and load precedence. By applying the energy management system on considered load types, the following outcomes are presented. The below simulation results show the graph of power consumption before DR and afterward DR, i.e., after applying control algorithm on load types (with and without optimization algorithms) for reducing peak load and load factor.

4.1 Before Demand Response (DR)

The maximum consumption of the power of domestic appliances occurs between 7.00 p.m. and 7.16 p.m., and it is of 14.652 kW. On the particular day, the total energy consumption is 87.0395 kWh. The average electricity energy consumption of domestic appliances is 3.6266 kWh, and the load factor is 24.75%. Daily load factor = Total KWh in 24 h / (peak load in KW * 24 h). Load factor means actual

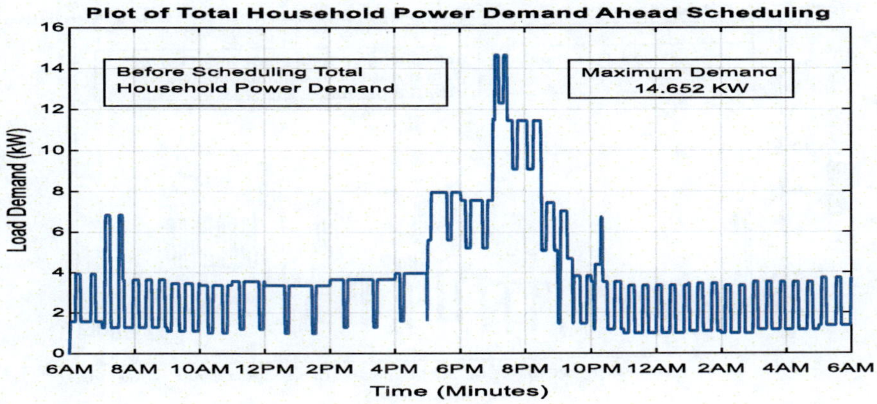

Fig. 2 Total home demand of power ahead scheduling

energy usage versus peak demand, i.e., one can improve the load factor by simply decreasing its peak demand and reducing the average cost per unit (KWh).

4.2 *Afterward Demand Response*

The scheduling of the home devices is along with optimization methods and excluding proposed optimization. The total home demand of power ahead scheduling is shown in Fig. 2. With the home energy management algorithm taking into account load precedence, user ease and demand threshold level and excluding optimization scheduling are done which is shown in Fig. 3. The peak load at the home decreases from 14.6520 to 8.6 kW. With the use of proposed optimization, scheduling of the total home devices with BSSA is shown in Fig. 4. With the proposed BSSA, the peak load at the home decreases from 14.6520 to 8.30 kW that gives additional reduction in peak load in comparison to other added methods.

4.3 *Comparison of BPSO and BSSA for Scheduling of Controlled Loads of the Home for Electricity Bill Minimizing*

Binary salp swarm algorithm (BSSA) actively exhibits better convergence in comparison to binary particle swarm optimization (BPSO). As it can be seen in Fig. 5 that the average fitness values are progressively declined. Due to BSSA exploration and exploitation stage, the fitness value gets better through the end of iterations in compare to BPSO. BSSA initiates converging at approximate 15th iteration so converging faster manifest the supremacy of stabilization and performance. BPSO

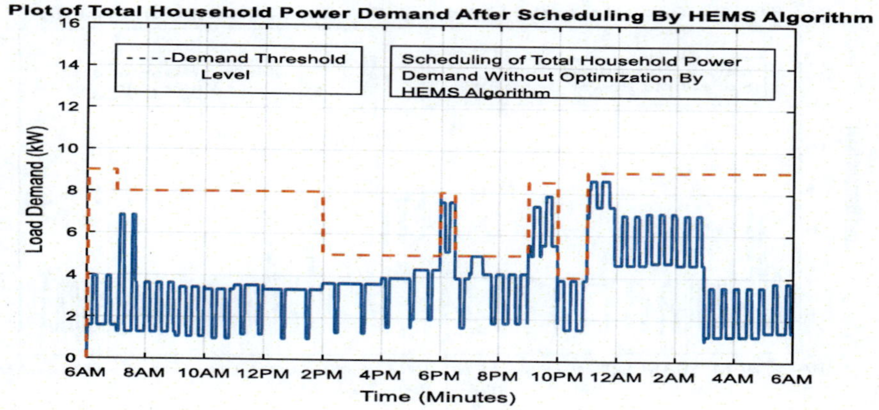

Fig. 3 Total home demand of power following HEMS algorithm for scheduling

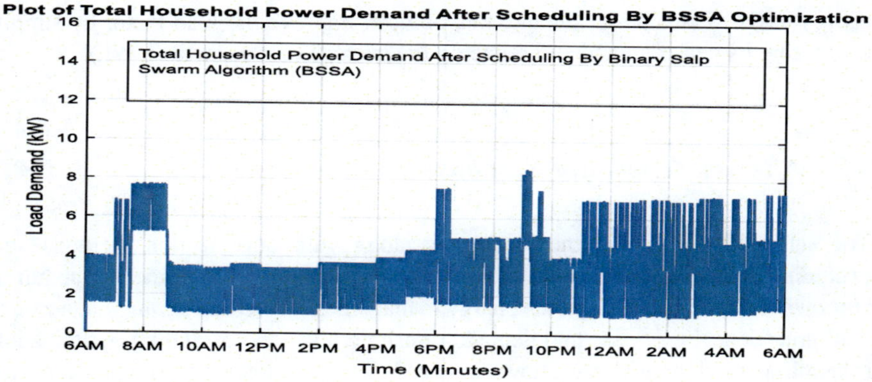

Fig. 4 BSSA scheduling of total home demand of power

initiates converging at approximate 54th iteration shown in Fig. 5. The convergence result of the two optimization is finest, but BSSA outdo as a result of revised sigmoidal transfer function. The departure due to variance in values of the BPSO and BSSA is 0.388%.

Thus, the above result clarifies that there is a reduction of 33% in daily energy usage with the proposed algorithm. The proposed salp swarm algorithm outperforms in a better way as compared to the other two added methods, and due to that, consumer saves on the electricity bill. Optimized and non-optimized HEM (Home Energy Management) algorithms efficiently transfer appliance uptime to non-productive periods and reduce per day energy costs to 398.311 ¢. Table 4 indicated that the proposed BSSA algorithm is superior in minimizing the everyday energy costs.

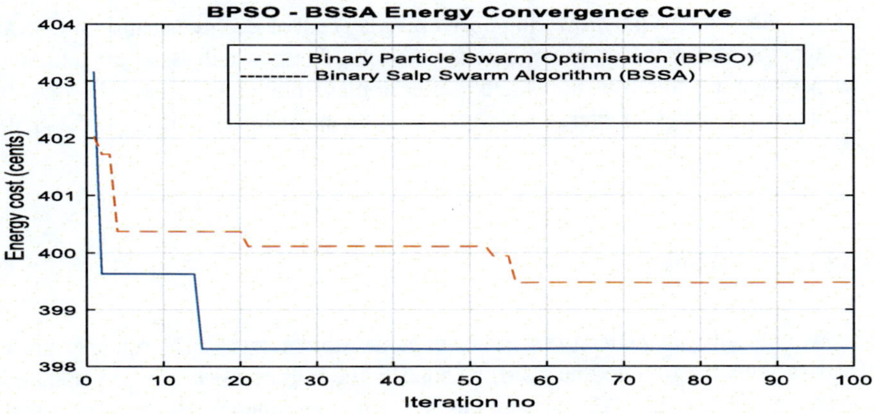

Fig. 5 Comparison convergence curve of energy cost

Table 4 Results of algorithms

Approaches	Cost of energy (cents)	Peak demand (KW)	% saving
Energy cost before applying energy managing algorithm	451.965	14.652	
Scheduling through energy managing algorithm	407.0324	8.60	34.48%
Scheduling by BPSO optimization	399.5	8.5	
Scheduling by BSSA optimization	398.311	8.30	33%

5 Conclusion

With the transition to the smart grid, the dependability and efficiency of the power grid have improved. From the deployment of smart grid technologies, advance metering infrastructure system, two-way communication, advance sensors, computing technologies supports in evaluation and modification in consumption pattern in relation to users response.

To examine and determine DR, it is essential to comprehend the physical framework of home appliances particularly for EWH, AC, CD, EV. In preparing the framework of non-indispensable load, the thermodynamic standards of home and other specified parameters have to be taken into consideration.

In the present study, proposed algorithm for optimal scheduling of the devices can handle non-indispensable loads which are EWH, AC, CD, EV unaffected consumer ease and predilections. With the incorporation of EV indoor DR program can be executed and avoid congestion problem in power network. The results attained from the proposed algorithm have shown that it is feasible to relocate the working time of the devices from peak demand period to lower demand period by taking into account pricing scheme. The scheduling algorithm results also indicated that user saves on electricity bill by lowering 33% everyday cost of energy because of BSSA optimization and make or become better the load factor which is 25% excluding energy managing algorithm and 42.75% after implementation of proposed BSSA. It diminish the demand for energy which is very high in peak period, i.e., 14.652–8.30 KW.

6 Future Works

This work feasibly extended to smart power generation premises. The future outlook of smart generating home one at which solar is the important supply of generated power. The devices like AC, heat pumps, EV charging, energy storage device that have some adaptability allow subsequent optimization of power usage with auto learning regulators. With the growth of PV and storage homes will have ability to turn into epicenters of energy and provide tractability to power grids.

References

1. Khodaei A, Shahidehpour M, Bahramirad S (2011) SCUC With hourly demand response considering intertemporal load characteristics. IEEE Trans Smart Grid 2(2):564–571
2. Logenthiran T, Srinivasan D, Shun TZ (2013) Demand side management in smart grid using heuristic optimization. IEEE Trans Smart Grid 3(3):1244–1252
3. Gellings CW (2009) The smart grid enabling energy efficiency and demand response. The Fairmont press Inc., Lilburn GA
4. Zhou Y, Chen Y, Xu G, Zhang Q, Krundel L (2014) Home energy management with PSO in smart grid. IEEE 23rd International Symposium on Industrial Electronics (ISIE), pp. 1666–1670. https://doi.org/10.1109/ISIE.2014.6864865
5. Pedrasa M, Spooner TD, MacGill IF (2009) Scheduling of demand side resources using binary particle swarm optimization. IEEE Trans Power Syst 24(3):1173–1181
6. Bie Z, Xie H, Hu G et al (2016) Optimal scheduling of power systems considering demand response. J Mod Power Syst Clean Energy 4:180–187
7. Shao S, Pipattanasomporn M, Rahman S (2012) Development of physical-based demand response-enabled residential load models. IEEE Trans Power Syst 28(2):607–614
8. Elyas SH, Sadeghian H, Alwan HO and Wang Z (2017) Optimized household demand management with local solar PV generation. *North American Power Symposium (NAPS)*, pp. 1–6. https://doi.org/10.1109/NAPS.2017.8107411
9. Aghaei J, Alizadeh MI (2013) Critical peak pricing with load control demand response program in unit commitment problem. IET Gener Transm Distrib 7(7):681–690
10. Rizk-Allah RM, Hassanien AE, Elhoseny M et al. (2019) A new binary salp swarm algorithm: development and application for optimization tasks. Neural Comput Appl 31, pp. 1641–1663. https://doi.org/10.1007/s00521-018-3613-z
11. Faris H, Mafarja MM, Heidari AA, Aljarah, Ala' MA-Z, Mirjalili S, Fujita H (2018) An efficient binary salp swarm algorithm with crossover scheme for feature selection problems. Knowl Based Syst 154:43–67. https://doi.org/10.1016/j.knosys.2018.05.009
12. Lokeshgupta B, Sivasubramani S (2018) Multi-objective dynamic economic and emission dispatch with demand side management. Int J Electr Power Energy Syst 97:334–343

Load Frequency Control for Microgrid Considering Small Hydro and Renewable Energy Sources

Hiramani Shukla and More Raju

Abstract The integration of distributed generation (DG) is very popular with the microgrid, but the challenge is to control frequency when it is integrated with the hydropower plant. Since high penetration of renewable energy sources (RES) may rise to stability issues, frequency deviation or power mismatch issues; hence, it is requisite to balance frequency deviation within the specified limits. This work studies two area microgrids under which one area is considered small hydro and another one is DG. In DG, RES includes solar photovoltaic (PV), aqua-electrolyzer (AE), fuel cell (FC), wind (WTS) and diesel (DEG). Load frequency is controlled under various loading conditions like step load disruption (SLD) in both the areas and random load disruption (RLD) in Area 1. The system is also considered the practical effect due to communication delay. The load frequency is controlled by integral (I) and proportional-integral controller (PI). To obtain the gains of these controllers, the prevalent particle swarm optimization (PSO) technique is used. The results are taken in MATLAB/Simulink platform, which shows that PI controller shows better results compared to I in terms of peak overshoot (PO), peak undershoots (PU) and setting time (ST).

Keywords Load frequency control (LFC) · Renewable energy sources (RES) · Particle swarm optimization (PSO) · Communication delay

1 Introduction

Sometimes it may be a challenge to provide electricity in the hilly and islanded areas where the source of water is abundantly available. In such areas, small hydropower plant plays a vital role for electricity, but it is not worthy to rely on it alone. The

H. Shukla (✉) · M. Raju
Department of Electrical Engineering, Maulana Azad National Institute of Technology Bhopal, Bhopal, India
e-mail: hiramanishukla143@gmail.com

M. Raju
e-mail: rajunitt1@gmail.com

K. Namrata et al. (eds.), *Smart Energy and Advancement in Power Technologies*, Lecture Notes in Electrical Engineering 926,
https://doi.org/10.1007/978-981-19-4971-5_50

addition of RES with a small hydropower plant may provide better dependency on electricity for rural areas. Moreover, it is a challenge to control the frequency when intermittent RES is penetrated into the grid.

Microgrids provide efficient solutions in a remote area with DG since this is reliable to provide electricity in such area [1]. Doolla and Bhatti in [2] talked about LFC for isolated small hydro plants with the reduction in the dump load from input side power control.

In [3], authors have discussed dual-mode multi-functional reconfigurable systems for small hydro (AC) and solar photovoltaic generation (DC). Kumari et al. [4] performed hydro turbine control with the help of a PID controller, which is a non-minimum phase system. To control grid frequency, authors in [5] have talked about grid support capability of variable speed hydropower with HVDC link. Author in [6] has considered DG in microgrid with thermal, wind and solar system for automatic load frequency control, while in [7] three-area system, wind, thermal and hydro are considered. In [8], three units of sources hydro-thermal-gas were considered, but their investigations were limited to a single area only. The works [9, 10] dealt with low-frequency deviation when hydropower plant implemented with wind turbine power while. In [9], the authors have also talked about multiple DGs for microgrid purposes. Bhuyan et al. [11] worked on demand-side management for the hydrothermal system. Authors in [12] have discussed grid frequency control for hydropower plants along with battery energy storage systems. In [13], a hydroelectric system in both areas is studied. The nonlinear active disturbance rejection control (ADRC) method of the LFC system has been made in [14].

It is very important to use a suitable controller along with a better optimization technique to get the proper gains of the controllers. Better the controller, better will be the control over the frequency at the load side, especially when we talk about the combination of a hydropower plant with penetration of RES. In the literature [10, 11, 13, 14] have implemented the particle swarm optimization (PSO) algorithm. From the literature, it is found that the PSO algorithm performs well to handle the frequency oscillation in a better manner [11]. Authors in [13] have implemented the PSO with intelligent fuzzy logic. The modified version of PSO so-called improved PSO considered for studies in [14]. It is seen from the various literature [13, 14] studies that PSO is able to suppress frequency oscillations, and it is performing better than GA and Zeigler-Nichols in terms of system dynamics. The above superior characteristics of the PSO algorithm inspired the authors to consider the PSO technique for the present study to optimize the parameters of discussed controllers. Instead of different inertia weights [15], in the present paper, constant inertia weights are used.

Summarizing the above discussion, the main motive of this paper is to.

(a) Develop the interconnected small hydro and RES for frequency control studies.
(b) Obverse the system dynamics with I and PI controllers.

(c) To obtain the dynamics in the system developed in (a) with the help of controllers considered in (b) and to compare the dynamics among I and PI controllers to decide the best one.
(d) To study the system responses for step and random load disturbances, i.e., SLD and RLD.

2 System Investigated

For the analysis, the two-area isolated microgrid system is considered. Two secondary controllers, namely I and PI controllers, are chosen to analyze the behavior of the system, whose gains are obtained by the PSO. The 1:1 ratio capacity system is assumed for studies. The system is tested under two different disturbance conditions: (a) The SLD of 1% is considered in both the areas of the system and (b) RLD in Area 1 only. The popularly known integral squared error (ISE) method is employed as cost function, J (1) with simulation time T (s).

$$J = \int_{0}^{T} \left(\Delta f_{\text{area-1}}^{2} + \Delta f_{\text{area-2}}^{2} + \Delta P_{\text{tie}}^{2}\right)\mathrm{d}t \tag{1}$$

2.1 Overall System Model

The overall system model under investigation is shown in Fig. 1. Area 1 includes a small hydropower plant, while Area 2 has DG.

2.2 Model of the Distributed Generator (DG)

Distributed generation (DG) is described as the generation of electricity for use on-site or near to consumers instead of transmitting the power over distances. The model of DG is shown in Fig. 2. The system is arranged in such a way that 60%PV power is directly supplying to the power system, and the remaining 40% is fed as input to the AE. In the transfer function of the PV system, K_{PV} and T_{PV} represent the gain and time constants. The aqua-electrolyzer (AE) utilizes a fraction of power (here 40%) from the PV system to make hydrogen (H_2) through water electrolysis using electricity, which acts as input to the FC. The fuel cell (FC) consumes H_2 for production of the electricity. Transfer functions of AE and FC have the terms K_{AE} and K_{FC}, which are the gains and T_{AE} and T_{FC} are time constants of the AE and FC, respectively.

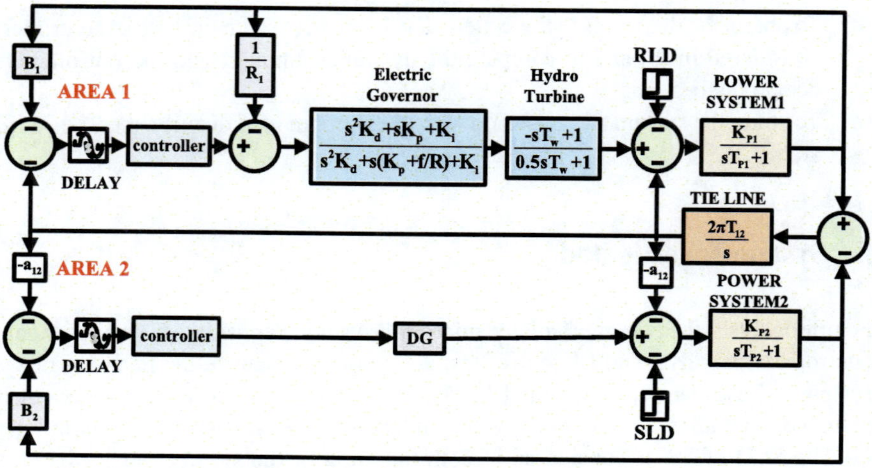

Fig. 1 Block diagram of the overall proposed system

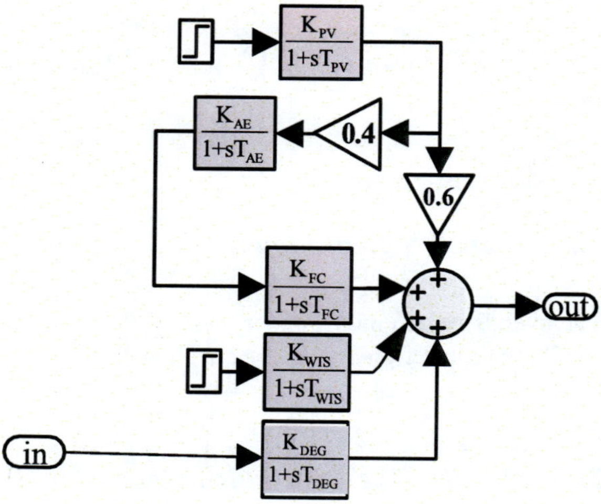

Fig. 2 Block diagram of a model of DG

The wind turbine system (WTS) and diesel engine generator (DEG) are also used in DG. The WTS converts kinetic energy into mechanical which leads to producing electrical energy through an induction generator. The relation between wind power (P_{WP}) and wind velocity (V_W) is $P_{WP} \propto V_W^3$. The first-order transfer function of WTS has K_{WTS} which is the gain, and T_{WTS} is the time constant of the WTS. The DEG uses liquid fuels as the primary fuel, which works based on air compression. The transfer function of DEG contains K_{DEG}, which is the gain, and T_{DEG} is the time constant of the WTS.

2.3 Proposed Controllers

The well-known PI and PID with filter are considered as the secondary controllers in the proposed system, whose equations are given by Eqs. (2, 3) and Fig. 3.

$$G(s)_{\text{I - CONTROLLER}} = \frac{K_I^i}{s} \tag{2}$$

$$G(s)_{\text{PI - CONTROLLER}} = K_P^i + \frac{K_I^i}{s} \tag{3}$$

$K_P{}^i$, $K_I{}^i$ and $K_D{}^i$ are the proportional, integral and derivative gains, and N is the derivative filter.

3 Particle Swarm Optimization (PSO)

The PSO technique was introduced by Kennedy and Eberhart [16] in 1995. The various steps in the PSO are explained by the authors in [11]. Here, for optimization of various SCs parameters, the PSO technique is utilized with the algorithm parameters mentioned below. The flowchart of the PSO technique is given in Fig. 4.

where

$i = 1, \ldots\ldots Z$,

Z	is the total number of particles,
M	is the current iteration number,
c_1 and c_2	are acceleration constants,

Fig. 3 I and PI controllers

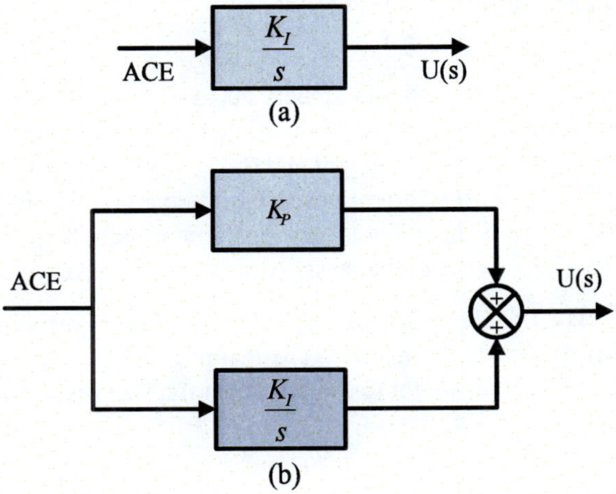

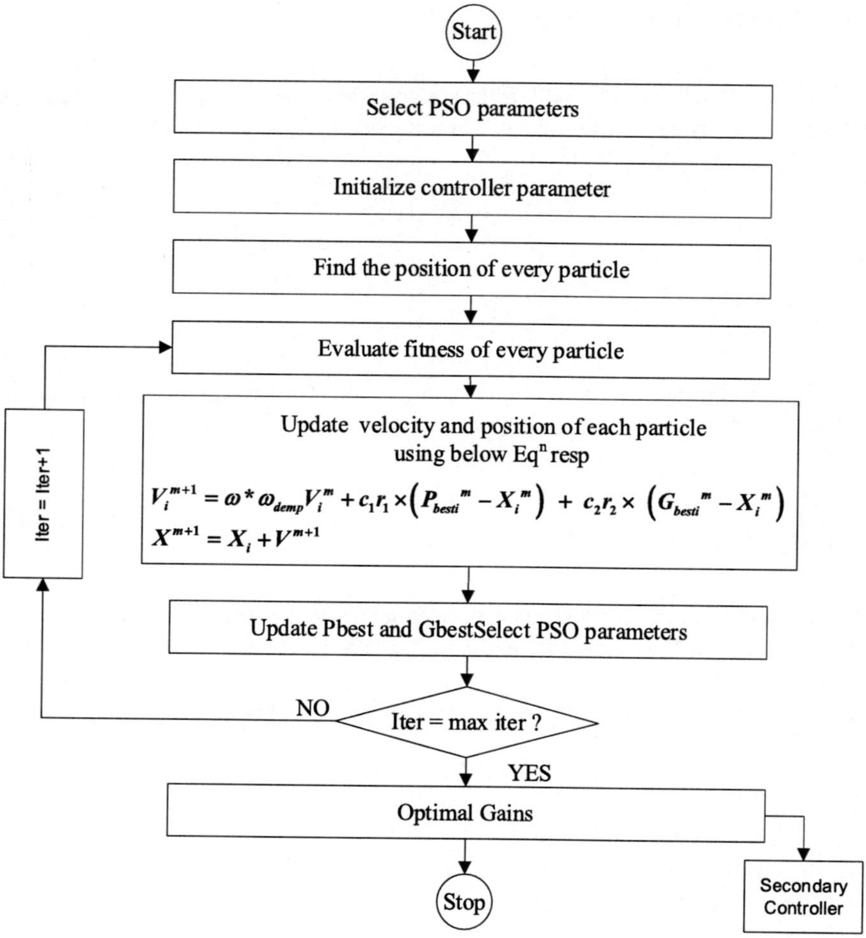

Fig. 4 Flowchart of PSO technique

r_1 and r_2	are random numbers range (0 1),
V_i^m	is the velocity of particle i at *m*th iteration,
(ω)	is inertia weight and (ω_{damp}) is damping factor,
X_i^m	is the current position of the particle i of iteration m,
$P_{\text{best}i}^m$	is the previous best position of particle i,
$G_{\text{best}i}^m$	is the global best position of particles.

The PSO technique is utilized to extract the various controllers' parameters. The PSO parameters considered are: learning rates $c_1 = c_2 = 1.5$, damping factor $(\omega_{\text{damp}}) = 0.7$, inertia weight $(\omega) = 1$, population size $(n) = 10$ and iterations (iter) $= 100$.

4 Results and Discussion

The simulation results are carried out and analyzed in two parts. The first one is with step load disturbance in both the areas considering I as well as PI controllers. Another part is with random load disturbance in the first area only along with I and PI controllers.

4.1 SLD in Both the Areas

See Fig. 5.

Table 1 represents the gains of the controllers obtained by the PSO. The settling time, peak overshoots, peak undershoots of the frequency deviation and tie-line

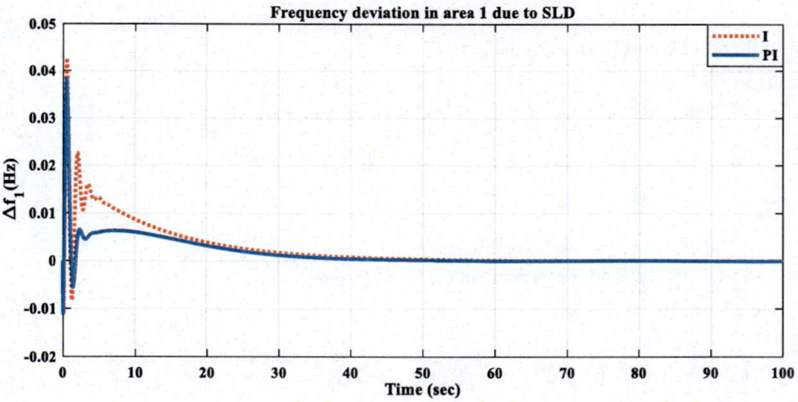

(a)

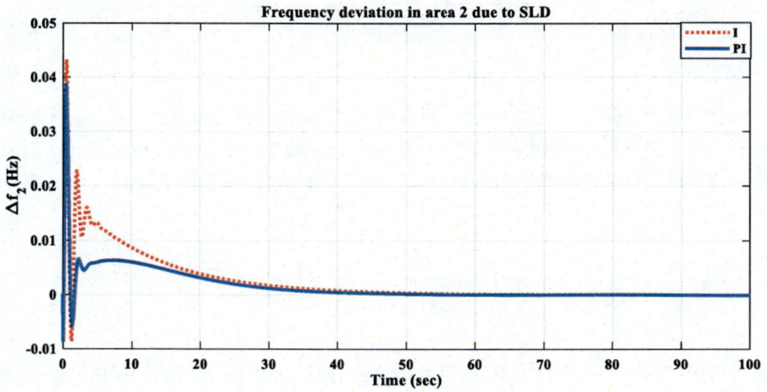

(b)

Fig. 5 Comparison of the dynamic response of the system due to SLD

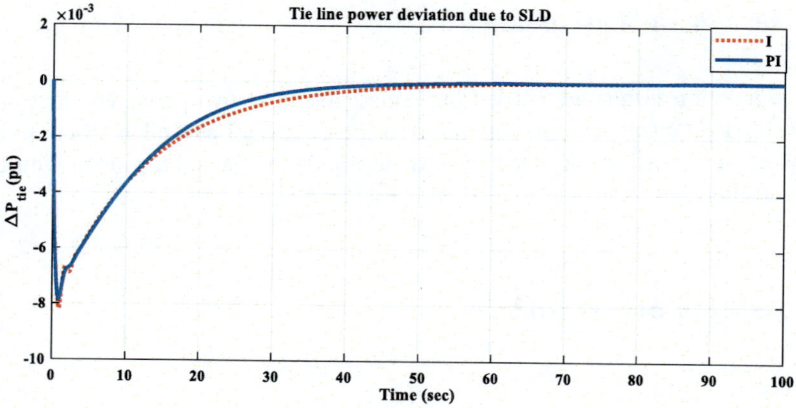

(c)

Fig. 5 (continued)

Table 1 I and PI control variables using PSO

Parameters (PSO)	(I-Controller K_I); (PI-Controller K_P, K_I)
Area 1	0.0089; 0.153,0.0888
Area 2	0.8763; 0.9739,0.6737

Table 2 ST (s), POs and PUs comparison for I and PI controllers

Dynamic response	Δf_1			Δf_2			ΔP_{tie}		
Controller	ST (s)	PO (in mili pu)	PU	ST (s)	PO (in mili pu)	PU	ST (s)	PO	PU (in mili pu)
I	60.2	15.6		61.7	15.5		64.8		7.2
PI	52.1	6.1		51.4	6.2		55		6.6

powers are mentioned in Table 2. The cost functions obtained are 0.005299 and 0.002663 for I and PI, respectively, which is represented by the bar graph in Fig. 6, and the convergence curve for I and PI controller is shown in Fig. 7.

4.2 RLD in Area 1

Figure 8 represents the random load pattern, this load is applied to the system, and the dynamic results obtained are shown in Fig. 9. The results include the comparative performance of I and PI controllers.

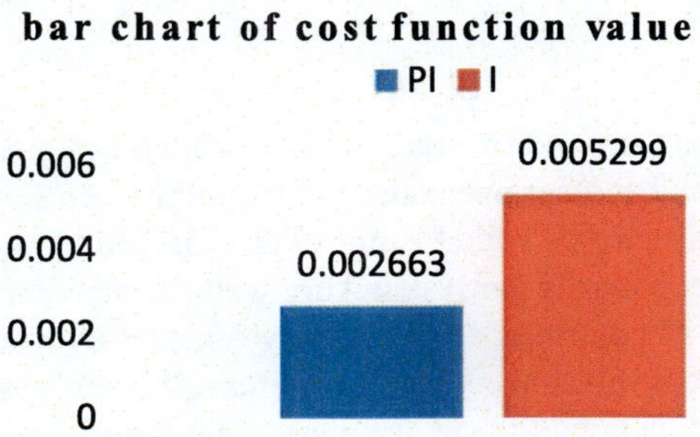

Fig. 6 Cost functions of the controllers

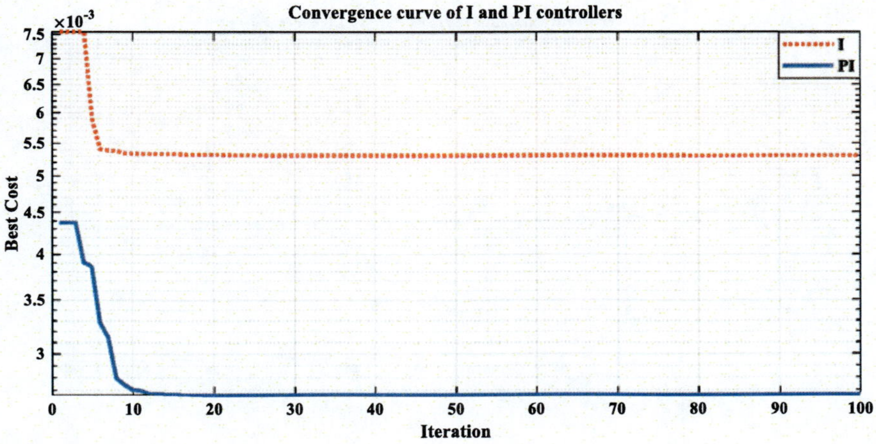

Fig. 7 Convergence curve of the controllers

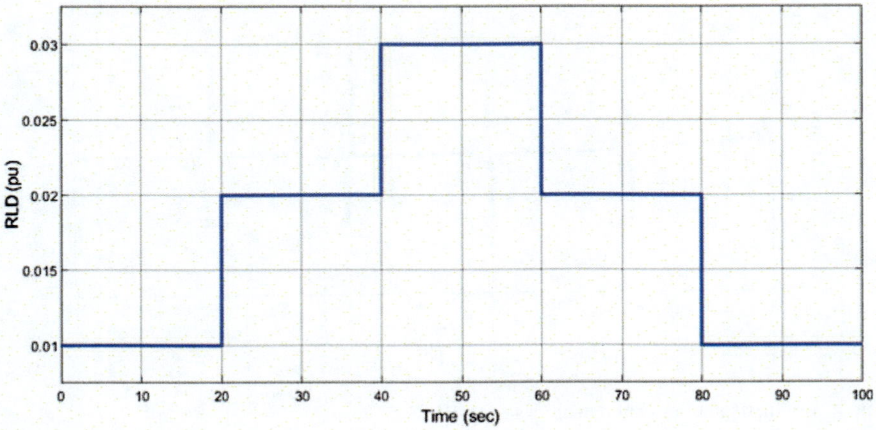

Fig. 8 Random load pattern

5 Conclusion

This paper describes the two-area microgrids in which one area belongs to a small hydropower plant, and another one is the combination of RES, which is termed DG. The whole system is tested by I and PI controllers. The Simulink results clearly show that the PI controller controls the system very well as compared to I. Also, in terms of ST, PO and PU, PI controller is giving very satisfactory results than I. This can also be observed from the convergence curve that PI is the better controller. Since the transfer turbine function of the hydropower plant is non-minimum phase system, this PSO is sufficient to control the frequency deviation.

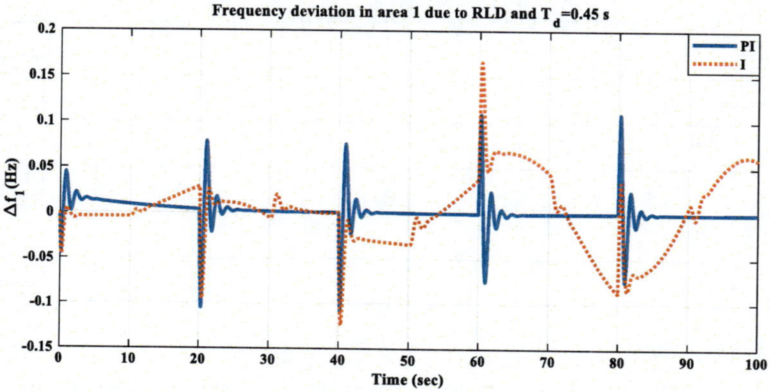

(a)

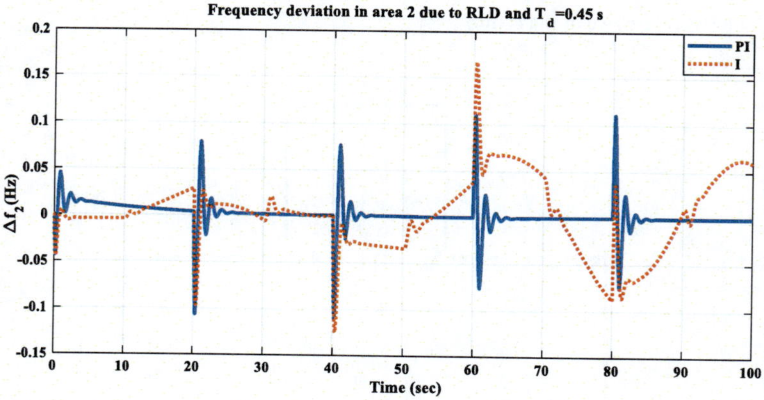

(b)

Fig. 9 Comparison of system dynamics due to RLD

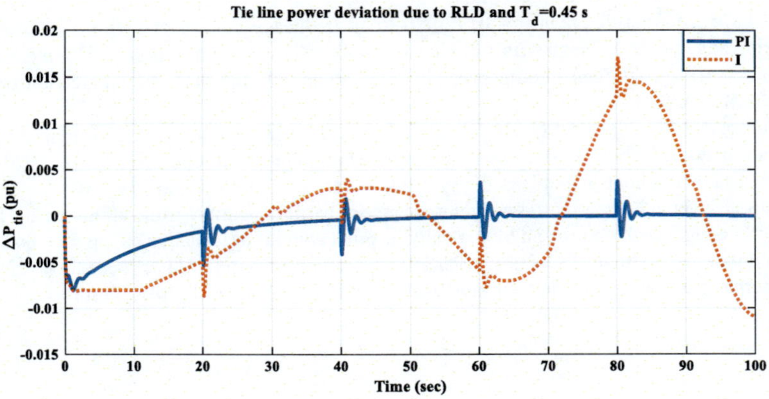

(c)

Fig. 9 (continued)

References

1. Farrokhabadi M, Cañizares CA, Bhattacharya K (2017) Frequency control in isolated/islanded microgrids through voltage regulation. IEEE Trans Smart Grid 8(3):1185–1194
2. Doolla S, Bhatti TS (2016) Load frequency control of an isolated small-hydro power plant with reduced dump load. IEEE Trans Power Syst 21(4):1912–1919
3. Reddy KM, Singh B (2018) Multi-objective control algorithm for small hydro and SPV generation-based dual mode reconfigurable system. IEEE Trans Smart Grid 9(5):4942–4952
4. Kumari R et al (2021) Improved hydroturbine control and future prospects of variable speed hydropower plant. IEEE Trans Ind Appl 57(1):941–952
5. Reigstad TI, Uhlen K (2021) Variable speed hydropower for provision of fast frequency reserves in the nordic grid. IEEE Trans Power Syst. https://doi.org/10.1109/TPWRS.2021.3078098
6. Ray PK, Puhan PS, Priyadarshini P (2021) Automatic load frequency control of distributed generations in a microgrid. In: 2020 3rd international conference on ener-gy, power and environment: towards clean energy technologies, pp 1–6
7. Shaker HK, Zoghby HE, Bahgat ME, Abdel-Ghany AM (2019) Advanced control techniques for an interconnected multi area power system for load frequency control. In: 2019 21st international middle east power systems conference (MEPCON), pp 710–715
8. Morsali J, Esmaeili Z (2020) Proposing a new hybrid model for LFC and AVR loops to improve effectively frequency stability using coordinative CPSS. In: 2020 28th Iranian conference on electrical engineering (ICEE), pp 1–7
9. Zhang J, Shao X, Li Y, Lin J, Li F, Zhang Z (2020) Research on frequency regulation strategy based on model predictive control for wind-hydro-storage complementary microgrid. In: 2020 4th international conference on HVDC (HVDC), pp 1031–1036
10. Niasse M, Zheng Q, Xin A, Quan FAF, Simiyu P (2020) Mitigating ultra-low-frequency oscillations in wind-penetrated hydro-dominant power grid through virtual inertia emulation strategy. In: 2020 4th international conference on power and energy engineering (ICPEE), pp 130–137
11. Bhuyan S, nee Dey SH, Paul S, Chaine S (2021) Analysis of frequency regulation for a hydro-thermal system with ALFC-DR model. In: 2021 1st international conference on power electronics and energy (ICPEE), pp 1–5
12. Mäkinen T, Leinonen A, Ovaskainen M (2020) Modelling and benefits of combined operation of hydropower unit and battery energy storage system on grid primary frequency control. In: 2020 IEEE international conference on environment and electrical engineering and 2020 IEEE industrial and commercial power systems Europe (EEEIC/I&CPS Europe), pp 1–6

13. Joshi M, Sharma G, Davidson IE (2020) Load frequency control of hydro electric system using application of fuzzy with particle swarm optimization algorithm. In: 2020 international conference on artificial intelligence, big data, computing and data communication systems (ICABCD), pp 1–6
14. Huang Z, Chen Z, Sun M, Sun Q (2020) Parameter optimization of load frequency active disturbance rejection control based on improved particle swarm optimization. In: 2020 IEEE 9th data driven control and learning systems conference (DDCLS), pp 331–336
15. Bansal JC et al (2011) Inertia weight strategies in particle swarm optimization. In: 2011 third world congress on nature and biologically inspired computing, Salamanca, pp 633–640
16. Kennedy J, Eberhart RC (1995) Particle swarm optimization. In: Proceedings of IEEE international conference on neutral networks, pp 1942–1948. IEEE Press, Perth

Current Status and Future Potential of Solar Energy Utilization in Rajasthan, India

Debajit Misra

Abstract Rajasthan requires a huge amount of power due to its own demand and to supply its nearby areas. High solar radiation and plenty of unoccupied land make the state in a position to run a variety of solar power plants and equipment. This paper describes the main ways of generating solar electricity in the state, like using solar photovoltaic (PV), concentrated solar power (CSP). The paper also highlights different solar schemes, and large numbers of solar projects have already been installed at different locations. Currently, Rajasthan possesses the world's largest solar park of 2245 MW capacity, named Bhadla Solar Park. Presently, total installed power capacity of Rajasthan is about 22.15 GW, while its total installed solar power capacity stands at 5.3 GW. At present, its share is about 15% of the total Indian solar installation. The state has already developed transmission and distribution network to supply power in the desert area. A strong power evacuation system is being developed, which could transmit 8000 MW of renewable energy to the state and national grid with the help of the Indian Solar Mission, the Asian Development Bank and the Clean Technology Fund. The state is involved in an aggregate solar power potential of 142.31 GW. Thus, in future, Rajasthan could be in a notable position towards solar energy deployment in India. The paper illustrates the current scenario, future potential and different routes of solar thermal energy utilization in Rajasthan.

Keywords Solar energy · Solar plant · Solar park · CSP plant · Rajasthan

1 Introduction

Renewable energy is becoming a more familiar part of the creation of a clean and green world. Among all renewable energy sources, solar energy is more abundant, environment friendly and the most reliable for long-term use [1–3]. There are so many ways to use this energy; it can be captured and converted to useful energy using photovoltaics (PV) or solar thermal collectors for clean energy production.

D. Misra (✉)
Department of Mechanical Engineering, Techno India Group, Kolkata 700091, India
e-mail: dmbesu@gmail.com

K. Namrata et al. (eds.), *Smart Energy and Advancement in Power Technologies*,
Lecture Notes in Electrical Engineering 926,
https://doi.org/10.1007/978-981-19-4971-5_51

Nowadays, these two technologies are extensively used all over the world for large-scale power generation. Besides power generation, solar energy can be used for other thermal projects like heating, cooling and ventilation [4–6]. Thus, solar energy technology happens to be a mature and promising option in the coming future than the other renewable energy technologies. Rajasthan is endowed with the maximum solar radiation intensity of 5.72 kWh/m^2/day and in a year 300–325 clear sunny days with a very low average rainfall [7, 8]. So its abundance is much more than the other Indian states. This state is the largest, hottest and driest Indian state by area (342,239 Km^2) which comprises 60% desert. The hottest desert means more free space for utilization of sunlight into useful energy. In Rajasthan, thermal power sources were the major options, despite the state's abundance solar radiation and waste land for renewable energy generation [9, 10]. The state often faces challenges to run traditional thermal power plants as there are less numbers of water sources and fossil fuels are brought from outside. Again, the state often imported electricity from other states when there was a deficit of electricity during the summer. The state is presently focusing on self-reliance power and launching various renewable energy projects. In order to minimize the above-mentioned issues, the promoting of the energy generation through wind and solar has been rapidly progressed in the state in the last few years. Though, previously the tapping of wind energy was the leading renewable energy in the state, recently, the solar energy is leading the chart in case of renewable energy generation.

2 Current Indian Scenario

At present, India's vast untapped renewable energy potential is being largely explored due to rising power demand and reducing consumption of fossil fuel. Thus, power generation by renewable sources is becoming an attractive option in India. India receives 4–7 kWh/m^2/day solar incident and 5000 trillion kWh/year energy in the most of the part of the country [11]. In 2015, the Government of India announced a target to generate 175 GW of renewable energy by 2022 [12, 13]. This target includes 100 GW of solar power, as country possesses abundance of solar energy. This 100 GW solar power comprises of 40 GW decentralized rooftop projects, 40 GW utility-scale solar plants and 20 GW ultra-mega solar parks. India's solar installation has been growing rapidly since 2014. In March, 2014 India's cumulative solar installation was just 2.6 GW [14, 15]. India crossed 20 GW in January 2018 and reached 30 GW in May 2019. In March, 2021 the country's renewable installation was 94.42GW, which is 24.7% of country's cumulative installation of all kinds of power generation. Of the 94.42GW renewable installations, 35.64 GW comes from solar, 39.24 GW from wind energy and additional 19.54 GW from other renewable energies. Solar installations were severely hampered in 2020 due to the pandemic and subsequent lockdowns.

According to the Central Electricity Authority (CEA), in FY 2020–21 solar energy generation was 59.63 terawatt hours (TWh) [16]. In between FY 2014–15 and FY

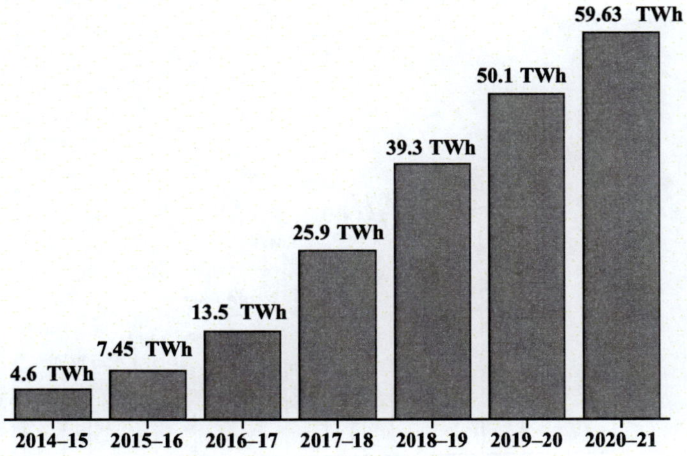

Fig. 1 Solar energy generation in India for the years [11, 14]

2020–21, in past seven years, the generation has been increasing rapidly with a compound annual growth rate (CAGR) of 46.6%. In 2017–18, the country first time made a historical change in terms of generation of solar energy (Fig. 1).

Presently, India's major solar plants are located in ten states of the country. Figure 2 shows the major locations and their installed capacity based on March 31, 2021 data [11]. Karnataka is on top in terms of solar installation, and its share is almost 20% to the total installation. Rajasthan is now in second in solar installation and shares 15% of the total installation. Other major states are Tamil Nadu, Andhra Pradesh, Telangana, Gujarat, etc.

3 Solar Energy Potential in Rajasthan

Rajasthan is located in a geographical and environmental advantageous position to deploy more solar projects. Among the Indian states, it has the highest solar irradiation and the state has vast tracts of unused, barren and affordable land. Thar Desert receives very high solar radiation, ranging from 5.85 to 6.44 kWh/m^2/day [7]. In this desert, yearly sunshine remains 345–355 days and rainfall is a scarce thing, average rainy days remains 10.4–20.5 days/year. The desert districts of the state like Jaisalmer, Bikaner, Barmer, Jodhpur, etc. are the prime hotspots to deploy wind mills and solar plants. Jodhpur, Jaisalmer and Bikaner are known as the desert triangle, and Jodhpur is named as Sun City as it is receiving maximum solar radiation. These determinant factors could be suitable for harnessing solar energy for a number of applications. To evacuate power from the solar plants, strong transmission network having capacities of 400 kV, 220 kV and 132 kV has already been developed in these locations [17].

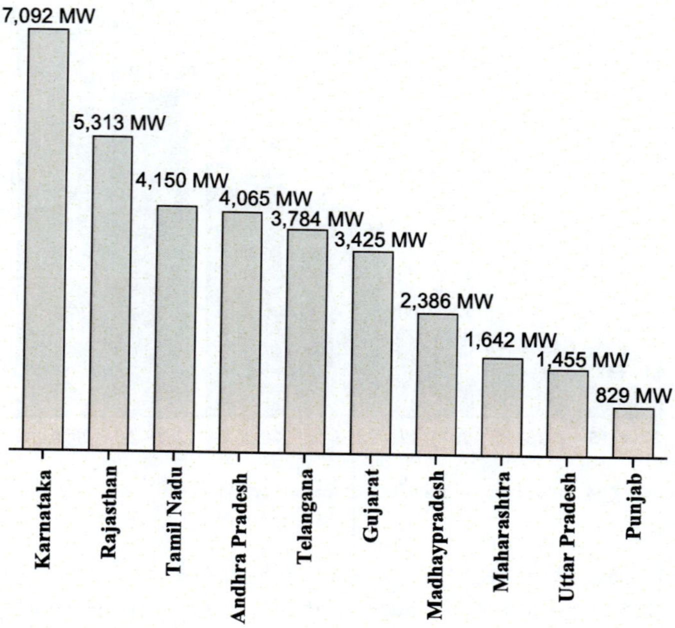

Fig. 2 Solar installation in the major Indian states [11]

The state is already providing incentive-based feed-in-tariff and inviting developers to establish PV-based projects. The firm policy and initiative of the state government have attracted attention of the manufacturers for the development of solar equipment. Many of the companies are involved in manufacturing and supplying solar modules with the effective range of varieties. Thus, the solar industry can become an economic engine to create jobs across the state and empower connected families in remote areas as well as generate billions of economic growth and tax revenue.

4 Current Scenario in Rajasthan

In Rajasthan, solar installation is high in the western parts like Barmer, Bikaner, Jaisalmer, Jodhpur owing to high solar irradiation. These parts already been connected with 400 kV network through Rajasthan Renewable Energy Transmission Investment Program (RRETIP). Recently, the Government of India, the Asian Development Bank (ADB) and the Clean Technology Fund are involving to develop the transmitting network which could transmit the maximum 8000 MW of renewable energy to the state as well as to the national grid with an estimated cost of Rs. 4813.89 Crore [16, 18]. Two transmission lines already have been established (i) 400 kV D/C Barmer-Bhinmal (PGCIL) line; (ii) 400 kV D/C Jaisalmer-2-Barmer line. In 2008, the Rajasthan Government first time made an initiative to develop

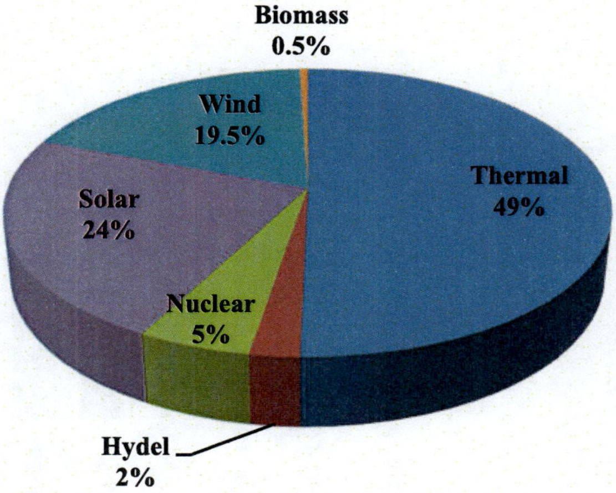

Fig. 3 Percentages of different types of power generation in Rajasthan [20, 21]

two solar projects having 5 MW of each under the Generation-Based Incentive scheme. Again, the Rajasthan Electricity Regulatory Commission, for the first time in India, imposed solar Renewable Procurement Obligation (RPO) for Distribution Companies (DISCOMs) in Rajasthan.

At present, the state comprises well-diversified power generating system. According to the report made by Solar Energy Corporation of India Ltd (SECI) in 2020 about 5 GW solar projects are under developing conditions in the state [19]. Following figure shows the split of generation capacities from different power generation sources up to March 2021. Thermal power generation, including gas turbine power plant, stands 49%. Solar power generation is highest in all other form of renewable energy generation, which is about 24% when total renewable energy generation is more than 43% (Fig. 3).

At present, total solar installation in Rajasthan up to March 2021 is 5313 MW on which grid-connected solar projects are 3226.79 MW [20, 21]. Figure 4 shows the capacities of solar installation for last ten years. It is clear that the growth rate was slow for first five years; then from 2016 to 17, momentum was started. The state's solar power installed capacity has gained pace over the past five years. In the fiscal year, 2019–20 more than 1800 MW solar power capacity was installed than the previous fiscal year (2018–19), which was the highest in the all-time in Rajasthan.

4.1 Solar Park in Rajasthan

Rajasthan Solar Park Development Company Limited (RSDCL) a state government undertaking company involving to set up solar park. Main activities of RSDCL are production, collection and distribution of electricity to the power grid. SECI and National Thermal Power Corporation (NTPC) are also engaged in the development

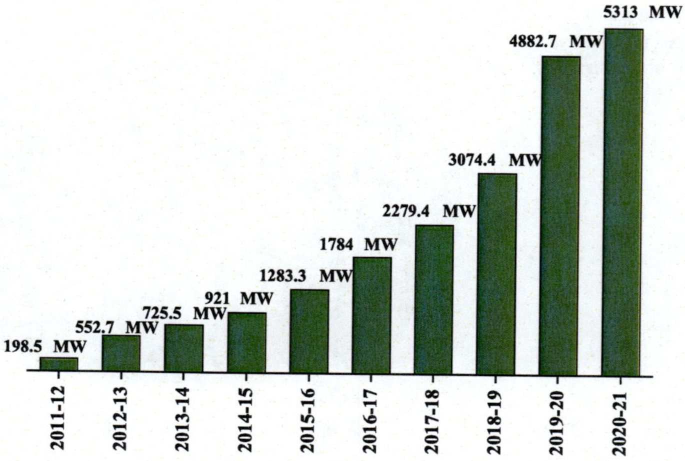

Fig. 4 Year-wise solar installation in Rajasthan [17, 20]

of solar park in the state. The PV technology has been utilized to establish solar park in the state.

India's first largest solar park was commissioned in Rajasthan, named Bhadla Solar Park. Bhadla Solar Park comprises phases I-IV having a total solar power generation capacity of 2245 MW. It is now the largest in the world, which is spreading over an area of 14,000 acres [22]. This solar park is significant in many aspects. It was completed at record-low tariffs (Rs. 4.34/kWh) and developed by more than 16 agencies through public–private partnership. Total estimated cost of the whole project is Rs. 985 crore [23]. Table 1 shows the different phases of the projects showing their capacities and developers.

Other three solar parks, naming as Phalodi-Pokaran (750 MW), Fatehgarh phase 1B (1500 MW) and Nokh (1000 MW), which are shown in Table 2, are in developing conditions [24, 25].

Table 1 Different phases of Bhadla Solar Park [23, 24]

Phase-I: (65 MW) Developer: Rajasthan Solar Park Development Company Ltd.	**Phase-II**: (680 MW) Developer: Rajasthan Solar Park Development Company Ltd.
Phase-III: (1000 MW) Developer: Surya Urja Company of Rajasthan Ltd.	**Phase-IV**: (500 MW) Developer: Adani Renewable Energy Park Rajasthan Ltd.

Table 2 Three still-developing solar parks [24, 25]

Phalodi-Pokaran Solar Park: 750 MW Developer: M/s Essel Surya Urja Company of Rajasthan Ltd Location: Ugraas 450 MW (Jodhpur) & Lavan&Purohitsar 350 MW (Jaisalmer)	Fatehgarh Phase-1B Solar Park: 421 MW through support of GoI out of 1500 MW Developer: Adani Renewable Energy Park Rajasthan Limited Location: Fatehgarh&Pokaran (Jaisalmer)	Nokh Solar Park: 980 MW Developer: Rajasthan Solar Park Development Company Ltd Location: Nokh, Jaisalmer

4.2 CSP Plant in Rajasthan

Concentrated Solar Power (CSP) plants began in Rajasthan to harness energy mainly from the desert. In a CSP plant, solar energy can be concentrated using by equipment to store heat that can be used to generate steam and eventually drive turbine to generate electricity. In 2011, country's first 2.5 MW CSP plant was commissioned at Bikaner, Rajasthan [26]. It is now under running condition. It was made of solar power tower technology, and in future, it will be scaled up to 10 MW. Presently, total 428.5 MW CSP plants are in running and construction phase in Rajasthan, which is more than 75% of the total CSP capacity in India. Now 177.5 MW plants are in operational phase [26]. India's first largest CSP plant of 125 MW capacity was developed at Dhusar in Rajasthan and commissioned under the government's National Solar Mission. It was made by the linear fresnel technology. The main benefit of that technology is that it requires less area to operate than other technologies for the same power output. Table 3 shows the CSP plants in all around Rajasthan showing their capacities, developers, locations, coverage areas and tariff rates.

4.3 Rooftop Solar Plant in Rajasthan

The Rajasthan state government has set a target of 1000 MW rooftop solar installation program by 2024–25, and it has reached 365.8 MW [29]. It is used mainly in commercial, domestic and government buildings. The Rajasthan Electrical Regulatory Commission (RERC) launches net metering facility for rooftops and small grid-connected solar systems, and under this system consumer can use the cheaper rooftop solar power instead of the costly DISCOM power. Currently in the state, many academic buildings, especially engineering institutes and universities, are also connected to the rooftop solar panels to generate their own energy. Table 4 shows the institutes that are generating solar power. In recent times, IIT Jodhpur initiated a plant to install 3 MW rooftop solar power plant, and relevant tender has been issued for commissioning the project [30]. A field survey has been carried out in Rajasthan

Table 3 The details of CSP plants in Rajasthan [26–28]

Project name and technology	Capacity (MW)	Developer	Location and coverage area (ha)	PPA (Rs./KWh)
(i) Dhusar (linear fresnel reflector)	125	Rajasthan Sun Technique Energy	Dhursar, 340	11.97
(ii) Godawari solar project (parabolic trough)	50	Godawari Green Energy Ltd.	Nokh, 150	12.2
(iii) ACME solar tower (solar power tower)	2.5	ACME Group eSolar Group	Bikaner, 4.85	13.45
(iv) Diwakar (parabolic trough)	100	Lanco Solar	Askandra	10.5
(v) KVK energy solar project (parabolic trough)	100	KVK Energy Ventures Ltd.	Askandra	11.2
(vi) Abhijit solar project (parabolic trough)	50	Corporate Ispat Alloys Ltd.	Phalodi, 157	12.24
(vii) Rajasthan solar one (parabolic trough)	1	Brahma Kumaris & World renewal spiritual trust	Mt Abu, 10.11	

and presented in Table 4 to show major rooftop solar projects that have already been installed in various institutes in the state.

Table 4 Major rooftop solar in various institutes in Rajasthan

S. No.	Institute	Capacity (KW)
(i)	Manipal University, Jaipur	800
(ii)	Arya Group of Colleges, Jaipur	500
(iii)	Malaviya National Institute of Technology,Jaipur	407
(iv)	Swami Keshvanand Institute of Technology Management, Jaipur	400
(v)	Sri Karan Narendra Agriculture University, Jobner- Jaipur	350
(vi)	Rajasthan Technical University, Kota	300
(vii)	Mamta Handicraft, Jaipur	180
(viii)	Maharaja Sawai Man Singh Vidyalaya, Jaipur	120
(ix)	Suresh Gyan Vihar University, Jaipur	110
(x)	Yagyavalkya Institute of Technology, Jaipur	100
(xi)	Poornima University, Jaipur	100

4.4 *Decentralized Off-Grid Solar Application in Rajasthan*

Another way of solar generation is decentralized and off-grid solar applications. The state government also considers incentives for promotion of decentralized and off-grid solar applications. The major applications are off-grid photovoltaic applications, which include solar PV home lighting, solar PV pumps, solar lantern and solar cooker. According to Energy Statistic Report 2019, India, total 48,175 numbers of off-grid solar pump, 6,852 numbers of street lighting systems, 187,968 numbers of home lighting systems, 225,851 numbers of solar lanterns have been installed in Rajasthan up to 31.03.2019 [31]. In that time off-grid solar plant was 30.34 MW.

In 2010–11, the state started solar pump along with microirrigation systems for farmers by giving solar pump subsidy. Nowadays, some farm-based PV powered system has been developed in the state. As an example, one 1 MW farm-based solar plant developed at Kotputli Bhaloji village, Jaipur, which could produce 17 lakh unit in a year [32].

The Rajasthan state government has strongly supported the utilization of solar thermal technology. A mandatory rule has been introduced in this state for utilizing solar water heating systems for new buildings if they require hot water and if plot area is equal to 500 m^2 or more [33]. This rule has been mandated in new hotels, industrial buildings, hospitals, swimming pools, sports complexes, hostels, barracks, public buildings and residential buildings, etc. [34]. An indirect subsidy scheme also has been introduced and granted for residential consumers in utilization of solar water heater. Now solar thermal energy is being also used in institutional cooking in Rajasthan. Brahma Kumaris World Spiritual University (BKWSU) developed parabolic community solar cooker using concentrated solar thermal power technology. It spreads over 25-acre area with 770 parabolic reflective dishes, each one with an area of 60 sq. meters [35]. It is the largest solar cooker in India with a capacity of 1 MW, generates steam with boiling water and can cook food for 3,800 people. A solar steam cooking system is using at Gargi hostel in Malaviya National Institute of Technology, Jaipur, for cooking 600 students' food [36].

5 Current Project Proposals

The Indian government has allocated a 25,000 MW Ultra Mega Renewable Energy Park in Rajasthan. The state government has identified 125,000 hectares of land in three districts Bikaner, Jaisalmer and Jodhpur for the park [19, 37–40].

Some of the project proposals are listed below:

(i) The Solar Energy Corporation of India (SECI) has invited proposals to set up 1,785 MW grid connected PV projects (Tranche-IV).
(ii) Grid-connected solar PV of 1,070 MW capacity (Tranche-III) has been commissioned for the state.

(iii) A tender was released for grid interactive solar rooftop project of 8 MW at various educational and training institutes in Rajasthan.
(iv) A project proposal with 2 MW rooftop solar has been released for Udaipur Smart City Limited.
(v) Jaipur Metro Rail Corporation received a 1 MW rooftop solar project.
(vi) Farm-based solar projects of 722 MW capacity have been commissioned on 623 farms in the state.
(vii) A tender has been released for installation and maintenance of 950 KW grid-connected rooftop solar plants at various buildings of Rajasthan University.
(viii) Rays Experts has successfully commissioned six solar projects of 600 megawatt (MW) capacity in Bikaner and Jodhpur.
(ix) ReNew Power developer has been launched 110 MW PV generation project in Jaisalmer.
(x) SECI has opened bids for developing 10 MW capacity grid-connected solar PV in Jaipur.
(xi) The tender for 3 MW roof PV solar installations has been issued for power supply at IIT Jodhpur.

6 Current Policies

In the budget of 2021–22, Indian government has decided to make Green Energy Corridor and allocated Rs. 300 crore for this purpose. Again, in this budget Rs. 5,753 crore has been allocated for the Ministry of New and Renewable Energy (MNRE) department [41, 42]. An additional fund of Rs. 1,000 crore has been allocated to SECI and Rs. 1,500 crore to Indian Renewable Energy Development Agency (IREDA). Presently, the central government has risen custom duties both for solar inverters and solar lanterns so that it can stay within the state. The central government has declared production-linked incentive (PLI) scheme providing Rs. 4,500 crore for five years in the manufacturing units to increase PV module efficacy [43]. Again, the plan of adding 30 GW of renewable energy in the desert areas of Rajasthan and Gujarat is initiated by the Indian government [44]. To boost the growth of renewable energy sectors, custom and excise duty benefits have been modified on rooftop solar projects, which could lower the installation and power generation cost. The central government has also unveiled a new scheme for rural farm development with the name "Prime Minister's Kisan Urja Suraksha Evum Utthan Mahabhiyan" (KUSUM).

On the other hand, in order to encourage the utilization of solar energy across the state, the Rajasthan government launched the new Rajasthan Solar Energy Policy, 2019, providing various facilities for developers. The goal of that policy was to attain 30,000 MW of solar power projects by 2025. Table 5 shows the distribution of solar projects.

According to the new policy, a maximum 1 MW solar rooftop systems can be set up under the net metering regulations and a maximum 50% surplus power can be supplied to a grid. Rajasthan state comprises 33 districts, and the state government

Table 5 Rajasthan's solar policy target by 2024–25 [41]

S. No.	Items	Capacity (MW)
(i)	Utility or grid-scale solar parks	24,000
(ii)	Distributed generation	4000
(iii)	Solar rooftop	1000
(iv)	Solar pumps	1000

is targeting to establish district's headquarters as "Green Energy Cities" by 300 MW rooftop solar installation by 2025. The state has also announced its Wind and Hybrid Energy Policy, 2019, targeting 3.5 GW of hybrid projects by 2024–25, where wind-solar hybrid projects will account for 2.7 GW.

The state will also promote setting up of decentralized solar projects. The state sets a goal to install 0.5 MW to 3 MW stand-alone solar projects near 33 kV gird sub-station for selling power to DISCOMs. Again, the state will promote solar power electric vehicle (EV) charging stations by the state or central public sector undertakings, private operators, or under the public–private partnership (PPP) models.

In 33rd meeting of the State Empowered Committee (SEC), private investors have shown their interest to invest in solar projects [45]. JSW group proposed to invest Rs. 48,500 crore in development of 10,000 MW solar projects. Adani group proposed Rs. 46,000 crore investment in 9700 MW solar and hybrid projects. ReNew Power, a solar and wind power developer, placed a proposal for establishing 10,000 MW solar projects with a Rs. 30,000 crore investment. Greenko group showed interest in investing Rs. 30,000 crore for setting up 4500 MW hybrid wind-solar projects. NTPC Ltd also taken lots of initiatives in the state for solar energy-based power generation.

7 Conclusions and Remarks

The present study shows how Rajasthan is expanding solar projects in the state with a strong determination. Both state and central governments are targeting desert land for utilization of free land and deployment of more solar projects. It is a great idea to use the land of Thar desert for installing more solar projects. However, only at certain regions solar plant can be installed where there is stable land, less sand storms and transmission line has already been developed. Lots of initiatives have been undertaken for infrastructure development in the desert regions to establish more solar projects. Presently, much more attention has been given on power evacuation and transmission system and 8000 MW of renewable power transmission system is almost ready for operation. The state is now in second largest position in terms of solar installation capacity in India. Presently, the state is satisfying India's renewable energy needs. However, a few years ago, the state was lagging behind in power generation and imported power from other states. The government solar policy 2019 could attract manufacturers and developers, leading to the development of the solar energy as well as employment opportunities for young generation. Many companies already been set up in the state to manufacture solar devices and equipment particularly in locations like Jaipur, Jodhpur, Udaipur, etc. Presently, more than seven hundred companies/agencies are involved from India and abroad in the development of solar projects in the state. It is found that educational institutions in Rajasthan have spread considerable awareness about solar energy utilization, which could encourage people to use solar equipment and accessories. Presently, the state provides facilities for setting up solar power projects on private land. The land acquisition and land conversion rule is very convenient for setting up solar plants. As the state has immense potential in the field of solar energy, the government is working towards providing green electricity to every village in the state. Solar energy is now in the leading position compared to wind energy towards clean energy generation in the state. In the last few years, the expense of solar energy has come down drastically, and it can be more affordable for the common people. Now, about 5 GW solar projects are under developing condition and some projects are in pipe line in the state. The state government set a target of 30 GW by 2024–25. The new investments have also arrived with new innovations in the sector. The government's strong policies and developers' interest towards greater investment indicate that the solar power generation in Rajasthan is likely to be in the leading position in the near future, beating the traditional power sources.

References

1. Misra D (2020) Floating photovoltaic plant in India: current status and future prospect. In: International conference on thermal engineering and management advances 2020, LNME. Springer, Singapore, pp 219–232
2. Misra D (2019) Design of a stand-alone rooftop PV system for electrification of an academic building. Int J Eng Adv Technol 9(2):3955–3964
3. Ganguly A, Misra D, Ghosh S (2010) Modeling and analysis of solar photovoltaic-electrolyzer-fuel cell hybrid power system integrated with a floriculture greenhouse. Energy Build 42(11):2036–2043
4. Misra D (2021) An experimental study on a portable SPV-integrated forced convective solar dryer. In: Advances in renewable energy and sustainable environment. Springer, Singapore, pp 233–244
5. Misra D, Ghosh S (2018) Evaporative cooling technologies for greenhouses: a comprehensive review. Agric Eng Int CIGR J 20(1):1–15
6. Misra D, Ghosh S (2019) Thermal modelling and performance assessment of a circular greenhouse with solar chimney assisted ventilation and fog Cooling. Agric Eng Int CIGR J 20(4):108–118
7. Singh H, Singh AK, Chaurasia PBL, Singh A (2005) Solar energy utilization: a key to employment generation in the Indian Thar Desert. Int J Sustain Energ 24(3):129–142
8. Pandey S, Singh VS, Gangwar NP, Vijayvergia MM, Prakash C, Pandey DN (2012) Determinants of success for promoting solar energy in Rajasthan, India. Renew Sustain Energy Rev 16(6):3593–3598
9. Meena RS, Sharma D, Rathore R (2014) The most promising solar hot spots in India development and policy: the Thar Desert of Rajasthan. Int J Eng Dev Res 3(1):74–79
10. Sukhatme SP, Nayak JK (1997) Solar energy in western Rajasthan. Curr Sci 62–68

11. Solar Energy, Ministry of New and Renewable Energy. Available at https://mnre.gov.in/solar/currentstatus/#:~:text=India%20is%20endowed%20with%20vast,providing%20huge%20scalability%20in%20India
12. Dawn S, Tiwari PK, Goswami AK, Mishra MK (2016) Recent developments of solar energy in India: perspectives, strategies and future goals. Renew Sustain Energy Rev 62:215–235
13. Report of the expert group on 175 GW RE by 2022. https://niti.gov.in/writereaddata/files/writereaddata/files/document_publication/report-175-GW-RE.pdf
14. Energy Statistics 2020. http://mospi.nic.in/sites/default/files/publication_reports/ES_2020_240420m.pdf
15. India becoming a global force in wind energy and solar power. https://www.evwind.es/2020/08/30/india-becoming-a-global-force-in-wind-energy-and-solar-power/76888
16. Central Electricity Authority of India. https://cea.nic.in/?lang=en
17. Rajasthan Rajya Vidyut Prasaran Nigam Ltd. https://energy.rajasthan.gov.in/content/dam/raj/energy/corporate-one-lines-vieer/pdf/Publications/T&C%20Manuals/Construction%20Manual%20For%20Lines1.pdf
18. Rajasthan Renewable Energy Transmission Investment Program. https://www.adb.org/sites/default/files/project-documents/45224/45224-002-dpta-en.pdf
19. Solar Energy Corporation of India Limited. https://www.seci.co.in/archives/data_archives
20. Rajasthan Renewables. https://www.renewablesindia.in/?region=29&selection_type=state
21. Rajasthan Renewable Energy Corporation Limited. https://energy.rajasthan.gov.in
22. Mercom India. https://mercomindia.com/world-largest-solar-park-bhadla/.
23. NS Energy. https://www.nsenergybusiness.com/projects/bhadla-solar-park-rajasthan/
24. Solar Parks. https://mnre.gov.in/img/documents/uploads/bcf7e95e88ae4f8dbfa8bd25d21e5e12.pdf
25. Government of India Ministry of New and Renewable Energy Lok Sabha Unstarred Question no-1538. http://164.100.24.220/loksabhaquestions/annex/15/AU1538.pdf
26. Concentrating solar power projects in India. https://solarpaces.nrel.gov/by-country/IN
27. CSP Focus. http://www.cspfocus.cn/en/study/detail_84.htm
28. India One Solar Thermal Power Plant. https://india-one.net/
29. Rajasthan to set up 30,000 MW solar power plants by 2024–25. https://www.eqmagpro.com/rajasthan-to-set-up-30000-mw-solar-power-plants-by-2024-25/
30. NTPC Subsidiary Tenders for 3 MW Rooftop Solar Plant at IIT Jodhpur. https://www.saurenergy.com/solar-energy-news/ntpc-subsidiary-tenders-for-3-mw-rooftop-solar-plant-at-iit-jodhpur
31. Energy Statistics 2019. http://mospi.nic.in/publication/energy-statistics-2019
32. First farm-based solar power plant comes up in Rajasthan. https://www.thehindu.com/news/national/other-states/first-farm-based-solar-power-plant-comes-up-in-rajasthan/article34226558.ece
33. Mandatory use of Solar Water Heating Systems in Rajasthan. https://www.iea.org/policies/321-mandatory-use-of-solar-water-heating-systems-in-rajasthan
34. Rajasthan Energy Conservation Building Rules & Code- 2018. https://energy.rajasthan.gov.in/content/dam/raj/energy/rrecl/pdf/Activities/EC/ECBC_RULES_codes_2018_%20pdf.pdf
35. Brahma Kumaris. http://www.brahmakumaris.org/hope/renewable-energy/india-one-solar-thermal-power-plant
36. Sun Focus, UNDP–GEF CSH Project. https://www.in.undp.org
37. Rajasthan sets a goal of 30,000 MW solar capacity in 5 years. https://www.sourceadvisory.in/forum/general-discussions/rajasthan-sets-a-goal-of-30-000-mw-solar-capacity-in-5-years
38. PV Magazine. https://www.pv-magazine-india.com/
39. Saur Energy. https://www.saurenergy.com/solar-energy-news
40. Back to Growth: Union budget 2021–22 announcement—renewable Energy. https://www.investindia.gov.in/team-india-blogs/back-growth-union-budget-2021-22-announcement-renewable-energy
41. Rajasthan Solar Policy 2019. https://jalore.rajasthan.gov.in/content/dam/doitassets/jalore/pdffiles/Rajasthan%20Solar%20Energy%20Policy2019.pdf

42. Rajasthan Budget Analysis 2020–21. https://prsindia.org/budgets/states/rajasthan-budget-analysis-2020-21
43. Steps to enhance domestic manufacturing of solar PV cells and modules. https://pib.gov.in/PressReleaseIframePage.aspx?PRID=1742795
44. Desert Power India 2050. https://www.powergrid.in/sites/default/files/footer/smartgrid/desert_power_india.pdf
45. Rajasthan has attracted investment proposals of over 1.5 lakh crore in renewable energy. https://government.economictimes.indiatimes.com/news/psu/with-1-5-lakh-crore-renewable-energy-investment-proposal-in-hand-rajasthan-aims-to-increase-installed-capacity-by-30000-mw-rrecl-cmd-subhodh-agarwal/82014059

Hydrokinetic as an Emerging Technology

Gaurav Saini and R. P. Saini

Abstract Due to ongoing environmental concerns and increased energy demand, the inclination toward renewable energy sources has increased significantly in the last few decades. The increased share of renewable energy in total energy generation is a sign of sustainable development. Hydropower has been considered better with good predictability and baseload applications. The non-conventional way of harnessing hydropower, viz. hydrokinetic power, is an emerging area of research. The technology to harness hydrokinetic energy plays a vital role in the lives of people living in rural and remote locations for their energy security. To increase the conversion efficiency of hydrokinetic technologies, a lot of investigations are being carried out. In this paper, various technological advancements on hydrokinetic energy technologies have been discussed at different stages of harnessing hydrokinetic energy. Based on the analysis, it has been found that the installations and selection of optimum system parameters are crucial to tap the energy from the hydrokinetic potential site. The investigated results would be useful for the selection of hydrokinetic turbines and to opt for the system parameters corresponding to better performance.

Keywords Renewable energy · Hydropower · Hydrokinetic energy · Turbine · System parameters

1 Introduction

The socio-economic development of the society (at the community and national levels) essentially requires access to electricity [1]. The availability and low production cost of fossil fuel-based electricity generation have always been encouraged in

G. Saini (✉)
Department of Mechanical Engineering, Harcourt Butler Technical University Kanpur, Kanpur 208002, India
e-mail: gaurav161990@gmail.com

R. P. Saini
Department of Hydro and Renewable Energy, Indian Institute of Technology Roorkee, Roorkee 247667, India

K. Namrata et al. (eds.), *Smart Energy and Advancement in Power Technologies*, Lecture Notes in Electrical Engineering 926,
https://doi.org/10.1007/978-981-19-4971-5_52

the past. In the present era, the environmental concerns and continuous depletion of conventional (fossil) fuel have discouraged electricity generation from these sources [2]. India is endowed with a lot of renewable energy sources such as solar, hydro, geothermal, biomass, and wind energy sites [3, 4]. The technology to harness the energy from these renewable energy sources has also been progressive. The continuous development of technology paved the way for the competitive environment for electricity generation from renewable energy resources [5, 6].

Among all types of renewable energy sources, the energy generation from hydropower has been considered as the most predictable and reliable [7]. The electricity generation from hydropower plants has contributed at different scales, i.e., large hydropower plants and small hydropower plants which are classified based on plant capacity [8]. The large capacity-based hydro-potential sites have already been used; however, the potential sites for the small hydropower are still being harnessed [9].

The operation of hydropower plants can further be classified into two different categories, i.e., conventional and unconventional. Conventionally, the potential energy (natural or artificial) of water is provided for the conversion of water energy into electrical energy. Unconventionally, the water kinetic energy can be harnessed by using the hydrokinetic energy convertors into electrical energy [10]. Nowadays, the unconventional way of harnessing hydropower has been encouraged due to their negligible impacts on the natural flow of water. Hydrokinetic energy convertors can extract the energy from existing water-carrying infrastructure without constructing an artificial dam, barrage, and diversion across the natural flow of water [11].

The technology utilized to tap the hydrokinetic energy is generally known as hydrokinetic technology which can further be classified in two categories, i.e., turbine type and non-turbine type [12]. The turbine-type hydrokinetic technology is usually preferred due to their better conversion efficiency. The term hydrokinetic turbine is generally interchangeably used with the 'free-flow turbine,' 'zero-head turbine,' and 'current turbine' [13]. The conversion of kinetic energy of water is analogues to wind energy conversion by wind turbines [14]. The exposure of the working fluid, i.e., water and air, is the only difference between the two types of technology. Comparatively higher density of water allows the hydrokinetic conversion system to generate sufficient amount of power at very low velocity of water [15].

Conversion efficiency of hydrokinetic turbines is mainly dependent on the type of technology utilized and adopted installation strategy [16]. Out of these two, technology selection is of prime concern [17, 18]. To improve the conversion efficiency, various adopted numerical, experimental, and analytical methodologies have been carried to investigate the system parameters of hydrokinetic turbines. In the present paper, it is aimed to study the effect of system parameters on the conversion efficiency of hydrokinetic turbines. Further, the investigated parameters will be categorized based on the turbine technology.

2 System Parameters

The performance of hydrokinetic technology is dependent on the selection of system parameter used. A number of system parameters are available which influence the performance (output power, starting torque, flow field distribution, power coefficient, and torque coefficient) [19]. The system parameters of hydrokinetic turbines are discussed as;

2.1 *Aspect Ratio*

Aspect ratio is the ratio of rotor height (H) to the rotor diameter (D). Aspect ratio is significant to increase the frontal area of the turbine. For optimum performance of different turbine, aspect ratio should be properly selected. This ratio is associated with all types of cross-flow hydrokinetic turbines.

2.2 *Solidity Ratio*

Solidity ratio is associated with the volume covered by the rotor blades to the total volume occupied due to rotor rotation. Analytically, solidity ratio is the ratio volume covered by the rotor blades to the total area occupied by the rotor. In case of axial flow turbine, due to negligible transverse length, the volume of rotor and blades is replaced with rotor area. Solidity ratio is important to signify the participation and interaction of fluid (water) to rotor blades irrespective of rotor domain.

2.3 *Overlap Ratio*

Overlap ratio is associated with the drag-type rotor of hydrokinetic technology, i.e., Savonius rotor. This parameter is related to placement of two blades which may help to improve the flow characteristics across the turbine. It is the ratio between the overlap of two blades and Savonius rotor diameter.

2.4 *Angle of Attack*

Angle of attack is related to the interaction of water with turbine blades. This parameter decides the energy transferring behavior of water corresponding to the positioning of rotor blade and direction of incoming water.

2.5 Number of Blades

The hydrokinetic turbine can accommodate a number of blades ranging from 2 to 4. This parameter of hydrokinetic turbine is a governing factor for the power output/efficiency, and it is associated with the solidity of the turbine also. More number of blades are significant for improving the starting torque behavior of the turbine.

2.6 Blade Profile

The generation of forces (lift and drag) over the turbine blades is dependent on the type of blade profile used for the rotor. The blade profiles (hydrofoil shaped) for lift force-based rotor generating better lift forces are generally considered, whereas for drag force-based rotors, blade shape-/profile-generating good drag forces are usually preferred.

The above-discussed parameters are crucial for harnessing the hydrokinetic energy from various hydrokinetic potential sites. A lot of investigations have been carried out for the performance improvement of hydrokinetic turbines by optimally selecting the system parameters. Based on the turbine type, various numerical, experimental, and analytical investigations on the different hydrokinetic turbines are discussed in the subsequent section.

3 Investigations on the Hydrokinetic Technology

There are different kinds of hydrokinetic turbine which are available for harnessing the hydrokinetic energy. Based on the site and turbine characteristics, the selection of the turbine is usually made. By keeping in mind the turbine technology, different types of investigations for performance improvement have been discussed.

3.1 Investigations on the Axial Flow Hydrokinetic Turbines

For axial flow hydrokinetic technology, the turbine rotor is of propeller type [20]. These types of turbine technology are suitable for high-depth potential sites such as ocean and tidal applications [14, 21, 22]. A lot of investigations have been performed to improve the conversion efficiency and to resolve the issues regarding installations/operation along with other associated drawbacks.

Ristic et al. [23] proposed the application of propeller rotors for the hydrokinetic applications. The propeller rotor poses less initial cost along with the satisfactory energy conversion efficiency. Further, it was recommended to analyze the rotor parameters for optimization. Lee et al. [24] experimentally investigated the wake recovery distance for axial flow hydrokinetic turbines under different TSR values. The study proposed the installation of hydrokinetic turbine between 6D-10D in stream-wise direction. It has also been found that rotor structure plays a significant role for wake distance and contributes approximately 30% deficit in velocity.

Chime and Malte [25] numerically investigated the axial flow hydrokinetic turbines under high blockage conditions. The results of the study revealed that as the blockage ratio increases with the presence of free surface and channel walls the energy extraction capability of the turbine increases. It has been recommended that free surface should be considered (using VOF model) for the prediction of hydrokinetic power. The enhanced mixing of fast and slow mixing of flow in high blockage conditions enable the early recovery of velocity in the downstream. In the similar lines, Kinsey and Dumas [26] investigated the effect of channel blockage on the conversion efficiency of axial as well as cross-flow hydrokinetic turbine. It has been observed that during dynamic stall conditions the generated power is insensitive to the blockage factor. Further, the correction factor is necessary for drag force and velocity field in order to predict the performance of turbine under a given confined conditions.

Anyi and Kirke [27] experimentally investigated a non-clogging axial flow hydrokinetic turbine in the field applications. The turbine was designed to overcome the issues raised due to the presence of debris. A clog-free design of the rotor was made which eliminates the clogging of debris around the rotor blade. The turbine was mainly design to operate in tropical rivers and to supply the energy to remote and rural areas. In an another study, Kirke [21] proposed the hydrokinetic turbine for moderate-sized rivers. Among different hydrokinetic rotors, axial hydrokinetic rotor requires high depth which is generally discouraged for shallow depth applications.

3.2 *Investigations on the Cross-Flow Hydrokinetic Turbines*

The cross-flow hydrokinetic turbine configurations mainly have two types of rotors, i.e., Savonius and Darrieus [28, 29]. These turbines were analyzed by various researchers under vertical and horizontal orientations. Further, very few researches studied the combinations of the two rotor in hybrid mode which has also been found in the literature.

Khan et al. [30] experimentally investigated the two-bladed three types of Savonius rotor having different stages. The study revealed that the double-stage Savonius rotor yields better power and torque characteristics. The toque fluctuation peaks were also eliminated in case of double-stage rotor which further may smoothen the power output. Saini and Saini [31] comparatively analyzed the multistage Savonius rotor which has utilized the vertical velocity gradient of flow. It has been found that multistage variable diameter rotor is applicable for the flow having fluctuations in discharge. However, the maximum performance was obtained corresponding to the single-stage constant diameter rotor.

Shashikumar et al. [32] numerically investigated the blade profile of Savonius rotor in order to improve the drag force. The conventional semicircular blade profile was tapered. The results of the study revealed that tapered profile of rotor reduces the blade surface which results in the 5% reduction in comparison with conventional semicircular blade profile. The variation of power and torque coefficient versus TSR is shown in Fig. 1.

Kumar and Saini [33] performed the numerical investigations on a modified twisted blade Savonius hydrokinetic turbine rotor. It has been found that the optimal twisting of turbine blade improves the torque generation and a maximum of 0.39 value of power coefficient was found corresponding to 12.5° twist angle. Jaohindy et al. [34] numerically analyzed the generation of transient force over the blades of Savonius rotor under different aspect ratios. Based on the investigations, it has been observed that aspect ratio is significant for lateral lift and longitudinal drag force generation. The results of the study revealed that forces on the Savonius rotor are the resultant of the lift and drag. Therefore, the Savonius rotor cannot solely be considered as drag-driven device. However, the fraction of lateral lift force is less. The generation of lateral lift and longitudinal drag force is represented by the schematic shown in Fig. 2.

Elbatran et al. [35] numerically analyzed the performance of Savonius hydrokinetic turbine under confined conditions. The results were comparatively analyzed with the results of rotor under normal operating (open conditions). The proposed installation configuration for the Savonius rotor is found to have less negative drag induced which increased the power coefficient by 78%.

Joo et al. [36] investigated the aerodynamics characteristics of a two-blade Darrieus rotor under different solidities conditions. Rotor solidity is found to have effect on the generation of peak torque and interaction of fluid with blades. The

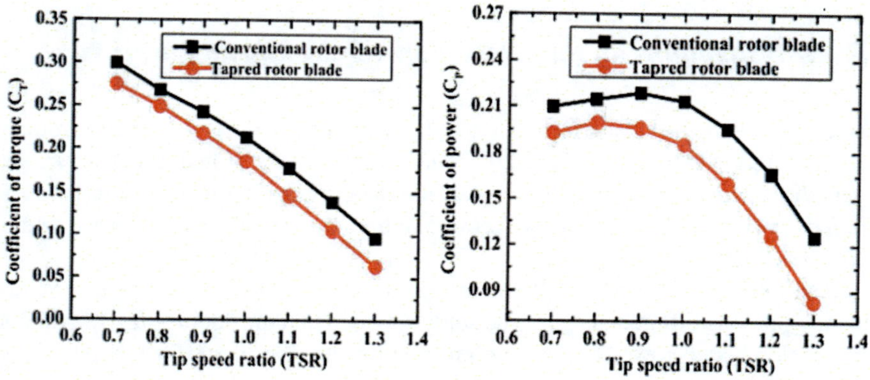

Fig. 1 Variations of toque and power coefficient corresponding to the conventional and tapered hydrokinetic turbine rotor [32]

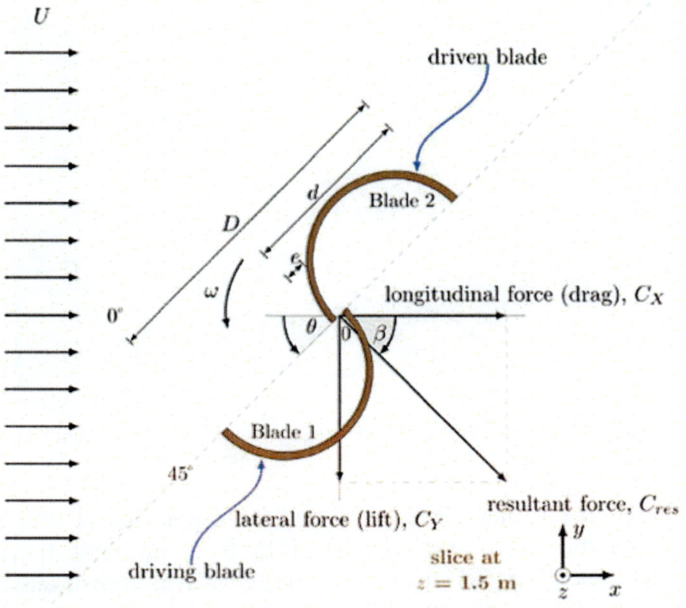

Fig. 2 Schematic for generation of forces on the Savonius rotor [34]

increment in the rotor solidity increases the fluid interaction and peak torque and vice versa. Further, the increased solidity affects the blockage which in turn increases the incidence velocity of the fluid.

Dai et al. [37] analytically investigated the hydrodynamics behavior of Darrieus rotor-based marine current turbine. It has been reported that parameters, i.e., blade profile, rotor solidity, and configuration of the rotor, are important for the performance analysis of rotor. It was proposed to install the turbine under vertical orientation. The application of CFD was recommended to effectively analyze the performance of rotor in comparison with analytical approach. Saini and Saini [38] investigated the Darrieus rotor under variable operating conditions and comparatively analyzed the performance of Darrieus rotor with hybrid rotor (combination of Savonius and Darrieus). The hybrid rotor was found to have better peak in the power and torque coefficient; however, the operating range (maximum TSR value) of the single Darrieus was more.

Kirke and Lazauskas [39] compared the performance of variable and fixed pitch-based Darrieus hydrokinetic rotor. In order to overcome the associated drawbacks of fixed pitch hydrokinetic turbine such as fluctuating torque, low efficiency, and poor starting torque, the application of variable pitch blades was suggested. Variable pitch blades were found to generate better starting torque with comparatively good conversion efficiency. Further, the vibration produced with the fixed blade turbine was also found to be eliminated in case of variable pitch-based Darrieus hydrokinetic turbine.

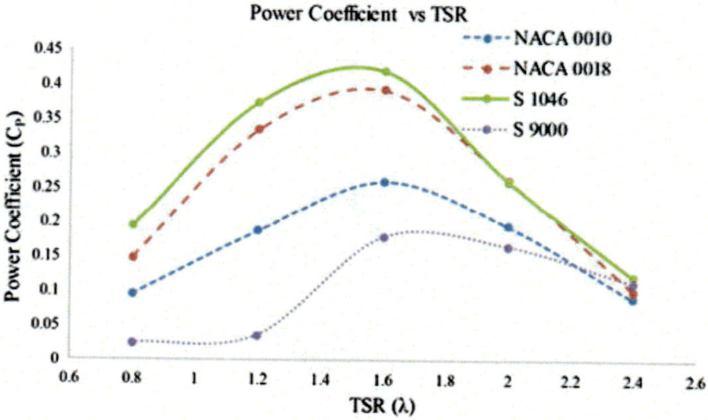

Fig. 3 Comparative performance of Darrieus hydrokinetic turbine with different hydrofoils

Saini and Saini [40] numerically analyzed the application of different blade profiles and their impact on the performance and the flow field distribution across the flow domain and inside the rotor structure. Four different hydrofoils of variable grades having symmetrical and un-symmetrical shape were used. The un-symmetrical shape S-1046 hydrofoil was found to have better hydrodynamic performance corresponding to variable operating conditions. Maximum coefficient of power as 0.46 was observed at 1.6 value of TSR for the Darrieus hydrokinetic turbine having the S-1046 hydrofoil. The comparative performance of Darrieus rotor with different hydrofoils is shown in Fig. 3.

Patel et al. [41] experimentally investigated the effect of rotor solidity, hydrofoil type, number of blades and turbine positioning in hydro-farm applications. General layout for positing of hydrokinetic turbine is shown in Fig. 4. The results revealed that three-bladed Darrieus rotor having symmetric hydrofoil (NACA 0018) generates maximum coefficient of power as 0.16 at 0.382 solidity value. Further, the optimized stream-wise and span-wise position for maximum energy extraction from a flow channel was reported as X/D = 7 and X/D = 3, respectively.

4 Conclusions

Under the present paper, the discussions on hydrokinetic technologies have been outlined to increase their applications in rivers and marine applications. Various latest investigations have been analyzed to find the technological as well as installation barriers. Different critical systems parameters have been discussed to justify their significance. The investigations results revealed that the blockage or confinement in the turbine placement significantly increases the flow velocity, which in turn increases the power density. The axial flow hydrokinetic turbine was suggested to deploy under high depth applications having low flow debris and silt conditions. In the case of

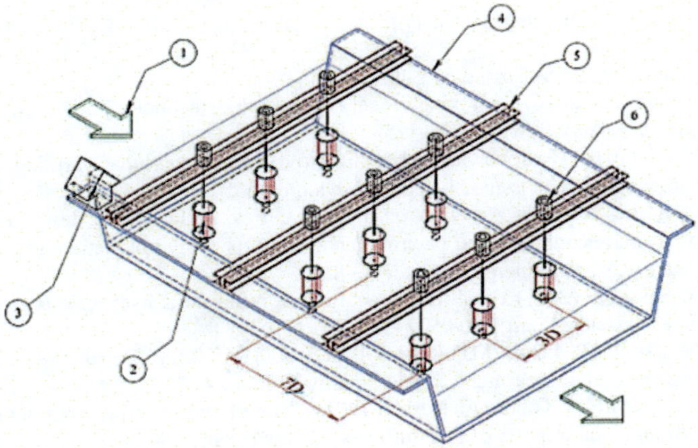

Fig. 4 General layout of Darrieus hydrokinetic turbine for hydro-farm applications [41]

shallow water flowing conditions, the cross-flow turbine was recommended. The placement of a turbine in a hydro-farm application is a tedious task that should be decided based on the wake characteristics. In stream-wise direction, hydrokinetic turbine was suggested to be deployed at 90% recovery of the flow velocity. Further, the turbine performance was found to be better, having good hydrodynamic blade profile. The study revealed that most investigations were performed to improve the energy conversion capability of turbine by utilizing the optimum system parameters. The present study will develop awareness about the resources, technology, and installation of hydrokinetic energy.

References

1. Anyi M, Kirke B (2010) Evaluation of small axial flow hydrokinetic turbines for remote communities. Energy Sustain Dev 14:110–116
2. Saini G, Saini RP (2018) Numerical investigations on hybrid hydrokinetic turbine for electrification in remote area. All India Semin Renew Energy Sustain Dev (Institution Eng July 27–28, 2018)
3. Deshpande P, Li X (2013) numerical study of giromill-type wind turbines with symmetrical and non-symmetrical airfoils. Eur Int J Sci Technol. 2:2304–9693
4. IRENA (2018) Power generation costs in 2017. Int Renew Energy Agency 2018 (Abu Dhabi), ISBN 978-92-9260-040-2, pp 1–160
5. Badrul Salleh M, Kamaruddin NM, Mohamed-Kassim Z (2019) Savonius hydrokinetic turbines for a sustainable river-based energy extraction: a review of the technology and potential applications in Malaysia. Sustain Energy Technol Assess 36:100554
6. National Research Council (2013) An evaluation of the U.S. Department of Energy's marine and hydrokinetic resource assessments. National Academies Press, Washington, DC.

7. Bahaj AS (2011) Generating electricity from the oceans. Renew Sustain Energy Rev 15:3399–3416
8. MNRE. mnre.gov.in/#. Assessed on 2 July 2021
9. Laws ND, Epps BP (2016) Hydrokinetic energy conversion: technology, research, and outlook. Renew Sustain Energy Rev 57:1245–1259
10. Killingtveit Å (2018) Hydropower. In: Letcher T (ed) Managing global warming. An interface of technology and human issues, 1st edn. Elsevier, pp 265–315
11. Kumar D, Sarkar S (2016) A review on the technology, performance, design optimization, reliability, techno-economics and environmental impacts of hydrokinetic energy conversion systems. Renew Sustain Energy Rev 58:796–813
12. Kumar A, Saini RP (2017) Performance analysis of a Savonius hydrokinetic turbine having twisted blades. Renew Energy 108:502–522
13. Ridgill M, Neill SP, Lewis MJ, Robins PE, Patil SD (2021) Global riverine theoretical hydrokinetic resource assessment. Renew Energy 174:654–665
14. Fallon D, Hartnett M, Olbert A, Nash S (2014) The effects of array configuration on the hydro-environmental impacts of tidal turbines. Renew Energy 10–25
15. Saini G, Saini RP (2019) A review on technology, configurations, and performance of cross-flow hydrokinetic turbines. Int J Energy Res 43:6639–6679
16. Saini G, Saini RP (2020) Study of installations of hydrokinetic turbines and their environmental effects. In: AIP conference proceedings, vol 2273. AIP, Sikkim, p 050022
17. Güney MS, Kaygusuz K (2010) Hydrokinetic energy conversion systems: a technology status review. Renew Sustain Energy Rev 14:2996–3004
18. d'Auteuil S, Birjandi A, Bibeau E, Jordan S, Soviak J, Friesen D (2019) Riverine hydrokinetic resource assessment using low cost winter imagery. Renew Sustain Energy Rev 105:293–300
19. Kerikous E, Thévenin D (2019) Optimal shape of thick blades for a hydraulic Savonius turbine. Renew Energy 134:629–638
20. Gorelov DN, Krivospitsky VP (2008) Prospects for development of wind turbines with orthogonal rotor. Thermophys Aeromechanics. 15:153–157
21. Kirke B (2020) Hydrokinetic turbines for moderate sized rivers. Energy Sustain Dev [Internet]. Int Energy Initiative 58:182–195
22. Ahmadian R, Falconer R, Bockelmann-Evans B (2012) Far-field modelling of the hydro-environmental impact of tidal stream turbines. Renew Energy 38:107–116
23. Ristić B, Božić I, Simić A (2021) A marine propeller as a hydrokinetic turbine—CFD analysis of energy characteristics. IOP Conf Ser Earth Environ Sci 774:012063
24. Lee J, Kim Y, Khosronejad A, Kang S (2020) Experimental study of the wake characteristics of an axial flow hydrokinetic turbine at different tip speed ratios. Ocean Eng 196
25. Hoseyni-Chime A, Malte PC (2014) Hydrokinetic turbines at high blockage ratio. In: Proceedings of the 2nd Marine energy technology symposium, April15-18,2014,Seattle,WA, pp 1–11
26. Kinsey T, Dumas G (2017) Impact of channel blockage on the performance of axial and cross-flow hydrokinetic turbines. Renew Energy 103:239–254
27. Anyi M, Kirke B (2015) Tests on a non-clogging hydrokinetic turbine. Energy Sustain Dev Int Energy Initiative 25:50–55
28. Teja DP (2017) Studi Numerik Turbin Angin Darrieus–Savonius Dengan Penambahan Stage Rotor Darrieus. Faculty of Industrial Technology Sepuluh Nopember Institute of Technology Surabaya
29. Sahim K, Santoso D, Puspitasari D (2018) investigations on the effect of radius rotor in combined Darrieus-Savonius wind turbine. Int J Rotating Mach 3568542:1–7
30. Khan MNI, Iqbal MT, Hinchey M (2008) Submerged water current turbines. IEEE, pp 1–6
31. Saini G, Saini RP (2018) Numerical investigations for performance comparison of savonius hydrokinetic turbine. In: 1st international conference on New Frontiers in Engineering, Science and Technology (NFEST/2018/R–100), New Delhi, India, January 8–12, 2018, pp 395–400
32. Shashikumar CM, Vijaykumar H, Vasudeva M (2021) Numerical investigation of conventional and tapered Savonius hydrokinetic turbines for low-velocity hydropower application in an irrigation channel. Sustain Energy Technol Assessments 43:100871

33. Kumar A, Saini RP (2017) Performance analysis of a single stage modified Savonius hydrokinetic turbine having twisted blades. Renew Energy 113:461–478
34. Jaohindy P, McTavish S, Garde F, Bastide A (2013) An analysis of the transient forces acting on Savonius rotors with different aspect ratios. Renew Energy 55:286–295
35. Elbatran AH, Ahmed YM, Shehata AS (2017) Performance study of ducted nozzle Savonius water turbine, comparison with conventional Savonius turbine. Energy 134:566–584
36. Joo S, Choi H, Lee J (2015) Aerodynamic characteristics of two-bladed H-Darrieus at various solidities and rotating speeds. Energy 90:439–451
37. Dai YM, Gardiner N, Sutton R, Dyson PK (2011) Hydrodynamic analysis models for the design of Darrieus-type vertical-axis marine current turbines. Proc Inst Mech Eng Part M J Eng Marit Environ 225:295–307
38. Saini G, Saini RP (2020) Comparative investigations for performance and self-starting characteristics of hybrid and single Darrieus hydrokinetic turbine. Energy Rep 6:96–100
39. Kirke BK, Lazauskas L (2011) Limitations of fixed pitch Darrieus hydrokinetic turbines and the challenge of variable pitch. Renew Energy 36:893–897
40. Saini G, Saini RP (2019) Numerical investigation of the effect of blade profile of a Darrieus hydrokinetic turbine. In: 2018 5th IEEE Uttar Pradesh Section International Conference on Electrical, Electronics and Computer Engineering (UPCON). IEEE, Gorakhpur, India, pp 1–6
41. Patel V, Eldho TI, Prabhu SV (2017) Experimental investigations on Darrieus straight blade turbine for tidal current application and parametric optimization for hydro farm arrangement. Int J Mar Energy. 17:110–135

3 KW PMSG Wind Turbine Stand Alone System for Residential Application

Sumit Sharan, Kumari Namrata, and Nishant Kumar

Abstract This paper summarizes operation of power conversion through small wind turbine conversion system with permanent magnet synchronous generator (PMSG) coupled to the turbine shaft and lead-acid batteries (LABs) for storage of converted energy in DC form and from battery the energy can be used for residential applications. The aim is to deliver required power to a three-phase 110–160 V/50 Hz domestic appliances through charged LAB. The delivered power to the load is continuous and constant, supplied by LAB. A simple wind energy conversion system is modelled to charge the LABs in an efficient way. To obtain maximum power characteristic of wind turbine, boost converter with MPPT control algorithm is modelled. Energy storage devices such as lead-acid batteries provide a continuous and constant power supply and make the wind conversion system to a stand-alone wind energy conversion system.

Keywords Small wind turbine · PMSG · VSI inverter · LABs · MPPT

1 Introduction

Electrical energy is the fundamental form of energy for human beings to sustain their life on earth. The demand for electrical energy is increasing per day, however since our generation is almost depends on conventional sources to fulfil the electricity demand that is limited, so we have to put our concentration on nonconventional sources of energy.

The gross electricity generation worldwide in the year 2018 was 3.9% higher that of in the year 2017 (from the report of, key electricity trends 2019, International Energy Agency or IEA) and is continuously increasing year by year. Sources such

S. Sharan (✉) · K. Namrata · N. Kumar
National Institute of Technology, Jamshedpur, Jharkhand, India
e-mail: sharansumit57@gmail.com

K. Namrata
e-mail: namrata.ee@nitjsr.ac.in

K. Namrata et al. (eds.), *Smart Energy and Advancement in Power Technologies*,
Lecture Notes in Electrical Engineering 926,
https://doi.org/10.1007/978-981-19-4971-5_53

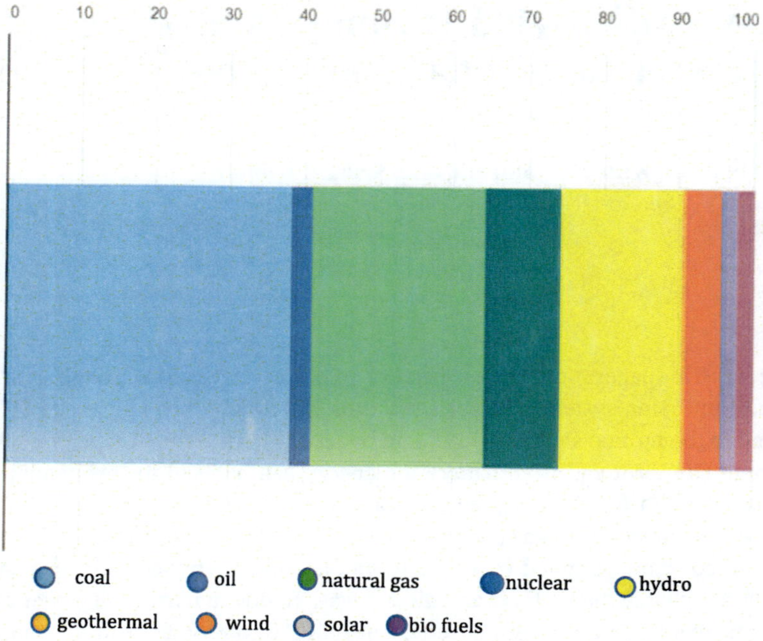

Fig. 1 Electricity production by source worldwide, in 2018 (IEA data)

as coal and coal products, natural gasses and biofuels are used as majority sources for electricity generation worldwide rather than renewable sources of energy.

Globally, the generation of electricity from renewable sources in the year 2018 was 12.4% from wind and 24.3% from solar energy. If we compare the contribution of renewable sources in electricity generation for last ten years, green energy has acquired a fast growth in the year 2018.

As our daily consumption of electrical energy is increasing per day and our majority of energy sources that we use to convert into this form (electricity) are conventional and are limited, and this conversion is directly related to pollution in some forms like greenhouse gasses emission. The only way to control pollution limit and balance electricity demand is to switch some part of conventional resources to green energy (Fig. 1).

1.1 System Architecture

In the architecture of the proposed model, a small turbine of 3 KW capacity is directly connected to PMSG generator. The output of PMSG is a 3-phase AC power, this AC form of generated power is then converted into DC with diode bridge rectifier, this rectified DC is boosted and maintained at a constant level with DC–DC converter

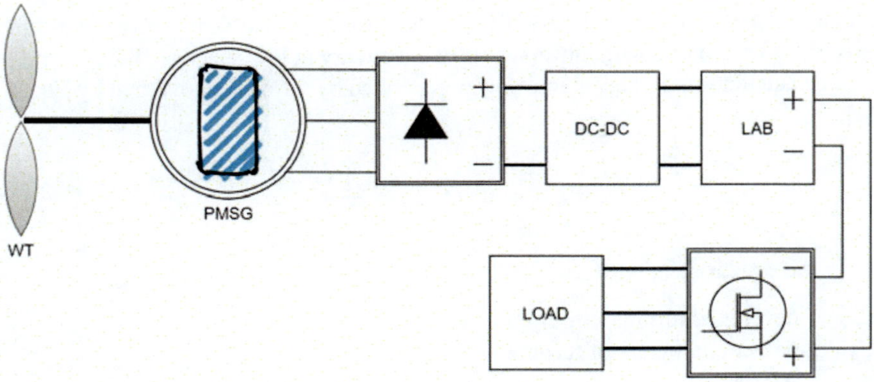

Fig. 2 System configuration

block and is supplied to LAB battery storage bank, then the dc constant voltage from the battery terminal is converted to three-phase AC from inverter and then feed to AC loads. The configuration is shown in Fig. 2 [1].

The design for this kind of energy conversion system is the result of various literature studies along with the new technologies and contributions that are developed by scientists to became self-dependent in power sector. While reviewing and studying various literatures on wind energy, author found an impressive literature on this topic that was by an author named "Luminita Barote", how she defined a energy conversion technique in an easiest and economical way that was the source of motivation. She has already published more than 15 papers on small scale wind energy conversion system with storage capability. When someone itself can generate and use and store power according to their uses and to be self-dependent in power sector. is the most motivational thing that brought interest of the author towards this field.

This section explores mathematical modelling of the proposed wind energy conversion system (Fig. 2).

1.2 Wind Turbine

Kinetic energy available in wind when it flows with some velocity, strikes the blades of wind turbine and lift and drag force together cause blades to rotate. Rotation of blades results in rotation of rotor and to rotor shaft attached to turbine and hence rotor of generator coupled with shaft starts rotating to generate electric power. For small turbines, the best thing associated with it is that they are mainly designed to work in low wind speeds, horizontal axis wind turbine (HAWT) is best choice for standalone wind conversion system. A 3KW capacity small wind turbine or three blade HAWT with rotor diameter of 2 m can work efficiently in low wind speeds. SD7080 model of airfoils may be the most suitable airfoil selection for blade designing to produce maximum power from small HAWT [2]. Also, utilizing a flexible bladed rotor turbine give rise to more efficient conversion of wind energy at low speeds [3].

The calculation of mechanical power generated by turbine is governed by Eq. (1) [4],

$$P_{\text{mech}} = C_p(\lambda, \beta)\rho\left(\frac{A}{2}\right)V^3 \tag{1}$$

In the equation,

P_{mech} developed turbine power
C_p power coefficient of turbine
ρ Air density
A swept area by turbine blades (in m^2)
V speed of wind (in m/s)
λ tip speed ratio
β pitch angle of turbine blades

Power coefficient of turbine is a function of tip speed ratio (lambda) and blade pitch angle (beta), which is calculated as:

$$Cp(\boldsymbol{\lambda}, \boldsymbol{\beta}) = C1\left(\frac{C2}{\lambda i} - \frac{C3}{\beta} - C4\right) * e^{\frac{C5}{\lambda i}} + C6\lambda \tag{2}$$

With

$$\frac{1}{\lambda} = \frac{1}{(\lambda + 0.08\beta)} - \frac{0.035}{(\beta^3 + 1)}$$

In the Eq. (2), when we define constants $C1 = 0.5176$, $C2 = 116$, $C3 = 0.4$, $C4 = 5$, $C5 = 21$ and $C6 = 0.0068$ and taking the value of beta, that is pitch angle to zero and lambda (tip speed ratio) to 8.1 [5], we get the maximum power coefficient (C_p) of turbine which is 0.48 (Fig. 3).

1.3 PMSG Model

Small wind turbines are best suitable with direct driven PMSG and can work efficiently at low wind speeds. Fractional power generation turbines with output capacity up to 20KW generally have directly coupled PMSG and don't need complex gear trains equipment [6]. The PMSG model has assumed to be with sinusoidal back EMF waveform, that is the magnetic lines of flux in the stator winding of machine is sinusoidal, and hence the EMF generated in machine is sinusoidal in nature.

The modelling of PMSG is configured in its d-q axis equivalent circuit and in figure the equivalent circuit for PMSG in its d and q axis, respectively, is shown

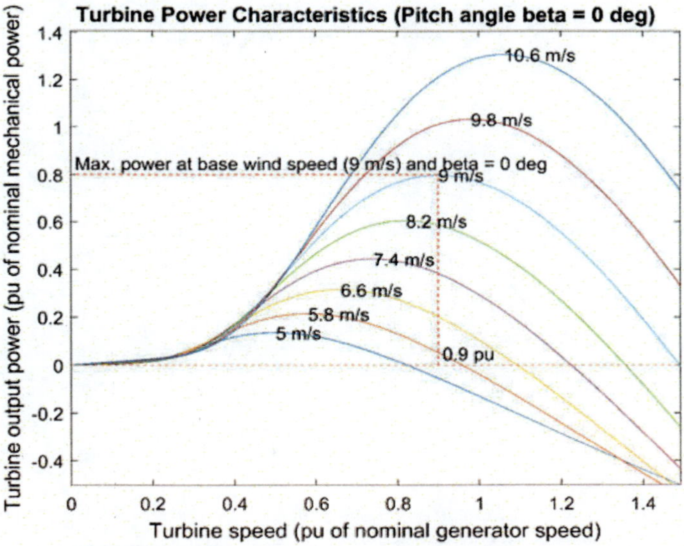

Fig. 3 Turbine power characteristics

in the d and q coordinates. The rotation is synchronous with an electrical angular velocity [7] (Fig. 4).

From figure above, *d*-axis voltage and *q*-axis voltage of the PMSG model in d-q coordinate can be expressed as [5],

$$V_d = -R_s i_d + \omega\varphi_q - \left(\frac{\mathrm{d}\varphi_d}{\mathrm{d}t}\right) \tag{3}$$

$$V_q = -R_s i_q - \omega\varphi_d - \left(\frac{\mathrm{d}\varphi_q}{\mathrm{d}t}\right) \tag{4}$$

The value of flux linkages in the above equations is given by,

$$\varphi_d = L_d i_d - \varphi_{\mathrm{PM}}$$

$$\varphi_q = L_q . i_q$$

And hence the resultant torque produced by the PMSG is,

$$T = \left(\frac{3}{2}\right)\left(\frac{P_n}{2}\right)\left[\left(L_{d\mathrm{axis}} - L_{q\mathrm{axis}}\right) i_{d\mathrm{axis}} i_{q\mathrm{axis}} - \varphi_{\mathrm{PM}} i_{q\mathrm{axis}}\right] \tag{5}$$

Φ_{PM} is the magnetic flux of permanent magnets. PMSG machine has salient poles.

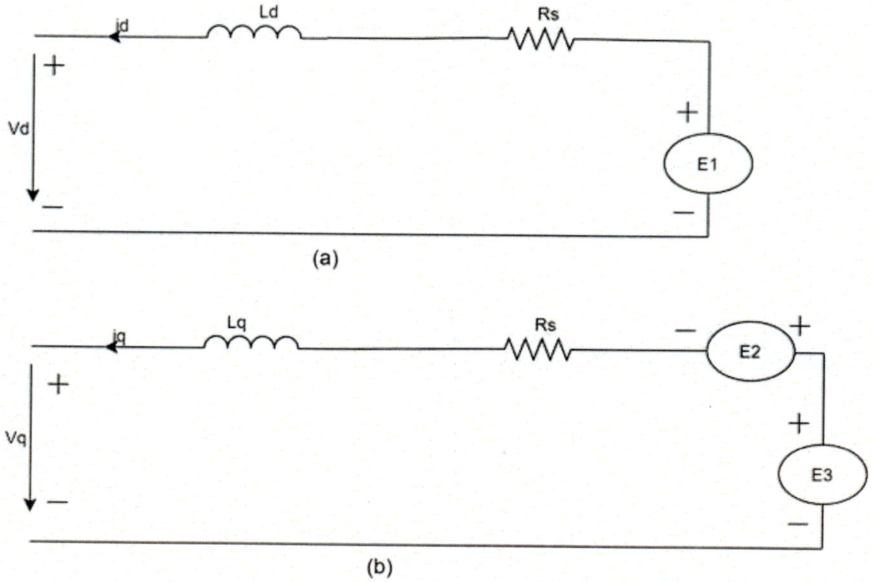

Fig. 4 **a** equivalent circuit of PMSG in d coordinate ($E1 = \omega L_q i_q$). **b** Equivalent circuit of PMSG in q coordinate ($E2 = \omega L_d i_d$) and ($E3 = \omega \varphi_{PM}$)

The poor voltage regulation performance of PMSG, and no any control to filed flux provided by permanent magnets cause the machine to work inefficiently during sudden changes in wind speed. Also fault tolerance of PMSG is of very low level so, for standalone system the use of PMSG is not reliable and satisfactory, it cannot achieve desired requirements. Selection of switched reluctance generator would may be an efficient choice for direct driven wind turbine system having advantages of low wind speed for starting, flexibility in control and operation, low cost [8].

1.3.1 DC–DC Converter (Boost Converter)

The unidirectional boost converter is an interface between the rectifier and the battery. The equivalent circuit model is shown in Fig. 5.

Output DC voltage at converter's terminal and calculation of the basic parameters *L* and *C* are developed with the help of general relationship:

$$V_{DC} = \frac{V_{in}}{(1 - D)}$$

$$L = \left\{ V_{in(min)} * \frac{\left(V_{out} - V_{in(min)}\right)}{(\Delta i * f_s * V_{out})} \right\}$$

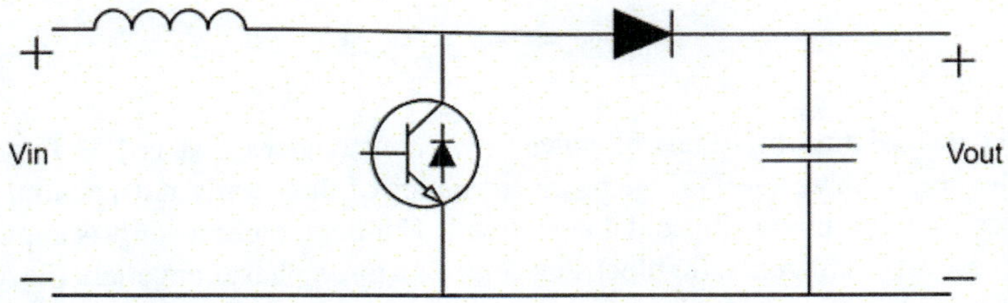

Fig. 5 Boost converter model

$$C = \frac{(I_{out*} D)}{f_s * \Delta V} \tag{6}$$

In the equation, *D* is defined as the duty cycle of the converter switch, which controls gate pulse of the switch.

Boost converter is employed with a MPPT control algorithm to obtain maximum power from the turbine at available wind speed. Output voltage and current of diode rectifier is used as reference to the MPPT function which controls the duty cycle D (that operates the DC–DC converter switch), when there is variation in wind speed or in output power of turbine [9].

1.4 The Voltage Source Inverter (VSI)

The equivalent circuit model for a three level VSI is shown in the following figure. It is implemented with IGBT-diode switches (Fig. 6).

The RMS value of fundamental component of voltage for each phase is given as:

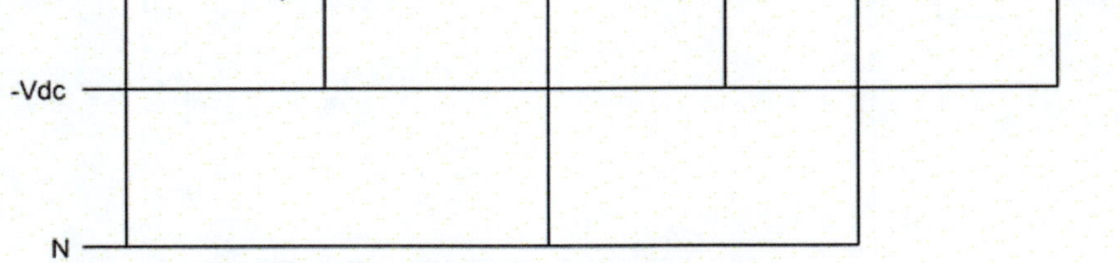

Fig. 6 Three arm inverter switching circuit

$$V_{\text{rms}} = \frac{V}{\sqrt{2}} = \frac{V_{\text{DC}}}{\sqrt{2}\pi}\sqrt{3} \tag{7}$$

Turn-on and turn-off times of power switching devices that is IGBT has not modelled in the three-level bridge block consisting of three arms and a neutral point clamped. A three-phase sinusoidal signal with 120-degree phase shift is applied to the three level PWM generator block that controls a three-phase converter. The output of converter has 12 pulses which is then passed through a LCL filter to smooth the sinusoidal output.

1.5 Lead-Acid Battery

Among other storage batteries lead-acid batteries are good choice to be implemented in renewable energy system with respect to their technical performances. Lead-acid batteries don't require more maintenance and are totally safe for residential applications consisting of non-flammable materials [10].

The equivalent circuit model for a LAB is modelled with the help of a controlled voltage source (E_b) in series with a resistance (Rs.). Rs. is considered as some resistances offered by the battery. The equivalent model circuit of LAB is shown in the Fig. 7. In the circuit, V_b is output terminal voltage of the battery.

The controlled voltage source Eb is a function of state of charge of the battery, battery constant (K) and temperature. E_b is expressed as:

$$E_b = E_{b0} - \frac{K \cdot Q}{\left(Q - \int_0^t i_b \mathrm{d}t\right)} \tag{8}$$

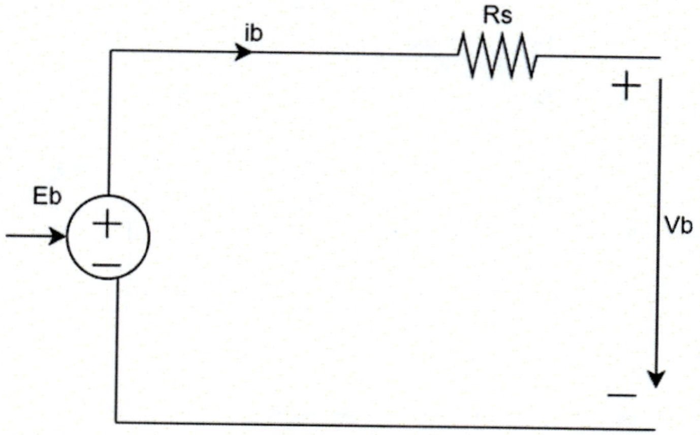

Fig. 7 Equivalent model of LAB

LABs are not recommended to get fully discharged. For efficient use of LAB, when battery is fully charged the state of charge (SOC) is considered as 1 or 100%, this is maximum SOC. And, the discharged SOC of battery is considered to be 0.2 or 20% and this is the minimum SOC of battery. This consideration may be applied for residential applications of LABs which results in longer battery life and safe use. The circuit for the LAB is considered as a practical control voltage source.

2 Simulation of Proposed Model

This section belongs to the Simulink model of the proposed system along with its parameters, and then the results of the simulation are discussed. Simulink block diagram of proposed model in MATLAB along with measurement blocks, is represented in Fig. 8.

The control strategy for wind turbine rotor is pitch angle controller [11]. By supplying a reference rotor speed to the pitch angle controller block, this reference value has set to compare the given input rotor speed. Whenever the rotor speed exceeds the reference value, differences in signals are generated in form of error. This error signal is followed by the angle limit block and rate limiter block which finally controls the pitch angle of the turbine, shown in Fig. 9. Model is designed for maximum wind speed of 9 m/s with base speed of rotor at 243.5544 rad/s.

The wind turbine block is found in Simulink library/Renewables in MATLAB. The block is modelled with the following parameters shown in Table 1,

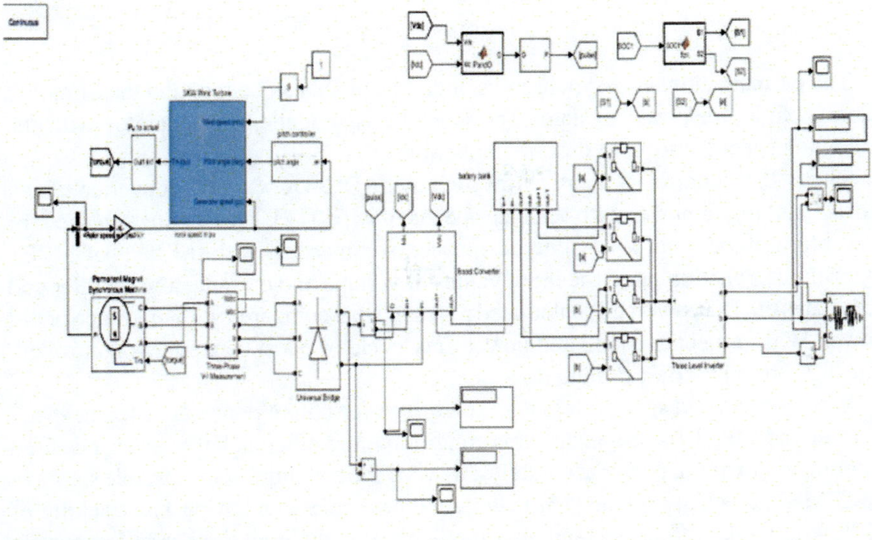

Fig. 8 Simulink block diagram

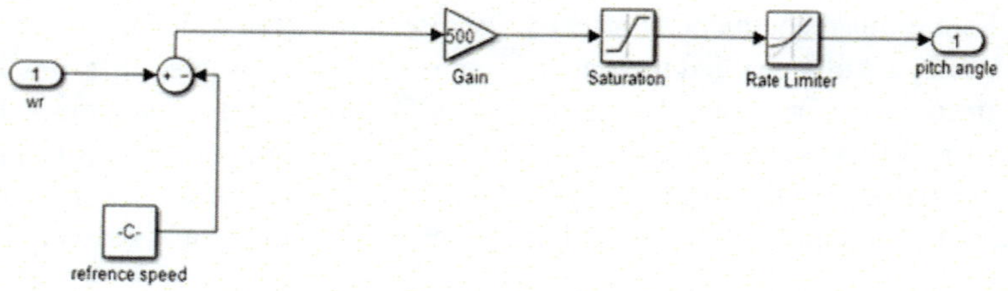

Fig. 9 Simulink model of pitch angle controller

Table 1 Wind turbine parameters

Parameters	Value
Power output	3 KW or 3e3 W
Base Power for PMSG	3e3/0.9
Base speed of wind	9 m/s
Maximum output power at base speed of wind	0.8

Table 2 Generator parameters

Parameters	Value
Rating	3KW
Base rotor speed	243.5544 rad/s
Per phase stator resistance	0.4578 Ω
Inductance of armature	0.003507 H
Flux Linkage	0.171
Number of Pole pairs	4

The permanent magnet synchronous machine block can be found in the Simulink Library/Simscape/power system. The machine is modelled as generator with the following specifications, as described in Table 2.

The MPPT control strategy is inserted with DC–DC (boost) converter which extract maximum power with variable speeds in wind [12]. MPPT function used for controlling maximum power output, may be summarized with the help of MPPT algorithms. Function takes the input values of voltage and current from rectifier and calculates the DC power, this calculated power is then compared with reference or old power. With respect to calculated power duty cycle of boost converter is controlled. The algorithm is shown in Fig. 10.

Variation in wind speed offers new conditions for output power at different speed of rotor and MPPT control makes this condition more efficient but to extract desired amount or maximum power accurately a new MPPT technique with variable step size P&O MPPT control scheme [13] may be the best approach having fast tracking of MPP with higher efficiency, and also another best technique is exponential moving average algorithm combining three control schemes DC-link voltage, rotor speed and pitch angle control working in a hierarchical control scheme [14]. Along with

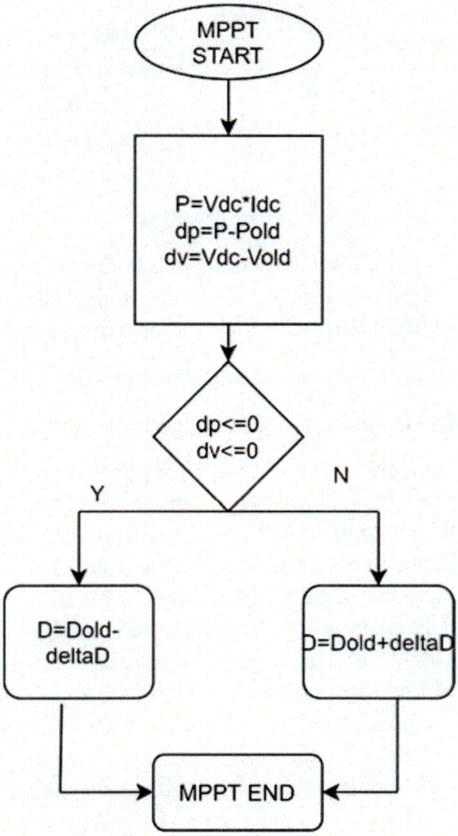

Fig. 10 P&O algorithm Flow chart

maximum power tracking methods, response of a turbine to changing or fluctuations in velocity of wind at the site of installation is also subject to improvement for best level of result from an installed turbine [15].

With the help of Eq. 6 and, the basic parameters of DC–DC converter (boost) are shown in Table 3.

Table 3 DC–DC converter parameters

Fixed parameters	Values
Input voltage	40–260 V
Output terminal voltage	200 V
Switching frequency (f_s)	100 kHz
Peak-peak ripple in voltage and current	0.01% and 0.4%, respectively
Calculated values	
Inductor L	1.44e−5 H
Capacitor C	7.04e−5 F

Table 4 Lead-acid battery parameter

Nominal voltage	200 V
Rated capacity (Ah)	720
Maximum Capacity (Ah)	750
Cut-off voltage	150 V
Nominal discharge current	144 A

LCL filter selected for this model for smoothing the output signal of inverter feeding a 3-phase load, the output of inverter is three-phase 50 Hz, with peak amplitude of 150–170 V approximately. The LCL filter has following parameters:

Inductance connected to inverter side in each three phases has value of 0.008H each.

Inductance connected to load side in each three phases has value of 0.006H each.

Capacitor connected in each phase in between inductors has value of 50μF each.

In the model, energy storage system is composed of two LABs (lead-acid battery). Connection of battery is controlled by a user defined MATLAB function and it connects to inverter circuit and DC–DC converter circuit with the help of circuit breakers. Control function is a function of SOC of one of the batteries and it is defined in such a way that when SOC of one of the specified batteries reaches to 80% or above, it gets disconnected from DC-DC converter circuit with the help of circuit breakers and simultaneously the other battery connects itself to output terminal of DC–DC circuit and starts charging. Batteries are also connected to the inverter with the help of circuit breakers, and are controlled by another user defined function in such a way that when one battery connects to DC–DC converter terminal for charging then the other charged battery is get connected to inverter input terminal and feeds the load. The LAB battery block is available in Simulink library/battery and its parameters are modelled as shown in Table 4. When simulation starts, SOC of charged battery is 82% and that of discharged battery is 40%.

2.1 Results

The simulation is run by supplying a constant wind speed of 5 m/s to the turbine's input with lead-acid battery (LAB) connected to the DC–DC converter terminal for charging, and this battery is acting as a load to the generator. Initial state of charge of battery was set at 40%, the results are as follows:

Figure 11a shows speed of rotor in rad/sec after simulation at wind speed of 5 m/s. after reaching to its steady state, rotor attains a constant speed of about 180 rad/sec and at this speed of rotation, machine is generating three phase power, when this AC power is converted to DC then output of diode rectifier is shown in Fig. 11b. In Fig. 11c showing output voltage at dc–dc converter's terminal that is approximately constant at 200v when there are fluctuations in rectifier output. At this constant level of output voltage, battery connected to converter's terminal is charging with increasing in SOC level showing through in Fig. 11d.

Now suddenly the wind speed input to the turbine is varied to 9 m/s from 5 m/s, at $t = 1.7$ s, and then the measured output comes out as shown in Fig. 12.

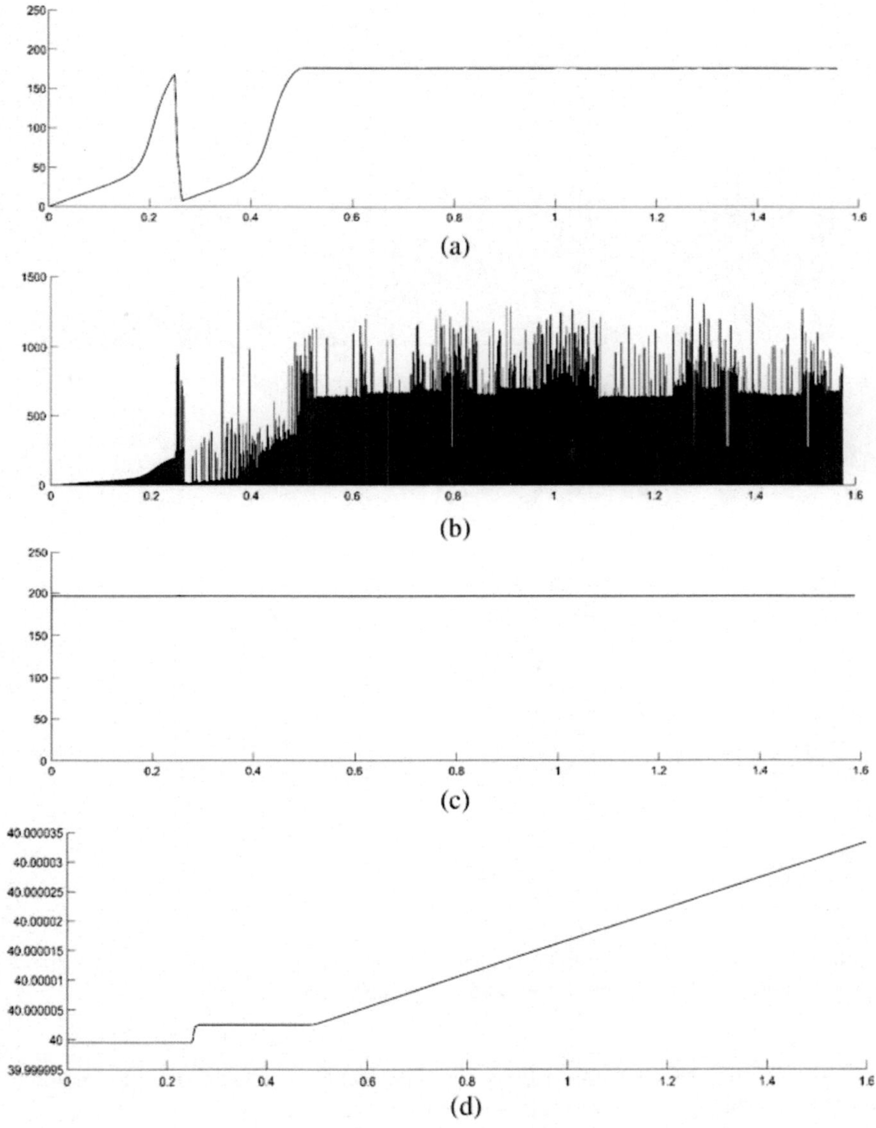

Fig. 11 Simulation result at 5 m/s wind speed: **a** rotor speed in rad/s **b** output voltage of rectifier bridge **c** dc–dc converter terminal voltage **d** SOC of battery

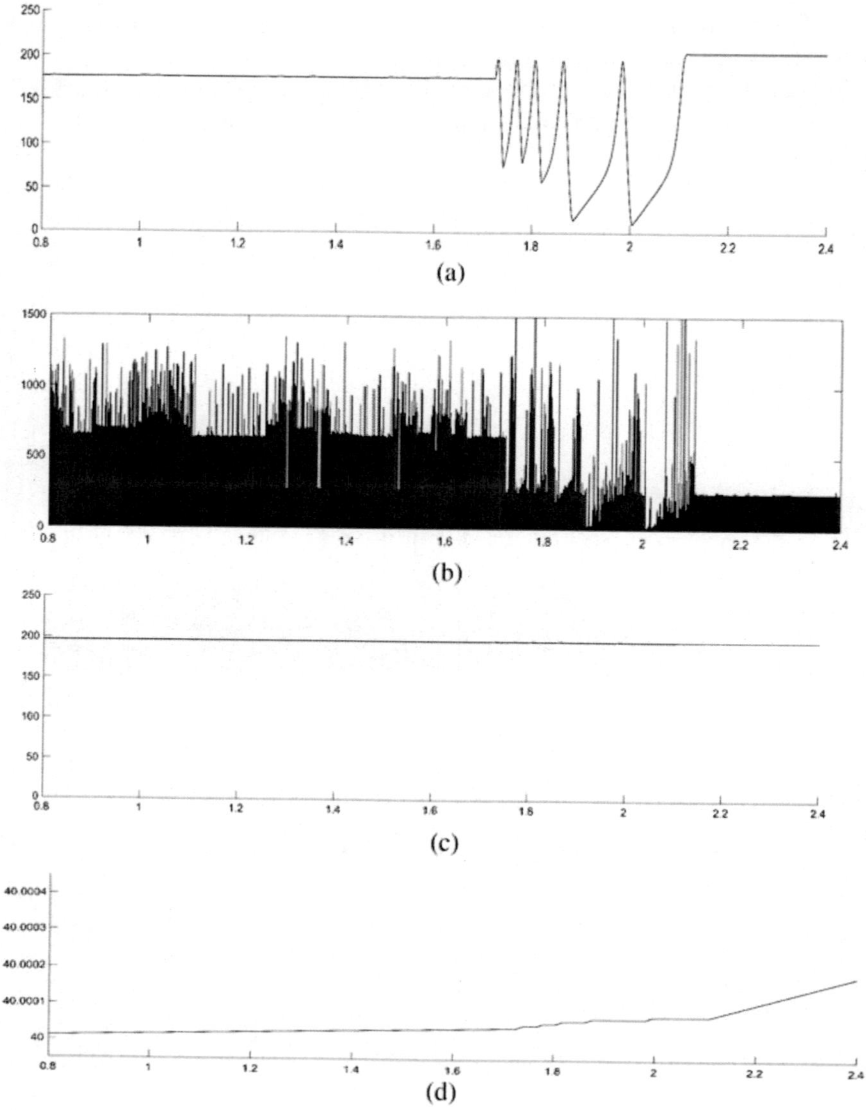

Fig. 12 Simulation results at sudden change in wind speed. **a** rotor speed in rad/sec **b** output voltage of rectifier bridge **c** DC–DC converter terminal voltage **d** SOC of battery

When wind speed of wind changes the disturbance in rotor speed of generator can be seen from Fig. 12a. From this fig, speed of generator get varied at $t = 1.7$ s and after 0.4 s generator gains a constant increased speed nearly to its base speed of 243.5544 rad/s. During this, disturbance of 0.4 s in speed of generator consequently output voltage fluctuations is more and voltage has reduced to zero many times, showing Fig. 12b and when generator runs at its base speed from $t = 2.1$ s onwards

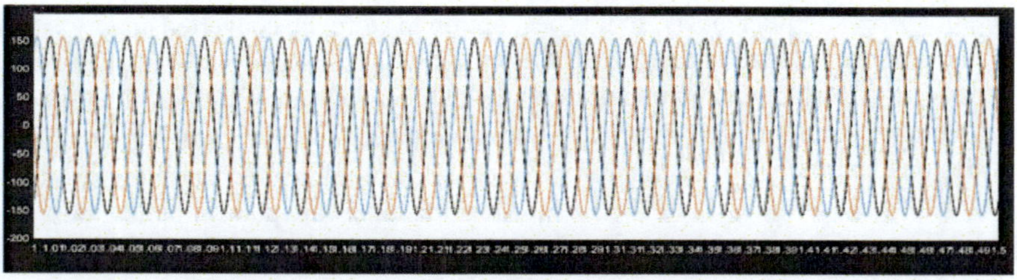

(a)

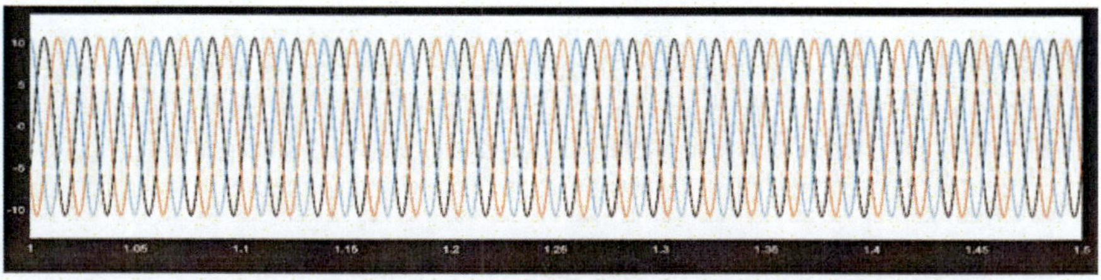

(b)

Fig. 13 VSI inverter output when supplied to load. **a** load voltage **b** load current

dc output becomes at a constant value of 250v nearly as can be seen from figure. Figure 12c shows output voltage of dc–dc converter, at this constant line of nearly 200v, still a constant value. In Fig. 12d instant SOC of battery that is charging has varied and deviated when the variation in generator speed occurs and then the slope increases due to increase in output current at base speed of rotor attained at maximum wind speed of 9 m/s.

The above results show behaviour of wind energy conversion system during charging the LAB. In this way, we can charge the battery and can store the converted energy, but simultaneously we can operate a load through this stored charge in battery. An inverter circuit connecting load and the charged LAB when runs in Simulink the load current and load voltage comes out as shown in Fig. 13. In 13a three-phase load voltage with peak-to-peak value of 150 V is shown and in 13b load current with peak-to-peak value of 10 amperes is shown, this value of load voltage and current results for a 2 KW resistive load. Since load power is driven by battery, it does not get affected with disturbances in wind speed or generated power.

3 Conclusion

Model demonstrates a small-scale wind turbine conversion system that may employed to use it for residential applications, and is analysed in Simulink environment. The results above clearly employ that use of storage system offers a constant supply to the loads and also charging of LABs in variable wind speed condition is not much get effected. And as we see variation in wind speed cause variation in output power simultaneously, so it is best option to store the converted energy rather to use directly. A different type of LAB with low voltage and power capacity may be used as a load to get charged at lower voltage levels, can be designed with buck converters maintaining a constant low output voltage. A simple charging circuit can be designed with low cost using uncontrolled diode rectifiers to charge low kilowatt-hour lead-acid batteries. It can be said for residential or low power applications, use of LABs as storage system is safe and is a good substitute to store converted energy.

References

1. Barote L, Marinescu C (2010) PMSG wind turbine system for residential applications. In: SPEEDAM 2010 international symposium on power electronics, electrical drives, automation and motion
2. Suresh A, Rajakumar S (2020) Design of small horizontal axis wind turbine for low wind speed rural applications. Mater Today Proc (2019 Elsevier Ltd., Peer-review under responsibility of the scientific committee of the Advanced Materials for Clean Energy and Health Applications (AMCEHA))
3. MacPhee DW, Beyene A (2018) Performance analysis of a small wind turbine equipped with flexible blades. Renew Energy. https://doi.org/10.1016/j.renene.2018.08.014
4. Ali RB, Schulte H, Mami A (2017) Modeling and simulation of a small wind turbine system based on PMSG generator. 2017 Evolving and adaptive intelligent systems (EAIS). IEEE, Solvenia
5. Peña Asensio A, Arnaltes Gómez S, Rodriguez-Amenedo JL, García Plaza M, Eloy-García Carrasco J, Alonso-Martínez de lasMorenas JM (2018) A voltage and frequency control strategy for stand-alone full converter wind energy conversion systems. Energies. www.mdpi.com/journal/energies
6. Hebala A, Ghoneim WAM, Ashour HA (2018) Detailed design procedures for low-speed, small-scale, PMSG direct-driven by wind turbines. IEEE
7. Barote L, Marinescu C, Cirstea MN (2013) Control structure for single-phase stand-alone wind-based energy sources. IEEE Trans Industr Electron 60(2)
8. Chen H, Xu D, Deng X (2020) Control for power converter of small-scale switched reluctance wind power generator. IEEE Trans Industr Electron
9. Putri RI, Pujiantara M, Priyadi A, Ise T, Purnomo MH (2018) Maximum power extraction improvement using sensorless controller based on adaptive perturb and observe algorithm for PMSG wind turbine application. IET Electr Power Appl 12:455–462
10. Maya GJ, Davidson A, Monahov B (2018) Lead batteries for utility energy storage: a review. J Energy Storage 15:145–157
11. Tang X, Yin M, Shen C, Xu Y, Dong ZY, Zou Y (2018) Active power control of wind turbine generators via coordinated rotor speed and pitch angle regulation. IEEE Trans Sustain Energy
12. Syahputra R, Soesanti I (2019) Performance improvement for small-scale wind turbine system based on maximum power point tracking control. Energies 12:3938. www.mdpi.com/journal/energies
13. Mousa HHH, Youssef A-R, Mohamed EEM (2019) Variable step size P&O MPPT algorithm for optimal power extraction of multi-phase PMSG based wind generation system. Electr Power Energy Syst 108:218–231
14. Lyu X, Zhao J, Jia Y, Xu Z, Wong KP (2018) Coordinated Control Strategies of PMSG-based wind turbine for smoothing power fluctuations. IEEE Trans Power Syst
15. Battisti L, Benini E, Brighenti A, Dell'Anna S, Raciti Castelli M (2018) Small wind turbine effectiveness in the urban environment. Renew Energy 129:102e113

Optimal Charging of Plug-in Electric Vehicle Using Interruptible Scheduling Algorithm

Sulthan Sheik Mohammed and Femin Titus

Abstract Various manufacturers are introducing electric vehicles (EVs) as an environmental-friendly alternative to automobiles with internal combustion engines (ICE), with numerous advantages. In comparison to ICE vehicles, EVs have lower running expenses and no carbon footprints. A plug-in electric vehicle (PEV) is a vehicle that uses rechargeable battery packs that can be charged from the grid. However, there are significant obstacles to EV adoption on a broad scale. The subject of EV charge scheduling is a large and difficult problem that has prompted lot of researches in recent years. In this paper, charge scheduling of EVs by interruptible scheduling algorithm is presented. Electric vehicle charging will be done in an interrupted manner so as to minimize the charging cost. The proposed scheduling algorithm is developed using MATLAB. To verify algorithm's efficiency and accuracy, tests were carried out under different cases. The algorithm is formulated based on ToUP tariff.

Keywords Plug-in electric vehicle (PEV) · Charging · Optimization · Interruptible scheduling · Time of use pricing (ToUP) · MATLAB

1 Introduction

Electric vehicles (EVs) were prevalent in the nineteenth century, but their popularity faded after the invention of the internal combustion engine. Automobile industries all around the world have worked to enhance the manufacture and use of electric vehicles in recent years. Electric vehicles (EVs) have gotten lot of attention recently as an environmentally safe and cost-effective option over the conventional vehicles driven by the ICEs [1]. Since EVs use batteries to store electric energy for powering the

F. Titus
Department of Electrical and Electronics Engineering, TKM College of Engineering Kollam, Kollam, India

S. Sheik Mohammed (✉)
Faculty of Engineering, Universiti Teknologi Brunei, Gadong A BE1410, Brunei Darussalam
e-mail: sheik.sulthan@utb.edu.bn

K. Namrata et al. (eds.), *Smart Energy and Advancement in Power Technologies*, Lecture Notes in Electrical Engineering 926,
https://doi.org/10.1007/978-981-19-4971-5_54

motor, it has been acknowledged that they will help to minimize the consumption of petroleum resources and the production of greenhouse gases. However, there are difficulties to the EV adoption on a wide scale. EVs are still more expensive to acquire than ICE vehicles, while having lower operating costs. In the present scenario, access to charging stations is very limited, and creating a public charging infrastructure requires significant capital expenditure [2]. Furthermore, connection of EVs to the grid is unpredictable, and they receive large amount of power from the grid while charging. As a result, uncoordinated charging of a large number of electric vehicles would result in negative influence on grid operations that includes reliable and quality power supply to the consumers. Several studies have looked into the influence of electric vehicles on the electricity grid [3]. The development in PEV-related load presents the electricity industry with new opportunities and problems. One way to reduce the grid's impact from EVs is to develop a controlled charging and discharging profiles. This can be accomplished by aggregating multiple sets of EVs for charging or discharging with varying start times and durations in order to maintain grid constraints. PEV charging has been studied and reported in several researches [4–6]. The impact of PEV penetration on the load profile of the distribution network is investigated in [7]. It has been demonstrated that depending on vehicle attributes and distribution grid type, peak load might increase by 10% to 35%. Another option is to encourage charging stations or aggregators to schedule charging during hours of low demand. Several scholars have taken on the problem of PEV charging scheduling in recent years. Real-time data is utilized to build and simulate the scheduling problem in [8]. Both charging and draining of PEVs are controlled using a robust and distributed algorithm. To model the energy demands and schedule the PEV charging, a linear program technique with a dynamic algorithm was applied [9]. Jing Huang suggested an algorithm for charging plug-in hybrid EVs (PHEVs) in shared parking lots in [10], with the purpose of managing demand and improving charging service quality. To discover the best schedule, several scenarios and techniques are tested. The technique of charging the earliest deadline first is proven as the most effective at balancing service quality and utilization. Using stochastic programming, Clement et al. suggest coordinated charging [11]. A clustering approach is utilized to group EVs in [12]. The purpose of the sliding window iterative approach used to schedule EVs online is to flatten the load profile and lower the cost to the distribution system operators (DSO) and customers.

Other research looks into demand response, and [13] proposes an optimization model for controlling the consumer's load level with real-time pricing (RTP)-based scheme. EV charge scheduling optimization has been studied extensively with a wide range of goals. The charging time for EVs is described in [14], with the goal of increasing feeder terminal load utilization while lowering power loss in the distribution network. The multi-agent strategy used in [15] coordinates EV charging in order to accomplish peak shaving and three-phase equilibrium. A real-time load management strategy for EV charging is proposed in [16] that reduces voltage drop and power loss. ToUP data is used to analyze the optimized charge scheduling mechanism in [17]. A global intelligent system is described in [18] for discovering the best charging and discharging techniques for electric vehicles. In [19], employing both particle swarm optimization and genetic algorithm, a centralized charging solution for EVs is introduced under the battery swapping concept. A day-ahead EV charge plan based on an aggregative game model is presented in [20]. The best charge scheduling of electric vehicles using a heuristic algorithm is studied in [21]. The charge scheduling problem is an optimization problem, and depending on the scheduling goals, several objective functions have been utilized in the problem formulations. Some studies have attempted to optimize the grid-side benefits such as lowering the financial cost of power supply [22], lowering grid operation costs, including the cost of wind and hydro-power availability for spinning reserves [23], lowering distribution system load variance [24], lowering distribution system loss and maximizing the profits of thermal and wind plants while minimizing energy trading risks.

In this paper, charge scheduling optimization of electric vehicle using interruptible scheduling algorithm is presented. The proposed works mainly aim at reducing the peak demand from grid by altering the charging pattern and thereby reduce the charging cost.

2 Development of Interruptible Charge Scheduling

2.1 *Problem Statement and Objective Function*

The explosion in the number of electric vehicles (EVs) has a significant impact on the energy systems and distribution network. Large-scale EVs inevitably increase the load on the grid, while uncoordinated PEV charging pose challenges to the stability and security of the grid. The uninterrupted charging schedules may cause huge amount of energy demand to the grid that result in the increased rate of EV charging cost. At the same time, coordinated charging can effectively improve the utilization rate of energy. Therefore, it is more important to study the economic and effective scheduling strategies to manage the charging behavior of EVs to enhance the beneficial effects, such as flattening the load curve, reducing the system cost and operation losses. To achieve the goal considering these factors, an appropriate scheduling strategy and an efficient optimization solution are essential. The main objective of the proposed work is to schedule the EV charging in an interrupted manner and thereby to reduce the charging cost to its minimum. The main factors considered for the algorithm development are as follows:

(i) Vehicle arrival time (A_t)
(ii) Vehicle departure time (D_t)
(iii) Length of charging duration (E)
(iv) Charging rate of vehicle (R_c)
(v) Tariff rate (C_T)
(vi) Number of off-peak, peak and shoulder hours (a_n)

2.2 Mathematical Modeling of Algorithm

In this case, scheduling PEV charging turns to a multistage decision-making problem, where during each interval, the aggregator has to decide whether to charge the PEV or not. The interruptible charge scheduling is formulated on the basis of ToUP tariff [17]. The 24-h time slot in a day is represented as T, where slot "1" is between 1:00 AM and 1:59 AM, likewise slot "16" is between 04:00 PM and 04:59 PM. The vehicle is allowed to charge only during the given time interval T, which is between A_t and D_t, according to its required charge duration (E). For example, consider the arrival time A_{ti} of ith vehicle to be 07:00 AM and its departure time D_{ti} to be 11:00 AM. Then, the timeline for charging the vehicle will be $T = [7, 8, 9, 10, 11]$.

At first, the charging cost for each hour in the timeline for ith vehicle has to find out using the expression given below,

$$\text{PR}_i(T) = \sum_{T=A_t}^{D_t} C_T \times R_{ci}$$

Since the ToUP tariff-pricing scheme includes off-peak hours, shoulder hours and peak hours, we get a set of three distinct costs from (1). Out of these set of costs, single value from each set of tariff has to be sorted out by using the expression given below,

$$\text{EC} = \text{unique}(\text{PR}(:))$$

where EC is the set of single values from off-peak hours, shoulder hours and peak hours arranged in ascending order. In order to schedule the charging for required time duration, it is necessary to find out the number peak hours, off-peak hours and shoulder hours during the particular time interval T. It can be found out by using the equation given below,

$$a_n = \text{sum}(\text{PR}(:) == n\text{th minimum value}) \tag{3}$$

where "n" ranges from 1 to 3 since we have three different tariff rates. a_1 gives the total number of off-peak hours during T, whereas a_2 and a_3 denote the total number of shoulder hours and peak hours, respectively. Then, schedule the charging according to the availability of off-peak hours using the equation given below,

$$\text{idPRMin} = \text{find}(\text{PR} == a_1, E_i) \tag{4}$$

If the number of off-peak hours is not enough to satisfy the required charging duration E_i, the remaining hours can be scheduled by taking shoulder hours using the equation below,

$$\text{idPRMin} = \text{find}(\text{PR} == a_2, E_i) \tag{5}$$

where $a_2 = E_i - a_1$. If both off-peak hours and shoulder hours are not enough to satisfy the required charging duration E_i, the remaining hours can be scheduled by taking peak hours using the equation below,

$$\text{idPRMin} = \text{find}(\text{PR} == a_3, E_i) \tag{6}$$

where $a_3 = E_i - a_1 - a_2$. The total charging cost during interrupted schedule can be calculated as given below,

$$\text{Total Cost}(\text{a}_\text{n}) = C_T(\text{idPRMin}) * R_{ci} \tag{7}$$

The charging status is denoted by U_i^T, where $U_i^T \in \{0, R_{ci}\}$. $U_i^T = R_{ci}$ implies that the ith PEV will be charging in the interval T, whereas $U_i^T = 0$ implies that it is not charging during the said interval. In addition to this, since the PEV has to charge between A_{ti} and D_{ti} $U_i^T = 0$, if $T < A_{ti}$ or $T > D_{ti}$. The stepwise development of the algorithm is shown in Table 1.

Table 1 Interruptible charge scheduling algorithm

Step 1	Load the vehicle data (A_t, D_t, R_c, and E) and ToUP data
Step 2	For $i = 1{:}n$
Step 3	Receive A_{ti}, D_{ti}, R_{ci}, and E_i and the tariff rate (C_T)
Step 4	Find the charging cost within the given interval (h), $\text{PR}_i(h) = \text{Tariff rate}(C_T) \times \text{Charging rate}(R_{ci})$
Step 5	Arrange the charging cost in ascending order using, EC = unique(PR)
Step 6	Find out the number of 1st, 2nd and 3rd minimum values using,$a_n = \text{sum}(\text{PR}(:) == n\text{th minimum value})$
Step 7	Check whether number of 1st minimum value ≥ Required charge duration Then, Schedule charging using, $\text{idPRMin} = \text{find}(\text{PR} == a_1, E_i)$ • If $a_1 < E_i$, Then $a_2 = E_i - a_1$ and Schedule charging using, $\text{idPRMin} = \text{find}(\text{PR} == a_2, E_i)$ • If $a_1 + a_2 < E_i$, Then $a_3 = E_i - a_1 - a_2$ and Schedule charging using, $\text{idPRMin} = \text{find}(\text{PR} == a_3, E_i)$
Step 8	Find the total charging cost during the charging hours using, $\text{Total Cost}(\text{a}_\text{n}) = C_T(\text{idPRMin}) * R_\text{ci}$
Step 9	Plot the schedule
Step 10	End

3 Simulation Results and Analysis

3.1 *Charge Scheduling of Electric Vehicle*

The charge scheduling of EVs is done for 24 h in order to examine the accuracy and efficiency of the algorithm. The ToUP tariff of Australian Capital Territory (ACT) is taken for this work [17]. The electricity cost for each hours is shown in Table 2.

From Table 2, it is evident that the off-peak hours are between 10:00 PM and 06:00 AM. During that time, electricity costs 10 cents. Peak hours are between 07:00 AM and 08:00 AM and 05:00 PM to 07:00 PM, where electricity cost is 21 cents. Rest of the period is shoulder hours, and the cost during shoulder period is 14 cents. The graphical representation of the ToUP tariff at ACT is shown in Fig. 1.

To test the algorithm, three electric vehicles with different power ratings and charging rates are chosen. Table 3 lists the parameters of EVs [25]. Since the algorithm is designed to calculate the charging cost on an hourly manner, the electric vehicle's rate of charging has been somewhat adjusted for the analysis. Table 4 shows the arrival and departure time, charge duration, charging rate and total energy of the selected vehicles.

According to the data given in Table 4, the charging cost for the selected EVs is calculated for unplanned, optimal uninterrupted and optimal interrupted charge scheduling. The algorithm is developed in MATLAB, and the scheduling status is generated for each case. The charging cost comparison is given in Table 5.

Table 2 Tariff rate for 24 h

Time slots	1	2	3	4	5	6	7	8	9	10	11	12
Tariff	10	10	10	10	10	10	21	21	14	14	14	14
	13	14	15	16	17	18	19	20	21	22	23	24
	14	14	14	14	21	21	21	14	14	10	10	10

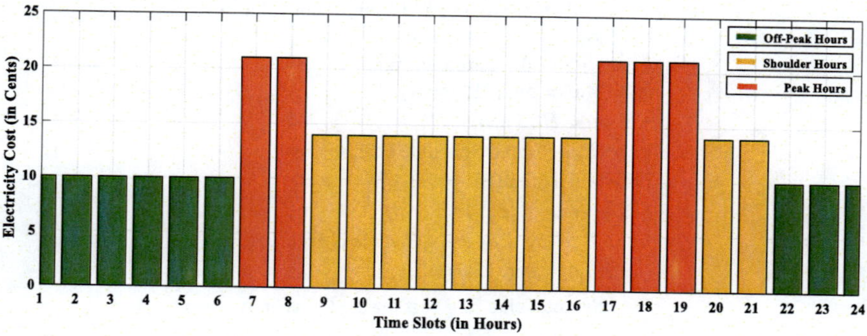

Fig. 1 ToUP tariff of Australian capital territory

Table 3 Parameters of EVs

Vehicle model	Vehicle's rate of charging	Battery capacity of vehicle (kWh)
Mercedes-EQA-250	7.4 kW (230 V, 32 A)	79.8
Skoda-Enyaq-iV-50	7.2 kW (230 V, 31 A)	58
Honda-e	3.7 kW (230 V, 16 A)	35.5

Table 4 Simulation details of EVs

Vehicle	Vehicle arrival time (A_t)	Vehicle departure time (D_t)	Length of charging duration (hr)	Charging rate of vehicle (kW)	Total energy (kWh)
Mercedes-EQA-250	11	23	10	7	77
Skoda-Enyaq-iV-50	10	21	8	7	56
Honda-e	7	22	11	3	33

Table 5 Charging cost comparison

Vehicle	Unplanned charge scheduling *(AUD$)	Uninterruptible optimal charge scheduling (AUD$)	Interruptible optimal charge scheduling (AUD$)
Mercedes-EQA-250	11.27	10.99	10.01
Skoda-Enyaq-iV-50	8.33	8.33	7.84
Honda-e	5.55	5.25	4.83
Total	**25.15**	**24.57**	**22.68**

* 100 Cents = 1 Australian Dollar

3.2 *Types of Charge Scheduling Methods*

Unplanned Charge Scheduling: In this case, the vehicle starts to charge from its arrival time itself. Then, it continues to charge until its required time duration. Since the charging is going on an uninterrupted manner, the vehicle is forced to charge during the peak hours also. It may results in an increased EV charging cost.

Uninterruptible Optimal Charge Scheduling: In this method, the optimal charging hours which incur the lowest charging cost are found out. Then, the vehicle is admitted to charge during that particular time period in an uninterrupted manner. In this scheduling also, peak hours may not be expelled out completely. This method is better efficient than unplanned scheduling. The reduction in the charging cost may not that appreciable when compared to interruptible optimal scheduling.

Interruptible Optimal Charge Scheduling: In interrupted charge scheduling, the charging of EVs is interrupted from charging during peak hours. The algorithm tries to charge the vehicle during time slots having minimum charging cost. This will result in the significant reduction of charging cost when compared to both unplanned and uninterruptible charge scheduling. The charging status of all the selected vehicles while applying the above-mentioned approaches is shown in Figs. 2, 3, and 4.

By analyzing Fig. 2, we can see that the required charge duration of the vehicle is 10 h. The vehicle starts its scheduling from 11th hour which is a shoulder period. The algorithm now rejects the peak hours between 17 and 19th slots and restarts its charging till the departure time. But, these hours were not sufficient to satisfy the vehicle's required charge duration. Therefore, the algorithm picks up one peak hour, i.e., the 17th hour.

In Fig. 3, by analyzing the first two plots, i.e., in the case of unplanned and uninterruptible charge scheduling, both the graphs are same. This is because minimum charging cost for EV is obtained on the first slot itself, i.e., the time period starting from the arrival time of vehicle to its required charging duration. Therefore, both the graphs are appeared to be same.

By analyzing Fig. 4, we can see that the required charge duration of the vehicle is 11 h. The vehicle actually starts its scheduling from 9th hour which is a shoulder period and continues to charge till 17th hour. The algorithm now rejects the peak hours between 17 and 19th slots and restarts its charging till the departure time. But, these hours were not sufficient to satisfy the vehicle's required charge duration. Therefore, the algorithm picks up one peak hour, i.e., the 7th hour. That is why, the vehicle is initiating its charging from 7th hour.

From Table 5, it is inferred that the total charging cost is significantly reduced during interruptible charge scheduling. Uninterruptible charging provides a reduction in cost, when compared to unplanned scheduling, but not to a great extent since the charging is going on without any interruption by taking the peak hours also. The interruptible scheduling algorithm always tries to exclude the peak hours and attains the required charge duration by taking the off-peak and shoulder hours. If the length of charge duration still demands charging hours, then only the algorithm considers the peak time slots. It will always result in the reduced rate of charging cost for EVs. Thus, the proposed method is simple and cost-effective for EV charging applications.

4 Conclusion

Charge scheduling of electric vehicles using interruptible scheduling algorithm is presented in this work. The algorithm is developed based on the ToUP tariff structure. The charging of EVs will get automatically interrupted during peak hours in ToUP. Simulations are carried out on MATLAB and plotted the graphs of scheduling for the case of unplanned charge scheduling, uninterruptible charge scheduling and for interruptible charge scheduling. It is found that scheduling along with interruptible algorithm can significantly reduce the charging cost of EV's to a great extent. The algorithm continuously checks for the availability of minimum charging cost so as to assign the vehicle to that particular time slot having least cost. If the algorithm

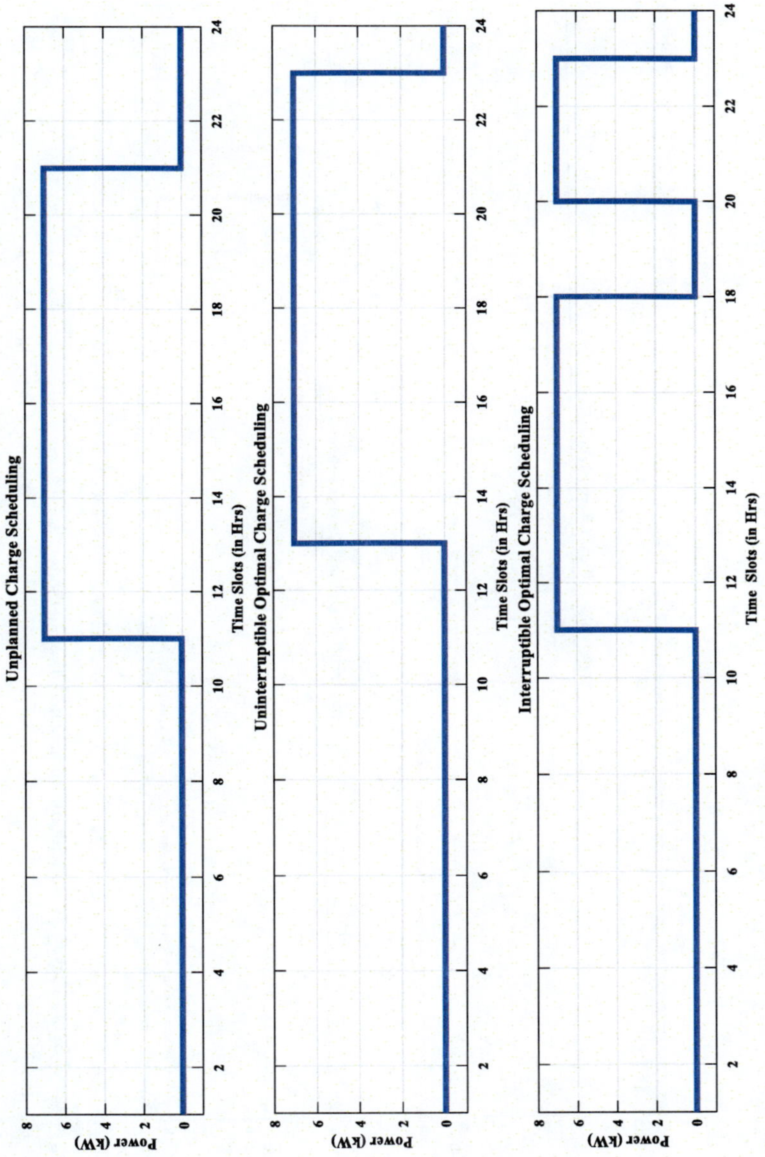

Fig. 2 Charging status of Mercedes-EQA-250 with different scheduling methods

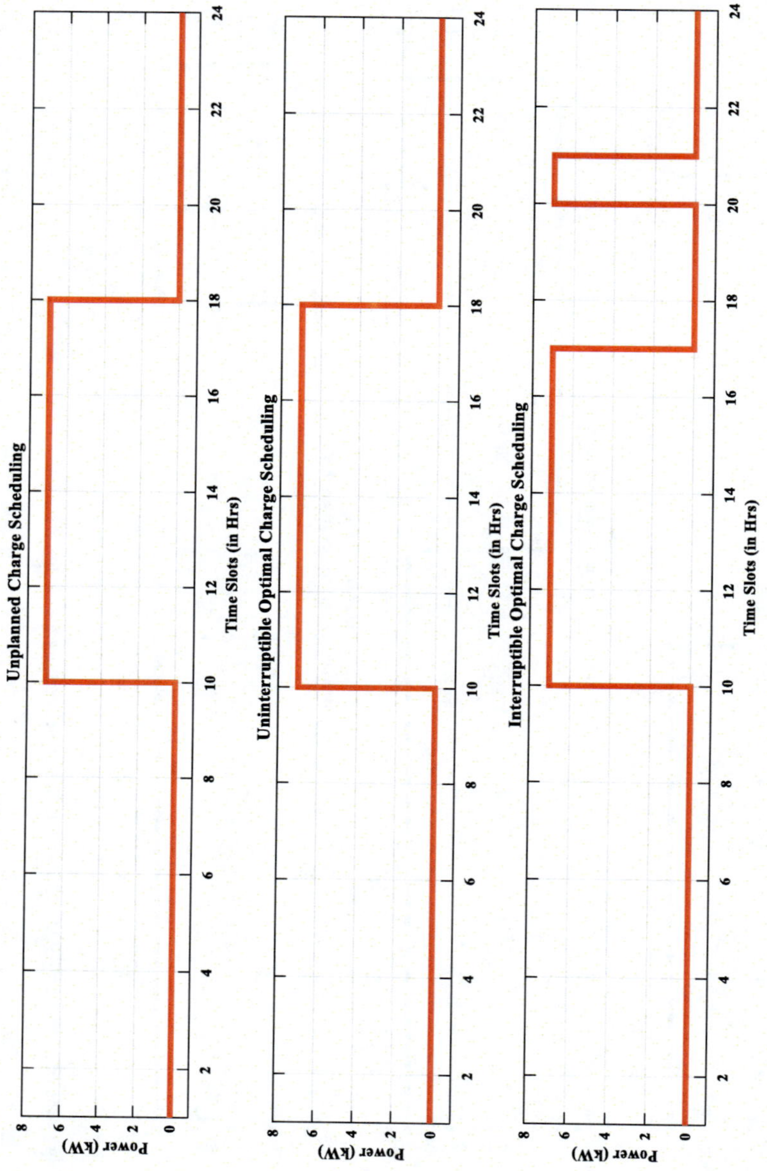

Fig. 3 Charging status of Skoda-Enyaq-iV-50 with different scheduling methods

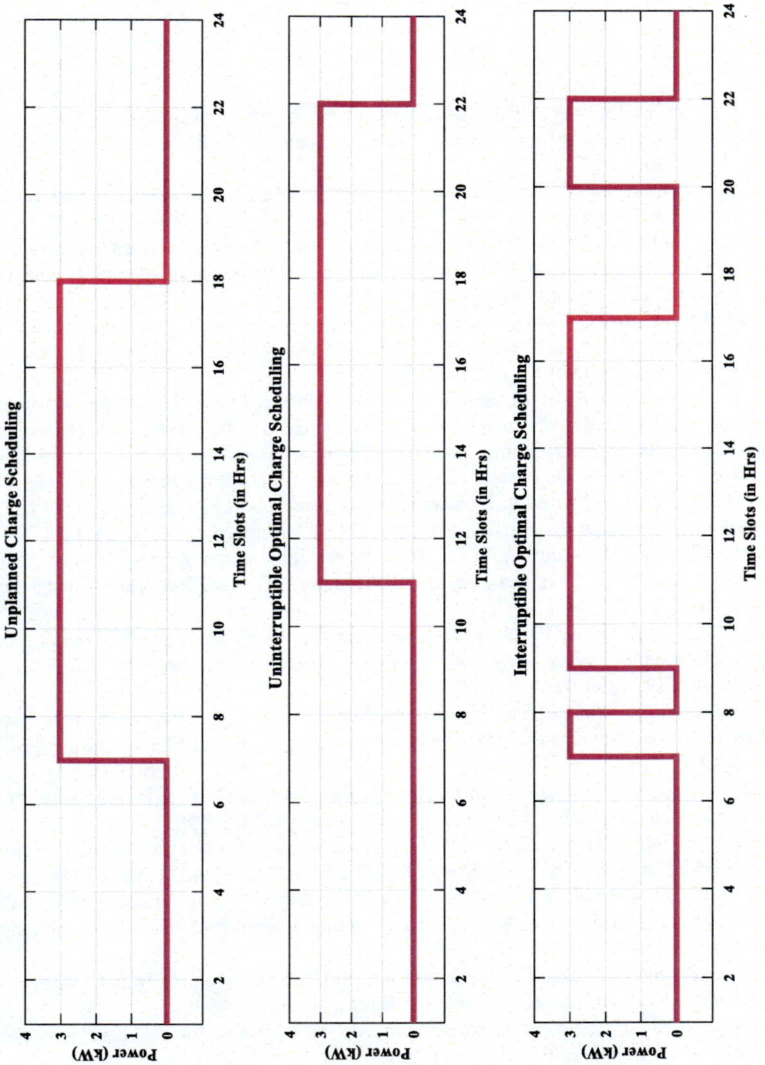

Fig. 4 Charging status of Honda-e with different scheduling methods

met the required charge duration, it automatically stops and generates the interrupted schedule of charging of the selected EV. The accurate charging slot for each vehicle is obtained from the algorithm at which minimum cost can be achieved among all the possible slots.

References

1. Jabalameli N, Su X, Ghosh A (2019) Online centralized charging coordination of PEVs with decentralized Var discharging for mitigation of voltage unbalance. IEEE Power Energy Technol Syst J 6(3):152–161
2. Li M, Gao J, Chen N, Zhao L, Shen X (2020) Decentralized PEV power allocation with power distribution and transportation constraints. IEEE J Sel Areas Commun 38(1):229–243
3. Konara MSY, Kolhe ML (2020) Priority based coordinated electric vehicle charging system for heterogeneous traffic. In: 2020 5th international conference on smart and sustainable technologies (SpliTech), Split, Croatia, pp 1–6
4. Zhu X, Han H, Gao S, Shi Q, Cui H, Zu G (2018) A multi-stage optimization approach for active distribution network scheduling considering coordinated electrical vehicle charging strategy. In: IEEE Access, vol 6, pp 50117–50130
5. Falco M, Arrigo F, Mazza A, Bompard E, Chicco G (2019) Agent-based modelling to evaluate the impact of plug-in electric vehicles on distribution systems. In: 2019 international conference on smart energy systems and technologies (SEST), Porto, Portugal, pp 1–6
6. Wu Y, Ravey A, Chrenko D, Miraoui A (2018) A real time energy management for EV charging station integrated with local generations and energy storage system. In: 2018 IEEE transportation electrification conference and expo (ITEC), Long Beach, CA, USA, pp 1–6
7. Zhao J, Xu Z, Wang J, Wang C, Li J (2018) Robust distributed generation investment accommodating electric vehicle charging in a distribution network. IEEE Trans Power Syst 33(5):4654–4666
8. Şengör İ, Güner S, Erdinç O (2021) Real-Time algorithm based intelligent EV parking lot charging management strategy providing PLL type demand response program. IEEE Trans Sustain Energy 12(2):1256–1264
9. Liu Z, Wu Q, Huang S, Wang L, Shahidehpour M, Xue Y (2018) Optimal day-ahead charging scheduling of electric vehicles through an aggregative game model. IEEE Trans Smart Grid 9(5):5173–5184
10. Ko H, Pack S, Leung V (2022) An optimal battery charging algorithm in electric vehicle-assisted battery swapping environments. IEEE Trans Intell Transp Syst 23(5):3985–3994
11. Sun B, Huang Z, Tan X, Tsang DHK (2018) Optimal scheduling for electric vehicle charging with discrete charging levels in distribution grid. IEEE Trans Smart Grid 9(2):624–634
12. Das S, Acharjee P, Bhattacharya A (2021) Charging scheduling of electric vehicle incorporating grid-to-vehicle and vehicle-to-grid technology considering in smart grid. IEEE Trans Ind Appl 57(2):1688–1702
13. Wang S, Bi S, Zhang YA (2021) Reinforcement learning for real-time pricing and scheduling control in EV charging stations. IEEE Trans Ind Inform 17(2):849–859
14. Zhang Y, You P, Cai L (2019) Optimal charging scheduling by pricing for EV charging station with dual charging modes. IEEE Trans Intell Transp Syst 20(9):3386–3396
15. Liu D, Wang W, Wang L, Jia H, Shi M (2021) Dynamic pricing strategy of electric vehicle aggregators based on DDPG reinforcement learning algorithm. IEEE Access 9:21556–21566
16. Elma O (2020) A dynamic charging strategy with hybrid fast charging station for electric vehicles. Elsevier J Energy 24(6):225–236
17. Sheik Mohammed S, Imthias Ahamed TP, Devaraj D (2019) Optimized charge scheduling of plug-in electric vehicles using modified placement algorithm. In: 2019 international conference on computer communication and informatics (ICCCI), Coimbatore, Tamil Nadu, India, pp 1–5

18. Cai H, Chen Q, Guan Z, Huang J (2018) Day-ahead optimal charging/discharging scheduling for electric vehicles in microgrids. Prot Control Mod Power Syst 3:9
19. Ioakimidis CS, Thomas D, Rycerski P, Genikomsakis KN (2018) Peak shaving and valley filling of power consumption profile in non-residential buildings using an electric vehicle parking lot. Energy 148:148–158
20. Amirhosseini B, Hosseini SMH (2018) Scheduling charging of hybrid-electric vehicles according to supply and demand based on particle swarm optimization, imperialist competitive and training-learning algorithms. Sustain Cities Soc 43:339–349
21. Quddus MA, Shahvari O, Marufuzzaman M, Usher JM, Jaradat R (2018) A collaborative energy sharing optimization model among electric vehicle charging stations, commercial buildings, and power grid. Appl Energy 229:841–857
22. Wu X, Hu X, Yin X, Moura SJ (2018) Stochastic optimal energy management of smart home with PEV energy storage. IEEE Trans Smart Grid 9:2065–2075
23. Aiswariya L, Imthias Ahamed TP, Mohammed SS (2020) Optimal microgrid battery scheduling using simulated annealing. In: 2020 international conference on Power Electronics and Renewable Energy Applications (PEREA), pp 1–6
24. Coordinated charging of electric vehicles connected to a net-metered PV parking lot. In: Proceedings of 2017 IEEE PES innovation smart grid technologies. Conference Europe ISGT-Europe 2017. January 2018, pp 1–6
25. Compare electric vehicles-EV database. https://ev-database.org/. Last accessed 23 Mar 2021

Impact of Phase-Locked Loop on the Control of TCSC

Gaurav Kumar Singh, Jai Prakash Sharma, and Om Hari Gupta

Abstract A case study on the impact of phase-locked loop (PLL) on the control and behavior of TCSC during unbalanced three-phase conditions is discussed in this paper. For this study, PSCAD/EMTDC platform is used to simulate a TCSC-based compensation in the transmission line. One PLL is taken from the PSCAD/EMTDC library, and the other one is simulated using the available blocks in the PSCAD/EMTDC library. The comparison reveals that the simulated PLL based on a per-phase control scheme gives a promising result than that available in the library.

Keywords FACTS · TCSC · PLL · TCR · Firing angle · PSCAD

1 Introduction

In the last few decades, flexible alternating current transmission system (FACTS) devices have emerged as a boon for the power system sector. It not only enhances the power flow capability but also provides voltage, angle, and frequency stability. To be more precise, among all the FACTS devices, thyristor-controlled series compensators (TCSCs) have attracted the power sector due to their numerous advantages such as lower cost, easy control, series compensation, ability to damp out oscillations, and enhancing the power flow capability [1]. TCSCs have gained at most importance in the last 2 decades.

TCSC first proposed in 1986 by Vithayathil is a method for varying the impedance of the line. In 1991, the first project of TCSCs was installed at 345 kV substation in West Virginia U.S.A. Industries have made use of TCSCs unfairly due to their ability to enhance the power transfer capability and minimizing the sub-synchronous resonance (SSR) phenomenon of transmission line between the conventional generating station and distribution end [2]. Modeling a TCSCs is complex since it contains energy storing elements inductor, capacitor, and nonlinear power electronics devices

G. K. Singh (✉) · J. P. Sharma · O. H. Gupta
Department of Electrical Engineering, NIT Jamshedpur, Jamshedpur, India
e-mail: gk752670@gmail.com

K. Namrata et al. (eds.), *Smart Energy and Advancement in Power Technologies*, Lecture Notes in Electrical Engineering 926,
https://doi.org/10.1007/978-981-19-4971-5_55

[3]. During the control of wind farm, an oscillation similar to SSR is generated, and TCSCs is a one-stop solution to mitigate this.

With the help of TCSCs, we can impose a dynamic control over the series compensation of the transmission line. In addition to this, it also protects against overvoltage. Fixed series compensation gives rise to an undesirable effect of SSR. TCSC overcomes this drawback. For smooth working of TCSC, the firing angle of thyristor should be synchronized either with the voltage of capacitor or line current [4].

A phase-locked loop (PLL) is used to achieve this synchronization [4]. This study compares the impact of two different PLLs on the TCSC impedance in the case of an unbalanced network condition. For this work, one PLL available in PSCAD/EMTDC library is used, and the other PLL is simulated using the blocks available in PSCAD/EMTDC library. The comparison of both is discussed and shown graphically.

The paper is structured in the following way: Sect. 2 describes the concept related to TCSC, Sect. 3 focuses on the PLL and related concept, Sect. 4 covers the TCSC control scheme, and Sect. 5 describes the result and discussion followed by the conclusion and references.

2 Thyristor-Controlled Series Compensator (TCSC)

TCSC is a FACTS device that helps in series compensation by decreasing the line reactance, thereby increasing power transfer capability [5]. TCSC helps in enabling dynamic control over the series compensation of the transmission line. TCSC is the combination of thyristor and TCR connected in parallel. The TCR is nothing but a series combination of a reactor and anti-parallel thyristors [6]. Figure 1 shows the circuit of TCSC.

The firing angle (β) is used to control the effective reactance of the TCR branch which, in turn, controls the overall impedance of TCSC. A variable reactance of TCSC is illustrated in Fig. 2.

A typical characteristic of TCSC [reactance Vs firing angle (β)] is illustrated in Fig. 3. It is clear from the characteristic that as the β is varied, TCSC reactance (Xtcsc) also varies and can become positive (inductive) and negative (capacitive) [7]. The

Fig. 1 TCSC

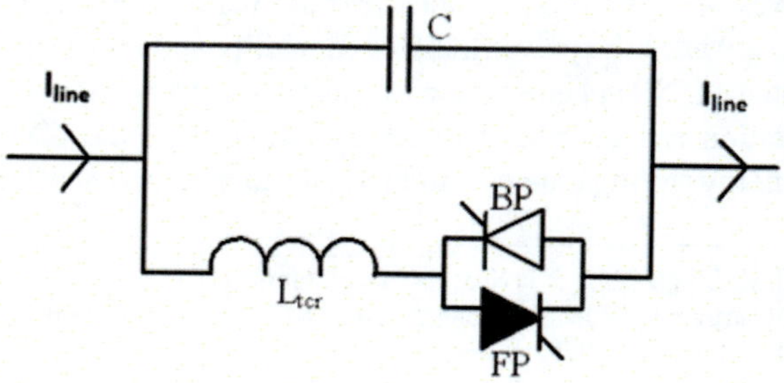

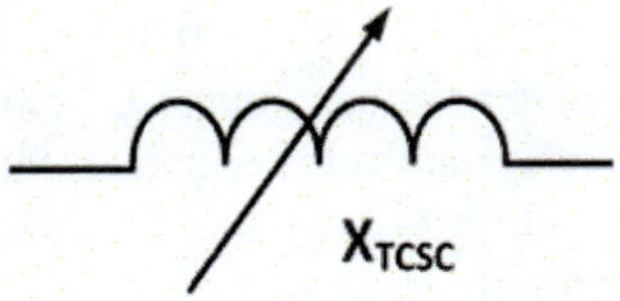

Fig. 2 Variable reactance of TCSC

principle of operation of TCSC is that increase the fundamental frequency voltage across the series compensated capacitor by changing the firing angle of anti-parallel thyristors and this enhanced voltage will change the reactance of the line.

For understanding the functioning of TCSC, it is to be understood that in TCSC, the voltage and current in fixed capacitor and TCR are not sinusoidal because of the switching action of the thyristor [8] LC circuit is shown in Fig. 4.

Thus, its equivalent impedance or reactance is given by

$$Z_{eq} = j\omega L \left\| \frac{1}{j\omega C} = -j\frac{1}{\omega C - \frac{1}{\omega L}} \right. \tag{1}$$

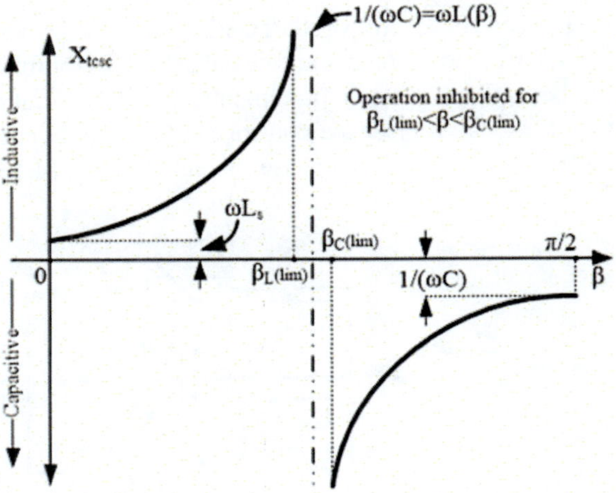

Fig. 3 Characteristic of TCSC

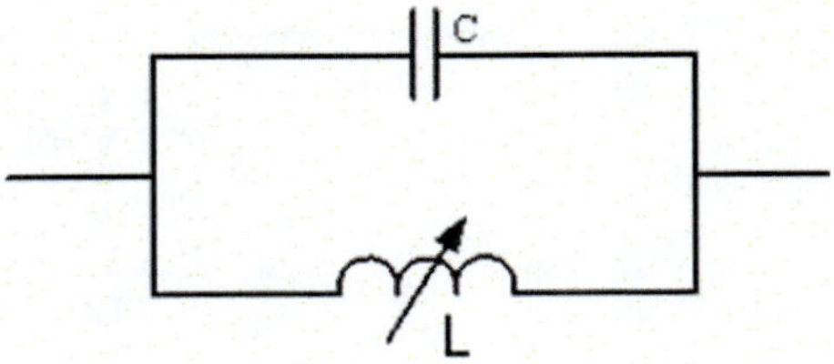

Fig. 4 LC circuit

If $(\omega C - \frac{1}{\omega L}) > 0$, which means that inductive reactance is greater than the capacitive reactance, then LC combination offers capacitive reactance, and this equivalent reactance value is greater than that of a fixed capacitor.

If $(\omega C - \frac{1}{\omega L}) = 0$, this is the case of resonance which is not feasible as it offers infinite reactance.

If $(\omega C - \frac{1}{\omega L}) < 0$, which implies that inductive reactance is greater than the capacitive reactance, then LC combination offers inductive reactance whose value is greater than fixed inductor.

Measurement of firing angle "β" is done with respect to V_{tcsc} and to change the firing angle "β", and the phase of V_{tcsc} is to be obtained. For this, the phase-locked loop (PLL) is required [9]. Therefore, a PLL structure has a direct impact on the control of TCSC impedance or reactance.

3 TCSC Control Scheme

As stated earlier in Sect. 2, TCSC is a combination of TCR and fixed capacitors. Overvoltage protection is provided by metal oxide varistor (MOV) [10]. MOV is illustrated in Fig. 5.

The control scheme of the TCSC used in the present work is illustrated in Fig. 5. Different control schemes are available for the control of TCSC [11]. However, based on the application and requirements, the TCSC control scheme is adopted [11]. In this work, the impedance control mode or phase control mode is used. Based on the TCSC voltage (V_{tcsc}) phase (obtained from PLL) and firing angle (β), the firing

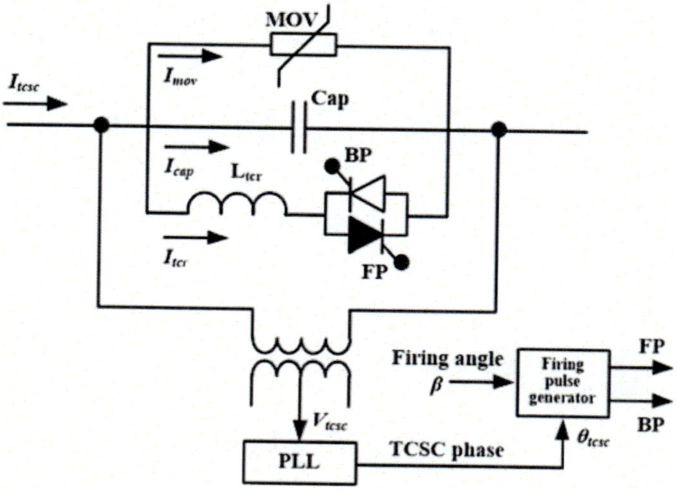

Fig. 5 Control scheme of TCSC with PLL

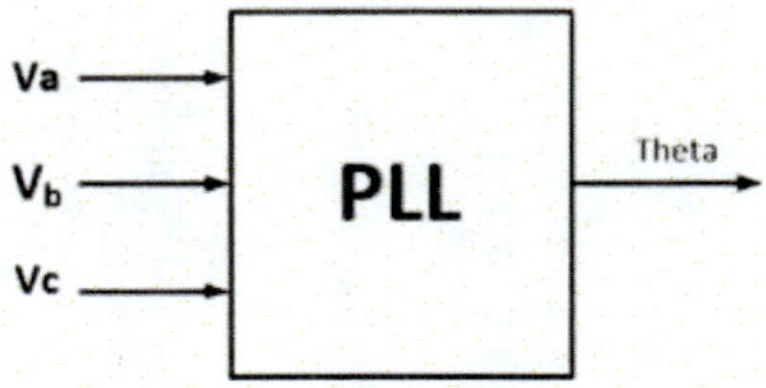

Fig. 6 PLL

pulses are generated. These pulses are given to the thyristors connected in an anti-parallel way. The firing pulse generator provides pulses for the forward thyristor (FP) and 180° shifted pulse for the backward thyristor (BP) [12].

4 Phase-Locked Loop (PLL)

The phase-locked loop or PLL was first invented in 1932. It is one of the most useful ways to recover and utilize the phase and frequency of the electrical system. A PLL generates an output waveform whose phase is exactly equal to the input waveform or we say phase is locked [13].

4.1 PLL from the PSCAD/EMTDC Library

Figure 6 shown below is a three-phase PLL, whose task is to generate a ramp signal "theta" that varies from 0° to 360°, which is locked in phase w.r.t phase voltage Va. Ramp "theta" is generated with the help of phase vector technique [14]. In this technique, trigonometric multiplication identities generate an error signal which varies the speed of the phase-locked oscillator, until the phase is exactly equal to input [15].

4.2 Proposed PLL Structure for an Unbalanced Network

The proposed PLL has been simulated using the blocks available in the PSCAD library as shown in Fig. 7.

The single line diagram of TCSC with compensation in a transmission system is illustrated in Fig. 8, and its PSCAD/EMTDC simulation diagram is shown in Fig. 9.

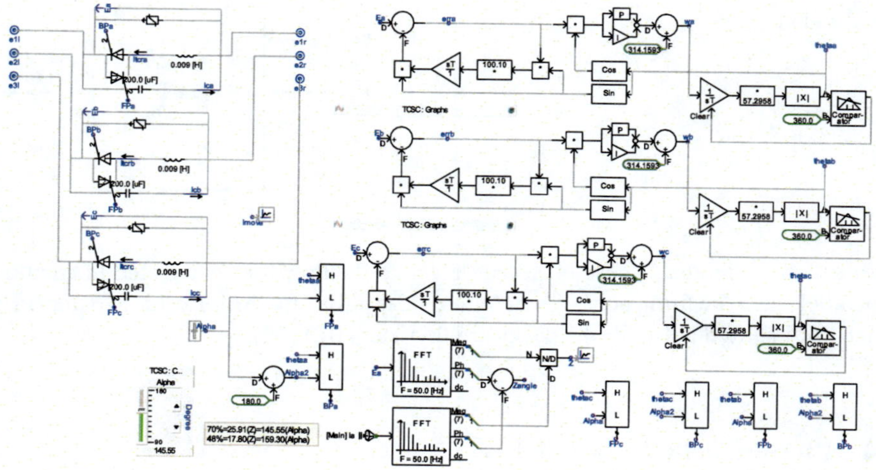

Fig. 7 TCSC and PLL simulated in PSCAD

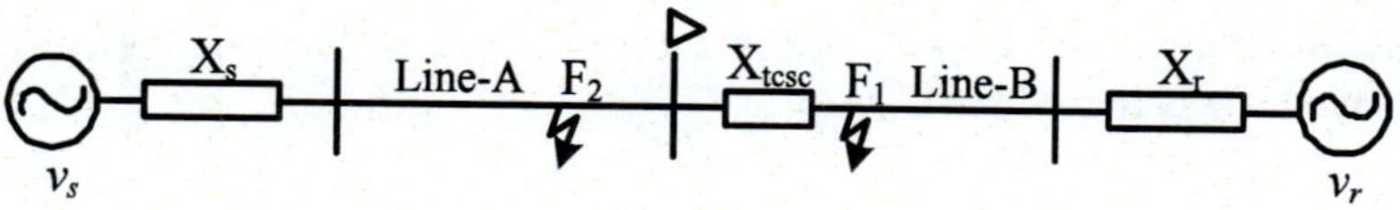

Fig. 8 Single line diagram of TCSC with compensation in the transmission system

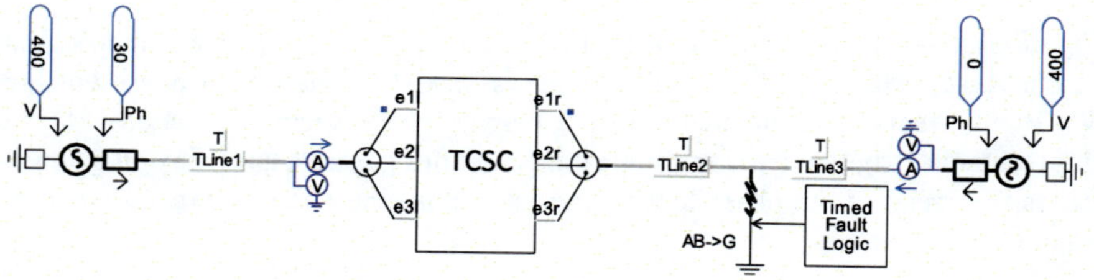

Fig. 9 PSCAD diagram of TCSC with compensation in the transmission system

5 Results and Discussion

This proposed PLL simulation has been carried out in PSCAD/EMTDC. Here, the prime objective was to compare the impact of PLL on the control of TCSC in an unbalanced condition. One PLL available in the PSCAD library is considered, and the other is simulated using the blocks available in the library. For the analysis purpose A-G fault (A-phase) has been considered at 50 km away from the relaying bus, and compensation level of TCSC was kept same in both the cases (30%). Practically it is expected that per-phase impedance should be the same before and after the fault. The run time of the simulation is 4 s, and the fault is created at 2 s (Fig. 10).

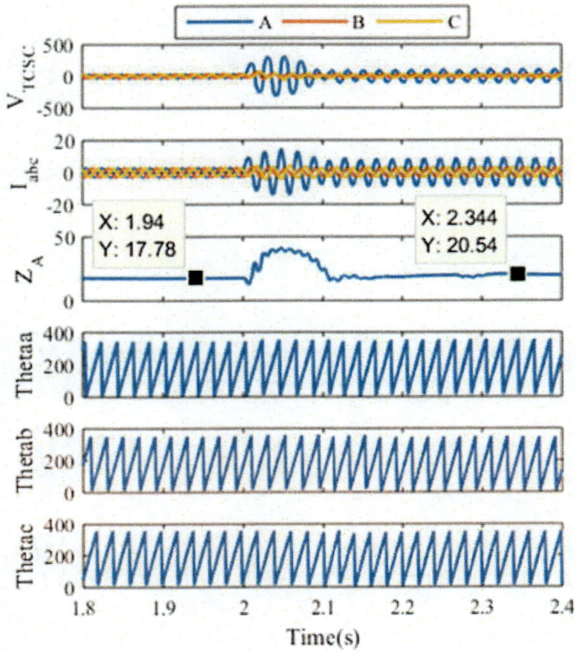

Fig. 10 Plot of V_{tcsc}, I_{abc}, Z_a, and theta (*a*, *b*, *c*) Vs time using PLL from library

5.1 *PLL Used from the PSCAD/EMTDC Library*

In this case, the PLL is used from the PSCAD library, and Fig. 10 shows the graphical plotting of V_{tcsc}, I_{abc}, Z_a, and theta (*a*, *b*, *c*) Vs time. It is clear from the figure that even after a fault at 2 s there is no deviation in theta (*a*, *b*, *c*). Also, the value of per-phase impedance (Z_a) before and after the fault gets changed. Pre-fault impedance value is 17.78 Ω, whereas post fault impedance is 20.54 Ω, but practically, it was expected to be the same. This shows that PLL used from the PSCAD library does not include practical scenarios.

5.2 *Proposed PLL Using Blocks from the PSCAD/EMTDC Library*

In this case, the PLL is simulated using blocks from the PSCAD library, and Fig. 11 shows the graphical plotting of V_{tcsc}, I_{abc}, Z_a, and theta (*a*, *b*, *c*) Vs time. It is clear from the figure that after a fault at 2 s, there is a deviation in theta (*a*, *b*, *c*). Also, the value of per-phase impedance (Z_a) before and after the fault remains the same (which we want practically). During pre-fault, the TCSC impedance value is 17.78

Fig. 11 Plot of *Vtcsc*, *Iabc*, Z_a, and theta (*a*, *b*, *c*) Vs time using proposed PLL

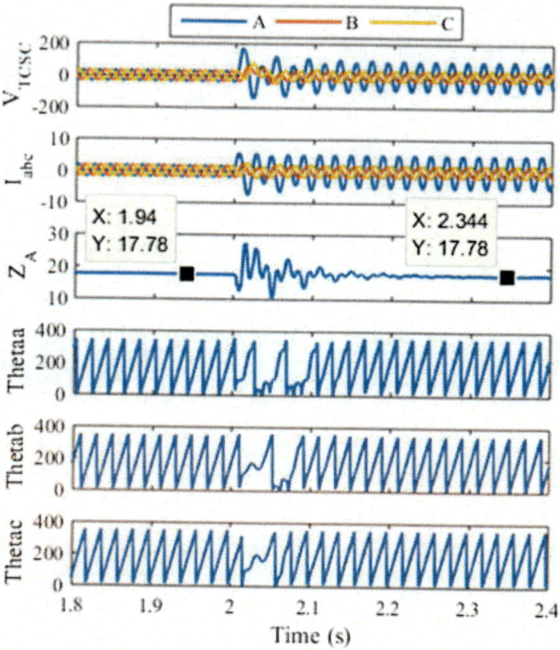

Table 1 Comparison of impedance for both PLLs

Type of PLL	Impedance before fault (Ω)	Impedance after fault (Ω)
PLL from PSCAD library	17.78	20.54
Proposed PLL	17.78	20.54

Ω, and during the fault, the TCSC impedance value is again 17.58 Ω. This shows that PLL simulated using blocks from the PSCAD library includes practicalities (Table 1).

6 Conclusion

Two different PLLs were simulated in PSCAD/EMTDC platform, and it was observed that the one, which was proposed using the blocks of the PSCAD library, has an impact on the control of TCSC in an unbalanced condition. The fault was created at 2 s to observe the changes in the system, and it is found that in the case of simulated PLL per-phase impedance remains the same before and after the fault, which is not found in one which is used from the library. Simulated PLL responds to the disturbance in the system. So, proposed PLL shows a better result as compared to the one available in the library. Thereby, the proposed method may help in locating the fault quite accurately than that of the available one.

References

1. Jovicic D, Pillai G (2005) Analytical modeling of TCSC dynamics. IEEE Trans Power Delivery 20(2):1097–1104
2. Mattarella P, Verghese GC, Stankovic A (1997) Phasor dynamics of thyristor-controlled series capacitor systems. IEEE Trans Power Syst 12(3):1259–1267
3. Jalali SG, Lasseter RH, Dobson I (1994) Dynamic response of a thyristor controlled switched capacitor. IEEE Trans Power Del 9(3):1609–1615
4. Raju R, Shubhanga KN (2015) Laboratory implementation of a thyristor-controlled series capacitor. In: 2015 IEEE international conference on signal processing, informatics, communication and energy systems (SPICES), pp 1–5. https://doi.org/10.1109/SPICES.2015.7091402
5. Meikandasivam S, Nema RK, Jain SK (2010) Performance of installed TCSC projects. In: India international conference on power electronics 2010 (IICPE2010), pp 1–8. https://doi.org/10.1109/IICPE.2011.5728103
6. Siddique A, Xu Y, Aslam W, Rasheed M (2019) A comprehensive study on FACTS devices to improve the stability and power flow capability in power system. In: 2019 IEEE Asia power and energy engineering conference (APEEC), pp 199-205. https://doi.org/10.1109/APEEC.2019.8720685
7. Okeke TU, Zaher RG (2013) Flexible AC transmission systems (FACTS). In: 2013 international conference on new concepts in smart cities: fostering public and private ALLIANCES (SmartMILE), pp 1–4. https://doi.org/10.1109/SmartMILE.2013.6708208
8. Asawa S, Al-Attiyah S (2016) Impact of FACTS device in electrical power system. In: 2016 international conference on electrical, electronics, and optimization techniques (ICEEOT), pp 2488–2495. https://doi.org/10.1109/ICEEOT.2016.7755141
9. Varma S (2015) FACTS devices for stability enhancements. In: 2015 international conference on green computing and Internet of Things (ICGCIoT), pp 69–74. https://doi.org/10.1109/ICGCIoT.2015.7380430
10. Hsieh G-C, Hung JC (1996) Phase-locked loop techniques. A survey. IEEE Trans Ind Electron 43(6):609–615. https://doi.org/10.1109/41.544547
11. Zhou X, Liang J (1999) Overview of control schemes for TCSC to enhance the stability of power systems. IEE Proc Gener Transm Distrib 146(2):125
12. Xie D, Zhang D, Gao P (2016) Research on phase-locked loop control and its application. In: 2016 IEEE information technology, networking, electronic and automation control conference, pp 818–821. https://doi.org/10.1109/ITNEC.2016.7560475
13. Singhal A, Madhu C, Kumar V (2014) Designs of all digital phase locked loop. In: 2014 recent advances in engineering and computational sciences (RAECS), pp 1–5. https://doi.org/10.1109/RAECS.2014.6799523
14. Bizzarri F, Brambilla A, Milano F (2018) Analytic and numerical study of TCSC devices: unveiling the crucial role of phase-locked loops. IEEE Trans Circ Syst I: Regul Pap 65(6):1840–1849. https://doi.org/10.1109/TCSI.2017.2768220
15. Canlzares CA, Faur ZT (1999) Analysis SVC and TCSC controllers in voltage collapse. IEEE Trans Power Syst 14(1):158–165

A Data-Driven Machine Learning Model for Transmission Line Faults Detection and Classification for the Smart Grid

Ani Harish, A. Prince, and M. V. Jayan

Abstract The smart grid is an intelligent power system network that should be reliable and resilient for sustainable operation. Wide area measurements systems are deployed in the power grid to provide real-time situational awareness to the power grid operators. Deriving meaningful insights from the growing voluminous data will be an excellent approach toward effectively using data being captured. This paper proposes an ensemble machine learning model for fault detection and fault type classification. The model is trained with features derived from data using an Ensemble feature extraction method. The ensemble feature extraction method's efficacy is compared with state-of-the-art feature extraction methods. The proposed model gives an accuracy of 99.9% for fault detection and above 90% for fault classification. A considerable decrease in model training time is also a beneficial characteristic of this model. The model is trained and validated by data from IEEE 39 bus system.

Keywords Fault detection · Fault classification · Transmission line · Machine learning · Smart grid

1 Introduction

A smart grid is an intelligent grid with distributed generation. Transmission line protection is a vital aspect of a resilient and self-healing smart grid. Intelligent methods for detecting, classifying, and localization of faults on transmission lines are critical for achieving a reliable smart grid. Power system operators can draw valuable insights into the system's state from wide area measurement systems (WAMS) signals data. Wide area situational awareness plays a pivotal role in improving the

Ani Harish (✉) · A. Prince
Department of Electrical Engineering, Rajiv Gandhi Institute of Technology Kottayam, APJ Abdul Kalam Technological University, Thiruvananthapuram, Kerala, India
e-mail: ani.ramachandran@gmail.com

M. V. Jayan
APJ Abdul Kalam Technological University, Thiruvananthapuram, Kerala, India

K. Namrata et al. (eds.), *Smart Energy and Advancement in Power Technologies*, Lecture Notes in Electrical Engineering 926,
https://doi.org/10.1007/978-981-19-4971-5_56

resiliency and reliability of the grid. Data mining techniques on WAMS data can be used to detect, localize, and classify short circuit faults on transmission lines. Machine learning algorithms based on pattern matching are efficient algorithms to detect and classify transmission line faults. The transmission line faults detection, classification, and localization methods found in the literature are generally of two types, the physics model-based and data-driven model-based [1]. The data-driven model-based approach does not require the knowledge of line and system parameters, a complex system model, and relies mainly on measurement data.

Jnaneshwar et al. [2] propose a transmission line protection scheme using discrete wavelet transform multiresolution analysis (DWT-MRA) [3] of current signals. Faults are classified using the energy values of the approximation coefficients. Here, the current signals from both ends of the transmission line are considered. [4] proposes wavelet entropy values of the voltage signals as inputs to an artificial neural network (ANN) for fault type classification. Hasabe et al. [5] propose transmission line fault classification using DWT feature extraction. DWT coefficients using db6 mother wavelet with level 6 decompositions of three-phase current signals form the features. An ANN with the sum of the level 6 detail coefficients as inputs does the fault type classification. In [6], the authors propose a feature extraction using wavelet packet entropy and ANN for faults classification. The mother wavelet chosen is db6 with three levels of decomposition. Features are the wavelet packet decompositions of three-phase fault current signals from the faulted line. [7] proposes a similar approach as [5] for feature extraction. [8, 9] present maximum overlap discrete wavelet transform (MODWT) [3] for feature extraction. The energy values of the MODWT coefficients form the features. Fast Fourier transform and S-transform are other data transformation methods used for feature extraction [10–11]. In [12] and [11], decision tree and ensemble tree classifier algorithms are used for classification [13, 14]. [15] presents fault classification by K nearest neighbor (KNN) algorithm. The instantaneous values of three-phase currents are the inputs to the classification algorithm. Gopakumar et al. in [16] propose extracting features from voltage signals using Fourier transform and Clarke's transformation and classifying using the Support Vector Machine (SVM) [17] classification algorithm.

From a wide area situation awareness perspective, we consider analyzing voltage, voltage angle, and frequency signals from all the buses of a scaled-down test power grid (IEEE 39 bus test system). This paper compares the different feature extraction methods to extract appropriate features from the power signal data. This work proposes an ensemble feature extraction method using wavelet transform. It identifies a machine learning classifier algorithm with high accuracy and good generalization capability for transmission line faults detection and classification with this feature data.

The structure of this paper is as follows. Section 2 describes the ensemble feature extraction, detection, and classification method. Section 3 has the results and discussion, and Sect. 4 has the concluding remarks.

2 Proposed Ensemble Feature Extraction, Detection, and Classification of Faults

The process flow diagram for detecting and classifying faults in a power system is shown in Fig. 1.

2.1 Data Acquisition

For data acquisition, we simulated an IEEE 39 bus system in the PowerWorld simulator, with the model downloaded from (Electric Grid Test Cases Repository, n.d.) [18]. The IEEE 39 bus New England Test System consists of 39 buses, ten generators, 19 loads, 37 lines, and 12 transformers. The nominal frequency is 60 Hz, and the nominal voltage is 345 kV. The parameters for the IEEE test systems can be obtained from [19].

Single line to ground faults (SLG), line to line faults (LL), double line to ground faults (LLG), and three-phase balanced faults (3PBF) were inserted at different times on different transmission lines of the 39 bus system. Each fault was simulated for five seconds, and then the signal dataset with the phasor values of voltage and frequency signals from all the buses were exported to .csv files.

The dataset comprises the pre-fault, fault, and post-fault signal values for different faults. The training and testing datasets are generated by varying fault types, fault distances from 0 to 90% and fault resistance from 0-50Ω. The training dataset has 40,970 records of 117 voltage, voltage angle, and frequency signals from the 39 buses. The testing dataset has 12,610 records of 117 signals. The number of records for the different classes for the training and test datasets is in Table 1.

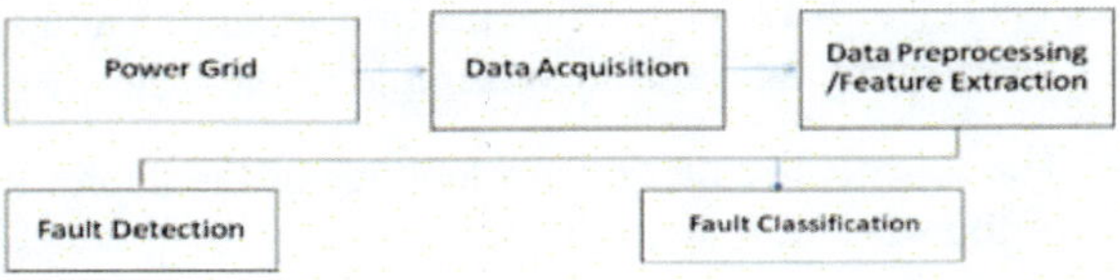

Fig. 1 Process flow diagram for fault detection and classification

Table 1 Dataset description

Dataset	NF	SLG	LL	LLG	3PBF
Training	23,033	2048	4447	5790	5652
Testing	7082	2004	1592	1028	904

2.2 Ensemble Feature Extraction

We imported the dataset from the .csv file into the MATLAB workspace. Decomposition of the signals into six levels of coefficients, was done using Discrete wavelet transform with Debauchies Db10 mother wavelet. Standard deviation, kurtosis, skewness, root mean square value (RMS), peak to RMS ratio, the statistical time-domain features of the DWT approximation, and detail coefficients were derived.

Entropy values of the wavelet coefficients of the signal dataset were calculated using (4). Shannon entropy quantifies the amount of information in a variable. The Shannon entropy of the wavelet coefficients at level j is as

$$\mathrm{SE}_j = \sum_{k=1}^{N} p_{jk} * \log p_{jk} \tag{1}$$

where N is the number of coefficients in the jth level and p_{jk} normalized squares of the wavelet coefficients at the jth level.

The energy (L2 norm) of each decomposition was calculated. The percentage of the energy of each wavelet approximation and detail component of each signal forms the wavelet energy feature.

$$\text{Wavelet Energy} = ||\mathrm{ca}||^2 + ||\mathrm{cd}||^2 \tag{2}$$

where ca is the approximation coefficient, and cd is the detail coefficient of the wavelet decomposition. Wavelet multiresolution analysis helps decompose the signal and provides information about the signal's transient changes and trends. Multiresolution analysis components are found by decomposing the signals into different components using empirical wavelet transform (EWT). The signals are broken into meaningful components, giving a better understanding of the signal variations. EWT partitions the energy of wavelet coefficients into separate passbands (MATLAB, n.d.) [20].

The feature dataset is an ensemble of the following.

- The statistical time-domain features of the approximation and detail coefficients of the wavelet transform.
- The wavelet energy values of the decomposition coefficients.
- The wavelet entropy values of the energy decompositions.
- The MRA decompositions of the signals with EWT.

Figure 2 shows the ensemble feature extraction module, fault detection, and the fault classification module.

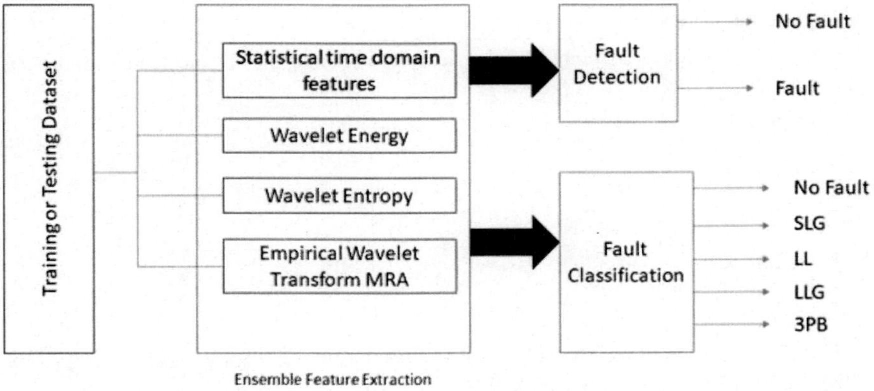

Fig. 2 Ensemble feature extraction for fault detection and classification

2.3 Fault Detection

Fault detection is a binary classification problem. The fault detection module has a binary classifier that classifies the feature dataset into 'no fault' and 'fault' categories. Boosting tree ensemble classifier (BTEC) is used for fault detection. The hyperparameters are chosen by minimizing the misclassification error Bayesian optimization. The hyperparameters for the fault detection module are as below.

- Ensemble method: AdaBoost
- Maximum number of splits: 9
- Number of learners: 18
- Learning rate: 0.0013928

2.4 Fault Classification

The steps for classification using the ensemble method are.

1. The ensemble feature extraction method extracted the feature dataset, from the training and test datasets, as the predictor data.
2. The response data is a categorical array of class labels. Class labels for the different classes are 1(No Fault), 2(SLG), 3(LL), 4(LLG), and 5 (3BPF). The training and testing datasets were labeled accordingly.
3. To overcome the data overfitting problem, a k-fold cross-validation approach is usually used while training. When overfitting occurs, the trained models lose the ability to generalize to new data sets. For the k-fold cross-validation, the training data is randomly shuffled and divided into k partitions. For each training-validation, a different partition is used for validation, and the remaining is used for training. Thus, each partition is used once for validation and (k − 1) times for training.

Table 2 Performance of classifier algorithms

Classification algorithm	Training/validation accuracy (%)	Training time (s) including optimization time
Decision tree	95.8	100.8
ANN	83.10	6300
KNN	93.3	2085.2
SVM	96.8	30,631
BTEC (RUSBoost)	98.3	2814

The performance of the five-fold cross-validated models for the DT, ANN, KNN, SVM, and BTEC is shown in Table 2.

The Ensemble tree classifier with the random undersampling (RUSBoost) ensemble method gave an accuracy of 98.3% for classifying the faults with the training dataset.

- The misclassification error was minimized by Bayesian optimization. The RUSBoost ensemble method had the minimum misclassification error. The optimized value of hyperparameters for RUSBoost was the maximum number of splits:1317, the number of learners: 10, and the learning rate: 0.8161.

3 Results and Discussion

The voltage phasor and frequency signals from 39 buses of the IEEE 39 bus test system are analyzed for transmission line faults detection and classification. A fault induced in a line can change the values of voltages in the nearby lines also. The features extracted by the ensemble feature extraction method are given as input to the fault detection and fault classification modules. The fault detection module is an AdaBoost ensemble tree classifier.

Ensemble AdaBoost tree binary classifier correctly classified 99.99% of the training features dataset and 99.90% of the testing features dataset records. The confusion matrix of a classification model depicts the number of correctly classified records and incorrectly classified for a class. The confusion matrix for the training and testing features dataset for fault detection is shown in Fig. 3.

Ensemble RUSBoost tree classifier correctly classified 98.3% of the training dataset records and 87.25% of the testing dataset records with features extracted by the proposed method. Figures 4 and 5 show the confusion matrix for the training and testing datasets.

The five-fold cross-validated efficiency of the classifier models with wavelet features is above 90%. Fourier Transform and S-transform models do not perform as well as the wavelet transform models.

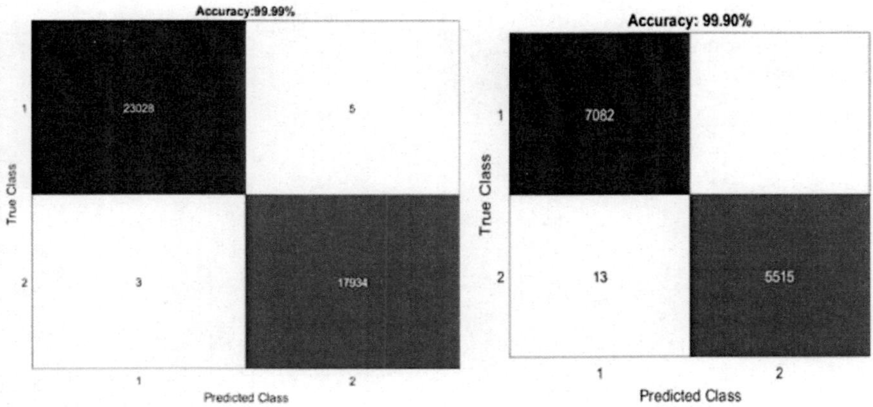

Fig. 3 Confusion matrix for fault detection—training and testing features data

Fig. 4 Confusion matrix for training dataset with ensemble feature dataset

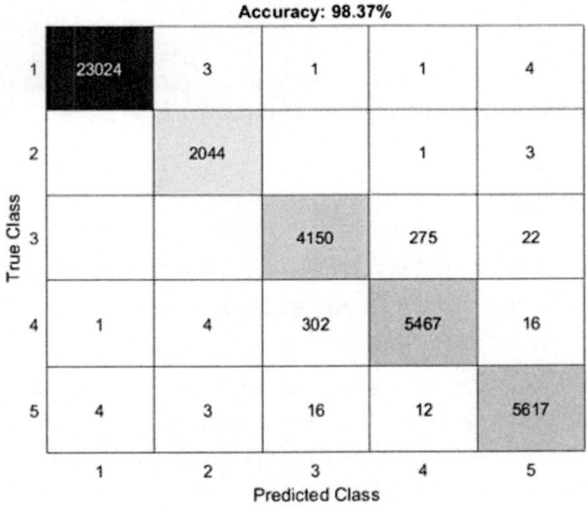

	1	2	3	4	5
1	23024	3	1	1	4
2		2044		1	3
3			4150	275	22
4	1	4	302	5467	16
5	4	3	16	12	5617

For some classification model's, the generalization capability with an unknown test dataset is weak. The classifier model with energy values of MODWT coefficients as input gave a 42.8% accuracy classifying an anonymous dataset, whereas the trained model gave a cross-validated accuracy of 99.22%. The classifier model for the WPT entropy feature dataset had a testing accuracy of 51.82% only.

When testing with a new dataset, the classifiers' accuracy is comparatively lesser which could be because of overfitting. The Confusion matrix of classifying testing features dataset is as in Fig. 5, which shows that the classifier incorrectly classified 727 class 3 records as class 4 and 362 class 4 records as class 3. This indicates that the separation boundary of these two classes is not well defined.

The algorithms were run on an Intel Core i7 9th gen processor machine with 32 GB RAM. The time taken to train the classifier models is high as the models are five-fold

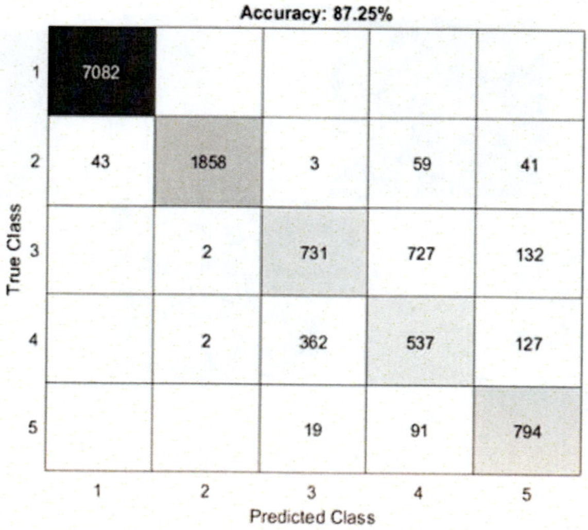

Fig. 5 Confusion matrix for testing dataset with ensemble feature dataset

cross-validated. Cross-validation improves the classifier model's performance and decreases the effect of overfitting.

3.1 Comparison of Performance of Ensemble Feature Extraction Method with Feature Extraction Methods Proposed in the Literature

We compare the feature extraction methods proposed in [2, 4, 6, 8, 10] with the proposed ensemble feature extraction method. The datasets assumed in the cases are very different from the datasets in this paper. In [2, 4, 6, 8, 10], the dataset are the signals from one end or both the ends of a single transmission line. For comparison, the features are extracted from the simulated IEEE 39 bus system training and testing datasets by the feature extraction methods proposed in [2, 4, 6, 8, 10] for faults classification.

Table 3 gives the training or the validation accuracy, testing accuracy, and training time of the RUSBoost ensemble tree classifier with the different feature extraction method. The Proposed ensemble features data set was classified with the highest accuracy of 87.25% in 1.75s.

Table 3 Comparison of performance measures

Feature extraction method	Training/cross-validation accuracy (%)	Testing accuracy (%)	Training time (s)	Testing time (s)
Fourier transform	56.2	53.39	824.94	14.30
Wavelet energy [2]	95.98	86.23	745	1.44
Wavelet entropy [4]	96.08	85.82	1552.7	1.94
MODWT energy [8]	99.22	42.48	421.4	1.761
WPT entropy [6]	89.54	51.82	3844	11.45
S-transform [10]	67.68	46.83	1500	9.18
Proposed ensemble feature extraction method	**98.3**	**87.25**	**1487.5**	**1.75**

4 Conclusion

The transient signal data analysis can provide insights into the different fault conditions. Machine learning algorithms can efficiently analyze and specify the type of fault from the signal data. The performance of a machine learning classifier is predominantly dependent on the input data. Appropriate features identified as inputs will improve the performance of the classifier. The proposed ensemble feature extraction method using statistical and time-domain features of wavelet coefficients, wavelet energy, wavelet entropy, and wavelet MRA gave an accuracy of 99.95% for fault detection and 87.25% for faults classification. The model's classification accuracy can be improved further by tuning the features dataset and the classification algorithm's parameters. Fault analysis methods that can provide a wide area situation awareness perspective to power system operators are desirable for a resilient smart grid with distributed generation. The primary protection methods employed in relays and other protection equipments deployed in the power grid are adequate but not flawless. A backup protection using the wide area measurement systems can be a suitable early warning system for any outages or blackouts. The fault detection and classification BTEC models trained and tested on the proposed feature extraction method can be deployed quickly for wide area situational awareness and power system protection. The testing time taken by the ensemble tree model is of the order of few seconds, and this model can be deployed for event detection and classification for a sustainable smart grid.

References

1. Tîrnovan RA, Cristea M (2019) Advanced techniques for fault detection and classification in electrical power transmission systems: an overview. In: Proceedings of 2019 8th international conference on modern power systems, MPS 2019, pp 1–10. https://doi.org/10.1109/MPS.2019.8759695
2. Jnaneswar K, Mallikarjuna B, Devaraj S, Roy DS, Reddy MJB, Mohanta DK (2020) A real-time DWT and traveling waves-based multi-functional scheme for transmission line protection reinforcement. Electr Eng. https://doi.org/10.1007/s00202-020-01117-0
3. Soman KP, Ramachandran K (2004) Insight into wavelets from theory and practice. PHI Learning Pvt. Ltd. ISBN :8120326504, 9788120326507. pp 447
4. Dasgupta A, Nath S, Das A (2012) Transmission line fault classification and location using wavelet entropy and neural network. Electr Power Compon Syst 40:1676–1689. https://doi.org/10.1080/15325008.2012.716495
5. Hasabe RP, Vaidya AP (2014) Detection and classification of faults on 220 KV transmission line using wavelet transform and neural network. Int J Smart Grid Clean Energy 2:2389–2393. https://doi.org/10.12720/sgce.3.3.283-290
6. Patel B, Bera P, Saha B (2018) Wavelet packet entropy and RBFNN based fault detection, classification and localization on HVAC transmission line. Electr Power Compon Syst 5008:1–12. https://doi.org/10.1080/15325008.2018.1431817
7. Ananthan SN, Padmanabhan R, Meyur R, Mallikarjuna B, Reddy MJB, Mohanta DK (2016) Real-time fault analysis of transmission lines using Wavelet multiresolution analysis based frequency-domain approach. IET Sci Meas Technol 10:693–703. https://doi.org/10.1049/iet-smt.2016.0038
8. Costa FB, Souza BA, Brito NSD (2012) Real-time classification of transmission line faults based on maximal overlap discrete wavelet transform. In: Proceedings of the IEEE power engineering society transmission and distribution conference, pp 1–8. https://doi.org/10.1109/TDC.2012.6281684
9. Ashok V, Yadav A, Abdelaziz AY (2019) MODWT-based fault detection and classification scheme for cross-country and evolving faults. Electr Power Syst Res 175:105897. https://doi.org/10.1016/j.epsr.2019.105897
10. Roy N, Bhattacharya K (2015) Detection, classification, and estimation of fault location on an overhead transmission line using s-transform and neural network. Electr Power Compon Syst 43:461–472. https://doi.org/10.1080/15325008.2014.986776
11. Jamehbozorg A, Shahrtash SM (2010) A decision tree-based method for fault classification in double-circuit transmission lines. IEEE Trans Power Deliv 25:2184–2189. https://doi.org/10.1109/TPWRD.2010.2050911
12. Mishra PK, Yadav A, Pazoki M (2018) A novel fault classification scheme for series capacitor compensated transmission line based on bagged tree ensemble classifier. IEEE Access 6:27373–27382. https://doi.org/10.1109/ACCESS.2018.2836401
13. Kattan M, Ishwaran H, Rao JS (2012) Decision tree: introduction. Encycl Med Decis Making 323–328. https://doi.org/10.4135/9781412971980.n97
14. Rocca J (2019) Ensemble methods: bagging, boosting and stacking. https://towardsdatascience.com/ensemble-methods-bagging-boosting-and-stacking-c9214a10a205
15. Asadi Majd A, Samet H, Ghanbari T (2017) k-NN based fault detection and classification methods for power transmission systems. Prot Control Mod Power Syst 2. https://doi.org/10.1186/s41601-017-0063-z
16. Gopakumar P, Reddy MJB, Mohanta DK (2015) Adaptive fault identification and classification methodology for smart power grids using synchronous phasor angle measurements. IET Gener Transm Distrib 9:133–145. https://doi.org/10.1049/iet-gtd.2014.0024
17. Ng A (2000) CS229 lecture notes margins: SVM. Intelligent systems and their applications IEEE. pt. 1, 1–25. https://doi.org/10.1016/j.aca.2011.07.027

18. Electric grid test cases: https://electricgrids.engr.tamu.edu/electric-grid-test-cases/ieee-39-bus-system/
19. Demetriou P, Asprou M, Quiros-Tortos J, Kyriakides E (2015) Dynamic IEEE test systems for transient analysis. IEEE Syst J 11:2108–2117. https://doi.org/10.1109/jsyst.2015.2444893
20. MATLAB Documentation :https://in.mathworks.com/help/wavelet/ref/ewt.html?s_tid=doc_ta

Optimal Sizing of DG Solar Wind Hybrid System Using HOMER

Abhishek Kumar Gupta, Pankaj Tripathi, and Shashank Dadhich

Abstract Distributed generation (DG) is a term describing the production of electrical energy by utilities that are suitably minor than primarily generating stations. It permits interconnection at closely any point in a power system according to IEEE. It is sited near the loads to avoid losses in transmission lines and hence increases the energy efficiency. This paper describes an optimal sizing of distributed generation solar wind system that fulfils the energy demand of an institute situated in Jaipur Rajasthan. This system can be designed and implemented for households, any institute, hospital, or banking sector for reliable and uninterrupted power supply. Sensitivity analysis is also done using hybrid optimization model for electric renewable (HOMER) Pro Software with 0, 10, and 20% capacitive shortage.

Keywords DG · HOMER · Microgrid · Photovoltaic (PV) · Wind energy · Cost of energy (COE) · Net present cost (NPC)

1 Introduction

From last five decades, centralized generation power system is reliable and irrefragable. Generally, the centralized power plants have large power generation capacity [1]. The primary cause for this was financially viable by making bigger power generation station offset the extra amounts of carrying the electrical energy to end user, and DG was virtually unreal in the 1990s [2]. However, with the new policy of government focussed at dipping the environment pollution by decreasing carbon emission and this cause more attention towards DG [3]. DG has low-transportation costs as well as it can be designed with the using renewable energy sources which reduce tremendous amount of carbon emission in comparison with

A. K. Gupta (✉) · S. Dadhich
Department of Electrical Engineering, Jaipur National University, Jaipur 302017, India
e-mail: abhi.kr.gupta@gmail.com

P. Tripathi
Department of Electrical Engineering, Invertis University, Bareilly 243123, India

K. Namrata et al. (eds.), *Smart Energy and Advancement in Power Technologies*,
Lecture Notes in Electrical Engineering 926,
https://doi.org/10.1007/978-981-19-4971-5_57

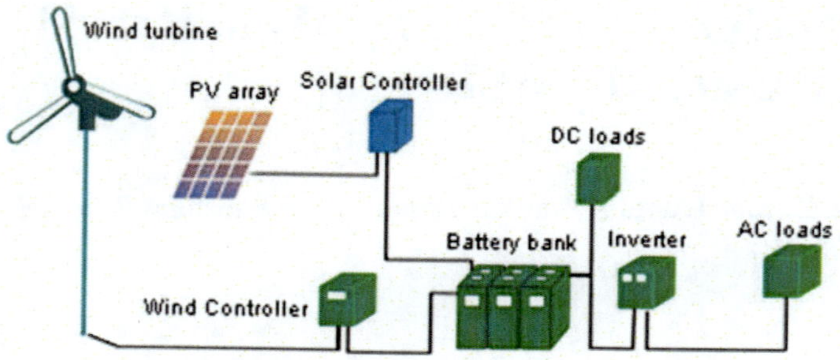

Fig. 1 General diagram of DG system with PV and wind energy sources [13]

coal-based conventional power plant [4]. Figure 1 shows the general diagram for DG system with PV and wind energy sources.

A DG hybrid energy generation model is designed and proposed for institutional area. Hybrid energy system consists PV cells and wind turbine as main source of energy [5–8]. It also consist a lead acid battery for power backup and off grid bidirectional converter for AC to DC and DC to AC conversion [9–12].

2 Input Data

Data of electrical engineering department has collected manually at Arya College of Engineering and Information Technology, Jaipur (ACEIT).

2.1 Flow Chart of Proposed Model

See Fig. 2.

2.2 Load Calculation

The following electrical loads are considered:

1. Electric fan = 174 nos. * 60 W/Fan = 10,440 W
2. Tube-light = 169 nos. * 40 W/Fan = 6760 W
3. Computers = 42 nos. * 150 W/PC = 6300 W
4. AC = 2 nos. * 1500 W/AC = 3000 W
5. Projector = 2nos. * 200 W/Projector = 400 W
6. Miscellaneous load = 1000 W.

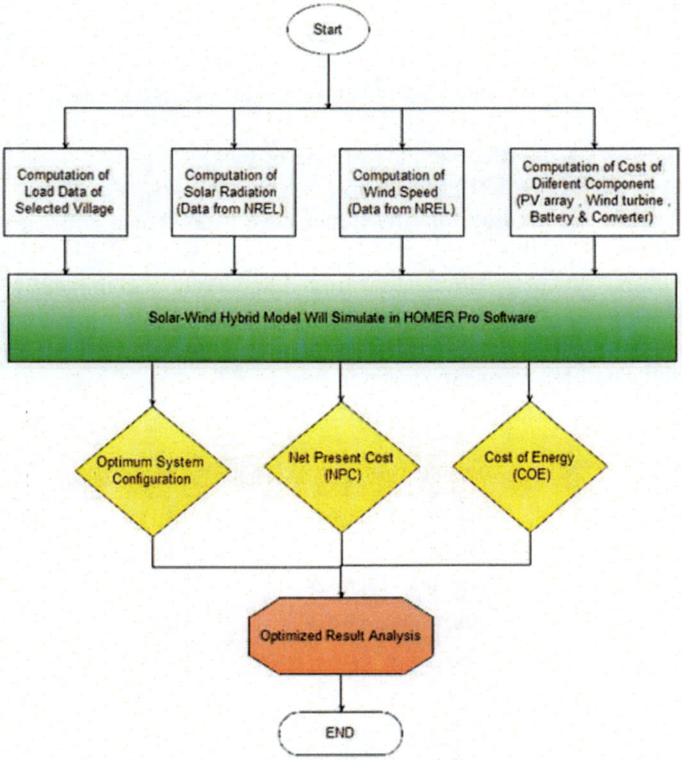

Fig. 2 Flow chart

The hourly load profile of Arya college of Engineering and I.T, Jaipur has been evaluated for the entire year which is shown in Fig. 3.

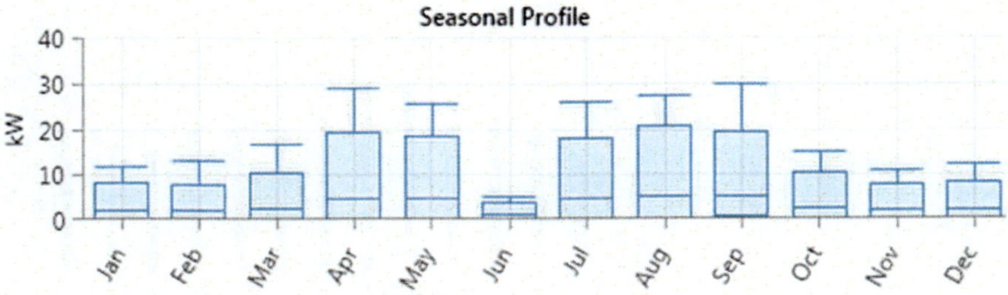

Fig. 3 Seasonal load profile of ACEIT

2.3 Solar Radiation

The latitude and longitude data for an institutional area located in Jaipur (Rajasthan) are,

Latitude—27°1.8' North, longitude –75°53.6' East,

Time zone—(GMT + 5:30) India.

Here in Table 1, base line solar resources have been shown with clearness and daily radiations.

The monthly average radiation varies from 3.866 kWh/m^2/d to 6.830 kWh/m^2/d, and the annual average radiation is found to be 5.42833 kWh/m^2/d at the site with average clearness index of 0.613083. The monthly average global horizontal solar radiation and clearness index are shown in Fig. 4.

Table 1 Base line data of solar resources

Months	Clearness index	Daily radiation(kWh/m^2/d)
January	0.645	4.128
February	0.643	4.882
March	0.628	5.717
April	0.619	6.427
May	0.614	6.812
June	0.604	6.830
July	0.515	5.748
August	0.500	5.269
September	0.611	5.756
October	0.672	5.329
November	0.663	4.382
December	0.643	3.866
Average	**0.613083**	**5.428833**

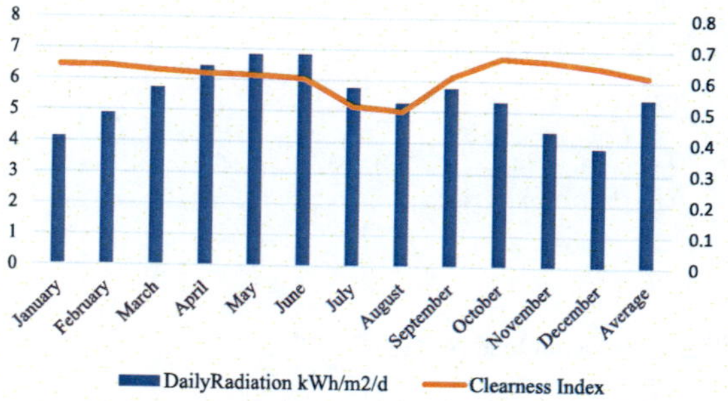

Fig. 4 Global solar radiation and clearness index

2.4 Wind Resources

Here in Table 2, wind speeds monthly have been given according to wind resources.

Figure 5 shows the average monthly speed of wind speeds at selected site.

Table 2 Base line data of wind resources

Months	Wind speed (m/s)
January	2.730
February	3.130
March	3.110
April	3.800
May	4.390
June	4.670
July	4.030
August	3.310
September	3.130
October	2.320
November	2.160
December	2.470
Annual average	**3.27**

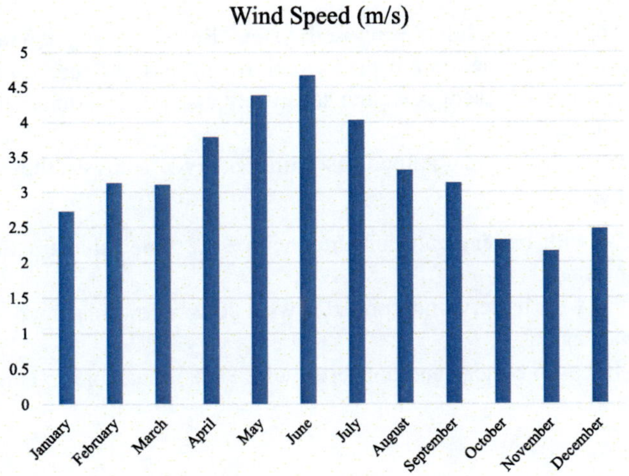

Fig. 5 Average monthly wind speed at ACEIT

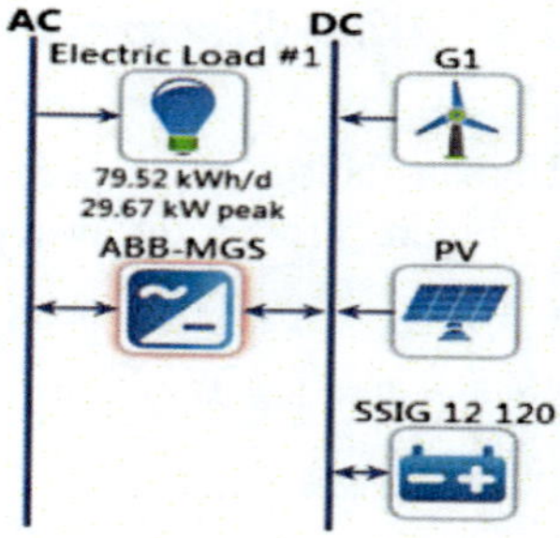

Fig. 6 System architecture of proposed model

3 System Architecture

Figure 6 shows the system architecture of proposed model. It is designed in HOMER Pro Software. It has both AC and DC buses.

It consists 79.52 kW PV panel as per load demand, and size of PV array is calculated by HOMER Pro Software. It also has battery bank for power back up rating of single battery is 12 V, 120 AH rating total 179 number of batteries needed suggested by simulation software. For making more reliable, 1 kW of wind turbine is connected. Rating of converter is 30 kW.

4 Result and Discussion

Table 3 shows optimized result of proposed system. Based on PV and wind recourses on selected site, the simulation has performed. Result has obtained on the basis of cost of energy and net present cost, and sensitivity analysis has done with different capacitive shortage.

Three primary cases have obtained based on sensitivity analysis, these three cases are as following:

- Case I—PV with battery and converter with 90% renewable fraction and 0% capacity shortage.
- Case II—PV with battery and converter with 90% renewable fraction and 10% capacity shortage.
- Case III—PV with battery and converter with 90% renewable fraction and 20% capacity shortage.

Table 3 Optimization result based on cost of energy (COE) and net present cost (NPC)

Sensitivity	Renewable fraction (%)	90	90	90
	Capacitive shortage (%)	0	10	20
Architecture	 PV (kW)	79.5	43.5	37.8
	 Wind Gen. (kW)	0	0	0
	 Battery 12 V, 120AH	179 Nos	39 Nos	37 Nos
	 ABB-MGS (kW)	30	19.9	14.3
Cost	NPC (INR)	4.96 M	2.60 M	2.29 M
	COE (INR)	13.31	7.20	6.84
	Operating cost (INR)	21,944	68,626	61,143
	Initial capital (INR)	4.68 M	1.72 M	1.51 M

4.1 Case I- PV with Battery and Converter with 90% Renewable Fraction and 0% Capacity Shortage

The first combination is solar with battery and converter, using 90% renewable fraction and 0% capacity shortage by HOMER. The system has PV array of 79.5 kW, converter of 30 kW, and single parallel string with 179 lead acid batteries bank. It has total net present cost (NPC) of ₹4.96 Million with operating cost of ₹21,944/year. It has highest COE, i.e. ₹13.31/kWh. Total electrical energy is produced by the system is 1,38,760 kWh/year. The consumption of AC primary load is 29,025 kWh/year, and excess electricity generated is 1,08,512 kWh/year (78.2%).

4.2 Case II- PV with Battery and Converter with 90% Renewable Fraction and 10% Capacity Shortage

The second combination is solar with battery and converter, using 90% renewable fraction and 10% capacity shortage by HOMER. The system has PV array of 43.5 kW, converter of 19.9 kW, and single parallel string with 39 lead acid batteries bank. It has total net present cost (NPC) of ₹2.60 Million with operating cost of ₹68,625/year. It has moderate COE, i.e. ₹7.20/kWh. Total electrical energy is produced by the system is 75,822 kWh/year. The consumption of AC primary load is 28,155 kWh/year, and excess electricity generated is 46,312 kWh/year (61.1%).

4.3 Case III- PV with Battery and Converter with 90% Renewable Fraction and 20% Capacity Shortage

The third combination is solar with battery and converter, using 90% renewable fraction and 20% capacity shortage by HOMER. The system has PV array of 37.8 kW, converter of 14.3 kW, and single parallel string with 37 lead acid batteries bank. It has total net present cost (NPC) of ₹2.29 Million with operating cost of ₹61,143 / year. It has least COE, i.e. ₹6.84/kWh. Total electrical energy is produced by the system is 66,027 kWh/year. The consumption of AC primary load is 26,117 kWh/year, and excess electricity generated is 36,649 kWh/year (58.5%).

5 Conclusion

Based on sensitivity analysis, three different optimal combination is obtained by HOMER as shown in Table 3. Each combination has two different power generation source, i.e. only PV as power generation source or PV and wind hybrid energy system as a power generation source.

Different component of cost, i.e. capital cost, replacement cost, operation and maintenance cost, fuel cost, salvages cost, and total cost for different cases is shown in Table 3. All costs shown in Fig. 7 are in Indian rupees (INR).

Based on above analysis as shown in Fig. 7, the following observations are as follows:

- PV and wind hybrid energy system are expensive in all the three cases.

  System having 20% capacitive shortage is most economical system but it is least reliable.
- System having 0% capacitive shortage is least economical system but it is most reliable. This system has complete power backup but it has tremendous excess amount of energy which is unused.
- System having 10% capacitive shortage is economical system as well as it is most reliable and has less excess energy.

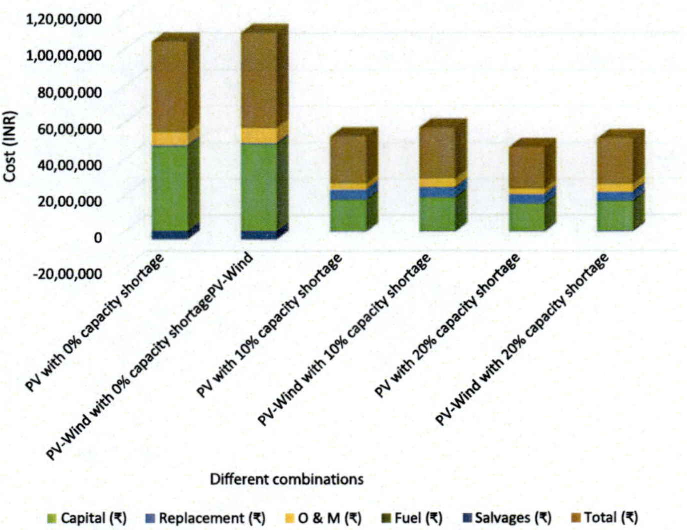

Fig. 7 Comparison of system architecture with sensitivity analysis

References

1. IEEE, "Institut of electrical and electronics engineers," [Online]. Available: http://www.ieee.org
2. IET (2006) Distributed generation. The Institution of engineering and technology. Tech Rep Sep
3. Government of India (2005) National electricity policy. Government of India
4. SIEMENS (2011) Microgrids. White paper
5. Sinha A, Ranjan R, Gupta AK, Jain VK (2020) Techno-economic feasibility analysis of off-grid electrification for remote areas: a review. In: 2020 9th international conference system modeling and advancement in research trends (SMART), pp 463–467. https://doi.org/10.1109/SMART50582.2020.9337112
6. Kumar and Singh, Mayengbam Sashilal (2020) Assessment and scope of decentralised power generation using renewable energy resources. Available at SSRN: https://ssrn.com/abstract=3554854 or https://doi.org/10.2139/ssrn.3554854
7. Dadhich S, Meena P, Singh S, Gupta AK (2019) A feasibility study of microgrids in India. In: 2019 8th international conference system modeling and advancement in research trends (SMART), pp 343–347. https://doi.org/10.1109/SMART46866.2019.9117339
8. Shekhawat K, Doda DK, Gupta AK, Bundele M (2019) Decentralised power generation using renewable energy resources: scope, relevance and application. Int J Innovative Technol Explor Eng. ISSN: 2278–3075 8(9)
9. Madhav D, Gupta AK, Sashilal M (2018) A review on renewable energy sources for hybrid power generation. Int J Creative Res Thoughts 6(1). ISSN: 2320–2882
10. Madhav D, Gupta AK, Ranjan R (2017) A review on assessment of hybrid renewable energy system. Int J Comput Appl (0975–8887), 179(8)
11. Kumar RR, Gupta AK, Kumar AR (2017) Design and simulate the solar-wind-diesel stand-alone systems for an institutional area. Int J Comput Appl (0975–8887) 165(12)

12. Kumar RR, Gupta AK, Ranjan R, Shrivastava S (2017) Off-grid and on-grid connected power generation: a review. Int J Comput Appl (0975–8887), 164(9)
13. https://www.mahaurja.com/meda/en/off_grid_power/small_wind_solar_hybrid

Automatic Load Frequency Control of Islanded Microgrid Using Social Group Optimization Technique

Anupam Shukla and Hiramani Shukla

Abstract Microgrids are a smaller form of main utility grid considering diverse energy sources and loads. This paper focuses on the automatic load frequency control (ALFC) of an islanded microgrid which consists of solar photovoltaic (SPV), wind turbine generator (WTG), aqua electrolyzer (AE), fuel cell (FC), diesel engine generator (DEG), and plug-in hybrid electric vehicle (PHEV). Storage elements such as flywheel energy storage system (FESS) and battery energy storage system (BESS) are also included. Proportional integral (PI) and proportional integral derivative (PID) controllers are connected in a feedback loop to ensure the stability of a microgrid system. A novel optimization technique, social group optimization (SGO), is used to optimize the controller parameters. MATLAB/Simulink results show that PID controller is showing better frequency deviation characteristics than PI controllers, and also frequency response shows that system considering storage elements is giving better frequency characteristics.

Keywords Automatic load frequency control (ALFC) · Social group optimization technique (SGO) · Proportional integral derivative (PID) · Islanded microgrid

1 Introduction

In isolated places such as villages or hilly areas, there is an unavailability of electric power. It is almost impractical to connect these areas with the existing grid; hence, we have to go with microgrids. A microgrid is a small power plant with diverse generating units and localized loads. As of now, the availability of fossil fuels is declining at a very rapid pace; hence, it is a wise choice that we should penetrate renewable energy sources (RES) as generating units for microgrids. RES are environment-friendly and help in the reduction of carbon footprint [1].

A. Shukla (✉) · H. Shukla
Electrical Engineering Department, Maulana Azad National Institute of Technology Bhopal, Bhopal, India
e-mail: anupamshukla580@gmail.com

K. Namrata et al. (eds.), *Smart Energy and Advancement in Power Technologies*, Lecture Notes in Electrical Engineering 926,
https://doi.org/10.1007/978-981-19-4971-5_58

Because of unpredictable solar irradiance and wind speed, output power from renewable energy sources is highly fluctuating in nature. Fluctuating output power from renewable energy sources consequently produces disturbances in the frequency profile of a microgrid system which leads to an unstable system. To avoid these disturbances, generally, the system is integrated with FC, DEG, and also some storage elements like BESS and FESS. However, batteries have certain limitations such as high maintenance cost and limited life span [1, 2]. Generally, microgrids are operated under two operating modes: (1) Grid-connected and (2) islanded mode [3]. In grid-connected mode, because of the larger inertia of a system, chances of instability of a system due to disturbances are comparatively lesser than islanded mode [4, 5]. In the islanded mode of operation, many ancillary services need to be addressed, and ALFC is one of those important issues in islanded mode [3, 6]. The basic idea of ALFC can be deduced from the swing equation [7], which is given by

$$2H_{SYS}\frac{d\Delta\omega}{dt} = P_m - P_e \tag{1}$$

where H_{SYS}, P_m, P_e, and $\Delta\omega$ are system inertia, mechanical power, electrical power, and change of the angular velocity, respectively [7]. From Eq. (1), it can be observed that power mismatch between demand and supply is the main reason for frequency deviation.

A plethora of researches has already available on the ALFC of an islanded mode of operation, but the problem is that many authors have applied outdated optimization techniques. Paper [1] investigates the frequency profile of microgrid systems considering PV, FC, and DEG as generating units. Particle swarm optimization (PSO) and firefly algorithm (FFA) are used to tune the PID controller parameters. Authors [2] introduce relatively advanced, fuzzy logic-based controllers for the ALFC of a microgrid system. But the problem is those fuzzy controllers take more storage; hence, its convergence time is more. In paper [3, 4], frequency profile of islanded mode is present by using terminal sliding mode control (TSMC) with the help of ant bee colony (ABC) optimization technique. Paper [8] compares the frequency characteristics obtain from cuckoo search algorithm (CSA) and the FFA optimization technique by using the cost function integral square error (ISE) [1, 8]. In the paper [9], authors have uses a novel, whale optimization algorithm (WOA) technique. Paper [10] investigates the frequency profile in the presence of a plug-in hybrid electric vehicle (PHEV). Many kinds of literature emphasize the mere reliance on storage systems such as BESS, FESS, and ultra-capacitors, to balance the power mismatch between demand and supply [7, 11, 12]. The frequency deviation of a microgrid system is also analyzed by using real-time data and power line communication [13–15].

After going through numerous articles on the ALFC of islanded microgrid, I have observed that many authors have used relatively outdated optimization techniques, such as PSO and CSO. In this work frequency profile is observed in an islanded mode of a microgrid system, consisting of PV, WTG, AE, FC, HP, and DEG as generating units along with storage elements such as BESS and FESS. PI/PID controllers are applied which are further tuned by novel optimization techniques namely the social group optimization (SGO) technique. SGO is based on the learning behavior of humans [16]. Since SGO is a new technique, its convergence and accuracy are far better than its relevant techniques such as PSO, FFA, CSO, WSO, and ABC [1–8]. This is the first time SGO is used for ALFC analysis of a microgrid system, though it is a well-established technique in the field of medical science [16].

After thorough review of available manuscripts on the topic ALFC of a microgrid system, it is observed that most of the work is based on the outdated optimization technique. In this proposed system, a novel technique which is implemented in ALFC of a microgrid system, name as social group optimization (SGO), is used to tune the controller gains. In the field of medical sciences, it is a well-established technique [16]. Its fast response and accuracy are the main characteristics that motivated the authors to implement this technique for ALFC.

The main objectives of this paper are:

(a) To design a microgrid system consisting of WTG, PV, DEG, AE, FC, PHEV, BESS, and FESS.
(b) To apply PI and PID controllers for frequency regulation of the above microgrid system and to compare the results of frequency deviation characteristics.
(c) To tune the gains of PI and PID controllers with the help of a newly introduced SGO technique.
(d) To compare the frequency profile of a system considering with and without energy storage elements.

2 Microgrid System's Components and Its Modeling

The proposed microgrid system is shown in Fig. 1. It consists of generating units such as PV, WTG, AE, FC, DEG, and HP. For better control, it is also equipped with energy storage elements such as BESS and FESS [1, 3]. In the succeeding subsections, each unit and the overall system is defined by the use of first-order transfer functions.

2.1 *Wind Turbine Generator (WTG)*

Wind turbine generation (WTG) is one of the most economical and feasible generating units. We modeled the WTG system as a first-order transfer function which is given by [1, 3].

$$G_{\mathrm{WTG}}(s) = \frac{K_{\mathrm{WTG}}}{sT_{\mathrm{WTG}} + 1} = \frac{\Delta P_{\mathrm{WTG}}}{\Delta f} \tag{2}$$

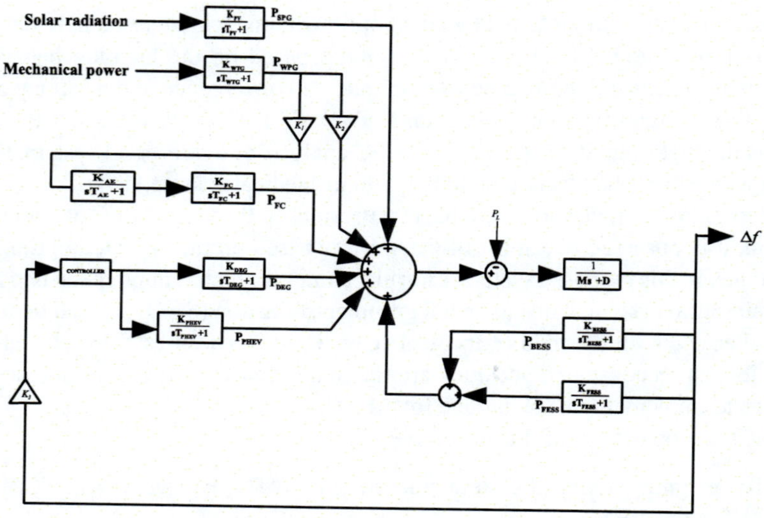

Fig. 1 Proposed microgrid system

2.2 Photovoltaic System (PV)

It is a series and parallel combinations of numerous solar cells which produce DC voltage. The first-order transfer function of the PV system can be represented by [1, 5].

$$G_{\text{PV}}(s) = \frac{K_{\text{PV}}}{sT_{\text{PV}} + 1} = \frac{\Delta P_{\text{PV}}}{\Delta f} \tag{3}$$

2.3 Aqua Electrolyzer (AE)

Some part of the power from WTG and PV is taken by AE for the production of hydrogen that is necessary for fuel cell (FC) units. For a representation of its power absorbing nature, we use a negative sign in the total power equation of the microgrid.

The first-order transfer function of the AE system can be represented by [5, 9].

$$G_{\text{AE}}(s) = \frac{K_{\text{AE}}}{sT_{\text{AE}} + 1} = \frac{\Delta P_{\text{AE}}}{\Delta f} \tag{4}$$

2.4 Fuel Cell (FC)

Generated hydrogen gas is utilized by FC, and it produces power with the help of a DC-AC converter during peak load demand. The first-order transfer function is given by [5, 9].

$$G_{\mathrm{FC}}(s) = \frac{K_{\mathrm{FC}}}{sT_{\mathrm{FC}} + 1} = \frac{\Delta P_{\mathrm{FC}}}{\Delta f} \tag{5}$$

2.5 Diesel Engine Generator (DEG)

During the unavailability of solar and wind power, DEG is used for backup purposes. It is an asynchronous generator with the advantages of higher efficiency and high starting torque. In a microgrid system, its representation is shown by a first-order transfer function which is given by [1].

$$G_{\mathrm{DEG}}(s) = \frac{K_{\mathrm{DEG}}}{sT_{\mathrm{DEG}} + 1} = \frac{\Delta P_{\mathrm{DEG}}}{\Delta f} \tag{6}$$

2.6 Plug-In Hybrid Electric Vehicle (PHEV)

It is a specific type of vehicle that can work on both fuel and battery; it can also be charged with the help of regenerative braking. A first-order transfer function is given by [8, 10].

$$G_{\mathrm{PHEV}}(s) = \frac{K_{\mathrm{PHEV}}}{sT_{\mathrm{PHEV}} + 1} = \frac{\Delta P_{\mathrm{PHEV}}}{\Delta f} \tag{7}$$

2.7 Battery Energy Storage System (BESS)

During less power demand at the load center, additional power is generated which is stored in BESS. This stored energy is utilized during peak hours or when renewable energy sources are down [9]. A first-order transfer function is given by [1].

$$G_{\mathrm{BESS}}(s) = \frac{K_{\mathrm{BESS}}}{sT_{\mathrm{BESS}} + 1} = \frac{\Delta P_{\mathrm{BESS}}}{\Delta f} \tag{8}$$

2.8 Flywheel Energy Storage System (FESS)

Its behavior is the same as BESS; their excellent desired expeditious response is very effective to mitigate the frequency deviation. BESS stored the electrical energy in the form of rotational energy [11]. The first-order transfer function is given by [1].

$$G_{\text{FESS}}(s) = \frac{K_{\text{FESS}}}{sT_{\text{FESS}} + 1} = \frac{\Delta P_{\text{FESS}}}{\Delta f} \tag{9}$$

2.9 Power and Frequency Dissimilarity Analyzing Transfer Function

The difference between actual and base frequency gives deviated frequency (Δf). Similarly; the difference between input mechanical power (P_m) and output electrical power (P_e) gives power deviation (ΔP_e). This can be mathematically represented by a transfer function [5].

$$G_{\text{SYS}}(s) = \frac{K_{\text{SYS}}}{sT_{\text{SYS}} + 1} = \frac{\Delta f}{\Delta P_e} = \frac{1}{K_{\text{SYS}}(1 + sT_{\text{SYS}})} = \frac{1}{D + \text{Ms}} \tag{10}$$

where M is the inertia constant, and D is the damping ratio of a microgrid system.

2.10 Controller Used

Due to the low inertia of islanded microgrid system, even small disturbances may produce larger frequency turbulence. Storage elements are not sufficient to mitigate these disturbances. The additional control action is required to make the system stable irrespective of power mismatch. In this work, PI and PID controllers are used in a feedback loop.

Proportional Integral (PI) Controller

It is a combination of two different controllers proportional and integral; the transfer function is given by [5, 8]. It relies on two parameters, proportional gain (K_P) and integral gain (K_I). The block diagram of the PI controller is shown in Fig. 2.

$$G_{\text{PI}} = K_P + \frac{K_I}{s} \tag{11}$$

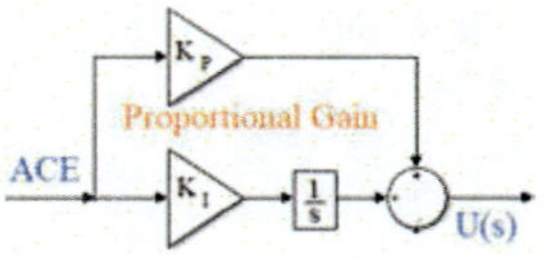

Fig. 2 PI controller

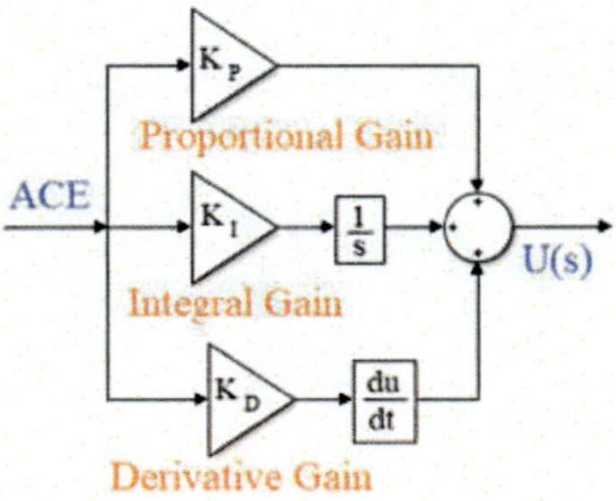

Fig. 3 PID controller

Proportional Integral Derivative (PID) Controller

It is one of the most versatile controllers; dynamic performance is excellent with good robust characteristics. It relies on three parameters, proportional gain (K_P), integral gain (K_I), and derivative gain (K_D) [1, 8]. The block diagram of the PID controller is shown in Figs. 3 and 4.

$$G_{\text{PID}} = K_P + \frac{K_I}{s} + K_D \tag{12}$$

2.11 *Objective Functions*

Performance of PI and PID controllers is analyzed through a cost function by calculating integral absolute error (IAE) of the frequency deviation (Δf). The detailed objective function and constraint have been given in sequence [1].

$$J = \text{IAE} = \int_0^t |\Delta f| \, \text{dt} \tag{13}$$

Subjected to:

$$K_P^{\min} \leq K_P \leq K_p^{\max}$$

$$K_i^{\min} \leq K_i \leq K_i^{\max}$$

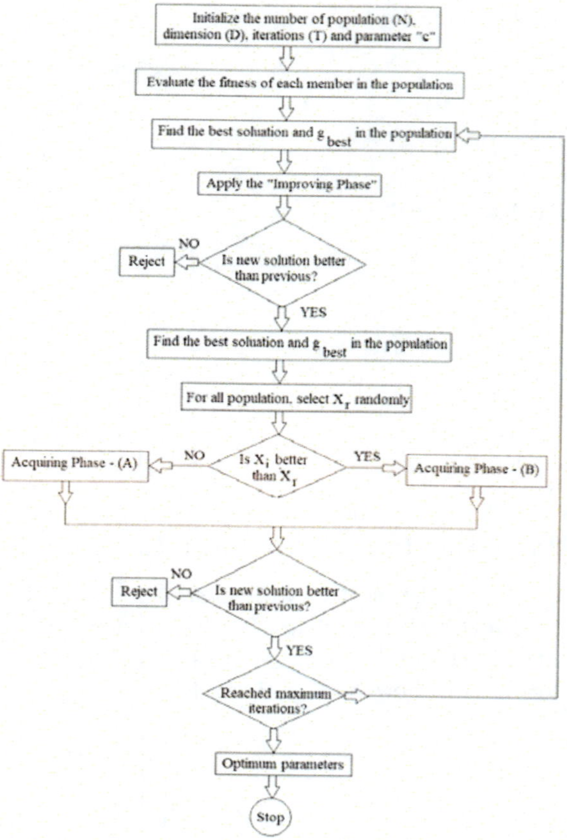

Fig. 4 Flowchart of SGO technique

$$K_d^{\min} \leq K_d \leq K_d^{\max}$$

Controller's parameters are optimized by minimizing the IAE. The range for all the gain parameters belongs to [8, 10].

3 Social Group Optimization (SGO) Technique

Social group optimization technique (SGO) is a metaheuristic approach that has been recently developed technique in the horizon of optimization techniques. Its principle is based on the learning behavior of humans in a group [15]. SGO technique consists of mainly two steps, the first one is the improving phase and the second one is acquiring phase. In the improving phase, knowledge of each person is upgraded with the influence of the best person. A person with the highest capability to solve a problem is known as the best person. In acquiring phase, each person upgrades his/her knowledge with mutual interactions [15]. Self-introspection parameter (c) is taken as 0.2 [16].

4 Results and Discussion

Figure 1 depicts the proposed microgrid system. Total output power in this microgrid system is given by [1],

$$P_{\mathrm{MG}} = P_{\mathrm{PV}} + P_{\mathrm{WTG}} + P_{\mathrm{DEG}} - P_{\mathrm{AE}} + P_{\mathrm{FC}} + P_{\mathrm{PHEV}} \pm P_{\mathrm{BESS}} \pm P_{\mathrm{FESS}} \tag{14}$$

Since power output from solar and wind generations are intermittent, output power is highly fluctuating. The mismatch in demand and supply causes system frequency deviation. To balance this situation, output power from BESS, FESS, DEG, PHEV, AE and, FC is adjusted with the help of controllers. Mainly two different cases of an islanded mode of microgrid are analyzed. Model is simulated in Simulink/MATLAB environment with a run time of 100 s.

4.1 Constant Wind and Solar Output with Constant Load

Case 1

The performance of the proposed microgrid system is analyzed considering the constant output of WTG and PV at 0.5 p.u and 0.3 p.u, respectively. A constant load of 1p.u (100%) has been taken. WTG and PV are not sufficient to meet the load demand hence auxiliary sources such as DEG, PHEV, and FESS start to operate to meet the demand. Controllers gain which is tuned with the help of the SGO technique taking population size as 10 and run up to 100 iteration counts is shown in Table 2.

Table 1 shows the parameters of the frequency response model of a microgrid system. System operation is further divided into two categories: first one is considering energy storage elements and the second one is without considering energy storage systems. In Fig. 5, system responses without having storage elements with PI and PID controllers are presented.

From Fig. 5, it evident that the PID controller is outperforming the PI controller in all the desired performance indices such as peak overshoot, peak undershoot, and settling times. Numerical analyses of these values are presented in Table 3.

System response is also analyzed with step load and constant output power from generating units. Figure 7 is representing the response of a system considering with and without storage elements in the presence of a PID controller. A system having storage elements shows better convergence characteristics which is evident from Fig. 6.

Table 1 Microgrid systems parameters

Generating units	Gains (K)	Time constants (T)
WTG	$K_{WTG} = 1$	$T_{WTG} = 1.5$
PV	$K_{PV} = 1$	$T_{PV} = 1.8$
AE	$K_{AE} = 1/500$	$T_{AE} = 0.5$
FC	$K_{FC} = 1/100$	$T_{FC} = 4$
DEG	$K_{DEG} = 1/300$	$T_{DEG} = 2$
PHEV	$K_{PHEV} = 1$	$T_{PHEV} = 0.15$
BESS	$K_{BESS} = -1/300$	$T_{BESS} = 0.1$
FESS	$K_{FESS} = -1/100$	$T_{FESS} = 0.1$
Damping constant	$M = 0.012$	–
Inertia constant	$D = 0.2$	–

Table 2 PI and PID control variables using SGO

Parameters (SGO)	(PI-Controller K_P, K_I); (PID-Controller K_P, K_I, K_D)
Without storage(PI; PID)	19.9999,20; 18.9987,19.7895,14.616
With storage (PID)	20,19.5687,10.2154

Table 3 ST (s), POs, and PUs comparison for PI and PID controllers without storage elements

Dynamic response	Δf		
Controller without storage	ST (sec)	PO (in mm pu)	PU (in mm pu)
PI	6.9	404.3	664.9
PID	6.1	6.93	24.34

Figure 7 tells that system with energy storage elements is way better than, a system having no storage elements in terms of frequency fluctuations.

4.2 *Constant Wind and Solar Output with Random Load*

Case 2

Similar to case 1, constant wind and solar power generation is considered. To make a system more realistic, random load behavior is taken. Figure 8 is showing the loading pattern for 100 s.

Again the system is analyzed under two different operating conditions in the previous fashion. Figure 9 shows the system response without energy storing elements in the presence of PI and PID controllers.

From Fig. 9, it is evident that the PID controller is outperforming PI controllers. From the zoomed plot in Fig. 4, it can be concluded that even in case of a large disturbance in loading, PID stabilizes better than the PI controller.

Similar to case 1, comparison between frequency responses of a system considering, with and without storage elements, is presented in Fig. 5, in the presence of a PID controller.

Form Fig. 10, it is evident that in the event of a large disturbance, the system considering storage elements stabilizes the deviation better than a system having no energy storage elements.

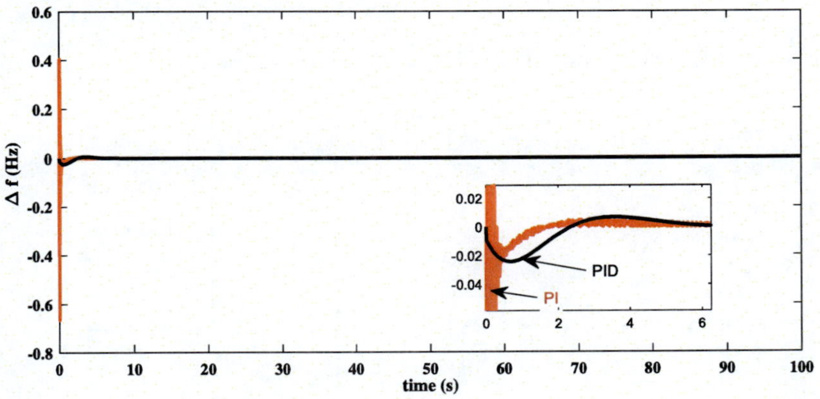

Fig. 5 Frequency deviation due to PI and PID without storage (step load)

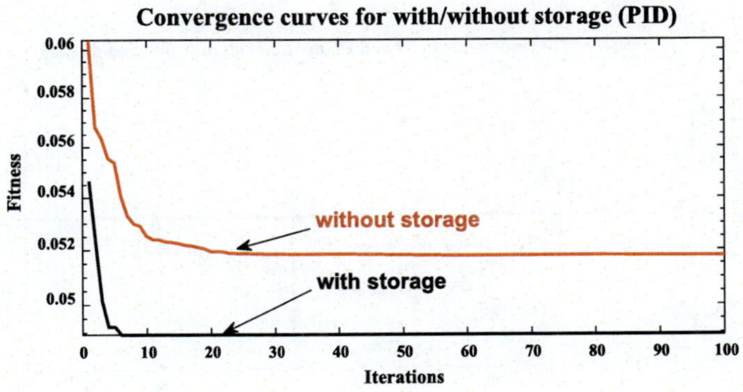

Fig. 6 Convergence curves

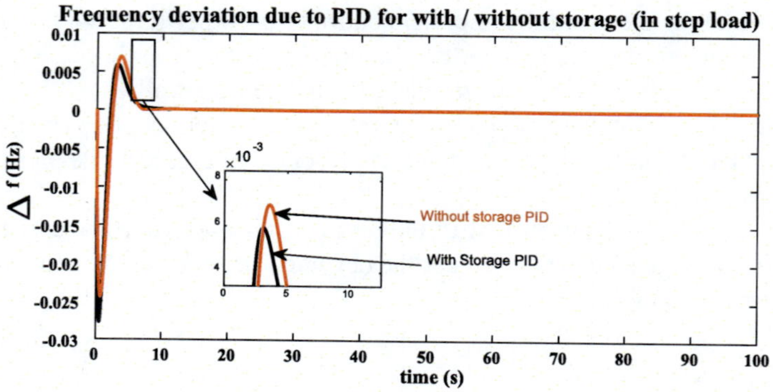

Fig. 7 Frequency deviation due to PID with/without storage (step load)

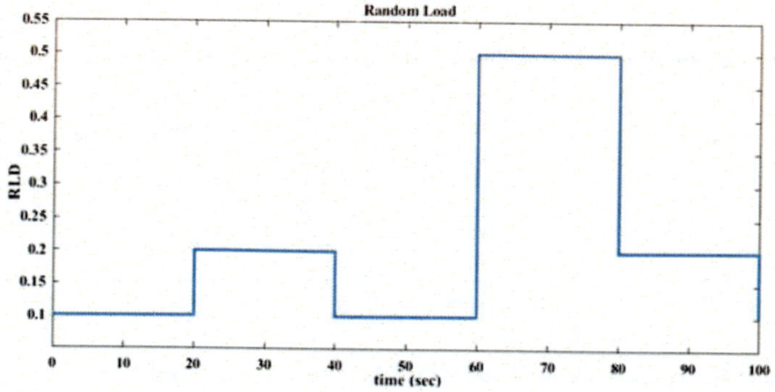

Fig. 8 Random loading

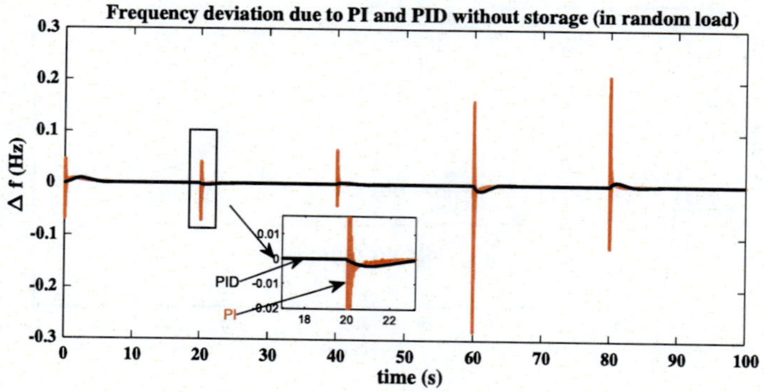

Fig. 9 Frequency deviation due to PI and PID without storage (random load)

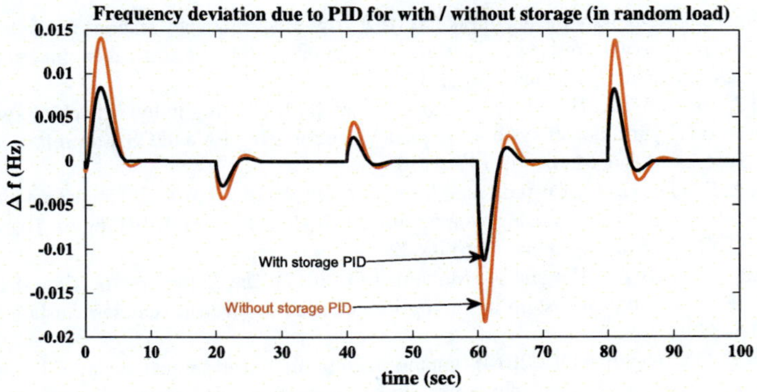

Fig. 10 Frequency deviation due to PID with/without storage (random load)

5 Conclusions

In this work, frequency deviation of microgrid system considering PV, WTG, AE, FC, DEG, FESS, and BESS is analyzed with constant and random loadings. Maiden application of the SGO technique in the domain of ALFC is a unique work. Controller gains are tuned under different operating conditions such as with storage elements, without storage, and random loading. Comparative analysis shows that PID outperforms PI controller in the context of peak overshoot, peak undershoots, and settling times. Also, a system considering energy storage elements mitigates frequency fluctuations better than a system without having any storage elements. It can be concluded that SGO tuned control technique with the help of PI and PID controller gives satisfactory results.

References

1. Ray P, Mohanty A (2019) A robust firefly–swarm hybrid optimization for frequency control in wind/PV/FC based micro-grid. Appl Soft Comput 85
2. Mohammadzadeh A, Kayacan, E (2020) A novel fractional-order type-2 fuzzy control method for online frequency regulation in ac micro-grid. Eng Appl Artific Intell 90
3. Abazari H, Monsef H, Wu B (2019) Coordination strategies of distributed energy resources including FESS, DEG, FC and WTG in load frequency control (LFC) scheme of hybrid isolated micro-grid. Int J Electric Power Energy Syst 109
4. Bagheri A, Jabbari A, Mobayen S (2021) An intelligent ABC-based terminal sliding mode controller for load-frequency control of islanded micro-grids. Sustain Cities Soc 64
5. Khan R, Gogoi N, Barman J, Latif A, Das DC (2019) Virtual power plant enabled co-ordinated frequency control of a grid connected independent hybrid microgrid using firefly algorithm. In: 2019 IEEE region 10 symposium (TENSYMP), Kolkata, India, pp 795–800
6. Mandal R, Chatterjee K (2020) Frequency control and sensitivity analysis of an isolated micro-grid incorporating fuel cell and diverse distributed energy sources. Int J Hydrogen Energy, 13009–13024

7. Teawnarong A, Chirapongsananurak P (2020) Providing frequency response in isolated micro-grids using battery energy storage systems. In: 8th International electrical engineering congress (iEECON), Chiang Mai, Thailand, pp 1–4
8. Latif A, Pramanik A, Das DC, Hussain I, Ranjan S (2018) Plug in hybrid vehicle-wind-diesel autonomous hybrid power system: frequency control using FA and CSA optimized controller. Int J Syst Assur Eng Manag 9:1147–1158
9. Latif A, Das DC (2019) Co-ordinated frequency support of solar-field/wind based independent hybrid micro-grid system using woa optimized single controller. IEEE region 10 symposium ((TENSYMP), Kolkata, India, pp 617–622
10. Khalil A, Rajab Z, Alfergani A, Mohamed O (2017) The impact of the time delay on the load frequency control system in microgrid with plug-in-electric vehicles. Sustain Cities Soc 35:365–377
11. Arani K, Gharehpetian GB (2014) Enhancement of microgrid frequency control subsequent to islanding process using flywheel energy storage system. In: 2014 smart grid conference (SGC), Tehran, Iran, pp 1–6
12. Sekhar PC, Mishra S (2016) Storage free smart energy management for frequency control in a diesel-PV-fuel cell-based hybrid AC microgrid. IEEE Trans Neural Netw Learn Syst 27(8):1657–1671
13. Taheri H, Taheri S, Pouresmaeil E (2017) A seamless control scheme for a microgrid with renewable energy sources. In: IEEE 30th Canadian conference on electrical and computer engineering (CCECE), Windsor, ON, pp 1–5
14. Elbaz M, Feliachi A (2012) Real time load frequency control for an isolated microgrid system. In: 2012 North American power symposium (NAPS), Champaign, IL, USA, pp 1–6
15. Satapathy S, Naik A (2015) Social group optimization (SGO): a new population evolutionary optimization technique
16. Dey N, Rajinikanth V, Shi F, Manuel RS, Tavares, Moraru L, Karthik A, Hong L, Kamalanand K, Emmanuel C (2019) Social-Group-optimization based tumor evaluation tool for clinical brain MRI of Flair/diffusion-weighted modality. Biocybernetics Biomed Eng 39(3):843–856

Power Generation for Street Lights Using Smart Tiles, Floor and Piezoelectric Shoes for Mobile Battery Charging Along with GPS Tracker Shoes

P. Srinivasan, D. Sivakumar, and V. N. Ganesh

Abstract In the current situation, the necessity for energy has been expanding in an upsetting rate and the accessibility of energy assets is not plentiful for maintainable turn of events and the need of great importance is to build up a prudent contamination-free endless energy asset to repay the expanding requests. Since walking is the common movement in human everyday life, a person walks, the energy from his feet is wasted on the ground. To save and use this power we are converting mechanical stress/ pressure which creates a strain in the loadcell thus converting it into electrical energy. The smart tile would be made out of 2 layers of plywood with footstep, layer being made extra hard to handle a high amount of pressure and is expected to handle higher mechanical stress. At the point when an individual strolls as well as runs, the pressing factor applied on the piezo-electric sensor that is inserted in the piezoelectric shoe can be changed over into electrical energy. To create energy to an enormous degree, in this Model, Execution of the smart tile layout along the pathway is completed which is then given to the path of streetlamps &alongside this the piezo shoe is carried out that would charge our mobile battery while walking and there is also a standalone GPS tracking shoe that would automatically receive data from satellite such as latitude, longitude, date and time.

Keywords Loadcell · Piezoelectric · GPS tracking · Smart tiles

1 Introduction

The urge for energy has been expanding in an upsetting rate and the feasibility of energy assets are not plentiful for the reasonable turn of events and it is the need of great importance to set up a practical contamination-free, limitless energy asset to repay the expanding requests. Since, walking is the most common action in everyday

P. Srinivasan (✉) · D. Sivakumar · V. N. Ganesh
Department of Electrical & Electronics Engineering, SRM Institute of Science and Technology, Ramapuram, Chennai, India
e-mail: srinivasp808@gmail.com

K. Namrata et al. (eds.), *Smart Energy and Advancement in Power Technologies*, Lecture Notes in Electrical Engineering 926,
https://doi.org/10.1007/978-981-19-4971-5_59

routine, at whatever point an individual walks, the person dissipates energy that is wasted on the surface.

The motive of this prototype is to produce smarter, greener renewable energy. In this model implementation, smart tile system along the pathway is done which is given to the path of streetlamps for alongside it the smart shoe is carried out that would charge our mobile battery while walking and has a standalone GPS tracking shoe so, a person would walk on smart floor with his piezoelectric shoes that would make the street lights glow as well as charge his mobile battery while walking and the location of the person would be known via GPS tracker shoes [6]. The major advantage is that these smart tiles is that it can also be used as smart steps [1, 7]. These tiles are easily portable and installable. These user-friendly tiles can also be easily removed.

In this model for operation, smart tile network along the footpath is carried out which is given to the lane of streetlights for a minor construction along with it the smart shoe is applied that would charge our mobile battery while walking and has a standalone GPS tracking shoe. So, a person would walk on smart floor with his piezoelectric shoes that would make the street lights glow as well as charge his mobile battery while walking and the location of the person would be known via GPS tracker shoes [13, 15]. The major advantage is that these smart tiles is that it can also be used as smart steps. These tiles are easily portable and installable. These user-friendly tiles can also be easily removed.

We will be using load cells for manufacturing power tiles. A load cell is a physical transducer that converts force or pressure into a quantifiable electrical yield. Load cells come in various shapes and sizes so they can be added to various hardware and gauging gear.

We are using the single point load cell that has off center loading on both its sides which can be associated with the smart tiles via M4 or M5 size screw. It has strain gauges in its middle such that when we exert pressure on the tile an unbalance would be created creating strain, and converting it into electrical energy.

2 Literature Review

Before beginning our research regarding the smart bin we decided to check the current advancements and innovations regarding this topic. We checked the research papers that came out recently and the following were the current developments. The paper discusses the different methodologies used to manage the GPS tracking shoe and describes the detailed functionality of producing electricity for lighting up LEDs, thus giving an overall idea of preparing applications related to information.

2.1 *Problem Definition*

In day-to-day human life, when people walk he/she exhausts their energy in the ground which goes to waste, so this energy which people hardly care about could be made into good use for generating some amount of electricity which can be used to light up LEDs (here we are lighting 93 LEDs which are present in the form of strips). In order to reserve and make use of this energy, we are converting mechanical stress/ pressure which creates a strain in load cell thus converting it into electrical energy. Moreover, the energy produced from some conventional sources like coal, oil, etc. is not clean energy as there is pollution along with some waste as an end product. So the aim should be economical, pollution free inexhaustible energy.

2.2 *Scope and Motivation*

2.2.1 Scope

In the existing system, a general prototype model of the piezoelectric floor would be made where a lot of piezoelectric crystals [14] would be embedded thus leading to a lot of clumsy wires and a small stick would be used to apply pressure on the crystal and an ampere value would be shown in ammeter [3]. Else a piezoelectric tile would be made with a lot of piezo sensors that really can't handle much weight and an accurate budget of INR 51,255 would be required. So, a highly efficient smart tile even as a prototype could handle less weight, look clumsier, be less efficient, and is obviously expensive. 30 piezo sensors need to be soldered together along with various wires and its cost for 10*10 tiles would sum up to INR 600. The prototype to make one piezo tile would cost INR 600. So, the cost of making 24 tiles would be INR 14,400. And additional costs of wires, controllers, battery, LCD, and all other electronic equipment would add up. Piezoelectric shoes for mobile battery charging could only be made to charge mobile battery which isn't very effective. In GPS Tracker shoes maybe only, the location would be shown and it might or might not get updated as the person moves. It would not show the exact latitude and longitude plots as the person keeps moving and there is no guarantee of server getting updated as the person keeps on moving.

As we can (see Fig. 1), different piezo sensors and transducers are embedded and soldered with each other. There are separate bushes on four sides and all sensors are attached on 4–5 different layers. There are four separate large bush stands on the corner of every tile. Top and bottom are fixed with 30*30 tile wood(example). Apart from these, there are several wires which make a single tile look clumsy. The weight it would handle would also be less and on calculating its cost of manufacture it would sum up to:

Piezo sensors: $10 * 5 * 20 = 1000$ INR.
Bush(4S + 4B): $50 + 250 = 300$INR.

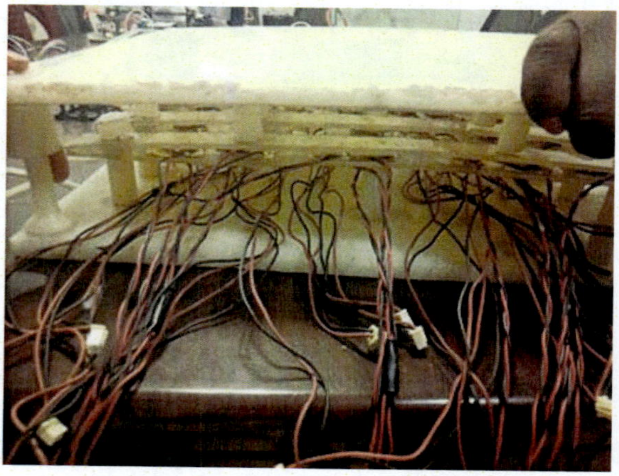

Fig. 1 Shows the proposed system

Tiles and wire(2 + n): 600 + 30 = 630INR.
Soldering and female pin connection: 330 INR.

So Approximately budget of INR 2500 would be required for making a single tile of 30*30 cm. Budget of **INR. 2500** is the honest and least possible budget with these electronic devices. This piezoelectric tile could handle a maximum footstep weight of 14–15 kg and subsequently, lower pressure and another main problem of using piezo sensors is that it may produce a vibration but not always [8, 9]. The vibration piezo sensors produce is not enough to produce electrical energy without the help of battery because the bottom line of sensors is that they aren't voltage sources or power sources, instead they can just sense a movement but that there's no guarantee that they would sense it every time [2].

2.2.2 Proposed System

In the proposed system, power tiles with 5 kg load cell would be soldered with HX711 Amplifier module. 24 power tiles would be employed front and back. Manufacture of 10 cm *10 cm power tiles with 15 mm*15 mm square shaped bush with countersunk M4 or M5 size screw hole for an unbalance to happen in load cell as shown in (see Fig. 2) that causes strain and this load cell being an electric transducer converts this into subsequent electric energy. A load cell would have a Wheatstone bridge circuit inside that, which would produce 2 ohms when subjected to forced pressure on top of the tiles. Higher the force applied on tiles, higher is the amount of resistance produced. This electrical signal is increased in order to transfer data efficiently via Arduino –IDE and Arduino MEGA. A Rechargeable Lithium Polymer battery with ratings of 11 V and 2.2 Ah is used to power up LED strips. An LCD Display would be attached on the top of floor with a long stick to show values such as force exerted,

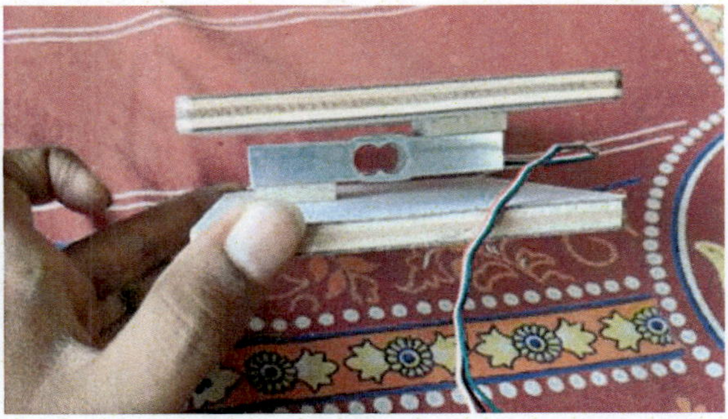

Fig. 2 Typical load cell module

pressure produced, and voltage harnessed. 8 LED lights would be stuck around each tile for clean, clear and perfect presentation. The prototype of proposed system is shown (see Fig. 4).

2.2.3 Description

In the described project just from a single activity of walking, three different benefits could be enriched. As you walk on our smart floor electrical energy for lightening up street lights would be generated [7]. Simultaneously As you walk on our smart floor with your smart piezo shoes charging current for your mobile battery would be made and a small LED at the beginning of our shoes would pave the way for those who walk in the dark. The block diagram of smart floor clearly depicts the processes of power generation from loadcell (see Fig. 5).

The circuit diagram in (see Fig. 6) shows us the connections in detail which are explained further. The load cell has a Wheatstone bridge circuit inside that, which would produce 2 ohms when subjected to forced pressure on top of the tiles. Higher the force applied on tiles, higher is the amount of resistance produced. When we stand on the tiles (force is exerted) pressure is created which is the form of electrical analog signal. This analog signal is then forwarded to the HX711 amplifier. The HX711 amplifier takes in measurable form of data and converts the analog signal to digital so, the HX711 amplifier is basically an analog to digital converter as shown (see Fig. 3).

Here the main unit (or in other terms the brain cell) is the Arduino Mega 2560 which has around 32 pins. It is nothing but a microcontroller that gives Input/output instructions to various parts connected to it.

A DC rechargeable Lithium Polymer battery is connected to Vin pin of the microcontroller. This battery would act as a voltage source for this huge combination of power tiles/smart floor. The ground pin is ported with the negative end of the battery. Pins A0 to A8 are connected to the LEDs. The LED here is connected in series and is in form of strips. Here 4 V LED Strips of 93 in number are used. These strips have positive and negative terminals where phase and neutral are soldered. A single tile would be having 8 led lights (pair of 4) which are arranged on its circumference. On

Fig. 3 Load cell soldered with HX711 amplifier on a dot board

Fig. 4 Prototype of proposed system

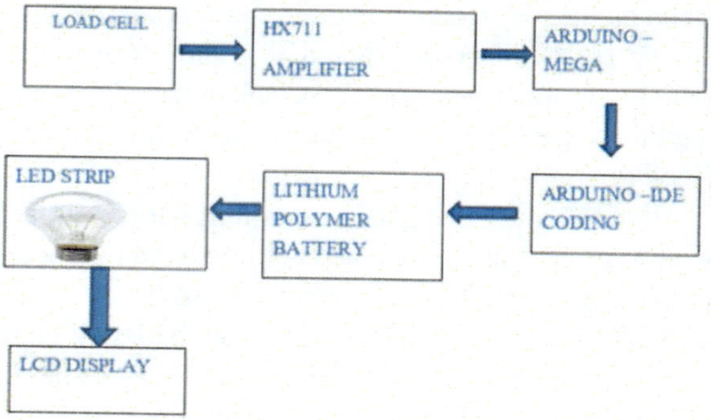

Fig. 5 Block diagram of smart floor

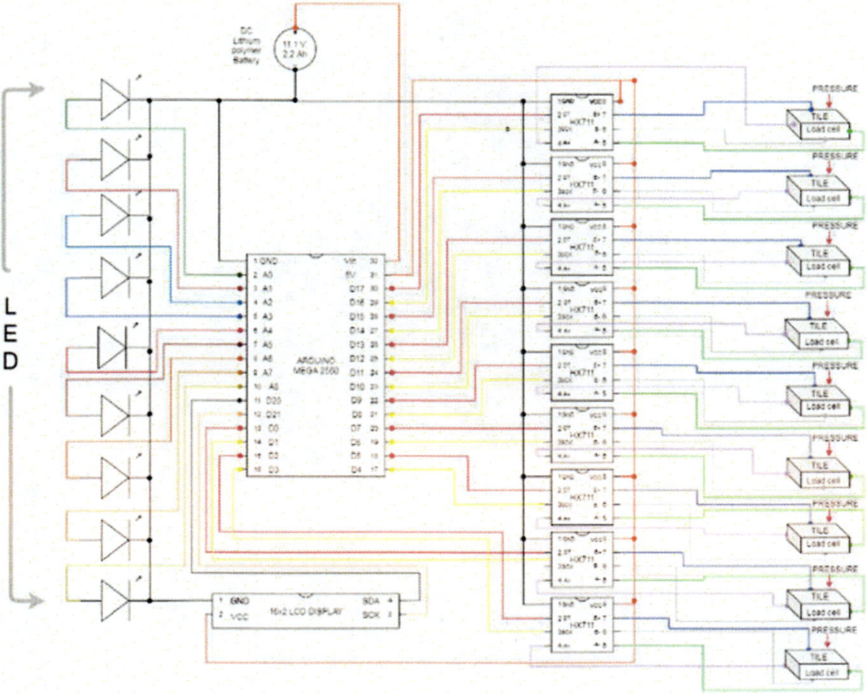

Fig. 6 Circuit diagram of smart tile

receiving ample amount of power, the LEDs glow. The rear end of the LED strip is connected to the ground port of the LCD screen to complete the circuit. Thus the working and connections of the smart tile is clearly illustrated from the above description.

2.2.4 GPS Tracker Shoes

Tracker shoes would be made with Node MCU ESP 2866, Ublox Neo 6 M GPS Module, LC7805 IC, dot board, female pin, male–female pin, and a 9 V Battery. Pins would be soldered on dot board. Dot board would act as a common point of connection. 9 V from battery would be converted to 5 V via LC7805 which would subsequently drop to 3.3 V for powering up Node MCU. Vin, G, D0, D1 pins of Node MCU would be connected to standalone GPS Module. Connections of male–female pin and female-female pin are carefully soldered from the backside of the dot board. The complete prototype for the Tracker Shoes is as shown (see Fig. 7).

The block diagram of GPS Tracker shoes (see Fig. 8) gives a clear idea and detailed steps for its setup.

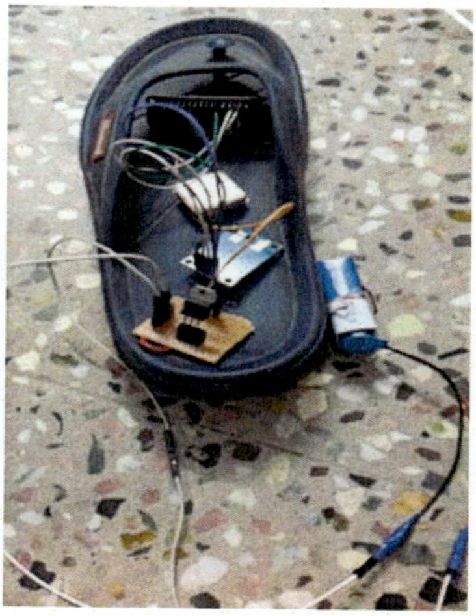

Fig. 7 Tracker shoes

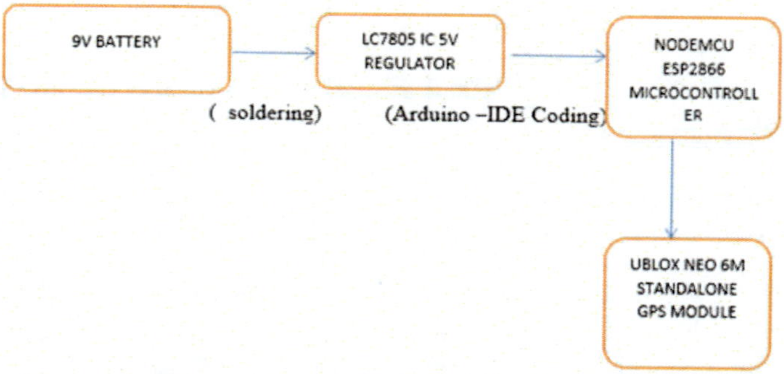

Fig. 8 Block diagram of GPS tracker shoes

The GPS Tracker shoes would show the location of the place. The latitude, longitude, time, date, and the link for Google maps is shown in the output. By clicking on the link, we get to know the precise location as shown in (see Figs. 9 and 10).

Fig. 9 Precise location details

Fig. 10 Precise location—google map

2.2.5 Piezoelectric Shoes

Piezoelectric shoes for mobile battery charging would be made with help of piezoelectric sensors, bridge rectifier (connection of 4 single IN4007 Diodes), small Li po battery, charging cable in a cost-effective way along with a small LED at the beginning of our shoe [4]. As we exert pressure on piezo sensors (they would be placed on heel of shoes) it would produce power for both mobile batteries charging and led light. This led light can be used to find way while walking in a dark room. So, the components and the cost of manufacturing this product are minimal and prove to be highly efficient for everyday life [10–12].

From (see Fig. 11) we can see that when we apply pressure on piezo embedded thermocol surface which is placed below slippers a spike current is produced by sensors which is then regulated by diode to dc current and it is then connected to capacitor that stores and discharges charge for backup. This dc along with 9 V DC battery is regulated to 5 V DC which is then sent to resistor for current division rule thus enabling both mobile battery and LED to function simultaneously. In the block diagram (see Fig. 12) gives a vivid picture of the components required for piezoelectric shoes and its connections [5].

Apart from the above components, there are additional components also required namely:

Plywood tiles- 10*10 cm in dimensions, 24 tiles (top and bottom).
Wooden Bush- 15 mm*15 mm in dimensions.
Jumper Wires- male-male, male–female, female-female.
LED Strips- 4 V led strips (93 in numbers required).
Piezoelectric sensors- O/P voltage- 30p-p, impedance- 500 Ω.

Glue Gun, Soldering Iron, Bread Board, Multimeter & Dot Board, Bridge rectifier, and LC7805 IC.

Combined Picture of Smart tile, GPS Tracker shoes, and Piezoelectric shoe is shown (Table 1):

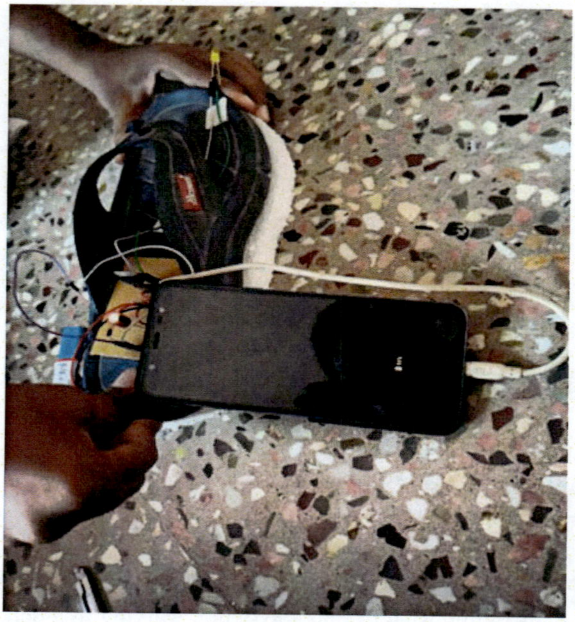

Fig. 11 Prototype of piezoelectric shoes

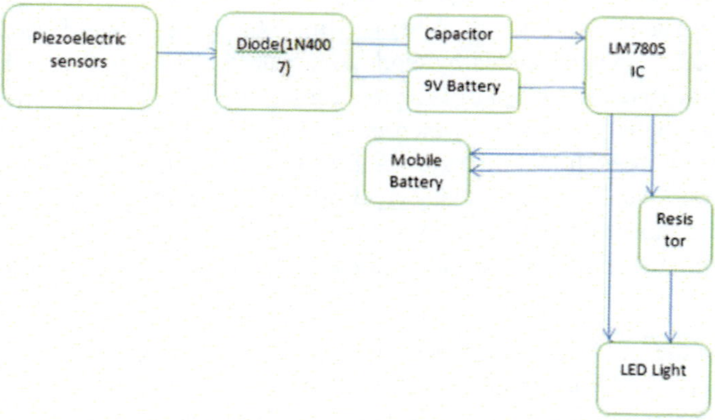

Fig. 12 Block diagram of piezoelectric shoes

2.3 Practical Implementation

We will be practically implementing this by making 10 power tiles that can handle weight up to 7 kg (kg), whereas a normal piezo tile can handle up to 500 g. So, in this regard, our power tiles can with stand more weight and pressure as shown (see Fig. 13).

Table 1 Components required

S. No.	Components	Type	Specifications
1	Amplifier module	HX711	Power Supply Voltage-(2.6–5.5)v Digital Supply Current 100 μA
2	Voltage regulator	LC7085	O/P voltage-(4.8–5.2)v Peak O/P current -2A
3	Load Cell	Strain Guage	Capacity -5 kg Dimensions-55.25*12.7*12.7 Materials-Aluminum Alloy
4	Node MCU	ESP 2866	Analog Voltage-(0–3.3)v
5	GPD module	UBLOX NEO 6 M	VCC—(0–2)v
6	Microcontroller	Arduino UNO ATmege 328p	Vin-(3.3–5)V Frequency-16 MHz
7	Battery	1. Lithium Polymer 2. Small Lithium Polymer	1.Voltage-11.2v Capacity- 2.2Ah 2.Voltage -3.3v Capacity- 0.5Ah
8	Display	LCD	Voltage-(4.7–5.3)v

Fig. 13 Typical prototype of the complete system

The current generated through these tiles would be used to light up 5 LED strips which consist of 93 LED lights attached on the circumference of every smart tile.

Piezoelectric shoes will also be made such that as while you put pressure on sensors fitted at heel of shoes, you will get your mobile battery charged. A small LED would be connected at the front of shoe for increased visibility which would also start glowing as you simultaneously put pressure on piezo sensors. Tracker shoes with ESP 2866(Node MCU), Ublox Neo 6 M Standalone GPS Module would be made through which GPRS would track our latitude, longitude, date, and time values and also show location via Google maps.

At the point when individuals walk or put pressure/power on the power tiles, it will produce little electrical energy. In this undertaking, power tiles are entirely executed as a wellspring of efficient power energy.

3 Result

In this prototype, the principal objective was to discover a way to deal with producing electrical energy from stress/force/pressure applied on a tile. It was seen that when the load cell was hit by a force/pressure, the load cell creates an unbalance which leads strain and is thus converted to electrical current of transferrable form but the amount of power generated was less, and output current from load cell was pretty decent and much better than piezo sensors.

Thorough calculations were done for each part of the project but indeed it can very well be changed and is still as useful as a green wellspring of energy. Along with this piezoelectric shoe (with LED) for mobile battery shoe and GPS Tracker shoes was made. The LCD screen mounted on wooden stick would display values such as force, pressure, and voltage generated. So, the objective was to harness 3 different benefits with a single activity of walking was thus achieved.

4 Future Scope

The undertaking can have heaps of advancement:

1. Additional improvements could be made on reducing the complexity of soldering in the power tiles project. For piezoelectric shoes, additional improvements could be made on charging the mobile battery more efficiently with pressure on piezo sensors.
2. The website link for GPS Tracker shoes could be sent as a message to user instead of displaying in serial monitor.

5 Conclusion

So, to conclude with, this project is a sub-classification under Waste to Energy Conversion (WtE). The energy that we send as waste to the ground, the pressure that we send as waste to ground is being converted to electrical energy. Along with these two different shoes were made. Rather than using just piezo sensors, we used different new materials in the field of electronics and embedded engineering, thus making our project a valuable success. This project was a great learning, exposure, and hands-on knowledge to all of us in a new field of electronics engineering, embedded systems, and embedded programming. We are hopeful to do even more projects in the field of automation, electrical, electronics, and embedded engineering.

Acknowledgements The Authors would like to extend their gratitude to the students K Prasanna Kumar, Manish Kumar Pandey, and Aggu Mallesh, who helped to bring this project to fulfillment. Authors would like to thank the management for the academic and technical support which they have given all the time during our work.

References

1. Ruman MR, Das M, Mahmud SMI (2019) Human footsteps for energy generation by using piezoelectric tiles, Dhaka,American International University, Bangladesh
2. Prasad PR, Bhanuja A, Bhoomika N (2019) Power generation through footsteps using piezoelectric sensors along with GPS tracking. SVCE Bangalore, India
3. Bhuvani G, Senthil M (2017) Piezoelectric method based wireless energy transfer, Kongu Engineering College Perundurai, Erode, Tamil Nadu
4. Tichý J, Erhart J, Kittinger E, Prívratská J (2014) Fundamentals of piezoelectric sensorics mechanical, dielectric, and thermodynamical properties of piezoelectric materials. Berlin, Springer
5. Tadigadapa S, Mateti K (2009) Piezoelectric MEMS sensors: state-ofthe- art and perspectives. Meas Sci Technol 20(9):092001
6. Piezo Systems: History of Piezoelectricity. [Online]. Available: http://piezo.com/tech4history.html
7. Piezo Tiles-Generate Electricity by Walking. [Online]. Available: http://www.instructables.com/id/Piezo-Tiles-Generate-Electricity-by-Walking
8. Mugali S (2018) Footstep power generation using piezo-electric transducers. Int J Latest Technol Eng Manage Appl Sci 7:2–5
9. Mahmud I (2018) Electrical power generation using footsteps. Euro Sci J
10. Singh M, Abdullah NAW, Singh B (2017) A review of walking energy harvesting using piezoelectric materials. Mater Sci Eng 1–8
11. Marshiana D (2016) Footstep power production using piezoelectric sensors. Res J Pharm Technol 1–5
12. Jahangir MP (2016) A study of human footsteps for power generation. IEEE
13. Foxlin E Pedestrian tracking with shoe-mounted inertial sensors. IEEE Comput Graph Appl 25(6):38–46. https://doi.org/10.1109/mcg.2005.140
14. Nye F (1985) Physical properties of crystals. Oxford Sci Publ 110–114
15. Foxlin E (2002) Motion tracking technologies and requirements. Handbook Virtual Environ Technol 182

A Review on Energy Storage Systems in Electric Vehicle Charging Station

Gaurav, Nakka Jayaram, Jami Rajesh, Satya Venkata Kishore Pulavarthi, and Jayachandra

Abstract The growth of electric vehicles (EVs) is very fast and will continue to grow exponentially in the coming days. Due to its high cost, the degree to which electric vehicles can be used is not up to the range in all the developing countries. The infrastructure of the charging stations in developing countries has not yet built to the level required to effectively operate an electric car in a developing world. Despite this, DC fast charging stations placed a heavy burden on the grid, resulting in high maintenance costs. Adding multiple storage systems to the DC fast charging station would help to mitigate these problems because it will act as a buffer between grid and vehicle. This review paper goes into the basics of energy storage systems in DC fast charging station, including power electronic converters, its cost assessment analysis of various energy storing devices for a range of charging scenarios.

Keywords Energy storages system (ESS) · Battery chargers · Batteries · Flywheel · Hydrogen · Vehicles · Battery energy storage system (BESS) · Flywheel energy storage system (FESS)

1 Introduction

The benefits of electrical vehicle are to decrease greenhouse gas emissions, as well as other controlled emissions that contribute to air pollution and other types of pollution done by human beings which has a negative health effect. High vehicle prices and charging technology are the key roadblocks to EVs adoption. While, the range can be further improved with larger and more energy density batteries, which results in change in vehicle mass, volume and cost [1, 2]. Design and implementation of large quantity of EVs, rapid fast charging station will help to ease range, charging issues on longer intercity drives without necessitating the purchase of extremely expensive EVs with big batteries [3, 4]. Because electric vehicle has high-efficiency,

Gaurav (✉) · N. Jayaram · J. Rajesh · S. V. K. Pulavarthi · Jayachandra
Electrical Engineering Department, National Institute of Technology Andhra Pradesh, West Godavari, India
e-mail: gaurav.sclr@nitandhra.ac.in

K. Namrata et al. (eds.), *Smart Energy and Advancement in Power Technologies*, Lecture Notes in Electrical Engineering 926,
https://doi.org/10.1007/978-981-19-4971-5_60

low-emission transportation system. Therefore, this target will only be achieved if the energy used to charge them is generated by renewable sources including solar and wind. One of the most important aspects of achieving this target is to balance the charging demand with irregular or erratic renewable energy availability. But green and clean energy storage without any pollution is very much required in the modern world, and electrical vehicle and DC fast charging station without any pollution are very useful in achieving the aim with less CO_2 emission [5, 6].

1.1 The Need for Energy Storage

Energy storage will greatly change how it will generate, transmit, and distribute, and the consumer pay for electricity tariff, according to the response. Energy storage facilities can integrate intermittent energy services, reduce net load issues, and provide electricity during periods of heavy demand [7]. Emergency outages in power system include storm-related outages, sudden device outages, and power supply transients. Accidents, as well as unintentional delays, are all the possibilities which can come. So, energy storage makes the power system more stable by compensating the fluctuation occurring in power system network in very less time interval, and it makes the Indian grid more resilient, efficient, and secure for all devices connected to it [8, 9].

1.2 Requirement of Energy Storage at DC Fast Charging Station

The direct connection between electric vehicles to a reliable grid is not always possible along highways and country roads, despite the fact that these are the locations where DCFC stations are most needed. On the other hand, drivers that need quick charging often need high-power charging as soon as the electrical vehicle is plugged in. A simpler alternative to DCFC stations is storage of energy in energy storage system such that the amount of power available from the grid is minimal enough to reduce or even eliminate the need for grid maintenance. When electricity demand and price are not up to the desired expectations, the ESS can be charged, so EVs can give more power from the energy storage system without affecting the grid and at a flat amount [10, 11].

2 Energy Storage Ways in DC Fast Charging Station

- Hydrogen energy storage
- Flywheel energy storage

- Battery energy storage
- Flywheel and battery hybrid energy storage.

2.1 Battery ESS Architecture

A battery energy storage system design with common dc bus must provide rectification circuit, which include AC/DC converter, power factor improvement, devices and voltage balance and control, and separation devices between the battery and the grid are all needed in a battery ESS DC fast charging architecture with a typical DC bus, which is done to reduce the effect of any ground faults coming into system if the battery is degraded or damaged battery shell. This need for grid-to-storage battery separation is a new limitation for DC fast charging station without energy storage, where isolation is needed between the grid and the electric vehicle. There are three strategies for isolating the grid from the storage battery. A low-frequency transformer, diode rectifier, power factor correction device, and DC-to-DC converter are used in [12, 13]. A large low-frequency transformer, on the other hand, will increase the system's cost and volume (Figs. 1 and 2).

$$\mathrm{SOC}_{\mathrm{battery}} = \mathrm{SOC}_{\mathrm{initial}} + 1/\left(C_{\mathrm{battery}} * 3600\right) \int_{t_{\mathrm{start}}}^{t_{\mathrm{end}}} i_{\mathrm{charging}}\, \mathrm{d}t \quad (1)$$

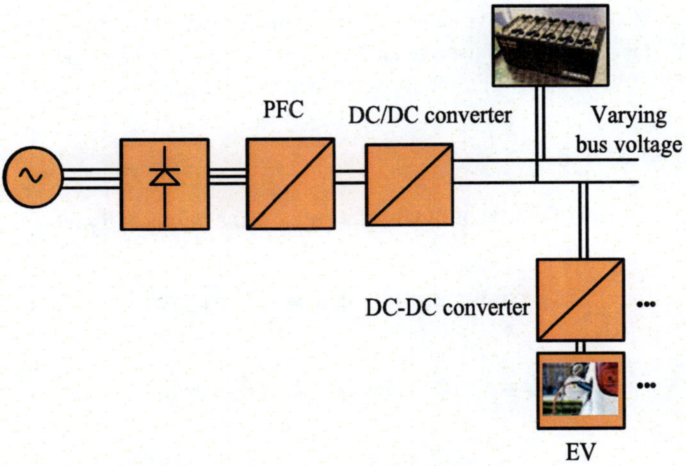

Fig. 1 Battery energy storage with dedicated low-frequency transformer

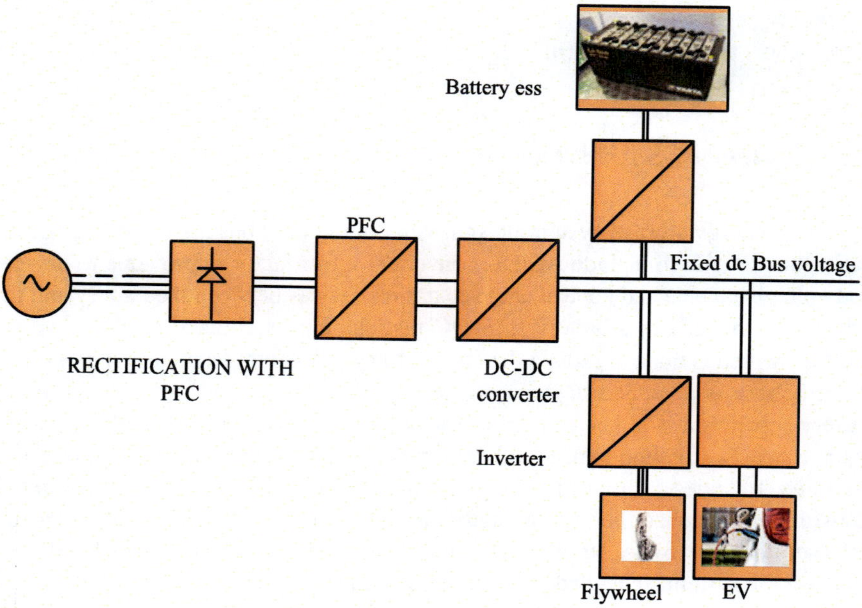

Fig. 2 Battery energy storage with isolated DC–DC converter varying bus voltage

2.2 *Flywheel ESS Design*

In fly wheel energy storage system design, there is an inner connection which connects a DC bus with a voltage spectrum of around 550–750 V [14, 15]. An external AC/DC converter is used in an AC input flywheel system so that power can flow in both direction between the flywheel and the AC grid (Fig. 3).

$$I = \int R^2 \, \mathrm{d}m$$

$$\text{moment of inertia of flywheel where } r \text{ is the radius of fly wheel.} \tag{2}$$

$$\text{Maximum speed} \quad V_{\max} = \sqrt{2k\sigma_{\max}/\rho}. \tag{3}$$

$$E = \frac{1}{2} I \omega^2 \text{energy stored by flywheel.} \tag{4}$$

$$\text{State of charge, flywheel} = \text{SOC}_{\text{initial}} + \text{SOC} \tag{5}$$

$$1/(E_{\text{CAP}} * 36000) \int_0^{\text{inf}} p_{\text{flywheel}} \, \mathrm{d}t \tag{6}$$

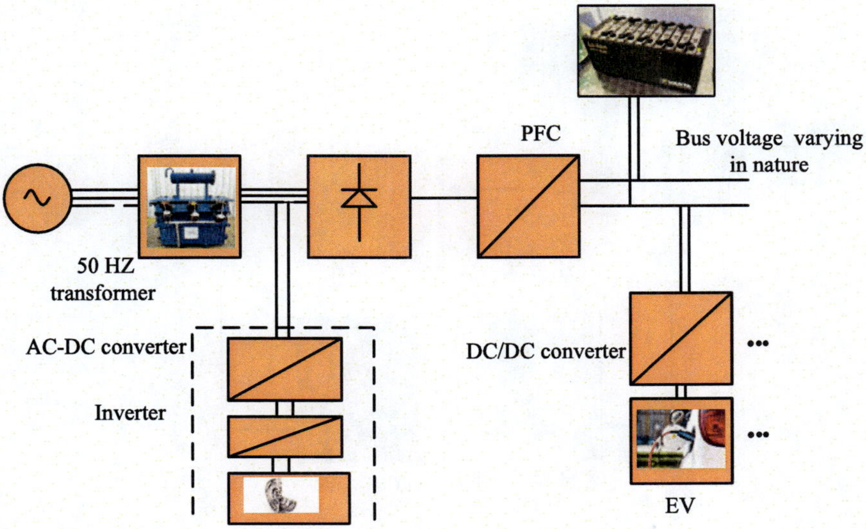

Fig. 3 Flywheel energy storage diagram with low-frequency transformer

$$\omega_{\text{flywheel}} = \sqrt{\text{SOC} * \left(\omega_{\max}^2 - \omega_{\min}^2\right) + \omega_{\min}^2} \tag{7}$$

2.3 Architecture of Hydrogen Energy Storage

The electrolyze does not require galvanic separation from the grid in a DC charging station, as it does for hydrogen energy storage, because the fuel cell is a generating source that can be isolated from the grid. It is standard procedure to employ a DC-to-DC converter at the fuel cell output side since fuel cells have more varied output voltages dependent on load and can function like current source with high voltage and low current [16, 17] (Figs. 4 and 5).

2.4 Flywheel–Battery Hybrid ESS Design

For flywheel battery hybrid energy storage system, there is separation between the grid and ESS components shown in Fig. 6. A hybrid design with a DC input flywheel is presented below with the DC bus allowing the battery voltage to differ. At the grid line, a DC-to-DC isolated converter provides separation DC-to-DC converter attached to the charge storing device such as battery has a constant DC bus voltage,

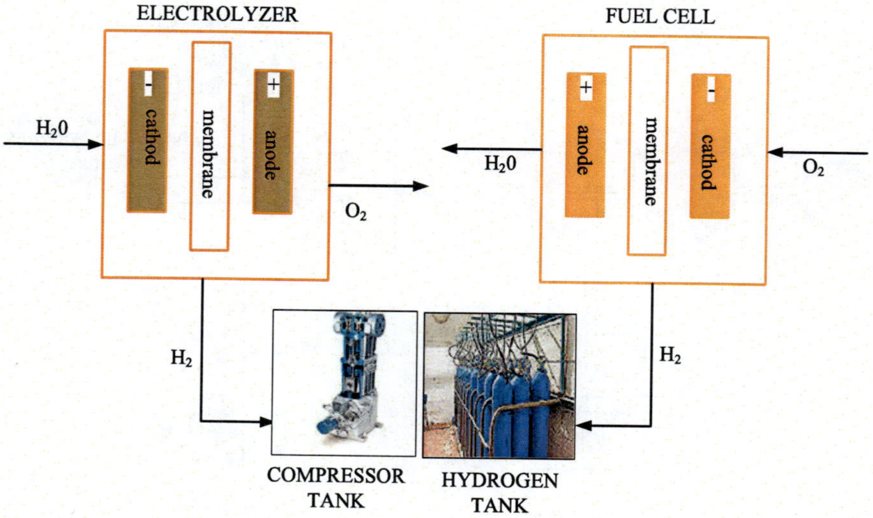

Fig. 4 Hydrogen storage system components

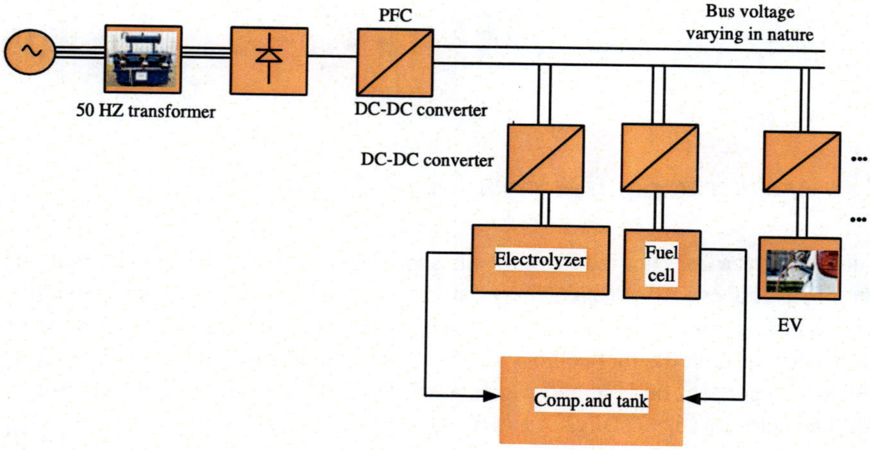

Fig. 5 Hydrogen energy storage isolated DC–DC at the fuel cell terminal

increases losses into and out of the battery while reducing the complexity of the DC-to-DC converter designs [18, 19]. Grid separation can also be accomplished with a transformer of low frequency, but this reduces device volume.

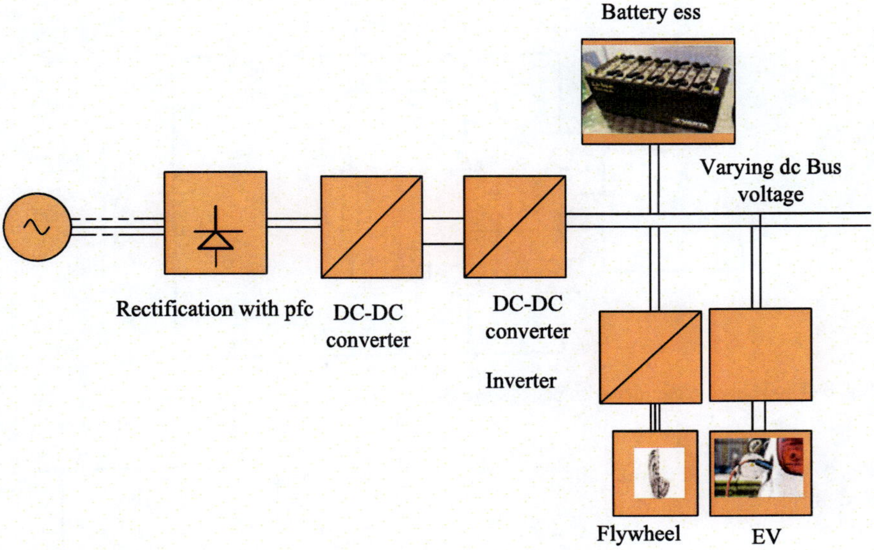

Fig. 6 Flywheel–battery hybrid energy storage with varying bus voltage

3 Different Types of Bridge Rectifiers

- Diode bridge rectifier with boost PFC
- Diode bridge rectifier with buck–boost PFC
- Totem-pole PFC with GAN switches.

In diode bridge rectifier, conduction loss is more; therefore, we prefer to go for bridgeless topology. Soft switching of power electronics converters in discontinuous mode can minimize losses in switches [20, 21] (Figs. 7, 8 and 9).

Fig. 7 Diode bridge rectifier with boost PFC

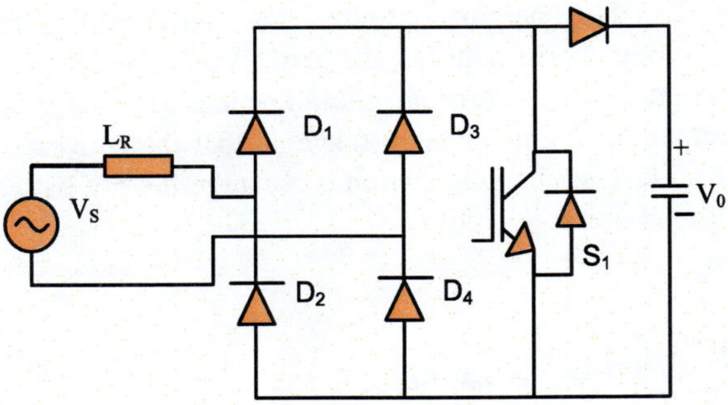

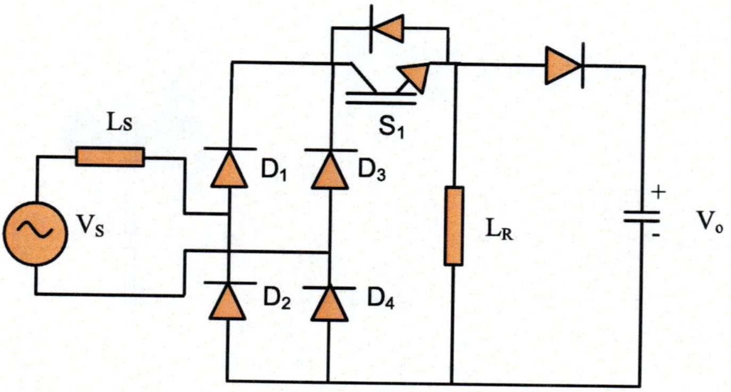

Fig. 8 Diode bridge rectifier with buck–boost PFC

Fig. 9 Totem-pole PFC with GAN switches

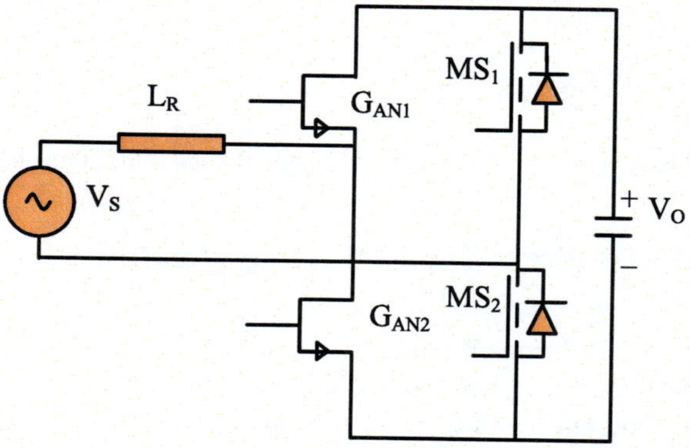

4 DC-to-DC Converter With Isolated Topologies

DC-to-DC converters with isolated topology used in the AC-to-DC conversion stage. At DC, fast charging station at low-frequency transformer is not available in this topology. As the battery voltage of EV is varying very fast and depends on the SOC of batteries, this converter should be made such that their voltage level in input and output should cover more range so that the converter works at its rated power so that DC fast charging station can charge the EV battery in minimum time [22, 23] (Figs. 10 and 11).

5 Vienna Rectifier

To obtain a high-power factor and to improve the harmonic profile this type of topology is used, each process has only one active turn, making the rectifier easier to

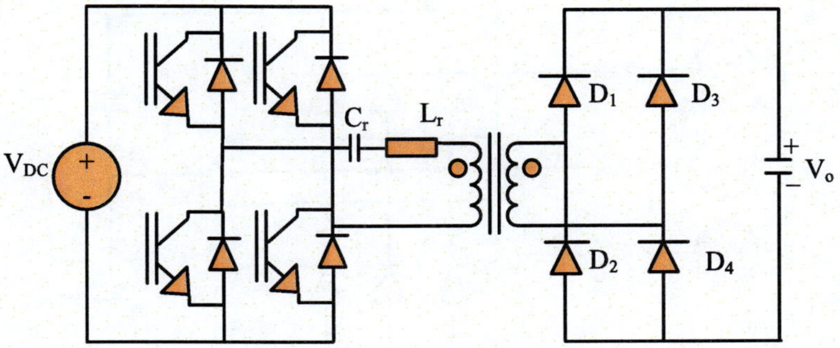

Fig. 10 Unidirectional full-bridge LLC converter

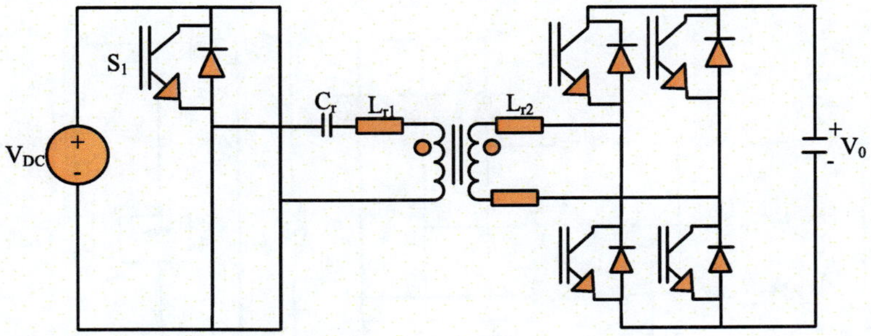

Fig. 11 Bidirectional full-bridge LLC converter

handle and more efficient. This is a pulse width modulated converter in which power factor change is determined by boost inductors at the input. The energy stored by the inductor when the switch is off is attached to the load through the diodes which are in series when the switch is turned on [24, 25]. The voltage stress on switches is $V_{DC}/2$; the general equation of Vienna rectifier is given as

$$V_{\text{bus}}/2 - V_{\text{rms}} = L * \Delta I/(D * T_s) \tag{8}$$

From this Eq. (8), the inductor can be designed as

$$L_i = (V_{\text{bus}}/2)/(4 * f_{\text{sw}} * \Delta I_{\text{pp}_{\max}}) \tag{9}$$

And for design of capacitor used in Vienna rectifier, the equation is given,

$$C = (1/3) * P_{\text{ac}}/(4 * f * (V^2 - (V - \Delta V)^2) \tag{10}$$

See Fig. 12.

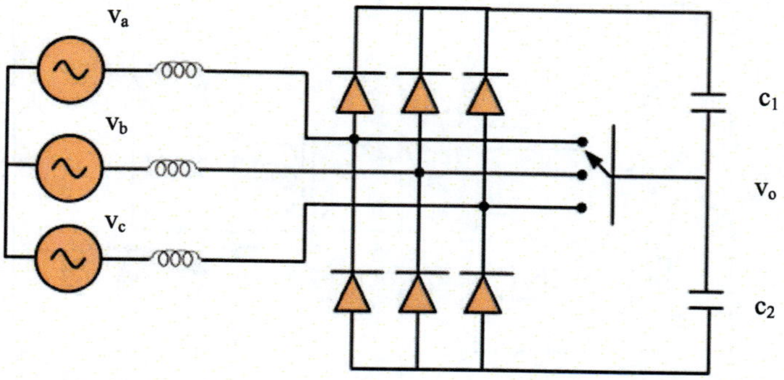

Fig. 12 Vienna recctifier

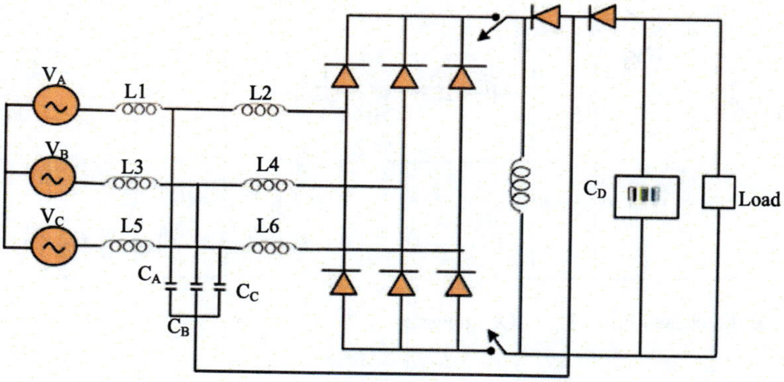

Fig. 13 AC-to-DC reduced switch topology

6 AC-to-DC Reduced Switch Topology

In this topology, the main advantages include its low cost, a fewer number of switches, and most important, buck–boost converter efficiency is high as compared to another topology. The voltage can be adjusted over a wide range of values. Though rectifiers have which used to convert AC–DC, most of them are converters used to increase the voltage level according to the input voltage that do not allow for voltage variation over a wide range [26, 27] (Fig. 13).

7 Boost Converters (Unidirectional)

Boost converters of unidirectional property are required when the output voltage needs to boost according to a set of conditions for loads that require higher voltage [28, 29]. When a unidirectional boost converter is used instead of a traditional diode

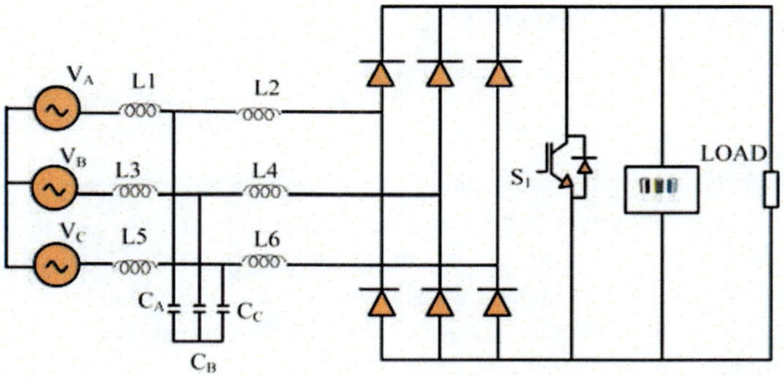

Fig. 14 Boost converter

bridge rectifier, power quality was improved and harmonics present in the system are eliminated at the input end, and a constant DC voltage is obtained at the output if undesirable AC instability occurred in the system [30, 31] (Fig. 14).

8 Comparative Study of Converter Topologies and Their Performance in DCFC

See Table 1.

The Vienna rectifier is the most optimal converter topology for DC fast charging stations among the observed topologies for the reasons mentioned above [30, 31].

Table 1 Comparison of converter topologies

Converter used	Operating mode	Current	Observation
Unidirectional boost converter	Boost	>5%	Design of this converter topology is easy.
Vienna rectifier	Boost	<5%	Power factor is high as compared to other converter topologies. The other advantages of Vienna rectifier include low THD, high-power density.
Reduced switch Buck–boost converter topology	Both buck and boost modes	Approx.15%	Efficiency is more as compared to other topologies

9 Energy Storage System Capital Cost Estimation at Time of Installation

See Table 2.

For flywheel energy storage system, we examined from the various literature studies that it may be cost efficient and can be used in medium DC fast charging station in the developing countries, data taken from various manufacturing company making the components [45]. So, we should install the fly wheel energy storage according to our requirement in the above for medium flywheel energy storage which we had used all the component cost fall in same range as in Tables 3 and 4.

For hydrogen energy storage, the specification is as given below so that it would fall under the range given in Table 2. In Table 4, we have considered the hydrogen energy storage for medium DC fast charging station [46].

For the battery used, the specifications are as given below [47] so that it would fall under the range mentioned in Tables 2 and 5.

For DC fast charging station, the DC charger specifications given below can charge the EV in rage of 50–150 kw (Table 6).

10 Storage Technologies Which are Suitable for DC Fast Charging Station

The initial cost of flywheel energy storage is high as conclude from various sources and from the literature study; therefore, it is not appropriate for small and medium DC fast charging stations. The advantage of FESS is its high-power capacity, and it can store large amount of electrical energy in less size. However, in the various circumstances considered here, more energy is needed, so FESS power is not completely utilized; as a result, flywheel energy storage can only be considered for DC fast charging stations that will provide stable grid that demand high-power capability and long-life cycles, as this would maximize the economic payback compared to the DC fast charging station. From one of the key parameters examined and written above in observation Table 2, battery energy storage has a high density but a low quality than hydrogen storage. So, the cost of hydrogen energy storage is only slightly higher than that of battery storage.

11 Conclusion

This paper includes a description of storage of energy with hydrogen, flywheel, battery, hybrid infrastructure of devices, and different power converters appropriate for each condition. The cost estimation for different energy storage gives information that the flywheels should be considered only if high-power grid services are also

Table 2 Cost estimation at time of installments

Component	Component details	Capital cost	Paper	Observations
EV charger	Isolated DC-to-DC converter	25,861.05/kW	[32]	There is a 40% reduction in front-end costs. Based on the total expense of the EV charger. Initially take an assumption. A small station has one outlet, a medium station has two, and a large station has five
Battery	Battery pack design	₹ 29,141.89/kWH	[33]	A total life of battery pack is 12 years
	Balance plant	₹ 45,907.91/kW	[34, 35]	For a balanced plant total life is 15 years. Calculation is based on energy efficiency. Every year cost is multiplied by 20
	Installation	₹11,249.91 /kWH	[36]	Installation includes different power converter which are suitable as requirement
Hydrogen	Tank	₹ 51,780.07/kW	[37]	The lifespan of hydrogen tank is ten years which includes the use of an AC-to-DC converter
	Electrolyzer	₹ 28032.19/kW	[38]	The total life of electrolyzer used in hydrogen energy storage is 18 years
	Compressor	₹ 1495.05kWH	[39]	Total life time of compressor used in hydrogen ESS is 20 years; the information is generally needed at the time of maintenance
	Fuel cell	₹ 19809.41/kW	[40, 41]	The total lifespan of fuel cell is ten years, and in this, DC-to-DC converter is used
Flywheel	Flywheel designed for power	₹ 84844.09/kw	[42]	The cost per kilowatt hour ranges from ₹48589.13 to 119604.00; the middle value is generally taken in this analysis

(continued)

Table 2 (continued)

Component	Component details	Capital cost	Paper	Observations
	Flywheel design for energy	₹336386.25/kwH	[43]	A lifetime of nearly 1 lack cycles and an 80% discharge depth centered on a 100 kw/125kwh flywheel
	Design cost	₹ 20930.70/kw	[44]	The range is 30% to 40% of the equipment cost in this case; we should assume the middle value

Table 3 Specification of flywheel energy storage in medium DC fast charging station

Specifications	Values
Power of flywheel	10 (kw)
Energy	33 (kwh)
Size of flywheel	1.75 (m^3)
Energy density	18.40 (kwh/m^3)
Power density	4.59 (kw/m^3)

Table 4 Hydrogen ESS specification for medium DC fast charging station

Grid power	EV charging rate	Quantity of Hydrogen requirement and equivalent energy
Connection to grid total power 50 kw	Electric vehicle charging rate 48 kwh	Hydrogen system requirement 0.217$m^3$and the energy equivalent in kwh is 98 kwh approximately
Connection to grid total power 50 kw	Electric vehicle charging rate 100 kwh	Hydrogen system requirement 0.170$m^3$and the energy equivalent in kwh is 150 kwh approximately
Connection to grid total power 50 kw	Electric vehicle charging rate 145 kwh	Hydrogen system requirement 0.135 $m^3$and the energy equivalent in kwh is 165 kwh approximately

Table 5 Battery energy storage system parameter for medium DC fast charging station

Battery parameters	Value
Battery capacity	225.6Ah
Battery charging resistance	1.2mohm
Battery nominal energy	85 kwh
Battery nominal voltage	370V
C rate	2.5

Table 6 DC fast charger specifications

Charger specifications	Values
Input voltage	200V
Output power	45,55kW
Maximum voltage	700V
Maximum current	750A

needed, and hydrogen storage should only be considered if fuel cell vehicle refueling is also required. Batteries, on the other hand, are simple to design and have the lowest number components, and their efficiency is more to that of other storage methods. Grid connectivity capability and EV charging efficiency have a huge effect on the optimum battery design, according to a study of battery designs in our country's industry. The Vienna rectifier can also be used to increase the performance of a DC fast charging station since it has the largest power factor of as compared to another converter topology, and nowadays, it is also used in wind energy applications as converter in front end.

References

1. Muratori M, Alexander M, Arent D, Bazilian M, Dede EM, Farrell J, Ward J (2021) The rise of electric vehicles—2020 status and future expectations. Progress Energy
2. Zhang S, Wang B, Wang S, Hu W, Wen X, Shao P, Fan J (2021) Influence of air pollution on human comfort in five typical. Chinese Cities Environ Res
3. Mowry A, Mallapragada D (2021) Grid impacts of highway electric vehicle charging and the role for mitigation via energy storage. MIT CEEPR Working Paper
4. Guzmán CP, Arias NB, Franco JF, Soares J, Vale Z, Romero R (2021) Boosting the usage of green energy for EV charging in smart buildings managed by an aggregator through a novel renewable usage index. IEEE Access
5. Soldo J, Škugor B, Deur J (2021) Analysis of optimal battery state-of-charge trajectory patterns for blended mode of a parallel plug-in hybrid electric vehicle and a wide range of driving conditions
6. Richard L, Petit M (2018) Fast charging station with battery storage system for EV optimal integration into the grid in 2018. IEEE Power Energy Soc General Meet (PESGM), pp 1,5
7. Knupfer S, Noffsinger J, Sahdev S (2019) How battery storage can help charge the electric-vehicle market. McKinsey & Company
8. Gallinaro S (2020) Energy storage systems boost electric vehicles' fast charger infrastructure. Analog Devices, pp 1–4
9. Baatar B, Heckmann K, Hoang T, Jarvis R, Sakhiya P (2019) Preparing rural America for the electric vehicle revolution
10. Baumgarte F, Kaiser M, Keller R (2021) Policy support measures for widespread expansion of fast charging infrastructure for electric vehicles. Energy Policy 156:112372
11. Nicholas M, Hall D (2018) Lessons learned on early fast electric vehicle charging systems
12. Francfort J, Salisbury S, Smart J, Garetson T, Karner D (2017) Considerations for corridor and community DC fast charging complex system design
13. Yilmaz M, Krein PT (2018) Review of battery charger topologies, charging power levels, and infrastructure for plug-in electric and hybrid vehicles IEEE Trans. Power Electron 28(5):2151–2169

14. Tu H, Feng H, Srdic S, Lukic S (2019) Extreme fast charging of electric vehicles: a technology overview. IEEE Trans Transp Electrif 5(4):861–878
15. Ronanki D, Kelkar A, Williamson SS (2019) Extreme fast charging technology:prospects to enhance sustainable electric transportation. Energies 12(19):3721
16. Ozbek E, Yalin G, Karaoglan MU, Ekici S, Colpan CO, Karakoc TH (2021) Architecture design and performance analysis of a hybrid hydrogen fuel cell system for unmanned aerial vehicle. Int J Hydrogen Energy
17. Buchroithner A, Wegleiter H, Schweighofer B (2018) Flywheel energy storage systems compared to competing technologies for grid load mitigation in EV fast-charging applications. In: IEEE 27th International Symposium on Industrial Electronics (ISIE), pp 508–514
18. Torres J, Moreno-Torres P, Navarro G, Blanco M, Lafoz M (2018) Fast energy storage systems comparison in terms of energy efficiency for a specific application. IEEE Access 6
19. He L, Xu X (2021) Soft-switching bidirectional switched-capacitor DC-DC converter with multiple phase-shift control methods. IEEE J Emerg Select Topics Power Electron
20. Faraji R, Ding L, Rahimi T, Kheshti M (2021) Application of soft-switching cell with inherent redundancy properties for enhancing the reliability of boost-based DC-DC converters. IEEE Trans Power Electron
21. Molnar C (2040) Norway needs $1.6-billion power grid upgrade to support EVs study Driving
22. Veldman E, Verzijlbergh RA (2015) Distribution grid impacts of smart electric vehicle charging from different perspectives. IEEE Trans Smart Grid 6(1):333–342
23. Khan AA, Cha H, Ahmed HF (2018) A new reliable three-phase buck-boost ac–ac converter. IEEE Trans Ind Electron 65(2)
24. Rąbkowski J, Zdanowski M, Wrona G (2021) Single two-and three-phase operation of the EV charger based on four-wire Vienna rectifier and two isolated DC-DC converters
25. Selva Kumar R, Gayathri Deivanayaki VP, Vignesh CJ, Naveena P (2019) Design and comparison of quadratic boost converter with boost converter. IJERT 5(01)
26. Dhananjaya M, Pattnaik S (2021) Review on multi-port DC–DC converters. IETE Tech Re 1–14
27. Di Capua G, Shirasavar SA, Hallworth MA, Femia N (2020) An enhanced model for small-signal analysis of the phase-shifted full-bridge converter. IEEE Trans Power Electron 30(3):1567–1576
28. Wang Y, Liu W, Ma H, Chen L (2015) Resonance analysis and soft-switching design of isolated boost converter with coupled inductors for vehicle inverter application. IEEE Trans Power Electron 30(3):1383–1392
29. Pramanik R, Pati BB (2021) Modelling and control of a non-isolated half-bridge bidirectional DC-DC converter with an energy management topology applicable with EV/HEV. J King Saud Univ Eng Sci
30. Bragard M, Soltau N, Thomas S, De Doncker RW (2019) the balance of renewable sources and user demands in grids: Power electronics for modular battery energy storage systems. IEEE Trans Power Electron 25(12):3049–3056
31. Chowdhury SM, Haque ME, Elrayyah A, Sozer Y, De Abreu-Garcia J (2018) An integrated control strategy for state of charge balancing with output voltage control of a series connected battery management system. In: 2018 IEEE energy conversion congress and exposition (ECCE)
32. Durgam P (2021) Traction battery pack design for BEV: a comprehensive approach no. 2021-26-0176 SAE Technical Paper
33. Chen Y, Ping Dai W, Zhou J, Hu E (2014) Study and design of a novel three-phase bridgeless boost power factor correction. IET Power Electron 7
34. Aanstoos T, Kajs JP, Brinkman W, Liu HP, Ouroua A, Hayes RJ (2015) High voltage stator for a flywheel energy storage system IEEE Trans. Magazine 37(1):242–247
35. Vajed I, Kohari Z, Benko L, Meerovich V, Gawalek W (2013) Investigation of joint operation of a superconducting kinetic energy storage (Flywheel) and solar cells. IEEE Trans Appl Supercond 13(2)
36. Olabi AG, Wilberforce T, Abdelkareem MA, Ramadan M (2021) Critical review of flywheel energy storage system. Energies

37. Thormann, B., Puchbauer, P., & Kienberger, T: Analyzing the suitability of flywheel energy storage systems for supplying high-power charging e-mobility use cases. Journal of Energy Storage, (2021).
38. Chen H, Cong TN, Yang W, Tan C, Li Y, Ding Y (2020) Progress in electrical energy storage system: a critical review. Prog Nat Sci
39. Ibrahima H, Ilincaa A, Perron J (2019) Energy storage systems characteristics and comparisons. Renew Sustain Energy Rev 12:1221–1250
40. Electric energy storage technology options: a white paper primer on applications, costs and benefits. Palo Alto, CA, Electric Power Research Institute (EPRI) (2010), p 1020676
41. Yue M, Lambert H, Pahon E, Roche R, Jemei S, Hissel D (2021) Hydrogen energy systems: a critical review of technologies, applications, trends and challenges. Renew Sustain Energy Rev 146:111180
42. Li X, Palazzolo A (2021) A review of flywheel energy storage systems: state of the art and opportunities
43. Rahman MM, Gemechu E, Oni AO, Kumar A (2021) The development of a techno-economic model for the assessment of the cost of flywheel energy storage systems for utility-scale stationary applications. Sustain Energy Technol Assessments 47:101382
44. Dache V, Sgarciu V (2021) Performance analysis of a low-cost small-scale flywheel energy storage system. In: 2021 23rd international conference on control systems and computer science IEEE
45. Hedlund M, Lundin J, De Santiago J, Abrahamsson J, Bernhoff H (2015) Flywheel energy storage for automotive applications. Energies 8(10):10636–10663
46. Ahsan N, Al Rashid A, Zaidi AA, Imran R, Qadir S (2021) A performance analysis of hydrogen fuel cell with two-stage turbo compressor for automotive applications. Energy Rep
47. Ghoulam Y, Mesbahi T, Durand S, Lallement C (2021) Electro-thermal battery model for automotive applications. in: pcim europe digital days 2021; international exhibition and conference for power electronics, intelligent motion, renewable energy and energy management

Modelling and Implementation of MPPT Controller for Off-Grid SPV System

Supriya Kumari, Prabhat Ranjan Sarkar, Akhilesh Kumar Pandey, Ahmad Faiz Minai, and Sandeep Kumar Singh

Abstract In the present scenario, green energy resources are widely available. Solar photovoltaic (SPV) energy is one of them, especially in northern regions of India. SPV system has nonlinear P–V and I-V characteristics so it becomes difficult to calculate and achieve maximum power at load line characteristics. Nowadays, the major concern of researchers to enhance the overall efficiency of SPV system by tracking maximum possible power. To abstract maximum power from SPV array, various maximum power point tracking (MPPT) methods are presented in literature. In the present paper, for achieving maximum power point, Mamdani's-based constant voltage (CV) MPPT method is used to automatically adjust the reference voltage at various environmental conditions. To calculate the open circuit voltage (Voc), fuzzy logic controller toolbox is simulated in MATLAB/Simulink environment.

Keywords Photo voltaic (PV) · Buck-boost converter · MPPT controller · Fuzzy logic tool box · PMDC motor

1 Introduction

In the current competitive world, energy security is an important issue. There is a need to generate and supply energy at a reasonable price. Sun energy is the most widely used non-conventional energy source. With the increasing demand for energy, solar PV is one of the fastest-growing technologies. It will play a vital role in global

S. Kumari · P. R. Sarkar (✉) · A. F. Minai · S. K. Singh
Integral University, Lucknow, India
e-mail: sarkarprabhat@gmail.com

S. Kumari
e-mail: supriyakm@iul.ac.in

A. K. Pandey
Lucknow University, Lucknow, India

K. Namrata et al. (eds.), *Smart Energy and Advancement in Power Technologies*, Lecture Notes in Electrical Engineering 926,
https://doi.org/10.1007/978-981-19-4971-5_61

electricity production. It reduces the effect of greenhouse gases and makes the environment pollution-free. There is two way of generating solar energy, one is solar thermal, and another is solar photovoltaic [1].

The SPV panel directly transforms sun energy into electricity through photovoltaic principle. During on-field operations, the performance of PV panels get affected by many climatic parameters such as temperature, humidity, wind velocity, soiling and cloud conditions. The results of these different parameters may have nonlinear characteristics [4]. It is essential to work SPV panel at its MPP condition to achieve the maximum yield. The efficiency of SPV panel is about 20–25%. To enhance the efficiency of SPV panel, maximum power point tracker (MPPT) is used.

In literature, various maximum power point tracking (MPPT) techniques/algorithms are available. Some of them are like constant voltage (CV) method, short circuit current (SCC) pulse method, fuzzy logic control (FLC), artificial neural network (ANN), particle swarm optimization (PSO), perturb and observe (P&O), etc., have been discovered to extract the maximum possible power from SPV system. These all methods have various applications which are based on a variety of features including the types of sensors required, range of effectiveness, convergence speed, implementation hardware requirements, cost, popularity.

In this, CV algorithm is used which is simplest MPPT techniques. Vmpp is achieved by calculating open circuit voltage and constant K by Vmpp = K.Voc $K.Voc$. The value of K (0.72–0.82) depends on the material used and solar radiation, it varies from experimental analysis [2, 3]. The value of Voc depends upon the ambient condition. The panel voltage is compared with the reference voltage Vref and Vmpp is achieved corresponding to the reference signal. In CV techniques [5], PI controller is used to adjust the duty cycle of MPPT converter. It keeps the SPV voltage close to the maximum possible power, and no extra input signal is needed for achieving MPP [6, 7].

Mamdani's-based CV method has been developed for achieving the MPP for SPV system. The Voc is calculated by the fuzzy logic toolbox and by this Voc, Vmpp is achieved as we discussed the above formula Vmp = K. Voc. It is easy to implement, less costly, and more flexible [8]. A comparative analysis is done between the panel is directly connected with load and panel with MPPT connect to the load.

2 Solar Cell Model

The output of the solar panel depends upon factors such as solar isolation and temperature [13]. Maximum current(Impp) and maximum voltage (Vmpp) are two points of the P–V and I-V curve on which we achieve maximum power, and these two are taken as references which can be used to track MPP shown in Fig. 1.

In the present paper, single diode model is used to design the SPV cell. The diode current (I_D), series resistance (R_s), and current source (I_{ph}) depend on the solar radiation which is falling on the SPV panel. By using Kirchhoff's current law, developed current of SPV cell can be written as [7, 9]:

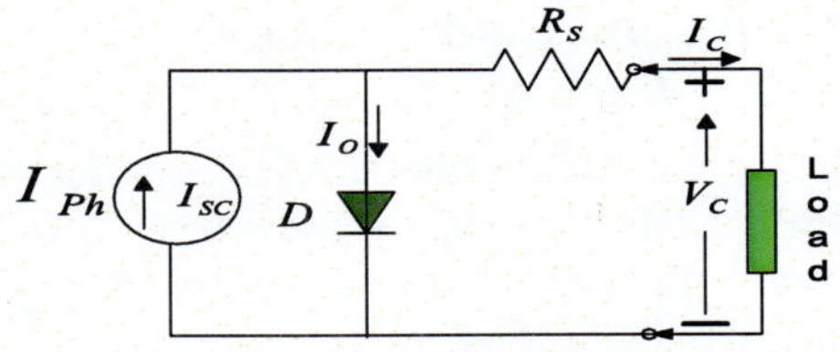

(a) Single diode equivalent model of SPV cell.

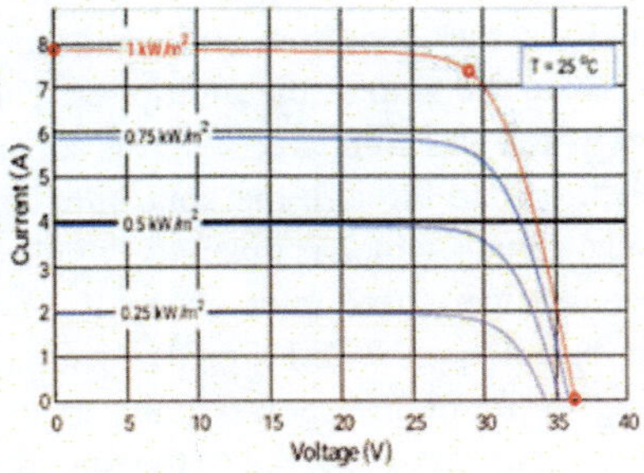

(b) I-V characteristics at 25°C & variable solar radiation

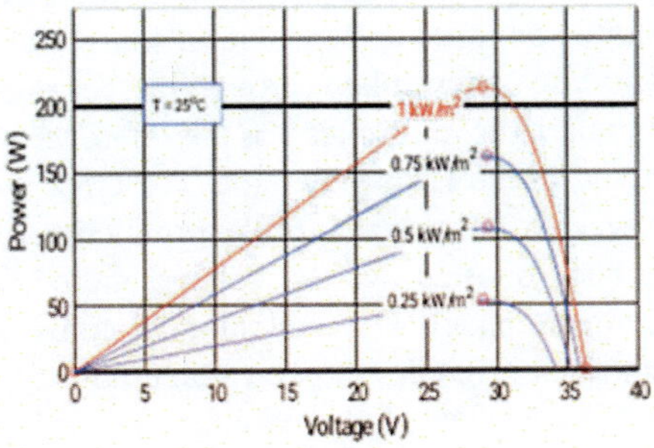

(c) P-V characteristics at 25°C & variable solar radiation.

Fig. 1 I-V and P–V characteristics at different conditions

$$I_C = I_{\text{Ph}} - I_D \tag{1}$$

In Eq. (1), diode current (I_D) is given by Shockley equation [11].

$$I_C = I_{\text{Ph}} - I_o \exp\left[\frac{(V_C + R_S T_C)}{\text{Ak}T_C} - 1\right] \tag{2}$$

Output voltage V_C of the SPV module that can be written as

$$V_C = \frac{\text{Ak}T_C}{q} \ln\left(\frac{I_{\text{Ph}} + I_O - I_C}{I_O}\right) - I_C R_S \tag{3}$$

where series resistance is denoted by $R_S = 0.001\ \Omega$, reverse saturation current of the diode is denoted by $I_o = 0.02$ mA. A is an identity factor of the diode, and charge of the electron $q = 1.6 \times 10^{-19}$ C, Boltzman constant $k = 1.38 \times 10^{-23}$ J/K, and cell temperature is T_C. The unit of T_C and k should be same, either in Celsius or Kelvin [10]. Current source value I_{Ph} was taken as 5A [11].

3 System Structure and Configuration

Figure 2 depicts the various components that are used for realizing the improved PV system with MPPT. Solar PV array supplies power to DC-DC converter which in turn supplies power to the load [13, 14]. The output voltage is measured through voltmeter. The measured output voltage of SPV, which is the actual voltage in actual weather conditions, is given as the feedback in the error generator [18, 19]. For generating reference value, fuzzy logic controller (FLC) first reads the value from the temperature and insolation sensor for actual weather condition and gives Voc. The Voc is multiplied through the constant "k" with the help of the multiplier. The output of the multiplier was the reference signal which was compared with the actual signal in the error generator. The error was then passed through the controller for corrective actions and for the production of PWM signals which is given to switches of buck-boost converter. Voltage and current sensor were used to calculate V_c and I_c of the panel. The switching of converter is based upon PWM signals received and thus gives an output voltage corresponding to the highest power point of the SPV panel. The generated output of the buck-boost converter is given to the load for utilization of electrical energy [8, 12].

In MATLAB/Simulink, FLC toolbox is programmed and simulated with SPV array, buck-boost converter, and PMDC motor [15, 16]. In this paper, SPV array consists of total 36 cells in each module in which series module = 4 and parallel = 5. At different solar isolation and temperature, open circuit voltage is calculated by FLC, and this is multiplied by 0.8 to achieved reference voltage for buck-boost converter [17]. In the present work, at constant solar isolation and different solar

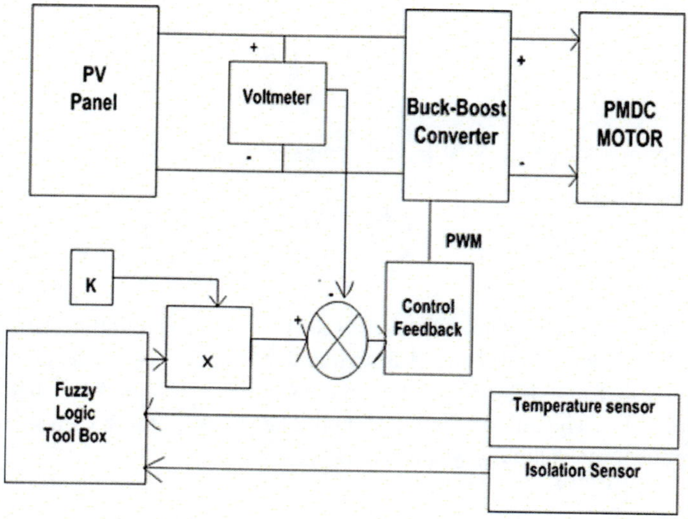

Fig. 2 System architecture with panels, FLC, DC-DC converter, and PMDC load

temperature or vice versa, the value of maximum power (P_{max}) is designed. The outcome of this can be easily seen in the output.

4 Mamdani's Techniques-Based CV MPPT Controller

Fuzzy logic (FL)-based tracking is a smarter way to achieve maximum power point (MPP) over the conventional tracking techniques. Without using accurate mathematical models, it tracks MPP by using fuzzy inputs as shown in Fig. 3. Fuzzy logic provides faster result as compared to other algorithm. In FL, knowledge of accurate model is not needed. It works on three different steps which are discussed as follows:

4.1 Fuzzification

This is the first step of this function. In this membership, functions are designed in the form of set value. Value is decided by the human.

IF–THEN method is used to calculate the value of input and output variable from the triangular type membership function.

Two inputs variables are used to in this paper, i.e. temperature and irradiance. The value of temperature is calculated at different conditions like very low, low, medium, high, and very high. Like the temperature value of irradiance also designed different conditions like very light, light, medium, bright, and very bright.

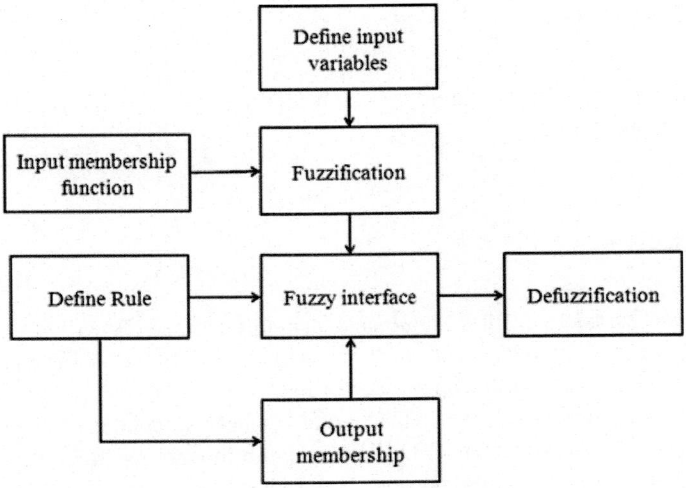

Fig. 3 Block diagram used to develop the algorithm for MPPT

Table 1 Rule base of Mamdani's algorithm

Sx / Tx	V light	Light	Medium	Bright	V bright
V low	VL	L	M	H	VH
Low	VL	L	L	H	VH
Medium	VL	L	L	H	H
High	VL	L	L	M	H
V high	VL	L	L	M	H

4.2 Rule Base Theory

This is the intermediate process of fuzzy logic controller. It related the behaviour of input variables with output variables by the set rule theory. All the set rules are shown in Table 1.

The first input is temperature which has five fuzzy levels as very low(VL), low(L), medium(M), high(H), very high(VH), and a another input is irradiance (S_x) which has five level VL (very) light, light, medium, bright, V (very) bright, and the outputs are very low(VL), low(L), medium(M), high(H), very high(VH). It can be seen easily that by using MPPT, it increase the output and efficiency of the overall system approx 30–35%. The simulation results are also shows the characteristics between maximum power versus time, motor speed versus time, and motor power versus time with MPPT and without MPPT as shown in Figs. 4 and 5.

4.3 Defuzzification

This is the last step of fuzzy logic in which the linguistic variable converted into human language.

5 Results

The study was carried out using Mamdani's techniques-based CV MPPT controller. A comparative study between with (W) and without (WO) MPPT has been done. The effect of simulation result on the performance of PMDC motor can be easily seen. Maximum power deliver to PMDC motor has been calculated at varying solar insolation from 100 mW/cm^2 to 40 mW/cm^2 and temperature 20–40 °C. Table 2 shows the variation of maximum power at different solar temperature and insolation by using with and without MPPT, respectively. It can be seen from Table 2 at a low-solar insolation, power decreases significantly, and the same effect is vice versa with

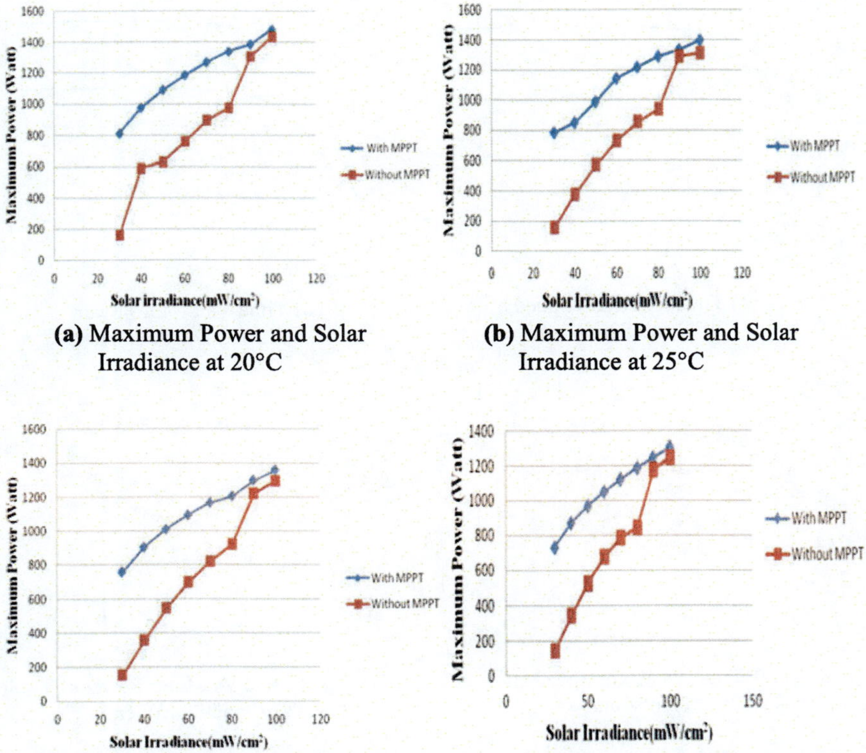

(a) Maximum Power and Solar Irradiance at 20°C

(b) Maximum Power and Solar Irradiance at 25°C

(c) Maximum Power and Solar Irradiance at 30°C

(d) Maximum Power and Solar Irradiance at 35°C

Fig. 4 Maximum power at different conditions of irradiance and temperature

the temperature. The variation of maximum power with directly couple and using buck-boost converter can be easily seen in Fig. 4a–e.

The effect on motor power and motor speed at different temperature and solar irradiance is shown in Fig. 5a–j.

6 Conclusion

Renewable energy is the possible solution amongst which solar leads in India. PV finds many applications in the standalone and tie-line connected system. The efficiency of the solar photovoltaic system is generally poor. To ameliorate the efficiency of SPV system, numerous MPPT algorithm has been developed and implemented. CV voltage MPPT method is the most economical and easy to implement. In this paper, Voc is calculated by FLC. And then find the maximum power using formula

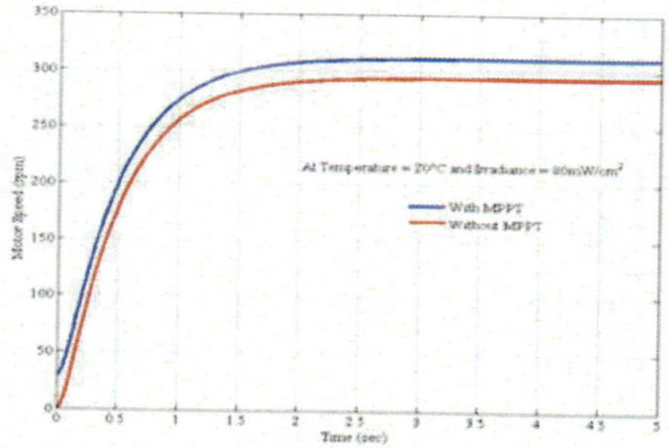

(a) Time Vs Motor Speed at 20°C and 80mW/cm^2

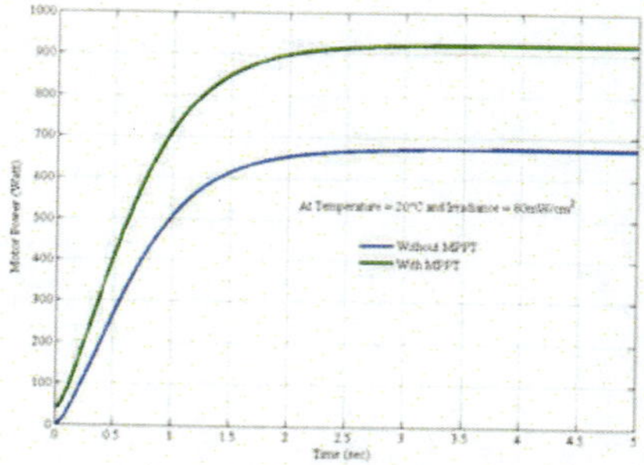

(b) Time Vs Motor Power at 20°C and 80mW/cm^2

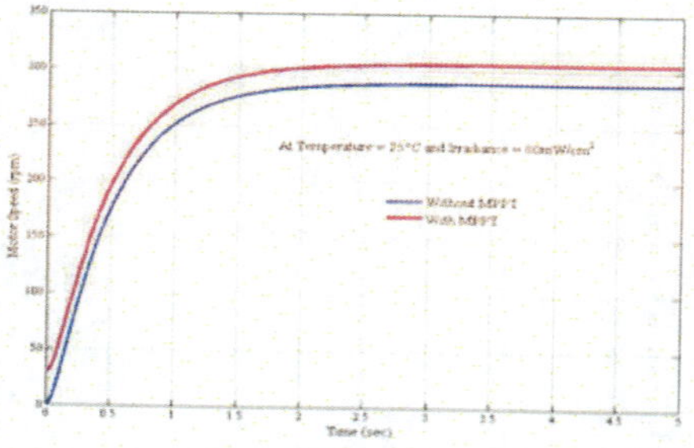

(c) Time Vs Motor Speed at 25°C and 80mW/cm^2

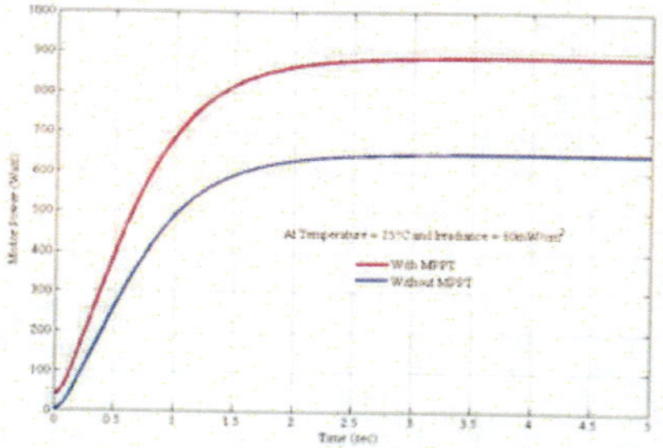

(d) Time Vs Motor Power at 25°C and 80mW/cm^2

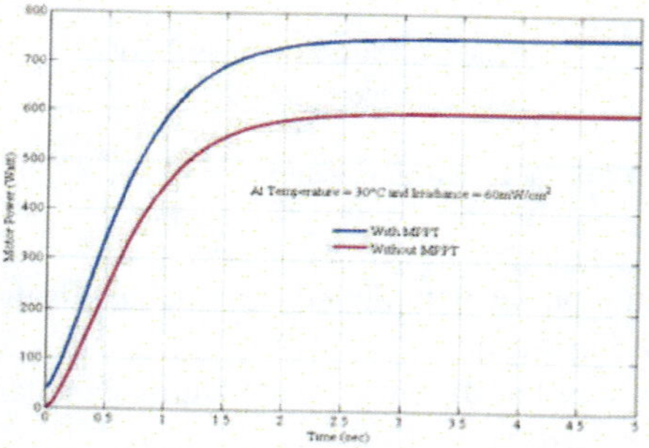

(e) Time Vs Motor Power at 30°C and 60mW/cm^2

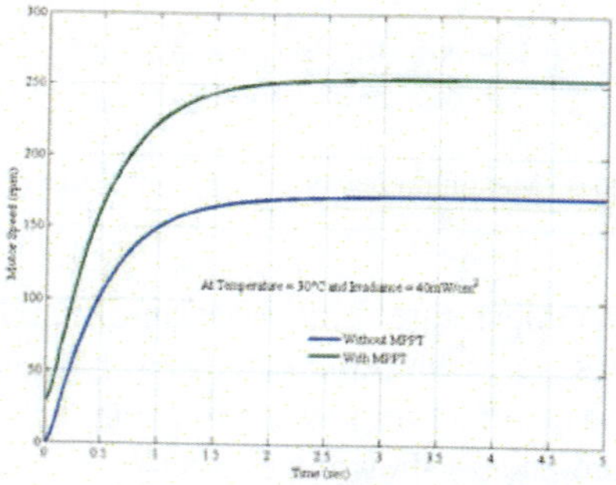

(f) Time Vs Motor Speed at 30°C and 40mW/cm^2

Fig.5 Characteristic of time versus motor speed at different conditions

Vmpp = *K*. Voc. Fuzzy logic-based tracking is a smarter way to calculate maximum possible power point than conventional tracking methods. Without using accurate mathematical models, it tracks MPP by using fuzzy inputs. A comparative analysis has been done by using with and without MPPT. Mamdani's techniques-based CV MPPT was developed which was made to run with PMDC motor as load. By using fuzzy logic, the performance of the system improved as it can be seen in above

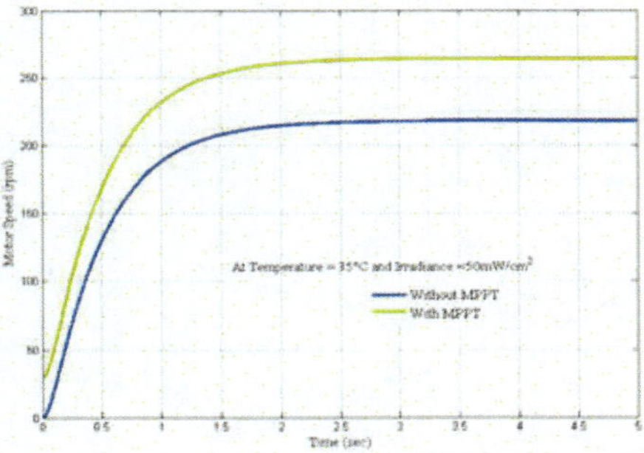

(g) Time Vs Motor Speed at 35°C and 50mW/cm^2

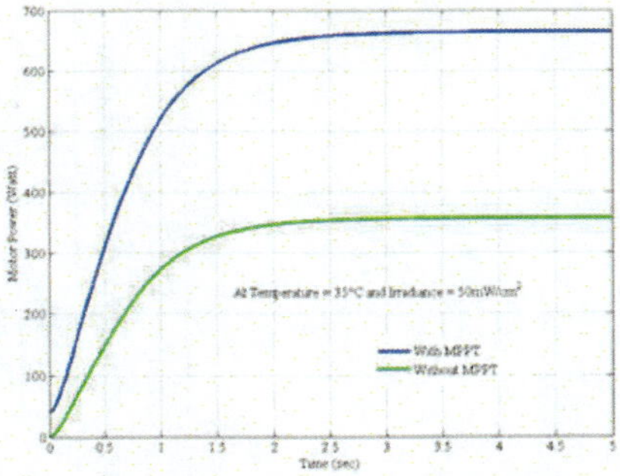

(h) Time Vs Motor Power at 35°C and 50mW/cm^2

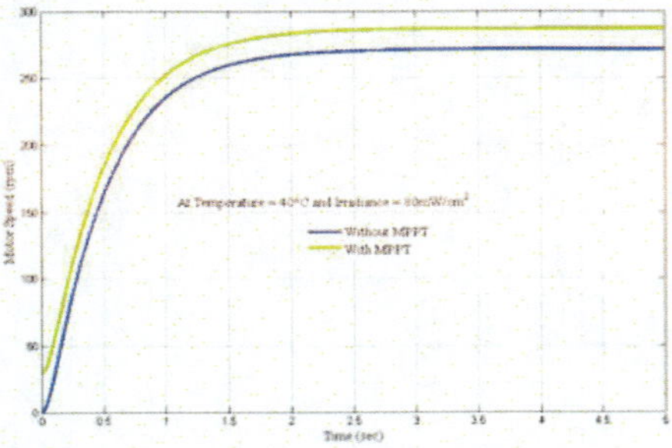

(i) Time Vs Motor Speed at 40°C and 80mW/cm^2

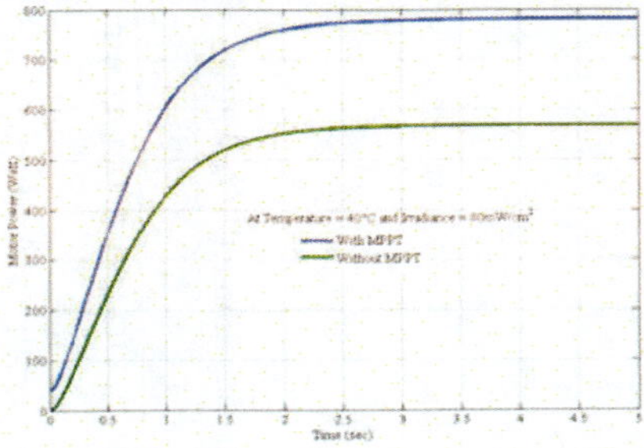

(j) Time Vs Motor Power at 40°C and 80mW/cm^2

Fig.5 (continued)

graph Fig. 5. This can be used in agrivoltaics system. The overall performance of the system has been simulated in MATLAB/Simulink. The advantage of fuzzy logic-based MPPT being fewer data required has less memory storage and provides faster result and less time required.

Table 2 Maximum power with different solar insolation and temperature with (W) and without (WO) MPPT

Sx (mWc/m^2)	T = 20 °C		T = 25 °C		T = 30 °C		T = 35 °C		T = 40 °C	
	W/MPPT	WO/MPPT	W/MPPT	WO/MPPT	W/MPPT	WO/MPPT	W/MPPT	WO/MPPT	W/MPPT	WO/MPPT
100	1485	1435	1395	1310	1358	1295	1301	1245	1247	1175
90	1385	1305	1330	1285	1298	1223	1244	1174	1192	1125
80	1339	975.5	1286	937.3	1205	920.5	1184	845.6	1135	828.4
70	1265	895.7	1215	851.7	1167	818.2	1120	785.5	1047	753.7
60	1183	760.3	1138	731.2	1093	695.8	1049	675.1	1007	648.1
50	1090	630.4	985.2	568.5	1008	546.8	969	525.6	930.6	485
40	976	590.8	850.3	373.6	904.7	359.7	870.3	346.2	830	333

References

1. Ossai CI (2017) Optimal renewable energy generation–approaches for managing ageing assets mechanisms. Renew Sustain Energy Rev 72:269–280
2. Ahmad J (2012) A fractional open circuit voltage based maximum power tracker for photovoltaic arrays. Int Conf Softw Technol Eng 247–250
3. Minai AF, Usmani T, Mallick MA (2015) Optimum sizing and estimation of a 30kWp hybrid solar photovoltaic system with multilevel inverter. Int J Res Sci Innov (IJRSI) 31–36. ISSN 912321–2705
4. Femia N, Granozio D, Petrone G, Spagnuolo G, Vitelli M (2007) Predictive and adaptive MPPT perturb and observe method. IEEE Trans Aerosp Electron Syst 43(3):934–950
5. Kumari S, Sarkar PR (2017) Maximum power point tracking method based on constant voltage for solar PV system. J Electric Power Syst Eng 3(2)
6. Esram T, Chapman PL (2007) Comparison of photovoltaic array maximum power point tracking techniques. IEEE Trans Energy Convers 22(2):439–449
7. Subudhi B, Pradhan R (2011) Characteristics evaluation and parameter extraction of a solar array based on experimental analysis. In: 2011 IEEE ninth international conference on power electronics and drive systems. IEEE
8. Elgendy MA, Zahawi B, Atkinson DJ (2010) Comparison of directly connected and constant voltage controlled photovoltaic pumping systems. IEEE Trans Sustain Energy 1(3):184–192
9. Minai AF, Usmani T, Iqbal A (2021) Performance evaluation of a 500 kWp rooftop grid-interactive SPV system at integral university, Lucknow: a feasible study under adverse weather condition. Stud Big Data 86. Springer, Singapore. https://doi.org/10.1007/978-981-33-4412-9_24
10. Ramli MAM, Twaha S, Ishaque K, Al-Turki YA (2017) A review on maximum power point tracking for photovoltaic systems with and without shading conditions. Renew Sustain Energy Rev 67(January):144–159
11. Asif M, Muneer T (2007) Energy supply, its demand and security issues for developed and emerging economies. Renew Sustain Energy Rev 11(7):1388–1413
12. Minai AF, Tariq A, Alam Q (2011) Theoretical and experimental analysis of photovoltaic water pumping system. India international conference on power electronics 2010 (IICPE2010), pp. 1–8. https://doi.org/10.1109/IICPE.2011.5728130
13. Yongxiu H, Xu Y, Pang Y, Tian H, Wu R (2016) A regulatory policy to promote renewable energy consumption in China: review and future evolutionary path. Renew Energy 89:695–705
14. Tariq, Asghar MSJ (2005) Development of an analog maximum power point tracker for photovoltaic panel. In: Proceedings of "the sixth international conference on power electronics and drive systems-PEDS 2005, Kuala Lumpur, Malaysia
15. Dubey S, Sarvaiya JN, Seshadri B (2013) Temperature dependent photovoltaic (PV) efficiency and its effect on PV production in the world–a review. Energy Procedia 33:311–321
16. Khare V, Nema S, Baredar P (2016) Solar–wind hybrid renewable energy ssstem: a review. Renew Sustain Energy Rev 58:23–33
17. Chenni R, Makhlouf M, Kerbache T, Bouzid A (2007) A detailed modeling method for photovoltaic cells. Energy 32(9):1724–1730
18. Naseem M, Husain MA, Minai AF et al (2021) Assessment of meta-heuristic and classical methods for GMPPT of PV system. Trans Electr Electron Mater. https://doi.org/10.1007/s42341-021-00306-3
19. Minai AF, Husain MA, Naseem M, Khan AA (2021) Electricity demand modelling techniques for hybrid solar PV system. Int J Emerg Electric Power Syst. https://doi.org/10.1515/ijeeps-2021-0085

Optimization-Based Speed Control Strategies for Induction Motor Drives in Plug-In Hybrid Electric Vehicle Using Quasi-Opposition Harmony Search Algorithm

Anish Kumar, Niranjan Kumar, and Amitesh Prakash

Abstract The optimization-based speed control schemes are extraordinarily utilized and applied in induction machine drives rather than vector and scalar control for their dependability and long-lasting capability, and low support cost. Considering reaction time and productivity, scalar control plot is second rate compared to present-day vector control plans, however, it requires lesser equipment assets and in this manner diminishing the expense. High performance is achieved in the drive system by decoupling of torque and flux-producing components. In this paper, the induction motor drive has been analyzed by scalar method and vector control methods and compared with a novel optimization technique, i.e., Quasi-Opposition Harmony Search Algorithm. The results obtained for the speed control by vector control technique can be further enhanced by using optimization techniques. It is expected that the results obtained by e Quasi –Opposition Harmony Search Algorithm will be better as compared to scalar and vector control techniques. To reproduce and approve the outcomes MATLAB/Simulink was used. Similar assessment of the Quasi –Opposition Harmony Search Algorithm and its comparison with vector as well as scalar control plans were plotted.

Keywords Plug-in hybrid electric vehicle · Quasi—opposition harmony search algorithm · Speed control

1 Introduction

Conventional vehicles that are being used now cause sound contamination, air contamination, a dangerous atmospheric deviation, and furthermore it causes a quick

---

A. Kumar (✉) · A. Prakash
Department of Electrical Engineering, BCE, Bhagalpur, Bihar, India
e-mail: nkumar.ee@nitjsr.ac.in; anishjsr.ak@gmail.com; anishjsr.ak@gmail.com

N. Kumar
Department of Electrical Engineering, NIT Jamshedpur, Jamshedpur, India

K. Namrata et al. (eds.), *Smart Energy and Advancement in Power Technologies*,
Lecture Notes in Electrical Engineering 926,
https://doi.org/10.1007/978-981-19-4971-5_62

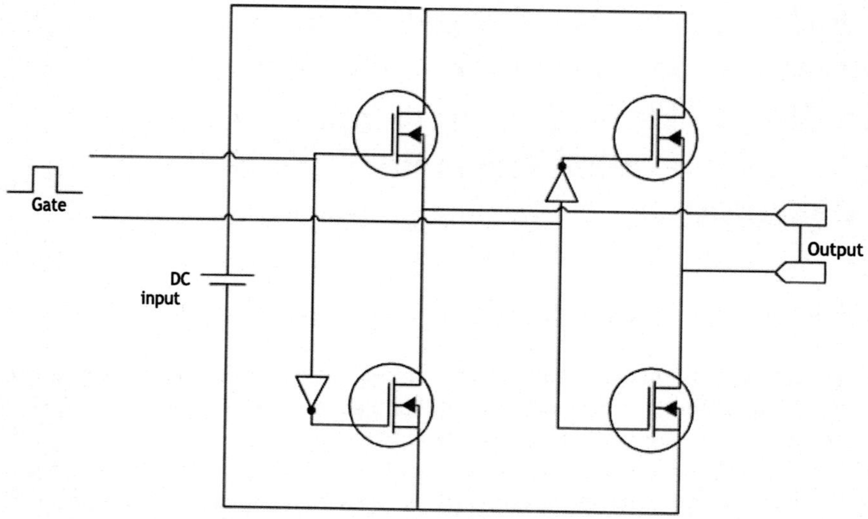

Fig. 1 Basic diagram of voltage source inverter

decrement of earth's normal assets. All these shortcomings of the conventional vehicles can be overcome by the use of electric vehicles. Recent research has also shown that electric vehicles can eliminate the shortcomings of the normally used vehicles.

The induction motor is readily available that's why it finds its application in numerous places. Due to availability, simplicity and easy control induction motor has become the best choice for electric vehicles. Many speed control techniques have been applied on induction motors and the results have shown that speed control strategies work well on the inductor motors. So, in electric vehicles also the induction motors are preferred over other motors.

DC input received from the batteries is changed to AC output of required frequency by the help of an inverter. The yield got from the inverter simply relies upon the exchanging activity performing by the circuit. The inverter it is constantly encouraged to lessen the exchanging or sounds by the switches to achieve better outcomes. Pure sinusoidal waveform is expected to be the output of the inverter.

There are basically two sorts of inverters:

"(1) voltage source inverter and (2) current source inverter" [4] (Fig. 1).

2 Strategies for the control of Induction Machine

2.1 Scalar Control

Only magnitude can be controlled by the scalar control method [1]. In order to obtain a steady force process, the functioning reach v/f proportion ought to be kept up with

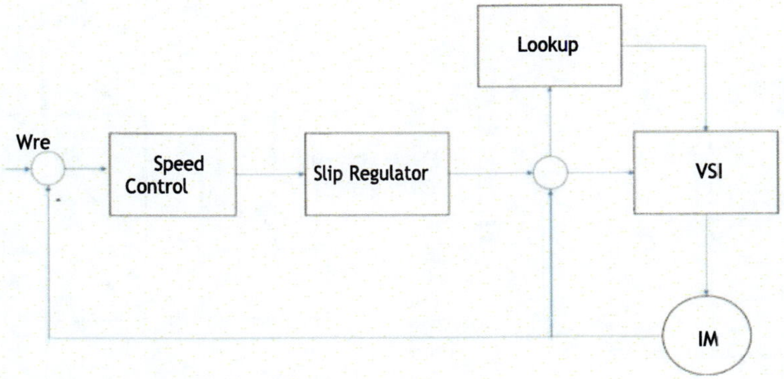

Fig. 2 Block diagram of Inverter fed IM with scalar control

consistent. An open circle control technique is followed in scalar control [18] so the expenditure is less when comparison is done with some other closed-loop framework technique in light of its improved on construction and plan. No criticism way makes the useful execution is simple compared to closed-loop framework. Figure 2 shows scalar control relies just upon extent of control values and doesn't rely upon the coupling impact of machine.

2.2 *Vector Control*

Most popular method for the speed control is the vector control technique. It is also known as called flux-oriented control [6]. Decoupling of motion and force is the main objective of vector control strategy. The vector control technique is used to obtain high performance of induction engine drive [18] (Fig. 3)

3 Need for Optimization

Optimization has become a very integral part of the research work. Nowadays it is being utilized in almost all the field of engineering. The term ideal might be perceived as most extreme value or least value. This least value could be set as per the need of the program of chosen by nature. A numerical stance of the issue should have been planned, and it's anything but a fundamental piece of the enhancement assignments to tackle the given issue. Nonetheless, in the majority of the cases, the actual acknowledgment of the model is too hard to even consider displaying because of different vulnerability, framework limitations, and natural conditions.

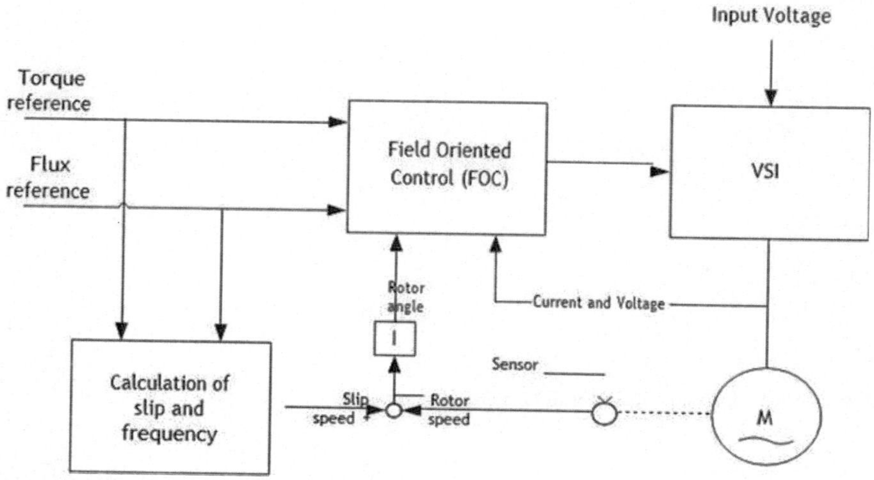

Fig. 3 Vector control scheme

3.1 *Quasi-Opposition-Based Learning: A Concept*

For optimization calculation, initiation of population is a very vital part. A decent optimization strategy always helps a great deal in discovering more good arrangements and further developing convergence rate. Rather than opposite numbers, semi-inverse focuses are utilized in the populace instatement of the fundamental HSA. The idea of semi-oppositional age hopping is used in the HSA. To produce semi-inverse populace from the current arrangement, semi-inverse hopping rate is used.

Quasi-opposite Number is shown by the equation below.

$$\mathrm{QOX}_0 = \left(\frac{\mathrm{para}^{\min} + \mathrm{para}^{\max}}{2} + \mathrm{para}^{\min} + \mathrm{para}^{\max} - X_0\right) \quad (1)$$

Quasi—Opposite Point is shown by the equation below.

$$\mathrm{QOP}_{0_i} = \mathrm{rand}\left(\frac{\mathrm{para}_i^{\min} + \mathrm{para}_i^{\max}}{2} + \mathrm{para}_i^{\min} + \mathrm{para}_i^{\max} - X_i\right), i = 1, 2, \ldots\ldots d \quad (2)$$

where d is the dimension of the problem.

3.2 *Quasi-Oppositional Population Initialization*

A haphazardly produced populace vector might be made do by using the semi-oppositional idea in the HSA. At first, a bunch of starting populace is produced within the arrangement space. Compared to this, wellness work esteem is assessed for the individual arrangement.

3.3 *Quasi-Oppositional Generation Jumping*

The interaction of streamlining acquired its viability to another applicant arrangement with some systems. In the QOHS calculation, in light of the idea of bouncing rate (Jr), another populace vector is created [19]. The pseudocode for QOHS calculation might be displayed in Algorithm 1.

4 Simulation and Results

MATLAB/Simulink MATLAB/Simulink was used to simulate the change of speed. The novel quasi-opposition harmony search algorithm was compared with scalar and vector control responses. Responses obtained from vector control method were very quick than the scalar-controlled method but the results obtained by the novel quasi-opposition harmony search algorithm were better than both the methods. Figures 6–8 show the speed responses of scalar, vector-controlled drives [11, 12] and with novel quasi-opposition harmony search algorithm (Figs. 4 and 5).

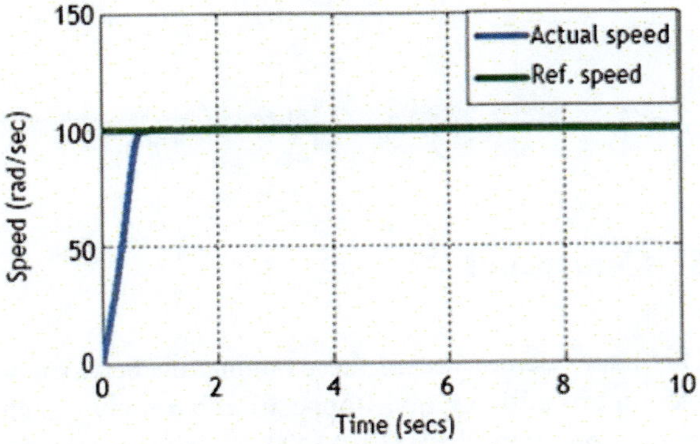

Fig. 4 Speed Vs Time Plot of a scalar-controlled drive

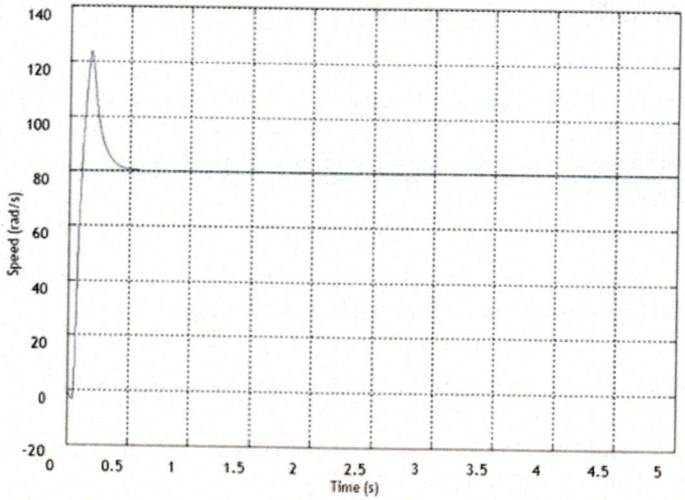

Fig. 5 Speed Vs Time Plot of vector-controlled drive

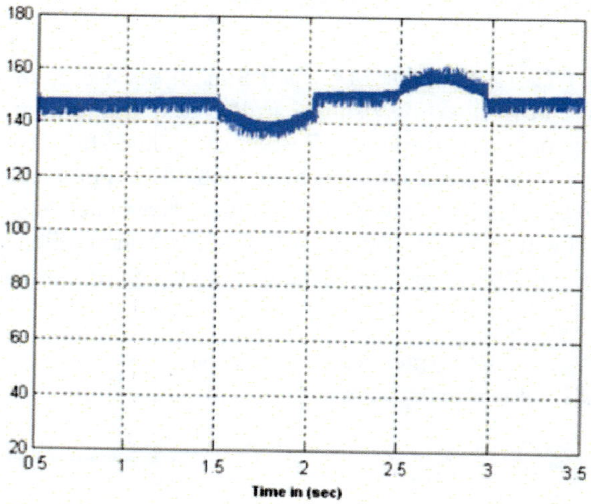

Fig. 6 Speed Vs Time Plot of QOHS Algorithm

5 Conclusion

This paper demonstrated the examination between scalar control, vector control, and control utilizing quasi-opposition harmony search algorithm techniques for an in- duction motor drive [6]. MATLAB/Simulink was used to simulate and examine the scalar, vector control, and quasi-opposition harmony search algorithm and to generate the speed waveforms of induction motor. Choice of controlling plans of drive depends on the exhibition of the speed and force attributes. The results of the simulation showed that vector control scheme has an edge over the scalar control scheme for the induction motor drive [14, 16] but better results were obtained by quasi-opposition harmony search algorithm. Even though vector control and quasi-opposition harmony search algorithm include complex calculations, they provide a better control over the electric vehicle.

References

1. De Doncker R, Pulle DWJ, Veltman A (2011) Control of induction machine drives. In: Advanced electrical drives. Power systems, Springer, Dordrecht
2. Bose, B.K., Chapter 8—Control and Estimation of Induction Motor Drives, in: Modern Power Electronics and AC Drives, Pearson Education, Inc, USA, 2002.
3. Nandi S, Ahmed S, Toliyat HA, Bharadwaj RM (2003) Selection criteria of induction machines for speed-sensorless drive applications. IEEE Trans Ind Appl 39(3):704–712
4. Richu SC, Rajeevan PP (2020) A Load Commutated multilevel current source inverter fed open-end winding induction motor drive with regeneration capability. IEEE Trans Power Electron 35(1):816–825
5. Bimbhra, P.S., Chapter 12–Electric Drives, in: *Power Electronics*, Khanna Publishers, India, 2008.
6. Zerdali E, Barut M (2017) The comparisons of optimized extended kalman filters for speed-sensorless control of induction motors. IEEE Trans Ind Electron 64:4340–4351
7. Smith A, Gadoue S, Armstrong M, Finch J (2013) Improved method for the scalar control of induction motor drives. IET Electr Power Appl 7(6):487–498
8. Fan L, Zhang L (2011) An improved vector control of an induction motor based on flatness. Procedia Eng 15:624–628
9. Jnayah, S. and Khedher, A., DTC of Induction Motor Drives Fed By Two and Three-Level Inverter: Modeling and Simulation, in: *2019 19th International Conference on Sciences and Techniques of Automatic Control and Computer Engineering (STA)*, pp. 376–381, IEEE, Sousse, Tunisia, 2019.
10. Sayed-Ahmed A, Demerdash NAO (2012) Fault-tolerant operation of delta-connected scalar- and vector-controlled AC motor drives. IEEE T. Power Electron. 27(6):3041–3049
11. Soufi Y, Bahi T, Lekhchine S, Dib D (2013) Performance analysis of DFIM fed by matrix converter and multilevel inverter. Energy Convers Manag 72:187–193
12. Sivakumar, Das, K., Ramchand, A., Patel, R., Gopakumar, C., A Five-Level Inverter Scheme for a Four-Pole Induction Motor Drive by Feeding the Identical Voltage-Profile Windings From Both Sides. *IEEE T. Ind. Electron.*, 57, 8, 2776–2784, 2010.
13. Zhang Y, Zhao Z, Zhu J (2011) A hybrid PWM applied to high-power three- level inverter-fed induction-motor drives. IEEE T Ind Electron 58(8):3409–3420
14. Akin, B. and Garg, N., *Scalar(V/f) Control Of 3-Phase Induction Motors, Texas Instruments, Inc*, pp. 1071–1080, C200 Systems and Applications Modelling Practice and Theory 17 Science Direct, Dallas, Texas, 2009.
15. Lekhchine S, Bahi T, Soufi Y (2014) Indirect rotor field oriented control based on fuzzy logic controlled double star induction machine. Int J Electr Power Energ Syst 57:206–211
16. Feroura, H., Krim, F., Talbi, B., Laib, A., Belaout, A., Sensorless Field Oriented Control of Current Source Inverter Fed Induction Motor Drive. *Rev. Roum. Sci. Techn.—Électrotechn. Et Énerg.*, 63, 1, 100–105, 2018. Bucarest.

17. de Rossiter Correa, M.B., Jacobina, C.B., da Silva, E.R.C., Lima, A.M.N.A., General PWM Strategy for Four-Switch Three-Phase Inverters. *IEEE T. Power Electron.*, 21, 6, 1618–1627, 2006.
18. Bose, B.K., Scalar Decoupled Control of Induction Motor. *IEEE T. Ind. Appl.*, IA-20, 1, 216–225, 1984.
19. Shiva CK, Mukherjee V (2015) A novel quasi-oppositional harmony search algorithm for automatic generation control of power system. Appl Soft Comput 35:749–765

Load Forecasting and Analysis of Power Scenario in Bihar Using Time Series Prediction and Machine Learning

Amitesh Prakash, Anish Kumar, Aduitya Kaushal, Kumari Namrata, and Niranjan Kumar

Abstract The present paper is regarding the reliable forecasting of electrical load demands of the Indian state of Bihar using the long short-term memory (LSTM) technique of machine learning. Bihar is an energy deficit state mainly owing to the lack of resources is now progressing at a good pace which is evident in its growth rate. The prediction of electricity demand, which is crucial for power generation unit planning and monitoring, is hence the need of the hour as this will enable higher growth for the state and prosperity for the people. The dataset of electrical load demand used covering the period of 2019–2020 available on kaggle has been taken from the weekly energy reports of POSOCO, a Government of India enterprise under the Ministry of Power. In the recent times, the transparency of the organizations as well as the development of technologies such as IoT, smart grid and smart energy metres has led to the availability of huge amount of data in the public domain. This involves power consumption as well as generation data of various entities. These developments have led to the need of electrical load forecasting which is useful in the financial planning as well as capacity addition in the generating stations. In this work, we have used deep learning technique called the long short-term memory (LSTM) technique in order to find patterns from a time series data. Results corresponding to the electricity consumptions for years 2019 and 2020 for the state of Bihar are presented and discussed. Finally, the future scope of time series prediction using big data techniques is presented.

A. Prakash (✉) · A. Kumar
BCE Bhagalpur, Bihar 813210, India
e-mail: amitesh.nitjsr1@gmail.com

A. Kaushal
BCE Bhagalpur, Bhagalpur 813210, India

K. Namrata · N. Kumar
NIT Jamshedpur, Jharkhand 831014, India
e-mail: namrata.ee@nitjsr.ac.in

N. Kumar
e-mail: nkumar.ee@nitjsr.ac.in

K. Namrata et al. (eds.), *Smart Energy and Advancement in Power Technologies*,
Lecture Notes in Electrical Engineering 926,
https://doi.org/10.1007/978-981-19-4971-5_63

Keywords Electrical load forecasting · Prediction · Long short-term memory (LSTM)

1 Introduction

Electricity demand forecasting is the backbone of power generation and planning. Forecast of electricity demands with higher accuracy enables better vision for the transmission line expansion and generation which finally helps in achieving the goal of sustainable growth [1–3].

Load forecasting is significant to fulfil the ever rising electrical power demand thereby helping to meet the growing electrical needs of the consumers [4]. It is also an integral part of electrical engineering especially power systems as it helps in planning for capacity addition as well as financial planning. It is important to predict energy demand as well as peak power demand, and the correct forecasting of power will ensure the reliability of the power system.

Load forecasting is nothing but the prediction of future load demand and their effect on power generation based on past data [4]. It is the backbone of the whole decision-making process in electricity production and distribution. In order to do a load forecasting, input variables are the past load demand data, whereas the target variable is the future load demand that is to be predicted.

The use of load forecasting can be in both the situations, first where the demand is ever increasing, and hence, we need to increase the generation by commissioning new power plants or by capacity addition to the existing ones, second where the load demand is decreasing due to some reasons (e.g., solar rooftop plants installed by the citizens) and some generators are hence needed to be turned off as lesser power is needed to be supplied. Both of these are determined by the generation control strategies like unit commitment [5], economic load dispatch, scheduling, etc., [6, 7]. Proper load forecasting also improves the reliability of the power system and facilitates proper planning of finances and capacity addition.

The energy forecast problems can be distinguished in the following categories as per the time span range of the prediction [8]:

1. Short-term load forecasting (STLF) used to predict up to 1 day ahead.
2. Medium-term load forecasting (MTLF) to predict for 1 day to 1 year ahead.
3. Long-term load forecasting (LTLF) used to predict up to 1–10 years ahead.

This paper discusses forecasting Electricity demand in one of the most populous state of India, Bihar. Historically, Bihar has been a power deficit state due to poor generation and some policy lapses but in the recent years, the power scenario in Bihar has improved a lot. As per the data available in public domain, Bihar had a meagre power consumption of just 700 MW in the year 2005 which gradually improved to 5932 MW in 2020–21. This has been possible due to positive steps taken by the government as almost 1900 MW power is being purchased by the state government to ensure uninterrupted power supply to the public. This has also led to increase in per capita electrical energy consumption of Bihar from 145 units in 2012–13 to 400 units in 2020–21.

This case study involves the electrical load demand of the state of Bihar from 02/01/2020 to 05/12/2020 and is useful in predicting the future load demand as well as capacity addition in the generation plants. The present paper is written in such a way so as to discuss the methodology, results, and conclusion with the future scope using big data analytics, respectively.

2 Methodology

This section explains how to forecast time series data using a long short-term memory (LSTM) network. Deep learning toolbox of MATLAB has been used for the time series forecasting using LSTM.

Recurrent neural network (RNN) is a type of artificial neural networks where connections between nodes form a directed graph along a temporal sequence which allows it to exhibit temporal dynamic behaviour. Derived from feedforward neural networks, RNNs can use their internal state (memory) to process variable length sequences of inputs [10, 11]. This enables RNNs to be applied in fields such as speech recognition [12, 13] or handwriting recognition [14].

Figure 1 shows a fundamental block diagram of an RNN which includes a loop structure and how it unfolds with time to form a chain-like structure where each unit A is alike and is connected to the next unit hence passing the information. Still, in practice, RNNs have a major drawback known as vanishing or exploding gradient problems. These drawbacks inhibit RNNs from learning temporal dependencies efficiently. In order to overcome this, long short-term memory (LSTM) networks were suggested by Hochreiter and Schmidhuber [15] which have dedicated memory blocks to store or forget information that is not modified during the learning process, and this allows it to remember values for a longer time. Figure 2 shows the fundamental block of LSTM network including the gates, an input block and also explains their operation. Long short-term memory (LSTM) is an RNN network [10] extensively employed in deep learning. Unlike conventional neural networks that have feed-forward connections, LSTM employs feedback connections. Apart from processing single data points (such as images), it can also process sequential data like speech, video or a time series data. This is the reason why we have used LSTM for this proposed paper.

In our proposed methodology, the dataset of electricity consumption in Bihar 'bihardata' is loaded which contains a single time series, with the time steps corresponding to days and values showing the power consumption in megaunits (MU). The output is a cell array, where each element is a single time step. The data is reshaped into a row vector.

The partition of dataset is then done, and the data is bifurcated into training data (first 90% of dataset) and test data (last 10% of the dataset). Training is done on the first 90% of the sequence and test on the last 10%. Ideally, we need to have zero

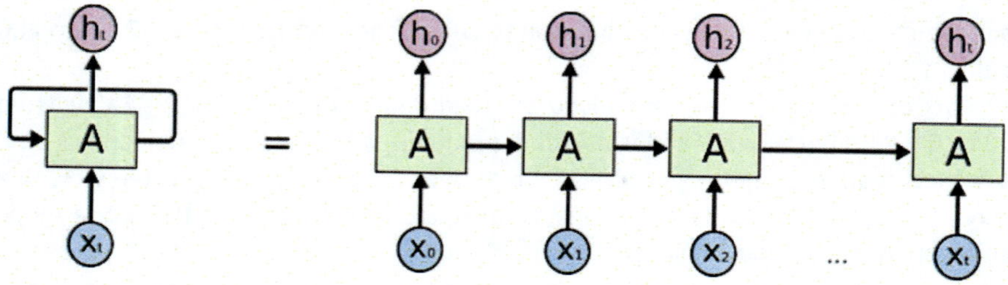

Fig. 1 Block diagram showing a simple RNN that unfolds with time forming a chain-like structure. A refers to one unit of the RNN, *x* and *h* refer to the input and the output at a given time step. The hidden state values of one unit are passed to the next one sequentially and are represented by the arrow. *Source* [9]

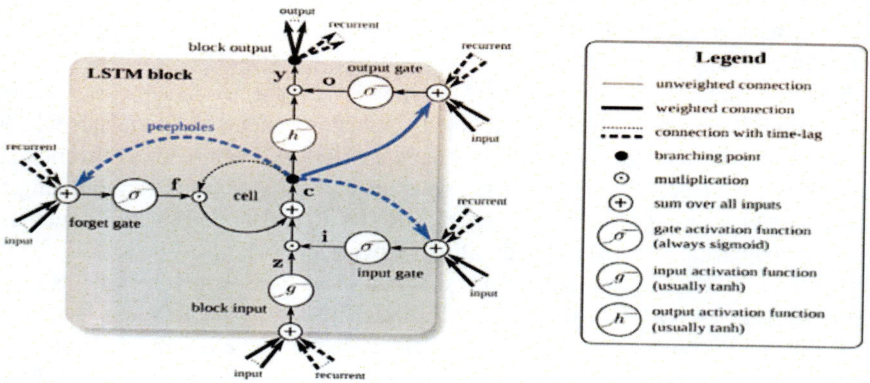

Fig. 2 Fundamental block of an LSTM network unit showing the three gates input, output and forget, the input as well as the output. *Source* [16]

mean and unit variance as well as no divergence in the data in order to ensure a better fit, and this is taken care of by the standardization of the training data. While doing the prediction, precaution must be taken such that the standardization of the test data is done using the parameters that are same as the training data.

The following were the steps followed for the time series prediction using LSTM:

Figure 3 shows the normal forecasting of future values of electrical power consumption in which the network state updating is done using the predicted values. Figure 4 shows the forecasting of future values of power consumption in the state of Bihar by updating the network state using the values observed in the previous step of prediction that we have termed here as observed values. Finally, it is generally observed that the predictions in this case have better accuracy when updating the network state with the observed values instead of the predicted values.

In this manner, we have successfully employed LSTM for the time series prediction of load forecasting.

Fig. 3 Flowchart for prediction of load demand by updating network state with predicted values

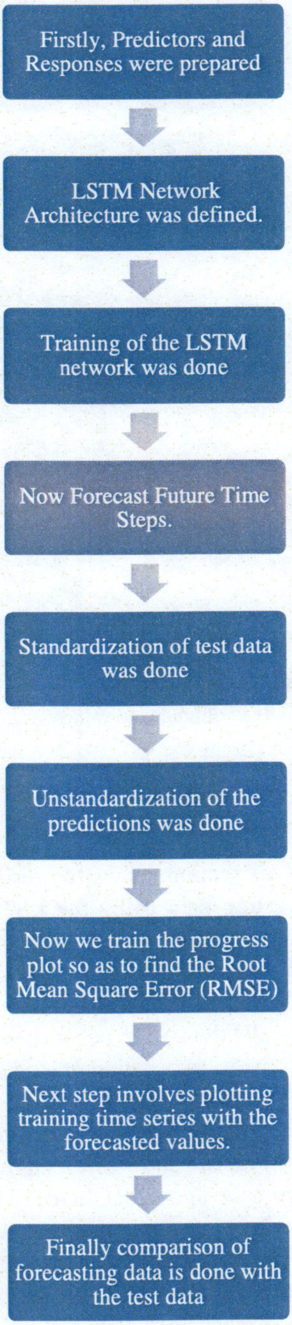

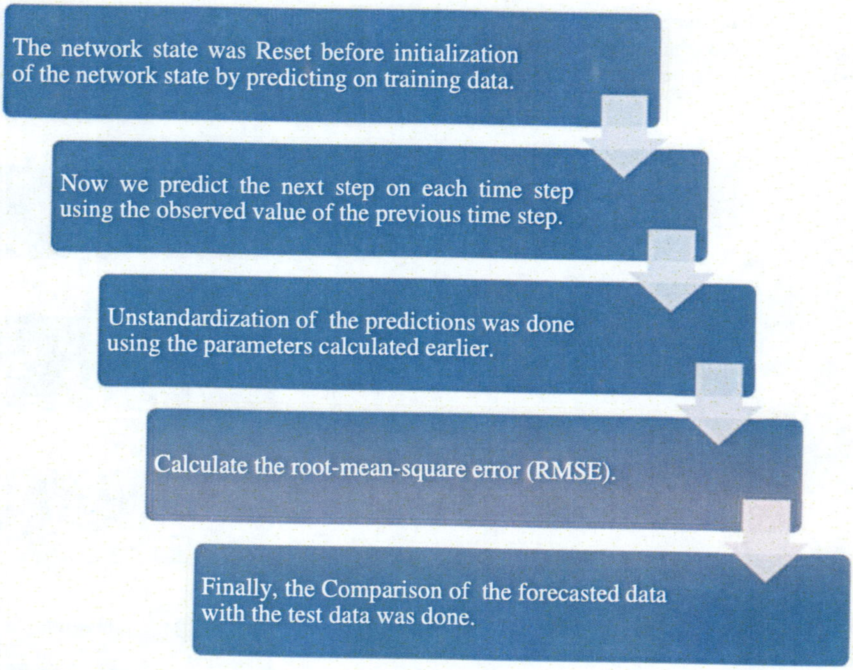

Fig. 4 Flowchart for prediction of load demand by updating network state with observed values instead of predicted values

3 Results

This section shows the results obtained in the time series prediction of load forecasting data using the LSTM network technique. Figure 5 shows the electricity consumption pattern in the state of Bihar during January, 2019 to December, 2020. This period is quite interesting as this includes the COVID-19 period as well. The dataset taken is partitioned into training and testing data as explained above and finally the LSTM network is trained using the training data which is shown in Fig. 6.

Now, forecasting of load demand is done using machine learning, and it is plotted as shown in Fig. 7. The blue lines show the observed data, whereas the red line in the graph shows the forecasted values.

Figure 8 shows the load forecast pattern along with the RMSE. It is clearly evident from the figure that there is a lot of difference between observed and predicted values.

To overcome this, we have one more method that involves forecasting the load demand by updating the network state with observed values instead of predicted values shown in Fig. 9. This ensures better prediction of electrical load demand.

It can be observed from these results that the future values of the electricity consumption can be easily predicted using LSTM deep learning algorithm. This

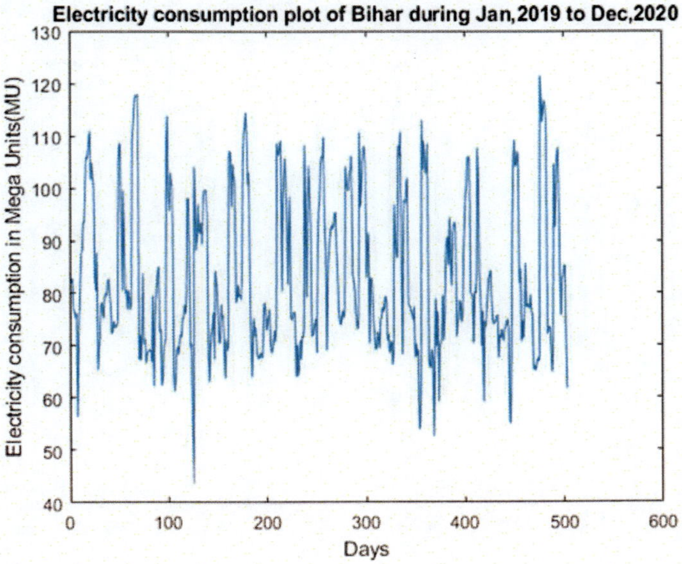

Fig. 5 Plot showing electricity consumption in Bihar during Jan, 2019 to Dec 2020

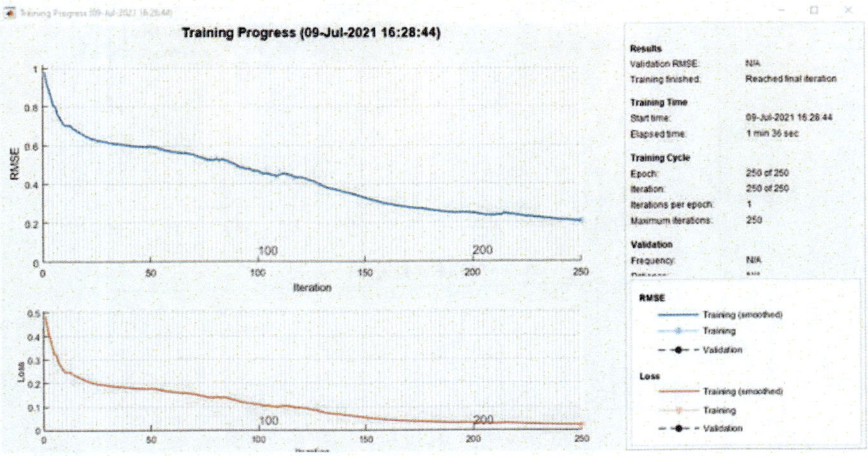

Fig. 6 Plot showing training progress of LSTM network

method is very much useful to predict any kind of time series data, and hence, we have successfully used it to forecast the electrical load in the state of Bihar.

It is evident from the forecast plot that there is still a lot of difference between the observed plot and forecasted plot. This is mainly as the load demand is based on many environmental factors like temperature, humidity as well as demography of the area under consideration which will be considered in future works. This forecasting

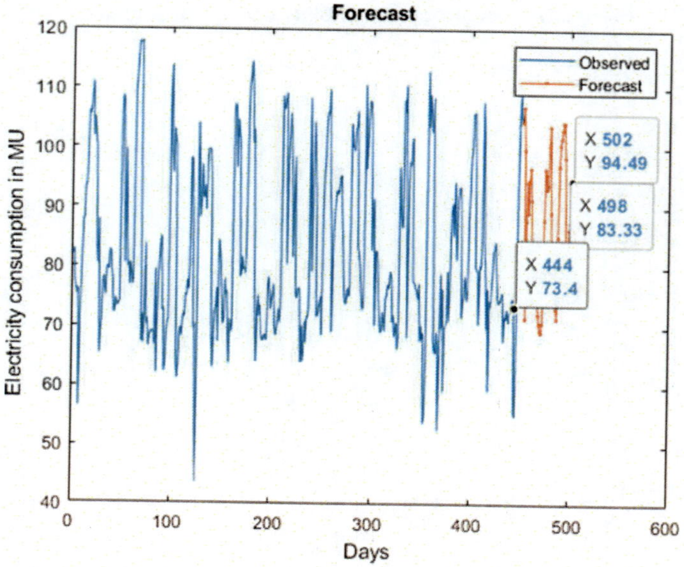

Fig. 7 Forecasting plot of electrical load based on the past data

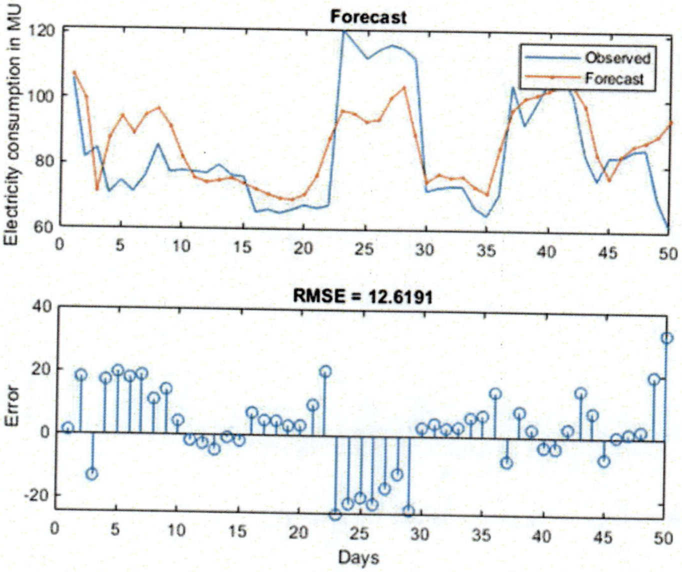

Fig. 8 Plot of forecasting data along with RMSE

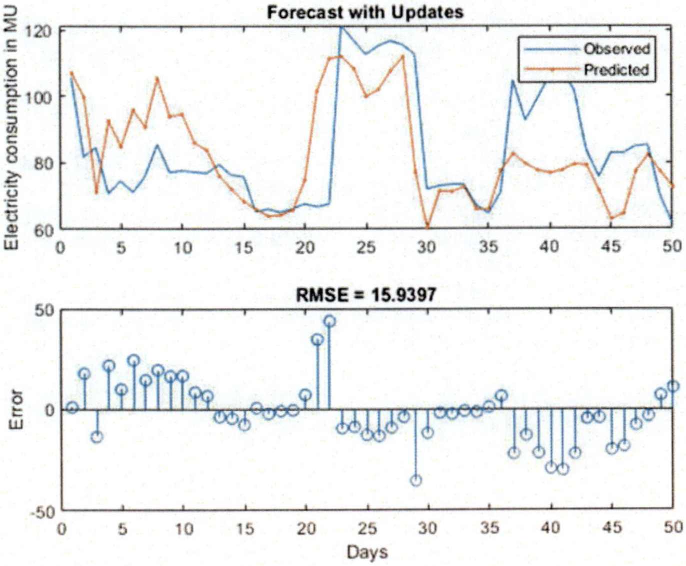

Fig. 9 Plot showing forecast by updating network state with observed values

can also be done for a large amount of data that is termed as big data using big data analytics, and we will consider this in our future works.

References

1. Zafer ZD, Hunt LC (2011) Industrial electricity demand for Turkey: a structural time series analysis. Energy Econ 33
2. Erdogdu E (2007) Electricity demand analysis using cointegration and ARIMA modelling: a case study of Turkey. Energy Policy 35(2):1129–1146
3. Pessanha JFM, Leon N (2015) Forecasting long-term electricity demand in the residential sector. Procedia Comput Sci 55:529–538
4. Keles D, Scelle J, Paraschiv F, Fichtner W (2016) Extended forecast methods for day-ahead electricity spot prices applying artificial neural networks. Appl Energy 162:218–230
5. Sharma V, Tung N, Prakash A (2013) Genetic algorithm optimization for thermal generating unit scheduling and economic dispatch evaluation. Int J Electron Commun Comput Eng 4(5):1575–1580
6. Hu Z, Ma J, Yang L, Yao L, Pang M (2019) Monthly electricity demand forecasting using empirical mode decomposition-based state space model. Energy Environ 30(7):1236–1254
7. Lee YW, Tay KG, Choy YY (2018) Forecasting electricity consumption using time series model. Int J Eng Technol 7(4):218
8. Román-Portabales A, López-Nores M, Pazos-Arias JJ (2021) Systematic review of electricity demand forecast using ANN-based machine learning algorithms. Sensors 21:4544
9. Example image of a recurrent neural network (rnn). URL http://colah.github.io/posts/2015-08-Understanding-LSTMs/img/RNN-unrolled.png.!p.ix, 17
10. Hochreiter S, Schmidhuber J Long short-term memory. Neural Comput 9(8):1735–1780

11. Graves A, Liwicki M, Fernández S, Bertolami R, Bunke H, Schmidhuber J (2009) A novel connectionist system for unconstrained handwriting recognition. IEEE Trans Pattern Anal Mach Intell 31(5):855–868
12. Gers FA, Schmidhuber J, Cummins F (2000) Learning to forget: continual prediction with LSTM. Neural Comput 12(10):2451–2471
13. Gers FA, Schmidhuber E (2001) LSTM recurrent networks learn simple context-free and context-sensitive languages. IEEE Trans Neural Networks 12(6):1333–1340
14. Li X, Wu X (2015) Constructing long short-term memory based deep recurrent neural networks for large vocabulary speech recognition. In: 2015 IEEE international conference on acoustics, speech and signal processing (ICASSP), pp 4520–4524
15. Hochreiter S, Schmidhuber J (1997) Long short-term memory. Neural Comput 9(8):1735–1780
16. Example image of a long short term memory network (lstm). https://devblogs.nvidia.com/parallelforall/wp-content/uploads/2016/03/LSTM.png

Fault Detection in Series Compensated Lines in the Presence of Power Swings Using Sum of Negative and Zero-Sequence Voltages

Sanjay Kumar Sena, Durgesh Himanshu, Ajay Agarwal, and Jitendra Kumar

Abstract The distance relays misinterpret power swings as fault like situation and produce undesirable tripping signals. So, the action of these relays is bypassed for power swing conditions. Distance relays are equipped with features like power swing blocking and out-of-step operation functions to deal with such situations. Although, the fault in the presence of the power swing can be identified, but to detect the faults when power swing is present in a series compensated line is quite problematic because of the issues related to effects generated by the series capacitor and its protection aspects. This paper introduces a technique that uses the sum of negative sequence and zero-sequence voltages to identify the faults in the presence of the power swings in a series compensated line. Various types of faults like fault involving single phase or two phases or all the phases incorporating high fault resistance are simulated using this technique in MATLAB from the data generated by simulating the system in EMTDC/PSCAD to test the proposed technique.

Keywords Power swings · Compensation · Sequence components

1 Introduction

To meet the vast increase in power demand while satisfying the economic and environmental constraints for building new lines the use of capacitors in series in long

S. K. Sena (✉) · D. Himanshu · A. Agarwal · J. Kumar
Department of Electrical Engineering, National Institute of Technology, Jamshedpur 831014, India
e-mail: 2020pgeeps12@nitjsr.ac.in

D. Himanshu
e-mail: 2020pgeeps09@nitjsr.ac.in

A. Agarwal
e-mail: 2020pgeeps17@nitjsr.ac.in

J. Kumar
e-mail: jitendra.ee@nitjsr.ac.in

K. Namrata et al. (eds.), *Smart Energy and Advancement in Power Technologies*,
Lecture Notes in Electrical Engineering 926,
https://doi.org/10.1007/978-981-19-4971-5_64

transmission lines is more prominent to enhance the line loading and optimize the performance and operation of the transmission networks. Though the series capacitor helps in achieving the power transfer issue, it poses problems and challenges in protection of such lines in the network [1].

The centre of inertia motion of the power system is maintained by the load generation balance in the system while some disturbances like line switching after a fault, sudden throw-off of large loads, generator disconnection, etc. disturbs the relative motion of the set of generators which can trigger severe power swings in the interconnected system. As a consequence, the measured impedance by the distance relays may change and fall into their operating characteristics, as a result, the distance relays maloperates and may result into uncontrolled tripping of the lines. To prevent this situation present-day distance relays are equipped with power swing blocking (PSB) function. Also, if a fault takes place when power swing is occurring the relay must identify the fault and operate as quickly as possible which is taken care of by the out-of-step operation (OST) feature [2].

However, to identify the fault when power swings are taking place in series of compensated lines is more tedious because fault signal contains components of various frequencies which also depends on the distance of the fault from the relay and type, compensation provided and operation of the protection system used for protecting series capacitor [2], the issue of voltage and current inversion [3]. Due to all these the impedance measured by the relay oscillates and relay cannot differentiate between a fault and the power swings.

Many techniques are present to identify faults in the presence of the power swings for the transmission lines not provided with compensation in series. A method that uses superimposed current components is introduced in [4]. A technique that uses the superimposed apparent power is proposed in [5]. A technique utilizing the rate of change of swing-centre voltage (SCV) is presented in [6]. A technique that monitors the phase angle of the voltage at the location of relay for the detection fault with high impedance is presented in [7]. A cross-linking scheme using the rate of change of the three-phase active and reactive power to detect faults involving all the phases when power swing is present is introduced in [8]. A method based on adaptive neuro-fuzzy inference system is presented in [9]. A method based on covariance indices of current signals is proposed in [10]. A fast detection of fault in series compensated line is presented in [11]. A symmetrical fault detection method based on phase space is discussed in [12]. A technique utilizing mean value of sampled data is presented in [13]. A method based on compressed sensing theory is presented in [14].

2 Issues in Detecting Faults When Power Swing is Present in a Series Compensated Line

2.1 Issues in Relaying

Due to transients during power swings and transients created by the series capacitor makes it is difficult for the relay to differentiate the faults from the power swings. These problems will vary with the amount of compensation, line length and configuration, type of capacitor and capacitor control, etc. Therefore, better transient response of the transmission line, as well as the relay, needs to be investigated [15].

2.2 System Transients

The transient component of current has a frequency approximated as

$$f_t = f_n\left(\frac{X_C}{X_L}\right)^{\frac{1}{2}} \tag{1}$$

where f_n is the fundamental frequency, X_C is the total series capacitive impedance and X_L is the total inductive line impedance. Assuming $X_L > X_C$ if the voltage drop due to load current is small than the voltage drop due to fault current, then after the inception of the fault the voltage drop in the inductor is of the same polarity as the voltage drop in the capacitor in the first cycle of the voltage. So, the capacitor reduces the fault current initially and when these two voltages become out of phase the current becomes larger. So, relaying and fault detection become difficult [15].

2.3 Voltage Inversion

If the measured impedance between the bus and the point of fault is inductive in nature, in addition, if the nature of the impedance measured between the relay location and the point of fault is capacitive leads to a condition called voltage inversion. Due to voltage inversion, the relay identifies the fault in the reverse direction for the fault occurring in the forward direction. And current inversion which is a rare phenomenon occurs when the voltage at the location of the relay remains in phase with the source voltage and current at the location of the relay due to high capacitance in the fault path leads source voltage by 90°. Both these problems create issues in directional relaying [15].

2.4 Higher Load Flows

Although series capacitors are employed to increase the line flow along with enhancing the stability of the system, higher line flows try to reduce the sensitivity of phase comparison schemes as well as some directional overcurrent comparison schemes. These problems become more prominent in long lines, use of distance relaying in sub-transmission systems and use of high-speed auto reclosing [15].

2.5 Very Long Series Compensated Line

For very long lines series compensated lines pilot schemes may become unsuitable for faults near one end because the magnitude and direction of fault current at remote end may not be determined.

2.6 Unbalance Line Impedance

Line impedances become unbalanced due to un-transposed or partially transposed lines, but the series capacitors put in phases are equal in all the phases. Thus, more impedance unbalancing may occur with the amount of compensation which may result in higher value to negative and zero-sequence currents and hence relay settings may require changes.

Also, there are issues related to sub-synchronous oscillations caused by series capacitors, transients produced by the action of MOV in the presence of the fault, etc. which pose challenges for the available relaying schemes.

Because of the issues listed above, the schemes present to identify the fault in the presence of the power swings in lines without series compensation become unsuitable for the series compensated lines. So, to differentiate faults when the power swings are present in a series compensated line, a technique taking the cumulative sum of difference in the value of sum of negative and zero-sequence voltages is proposed in the chapter.

The single-line diagram of the circuit used for the study is shown in Fig. 1.

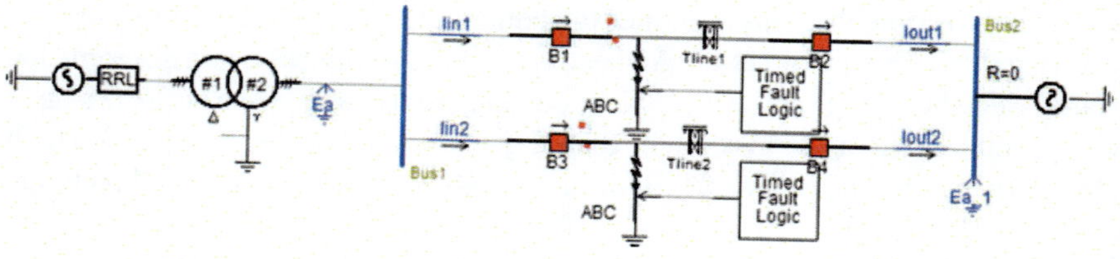

Fig. 1 400 kV single machine infinite bus system

3 Proposed Technique to Detect Faults

Though power swings are balanced in nature, a small amount of unbalance is present. Due to which sequence components of currents and voltages appear. Unbalanced faults produce a large value of negative sequence component of voltages and currents. For unbalanced faults involving ground, a zero-sequence component voltage is observed. Three-phase fault being balanced in nature, in the presence of the power swings some amount of negative sequence current is observed, due to which corresponding amount of negative sequence voltage is also present and if three-phase fault involves ground then zero-sequence current and voltage is present.

The presence of these sequence components during the power swings allows us to use negative sequence voltage in our algorithm. Also, during faults involving ground, a zero-sequence component of current and voltage is present. So, in the algorithm, a change in magnitude of sum of negative sequence and zero-sequence voltage (ΔV) is introduced to identify the fault. Therefore, a technique employing the cumulative sum (CUSUM) of (ΔV) with appropriate threshold is used in the technique. The steps for the computation for the technique are

$$V_0 = \frac{V_a + V_b + V_c}{3} \tag{2}$$

$$V_2 = \frac{V_a + \alpha^2 V_b + \alpha V_c}{3} \tag{3}$$

where V_0 and V_2 are the zero and negative sequence components of voltage, respectively; V_a, V_b and V_c are the phase voltages; $\alpha = e^{j2\pi/3}$.

$$V = V_0 + V_2 \tag{4}$$

A signal m_t is calculated as

$$m_t = \Delta\left|V_{2_t}\right| = \left|V_{2_t}\right| - \left|V_{2_{t-1}}\right| \tag{5}$$

For $m_t > \mu$, the CUSUM test is written as

$$g_t = \max(g_{t-1} + m_t - \mu, 0) \tag{6}$$

where g_t is the parameter which indicates the test results and μ is the parameter that nullifies the value of m_t for power swing during non-fault conditions.

A fault is detected when

$$g_t > l \tag{7}$$

where l is a constant and is zero ideally. μ affects the performance of the detector as it provides low pass filtering in (6). The value of g_t becomes more and more as the difference between m_t and μ increases, i.e., when $m_t > \mu$. With more samples of voltage, the CUSUM technique decides on the fault situation by using (7). m_k is again set to zero for the power swing condition as m_k will be zero during power swing because $\Delta V < \mu$.

The performance of the algorithm depends on the value of μ and l. In the proposed technique of fault detection based on CUSUM method, the value of μ is set to make $m_t = 0$ when power swing is present. In the chapter the value of $\mu = 0.8$ for all the cases except for the single pole tripping for which $\mu = 3$ which is a special case and as the technique is based on CUSUM method, a higher value of the index m in obtained during the fault. The value of $l = 10$ in this chapter. The value of threshold is selected such that no trip signal is generated during the stable power swing.

4 Results

The algorithm is examined for faults at different locations in the line length and for several fault situations like balanced and unbalanced faults incorporating low resistance and high resistance faults. MATLAB has been used to simulate the data obtained by modelling the system in EMTDC/PSCAD.

4.1 Faults Involving All the Phases in Series Compensated Line

Three-phase faults in the present power swings is difficult to differentiate because both the phenomenon are balanced in nature. To check the algorithm a fault involving all the phases when power swing is present is simulated at 2.6 s with fault present at 200 km from the relaying point in line 1. From Fig. 2. it can be seen that a signal to trip the circuit breakers is generated and hence the fault can be identified after few seconds from the inception of the fault.

4.2 Line-To-Ground Fault in Series Compensated Line with High Fault Resistance

To check the algorithm an ag-type fault with faulted path having resistance of 200 Ω located at 150 km from the relaying point is created at 2.5 s. It is difficult to detect the ag-type fault with large resistance in the faulted path as the current when fault is taking place is comparable to the current when power swing is present. Hence

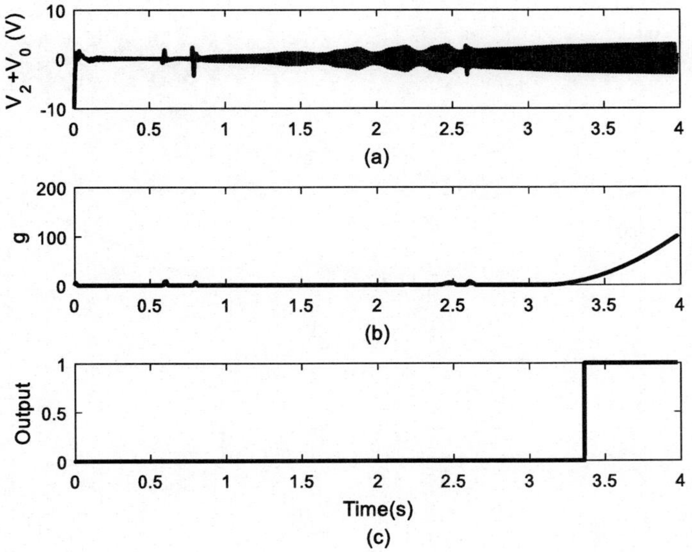

Fig. 2 Test results of three-phase-to-ground fault

the voltage variation will also be comparable to voltage when power swings taking place. Figure 3 depicts the result of the technique used. It detects the fault after a few milliseconds delay from the inception of the fault.

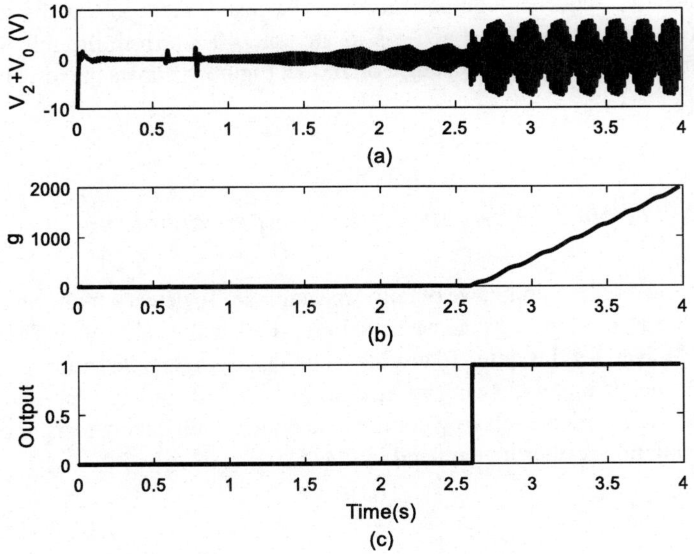

Fig. 3 Test results of line-to-ground fault

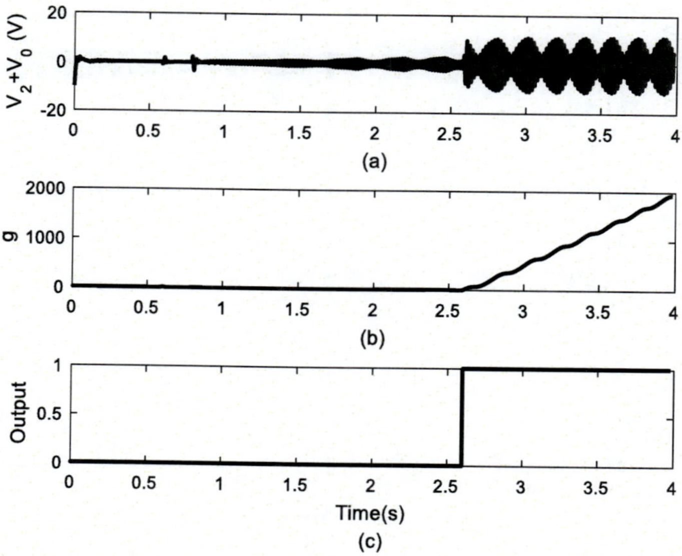

Fig. 4 Test results for double line fault

4.3 Double Line Fault in Series Compensated Line

Faults that involve the two phases is an unbalanced phenomenon but does not include zero-sequence component of voltage or current hence CUSUM technique applied only to negative sequence component gives the result. Here to test the technique a fault of bc-type is simulated at 2.5 s located at 240 km from the relaying point with the faulted path having resistance of 200 Ω. Figure 4. shows the test simulation results.

4.4 Single Pole Tripping in Series Compensated Line

In case of single pole tripping, we obtain a special case. To detect a fault, we first need to detect single pole tripping then we need to recalculate the value of μ. The value of $\mu = 3$ is obtained for this case. To test the technique a single pole tripping is created in phase A, hence a power swing is created. Then a fault of bcg-type is simulated located at 50 km from the relaying point with a faulted path having resistance of 200 Ω. The simulation results are depicted in Fig. 5

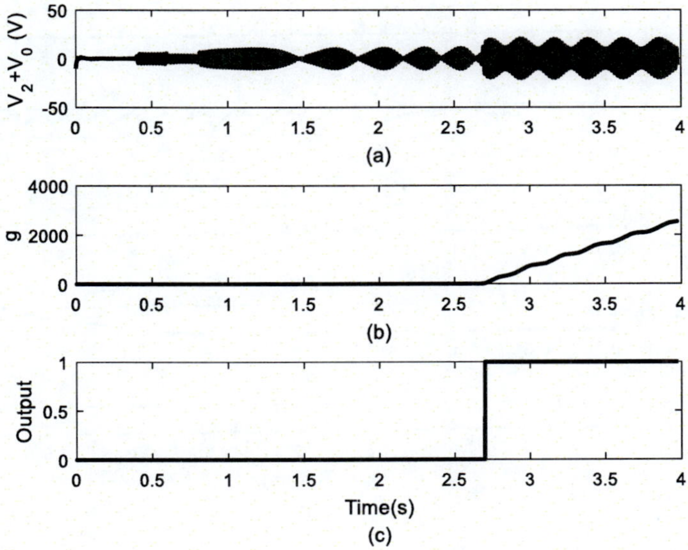

Fig. 5 Test results of single pole tripping

5 Comparison of Proposed Technique with Other Techniques

There are various techniques to identify the faults in the presence of the power swings for uncompensated lines but when applied to the compensated lines they do not provide satisfactory results. In this paper, we have considered three techniques: continuous impedance calculation, swing-centre voltage (SCV) and apparent resistance for comparison.

A double line-to-ground fault with faulted path having resistance of 200Ω located at 240 km from the relaying point at 2.5 s is simulated during power swing to compare the techniques. From Fig. 6a. it can be seen that continuous impedance calculation technique does not produce significant change in impedance value for fault conditions during power swing. Also, from Fig. 6b the value of SCV is high and it is oscillating which produces difficulty in identification of the fault when power swing is present in the system. Figure 6c shows that the value of apparent resistance calculated from the relay location oscillates which may produce delayed response. Figure 6d shows the value of index *g* which jumps to a large value from zero on the occurrence of the fault. This manifests how correctly the proposed technique can identify the faults when power swing is present when the line is series compensated.

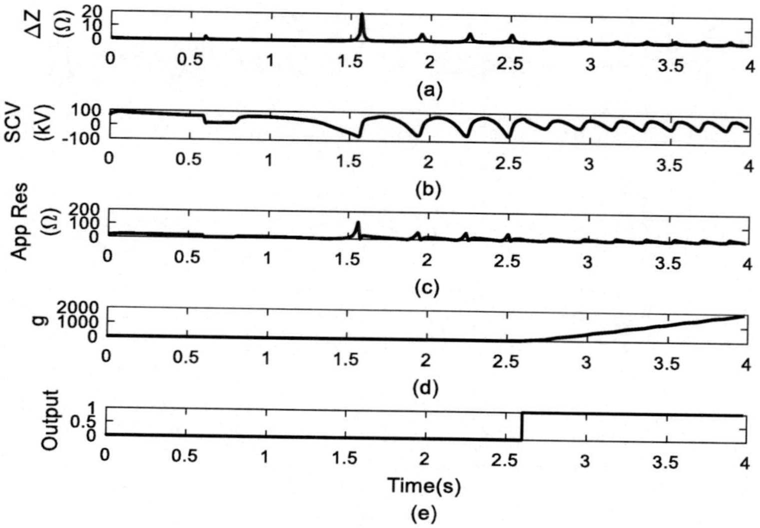

Fig. 6 Comparison of proposed technique with other techniques

6 Conclusion

The proposed technique provides satisfactory results for different fault conditions. A special case is obtained in case of single pole tripping where single pole tripping phenomenon is first detected and then the value of μ is recalculated to detect a fault arising further. The technique is compared with various other available techniques and result is found to be satisfactory.

References

1. Nayak PK, Pradhan AK, Bajpai P (2013) A fault detection technique for the series-compensated line during power swing. IEEE Trans Power Del 28(2):714–722
2. IEEE Power System Relaying Committee of the IEEE Power Eng. Soc., Power swing and out-of-step considerations on transmission line. Rep PSRC WG D6, (2005)
3. Jena P, Pradhan AK, Bajpai P (2010) A positive-sequence directional relaying algorithm for series-compensated line. IEEE Trans Power Del 25(4):2288–2298
4. Apostolov AP, Tholomier D, Richards SH (2004) Superimposed components based sub-cycle protection of transmission lines. Proc IEEE Power Eng Soc Power Syst Conf Expo 1:592–597
5. Kumar J, Jena P (2015) Fault detection in a series compensated line during power swing using superimposed apparent power. 2015 IEEE Power Energy Soc Gener Meet
6. Benmouyal G, Hou D, Tziouvaras D (2004) Zero-setting power-swing blocking protection, presented at the the 31st AnnualWestern Protective Relay Conf., Spokane, WA
7. Mechraoui A, Thomas DWP (1997) A new principle for high resistance earth fault detection during fast power swings for distance protection. IEEE Trans Power Del 12(4):1452–1457
8. Lin X, Gao Y, Liu P (2008) A novel scheme to identify symmetrical faults occurring during power swings. IEEE Trans Power Del 23(1):73–78

9. Zadeh HK, Li Z (2008) A novel power swing blocking scheme using adaptive neuro-fuzzy inference system. Elect Power Syst Res 78:1138–1146
10. Musa MHH, He Z, Fu L, Deng Y (2019) A convariance indices-based method for fault detection and classification in a power transmission system during power swing. Int J Electric Power Energy Syst 105:581–591
11. Daryalal M, Sarlak M (2017) Fast fault detection scheme for series compensated lines during power swing. Int J Electric Power Energy Syst 92:230–244
12. Dubey R, Samantaray SR, Panigrahi BK, Venkoparao VG (2016) Phase-space-based symmetrical fault detection during power swing. IET Gener Transm Distrib 10(8):1947–1956
13. Taheri B, Faghihlou M, Salehimehr S, Razavi F (2020) Symmetrical fault detection during power swing using mean value of sampled data from thecurrent signal. IETE J Res. https://doi.org/10.1080/03772063.2020.1798823
14. https://ietresearch.onlinelibrary.wiley.com/doi/https://doi.org/10.1049/iet-gtd.2020.0540
15. Wilkinson S Series compensated line protection issues, GE Power Management, GER 397

Home Energy Management System with Improved Binary PSO

Arshad Mohammad, Saeem Ansari, Faiz Ali, and Imtiaz Ashraf

Abstract Home energy management systems (HEMS) have been proposed to manage energy consumption in smart houses. To reduce the amount of power utilized by the homes, an optimal home energy management strategy (OHEM) has presented in this paper. The OHEM algorithm is applied to determine when electric tasks are performed to optimize customer satisfaction and electric cost. There are two disadvantages to the existing scheduling methods. Most of the current techniques do not take into account the user's comfort, and secondly, there is a lack of an effective optimization algorithm. In this paper, an improved binary particle swarm optimization (IPSO) is used to get accurate, optimal, and desirable solutions for the power consumption that is taking place in smart homes. The representation is done in a time slots manner. The main objective of the algorithms described here is to lower the cost of electricity and the conformity of the user. The mentioned IBPSO technique is then used to find the best working model that achieves the objectives mentioned above. Moreover, the improvement and progress range in minimizing the cost and consumer's comfort is regulated and managed using a weighting parameter.

Keywords Home energy management system · Electricity bill · Customer satisfaction · Improved binary particle swarm optimization

1 Introduction

In smart houses, the need for electrical energy has been increasing significantly. In the smart grid, the user platform is connected with numerous portals and assets in establishing a secure connection between the physical power source and the information. Many unique and innovative technologies like communication and distribution systems are combined together in making the method best in safety, security, and storage options thereby allowing for a significant increase in the dependability and resilience of the power network, lowering energy prices [1]. Demand response (DR)

A. Mohammad (✉) · S. Ansari · F. Ali · I. Ashraf
Aligarh Muslim University, Aligarh, India
e-mail: arshad.gb2140@gmail.com

K. Namrata et al. (eds.), *Smart Energy and Advancement in Power Technologies*, Lecture Notes in Electrical Engineering 926,
https://doi.org/10.1007/978-981-19-4971-5_65

helps to change the consumer's electricity usage overtime because there are variable hourly prices for power [2]. As a result, demand response may be handled effectively.

The smart meter is the fundamental component of SG, and it is through it, the electrical operator may trigger the directives for electricity control [3]. The electricity process is defined by the real-time price (RTP) for each hour of the day. The energy management for smart homes considers the lowest cost of electricity usage according to the RTP and the user comfort of electricity usage in terms of electrical device operation time slots [4]. The usage of HEMS is seen to be very popularly used for most smart houses over the last decade. In the case of dynamic electricity prices or demand response based on incentives programs, HEM systems, appliances schedules, and the storage of the battery units of the automobiles will reduce the electricity bill of the houses [5].

The power and working of certain appliances are being shifted to the range when the power is low or when incentives are given out, to lower the electricity bill [6]. This is done by the HEMS methodology. The scheduling is usually done on a day-ahead basis in HEMS. However, improving the consumer's preferences and electricity cost is conflicting goals. As a result, a utility function should be formed to integrate the two objectives. The optimum solution must be addressed for the better method of finding the scheduled price when the shifting of certain appliances take place. The problem of constrained optimization is related with the integer choice variables is a frequent formulation for this problem. Some methodologies for achieving this goal have been proposed in the literature.

The scientific community has proposed several techniques for scheduling in-home power. Generally, linear programming, mixed-integer programming, and metaheuristics optimization techniques have been employed for scheduling [4, 7–10]. Linear programming is used in [11] to optimize the power of EWHs which are known as electric water heaters to reduce electricity expenditure. Another technique called as MILP —mixed integer linear program—is been used in [12] for the purpose of better schedule of appliances, electric vehicles, and various other instruments that can be managed through thermostat and not. The reduction of the power bill and the enhancement of user comfort are considered as the problem statement, and the optimization is done accordingly. In [13], authors proposed a home energy management system that minimizes the electricity bill by full utilization of energy storage system. They find an optimal solution using binary PSO. A search algorithm that is based on the greedy approach HEMS has been used [14]. The calculation done by HEMS is done in the best manner for such home appliances like dishwashers and washing machines as they are relocated to times when there is a surplus of PV generation regarding the amount of energy they consume.

All the work, as mentioned above, didn't formulate the problem effectively. Moreover, there is a lack of an effective optimization algorithm. Therefore, in the proposed research, the method of OHEM, optimal home energy management system, is employed to find the best range of output in terms of the working and operation of the power in terms of time slot measurement. The electrical tasks have been completed in order to optimize both user's comfort and cost. The most effective algorithms for solving challenging engineering optimization issues are metaheuristics.

Improved binary particle swarm optimization (IPSO) is a swarm-based metaheuristic optimization technique that promises to deal with various engineering optimization issues. Improved binary particle swarm optimization (PSO) method is used for that solution.

The remaining portion of the research is mentioned as given below. In the second section, the allotment and schedule of the appliances used in the homes are given. The third section contains the results and the analysis involved. Finally, the conclusion are presented in the fourth section.

2 Problem Formulation

The proposed OHEM is aimed to schedule electrical tasks in houses by determining the start and finish times for each task in order to reduce overall energy costs while maintaining reasonable user comfort. The sole objective is to cut down the cost factor, bring about the lowest load of power, and minimize the range of units taken for consumption by the homes. The issue faced here is due to the optimization that has taken place. It is mostly based on the terms of decision variables which denote the ON and OFF values when taking the time intervals into consideration. According to the functioning of the electronic appliances, the problem said here is taken as multi-domain, and the optimization technique is carried out accordingly [15]. Billing system mentioned here has been altered so that the users must pay a surcharge that is equal to the amount consumed which leads to the rise of the voltage issue. The status of the ON and OFF switch is set by HEMS in a method that the user's electricity consumption amount is considerable reduced.

An optimization issue must be addressed in order to find the ideal scheduling of the pricing mechanism and the usage according to the time slots. A constrained issues that taken place with integer choice variables are a frequent formulation for this problem. Improvements in electricity costs and consumer preferences, on the other side, are at contradiction. As a result, a utility function should be created to integrate the two objectives. In addition, to solve the suggested problem, an optimization algorithm known as improved binary PSO is employed.

The method that is involved in estimating the sole purpose of the research is estimated using certain decision variables and terms as discussed in this section. Multi-domain problem with the choices of the values that have the power constraints is given here. The relation is expected to have no non-shiftable non-interruptible appliances. The optimization of such appliances is curated using the decision vector to settle the optimization issue, from the initiation of the appliances working as shown below.

$$X = [S_1 S_2 S_3 \ldots\ldots\ldots\ldots S_i \ldots\ldots\ldots\ldots S_{\mathrm{Na}}] \tag{1}$$

where S_i denotes the appliance's start time. The number of choice variables is clearly Na, and they are all integer variables. From beginning, the state of the home applications is estimated using the time slot calculation by using the decision vector basis.

$$\text{stat}_{i,t} = \begin{cases} 0 & t < S_i \\ 1 & S_i \leq t \leq S_i + N_i - 1 \\ 0 & t > S_i + N_i - 1 \end{cases} \tag{2}$$

I is the state of the application, and the time slot is denoted by the term t. It will be 1 if the appliance is turned ON and 0 if it is turned OFF. Indeed, each appliance i's operation begins at S_i and ends at $S_i + N_i - 1$.

The home's electricity bill is then calculated by using the equation below,

$$\text{bill} = \sum_{t=1}^{N_t} \sum_{i=1}^{N_a} \pi t * P_i * \text{Stat}_{i,t} * t * \text{d}t \tag{3}$$

The single variable represents the amount and range of the electricity cost. Here, the non-controllable applications are not included. The next variable denotes load curtailment incentives. The bill given does not comprise the tax amount, and other service fees that are associated with the current charge. πt stands for the price of electricity. P_i gives the power of the application taken for analysis which respect to the time interval t. int represents the incentive paid per kWh of load at time interval t for appliance i at time slot t. curtailment, dt stands for time resolution, and statbs is for metrics. The current stage of the applications is analyzed as per the time slot.

The index of discomfort is given on the underlying principle of the variation in the schedule baseline, and the schedule after the program is done in the DR to show the involvement of the factor according to the user preference. It takes place in the program of DR. Any variation and deviation between the normal schedule and the time interval of the appliance is defined according to the index value, which is given to the relation of the add-ons below.

$$\text{discomf} = \sum_{t=1}^{N_t} \sum_{i=1}^{N_a} \left| \text{stat}\frac{\text{bs}}{i.t} - \text{stat}_{i,t} \right| \tag{4}$$

where discomf denotes the level of discomfort.

Because the generated optimization issue has two objectives, the outcomes of the multi-domain issues of the involved one-time problem of the appliances taken are done with the LWS technique as presented. It is called as linear weighted sum. It is shown as,

$$\text{OF} = w * \text{bill} + (1 - w) * \text{Discomf} \tag{5}$$

Here, the term W shows the index's discomfort, and the value of OF is termed as the value of the total objective taken.

The initiation time of the considered applications must possess the following range values with accordance with the optimization problem's parameters.

$$\text{lb}_i < S_i < \text{ub}_i - N_i + 1(i = 1, 2, 3, \ldots\ldots\ldots, N_a) \tag{6}$$

where lb_i and ub_i signify the below and above range of operations that are present in the taken appliances, respectively. The above and below values have been chosen so that appliance i's operation is completed no further to ub_i.

The complete terms and requirements are that every application must have certain number of N_i terms each day as per the count.

$$\sum_{t=\text{lb}\quad i}^{\text{ub}\quad i} \text{stat}\, i, t \geq N_i(i = 1, 2, 3 \ldots\ldots\ldots\ldots, N_a) \tag{7}$$

3 Binary Particle Swarm Optimization

Binary particle swarm optimization (BPSO) is an evolving computational algorithm that models the behavior of a swarm of particles. The BPSO algorithm is commonly used to solve problems and issues involving continuous optimization. At the start of BPSO using an algorithm, a population of particles is generated and distributed at random placed in the problem's search space to be optimized. Each particle advances to a new position with each repetition to locate an effective solution inside the search space. The current position and velocity are used to calculate a new position. The following formula is used to calculate the new position $x_i(t+1)$ and velocity $v_i(t+1)$ of particle i at iteration $t+1$.

$$v_i(t+1) = \omega \cdot v_i(t) + Cl \cdot rl \cdot (\text{lb}_i(t) - x_i(t)) + Cg \cdot rg \cdot (gb(t) - x_i(t)) \tag{8}$$

$$x_i(t+1) = x_i(t) + v_i(t+1) \tag{9}$$

where ω is the weight of inertia and is a constant. The velocity of a particle at iteration t is $v_i(t)$.Cl is a constant that represents the acceleration coefficients for personal best. rl is a random number between 0 and 1 that represents a personal best. $\text{lb}_i(t)$ is the particle's personal best position at iteration t; $x_i(t)$ is the particle's position at iteration t; Cg stands for universal best acceleration coefficients and is a constant; rg stands for universal best random number spread from 0 to 1, and $gb(t)$ stands for universal best position at iteration t. All particles will move to the best place (best solution) of the problem after a number of repetitions. We must use Kennedy and

Eberhart's [11] binary particle swarm optimization (BPSO) for binary variables. At iteration $t + 1$, the formula for each particle's velocity in the BPSO algorithm is the same as (8). Instead of using velocity as (8), we utilize a sigmoid function $S(.)$ given by to update the value of particle $x_i(t + 1)$ at iteration $t + 1$ as given in (10).

$$S(v_i(t+1)) = \frac{1}{1 + e - v_i(t+1)} \tag{10}$$

and

$$x_i(t+1) = \begin{cases} 1 & S(v_i(t\ +\ 1)) > \text{rand}() \\ 0 & S(v_i(t+1)) \leq \text{rand}() \end{cases} \tag{11}$$

where rand() is a function that generates a pseudo-random number between in the range [0.0, 1.0]. The working flow of our BPSO algorithm is depicted in Fig. 1.

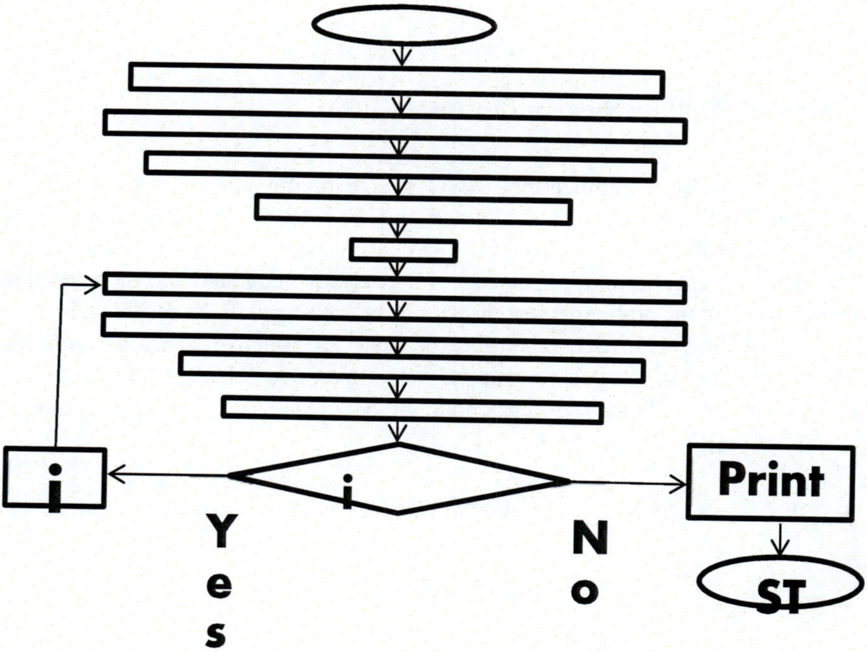

Fig. 1 Flowchart for BPSO algorithm

Table 1 Specifications of appliances used in smart home

Appliances name	Rated power (kwh)	Compulsory operating time duration (hrs.)	Baseline operating time range
Dishwasher	2.5	1	9–10
Washing machine	3	1	9–10
Spine dryer	2.5	0.5	14–15
Cooker hub	3	0.5	8–9
Electric vehicle	3.5	2	19–21

4 Result and Simulation

In this section, event-driven simulation is used to evaluate the proposed OHEM. The simulator is implemented using MATLAB. The simulator is run using a computer with 2.5 GHz processor and Intel Core i7 generation and 8 GB memory. The billing cycle has been set to ($n = 1$ h).

In the beginning, a set of household routines with a fixed power price for a single day have evaluated. Table 1 shows the household routines during one day. The properties for each household include the operation time, power consumption, and user-preferred baseline operating time range.

Figures 2, 3 and 4 show the power consumptions with scheduling (PC), objective function (F obj), and daily cost (DC) versus the iteration number for different w. For all figures, PC, Fobj, and DC are improved with iteration since improved binary PSO looks for a better solution for each iteration. For $w = 0.25$, the DC has 0.25, while customer comfort is 0.75. More attention is given to customer satisfaction. It is clear from Fig. 2 that devices operation has less deviated from the preferred time slot. Therefore, the DC is improved with iteration. On the other hand, PC is getting worse with iteration. For $w = 0.50$, both DC and PC are considered equally in the objective function (F obj), and hence, OHEM improves both of them. Thus, binary PSO searches for a solution that minimizes both DC and PC with an equal degree. The object function (Fobj) in Fig. 4 $w = 0.75$ means PC has given more concentration while DC has given less priority. The operation of devices in different time slots has changed according to the weight of the objective function shown in Figs. 2, 3 and 4. As seen, the weighting parameter w in objective function can control the degree of improvement for DC and PC.

5 Conclusion

In this paper, the issue of a home energy management system (HEMS) is discussed. The optimal house energy management (OHEM) algorithm is presented for scheduling the start and finish periods of home electrical tasks so that energy

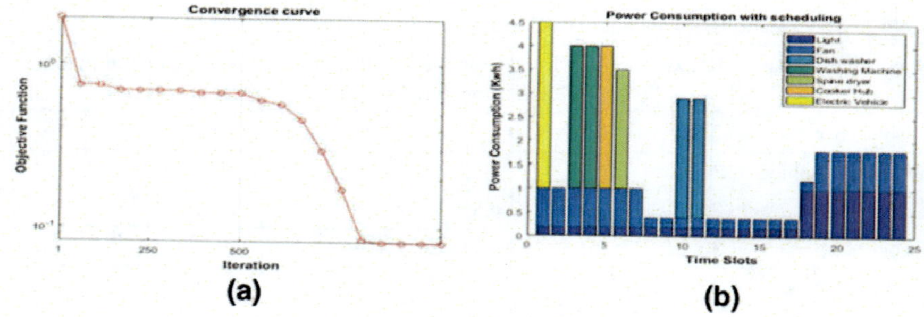

Fig. 2 **a** Convergence curve for $w = 0.25$ and **b** power consumption graph for $w = 0.25$

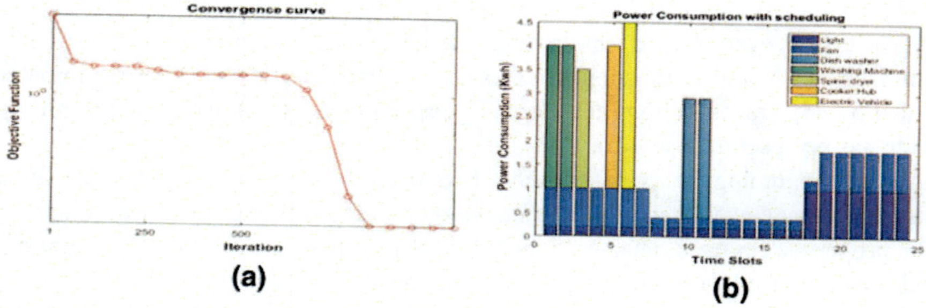

Fig. 3 **a** Convergence curve for $w = 0.5$ and **b** power consumption graph for $w = 0.5$

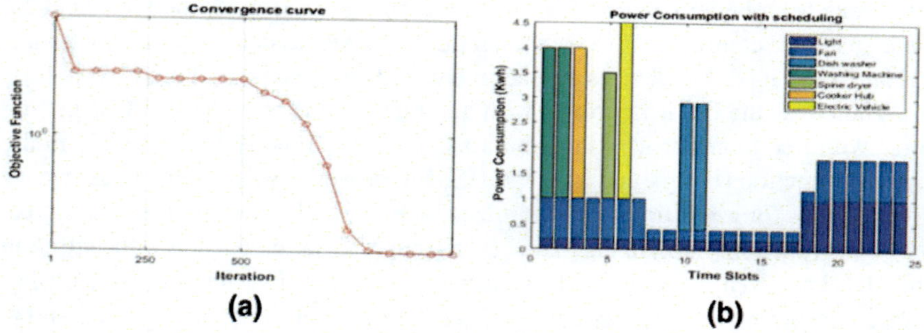

Fig. 4 **a** Convergence curve for $w = 0.75$ and **b** power consumption graph for $w = 0.75$

costs are mitigated, while user comfort is maintained. The OHEM algorithm specifies the hourly power price during the day using the real-time pricing (RTP) strategy. The improved binary particle swarm optimization (PSO) optimization mechanism, which is a local search-based algorithm, and the OHEM algorithm are built on this framework. The suggested OHEM algorithm is evaluated, and the simulation results show that the daily cost (DC), power consumption (PC), and objective function (F obj) are improved with the iteration time.

References

1. Leitao J, Gil P, Ribeiro B, Cardoso A (2020) A Survey on home energy management. IEEE Access 8:5699–5722
2. Jordehi AR (2019) Optimal scheduling of home appliances in home energy management systems using grey Wolf optimisation (Gwo) algorithm. In: 2019 IEEE Milan PowerTech, PowerTech 2019, pp 1–6
3. Alilou M, Tousi B, Shayeghi H (2020) Home energy management in a residential smart micro grid under stochastic penetration of solar panels and electric vehicles. Sol Energy 212(November):6–18
4. Dinh HT, Kim D (2021) An optimal energy-saving home energy management supporting user comfort and electricity selling with different prices. IEEE Access 9:9235–9249
5. Zhai S, Wang Z, Yan X, He G (2019) Appliance flexibility analysis considering user behavior in home energy management system using smart plugs. IEEE Trans Ind Electron 66(2)
6. Kaluthanthrige R, Rajapakse AD (2021) Demand response integrated day-ahead energy management strategy for remote off-grid hybrid renewable energy systems. Int J Electr Power Energy Syst 129:106731
7. Abdalla MAA, Min W, Mohammed OAA (2020) Two-stage energy management strategy of EV and PV integrated smart home to minimize electricity cost and flatten power load profile. Energies 13(23)
8. Aznavi S, Fajri P, Asrari A, Harirchi F (2020) Realistic and intelligent management of connected storage devices in future smart homes considering energy price tag. IEEE Trans Ind Appl 56(2):1679–1689
9. Ahmad A et al (2017) An optimized home energy management system with integrated renewable energy and storage resources. Energies 10(4):1–35
10. Javaid N et al (2017) An intelligent load management system with renewable energy integration for smart homes. IEEE Access 5:13587–13600
11. Melhem FY, Grunder O, Hammoudan Z, Moubayed N (2017) Optimization and energy management in smart home considering photovoltaic, wind, and battery storage system with integration of electric vehicles optimisation et gestion de l'énergie dans une maison intelligente en considérant le photovoltaïque, l'éolienn. Can J Electr Comput Eng 40(2):128–138
12. Melhem FY, Grunder O, Hammoudan Z, Moubayed N (2018) Energy management in electrical smart grid environment using robust optimization algorithm. IEEE Trans Ind Appl 54(3)
13. Erdinc O, Paterakis NG, Mendes TDP, Bakirtzis AG, Catalão JPS (2015) Smart household operation considering bi-directional EV and ESS utilization by real-time pricing-based DR. IEEE Trans Smart Grid 6(3):1281–1291
14. Rasouli V, Goncalves I, Antunes CH, Gomes A (2019) A comparison of MILP and metaheuristic approaches for implementation of a home energy management system under dynamic tariffs. In: 2019 International conference on smart energy systems and technologies (SEST), pp 1–6
15. Jordehi AR (2020) Enhanced leader particle swarm optimisation (ELPSO): a new algorithm for optimal scheduling of home appliances in demand response programs. Artif Intell Rev 53(3):2043–2073

Solar Power Generation and Utilization—Policies in India

Santu Hore, Raja Kumar Sakile, and Umesh Kumar Sinha

Abstract India is the fastest growing country, and Indian industries are growing faster than other developing countries to support Make in India initiative. With the technological advancement, every Indians are using multiple gadgets, applications to ensure their comfortable lifestyles resulted per capita consumption of electrical energy in India is increasing day by day. To meet the industrial, commercial, and residential power demand, India is increasing power generation capacity on every year, but the major challenges to generate the conventional power are sharp depletion of fossil fuels as country like India primarily depends on fossil fuels for power generation. Country like India can harvest solar energy in such that it can fulfill country's complete power requirements. Generation and use of power from renewable sources will enhance the energy security, economic growth, and stability of the Indian economy and reduce the dependency on other countries for fossil fuels, and therefore, vision, short-term and long-term planning, guidelines, policies, strategies are utmost important for significant growth in renewable sector in India. At present, the renewable sector is working under Central Government directly through State Government agencies, and therefore, common guidelines and policies are utmost important to ensure the maximum use of renewable sources across the country. This paper is focusing on the present policies and agencies available in India to promote generation, transmission, distribution, and utilization of solar and other renewable sources across the country.

Keywords Electricity act · Jawaharlal Nehru National Solar Mission · PM-KUSUM · Net and gross metering

---

S. Hore (✉) · R. K. Sakile · U. K. Sinha
Department of Electrical Engineering, NIT Jamshedpur, Jamshedpur, India
e-mail: horeaec@gmail.com

U. K. Sinha
e-mail: uksinha.ee@nitjsr.ac.in

K. Namrata et al. (eds.), *Smart Energy and Advancement in Power Technologies*, Lecture Notes in Electrical Engineering 926,
https://doi.org/10.1007/978-981-19-4971-5_66

1 Introduction

India and Indian industries are growing faster than other countries in the world, and to continue with this growth, per capita energy consumption is increasing day by day. Power sector is playing an important role for India's economic and industrial growth. In the optimistic scenario of 8% GDP growth, electrical energy requirement in India is projected to increase from 1152.4 BU in 2016–17 to 1905.5 BU in 2026–27, 2458.9 BU in 2031–32 and 3175.4 BU in 2036–37 [1].

Power system in India is primarily dominated by fossil fuels, and about 61% of India's total installed generation capacity depends on coal, lignite, natural gas, and diesel. Continuous growing of prices and depletion of fossil fuels are the major challenges to meet the rising energy demand. High dependency on fossil fuel import is giving additional burden, and therefore, the country has to frame how to reduce the utilization of fossil fuels and enhance the utilization of renewable energy sources (RESs) [2].

As on March-21, the country has an installed power capacity of 382.15 GW, out of which thermal: 234.73 GW, nuclear: 6.78 GW, hydro: 46.21 GW, and renewable energy sources (RESs): 94.43 GW. The solar power generation capacity is 40.085 GW; wind power generation capacity is 39.25 GW; capacity of bio-power and small hydropower plant is 10.32 GW and 4.79 GW, respectively [3] (Figs. 1 and 2).

In past few decades, Government of India (GOI) has introduced frameworks to reduce use of conventional energy and increase the use of renewable energies. Among all other renewable resources, solar energy is the primary nonconventional resource, and India has the tremendous potential to generate, use, and export the energy cultivated from solar energy. Land area of India receiving approx. 4–7 kWh/m^2/day with approx. radiation of 1200–1300 kWh/m^2/annum, and therefore, solar photovoltaic

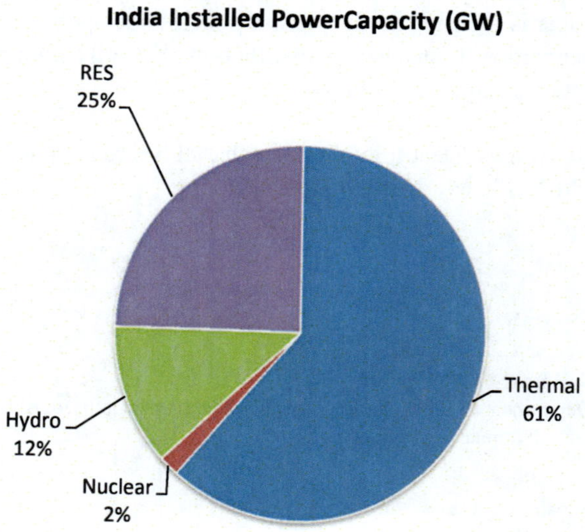

Fig. 1 Installed power capacity (GW) in India as on March-21 (*Source* CEA, 2021)

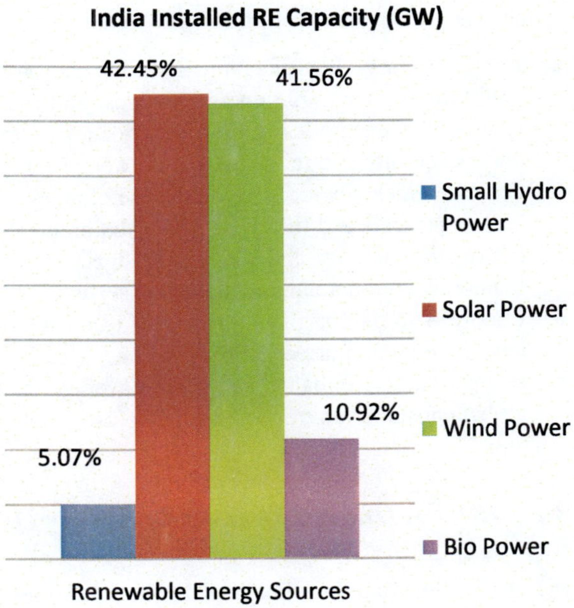

Fig. 2 Installed RE power capacity (GW) in India as on March-21 (*Source* CEA, 2021)

power can easily and effectively be generated using suitable infrastructures in India [4]. In India, solar power is sharing approx. 42.45% of the total renewable energy capacity till March 21.

2 Government Agencies

2.1 MNRE (Ministry of New and Renewable Energy)

After the severe energy crisis in 1970, a new commission, CASE has been formed in March 1981 under the Department of Science and Technology (India). The primary responsibility of Commission for Additional Sources of Energy (CASE) was introduction of policies toward development, formulation of new and nonconventional energy resources. In the year 1982, another new department was introduced named Department of Nonconventional Energy Sources (DNES), and then, CASE merged with DNES to continue the activities associated with renewable energy. After successful set up of an individual Ministry for nonconventional energy sources in 1992, the present name was adopted in the year 2006 [5].

The primary objectives of this Ministry are,

a. Energy Security: Identification, development, and implementation of use of alternative fuels, renewable energy sources, and reduction of use of conventional resources to ensure less import dependency on fossil fuel followed by increase energy security and fulfill the gap between supply and demand;
b. Promote green power: Increase the use of renewable resources against the fossil fuel to generate and fulfill the power requirement;
c. Energy Availability and Access: Ensure easy availability and use of RES in industrial, commercial, and domestic sectors;
d. Energy Affordability: Limit the tariff of renewable energy, promote subsidy to ensure maximum use of RES;
e. Energy Equity: Enhance and ensure average energy consumption in line with global consumption trend and suitable use of renewable energy sources.

2.2 IREDA (Indian Renewable Energy Development Agency)

IREDA is Non-Banking Financial Institution established in 1987 as a subsidiary company of MNRE. Major objectives of the agency are to promote, develop, and extend financial supports for implementation of projects related to new and renewable energy sources.

2.3 SECI (Solar Energy Corporation of India)

SECI is a non-profit public sector unit under the Companies Act, 1956, is established in 2011 under the MNRE administration. The purpose of this organization enhances support toward implementation of Jawaharlal Nehru National Solar Mission. In the present, SECI enhances implementation of a number of projects under MNRE, small and large-scale grid-connected installations, grid-connected ground mounted, rooftop solar PV plants, canal-top, etc.

2.4 NISE (National Institute of Solar Energy)

National Institute of Solar Energy is a research and development institute under MNRE mainly responsible for design, development, testing, certification, and standardization of solar components. This institute is also responsible for capacity planning and policy planning to ensure sustainable growth of solar energy. This institute also engaged in research and development projects on hybrid and storage systems (Fig. 3)

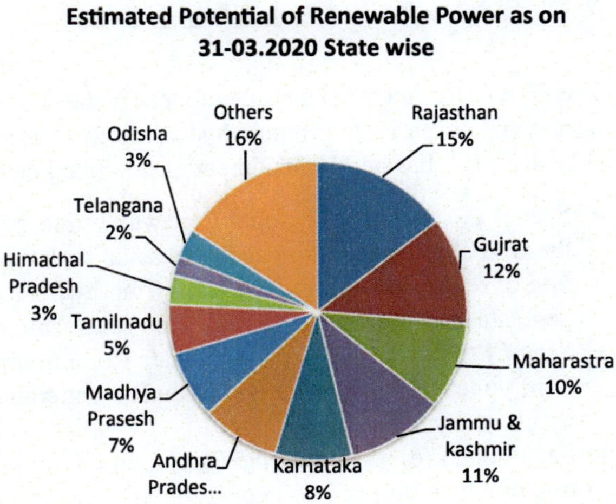

Fig. 3 State-wise estimated potential of renewable power (*Source* Energy Statistics India 2021)

3 Government Policies

Since 2000, Government of India (GOI) has introduced acts, regulations, and policies toward promotion of renewable energy. Primary objective of policy guidelines is to enhance the contribution of renewable energy sources in country's total energy requirement. In this connection, the major contributors are as follows.

3.1 Electricity Act 2003

- Sect. 3 states that National Electricity Policy shall be developed by the GOI in discussion with the State Governments and the power development authority, in line with optimum use of fossil fuels and renewable sources.
- Sect. 4 states that the GOI shall develop National Policy on standalone systems in connection with nonconventional energy systems for rural areas after consultation with the state governments.
- Sect. 61 (h) states that the Appropriate Commission shall formulate to determine tariff policy to promote generation and co-generation of electricity from renewable sources.
- Sect. 86 (1) (e) states that the State Commission shall empower toward promotion of generation and co-generation of electrical energy through grid-connected renewable resources, purchase electricity from renewable sources, and sale to any person with a defined percentage of total energy consumptions [6].

3.2 *Electricity Act Amendment 2020*

The Electricity Act 2003 has amendment in 2000. In the amendment bill, focus has been given primarily to promote hydro energy sources, and accordingly, Sect. 3, 61 (h) and 86 (1), has been amendment as described below:

- Sect. 3 A states that National Renewable Energy Policy shall be developed by the GOI after consultation and discussion with the state governments for promotion of renewable energy primarily from hydro sources and prescribe minimum percentage of purchase of power from renewable sources.
- Sect. 61 (h) states that the Appropriate Commission shall formulate to determine tariff policy to promote generation and co-generation of electricity primarily from hydro renewable sources.
- Sect. 86 (1) (e) states that the State Commission shall empower toward promotion of generation and co-generation of electrical energy through grid-connected renewable resources, purchase electricity from renewable sources primarily from hydro sources and sale to any person with a defined percentage of total energy consumptions.

3.3 *National Electricity Policy*

The policy states use of renewable energy sources shall be increased progressively for all households, energy, and demand shortages to be addressed and adequate reserve to be ensured. Reliable power supply also to be ensured in an efficient manner and at affordable rates [7]. The government has also initiated to revise the NEP released in 2021 to focus on optimum regulatory framework to make Indian Power sector future ready.

3.4 *National Tariff Policy*

The policy states financial feasibility of renewable energy sector and attract investments to ensure sufficient electricity to consumers at competitive rates. The policy also promotes transparency, consistency, operational efficiency to improve power supply quality [8].

3.5 National Action Plan on Climate Change

The NAPPC 2008 sets year-wise target for procurement of renewable energy percentage and also recommends strong regulatory measures and actions to achieve the targets.

3.6 National Rural Electrification Policies

The policies states the extension of reliable power supply to every consumers at reasonable tariff. The policy also explains the standalone process for electrification to households where grid connectivity is not available including plan and delivery mechanism.

3.7 Jawaharlal Nehru National Solar Mission (JNNSM)—2010

Modern technologies ensured about sunlight as renewable energy source can generate electricity. India has vast potential of solar energy provides to generate power on a connected and distributed basis. Off-grid or decentralized standalone applications are more advantageous in rural electrification process, and grid connectivity systems are more advantageous to industrial processes.

There has been a considerable influence of solar power to the power system of India as on date, and solar power dependent distributed and standalone systems have comforted people across the country by fulfilling their needs to illumination, heating, cooking, and other requirements in terms of energy uses. In order to support tail-end villages, JNNSM one of the global initiative was initiated in the year 2010. The target of the mission is of deployment 2 GW off-grid solar power and 20 GW (later increased by 100 GW) grid-connected solar power on or before 2022 [9].

3.8 Ujjwal Discom Assurance Yojna (UDAY) Scheme

UDAY is an initiative was taken by GOI in 2015 to find permanent solar power solution with financial aid which was seriously faced by the DISCOMs. The initiative was optional for states to join. The primary objectives of this initiative were deployment of renewable energy, power generation cost reduction, and improvement of energy efficiency.

3.9 *Pradhan Mantri Kisan Urja Suraksha Evam Utthaan Mahabhiyan (PM-KUSUM)—2019*

As a part of 40% enhancement of installed power capacity from nonconventional energy sources by 2030, the PM-KUSUM scheme launched in 2019 with three different categories:

- Component A: In this category, individual or group of farmers can installed renewable energy-based power plant of 500 KW to 2000 KW on barren land. To reduce the transmission losses, the power plant shall be installed within the 5000 m radius from the nearest sub-station.
- Component B: In this category, individual farmers can replace existing diesel-driven pumps used in irrigation purpose by standalone solar power-driven pumps where grid power is not available. The financial support shall be provided by the GOI limited to pump capacity of 7.5HP.
- Component C: In this category, individual farmers can feed extra power to the grid after use of power generated from grid-connected solarize pumps to local DISCOM with a pre-fixed tariff rate (Fig. 4).

4 Metering Arrangements

With the initiation and implementation of policies for generation and use of renewable sources from central and state government, residential, commercial, & industrial consumers are installing ground-mounted/rooftop solar PV plant to meet their daily energy requirement and reduce the manufacturing cost applicable to industries [10]. While the residential and commercial loads are stable in nature for working and off days, but industrial loads are fluctuating in nature in working and off days. Large rooftops are available in industries, and industrial consumers are installing grid-connected solar plants in large scales [11]. During working days, industrial consumers are consuming total power generated by the plant, while in off days, industrial consumers are unable to consume the total power generated by the solar plant, and excess power is back-feeding to the grid. Toward measuring and monitoring of the back-feeding power, net metering and gross metering arrangements come into system [12].

4.1 *Net Metering*

Industrial consumers are installing ground-mounted and rooftop grid-connected solar PV plants in large scales. After fulfilling the energy requirement of the respective consumer, the excess generation can be supplied to utility grid, and the supplied power can be monitored and measured by suitable metering system [13]. Net metering is the

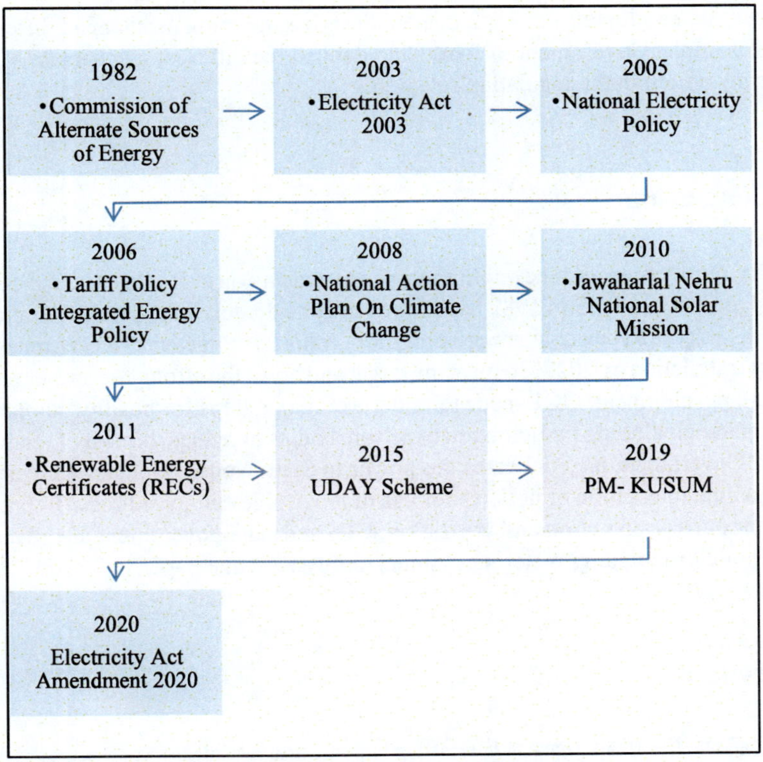

Fig. 4 Year-wise renewable energy policies implementation in India

arrangement through which consumers can sell extra generation from plant to utility grid. In this system, bi-directional energy meters are used, and consumers are using power from utility grid are recorded into import section, and consumers exporting excess electricity to utility grid are recorded into export section of the energy meter.

Every states adopted and implemented the net metering mechanism differently. As an example, Government of Jharkhand extended net metering system for the rooftop solar plant capacity maximum to 2 MWp [14].

4.2 Gross Metering

In gross metering system, total generation by the plant is injected to the utility, and consumer imports electricity from the grid. During the billing cycle, consumer is compensated for the electricity exported to the grid at decided tariff by the Regulatory Commission of the state [15].

Every states adopted and implemented the gross metering mechanism differently. As an example, Government of Jharkhand defined the different mechanism of gross metering system in the regulation released.

5 Conclusion

In India, requirement of renewable energy is increasing on every day, and the country started journey for RES in 1982 has improved considerably in this field. As a part of 'World's largest Renewable energy program', India has targeted 175 GW renewable energy installation by 2022. To meet the target and boost the efforts for further promotion and development, GOI needs to undertake major steps for healthy competition in the market including participation of private bodies to reduce the cost of renewable power. Government also delicates the power to State Nodal Agencies for easy and smooth implementation of policies for use of renewable energy sources. To promote the solar power generation and utilization, GOI and state government are governing by all policies including metering systems discussed in the paper.

References

1. Long Term Electricity Demand Forecasting Report, https://cea.nic.in
2. Shyam B, Kanakasabapathy P (2017) Renewable energy utilization in India—policies, opportunities and challenges. International conference on technological advancements in power and energy (TAP Energy)
3. Energy Statistics India (2021) Ministry of statistics and programme implementation (MOSPI): http://mospi.nic.in
4. Suman SK, Ahamad J (2018) Solar energy potential and future energy of India: an overview. Int J Eng Sci Comput. http://ijesc.org/
5. Upadhyay A, Chowdhury A (2014) Solar energy fundamentals and challenges in Indian restructured power sector. Int J Sci Res Publ 4(10). ISSN 2250–3153 www.ijsrp.org
6. The Electricity Act (2003)
7. National Electricity Policy; https://powermin.nic.in
8. Tariff Policy; www.powermin.gov.in
9. www.mnre.gov.in
10. Srivastava SP, Srivastava SP (2013) Solar energy and its future role in Indian economy. Int J Environ Sci: Develop Monit (IJESDM). ISSN No. 2231-1289, 4(3)
11. Mishra SS Design and installation of grid connected roof top solar Pv system. In: International conference on recent innovations in electrical, electronics and communication engineering—(ICRIEECE)
12. Omkar K, Srikanth MV, Swaroop KP, Rama Rao PVV4 (2015) Performance evaluation of 50 KWp rooftop solar PV plant. In: 2015 international conference on industrial instrumentation and control (ICIC), college of engineering Pune, India
13. Thombre S, Vetal S Net metering technology. State level conference on engineering innovations (SCEI-17). Int J Adv Eng Res Sci (IJAERS). https://doi.org/10.22161/ijaers/scei.2017.3 ISSN: 2349–6495(P) | 2456–1908(O)

14. The Jharkhand gazette on rooftop solar PV grid interactive systems and net/gross metering (1st Amendment) regulations (2019)
15. Kumar A, Hussain DMA (2018) A review paper on solar energy in India. Gyancity J Electron Comput Sci 3(1):1–10. ISSN: 2446–2918. https://doi.org/10.21058/gjecs.2018.31001